WITHDRAWN

**PHYSICAL METHODS
IN HETEROCYCLIC CHEMISTRY**

GENERAL HETEROCYCLIC CHEMISTRY SERIES

Edward C. Taylor and Arnold Weissberger, Editors

> MASS SPECTROMETRY OF HETEROCYCLIC COMPOUNDS
> by Q. N. Porter and J. Baldas*
>
> NMR SPECTRA OF SIMPLE HETEROCYCLES
> by T. J. Batterham*
>
> HETEROCYCLES IN ORGANIC SYNTHESIS
> by A. I. Meyers (out of print)
>
> PHOTOCHEMISTRY OF HETEROCYCLIC COMPOUNDS
> by Ole Buchardt (out of print)
>
> STEREOCHEMISTRY OF HETEROCYCLIC COMPOUNDS
> by W. L. F. Armarego*
> Part I, Nitrogen Heterocycles
> Part II, Oxygen; Sulfur; Mixed N, O, and S; and Phosphorus Heterocycles
>
> 1,3-DIPOLAR CYCLOADDITION CHEMISTRY
> edited by Albert Padwa
>
> PHYSICAL METHODS IN HETEROCYCLIC CHEMISTRY
> edited by R. R. Gupta

*Now available from   Krieger Publishing Co., Inc.
P.O. Box 9542
Melbourne, Florida 32901

PHYSICAL METHODS IN HETEROCYCLIC CHEMISTRY

Edited by

R. R. Gupta

*Department of Chemistry
University of Rajasthan
Jaipur, India*

A Wiley-Interscience Publication

JOHN WILEY & SONS
New York • Chichester • Brisbane • Toronto • Singapore

Copyright © 1984 by John Wiley & Sons, Inc.

All rights reserved. Published simultaneously in Canada.

Reproduction or translation of any part of this work
beyond that permitted by Section 107 or 108 of the
1976 United States Copyright Act without the permission
of the copyright owner is unlawful. Requests for
permission or further information should be addressed to
the Permissions Department, John Wiley & Sons, Inc.

Library of Congress Cataloging in Publication Data

Main entry under title:

Physical methods in heterocyclic chemistry.
    (General heterocyclic chemistry series, ISSN 0363-8626)
    "A Wiley-Interscience publication."
    Bibliography: p.
    Includes index.
    1. Heterocyclic compounds. 2. Chemistry, Physical organic.
I. Gupta, R. R. (Radha Raman), 1941– . II. Series.
QD400.P593 1983 547′.59045 83-14609

ISBN 0-471-09855-8

Printed in the United States of America

10 9 8 7 6 5 4 3 2 1

CONTRIBUTORS

J. ARMAND
  Laboratoire de Physicochimie des Solutions
  Université Pierre et Marie Curie
  Paris, France

ILIE I. BĂDILESCU
  Department of Chemistry
  Institute of Higher Education
  Pitești, Romania

SIMONA BĂDILESCU
  Spectroscopy Laboratory
  Petrochemical Works Pitești
  Pitești, Roumania

ALEXANDRU T. BALABAN
  Department of Organic Chemistry
  The Polytechnic
  Bucharest, Roumania

P. K. BASU
  Solid State and Structural Chemistry Unit
  Indian Institute of Science
  Bangalore, India

E. A. BOUDREAUX
  Department of Chemistry
  University of New Orleans
  New Orleans, Louisiana

M. D. FENN
  Medical Chemistry Group
  The John Curtin School of Medical Research
  Australian National University
  Canberra City, Australia

R. R. GUPTA
  Department of Chemistry
  University of Rajasthan
  Jaipur, India

SVEN E. HARNUNG
  Department of Inorganic Chemistry
  The H. C. Ørsted Institute
  The University of Copenhagen
  Universitetsparken, Denmark

ELLEN A. KERR
  School of Chemistry
  Georgia Institute of Technology
  Atlanta, Georgia

ERIK LARSEN
  Chemistry Department
  The Royal Veterinary and Agricultural University
  Thorvaldensvej, Denmark

HELEN C. MACKIN
  School of Chemistry,
  Georgia Institute of Technology
  Atlanta, Georgia

R. J. MAJESTE
  Department of Chemistry
  Southern University in New Orleans
  New Orleans, Louisiana

J. PINSON
  Laboratoire d'Electrochimie
  Université Paris 7
  Paris, France

C. N. R. RAO
Solid State and Structural Chemistry Unit
Indian Institute of Science
Bangalore, India

L. M. TREFONAS
Department of Chemistry
University of Central Florida
Orlando, Florida

WORTH E. VAUGHAN
Department of Chemistry
University of Wisconsin
Madison, Wisconsin

NAI-TENG YU
School of Chemistry
Georgia Institute of Technology
Atlanta, Georgia

INTRODUCTION TO THE SERIES

General Heterocyclic Chemistry

The series, "The Chemistry of Heterocyclic Compounds," published since 1950 by Wiley-Interscience, is organized according to classes of compounds. Each volume deals with syntheses, reactions, properties, structure, physical chemistry, etc., of compounds belonging to a specific class, such as pyridines, thiophenes, and pyrimidines, three-membered ring systems. This series has become the basic reference collection for information on heterocyclic compounds.

Many aspects of heterocyclic chemistry have been established as disciplines of *general* significance and application. Furthermore, many reactions, transformations, and uses of heterocyclic compounds have specific significance. We plan, therefore, to publish monographs that will treat such topics as nuclear magnetic resonance of heterocyclic compounds, mass spectra of heterocyclic compounds, photochemistry of heterocyclic compounds, X-Ray structure determination of heterocyclic compounds, UV and IR spectroscopy of heterocyclic compounds, and the utility of heterocyclic compounds in organic synthesis. These treatises should be of interest to *all* organic chemists as well as to those whose particular concern is heterocyclic chemistry. The new series, organized as described above, will survey under each title *the whole field of heterocyclic chemistry* and is entitled "General Heterocyclic Chemistry." The editors express their profound gratitude to Dr. D. J. Brown of Canberra for his invaluable help in establishing the new series.

Edward C. Taylor

Department of Chemistry
Princeton University
Princeton, New Jersey

Arnold Weissberger

Research Laboratories
Eastman Kodak Company
Rochester, New York

PREFACE

Heterocyclic chemistry is one of the most important branches of organic chemistry, one in which modern research would be impossible without the structural information provided by physical methods. Thus, physical methods and synthetic chemistry are complementary, and chemists working in the field of heterocyclic chemistry must be fully aware of this intimate relationship. The identification and characterization of heterocyclic compounds critically depend upon the use of the physical methods discussed in this volume. These include IR spectroscopy, NMR spectroscopy, photoelectron spectroscopy, diamagnetism, dipole moments, X-ray studies, Raman spectroscopy, electrochemistry, and natural and magnetic circular dichroism.

Efforts have been made to maintain a consistent style in all chapters. Each author has been given complete freedom to express his own viewpoints, and the authors alone are responsible for the quality of their chapters. I am indeed thankful to Professor E. C. Taylor and Dr. Arnold Weissberger for their constant encouragement, invaluable guidance, and full cooperation. I also extend my sincere thanks to the staff of John Wiley & Sons.

R. R. GUPTA

University of Rajasthan
Jaipur, India
January 1984

CONTENTS

1. **Infrared Spectra of Heterocyclic Compounds**   1
    A. T. Balaban, S. Bădilescu, and I. I. Bădilescu

2. **Nuclear Magnetic Resonance Spectroscopy**   141
    M. D. Fenn

3. **Ultraviolet Photoelectron Spectroscopy of Heterocyclic Compounds**   231
    C. N. R. Rao and P. K. Basu

4. **Diamagnetism of Heterocyclic Compounds**   281
    E. A. Boudreaux and R. R. Gupta

5. **X-Ray Studies of Small-Ring Nitrogen Heterocycles**   313
    L. M. Trefonas and R. J. Majeste

6. **Dipole Moments of Heterocyclic Compounds**   355
    W. E. Vaughan

7. **Raman Spectroscopy of Heterocyclic Compounds**   373
    H. C. Mackin, E. A. Kerr, and N.-T. Yu

8. **Electrochemical Behavior of Heterocyclic Compounds**   427
    J. Armand and J. Pinson

9. **Natural and Magnetically Induced Circular Dichroism of Heterocyclic Compounds**   627
    S. E. Harnung and E. Larsen

    Index   669

1 INFRARED SPECTRA OF HETEROCYCLIC COMPOUNDS

ALEXANDRU T. BALABAN

*Department of Organic Chemistry
The Polytechnic
Bucharest, Roumania*

SIMONA BĂDILESCU

*Spectroscopy Laboratory
Petrochemical Works Pitești
Pitești, Roumania*

ILIE I. BĂDILESCU

*Department of Chemistry
Institute of Higher Education
Pitești, Roumania*

1.1.	Introduction		3
1.2.	Small Rings		4
	1.2.1.	Three-Membered Rings, 4	
		1.2.1.1. Ethylene Oxide, 4	
		1.2.1.2. Aziridines, 5	
	1.2.2.	Four-Membered Cyclic Peroxides, 7	
		1.2.2.1. 1,2-Dioxetanes, 7	
1.3.	Five-Membered Rings with One Heteroatom		8
	1.3.1.	Nonaromatic Rings without Carbonyl Groups, 8	
		1.3.1.1. Tetrahydrofuran (THF), 8	
		1.3.1.2. 2,5-Dihydrothiophenes, 8	
	1.3.2.	Aromatic Rings without Carbonyl Groups, 8	
		1.3.2.1. Furan, Thiophene, Pyrrole, Selenophene, and Tellurophene, 8	
		1.3.2.2. Pyrrole Derivatives, 15	
		1.3.2.3. Thiophene Derivatives, 17	
		1.3.2.4. Furan Derivatives, 20	
	1.3.3.	Rings with Carbonyl Groups, 22	
		1.3.3.1. Imides and Anhydrides, 22	
		1.3.3.2. Lactams, 27	
		1.3.3.3. Lactones, 31	

2   Infrared Spectra of Heterocyclic Compounds

1.4. Five-Membered Rings with Two or More Heteroatoms   34
   1.4.1. Rings with Two Heteroatoms with Carbonyl or Thiocarbonyl
          Groups, 34
          1.4.1.1. Ethylene Carbonate, Ethylene Trithiocarbonate,
                   Oxazolidine-2-one, Oxazolidine-2-thione, Ethyleneurea,
                   Imidazoline-2-thione, Thiazolidine-2-one, and
                   Thiazolidine-2-thione, 34
          1.4.1.2. Dioxolan-2-ones, 37
          1.4.1.3. Methylene Oxalate, 38
          1.4.1.4. Rhodanine, 38
          1.4.1.5. Imidazoline Derivatives, 39
          1.4.1.6. Hydantoins, 41
          1.4.1.7. Thioamides, 41
          1.4.1.8. Thiazolidinediones, 42
          1.4.1.9. Saccharin, 42
   1.4.2. Aromatic Rings with Two Heteroatoms without Carbonyl Groups, 42
          1.4.2.1. Pyrazole, 42
          1.4.2.2. Imidazole and Derivatives, 43
          1.4.2.3. Thiazole, Isothiazole, and Derivatives, 45
          1.4.2.4. Oxazole and Derivatives, 45
          1.4.2.5. Dithiole and Derivatives, 46
   1.4.3. Aromatic Rings with Three or Four Heteroatoms without Carbonyl
          Groups, 46
          1.4.3.1. Triazoles, 46
          1.4.3.2. Oxadiazoles and Thiadiazoles, 47
          1.4.3.3. 1,2,3,4-Thiatriazoles, 48
          1.4.3.4. 1,2,3,4-Tetrazoles, 49

1.5. Six-Membered Saturated Rings   49
   1.5.1. Rings with One Heteroatom, 49
          1.5.1.1. Tetrahydropyran, Thiacyclohexane, Piperidine,
                   Thiopiperidones, and Thiomorpholides, 49
          1.5.1.2. Tetrahydroquinolines, 51
          1.5.1.3. Spiropyrans, 51
   1.5.2. Rings with Two Heteratoms in 1,4 Positions, 52
          1.5.2.1. 1,4-Dioxan, 1,4-Dithian, and 1,4-Oxathian, 52
          1.5.2.2. 1,4-Thioxan, 1,4-Selenoxan, 1,4-Telluroxan, and
                   1,4-Thioselenan, 55
          1.5.2.3. Morpholine, 56
          1.5.2.4. Other Six-Membered Heterocyclic Rings, 57
   1.5.3. Rings with Three or Four Heteroatoms, 58
          1.5.3.1. 1,3,5-Trithiane and Derivatives, 58
          1.5.3.2. 1,2,5-Oxadiazines, 59
          1.5.3.3. Bicyclic Compounds, 60
          1.5.3.4. Tetrathiaadamantane, 61
   1.5.4. Six-Membered Rings with Carbonyl Groups, 62
          1.5.4.1. Pyrones, 62
          1.5.4.2. Glutarimide, 63
          1.5.4.3. 2,5- and 2,3-Piperazinediones, 63
          1.5.4.4. 1,3-Oxazine-4,6-diones, 63
          1.5.4.5. Flavones and Isoflavones, 65
          1.5.4.6. Chromone and Thiochromone, 66

1.6. Aromatic Six-Membered Rings   68
   1.6.1. Pyridine and Related Systems, 68

        1.6.1.1.   Pyridine, 68
        1.6.1.2.   2,2′-Bipyridyl, 70
        1.6.1.3.   Substituted Pyridines, 71
        1.6.1.4.   Pyrylium Cation, 75
        1.6.1.5.   Phosphabenzene and Arsabenzene, 75
        1.6.1.6.   Quinolines, 78
        1.6.1.7.   Carbolines, 81
        1.6.1.8.   Naphthyridines and Other Fused Systems with One Nitrogen Heteroatom Per Ring, 81
    1.6.2. Diazines, 83
        1.6.2.1.   Pyridazines, 83
        1.6.2.2.   Pyrimidines, 83
        1.6.2.3.   Pyrazines, 89
        1.6.2.4.   Fused Diazines, 91
    1.6.3. Vibrational Spectra of Nuclei Acid Constituents, 97
        1.6.3.1.   Pyrimidines, 97
        1.6.3.2.   Purines, 105
        1.6.3.3.   Hydrogen Bonding Interactions of Nuclei Acid Constituents, 107
    1.6.4. Triazines, 107
        1.6.4.1.   s-Triazines, 107
        1.6.4.2.   1,2,4-Triazines, 109
    1.6.5. Tetrazines, 110

1.7. Seven-Membered Rings 111

1.8 Metallo-organic and Elemento-organic Chelate Heterocycles 114
    1.8.1. Metal Chelates, 114
    1.8.2. Boron and Silicon Heterocycles, 119

References 122

1.1. INTRODUCTION

Some organic chemists engaged mainly in synthetic work may think that infrared (IR) spectroscopy is an obsolete technique and that the advent of multiple-nuclei pulse-Fourier-transform (PFT) NMR spectroscopy and the various mass spectrometric techniques can deliver much more information about molecular structure than IR spectroscopy. While this is partly true with respect to routine spectra obtained with rudimentary instruments, IR spectroscopy has also evolved in recent years, resulting in the development of both sophisticated instrumentation (far-IR and computerized PFT-IR) and new auxiliary techniques.

A nonexhaustive list of recent advances follows:

Photoelectron spectroscopy provides the energy of the symmetric vibrations in different ionic states.

Matrix isolation vibrational spectroscopy allows differentiation between monomer and dimer bands.

Solid-state IR spectroscopy (oriented crystalline samples obtained by solidification of liquid films under a temperature gradient, examined in polarized light, i.e., dichroism measurements on single crystals) contributes to the understanding of lattice dynamics.

The measurement and identification of low-frequency IR vibrations have important consequences for the identification of conformational changes.

The increasing use of measured IR intensity allows assessment of the extent of intramolecular interaction in conjugated systems.

On the other hand, old and new methods for assigning IR bands continue to advance the frontiers of knowledge. Some of these methods are:

Multiple isotope labeling techniques for the complete vibrational assignment that allow force constant calculations to be checked against experiment

Measurement of the intensity and polarization of Raman lines, the rotational envelope contours, and *P–R* separations in gas-phase IR spectra

Comparison of the spectra of the neutral molecule and the corresponding radical anion

Studying the influence of protonation on vibrations

Complexation with Lewis acids (BF_3 for discriminating between C=O and C=C vibrations; metallic acceptors to lower the symmetry relative to the free ligands)

Adduct formation (e.g., between sulfoxides and silver or cadmium halides as a diagnostic tool to differentiate axial from equatorial S=O bands in cyclic six-membered compounds)

Assignment of torsion modes by far-IR–PFT spectroscopy (e.g., for the amino nitrogen inversion, a 20-m multiple reflection cell was used, and thousands of pulses were averaged)

This review is aimed at the reader who is familiar with the earlier reviews by Katritzky and coworkers (1) and discusses mainly the advances since the publication of those reviews. Since the field of heterocyclic chemistry is so vast, we have restricted the discussion to results that are significant to a class of compounds, rather than a single compound; omissions are therefore inevitable, and we would appreciate being notified about them.

1.2. SMALL RINGS

1.2.1. Three-Membered Rings

1.2.1.1. Ethylene Oxide

The infrared spectra of ethylene oxide have been studied extensively (2–6), and a number of force field calculations have been performed (7–9). Lord and Nolin's work (3) was based on a low-resolution study of ethylene oxide (gas and liquid), and some of their assignments were revised by Le Brumant (5, 6)

A reassignment of the fundamentals has been made (10) on the basis of a new study of the solid IR spectra of various deuterated derivatives: C_2H_3DO, C_2D_3HO, and $C_2D_2H_2O$ (*gem*, *cis*, and *trans* isomers). These data were used to derive a force field that reproduced all frequencies for these molecules with an average error of less than 1%. The observed frequencies for ethylene oxide and ethylene oxide-d_4 are assigned to vibrations of symmetry species in Table 1.1.

Good agreement was found between the observed and calculated frequencies for symmetry species A_2 and B_2 if rocking modes are assumed to have higher frequencies than twisting modes. Data reported for substituted epoxy rings (ring breathing and deformation modes) were collected by Katritzky and Ambler (1).

1.2.1.2. Aziridines

The vibrational assignment for aziridine has been reported by Thompson and Cave (2). The infrared absorption and the conformations of some aziridines and N-deuterated aziridines were studied, and there was evidence of ν_{NH} and ν_{ND} splitting (11). The invertomer populations for disymmetrically substituted aziridines have been studied (12) by absorbance measurements of NH or ND overtone IR bands. Thermodynamic parameters of *syn-anti* configurational equilibrium determined on the two ν_{0-2} (NH) bands (6480 and 6409 cm^{-1}) of *cis*-2,3-dimethylaziridine, are

$$\Delta H° = -0.45 \pm 0.20 \text{ kcal mol}^{-1}$$

$$\Delta G°_{25} = -1.1 \text{ kcal mol}^{-1}$$

$$\Delta S° = 2.3 \text{ cal mol}^{-1} \text{ K}^{-1}$$

The graphical resolution method is applicable only when the temperature variation of invertomer ratios is not too small. It was shown that steric factors predominate in controlling the equilibrium **1a ⇌ 1b**.

1a   **1b**

The $\nu_{C=O}$ band in *N*-substituted α-aziridinyl ketones also exhibits a splitting of conformational origin. The higher frequency band was ascribed to *cisoid* conformation (13). The low-temperature spectral data for 3,3-dimethyldiaziridine were not convincing with regard to the existence of more than a single form, and it was assumed that the *trans* structure is the more stable molecule (14). The vibrational assignments for some fundamentals (in terms of the C_2 point group), based upon isotopic shift ratios and on comparing the data with assignments for other three-membered ring compounds, are given in Table 1.2.

Table 1.1. Vibrational Assignments (cm^{-1}) of Ethylene Oxide and Ethylene Oxide-d_4

Symmetry species (C_{2v})	Vibration No.	C$_2$H$_4$O Obs. Lord (3)	C$_2$H$_4$O Obs. Cant (10)	Calc.[a]	C$_2$D$_4$O Obs. (10)	Description
A_1	1	3005	3018	3015	2200	ν_{CH}
	2	1490	1498	1505	1309	CH$_2$ scissoring
	3	1120	1270	1266	1012	CH$_2$ wagging
	4	1266	1148	1152	968	Ring breathing
	5	877	877	875	752	Ring β deformation
B_1	9	892	3006	3002	2176	ν_{CH}
	10	3079	1472	1462	1084	ν_{CH}
	11	1143	1151	1156	903	CH$_2$ twisting
	12	821	840	838	809	CH$_2$ rocking
A_2	6	3019	3063	3040	2250	ν_{CH}
	7	1470	1047	1060	862	
	8	1153	825	831	595	
B_2	13	3063	3065	3088	2318	ν_{CH}
	14	1345	1147	1159	896	CH$_2$ twisting
	15	807	794	800	588	CH$_2$ rocking

[a] Calculated (10) considering rocking vibrations to have higher frequencies than twisting vibrations.

Table 1.2. Vibrational Assignment for Some of the Fundamentals in 3,3-Dimethylaziridine

Symmetry Species	Vibration No.	cm^{-1}	Approximate Description
a	ν_1	3162	NH stretching
	ν_2	3001	CH$_3$ stretching
	ν_3	2951	CH$_3$ stretching
	ν_8	1192	NH bending
	ν_9	1132	Ring deformation
	ν_{14}	653	Ring deformation
b	ν_{18}	3193	NH stretching
	ν_{24}	1394	Ring deformation
	ν_{27}	1207	NH bending
	ν_{28}	1069	NH bending

The IR and Raman spectra of 1,1'-bisaziridyl in the gaseous, liquid, and crystalline states (15) showed that from all possible conformers only the *s-trans* form is present in each of the three states **2a–2c**. Since the inversion barrier is particularly

high in aziridines, the internal rotation must also be strongly hindered, and only centrosymmetric conformers may be stable.

1.2.2. Four-Membered Cyclic Peroxides

1.2.2.1. 1,2-Dioxetanes

For 1,2-dioxetanes **3** a weak O–O deformation band at 845–895 cm^{-1} has been reported (16, 17). A characteristic $\nu_{C=O}$ band (1870 cm^{-1}) in α-peroxylactone **4** has been evidenced (18).

1.3. FIVE-MEMBERED RINGS WITH ONE HETEROATOM

1.3.1. Nonaromatic Rings without Carbonyl Groups

1.3.1.1. Tetrahydrofuran (THF)

Earlier assignments for THF **5** were reconsidered (19) on the basis of its deuterated derivatives and new evidence on their conformation [puckered ring able to undergo pseudorotation (20)]. The new interpretation of the vibrational spectrum of THF together with Palm and Bisser's earlier assignment (21) are given in Table 1.3.

1.3.1.2. 2,5-Dihydrothiophenes

Raman and IR spectral data for 2,5-dihydrothiophenes **7** were reported by Green and Harvey (22). A vibrational assignment of the observed frequencies of sulfolane **6** and 3-sulfolene **7** was proposed (23, 24). Oriented crystalline samples of 3-sulfolene were obtained by solidification of liquid films under a temperature gradient. Their IR spectra were examined in polarized light, and the dichroic ratios

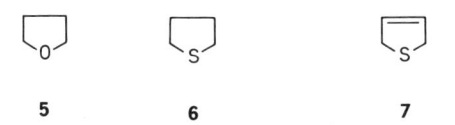

of the IR bands were measured. In the IR spectrum of the oriented solid the band at 1310 cm^{-1} disappeared when the electric vector was parallel to the direction of crystal growth; therefore this band was assigned to the asymmetrical stretching mode of the SO$_2$ group. A strong band at 1127–1122 cm^{-1} in the IR spectrum and at 1111 cm^{-1} in the Raman spectrum of the solid was assigned to the SO$_2$ symmetric stretching mode. It is interesting to note the unusual difference between the IR and Raman frequencies in the liquid-phase spectra (1139 and 1126 cm^{-1}, respectively). The difference was accounted for on the basis of strong dipole-dipole interactions, an effect that has been recognized in liquids having high dipole moments. This effect involves only very strong absorption bands corresponding to totally symmetric vibrations (25, 26). Ring stretching fundamentals were identified at 948 and 705 cm^{-1} as strongly polarized Raman bands and ring bending at 440 cm^{-1}. The depolarized Raman band at 390 cm^{-1} was assigned to the so-called SO$_2$ wagging modes.

1.3.2. Aromatic Rings without Carbonyl Groups

1.3.2.1. Furan, Thiophene, Pyrrole, Selenophene, and Tellurophene

Table 1.4 presents the generally accepted assignments of the IR and Raman bands for furan (27–29), thiophene (30, 31), and pyrrole (32–34), and also the new photoelectronic implications. Other, earlier IR and Raman studies on five-membered

Table 1.3. Vibrational Assignments for Tetrahydrofuran

Symmetry Species	ν_{calc}	ν_{obs} Vapor	ν_{obs} Liquid	ν_{obs} Solid	Assignments[a] Ref. 21	Assignments[a] Ref. 19
B_2	2977	2976	2977	2947	CH str. (a)	α-CH str.
A_2	2977	2964				
A_2	2942			2924	CH str. (a)	β-CH str.
B_2	2935					
A_1	2864	2861			CH str. (s)	α-CH str.
B_1	2861					
A_1	2853	2847		2849	CH str. (s)	β-CH str.
B_1	2852					β-CH₂ bend.
A_1	1491			1487	CH₂ bend.	β-CH₂ bend.
B_1	1490					
B_1	1466	1458	1461	1466		α-CH₂ bend.
A_1	1451			1441		α-CH₂ bend.
B_1	1376	1366	1364	1368	CH₂ wag.	α-CH₂ wag.
A_1	1352		1333	1339		α-CH₂ wag.
B_2	1318		1289	1323, 1307	CH₂ wag.	α-CH₂ twist.
A_2	1286					α-CH₂ twist.
B_1	1251	1238	1234	1241	CH₂ wag.	CH₂ wag.
B_2	1208					β-CH₂ twist.
B_1	1196	1177	1177	1179	Ring str.	CO str. (a)
A_1	1195					β-CH₂ wag.
A_2	1169					β-CH₂ twist.
A_1	1093	1076	1067	1058	Ring str.	Ring str.
A_2	1046		1030	1043	CH₂ rocking	β-CH₂ rocking
B_2	966			980		α-CH₂ rocking
B_1	957			954	CH₂ rocking	CO str. (a)

9

Table 1.3. (continued)

Symmetry Species	ν_{calc}	ν_{obs} Vapor	ν_{obs} Liquid	ν_{obs} Solid	Assignments[a] Ref. 21	Assignments[a] Ref. 19
A_1	954			921		CO str. (s)
A_1	904	912	908	908		Ring breathing
A_2	885			981, 871		α-CH$_2$ str.
B_2	827	821		838		β-CH$_2$ rocking
A_1	660	654	654	662		CH$_2$ rocking
B_1	604				In-plane ring bending	In-plane ring bending
A_2	277				Out-of-plane ring bending	Out-of-plane ring bending
B_2	214				Out-of-plane ring bending	Out-of-plane ring bending

[a] Abbreviations: str., stretching; (s), symmetrical; (a), asymmetrical; wag., wagging.

Table 1.4. Vibrational Assignments of Five-Membered Aromatic Rings

Vibration No. (32)	Furan IR, Raman cm^{-1}	Furan PES (31)* IP, eV cm^{-1}	Thiophene IR, Raman cm^{-1}	Thiophene PES (31)* IP, eV cm^{-1}	Pyrrole IR, Raman cm^{-1}	Pyrrole PES (31)* IP, eV cm^{-1}	Assignment
Symmetry Species A_1							
1	724 (27) 720 (28) 839 (29)	8.9(1a$_2$) 839 10.3(2b$_1$) 871	604 (30)	8.9(1a$_2$) 645	711 (35) 708 (29) 865 (31)	8.2(1a$_2$) 871 9.2(2b$_1$)	β_{ring} distortions
2	1061 (27) 1067 (28) 1138 (29)		1032 (30)		1072 (35) 1076 1235 (29) 1235 (31)		β_{CH}
3	986 (27) 995 (28)	8.9(1a$_2$) 952 10.3(2b$_1$) 1024	832 (31)		1144 (35) 1144 (29, 31)		Ring breathing
4	1140 (28) 1061 (27) 1061 (29)	8.9(1a$_2$) 1073	1079 (30) 1079 (31)	8.9(1a$_2$) 1137	1237 (35) 1072 (29)	8.2(1a$_2$)	β_{CH}
5	1380 (27) 1384 (28) 1380 (29)	10.3(2b$_1$) 1355	1358 (30) 1358 (31)		1384 (35) 1379 (29)	8.2(1a$_2$) 1371	ν_{ring}
6	1483 (27) 1490 (28) 1483 (29)	8.9(1a$_2$) 1420	1404 (30) 1404 (31)	8.9(1a$_2$) 1395	1467 (35) 1469 (29)	8.2(1a$_2$) 1468	ν_{ring}
7	3089 (27) 3124 (28) 3089 (29)		3084 (31) 2996 (30)		3103 3045 (28) 3100 (35)		ν_{CH}
8	3121 (27) 3124 (28) 3154 (29)	17.4(7a$_1$) 2900	3093 (30) 3108 (31)		3133 (35) 3138 (29)		ν_{CH}
9	—		—		3400 (35) 3383 (29)		ν_{NH}

Table 1.4. (continued)

Vibration No. (32)	Furan IR, Raman cm^{-1}	Furan PES (31)* IP, eV cm^{-1}	Thiophene IR, Raman cm^{-1}	Thiophene PES (31)* IP, eV cm^{-1}	Pyrrole IR, Raman cm^{-1}	Pyrrole PES (31)* IP, eV cm^{-1}	Assignment
Symmetry Species B$_1$							
10	874 (28)		872 (30)		867 (50, 51)		β_{ring}
11	1268 (28)		1290 (30)		1046 (35)		β_{CH}
12	1181 (28)		909 (30)		1015 (35)		β_{CH}
13	—		—		1074 (35)		β_{NH}
14	1460 (28)		1252 (30)		1418 (35)		ν_{ring}
15	1586 (28)		1590 (30)		1530 (35)		ν_{ring}
16	(3163) (28)		2996 (30)		3111 (35)		ν_{CH}
17	3163 (28)		3093 (30)		3133 (35)		ν_{CH}
Symmetry Species A$_2$							
18	550 (28)		565 (30)		618 (34)		γ_{ring}
19	660 (28)		686 (30)		711 (35)		γ_{CH}
20	1030 (28)		748 (30)		652 (50, 51)		γ_{CH}
Symmetry Species B$_2$							
21	—		—		565 (35)		γ_{NH}
22	744 (28)		710 (30)		735 (50, 51)		γ_{CH}
23	601 (28)		453 (30)		649 (34)		γ_{ring}
24	838 (28)		832 (30)		838 (34)		γ_{CH}

* IP – ionisation potential.

heterocycles were carried out on furan (35–39), thiophene (27, 32, 35, 38, 40–47), and pyrrole (35, 48–52).

In the IR spectra of the neat liquid, fewer bands are observed than the expected number of vibrations of the five-membered ring molecules 8. In the opinion of some authors (53), these molecules (especially pyrrole) have a shape closer to the higher symmetry point group D_{5h} than to C_{2v}. Thus some modes that are active under C_{2v} symmetry have low intensity and are not observed.

X = O, S, N, Se, Te

8

The agreement between different qualitative descriptions of some A_1 vibrations is poor, and in this situation the information derived from photoelectron spectroscopy (PES) is extremely valuable.

It has been shown (31) that photoelectron spectroscopy is a new method that is of considerable value in the interpretation of Raman and infrared spectra of molecules possessing aromatic character. Analysis of the vibrational structures of photoelectron bands provides the energy values of the vibrations in different ionic states, and Asbrink and coworkers (54) have observed that the energies of totally symmetric vibrations (A_1) in the neutral molecules closely resemble their energies in the ionic state. On the basis of photoelectron-spectroscopical results, new assignments are made for some vibrations, and uncertainties concerning the others are largely eliminated.

Table 1.4 presents the comparative Raman, infrared, and photoelectron spectral data for furan, thiophene, and pyrrole molecules. The discussion is limited to the A_1 vibrations, since only they are expected to be observed in the photoelectron spectra.

Photoelectron identifications indicated that the generally accepted assignments were partly incorrect. The information derived from photoelectron spectroscopy is extremely valuable when the depolarization factors, intensities, and linewidths of Raman bands are comparable and thus it is not possible to decide on this basis which of these lines represent A_1 vibrations. The photoelectronic data confirmed the correct assignment of the 839-cm^{-1} Raman line as an A_1 vibration of furan, contrary to the assignment of Thompson and Temple (27). The PES assignment is in agreement with the Raman data of Sidorov and Kalashnikova (29). These authors made careful measurements of intensities, depolarization factors, and widths of the Raman lines and found comparable intensities and linewidths for both bands at 724 and 839 cm^{-1}, so that is was not possible to decide from the Raman data alone which line represents A_1 vibration. Thompson and Temple (27) and Bak and coworkers (28) assigned the 839-cm^{-1} line to a B_2 vibration, but there is not sufficient evidence for this. Concerning the two CH bending vibrations (ν_2 and ν_4) it was not clear from earlier results which of the two lines (1061 and 1138 cm^{-1}) is ν_2 and which is ν_4. On the basis of photoelectron spectral results that indicated a correspondence between 1073 cm^{-1} and ν_4, the Raman line of

Table 1.5. Ring Vibrations of Five-Membered Compounds (cm^{-1})

Approximate Description	Thiophene	Selenophene	Tellurophene
ν_6 Ring stretching	1408	1419	1432
ν_5 Ring stretching	1360	1341	1316
ν_3 Ring stretching	833	758	687
ν_1 Ring in-plane deformation	606	456	380
ν_{18} Ring out-of-plane deformation	565	541	507
ν_{10} Ring (deformation + stretching	871	820	797
ν_{20} Ring in-plane deformation	750	623	552
ν_{23} Ring out-of-plane deformation	453	394	354

furan at 1061 cm^{-1} was assigned to ν_4. Other implications concerning the interpretation of the Raman and infrared spectra are mentioned in the last column of Table 1.4.

A complete theory of the connection between the molecular orbital from which the electron has been ionized and the vibrational structure of the bands has not yet been developed, but the photoelectron identification of symmetric vibrations assists considerably in the interpretation of the Raman and infrared spectra.

A complete analysis of the observed overtone and combination frequencies was carried out (55) by means of a computer program. The program permitted calculation of all binary combinations and corresponding symmetry species from the proposed fundamentals, or the calculation of fundamentals and their symmetry from the given combinations and assumed fundamentals.

It is of interest to compare the frequencies of ring vibrations in thiophene with those of other five-membered heterocyclic compounds like selenophene and tellurophene (56) in order to rationalize their vibrational behavior. The assignment proposed is based on experimental and theoretical criteria (such as the intensity and polarization of the Raman lines, the contours and P–R separations in gas-phase IR spectra) and is supported by the good correspondence found for the isotopic product and sum rules. The frequencies of the ring vibrations for the three compounds are compared in Table 1.5 (the description of the normal vibrations are those from ref. 48. The proposed vibrational assignments fit well those of Aleksanyan and coworkers (46). Except for the ν_5 ring stretching mode, ring vibrations are shifted to lower values (in the order thiophene > selenophene > tellurophene), and this is easily understood on the basis of the mass and geometric effects. Concerning the ν_5 mode (symmetric stretching of the double bond), Palianti and coworkers (56) explained the increase in its frequency by the decreasing aromaticity in the above sequence due to a greater localization of the double bonds. In this context the higher value of this vibration in furan (1483 cm^{-1}) is easily understandable.

Andrieu and coworkers (57) have studied the intensity and multiplicity of the $\nu_{C=O}$ and $\nu_{C=C}$ bonds in 2-formyl- and 2-acetyltellurophene. The higher frequency and intensity of the $\nu_{C=O}$ band in the formyl, compared with the acetyl, derivatives as in the corresponding congener derivatives, has been related to inductive and steric effects. It is hypothesized that the doublet in the region of the $\nu_{C=O}$ vibration is due to a conformational equilibrium of (Te, O)-*cis* and (Te, O)-*trans* forms, but no evidence is given to exclude Fermi resonance. The literature up to 1975 has been covered in a review on tellurophene by Fringuelli and coworkers (58).

1.3.2.2. Pyrrole Derivatives

The molecules containing five-membered rings may act as guest molecules (G) in Hofmann-type clathrates of the general formula M(NH$_3$)$_2$Ni(CN)$_4$. 2G; M = Mn, Fe, Co, Ni, Cu, Zn; G = benzene, thiophene, furan, pyrrole. The analysis of the vibrational spectra of the clathrates (53) indicated the presence of hydrogen bonding between the π electrons of the aromatic guest molecules and the host lattice ammonia. This interaction in the clathrates induces a shift (of about 20 cm^{-1} to higher frequency) of an out-of-plane hydrogen deformation mode, and simultaneously a general reduction in the intensity of the bands compared to the neat liquid or solid. This was explained in terms of changes in the π-electron density in the ring of the aromatic guest molecules. The IR spectra of the pyrrole clathrates indicated a weak hydrogen bond between the pyrrole N–H group and the host lattice.

Infrared spectral data on some pyrrole derivatives are given in references 59 and 60.

1.3.2.2.1. Indole and Indolizine. Lautié and coworkers (61) have investigated recently the IR and Raman spectra of liquid and crystalline indole, indolizine, and deuterated derivatives. These authors gave evidence for the autoassociation of indole in chains by NH···π-hydrogen bonding characterized by a relative NH stretching frequency shift of 3.5%. The ν_{NH} frequency has been reported to be 3523 cm^{-1} in the gas phase, 3493 cm^{-1} in CCl$_4$ solution, 3418 in neat liquid, and 3399 cm^{-1} in the crystalline state. The assignments given for the fundamental vibrations of the indole molecule are summarized in Table 1.6.

1.3.2.2.2. Isoindoles. IR data for isoindoles have been reported (62): a strong NH band in 3400–3500 cm^{-1} region and one or two bands in the 1600 cm^{-1} region (63, 64).

1.3.2.2.3. Isatin and Derivatives. Infrared vibrational frequency correlations of isatin **9** and substituted isatins have been reported (65, 66). (See also Sec. 1.3.3.1.2.) An infrared study of hydrogen bonding in isatin-3-oximes has been also reported (67). Infrared data of some aminochromes **10** have been reported by

Table 1.6. Vibrational Assignment (61) for Indole

Frequency, cm^{-1}	Assignment
1616	$R'\phi(8a)$
1576	$R'\phi(8b)$
1509	$R'\pi(15)$
1487	$R'\phi(19a)$
1458	$R'\phi(19b)$
1412	$R'\pi(6)$
1352	$R'\pi(14)$
1334	$R'\phi(14)$
1276	$R'\pi(5)$
1245	$\delta_{CH}\phi(3)$
1203	$R'\pi$
1191	$\delta_{CH}\phi(9a)$
1147	δ_{NH}
1119	$\delta_{CH}\phi(9b)$
1092	$\delta_{CH}\pi$ en 2
1064	$\delta_{CH}\pi$ en 3
1010	$\delta_{CH}\phi(18b)$
970	$\gamma_{CH}\phi$
930	$\gamma_{CH}\phi(17b)$
895	$R'\pi(d')$
873	$\gamma_{CH}\phi(10b)$
848	$\gamma_{CH}\pi$ en 2
767	$R'\pi$
758	$R'\phi(d)$ breathing
743	$\gamma_{CH}\phi(11)$
725	$\gamma_{CH}\pi$ en 3
608	R''
607	$R'\phi$
575	$R'\phi$
542	$R'\phi$
487	γ_{NH}
423	$R''\phi(16b)$
397	$R''\phi(18)$
254	R''
224	R''

Heacock and Scott (68) evidencing the complex absorption pattern in the "carbonyl" region. In this range a strong band at 1550–1600 cm^{-1} (ionized carbonyl) has been observed.

For isatogens 11 and indolones 12, carbonyl bands in the region 1700–1720 cm^{-1} have been evidenced (69). The $\nu_{N^+-O^-}$ stretching vibration of 11 was identified at 1175 cm^{-1}.

1.3.2.2.4. Isophosphindoles. For 1,2-dihydro-2-hydroxyisophosphindole-2-oxide **13**, $\nu_{P=O}$ at 1240 cm^{-1} and ν_{P-OH} at 975 cm^{-1} have been reported (70), whereas for isophosphindoline **14** a weak band at 2270 cm^{-1} has been evidenced and assigned to ν_{P-H}.

1.3.2.3. Thiophene Derivatives

1.3.2.3.1. Substituted Thiophenes. A comparison was made between the electronic effect of bromine and chlorine substitution on vibrational modes of the thiophene nucleus. The assignment of the fundamental frequencies of tetrachloro- and tetrabromothiophene based on the polarization of the Raman lines was assisted by assignments of the thiophene vibrational mode (47). Ring modes consist of four stretching bands of symmetry $3a_1 + b_2$, three in-plane deformations classified as $a_1 + 2b_2$, and two out-of-plane deformations each of a_2 and b_1 species. All should be active in the IR and Raman except the a_2 mode, which is inactive in the IR.

The frequencies of ring vibrations in thiophene (47), tetrachloro-, and tetrabromothiophene (71, 72) are compared in Table 1.7. The ring vibrations occur at higher frequencies in tetrachlorothiophene than in thiophene. In tetrabromothiophene the vibrations were more difficult to locate, owing to the instability of the carbon-bromine bond in the laser beam of the Raman spectrometer. It was established, however, that the deviation is less than in tetrachlorothiophene. This may be attributed to the difference in the electronegativity of the halogens showing that the resonance interaction between bromine and the thiophene nucleus is not as strong as was observed in tetrachlorothiophene.

In dichloro and dibromo derivatives of 2,5-dimethylthiophene the ring vibrations were observed (73) at lower wavenumbers than in the corresponding tetrahalo derivatives.

Vibrational analysis of four 2-fluorothiophene derivatives was performed by Rogstad (74). The fundamental modes were assigned on the basis of IR gas-phase contours, Raman polarization measurements, force constant calculations, and similarities between the spectra of these molecules and other substituted thiophenes. The developed force fields were used to calculate the mean amplitudes of vibrations (u). The u values 0.050 Å (C=C) and 0.052 Å (C-C) are slightly higher than the

Table 1.7. Ring Vibrations in Thiophene Derivatives

Symmetry Species	Vibration No.[a]	Thiophene (47)	Tetrachloro-thiophene (71)	Tetrabromo-thiophene (72)	Normal Mode Description
A_1	6	1409	1449	1410	⎫ Ring stretching
	5	1360	1436	1280	⎬
	3	834	1309	1275	⎭
	1	608	558	572	Ring in-plane deformation
A_2	18	568	635	528	⎫ Ring-out-of-plane deformation
B_1	23	452	508	450	⎭
B_2	15	1504	1518	1494	⎫ Ring stretching
	10	870	854	865	⎭
	20	748	711	740	Ring in-plane deformation

[a] The number and types of normal vibrations are those of reference 32.

benzene value (0.046 Å). The ring stretchings in 4-chloro-2-fluorothiophene are situated at 1558, 1463, 1353, and 885 cm^{-1}.

When the heterocyclic and benzenic systems coexist in the same molecules, for example in phenyl 2-thienyl ketone (75), the ring-stretching vibrations are shifted to higher frequencies in comparison with the corresponding frequencies in thiophene (1515, 1415, and 1355 compared with 1409, 1360, and 834 cm^{-1}).

Electronic spectra and infrared characteristics of *trans*-2-styrylthiophene and nitro derivatives have been studied (76). In *trans*-2-styrylthiophene, the 1430 cm^{-1} band of small or medium intensity was assigned to thiophene in-plane vibrations. Infrared data for 2-thienylmethacrylates have been reported by Hawkins (77).

In the last few years IR spectroscopy of 2-thienyl- and 2-furylcarbonyl compounds has provided information on the conjugative effect of the thiophene ring (78-80). The carbonyl stretching frequencies of diaryl ketones (Ar–CO–Ar' in which Ar and Ar' are furyl, thienyl, or phenyl) show a regular shift depending on the groups attached to the carbonyl group. The lower $\nu_{C=O}$ values of the 2-thienyl compounds in comparison with the corresponding 2-furyl derivatives are in agreement with the greater conjugative effect of the thiophene ring. The IR data suggest that the sequence of the conjugative effect is thiophene > furan > benzene.

A new, powerful method for vibrational assignment is the comparison between the spectra of the neutral molecules and the corresponding anion radicals. Recently, the IR spectra of anion radicals of tetracyanothiophene were recorded at −78°C, and the large difference observed between the neutral molecule and the anion radicals was interpreted by the change in the electronic structure (81, 82). The corresponding anion radicals were prepared by evaporating tetracyanothiophene on alkali halide films at −78°C and 10^{-6} torr in a low-temperature IR cell. The results showed that almost all the bands due to the stretching vibrations of the single bonds are shifted to higher frequencies, and those of the double and triple bonds to lower frequencies, in going from the neutral molecule to the anion radical. For example, the ν_{15}, ring-stretching vibration, 1512 cm^{-1} in the neutral molecule (TCNT), is shifted to 1450 cm^{-1} in TCNT$^-$Na$^+$, whereas ν_{18} is shifted from 770 to 802 cm^{-1}.

1.3.2.3.2. Thienothiophenes. The complete vibrational analysis of thienothiophenes 15 and 16 and of the isotopically substituted compounds has been performed by Kimel'feld and coworkers (83) assuming a planar structure of point group C_{2v} for 15 and C_{2h} symmetry for 16.

IR data on these compounds were reported also by Cozien and Saumagne (84).

15   16

1.3.2.3.3. Benzo[b]thiophene and Derivatives. Vibrational assignments of the frequencies of benzo[b]thiophene and benzo[b]furan have been proposed by Mille and coworkers (85), and the vibrational analysis of benzo[b]thiophene-2,3-

dione has been reported (86). The infrared spectrum of dibenzothiophene is related to that of tricyclic compounds such as anthracene and thianthrene (87). Zahradnik and Bocek (88) have determined the O=S=O stretching vibrations in dibenzothiophene-5,5-dioxide: 1309 cm^{-1} for $\nu_{asym}(SO_2)$ and two $\nu_{sym}(SO_2)$ bands at 1161 and 1171 cm^{-1}.

Many substituted benzo[b]thiophenes have been studied by Cagniant and coworkers (89–91) in order to correlate band position with the substitution pattern. The tautomeric equilibrium of 2-amino-benzo[b]thiophenes has been studied (92), and the existence of the amino tautomer has been evidenced. It was reported (93, 94) that 2- and 3-hydroxybenzo[b]thiophenes exist solely as the keto tautomers.

The polarized Raman spectra of the single crystals measured in all possible orientations, supplemented by the measured depolarization ratios of Raman lines in solution spectra, were used (95) for vibrational assignments of larger molecules such as dibenzothiophene. Approximate normal coordinate calculations were facilitated by transferring simplified force fields from similar molecules.

Details of the IR spectra of many benzothiophene derivatives are available (96–100).

1.3.2.4. Furan Derivatives

Correlations of substituent effects with positions and intensities of bands of the furan ring have been established (101, 102). The effect of substitution on the vibrational spectrum of the furan ring has been discussed by Volka and coworkers (103) in several 2- and 3-substituted derivatives for which normal coordinate analysis has been performed by these authors. The substituents have been divided into two groups, the first containing monoatomic substituents or substituents with single bonds ("σ substituents"), the second covering substituents involving multiple bonds ("π substituents"). The authors assumed that the spectral differences in the σ series are due primarily to the mass effect. In the π series the bond orders of the furyl system are altered as a result of the charged delocalization, the coupling of the substituent vibrations with those of the furan ring is greater, and the vibration mechanics is changed.

The assignment of the bands in 2- and 3-furonitrile is based (104) especially on the frequency shift accompanying the substitution of ^{14}N by the isotope ^{15}N. In 3-furonitrile the $\nu_{C\equiv N}$ band appears as a doublet owing to a combination tone, and the more intense component is shifted by -38 cm^{-1} in the liquid state on ^{15}N-labeling. Small variations were observed in the positions of the in-plane fundamentals of 2- and 3-furonitrile.

Conformational isomerism in molecules like bifuran, in which two rings are connected by a central bond, was evidenced by means of a vibrational study on different states of aggregation of the compound (105). To obtain conclusive experimental evidence of the conformation of the molecules in the solid state, IR spectra in polarized light have been obtained on oriented polycrystalline samples. Two different crystal faces were formed by rapid cooling of the liquid film between potassium bromide windows (face I) and by successive melting and cooling of the same sample (face II). The existence of a *trans* molecular configuration (**A**) in the crystal was established on the basis of a comparison of IR and Raman bands; the lack of coincidence between the bands indicates a center of symmetry, and this is possible only for a *trans* conformer. The same comparison indicated a loss of the center of symmetry in the liquid state and the presence of different configurations belonging to C_{2v} or a C_2 point group.

Infrared spectra of 3-furaldehyde and its deuterated analog were measured in the gaseous and liquid states (106). The fundamentals were identified by employing normal coordinate calculations, and the results obtained were compared with those of 2-furaldehyde.

An assignment of the absorption bands to vibrational modes of 2-nitrofuran was suggested (107), and a strong coupling of the vibrational motions of the individual bonds was evidenced by calculation of the vibrational potential energy distribution. The skeletal vibrations of 2-substituted furans (furan-2-carboxylates) are sensitive to conformational properties of these compounds (108). The results confirm that the investigated compounds exist as mixtures of two conformers, the *s-cis* and *s-trans* forms. From the investigated skeletal vibrations the most sensitive is ν_7 in the notation of Sénéchal and Saumagne (102). The difference between the two conformers (**B** and **C**) is 20 cm^{-1} for furfural. For the other ring vibrations ($\nu_5-\nu_{12}$) the splitting is no more than 10 cm^{-1} in CCl$_4$ solution. The difference between the relative enthalpy of the two conformers was studied following the effect of solvent polarity on their ratios. The results indicated the presence of both conformers in approximately equal amounts of polar media, while earlier results (109) concerning rotational isomerism of furfural had shown that the *s-trans* isomer (**B**) becomes favored. This topic was reviewed recently (109a). Results of NMR spectroscopy (109a, b) complement those of IR studies.

A **B** **C**

The 2-carboxylic acids of various five-membered rings were studied in $\nu_{C=O}$ region, in order to evidence the influence of the heteroatom on the carbonyl stretching frequency of monomers and hydrogen-bonded dimers, respectively (110). The results are summarized in Table 1.8. The results show the decrease of $\nu_{C=O}$ in the order $O > S > Se > Te$, the effect being more pronounced for the monomer carbonyl.

The IR spectra of 2-furaldehydes bearing substituted phenyl groups showed a linear correlation between the CO stretching frequency and the σ-substituent constants (111). The transmissive factor of substituent effects for the furan ring calculated from these correlations is 0.65 ± 0.12 in CCl$_4$ and 0.48 ± 0.06 in CHCl$_3$.

22 Infrared Spectra of Heterocyclic Compounds

Table 1.8. Values of $\nu_{C=O}$ in 2-Carboxylic Acids of Various Five-Membered Rings

	2-Carboxylic Acid (a)			Benzo-fused 2-Carboxylic Acid (b)		
X	pK_a	Monomer, cm^{-1}	Dimer, cm^{-1}	pK_a	Monomer, cm^{-1}	Dimer, cm^{-1}
O	4.54	1755	1700	4.40	1758	1704
S	5.05	1734	1682	4.67	1733	1685
Se	5.14	1728	1680	4.79	1731	1679
Te	5.48	1721	1673	5.13	1716	1671

1.3.3. Rings with Carbonyl Groups

1.3.3.1. Imides and Anhydrides

1.3.3.1.1. Maleimide. 17 and its N-deuterated counterpart have been the subject of a detailed study by Woldbaek and coworkers (112). Some of their assignments have been contradicted by two other studies (113, 114), especially concerning the NH(D) bending modes. Recently, Barnes and coworkers (115) have used matrix isolation vibrational spectroscopy to distinguish the monomer and dimer bands by varying the solute concentration. The vibrational assignments obtained from matrix spectra are summarized in Table 1.9 together with the vapor-phase and low-temperature solid-phase spectra.

17

Doping studies were also carried out with water and with hydrogen chloride to investigate the structure of hydrogen-bonded complexes (115). The data obtained showed that the dopant interacts with maleimide through one of the carbonyl groups. The structures of maleimide dimer, maleimide–hydrogen chloride complex, and maleimide-water complex are indicated by formulas **18, 19,** and **20,** respectively.

18   19   20

Table 1.9. Comparison of IR Spectra of Maleimide in the Vapor Phase, in Argon Matrices, and as a Solid[a]

Assignment	Symmetry Species	Vapor	Monomer[b] Ar (20 K)	Dimer Ar (20 K)	Solid (20 K)
ν_1 NH str.	A_1	3482 (s)	3486 (s)	3255 (sh) 3229 3212 (sh)	3219 (sh) 3169 (s)
ν_{13} CH str.	B_1	—	—	3110	3115 (sh)
ν_2 CH str.	A_1	3090 (vw)	—	3092	3100 (m)
$\nu_{14} + \nu_{15}$	A_1	—	—	3087	3076 (m)
$\nu_{10} + \nu_{18}$	B_2	1865 (w)	1870 (vw)	1867	—
$\nu_{16} + \nu_{20}$	A_1	—	1816 (w)	1816	1840 (w)
$2\nu_7$	A_1	1790 (sh)	1801 (ms)	1801	1804 (w)
ν_3 C=O str.	A_1	1775 (sh)	1770 (s)	1770	1775 (m)
$\nu_8 + \nu_{17}$	B_1	—	1762 (m)	—	—
$\nu_9 + \nu_{15}$	B_1	—	1757 (s)	—	1765 (vw)
ν_{14} C=O str.	B_1	1756 (vs)	1744 (vs)	1735 1730	1705 (vs, br)
ν_4 C=C str.	A_1	1580 (w)	1574[c]	—	1583 (w)
$\nu_{11} + \nu_{22}$	B_1	1396 (m)	1393 (w)	—	—
ν_{15} NH bend/CN str.	B_1	—	1346 (s)	1360 (br)	1371 (m, br) 1360
ν_5 CN str.	A_1	1330 (s)	1330 (s)	1333	1344 (s)
ν_{16} CH bend	B_1	1285 (w)	1285 (mw)	1285	1303 (m, w)
$2\nu_8$	A_1	1265 (w)	1262 (mw)	1262	1295 (w)
$\nu_{12} + \nu_{21}$	B_1	—	—	1155 (sh)	1160 (m) 1153 (m)
ν_{17} CN bend	B_1	1130 (s)	1132 (ms)	1149	1143 (s)
ν_6 CH bend	A_1	—	1123 (w) 1119 (mw)	1065	1065 (w)

23

Table 1.9. (continued)

Assignment	Symmetry Species	Vapor	Monomer[b] Ar (20 K)	Dimer Ar (20 K)	Solid (20 K)
$2\nu_{23}$	A_1	1015 (m)	1008 (m)	—	—
			1001 (w)		
ν_{10} CH bend	A_2	—	—	—	970[d]
ν_{18} CC str.	B_1	906 (s)	905 (ms)	925	948 (m)
					943 (m)
ν_7 CC str.	A_1	—	895 (w)	895	902 (vw)
ν_{21} CH bend	B_2	831 (s)	829 (s)	844	852 (s, br)
ν_{11} CO bend	A_2	—	—	—	775 (vw)
ν_{19} ring bend	B_1	668 (s)	670 (m)	678	683 (s)
ν_8 ring bend	A_1	637 (vw)	633	642	647 (w)
ν_{22} C=O bend	B_2	620 (m)	624 (w)	624	630 (w)
ν_{20} C=O bend	B_1	514 (sh)	536 (w)	542	550 (vw)
ν_{23} NH bend	B_2	504 (s)	510 (m)	751	744 (ms, br)
			505 (s)	729	736
				715	
ν_9 C=O bend	A_1	—	410 (m)	418	418 (w)
ν_{12} ring bend	A_2	—	—	—	291[d]
ν_{24} ring bend	B_2	—	—	—	172[d]

[a] Abbreviations: vs, very strong; s, strong; ms, medium strong; m, medium; mw, medium weak; w, weak; vw, very weak; br, broad; sh, shoulder.
[b] References 112–114.
[c] Frequency taken from Raman spectrum (very weak in IR).
[d] Not observed in this report (115).

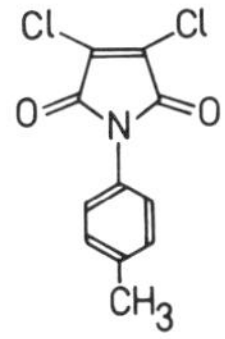

Five-Membered Rings with One Heteroatom 25

21

N-*p*-Tolyldichloromaleimide **21** exhibits a strong $\nu_{C=C}$ band at 1720 cm^{-1} (KBr) (116).

1.3.3.1.2. Phthalimide. Previous vibrational assignments reported in the literature on phthalimides were concerned mainly with the behavior of the NH and CO stretching bands (117, 118). Recently, polarized IR spectra of oriented single crystals of phthalimide **22** and isatin **9** have been also obtained (119). (See also Sec. 1.3.3.1.4.) On the basis of dichroism measurements on single crystals and Raman spectra of polycrystalline samples, an almost complete and reliable assignment of the vibrational modes of these two compounds has been proposed. The crystal spectra were interpreted on the basis of free molecules having C_{2v} symmetry. The 3187-cm^{-1} band was assigned to the ν_{NH} mode. This study evidenced the complex structure of this band and its shift of the absorption maximum to lower wavenumbers on cooling. On deuteration the band shifts to 2436 cm^{-1} in **22**, whereas in **9** the shift is larger (3188 to 2370 cm^{-1}). The four C=O oscillators of two isatin molecules arranged in a centrosymmetric structure in the unit cell of the single crystal give a strong Raman band at 1724 cm^{-1} and an IR absorption at 1750 cm^{-1}.

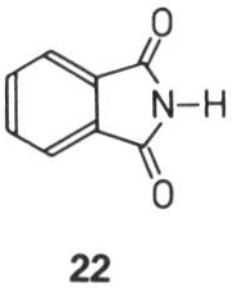

22

1.3.3.1.3. Thiophthalic Anhydride and Thionaphthenequinone. Bigotto and and Galasso (86) have performed a vibrational analysis of thiophthalic anhydride **23** and thionaphthenequinone **24**, for which only limited assignments of carbonyl stretching frequencies were reported previously (120). The isolated thiophthalic anhydride molecule was treated as having C_{2v} symmetry because the deviation from planarity is slight. The region around 1700 cm^{-1} appears complex in IR because of the presence of combination bands enhanced by resonance of the C=O stretching fundamentals. The symmetric C=O stretching mode was assigned to the strongly polarized Raman line at 1742 cm^{-1} with a weak counterpart in the IR spectrum.

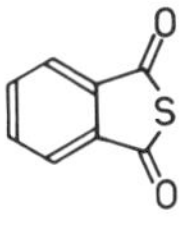

23 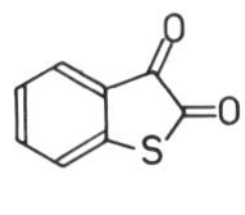   **24**

1.3.3.1.4. Phthalic Anhydride and Derivatives.

The carbonyl stretching vibrations of the O=C–X–C=O resonating system in several derivatives of phthalic anhydride **25** and phthalimide **22** have been discussed (121, 122). The Fermi resonance-corrected fundamental frequencies of the C=O stretching vibrations ν_1 (in-phase stretching) and ν_{27} (out-of-phase stretching) of $C_6H_4(CO)_2X$, $C_6Cl_4(CO)_2X$,

25

and $C_6Br_4(CO)_2X$ are summarized in Table 1.10 with C=O bond force constants $K_{C=O}$ obtained from the normal coordinate analysis. These remarkably large frequency separations between ν_1 and ν_{27} cannot be explained in terms of the mechanical interaction between two carbonyl bonds, since the two C=O bonds in the O–C–X–C=O system are not connected directly. The frequency separations were reproduced by using the interaction force constants to express the electronic effects arising from resonance stabilization by ionic structures. The effective

Table 1.10. Carbonyl Stretching Frequencies of

			$\nu_{C=O}$			
Y	X	Compound	ν_1	ν_{27}	$\Delta\nu$	$K_{C=O}$
H	O	Phthalic anhydride	1849	1763	86	11.10
H	NH	Phthalimide	1766	1736	30	9.85
H	ND	Phthalimide-d	1765	1717	48	
H	N⁻	K-Phthalimide	1706	1587	119	8.87
Cl	O	Tetrachlorophthalic anhydride	1845	1775	70	11.18
Cl	NH	Tetrachlorophthalimide	1774	1706	68	9.76
Cl	ND	Tetrachlorophthalimide-d	1773	1701	72	
Cl	N⁻	K-Tetrachlorophthalimide	1719	1607	112	8.76
Br	O	Tetrabromophthalic anhydride	1866	1773	93	11.12
Br	NH	Tetrabromophthalimide	1770	1728	42	9.71
Br	ND	Tetrabromophthalimide-d	1769	1731	38	
Br	N⁻	K-Tetrabromophthalimide	1711	1615	96	8.84
I	O	Tetraiodophthalic anhydride	1809	1787	22	
I	N⁻	K-Tetraiodophthalimide	1698	1601	97	
I	NH	Tetraiodophthalimide	1763	1719	44	
I	ND	Tetraiodophthalimide-d	1772	1719	53	

negative charges on the O=C–X–C=O skeleton increase in the order O=C–N̈–C=O > O=C–NH–C=O ≈ O=C–ND–C=O > O=C–O–C=O; in the same order the contribution of the ionic resonance structures to the electronic structure of the molecule increases and the carbonyl bond order decreases.

It is interesting to note the very large separation between the two carbonyl bands (1715 and 1640 cm^{-1}) in the IR spectrum (KBr pellet) of N-phenyl-3-fluorophthalimide (123).

1.3.3.1.5. Quinolinic Thioanhydride. Recently, Bigotto and Galasso (124) have assigned the fundamentals of quinolinic thioanhydride **26** and N-methylquinolinimine **27**. Infrared and Raman spectra of partly oriented polycrystalline films were obtained by slow cooling of melts between CsI plates. Carbonyl absorption bands in **26** were found at 1752 cm^{-1} (Raman) and 1702 cm^{-1} (IR) and skeletal vibrations at 1305 and 1296 cm^{-1}.

26   **27**

1.3.3.1.6. Thio- and Selenophthalides. These have been stuied (125), and the following data were reported.

28 $\nu_{C=N}$ 1685 cm^{-1}

29 $\nu_{C=O}$ 1660 cm^{-1}

30 $\nu_{C=O}$ 1695 cm^{-1}

1.3.3.2. Lactams

1.3.3.2.1. Introduction. The characteristic IR bands in lactams with five-, six-, and seven-membered rings have long been established, (126, 127), and their dependence on ring size was discussed by Hallam and coworkers (128, 129) and Hall and Zbinden (130).

It is well known that in lactams with the general formula **31**, the so-called amide II band appears at higher frequencies for the strained five-membered rings than for the seven- and eight-membered analogs. In benzooxalactams **32** with $n = 5-9$, $\nu_{C=O}$ is shifted from 1707 to 1662 cm^{-1} due to the lower conjugative interaction between the amidic nitrogen and the aromatic ring in the larger rings ($n = 8$ or 9).

(CH₂)ₙ₋₂ structure **31**   benzooxalactam structure **32**

Infrared studies on lactams, as *cis-trans* models for the peptide bond, have been performed (131). The IR spectra of a homologous series of lactams were studied as a function of temperature and concentration in various solvents. Three spectral regions show distinctive features for *cis*- and *trans*-peptide conformation in CCl$_4$: the free NH and associated NH stretching absorptions, the "Amide I" and "Amide II" bands. The bands around 1550 and 1510 cm^{-1} were assigned to the "Amide II" band of the associated NH group and the free NH group in *trans* amides, respectively. The lactams **31** ($n = 9$) were found to be of a critical dividing size for the *cis* and *trans* configurations, because they exist in both configurations. The ν_{NH} vibrations in thio- and selenolactams with $n = 4-12$ were used (129) as IR criteria for *cis* and *trans* conformations of the thio- and selenopeptide links in cyclic systems.

The near-infrared bands (overtone and combination bands) of γ-butyrolactam, δ-valerolactam, and ε-caprolactam have been elucidated (132). The $2\nu_{NH}$ overtone mode and the ν_{NH} + "Amide II", ν_{NH} + "Amide III", and $2\nu_{C=O}$ + "Amide III" combination modes were assigned with the aid of deuteration, methylation, and solvent-shift studies, and account for all amide group absorptions observed in the near-IR spectra of lactams. The calculated and observed combination modes for three *cis*-lactams are compared in Table 1.11.

A comparison of the mechanical anharmonicity constants for the N–H stretching vibration of the *cis*-lactams with secondary amides possessing *trans* configuration demonstrated that the magnitude of these constants is not affected by a change in the configuration of this amide group.

Table 1.11. The Near-Infrared Combination Band ($\nu_{NH} + \nu_{C=O}$) for *cis*-Lactams (132) (cm^{-1})

Lactam	Calculated	Observed	Anharmonicity
γ-Butyrolactam	5160	5155	5
δ-Valerolactam	5095	5075	20
ε-Caprolactam	5105	5090	15

Table 1.12. Infrared Data on Penems (cm^{-1})

Compound	Infrared Data, cm^{-1}	Ref.
33 COOR′ on penem skeleton R = CH$_3$; R = C$_6$H$_5$; R′ = p-NBz	1795, 1718, 1587, 1527, 1351, 1316, 1204 1802, 1718, 1527, 1351, 1307, 1198, 1183, 1093, 1015	134
34 (5R)-Penem-3-carboxylic acid	1802, 1695, 1550	135
35 (5S)-Penem-3-carboxylic acid	1802, 1695, 1550	135
36 Methyl-6(S)-Bromo-(5R)-penicillanate	1795, 1745, 1435, 1415, 1299, 1290–1247, 1212, 1181	136
37 threo-trans-6-(α-Hydroxyethyl)penem-3-carboxylic acid	3480, 1795, 1670, 1545, 1435, 1255, 1160	137

30 Infrared Spectra of Heterocyclic Compounds

1.3.3.2.2. β-Lactams. In penicillins, the β-lactam carbonyl absorbs at 1765 and 1800 cm^{-1} for penicillin-sulfones (133). Table 1.12 presents IR data for some derivatives of penems, a new class of β-lactam antibiotics recently prepared by Woodward and coworkers (134–137).

1.3.3.2.3. Seven-Membered Lactams. Infrared data on seven-membered lactams are given in Table 1.13. Owing to the conjugation effects in benzolactams, the carbonyl frequency is lowered by 5–20 cm^{-1} or more (Table 1.14).

Table 1.13. Infrared Data on Seven-Membered Lactams

Compound	Infrared Data, cm^{-1}	Ref.
38	1650–1655 (C=O) 1200 (C–O–C)	138
39	1650 (C=O) 1200 (C–O–C)	139
40	1710, 1675	140
41	1667–1647	141
42	1675	142
43	1653	143

Table 1.14. Infrared Data on Seven-Membered Benzolactams

Compound		Infrared Data, cm^{-1}	Ref.
44	R = H R = X	1670 1666, 1670	144 144
45	R = X	1650, 1656	144
46	R = H R = CH$_3$	1615 1630	145 145
47	R = H R = CH$_3$	1615 1630	145 145
48		1668, 1648	146

1.3.3.2.4. Eight-Membered Lactams. Infrared data on benzoxazocinones, benzothiazocinones, and benzodiazocinones are given in Table 1.15.

1.3.3.3. Lactones

Some data on five-membered lactones are summarized in Table 1.16.

32 Infrared Spectra of Heterocyclic Compounds

Table 1.15. Infrared Data on Some Benzoxazocinones, Benzothiazocinones, and Benzodiazocinones

Compound	Infrared Data cm^{-1}	Ref.
49	3322 (NH), 1635 (C=O)	147
	3300 (NH), 1655 (C=O)	148
50	3311 (NH), 1658 (C=O)	147
	3290 (NH), 1675 (C=O)	148
51	3294, 3180, 1673 (C=O)	149
52	3370 (NH), 1665 (C=O)	150
53	3200 (NH), 1660 (C=O)	150
54	3200 (NH), 1668 (C=O)	150
55	3175 (NH), 1672 (C=O)	151

Table 1.16. Infrared Data on Five-Membered Lactones

Compound	Infrared Data, cm^{-1}	Ref.
trans-α-(2-Pyrryl-methylidene)-γ-butyrolactone **56**	3330 (NH), 1717 (C=O), 1635 (C=C)	152
56'	1730 (C=O), 1645 (C=C)	
α,β-Dichloro-γ,γ-difluoro-γ-crotonolactone **57**	1821 (C=O), 1629 (C=C)	153
5-Methoxynaphtho[2,1-b]-furan-2-(1H)-one **58**	1810 (C=O)	154
3-(2-Isobutenyl)-5-methoxy-3-phenyl-2-benzofuranone **59**	1800 (C=O; CDCl$_3$)	155
1-Phenyl-2-hydroxymethyl-3-naphthoic acid lactone **60**	1631 (C=O)	155

1.4. FIVE-MEMBERED RINGS WITH TWO OR MORE HETEROATOMS

1.4.1. Rings with Two Heteroatoms with Carbonyl or Thiocarbonyl Groups

1.4.1.1. Ethylene Carbonate, Ethylene Trithiocarbonate, Oxazolidine-2-one, Oxazolidine-2-thione, Ethyleneurea, Imidazoline-2-thione, Thiazolidine-2-one, and Thiazolidine-2-thione

Considerable attention is centered on the study of five-membered heterocyclic molcules owing to their biological and pharmaceutical importance. These compounds are of great interest because they provide potential binding sites for metal ions in physiological systems.

A comparative IR reinvestigation of 1,3-thiazolidine-2-one **61** and -2-thione **62** and their oxazolidine analogs **63** and **64** has been reported (156). Infrared and Raman assignments of thiazolidine on an empirical basis were performed by Guiliano and coworkers (157).

Recently a complete vibrational assignment by normal coordinate treatments for eight five-membered rings **61–68** was reported (158). The vibrational assignments for these compounds are given in Table 1.17.

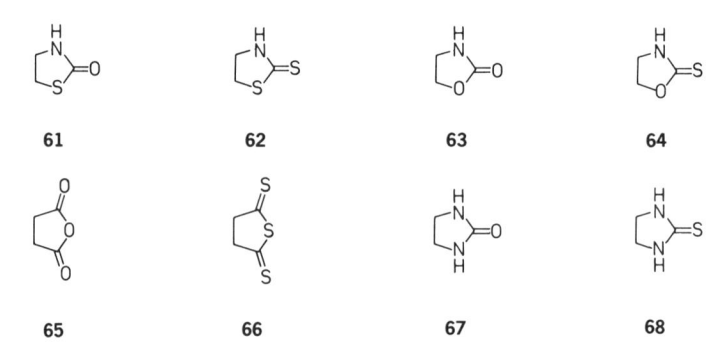

Devillanova and coworkers (159) investigated the IR spectra of **62** and its selenium analog, considering the selenium only as an exocyclic atom and proving that selenation is an effective method for assigning vibrations of the thioketonic group. The ν_{NH} band in the selenium compound was found at 3100 cm^{-1}, i.e., 30 cm^{-1} lower than in the thio analog, and this decrease can be attributed to a larger positive charge on the nitrogen in the former than in the latter according to the smaller force constant of the carbon-selenium bond. The selenium atom instead of the sulfur *in the ring* would produce a higher ring distortion in consequence of its bulk, and one would expect it to influence also the π system of the selenoamido group (160). It was concluded that the selenation in the ring also affects the vibrations related to the common structural elements, modifying the low region of the spectra where ν_{CS} vibrations and ring deformations give high contributions.

Table 1.17. Infrared Assignments for Five-Membered Heterocyclic Compounds[a]

61	62	63	64	65	66	67	68	Assignment
				In-plane a' or A_1, B_1				
3200	3130	3260	3205	—	—	3285	3275	ν_{NH}
							3250	
2865	2845	2980	2950	2925	2965	3000	2900	ν''_{CH}
					2922			
2930	2960	2840	2885	2925	—	2960	2900	ν'_{CH_2}
—	1510	1405	1523	—	—	1450	1528	$\nu_{CN} + \delta_{NH}$
							1508	
1435	1430	1479	1465	1483	1428	1502	1470	δ''_{CH_2}
					1418			
1455	1448	1445	1452	1483	—	1485	1480	δ'_{CH_2}
		1430						
1440	1345	1385	1395	—	—	1423	1376	$\delta_{NH} + \nu_{CN}$
1322	1290	1350	1282	1421	1279	1274	1312	ω_{CH_2}
							1286	
1260	1250	1325	1228	1386	1248	1207	1212	ω_{CH_2}
—	1085	1725	1107	1868	1062	1668	925	$\nu_{C=X}$
1047	1047	1019	1029	1125	—	1105	1050	$\nu_{C'N}$
675	655	1080	1107	1087	674	1041	1008	$\nu_{C''Y}$
853	933	964	902	960	983	933	1120	$\nu_{C''Y}$
555	585	1405	1395	881	—	991	—	ν_{CY}
		918	902					
675	655	620	630	715	—	660	678	RD[b]
	292	580	283	527	248	511	343	$\delta_{C=X}$
476	545	512	513	696	503	—	516	RD[b] + $\nu_{C=X}$
	434				454			

35

Table 1.17. (continued)

61	62	63	64	65	66	67	68	Assignment
				Out-of-plane a'' or A_2, B_2				
2955	2882	2980	2970	3004	2974	2980	2967	$\nu_{C''H_2}$
—	2998	2900	2935	3000	2922	2915	—	$\nu_{C'H_2}$
1226	1203	1228	1198	1218	1279	1207	1312	$t_{C''H_2}$
					1150			
1167	1160	1150	1165	1157	1107	—	1218	$t_{C'H_2}$
980	999	865	965	768	955	—	1108	$\rho_{C''H_2}$
930	850	845	845	660	882	—	925	$\rho_{C'H_2}$
					832			
—	434	766	350	620	478	768	200	$\pi_{C=X}$
832	700	700	690	—	—	703	598	π_{NH}
—	103	204	178	230	234	250	109	τ_{ring}
—	86	85	102	215	92	—	90	τ_{ring}

[a] All data in cm^{-1}. The structure is considered to be:

If Y = NH, C' denotes carbon bonded to nitrogen atom.
[b] RD = ring deformation.

1.4.1.2. Dioxolan-2-ones

Five-membered cyclic carbonates (1,3-dioxolan-2-ones) such as ethylene carbonate **69**, were studied by spectroscopic methods and assignments for IR and Raman bands have been proposed (161). Spectroscopic studies of ring systems have become of interest owing to the pseudorotation exhibited by these systems. The term *pseudorotation* describes the interconversion of conformations belonging to the chair and half-envelope structures, when the maximum out-of-plane amplitude moves around the ring. As a result of this motion the ring appears to rotate, although the actual motion of the ring atoms is perpendicular to the plane of the ring. A slightly distorted half-chair structure was assumed for these molecules.

69

The nine skeletal modes and fundamental vibrational frequencies in 4-methyl-1,3-dioxolan-2-one are:

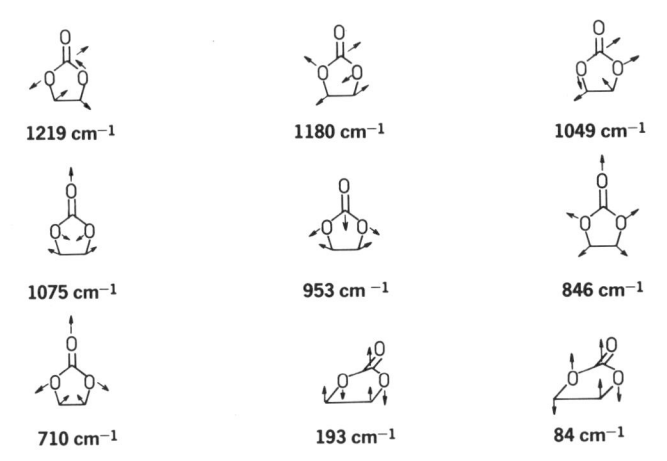

1219 cm⁻¹	1180 cm⁻¹	1049 cm⁻¹
1075 cm⁻¹	953 cm⁻¹	846 cm⁻¹
710 cm⁻¹	193 cm⁻¹	84 cm⁻¹

The high intensity and large width of the low-frequency modes were attributed to the coupling of pseudorotation with the deformation of the C=O bond. It was concluded that the cyclic carbonates exhibit reasonably rigid structures and undergo only limited dynamic changes associated with pseudorotational motion.

The methods of solvent variation and isotopic substitution (^{13}C- and d_4-labeled compounds) have been used (162) to assign the origin of the perturbing vibrational mode with the carbonyl group in the Fermi resonance of various five-membered rings. The nonlabeled and ^{13}C-labeled ethylene carbonates, for example, have a doublet carbonyl absorption (1809 and 1776 cm^{-1} for the nonlabeled and 1784 and 1756 cm^{-1} for the ^{13}CH$_2$–CO derivative in CHCl$_3$ solution), whereas the d_4 compound has only a single absorption. The origin of the perturbing vibrational mode responsible for Fermi resonance was shown to be due to the first overtone of the absorption at 897 cm^{-1}, which was assigned to CH$_2$ rocking. Infrared and acoustic studies in relation to the structure of several 1,3-dioxolan-2-ones have also been performed (163).

1.4.1.3. Methylene Oxalate

For methylene oxalate **70**, an interesting experimental technique was applied to measure the dichroic ratios on oriented films (164). Oriented films of solid methylene oxalate were obtained by allowing a liquid capillary layer to freeze between KBr windows under a temperature gradient. The spectra in polarized light were obtained with a beam condenser and an AgCl polarizer. The molecule belongs to the C_{2v} point group, and all polarized Raman bands correspond to perpendicular bands in the IR spectrum of the oriented solid. The in-phase carbonyl stretching fundamental appears as a strongly polarized Raman band at 1852 cm^{-1}, and the out-of-phase $\nu_{C=O}$ as a nearly depolarized band at 1797 cm^{-1}. Bands at the same frequencies appear also in the IR spectrum. The angle α between the directions of the two carbonyl bonds were determined to be about 62°, with the formula (165)

$$\frac{A_{\text{in-phase}}}{A_{\text{out-of-phase}}} = tg^2\left(\frac{\alpha}{2}\right)$$

in which the A's are the integrated intensity of the carbonyl in-phase and out-of-phase stretching vibrations, respectively.

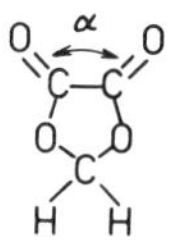

70

1.4.1.4. Rhodanine

Some of the five-membered rings act as ligands in various complexes, and IR spectra allow identification of the atom involved in the coordination. For example, for the rhodanine complexes **71** of zinc(II), cadmium(II), or mercury(II), it was shown (166) that the rhodanine anion has the tendency to act as an S,N-bonding ligand and that the CO group could not be involved in the coordination. Spectral studies on 3-substituted rhodanine and 5-substituted isorhodanines have been performed (167, 168).

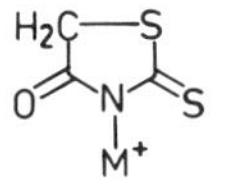

M = Zn(II), Cd(II), Hg(II)

71

1.4.1.5. Imidazoline Derivatives

4,5-Diphenyl-4-imidazoline-2-one **72**, its 2-thioxo analog, and all possible *N*- or *S*-substituted monomethyl- and *N*,*N*- or *N*,*S*-dimethyl derivatives (R^1 and R^2 = Me and H) were found (169) to exist exclusively in the symmetrical tautomeric form **73**.

The nonmethylated parent compound and the monomethyl derivatives form cyclic dimeric associations **74**.

Tautomeric equilibria of 4,4/5,5-disubstituted 4*H*/5*H* imidazoline-5/4-ones in the solid state and in polar solvents were studied by Edward and Lantos (170). The question of the correct tautomeric formula for these compounds has long been confused, chiefly because early evidence was based on the products obtained by methylation. Later workers (171) used more reliable spectroscopic methods and showed that in the crystalline solid the compound **76/77** (R = H) existed as the tautomer **76** and that the compound **76/77** (R = SMe) could be obtained in either tautomeric form, depending on the solvent used for crystallization (172).

The solid spiro compound **78/79** (R = H) appeared to exist as the "conjugated" tautomer **79** (R = H) (173), and the 4,4/5,5-dimethyl compound (**76/77**; R = Me) as the "unconjugated" tautomer (174). From these results it would seem that the energy differences between "conjugated" and "unconjugated" tautomers are small

Table 1.18. Infrared Data of Imidazoline Derivatives

Compound	Infrared Data, cm^{-1}	Ref.
Imidazolin-N-oxide **80**	3200 (ν_{NH}) 1593 ($\nu_{C=N}$) 1183 (ν_{N-O})	175
2-Oxo-Δ^3-imidazolin-3-oxide $R^1 = R^2 = C_2H_5$ **81**	1800 ($\nu_{C=O}$) 1600 ($\nu_{C=N}$) 1237 (ν_{N-O})	176
2,4-Dioxo-3,5,5-trimethylimidazoline **82**	1756 ($\nu_{C=O}$) 1698 ($\nu_{C=O}$)	176
2-Thioxo-5-methyl-4-phenyl-Δ^3-imidazoline-3-oxide $R^1 = CH_3$ $R^2 = C_6H_5$ **83**	3120 (ν_{NH}) 1646 ($\nu_{C=N}$) 1512 (ν_{NH}) 1253 (ν_{N-O}) 1208 ($\nu_{C=S}$)	177
2-Thioxo-5-methyl-4-phenyl-Δ^3-imidazoline **84**	3080 (ν_{NH}) 1640 ($\nu_{C=N}$) 1513 (ν_{NH})	177

and that the particular tautomer found in the solid reflects specific effects of crystal packing and is not necessarily the thermodynamically more stable tautomer in solution. Infrared data of Edward and Lantos indicated a greater stability of "unconjugated" systems **76** and **78** (R = H) in nonpolar solvents. Small amounts of the "conjugated" tautomers **77** and **79** (R = H) in more polar solvents were also evidenced. In compounds with an electron-donating substituent in the 2-position, the "conjugated" tautomer may become predominant in these solvents.

Some imidazoline derivatives have been studied by Gnichtel and coworkers (175–177), and IR data have been reported (Table 1.18).

The tautomeric equilibrium in some pyrazoline-5-ones has been studied by IR methods (178, 179)

1.4.1.6. Hydantoins

Seth Paul and Demoen (180) studied a series of substituted hydantoins **85** by IR spectroscopy and assigned the upper band in the carbonyl region (1790–1730 cm^{-1}) to C^2=O stretching vibration and the lower one (1735–1690 cm^{-1}) to C^4=O. Katritzky (1) questioned these views and explained the two bands as a result of a coupling effect between both vibrations as reported for anhydrides (181) and cyclic imides (182). A similar pattern of the carbonyl stretching region in the spectra of some hydantoin spiro derivatives **86** has been reported (183).

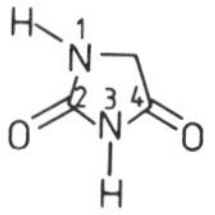

85

Hydantoin compounds with very different substitution patterns exhibit two strong carbonyl absorption bands (184, 185). Very low and broad ν_{NH} bands were found in the solid-state IR spectra due to very strong intermolecular hydrogen bonds between the weakly acidic N^3–H group of the hydantoin ring and the basic piperidine nitrogen atom **86**. A cyclic dimeric structure **87** was assigned (186) to *gem*-diphenyl-substituted hydantoin derivatives.

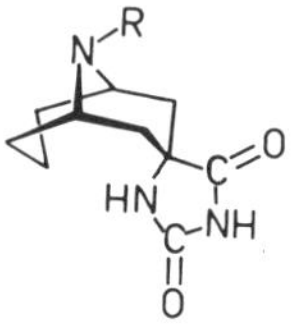

86 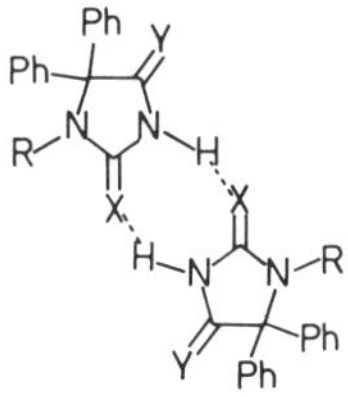   **87**

1.4.1.7. Thioamides

Hydrogen bonding in heterocyclic thioamides **88–90** was examined (187). The three compounds form cyclic dimers via cooperative proton transfer and resonance stabilization. Dimerization constants were determined: $88 \pm 9\ M^{-1}$ **88**; $570 \pm 60\ M^{-1}$ **89**, and $5540 \pm 500\ M^{-1}$ **90**, and the strength of the hydrogen bonds was estimated from the equation

$$\Delta \nu = \nu_{NH\ free} - \nu_{NH\cdots S}$$

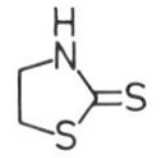

88 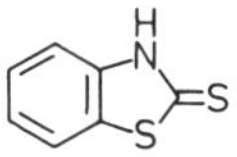   **89** 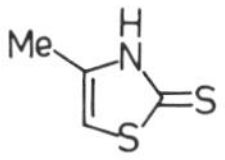   **90**

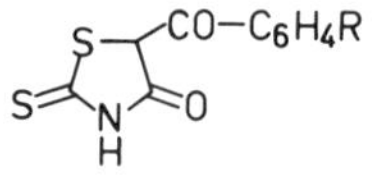

42 Infrared Spectra of Heterocyclic Compounds

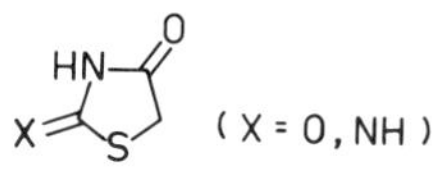

91   **92** (X = O, NH)

The strength of the hydrogen bonding increased in the order **88 < 89 < 90**. The stabilizing effect of intramolecular hydrogen bonding on enolization of **91** was also proved by IR spectroscopy (188).

1.4.1.8. Thiazolidinediones

Infrared spectral data concerning 2,4-thiazolidinedione **92** have been reported (189–191).

1.4.1.9. Saccharin

The infrared spectrum of saccharin (3-oxo-2,3-dihydrobenz[d]isothiazole-1,1-dioxide) **93** exhibits higher $\nu_{asym}SO_2$ than acyclic sulfonamides (192). Self-association and the formation of complexes with various compounds have been studied by IR spectroscopy (193).

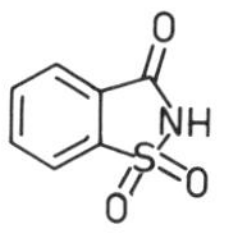

93

1.4.2. Aromatic Rings with Two Heteroatoms without Carbonyl Groups

1.4.2.1. Pyrazole

Infrared spectra of pyrazole **94** (crystalline form and concentrated solution) show a strong $\nu_{C=N}$ band (1592 cm^{-1}) and two weak $\nu_{C=C}$ bands at 1658 and 1552 cm^{-1} (194–196). Infrared data concerning pyrazole derivatives have been reported (197, 198).

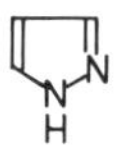

94

Low ν_{NH} frequencies have been found in heterocycles such as **94** (199). On the basis of the structure of the NH stretching band of **94**, it was concluded that the splitting is due to interaction of the stretching vibration with overtones and combinations of the lower-frequency intramolecular vibrations.

1.4.2.2. Imidazole and Derivatives

The spectra of solutions of 4(5)-methylimidazole in carbon tetrachloride showed numerous subbands in the NH stretching range (200). These bands were correlated with the binary combinations of fundamental vibrations between 1000 and 1600 cm^{-1}. The Fermi resonance of $\nu_{NH(D)}$ with overtones and recombinations of internal modes was illustrated recently by ^{15}N substitution in polycrystalline imidazole **95** (201). The observed ν_{NH} ^{15}N shift was from 10 to 15 cm^{-1}, which is more than expected for pure modes.

95

Far-infrared data (202) indicated that the intermolecular hydrogen bond strengths decrease in the order imidazole > pyrazole > pyrrole.

Some new band assignments have been made by matrix isolation of the monomeric species of **95** in an inert gas (203).

In protonated imidazole salts, hydrogen bonding occurs to various degrees depending on the anion (204). Comparison of the spectra of the imidazolium cation and **95** shows that the frequencies of the CH stretching and certain in-plane ring vibrations increase upon protonation. The influence of protonation on various vibrations of **95** is shown in Table 1.19.

The important shifts of skeletal vibrations on protonation have been explained by a strong vibrational coupling with the ν_{NH} mode. In the spectrum of 2-chloro-

Table 1.19. Influence of Protonation on IR Bands of Imidazole (cm^{-1})[a]

		Imidazolium	
Vibration	Imidazole	Cl$^-$ Salt	SnCl$_6^{2-}$ Salt
ν_{CH}	3145	−10	+23
ν_{CH}	3125	0	+15
R$_1$[b]	1541	+37	+39
R$_2$	1485	+50	+45
R$_3$	1448	−6	−8
R$_4$	1324	+84	+98
R$_5$	1142	+63	+48
R$_9$	620	+5	0
$\delta_{NH(ND)}$	1242	−60	−79
δ_{CH}	1262	−2	−17
δ_{CH}	1099	−8	−13
δ_{CH}	1055	−55	−9

[a] Solid state spectra.
[b] Skeletal vibrations.

2-amidazoline sulfate, strong bands have been assigned (205) to vibrations of the protonated nitrogen: 2300 ($\overset{+}{N}H$), 1620 (C=N⁺), and 1580 cm⁻¹ ($\overset{+}{N}H$, NH).

Stable iminoxyl radicals of 3-imidazolines **96** have been prepared and characterized by IR spectra (206). In biimidazole derivatives **97** the ν_{NH} band appears at 3440 cm⁻¹ in CHCl₃ solution (207).

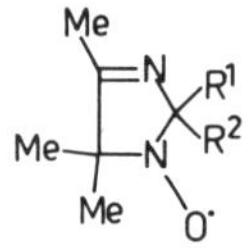

96 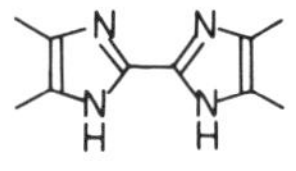   **97**

Infrared absorption data for imidazo[1,2]pyridine **98** and imidazo[1,2-*a*]pyrimidine **99** derivatives are given in Table 1.20 (208–211).

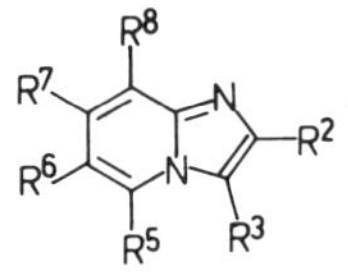

98 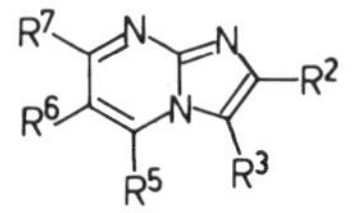   **99**

Table 1.20.   Infrared Data on Imidazo[1,2]pyridines (1-Aza-indolizines) and Imidazo[1,2-*a*]pyrimidines (1,8-Diaza-indolizines) (208)

\multicolumn{2}{c}{Imidazo[1,2]pyridines **98**}	\multicolumn{2}{c}{Imidazo[1,2-*a*]pyrimidines **99**}
cm⁻¹	Assignment	cm⁻¹	Assignment
3133 ± 15	Imidazole band (209)		
3044 ± 6	Pyridine band (210)	3072 ± 12	Imidazole band (209)
1638 ± 17		1619 ± 8	
1567 ± 7	Imidazole and Pyridine bands (209, 210)	1533 ± 5	
1455 ± 5	Imidazole band (209)	1373 ± 3	Pyrimidine band (211)
1391 ± 7		1349 ± 9	
1120 ± 10		1241 ± 3	
860 ± 5		1032 ± 10	Pyrimidine band (211)
783 ± 8	Pyridine band (210)		
756 ± 5	Imidazole and pyridine bands (209, 210)		
725 ± 4		768 ± 28	Imidazole and pyrimidine bands (209, 210)
695 ± 6		703 ± 6	
618 ± 8			
526 ± 6		491 ± 5	

Infrared data were reported for imidazopyrazole (212), benzimidazoles (213, 214), and naphthimidazoles (215).

1.4.2.3. Thiazole, Isothiazole, and Derivatives

Infrared data on thiazoles **100** and thiazolium salts were reported (216–219). Califano and coworkers (220) reported a full assignment of the fundamental vibrations of isothiazole **101**. A valence force field was determined from vibrational spectra of **101** and three deuterated species (221). Force constants have been calculated, and a potential energy distribution analysis has been performed. This analysis showed the complex nature of the vibrations in the 1000–1250 cm^{-1} region.

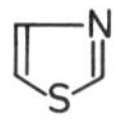

100 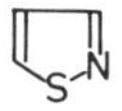   101

Some thiazole derivatives of pharmaceutical importance were prepared by Gudrinietze and coworkers (222–224). Characteristic IR bands for these thiazole derivatives are summarized in Table 1.21.

1.4.2.4. Oxazole and Derivatives

Oxazole **102** belongs to the C_s point group (assuming a planar structure). The vibrational assignment of the 18 normal modes has been made (225, 226).

Table 1.21. Infrared Data on Thiazole Derivatives 101A (222–224)

Compound	Infrared Bands, cm^{-1}
2-Amino-5,5-dimethyl-4,5,6,7-tetrahydrobenzothiazol-7-one 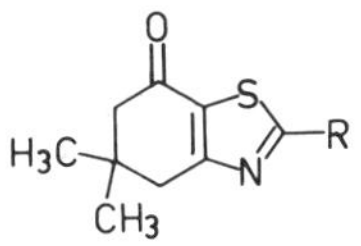 **101 a**	
R = NH$_2$	1606, 1511 (thiazole ring)
R = NH–CH$_2$CH$_2$CH$_3$	1639 (C=O); 1528, 1530 (ring vibrations)
R = N(CH$_2$CH$_2$CH$_3$)(COCH$_3$)	1623 (C=O), 1611, 1522
R = NHC$_6$H$_5$	1630 (C=O), 1500 (ring), 1554 (NH)
R = NH-(CH$_2$)$_4$NH-thiazole ring	1634, 1629 (C=O), 1570 (NH), 1538 (ring)

102

Borello and coworkers (227) have also assigned the absorption bands of 4-methyl- and 2,4-dimethyloxazole. Vibrations arising from the heterocyclic ring system (1660–1600 cm^{-1} and 1585–1500 cm^{-1}) have been evidenced (228).

Hayes and coworkers (229) have discussed the IR spectra of several phenyloxazoles.

1.4.2.5. Dithiole Derivatives

The IR spectra of α-(1,2-dithiol-3-ylidene) ketones and aldehydes **103** have been studied (230–232), and the carbonyl absorption has been assigned by ^{18}O substitution. Unusually low carbonyl bands were found in these compounds (1550–1570 cm^{-1}); on the basis of differences between observed and calculated isotopic displacements, the contribution of the carbonyl modes to this vibration was calculated.

103 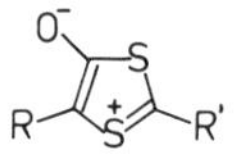 **104**

IR data on anhydro-1,3-dithiolium-4-oxides **104**, mesoionic aromatic compounds, have been reported (233). Dithiolium-4-oxides **104** show a strong absorption band in the range 1575–1610 cm^{-1} (234).

1.4.3. Aromatic Rings with Three or Four Heteroatoms without Carbonyl Groups

1.4.3.1. Triazoles

Characteristic absorptions have been observed in the spectra of 1,2,3-triazoles **105** (235, 236). The NH stretching band appears in carbon tetrachloride at 3470 cm^{-1}, in-plane and out-of-plane deformation CH bands at 1237 and 1076 cm^{-1}, and ring-breathing vibrations in the region 1150–900 cm^{-1}. On the basis of IR spectra it may be concluded that the 1H structure is present in dilute solution.

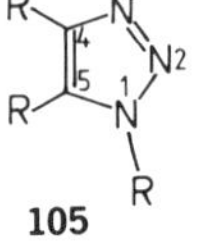

105 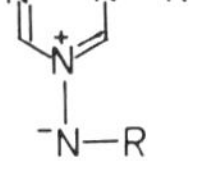 **106**

The wave numbers of the N-imine group in 1,2,4-triazole-N-imine derivatives **106** are shifted by 100 cm^{-1} to lower values than the unsubstituted ones (R = H), suggesting a mesomeric stabilization of the negative charge in the imino group (237, 238).

1.4.3.2. Oxadiazoles and Thiadiazoles

The oxadiazole ring **107** has been characterized by bands at about 970 and 1020–1030 cm^{-1} due to C–O vibration and at 1560–1640 cm^{-1} due to the C=N vibration (239–242). A high frequency $\nu_{C=O}$ (1740–1785 cm^{-1}) in 1,3,4-oxadiazolin-5-ones has been reported (243).

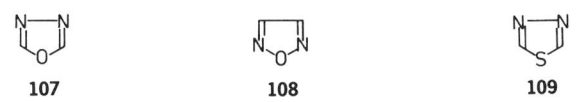

107   108   109

Vibrational spectra of 1,2,5-oxadiazole **108**, 1,3,4-oxadiazole **107**, 1,3,4-thiadiazole **109**, and some of their isotopic species were studied by Christensen and coworkers (244, 245), in order to obtain an unambiguous check on the earlier assignments of **108** (246, 247) and **109** (248–251). [D_2]-1,2,5-oxadiazole, [D]-1,3,4-oxadiazole, [D_2]-oxadiazole, and [2-D]-, [2,5-D_2]-, [2-^{13}C]-, and [3-^{15}N]-1,3,4-thiadiazole were prepared; the resulting complete assignment of the fundamental vibration frequencies of these compounds is presented in Table 1.22. The assignments for all the isotopic molecules fulfill the Teller-Redlich product rule and also the complete isotopic rule given by Brodersen and Langseth (252). The complete vibrational assignment of various isotopically substituted compounds is always extremely useful in force-constant calculations. Christensen and co-workers (244, 245) compared the frequencies of the two lowest fundamentals of **108**, $\nu_8 = 635$ cm^{-1} and $\nu_{15} = 631$ cm^{-1}, with the values calculated from the intensities of microwave lines of vibrationally excited states. The values found by Saegebart and Cox (253) are $\nu_8 = 694 \pm 30$ cm^{-1} and $\nu_{15} = 665 \pm 20$ cm^{-1}. By using the fundamental frequencies given in Table 1.22 and the rotational constants from the microwave work, several thermodynamic functions (heat content, free

Table 1.22. Vibrational Assignments for Oxadiazoles and Thiadiazole (cm^{-1}) (245)

Symmetry Species	Vibration	107	108	109
A_1	1	3169	3157	3115
	2	1534	1418	1403
	3	1272	1316	1391
	4	1092	1036	1224
	5	951	1005	962
	6	920	872	894
A_2	7	840	888	796
	8	657	635	616
B_1	9	3167	3144	3112
	10	1541	1541	1526
	11	1215	1175	1193
	12	1078	953	897
	13	925	820	743

energy, entropy, and heat capacity C_p^o) have been calculated. Large frequency shifts occur between the liquid and vapor phase, whereas smaller deviations were observed between the vapor spectrum and the spectra of dilute solutions of these compounds. These shifts were more pronounced for fundamentals involved mainly motion of the hydrogen atoms, but some of the ring vibrations were also influenced. As in similar molecules these shifts indicate association in the condensed phase. The assignments of the fundamental vibration frequencies of **109** and its deuterated derivatives provided sufficient frequency data to calculate the harmonic potential function for the out-of-plane vibrations and normal coordinates (248). Molecular motion in liquid **110** were studied by Raman and depolarized Rayleigh scattering (254).

A complete assignment of all the IR-active modes of 1,2,5-thiadiazoles **110** has been reported (255). On the basis of the rotational envelope contours of the gas-phase spectra and isotope shifts, the following ring vibrations have been assigned: 1350, 1251, 1417 cm^{-1} (ring stretching); 806, 688, 895, 780 cm^{-1} (ring in-plane bending vibration), and 500 and 520 cm^{-1} (ring out-of-plane bending vibration). These vibrations were found in accordance with the assignments reported for isomeric thiadiazoles (1).

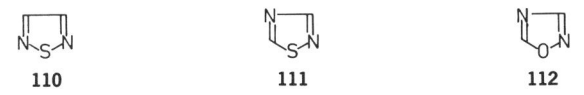

        110   **111**   **112**

Characteristic bands of several furoxan derivatives have been reported by Boyer and coworkers (256), and the following regions have been proposed for diagnostic purposes: 1625–1600 cm^{-1} (C=N$^+$–O$^-$), 1475–1410 cm^{-1} $\left(=N^+<^O_{O^-}\right)$, 1360–1300 cm^{-1} (N–O), and 1190–1150, 1030–1000, and 890–840 cm^{-1} (ring vibrations). Useful bands for diagnostic purposes were also reported for benzofuroxan at 1630, 1600, 1545, and 1500 cm^{-1} (257).

Infrared IR data concerning 1,2,4-thiadiazole **111** and derivatives have been reported (258–261). Zecchina and coworkers (262) have performed the vibrational analysis of IR spectra of 1,2,4-oxadiazoles **112**. The following assignments were suggested for the various ring modes: 1560, 1430, 1365, 1289 cm^{-1} (in-plane stretching); 1229 cm^{-1} (in-plane bending); 1125 cm^{-1} (ring breathing); 1093 cm^{-1} (in-plane bending; 956, 858, 941 cm^{-1} (CH out-of-plane bending), and 886, 649, 618 cm^{-1} (out-of-plane bending). IR data for Δ^2-oxadiazolines have been reported by Huisgen and coworkers (263) and Barraws and coworkers (264).

1.4.3.3. 1,2,3,4-Thiatriazoles

Kuhn and Mecke (265) have assigned the strong bands in the 1540–1590-cm^{-1} range to C=N and N=N stretching vibrations of the heteroaromatic system of 1,2,3,4-thiatriazoles **113**. In various 5-substituted thiatriazoles, ring-breathing absorptions were found at 1320–1260, 1238–1215, 1105–1080, and 1030–1000 cm^{-1} (266).

113

1.4.3.4. 1,2,3,4-Tetrazoles

The amino-imino tautomerism in 5-aminotetrazoles **114**, **115** has been studied by IR spectroscopy, and the presence of an imino form in the solid state has been suggested (267).

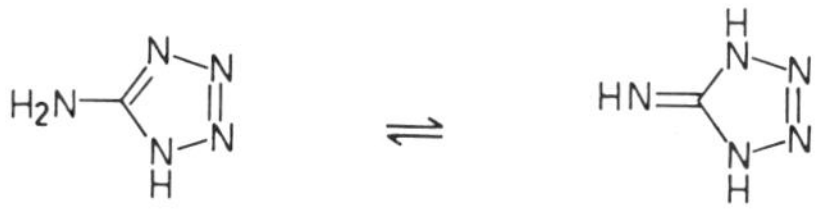

114 ⇌ **115**

1.5. SIX-MEMBERED SATURATED RINGS

1.5.1. Rings with One Heteroatom

1.5.1.1. Tetrahydropyran, Thiacyclohexane, Piperidine, Thiopiperidones, and Thiomorpholides

The close correspondence between the vibrational spectra of tetrahydropyran **116**, tetrahydrothiopyran or thiacyclohexane **117**, and piperidine **118** was reported by Vedal and coworkers (268–270). These molecules exist in the chair form. For **116** the coalescence temperature ($-80°$C) of the ^{1}H-NMR spectra indicates (270) a barrier of 10.7 kcal/mol for the ring inversion. The boat form is predicted in 2,2,6,6-tetramethyl-4-tetrahydropyranols **119**. IR data indicated a transannular hydrogen bond between OH and the heteroatom (271). Compound **118** exists in an equilibrium between the equatorial and axial NH conformers, and the conversion can take place as inversion of the ring or at the nitrogen atom. The inversion barriers for the two processes are expected to be of the same magnitude, approximately 10 kcal/mol.

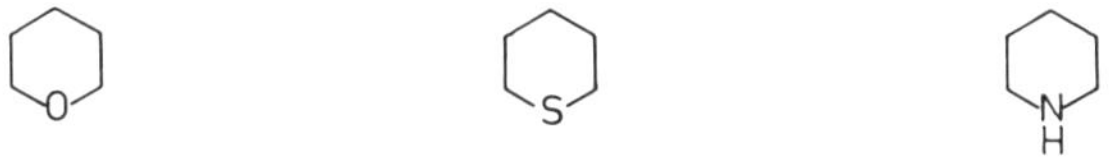

116   **117**   **118**

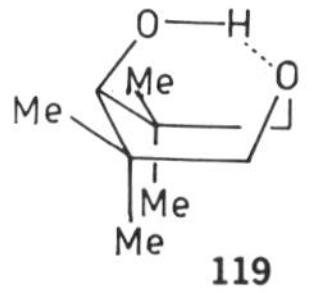

119

Katritzky and coworkers (272) have established a slight preference for the equatorial conformer of **118** in the vapor phase and in noninteracting solvents ($\Delta G = 0.4 \pm 0.2$ kcal/mol). Examination of **118** and cis-2,6-dimethylpiperidine in CCl$_4$ gives (273) $\Delta H = 0.6 \pm 0.2$ kcal/mol for the preference for equatorial NH.

50   Infrared Spectra of Heterocyclic Compounds

All the fundamentals ($23\,a' + 19\,a''$ belonging to the C_s symmetry group) are IR and Raman active, and the assignments were facilitated by Raman polarization data and IR vapor spectra. Since the largest and smallest moments of inertia lie in the symmetry plane of these molecules, vibrations of species a' are expected to have an A/C hybrid band contour and those of a'' type to have a B-type contour. Many irregular band contours were found, and rotational band progressions were detected for the bands at 1036, 1012, 564, and 402 cm^{-1} of **116**.

In the vapor spectra, ν_{NH} frequencies were assigned (274) for the two conformers: 3364 cm^{-1} for the equatorial and 3326 cm^{-1} for the axial conformer, in good agreement with the overtones at 6577 and 6499 cm^{-1}. Bands (IR and Raman) at 906 and 898 cm^{-1} were assigned to in-plane N–H bending modes of the equatorial and axial conformers, respectively. In the deuterated compound these bands are shifted to 657 and 678 cm^{-1}. The ring-deformation modes appeared below 500 cm^{-1}, as six fundamentals: four a' + two a''.

The IR spectrum of **118** as a solid at $-180°$C (obtained by condensing the vapor on a cooled CsI plate) was recorded. The presence of both conformers in the crystal at low temperature was evidenced, and the N–H frequencies were assigned as 3236 cm^{-1} ν_{N-H}(eq) and 3198 cm^{-1} ν_{N-H}(ax), which was explained by a ring packing that is not favored in other six-membered cyclics (two piperidine rings oriented in parallel planes, with the nitrogens facing each other). The frequencies appear shifted by about 100 cm^{-1} in comparison with the liquid spectra of selectively deuterated species of **118** (275). IR and Raman spectra of six-membered saturated molecules revealed splittings into doublets and triplets in the crystalline state, indicating two molecules in the unit cell. In **118** the splittings may be caused by fundamentals of the e and a conformers, which appear separated only in the low-temperature crystal spectra.

Nitrogen-hydrogen stretching vibrations and the conformation of various piperidine derivatives in carbon tetrachloride solution have been discussed (Table 1.23) (276).

In 3-hydroxypiperidine the existence of various forms **120–123** was predicted (276).

Table 1.23. Infrared Data for Piperidine Derivatives (cm^{-1})

Compound	ν_{NH}(eq)	ν_{NH}(ax)	E_e/E_a
Piperidine	3353	3330	2.0
2-Methylpiperidine	3345	3325	1.6
2-Ethylpiperidine	3350	3330	1.4
Conhydrine	3345	3328	0.8
3-Methylpiperidine	3358	3328	2.1
3-Hydroxypiperidine	3363	3333	3.3
3-Hydroxy-3-phenyl piperidine	3365	3330	7.0
4-Methylpiperidine	3355	3325	2.2
Morpholine	3358	3330	2.0
Camphidine	3360	3320	7.6
1,2,3,4-Tetrahydroisoquinoline	3363	3333	1.5

120 **121** **122** **123**

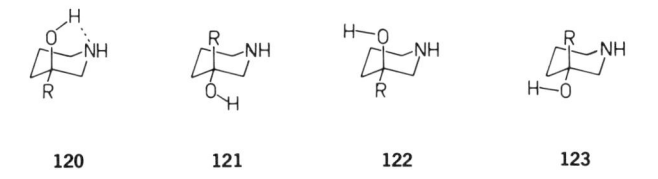

Two stable conformations were predicted (277) also for *N*-chloropiperidine on the basis of the two observed ν_{N-Cl} frequencies in the liquid state, 606 and 680 cm^{-1}. The stereochemistry of 2,6-dialkyl-3-piperidinols has been studied using IR and NMR methods by Brown and coworkers (278).

The vibrational spectrum of the persistent free radical 2,2,6,6-tetramethylpiperidine-1-oxyl has been interpreted (279). A model of the force field has been suggested whose validity was proved by coincidence of the experimentally obtained and predicted shift in the vibrational frequency of the ν_{N-O} band on substitution of ^{14}N by ^{15}N.

Infrared spectra of several thiopiperidones and thiomorpholines have been studied. Three characteristic thioamide bands were located and assigned on the basis of the behavior of these bands on complexing with iodine (280).

1.5.1.2. Tetrahydroquinolines

Cyclization reactions of peptides containing 1,2,3,4-tetrahydroquinoline-2-carboxylic acid have been performed (281, 282). The determination of absolute configurations for 1,2,3,4-tetrahydroquinoline-2-carboxylic acids and stereochemical correlations of some 2-substituted tetrahydroquinolines were reported. IR data concerning these compounds are summarized in Table 1.24. Pertinent IR data have been reported for 1,2,3,4,5,6,7,8-octahydroquinolines by Wittekind and Lazarus (283).

1.5.1.3. Spiropyrans

Two isomeric forms of spiropyrans **128**, **129** have been discussed on the basis of the O–C–O bands, and thermochromic behavior has been evidenced (284).

128
(*R* and *S*)

129 ⇌ **130**

Table 1.24 Infrared Data on Some Quinoline Derivatives

Compound	cm^{-1}
N-Benzyloxycarbonyl-(S)-phenylalanyl-(R)-1,2,3,4-tetrahydroquinoline-2-carboxylic acid **124**	KCl: 3520, 3380, 3290, 1715, 1680, 1630
(+)-(R)-2,3-Dihydro-2-methylquinolin-4-(1H)-one **125**	KBr: 3350–3330, 1650, 1610, 750
(−)-(S)-1,2,3,4-Tetrahydroquinoline-2-ylmethanol **126**	CCl$_4$: 3620, 3400
1H,3H,5H-Oxazolo[3,4-a]quinolin-3-one **127**	KBr: 3400, 1792, 1715, 1695

1.5.2. Rings with Two Heteroatoms in 1,4 Positions

1.5.2.1. 1,4-Dioxan, 1,4-Dithian, and 1,4-Oxathian

The earlier work on 1,4-dioxan **131** has been cited by Malherbe and Bernstein (285) and by Kirchner (286). More recent data on liquid **131** is included in references 287–289. The IR spectra of the completely deuterated molecule dioxan-d$_8$ were

Table 1.25. Vibrational Assignments (cm^{-1}) of Dioxan-h_8 and -d_8

Symmetry Species	Dioxan-h_8 Vibration	Assignment	Dioxan-d_8 Vibration	Assignment	Approximate Motion
a_g	ν_1	2968	ν_1	2242	CH$_2$ stretch.
	ν_2	2856	ν_2	2098	CH$_2$ stretch.
	ν_3	1444	ν_4	1108	CH$_2$ scissor
	ν_4	1397	ν_5	1008	CH$_2$ wagging
	ν_5	1305	ν_6	832	CH$_2$ twisting
	ν_6	1128	ν_3	1225	Ring stretch.
	ν_7	1015	ν_7	808	CH$_2$ rocking
	ν_8	837	ν_8	752	Ring stretch.
	ν_9	453	ν_9	490	Ring bending
a_u	ν_{11}	2970	ν_{11}	2235	CH$_2$ stretch.
	ν_{12}	2863	ν_{12}	2086	CH$_2$ stretch.
	ν_{13}	1449	ν_{15}	1030	CH$_2$ scissor
	ν_{14}	1369	ν_{13}	1191	CH$_2$ wagging
	ν_{15}	1256	ν_{16}	922	CH$_2$ twisting
	ν_{16}	1136	ν_{14}	1117	Ring stretch.
	ν_{17}	1086	ν_{17}	809	CH$_2$ rocking
	ν_{18}	881	ν_{18}	762	Ring stretch.
	ν_{19}	288	ν_{19}	254	Ring bending

Table 1.25. (continued)

Symmetry Species	Dioxan-h_8 Vibration	Assignment	Dioxan-d_8 Vibration	Assignment	Approximate motion
b_g	ν_{20}	2968	ν_{20}	2226	CH_2 stretch.
	ν_{21}	2856	ν_{21}	2088	CH_2 stretch.
	ν_{22}	1459	ν_{22}	1070	CH_2 scissor
	ν_{23}	1335	ν_{24}	956	CH_2 wagging
	ν_{24}	1217	ν_{25}	888	CH_2 twisting
	ν_{25}	1110	ν_{23}	1023	Ring stretch.
	ν_{26}	853	ν_{26}	711	CH_2 rocking
	ν_{27}	490	ν_{27}	422	Ring bending
b_u	ν_{28}	2970	ν_{28}	2232	CH_2 stretch.
	ν_{29}	2863	ν_{29}	2098	CH_2 stretch.
	ν_{30}	1457	ν_{32}	1042	CH_2 scissor
	ν_{31}	1378	ν_{30}	1153	CH_2 wagging
	ν_{32}	1291	ν_{31}	1087	CH_2 twisting
	ν_{33}	1052	ν_{33}	896	Ring stretch.
	ν_{34}	889	ν_{34}	732	CH_2 rocking
	ν_{35}	610	ν_{35}	490	Ring bending
	ν_{36}	274	ν_{36}	238	Ring bending

Table 1.26. ν_{CH} Vibrations in 1,4-Dioxan, 1,4-Dithian, and 1,4-Oxathian (cm^{-1})

Vibration	1,4-Dioxan 131 d_8	$d_4{}^a$	1,4-Dithian 132 d_7	1,4-Oxathian 133 d_8
$\nu_{CH(D)}$(ax)	2875.9 ± 0.2	2140	2930.6 ± 0.2	2879 ± 0.5
$\nu_{CH(D)}$(eq)	2961.9 ± 0.3	2200	2935	2950

^{a} 2,3,5,6-D$_4$.

published by Marsault (290). Infrared vapor contours and accurate Raman polarization data were obtained by Ellestad and coworkers (291). To obtain a satisfactory experimental basis for interpreting the vibrational spectra of **131**, vapor spectra were recorded at various temperatures using a multiple-reflection cell with a 1-meter optical path. Electron diffraction studies and Kerr constant measurements evidenced the chair conformation of nondeuterated **131** in the vapor state and in solution. The spectra were interpreted in terms of C_{2h} symmetry; the 36 fundamentals comprise ten a_g (Raman-active, polarized), nine a_u (IR-active), eight b_g (Raman-active, depolarized), and nine b_u (IR-active). The vibrational assignment for nondeuterated and deuterated **131** from Raman liquid and IR vapor spectra (checked by the Teller-Redlich product rule) are given in Table 1.25.

131   **132**   **133**   **134**

The presence of "hot bands" in the vapor spectra was also demonstrated. Recently, Caillod and coworkers (292) measured the IR spectra of partially deuterated species of 1,4-dioxan **131**, 1,4-dithian **132**, 1,4-oxathian or 1,4-thioxan **133**, and cyclohexane **134** and succeeded in distinguishing between ν_{CH} axial and ν_{CH} equatorial vibrations. The axial and equatorial CH frequencies for three of these molecules are given in Table 1.26. In all compounds the equatorial carbon-hydrogen frequency is higher than the axial.

It is interesting to note the IR study of a crystalline adduct formed between dioxan and hydrogen peroxide (293). The spectra are consistent with a chain structure of alternating dioxan and peroxide molecules linked by hydrogen bonds, both component molecules retaining their normal structure and configuration (chair form for *p*-dioxan and *gauche* configuration for hydrogen peroxide).

Infrared spectra and stabilities of Hofmann-type dioxan clathrates were reported (294). All the IR bands of the dioxan guest molecules in cadmium or nickel dioxan clathrates correspond in frequency and intensity to those in the IR spectra of liquid **131**.

1.5.2.2. *1,4-Thioxan, 1,4-Selenoxan, 1,4-Telluroxan, and 1,4-Thioselenan*

The vibrational spectra of several six-membered nonaromatic cyclic molecules with O, S, Se, and Te as heteroatoms in the 1,4 positions have been interpreted (295),

Table 1.27. Vibrational Assignments for 1,4-Thioxan, 1,4-Selenoxan, 1,4-Telluroxan, and 1,4-Thioselenan

133	136	137	138	Heteroatom[a]	Approximate Motion
2957	2968	2972	2932	X—CH$_2$	CH$_2$ asym. stretch
2942	2946	2942	2960	Y—CH$_2$	CH$_2$ asym. stretch
2911	2929	2928	2915	Y—CH$_2$	CH$_2$ sym. stretch
2853	2848	2847	2899	X—CH$_2$	CH$_2$ sym. stretch
1454	1450	1451	1411	X—CH$_2$	CH$_2$ scissor
1416	1411	1406	1401	Y—CH$_2$	CH$_2$ scissor
1383	1379	1377	1296	X—CH$_2$	CH$_2$ wagging
1315	1311	1307	1268	Y—CH$_2$	CH$_2$ wagging
1266	1238	1204	1143	X—CH$_2$	CH$_2$ twisting
1200	1180	1154	1127	Y—CH$_2$	CH$_2$ twisting
998	998	990	922	X—CH$_2$	CH$_2$ rocking
966	932	888	866	Y—CH$_2$	CH$_2$ rocking
1045	1027	1015	1000		C—C stretch.
824	811	791	645	X	C—X stretch.
663	580	504	560	Y	C—Y stretch.
555	529	538	441	X	CCX deform.
390	394	399	310	Y	CCY deform.
341	266	228	222	X	CXC deform.
206	170	145	135	Y	CYC deform.

[a] For thioxan X = O and Y = S; selenoxan X = O and Y = Se; telluroxan X = O and Y = Te.

and the spectral features compared with those of the symmetrical molecules 131 (291), 132 (296, 297), and diselenan 135 (298). Many similarities were detected in the spectra of 1,4-thioxane 133, 1,4-selenoxane 136, 1,4-telluroxane 137, and 1,4-thioselenane 138 as can be observed from Table 1.27. The most striking differences appeared in the spectra when the heteroatom changed from O to S, and they diminished for heavier heteroatoms.

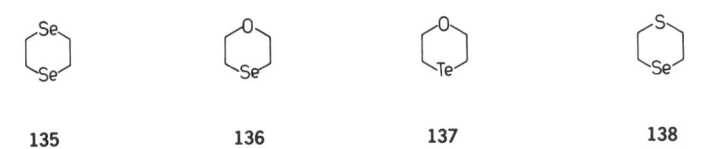

135   136   137   138

1.5.2.3. Morpholine

The IR spectrum of morpholine 139 is closely related to that of piperidine or dioxan, and many of the empirical frequency correlations derived for the fundamental vibrations of the previously studied molecules are applicable to morpholine as well. In the unit cell of crystalline morpholine, an antiparallel orientation of the rings is achieved, with a nitrogen facing an oxygen. The ν_{NH} frequencies are slightly lower, and the shift between the equatorial and axial conformers is somewhat larger than for piperidine. Force fields were derived for the N–H(D) equatorial bond in conformers of **118** and **139**, and the calculated frequencies of the parent molecules and the *N*-deuterated species were correlated with the experimental data Changes of the vibrational spectra of **118** and **139** and their *N*-methyl derivatives adsorbed on aerosil silica were discussed (299). The existence of an ND stretching band on the deuterated silica suggested that hydrogen bonds were formed between nitrogen atoms and silanols.

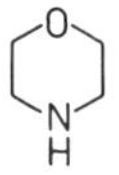

139

Morpholine represents the most interesting model for the consideration of competitive coordination of oxygen and nitrogen atoms (300). IR spectra of morpholine complexes of antimony(III) and bismuth(III) trihalides showed slight shifts of both ν_{C-O-C} and ν_{N-H} vibrations in comparison with those in the free ligand **139**.

1.5.2.4. Other Six-Membered Heterocyclic Rings

Stable nitroxide radicals derived from six-membered phosphorus heterocycles **140** have been prepared and characterized by their IR, ESR, and mass spectra (301).

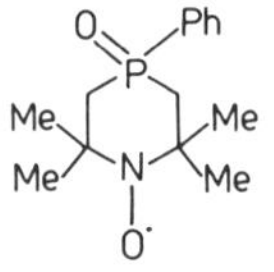

140

The IR, far-IR, and Raman spectra of tetrafluoro-1,3-dithietane were studied, and it was concluded that the molecule is planar, possessing D_{2h} symmetry. Force constants of the approximate Urey-Bradley force field were calculated (302).

Infrared data for the two conformers of **141** (X = O, S; R = H, 3-Cl, 4-Me) and its geometrically isomeric tautomers **142** and **143** were obtained in order to study the effects of concentration, temperature, and solvent on the structure. The dimer structures were discussed (303).

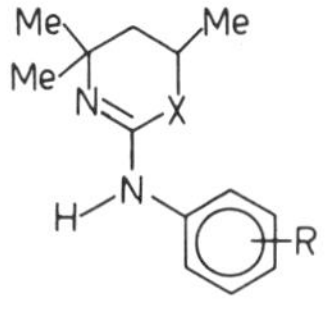

141 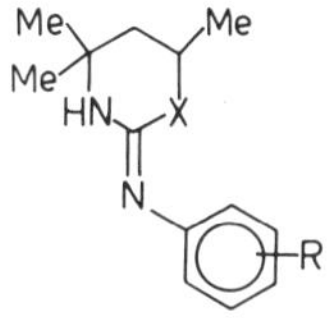   **142** 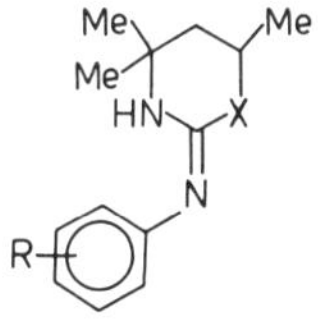   **143**

58   Infrared Spectra of Heterocyclic Compounds

N-Unsubstituted thiazine dioxides **144** show NH absorption in the 3360–3380 cm^{-1} region, indicating the existence of these compounds as 4*H* tautomers (304). Strong absorptions at 1130–1160 and 1310–1335 cm^{-1} were assigned to $\nu_{as}(SO_2)$ (305).

144

Infrared data have been reported for fused systems such as benzoxazine derivatives **145** (306) and benzothiazine derivatives **146** (307).

145
1565 cm^{-1} (C=N)

146
1686 cm^{-1} (C=O)
1333 and 1149 cm^{-1} (SO$_2$)

1.5.3. Rings with Three or Four Heteroatoms

1.5.3.1. *1,3,5-Trithiane and Derivatives*

Asai and Noda (308) have reported IR data of 1,3,5-trithiane **147** and 1,3,5-trithiane-1-oxide **148**. The normal coordinate analysis of these compounds has been carried out by assuming molecular symmetry C_s and two molecular models of the chair conformation with the sulfinyl group either equatorially or axially oriented. An SO stretching vibration was observed at 1036 cm^{-1} and two SO bending vibrations at 366 and 297 cm^{-1}. Ring vibrations in these compounds at 753, 738, 664, and 654 cm^{-1} were reported. The splitting of the SO band in 2-oxo-1,3,2-dioxathianes **148A** was evidenced (309) and was assigned to Fermi resonance by the method of solvent variation. The $\nu_{S=O}$ frequencies were classified according to the possible orientations of the S=O group on a six-membered cycle. The integral molar absorption coefficient of the $\nu_{S=O}$ bands was found to be nearly constant and independent of the S=O orientation. ν_{NH} vibrations for 1,2,4-benzothiadiazine-1,1-dioxides **149** have been reported (310).

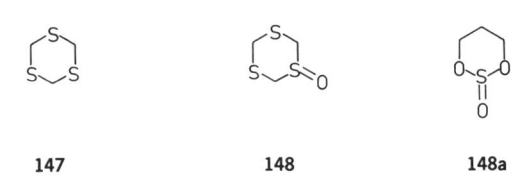

147   **148**   **148a**

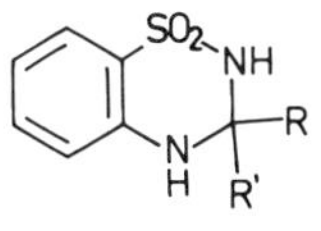

Six-Membered Saturated Rings 59

149

1.5.3.2. 1,2,5-Oxadiazines

Infrared spectra of 1,2,5-oxadiazines were studied by Gnichtel and coworkers (311–313). Pertinent IR bands of oxadiazines are summarized in Table 1.28. In compounds with high symmetry such as 1,4-dithia-2,3,5,6-tetrazine-1,4-dioxide **153**, $\nu_{S=O}$ appears as a strong band at 1190 cm^{-1} (314).

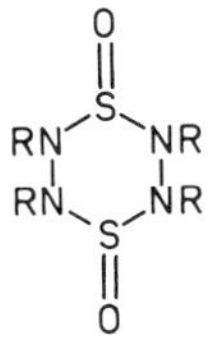

153

Table 1.28. Infrared Data on Oxadiazines

Compound	cm^{-1}	Ref.
6-Methyl-3-phenyl-5,6-dihydro-4H-1,2,5-oxadiazine 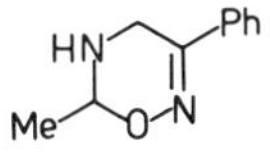 **150**	3280 (ν_{NH}) 1598 ($\nu_{C=N}$) 1068 (ν_{C-O}) 911 (ν_{N-O})	311
6-Thioxo-3-phenyl-5,6-dihydro-4H-1,2,5-oxadiazine 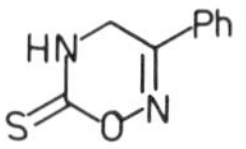 **151**	3200 (ν_{NH}) 1639 ($\nu_{C=N}$) 1572 (ν_{NH}) 1282 (ν_{C-N}) 1139 ($\nu_{C=S}$)	312
6-Oxo-3-phenyl-5,6-dihydro-4H-1,2,5-oxadiazine 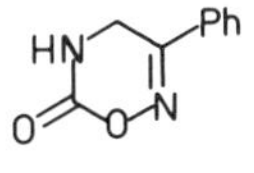 **152**	1750 ($\nu_{C=O}$) 1635 ($\nu_{C=N}$) 1122 (ν_{C-O}) 920 (ν_{N-O})	313

Table 1.29. Vibrational Assignment of 1,4-Diazabicyclo[2.2.2]octane (322)

Frequency No.	Symmetry Species	IR	Raman	Approximate Description
1	A_1'		2870 s	ν_s C–H
2	A_1'			ν_s CH$_2$
3	A_1'		1328 w	ω CH$_2$
4	A_1'		961 m	ν_{as} N–C–C
5	A_1'		805 m	ν_s C–C
6	A_1'		596 w	δ_s N=C$_3$
7	A_2''	2875 s		ν_s C–H
8	A_2''	1450 w		δ_s CH$_2$
9	A_2''	1350 w	1349 vw	ω CH$_2$
10	A_2''	897 m	897 w	ν_s NC$_3$
11	A_2''	755 m		ρ CH$_2$
12	E'	2950 s	2948 vs	ν_{as} C–H
13	E'	2875 s	2870 s	ν_{as} C–H
14	E'	1459 s	1455 mw	δ_{as} CH$_2$
15	E'	1315 m		ω CH$_2$
16	E'	1295 mw	1296 mw	τ CH$_2$
17	E'	1057 vs	1061 mw	ν_{as} N–C–C
18	E'	986 m		ν_s N–C–C
19	E'	830 m	805 m	ν_{as} NC$_3$
20	E'	425 vw	423 w	δ_{as} NC$_3$
21	E''			ν_{as} CH
22	E''			ν_{as} CH
23	E''		1448 w	δ_{as} CH$_2$
24	E''			δ_{as} CH$_2$
25	E''			ω CH$_2$
26	E''			τ CH$_2$
27	E''			ν_{as} C–C
28	E''		579 w	δ_{as} NC$_3$
29	E''		332 vw	ρ NC$_3$

1.5.3.3. Bicyclic Compounds

Spectroscopic studies of boron trihalide and borane complexes of 1,4-diazabicyclo[2.2.2]octane (DABCO) were used to determine the molecular symmetry of DABCO itself by correlating the frequency data obtained for the complexes (322). For an earlier study on DABCO, see reference 323. The assignment of observed frequencies for DABCO is given in Table 1.29. IR and Raman spectra of quinuclidine complexes with boron trihalides and borane were also investigated as an aid in the spectral interpretation of the DABCO complexes.

Recently, Grech and coworkers (323a) observed in the IR spectra of salts of DABCO the bands for free DABCO with minor frequency shifts and occasionally with splittings. The strong temperature effect on IR spectra was interpreted as involving a coupling of ν_{NH} vibrations with skeletal δ_s(N–C$_3$) and δ_{as}(N–C$_3$).

Six-Membered Saturated Rings 61

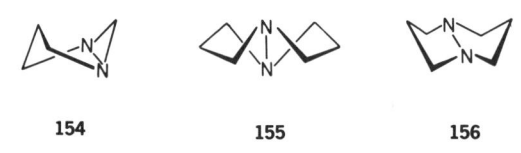

154   **155**   **156**

The molecular symmetry of five bicyclic hydrazines were determined from the number of observed bands and the frequency coincidences and Raman depolarization factors of their vibrational spectra (324). For 1,5-diazabicyclo[3.1.0]hexane **154**, the conformation **154** was assigned, but for 1,5-diazabicyclo[3.3.0]octane, on the basis of vibrational spectra and the analysis of vibrational spectra, it was impossible to distinguish between conformations **155** and **156**.

1.5.3.4. Tetrathiaadamantane

Interesting IR data on 1,3,5,7-tetramethyl-2,4,6,8-tetrathiaadamantane and 1,3,5,7-tetramethyl-2,4,6,8,9,10-hexathiaadamantane have been reported (315, 316). A previous IR study of the former compound (317) was restricted to the conventional IR region and no assignment was given. Barnes and coworkers (315, 316) performed a normal coordinate analysis of the skeletal vibrations assuming D_{2d} symmetry. The observed and calculated frequencies in tetramethyltetrathiaadamantane are summarized in Table 1.30. Assignments were similar to 1,3-dithian, the structural unit of which the skeleton is built. Fredga and Olsson (318) have obtained the IR

Table 1.30. Infrared Data on Tetramethyltetrathiaadamantane (cm^{-1}) (315)

Class	Description	Assignment	Calculated	Observed
A_1	C–S stretch	1	668	650
	C–C stretch	2	1033	1040
	SCS	3	418	442
	CCS	4	476	470
B_1	C–S stretch	5	709	712
	CCS bend	6	314	307
B_2	C–S stretch	7	620	593
	C–C stretch	8	886	886
	SCS bend	9	243	244
	CCS bend	10	508	523
E	C–S stretch	11	573	577
	C–S stretch	12	685	680
	C–C stretch	13	947	948/956
	SCS bend	14	269	264
	CCS bend	15	311	330
	CCS bend	16	367	369

62   Infrared Spectra of Heterocyclic Compounds

Table 1.31. Distribution of the Skeletal Modes Between Symmetry Classes for Hexathiaadamantane (316)

Type	Class	A_1	A_2	E	F_1	F_2
C–S	Stretching	1	0	1	1	2
SCS	Bending	1	0	1	1	2

spectrum of 2,4,6,8,10-hexathiaadamantane as a solid but did not assign any of the bands to specific vibrations. For the normal coordinate analysis of the skeletal vibrations, T_d symmetry is assumed, and the skeletal modes are distributed between symmetry classes as in Table 1.31.

Infrared data on N,O,S-adamantane (319) and on 1,3,5-triaza-7-phosphaadamantane-7-oxide have been reported (320, 321).

1.5.4. Six-Membered Rings with Carbonyl Groups

1.5.4.1. Pyrones

Pyran-4-ones and pyran-4-thiones have been investigated by Katritzky and Jones (325). They assigned the carbonyl frequencies and ring modes of the parent compound **157** and a great number of derivatives by making use of solvent shifts. Table 1.32 summarizes some of the IR data of the pyran-4-ones and related structures.

157

Table 1.32. Infrared Data on γ-Pyrones and Related Structures

Compound	Infrared Data	Ref.
Pyran-4-one (γ-pyrone)	1660 ($\nu_{C=O}$)	325
	1634, 1464, 1317, 919 (ring modes)	
2,6-Dimethyl-γ-pyrone	1639 ($\nu_{C=O}$)	
	1678 ($\nu_{C=C}$)	
Pyran-4-ones		
2-substituted	1650, 1619, 1385 (ring modes)	326
3-substituted	1648, 1618, 1395 (ring modes)	326
3,5-Diphenyl-4-pyrone	1638 ($\nu_{C=O}$)	327
	1600 ($\nu_{C=N}$)	
3,5-Diphenyl-4-thiopyrone	1100 ($\nu_{C=S}$)	327
2,6-Diphenyltetrahydroseleno-pyran-4-one	3335, 1707, 1641, 1496, 1459, 1386, 1245, 1192, 1055, 1027, 807, 762	328
Phenyl-2-selenochromone	1600–1620 ($\nu_{C=O}$)	329

1.5.4.2. Glutarimide

Thompson and coworkers (333) have performed a vibrational analysis of glutarimide, and several of the fundamental vibration modes have been assigned.

1.5.4.3. 2,5- and 2,3-Piperazinediones

Amino acid anhydrides (2,5-piperazinediones) **158A** are convenient compounds for the IR study of the *cis* peptide bond. A weak band at 3370 cm^{-1} was assigned to free N–H stretching, whereas two stronger bands at 3230 and 3190 cm^{-1} were assigned (334) to hydrogen-bonded N–H stretching of the *cis* peptide bond. The carbonyl stretching in 2,5-piperazinediones appears as a very strong band at 1701 cm^{-1}. Calculation of the potential energy distribution has shown that the N–H in-plane bending and ring skeletal vibration couple with each other. 2,3-Piperazinediones **158B** exhibit (335) three N–H bands in the range 3080–3350 cm^{-1} and two strong carbonyl bands between 1680 and 1740 cm^{-1}.

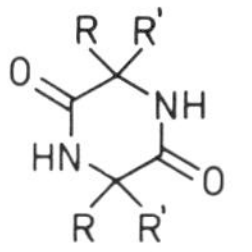

158a158b

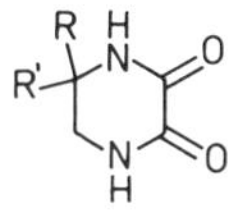

3-Benzylidene-2,5-dioxopiperazine **159** was studied by Dominy and Cawton (336).

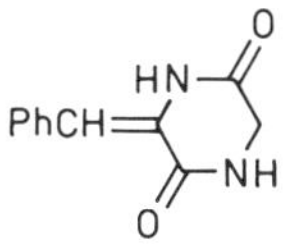

159

1.5.4.4. 1,3-Oxazine-4,6-diones

A conformational study of 1,3-oxazine-4,6-diones was performed (337) by following the temperature dependence of the characteristic IR and NMR absorptions. The study showed that 5-(R)-substituted 1,3-oxazine-4,6-diones **160** were in the oxo form at room temperature.

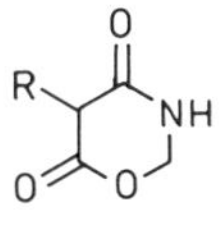

160

64   Infrared Spectra of Heterocyclic Compounds

Fused systems like tetrahydrocinnolinone **161** exhibit (338) strong carbonyl absorption at 1640 cm^{-1} and two NH bands at 3210 and 3160 cm^{-1}. In indazolo[2,1-a]cinnoline(7H)-6,13-dione **162** a single $\nu_{C=O}$ band appears at 1690 cm^{-1}.

161   **162**   **163**

Cyclic hydroxamic acids have been characterized by means of IR spectroscopy (339). In the solid state, simple derivatives of 3,4-dihydro-4-hydroxy-3-oxo-2H-1,4-benzoxazine **163**, X = O and related benzothiazines exist as mixtures of intermolecularly and intramolecularly hydrogen-bonded forms. The IR spectra of both classes of compounds showed one or two strong carbonyl bands within the range 1627–1689 cm^{-1}. The IR spectra of 2-arylidene-3,4-dihydro-4-hydroxy-3-oxo-1,4-benzothiazine are similar to the preceding ones except that the strong carbonyl bands are located within the range 1608–1649 cm^{-1}. 3,4-Dihydro-4-hydroxy-3-oxo-2H-1,4-benzothiazine-1,1-dioxide **163**, X = SO$_2$ exhibits dimorphism. One crystalline form shows intramolecular hydrogen bonding, and the second form shows intermolecular hydrogen bonding. Other benzothiazine-1,1-dioxides exist only in the intermolecularly bonded form. A single strong CO band (1665–1696 cm^{-1}) and a strong, relatively sharp OH stretching band (3245–3370 cm^{-1}) were observed in all spectra.

Quinoline-hydroxamic acids present one or two medium or strong CO bands (1603–1665 cm^{-1}) and a very broad OH stretching band of medium intensity. Intramolecular hydrogen bonding predominates in these compounds. A strong CO stretching band in the 1530–1565 cm^{-1} region was observed in the spectra of ferric chelates of several benzothiazine-hydroxamic acids.

Infrared absorption spectra of 7-mono- and 7,8-disubstituted theophylline and 1-substituted theobromine derivatives were studied (340). The 1-substituted theobromine **164**, for example, exhibits five bands in the 1750–1500 cm^{-1} range at 1710, 1668, 1635, 1587, and 1545 cm^{-1}. IR data on 1,2,3,4,5,10-hexahydrobenzo[g]quinoxaline-5,10-diones **165** were reported (341). These data are summarized in Table 1.33.

164   **165**

Table 1.33. Infrared Data on 1,2,3,4,5,10-Hexahydro[g]quinoxaline-5,10-diones 165

			Frequency (nujol)	
R	R'	Configuration	ν_{NH}	ν_{CO}
CH$_3$	H		3320	1612
C$_2$H$_5$	H		3300	1620
C$_6$H$_5$	H		3290	1610
CH$_3$	CH$_3$	Trans	3325	1615
C$_6$H$_5$	C$_2$H$_5$	Trans	3290	1610
C$_6$H$_5$	CH$_2$C$_6$H$_5$	Trans	3320	1640

1.5.4.5. Flavones and Isoflavones

Some of the earlier infrared data on naturally occurring flavones and isoflavones are summarized in Table 1.34. Although the IR spectra of flavones and isoflavones have been extensively studied, they have not been found so useful as their UV or ^{1}H-NMR spectra for structure elucidation. Analysis of the reported assignments of the bands in the double-bond region reveals that it was often assumed erroneously that the highest frequency band was the carbonyl absorption. To make an unambiguous assignment of carbonyl absorption in flavone and isoflavone, Jose and coworkers (345) used the method of complexation (346). This method, which consists in the addition of equimolar proportions of iodine to a solution of the carbonyl compounds in CCl$_4$, was extended by Jose and coworkers (345) to the effects of stronger complexing agents, such as boron trifluoride, which are known to give larger carbonyl shifts.

Table 1.34. Carbonyl Frequencies in Some Flavones and Isoflavones

Compound	$\nu_{C=O}$, cm^{-1}	Ref.
5-Hydroxyflavone	1652	342
5-Hydroxy-7-methoxyflavone	1649	342
Flavone	1695	342
5-Hydroxyflavanone	1648	342
5-Methoxyflavone	1645	343
Isoflavone	1640	344
5-Hydroxyisoflavone	1660	344
5-Methoxyisoflavone	1640	344
7-Hydroxyisoflavone	1620	344
5-Hydroxy-7-ethoxyisoflavone	1660	344

Table 1.35. Effect of Complexation on the IR Bands of Flavone Derivatives

Compound	IR Bands in the 1500–1700 cm^{-1} Region	BF$_3$ Adduct (In Nujol)
Flavone	1638 (C=O, C=C)	1618
	1610	1598
	1600	1582
	1562	1562
		1526
Flavonol (3-Hydroxyflavone)	1622 sh (C=C)	1618 (C=C)
	1605 (C=O)	1607
	1555	1555
		1495 (C=O)
5-Hydroxyflavone	1645 (C=C)	1635 (C=C)
	1612 (C=O)	1590
	1580	1550
		1525 (C=O)
7-Methoxyflavone	1652 (C=C, C=O)	1627 (C=C)
	1627	1597
	1606	1578, 1550
		1510 (C=O)

The addition of iodine to flavone, isoflavone, and 7-methoxyflavone yielded evidence that the band at 1663 cm^{-1} consists of both C=O and C=C absorptions. The spectra of 5-hydroxyflavone and 5-hydroxyisoflavone did not show any change upon addition of iodine, indicating the absence of complex formation between iodine and the C=O group. This is not unexpected in view of the strong intramolecular hydrogen bonding, which prevents the iodine molecule from forming the complex. Further evidence in support of the above assignments was obtained by studying the BF$_3$ complexes of flavones and isoflavones. The boron trifluoride adducts were obtained by introducing gaseous BF$_3$ into a solution of the compound in CCl$_4$. The adducts were then separated and studied as nujol mulls. The IR absorption bands in the double-bond region of flavone derivatives and their BF$_3$ adducts are given in Table 1.35. Frequency shifts of the order of 100 cm^{-1} were obtained for the carbonyl band. The IR data have been explained by taking into account the C=C absorption, which often overlaps with the C=O band, so that the complexation method may be a good diagnostic tool in a correct assignment.

1.5.4.6. Chromone and Thiochromone

Complex systems containing chromone and thiochromone were studied by Görlitzer (347–351). The IR data are given in Table 1.36.

Table 1.36. Infrared Data on Chromone and Thiochromone

Compound	IR Data, cm^{-1}	Ref.
Benzothieno[3,2-b][1]-chromone **166**	1630 (C=O) 1610, 1600, 1590 (C=C)	347
Benzo[b]-indeno[2,1-e]-pyrone(4) **167**	1637 (C=O) 1620, 1600 (C=C)	348
Benzo[b]indeno[2,1-e]-thiopyrone(4) **168**	1605 (C=O) 1580 (C=C)	349
Benzofuro[3,2-b]thiocarbonyl- thiochromone **169**	1600, 1585 1542, 1496	349
Benzofuro[3,2-b]thiochromon-S,S-dioxide **170**	1676 (C=O) 1600, 1580, 1567 (C=C) 1300, 1148 (SO$_2$)	349
11-Oxo-5,11-dihydrobenzofuro[3,2-b][1]- quinoline **171**	1632 (C=O) 1600, 1570 (C=C) 3065 (NH)	350

68 Infrared Spectra of Heterocyclic Compounds

Table 1.36. (continued)

Compound	IR Data, cm^{-1}	Ref.
11-Oxo-5,11-dihydrobenzothieno[3,2-*b*][1]-quinoline 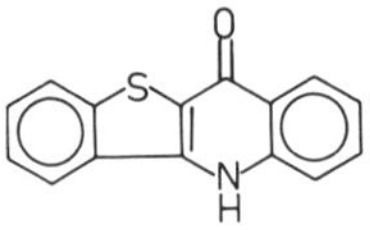 **172**	1620 (C=O) 1590, 1560 (C=C) 3060 (NH)	350
11-Oxo-5,11-dihydro-1,3-dihydroxybenzothieno[3,2-*b*][1]-chromone 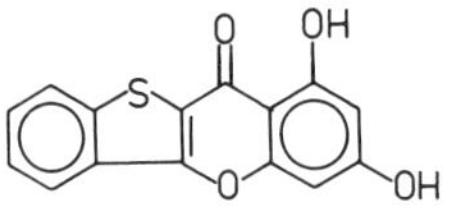 **173**	1635 (C=O) 1590 (C=C) 3310 (NH)	350
11-Oxo-5,11-dihydro-indeno[1,2-*b*]-quinoline 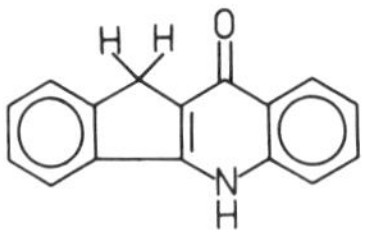 **174**	1628 (C=O) 1608 (C=C) 3420 (NH)	351

1.6. AROMATIC SIX-MEMBERED RINGS

1.6.1. Pyridine and Related Systems

1.6.1.1. *Pyridine*

The numerous spectral studies carried out on pyridine **175** are based on extensive investigations on the vibrational spectra of benzene (352–357). For a survey of early references, see reference 1.

175

The vibrational assignments (358) for pyridine and pyridine-d_5 are given in Table 1.37.

The distinction between the ring-stretching and C–H bending vibrations of pyridine was confirmed by an isotopic (deuterium) labeling study of pyridine

Table 1.37. Vibrational Assignments of Several Fundamental Vibrations of Pyridine and Pyridine-d_5[a]

C_{2v} Species	Vibration No.	Pyridine h_5	Pyridine d_5	Description
B_2	20b	3080 vs		Ring-H stretch
A_1	2, 20a	3054 s		
B_2	7b	3036 vs		
			2293 s	Ring-D stretch
			2285 s	
			2270 s	
			2254 vs	
A_1	8a	1583 vs	1530 s	Ring stretch (C=N and C=C)
B_2	8b	1572 s	1542 s	Ring stretch (C=N and C=C)
A_1	19a	1482 s		Ring stretch and ring-H bend
			1340 w	Ring stretch and ring-D bend
B_2	19b	1439 s		Ring stretch and ring-H bend
			1301 s or 1228 m	Ring stretch and ring-D bend
B_2	14	1375 m	1301 s or 1322 m	
B_2	3	1288 s		
A_1	9a	1218		Ring-H in-plane bend
B_2	15	1148 m		Ring-H in-plane bend
A_1	18a	1068 s	1043 m	Ring-D in-plane bend
			1006 m	Ring stretch and C–D bend
A_1	1	992 vs	962 s	Ring breathing
B_1	10b	886 m		Ring-D in-plane bend
B_2	6b	652 w	625 w	Ring bend
A_1	6a	605 s	582 s	(in-plane)

[a] See also Tables 1.41 and 1.42.

(359), which indicated that the C–H bending modes undergo a characteristic shift to lower frequency ($\Delta \bar{\nu}$ varies between 100 and 350 cm^{-1}), while the ring-stretching vibrations are less affected. The ratio between the frequencies of the deuterated and nondeuterated species is 0.75 for the C–H bending modes and 0.95 for the ring-stretching modes. Recently, application of multiple isotopic labeling techniques to the band assignments in the IR spectra of metallic complexes with C_{2v} axially coordinated pyridine provided firm assignments for the pyridine ligand vibrations (360, 361). The pyridine ligand vibrations are identified as those bands which are sensitive to both ^{15}N (Py) and Py-d_5 labeling. ^{15}N labeling of the pyridine nitrogen atom enabled a check on the assignments resulting from pyridine deuteration.

1.6.1.2. 2,2'-Bipyridyl

2,2'-Bipyridyl **176** is basically a 2-substituted pyridine, with pyridine being the substituent group. Strukl and coworkers (358) presented a complete assignment of the IR spectrum of **176**, based primarily on a comparison with its deuterated analog (to differentiate between the hydrogen-involving modes and ring modes) and secondarily on a comparison with the spectral assignments of **175**, 2-substituted pyridines and biphenyl.

The assignment was possible only for 2,2'-bipyridyl in the crystalline state in the *trans* coplanar conformation **176**, because it displays the simplest spectral geometry (C_{2h} with a center of symmetry). Aside from its pyridine character, this compound possesses certain inter-ring motions similar to those of biphenyl. The use of a far-infrared Fourier transform spectrophotometer allowed also the assignment of the low frequencies. The empirical assignments of these modes (crystalline **176** embedded in polyethylene matrix) are given in Table 1.38. For comparison, two of the low-frequency modes of biphenyl are also given (362–364).

Table 1.38. Infrared Frequencies (cm^{-1}) and Empirical Assignments of 2,2'-Bipyridyl-h_8 and -d_8 in the Low-Frequency Region

2,2'-Bipyridyl			
h_8	d_8	Biphenyl	Description
398 m	351 m	–	Ring torsion
169 m	–	140	In-plane scissoring
94 m	–	120	Out-of-plane scissoring
42 w	–	–	Torsion or crystal mode

Infrared studies (365) of 1:1 transition metal:bipyridyl chelates indicated the shift of the 398 cm^{-1} ring torsion band to 420 cm^{-1} in L·PdCl$_2$ (L = 2,2'-bipyridyl). Spectral data for metallic derivatives of **176** are given in references 366 and 367 and for 2,2'-bipyridyl-N,N'-dioxides and their complexes with cobalt(II), nickel(II), zinc(II), cadmium(II), and mercury(II) thiocyanates in reference 368.

Infrared and Raman spectral studies of zinc complexes of 2,2'-, 3,3'-, and 4,4'-bipyridyls and copper(II) complexes of 2,2'-bipyridyls were reported by Grzeskowiak and coworkers (369). For 6,6'-dibromo-2,2'-dipyridyl-1-oxides **177**, an N−O vibration at 1265 cm^{-1} has been reported (370).

1.6.1.3. Substituted Pyridines

1.6.1.3.1. Halopyridines. Published information on vibrational spectra of halopyridines is confined to several monosubstituted derivatives (371, 372), two pentasubstituted pyridines (373, 374), and 2,6-difluoropyridine (375). Recently Lui and coworkers (376), by selective deuteration, and using the results of microwave investigations (molecular geometry and the low-frequency vibrational states of the molecule) have performed a complete vibrational analysis of 2,6-difluoropyridine. The vibrational frequencies were calculated with the Urey-Bradley force field of the pyridine molecule as starting point and with carbon-fluorine force constants transferred from vinyl fluoride. The force field was then refined by assigning successively the observed frequencies to fundamental vibrations and including these in an iterative fitting procedure. The results have shown good agreement between the observed and calculated vibrational frequencies.

1.6.1.3.2. Alkyl Substituted Pyridines. Green and Harrison (377) reported a complete characterization by vibrational spectroscopy of five of the six possible isomeric trimethylpyridines (collidines), and a comparison with the analyzed spectra of the structurally related isoelectronic benzene analogs. For the identification of the fundamentals in these compounds, consideration of the IR and Raman spectra taken together was essential, especially for the isomers with lower symmetry (2,3,6-, 2,3,5-, and 2,4,5-trimethylpyridines). The methyl vibrations in all collidines are given in Table 1.39. For other alkyl-substituted pyridines, the earlier available data are summarized in reference 1. Characteristic absorption patterns for mono- and disubstituted pyridines in the overtone region are indicated, and correlations in the different spectral regions are discussed.

1.6.1.3.3. Acetylpyridines and Related Structures. Forrest and coworkers (378) illustrated the particular value of the symmetric ring breathing vibration at 992 cm^{-1} in characterizing pyridine substitution patterns. The symmetric ring breathing (Raman) frequencies of monosubstituted pyridines are given in Table 1.40. Forrest and coworkers (378) applied IR and Raman spectroscopy to cyclopropyl pyridyl ketones to test the presence of the rotational isomers by analogy with conformational isomerism suggested for acetylpyridines (379, 380) and pyridine carboxaldehydes (380–382). No significant spectral changes indicative of multiple pyridyl-carbonyl or cyclopropyl-carbonyl conformations were observed in either the liquid or low-temperature solid spectra.

Table 1.39. Methyl Vibrations in Collidines (cm^{-1})

Vibration	2,4,6-	2,3,6-	2,3,5-	2,4,5-	2,3,4-
$\nu_{as}(CH_3)$	2973 s	2972 s	2971 s	2968 s	2986 s
	2950 s	2946 s	2940 s	2945 s	2945 s
$\nu_s(CH_3)$	2920 s	2922 s	2919 s	2920 s	2920 s
$\nu_{as}(CH_3)$	1456 m	1437 sh	1450 s	1449 s	1443 s
$\delta_s(CH_3)$	1384 s	1389 s	1398 s	1384 s	1385 s
	1372 s	1377 s	1379 s	1372 s	1368 s
	1030 m		1038 s	1033 s	1044 m
CH$_3$ rock		1017 s	1017 s	1020 s	1007 s
		1000 s	997 s	1000 s	976 s
				985 sh	
		970 m			
$2\delta_a(CH_3)$	2850 m	2856 m	2863 m	2860 m	2864 m
$2\delta_s(CH_3)$	2724 w	2734 w	2733 w	2739 w	2730 w
		2723 w	2720 sh	2730 w	

The vibrational spectra and structure of crystalline dipicolinic acid (2,6-pyridinecarboxylic acid) and calcium dipicolinate trihydrate — compounds of biological interest — were investigated (383). A tentative assignment was given assuming that the vibrational spectra of these compounds were superpositions of bands corresponding to the COOH groups and pyridine. The hydrogen-bonded OH band was shown to have a complex structure. The center of the broad band was assumed to be around 2800 cm^{-1} and the main subbands (2835, 2662, and 2558 cm^{-1}) were interpreted in terms of Fermi resonances between the ν_{OH} and $2\delta(OH)$, $\delta(OH)$, $\delta(OH) + \nu(C-O)$, and $2\nu(C-O)$ overtone combinations. On the basis of spectral features, cyclic, centrosymmetric, and dimeric structures involving the carboxylic groups were proposed.

Alkyl 2-picolyl ketones exhibit (384) a ketoenolic tautomerism **178 ⇌ 179**.

Table 1.40. Symmetric Ring-Breathing Frequencies of Monosubstituted Pyridines (cm^{-1})

Compound	2	3	4	Technique
Pyridine	992	992	992	Raman
Cyclopropyl pyridyl ketones	993	1025	990	Raman
Pyridinecarboxaldehyde	993	1036	990	Raman
Acetylpyridine	993	1034	989	Raman
Pyridine-d	989	980	989	IR
Methylpyridine	994	1025	994	IR

Aromatic Six-Membered Rings 73

178 ⇌ **179**

The enol-keto ratio, estimated by the ratio of the absorbancies due to the carbon-carbon double bond and the carbonyl, depends both on the solvent polarity and on the nature of the alkyl R group. A similar tautomerism was evidenced for pyrylium derivatives **180** ⇌ **181** (385).

180 ⇌ **181**

1.6.1.3.4. Aminopyridines. Recently, aminopyridines have received much theoretical and experimental attention because they act as good model systems for the purine and pyrimidine bases. The amino-nitrogen inversion vibration was studied in the far-infrared spectra of various aminopyridines and their mono- and dideuterated derivatives (386). These are all solids with very low vapor pressure, and in order to obtain a good signal-to-noise ratio, 6000 scans were averaged on a Fourier transform (Digilab FTS-16) interferometer using a 20-meter multiple reflection cell. The inversion barriers (241 cm^{-1} for 2-aminopyridine, 295 cm^{-1} for 4-aminopyridines and 500 cm^{-1} for 3-aminopyridine) are shown to correlate extremely well with the calculated electron density on the amino nitrogen. The far-infrared spectra of charge-transfer complexes of 2-, 3-, and 4-aminopyridine were studied by Sasaki and Aida (387).

1.6.1.3.5. 2-Hydroxypyridines. These are also model compounds for hydroxypyrimidines (uracil and cytosine) and hydroxypurines (guanine, xanthine, and hypoxanthine) of biological importance. IR studies of the lactim-lactam tautomeric equilibrium of 2-hydroxypyridines (388) and of some of their chlorinated derivatives (389) illustrated the spectral features of the individual tautomers and also the influence of the carboxylic acid association upon this equilibrium. The lactam-lactim ratio in chlorinated 2-pyridinols isolated in low-temperature argon matrices (389) decreases in the order: 3,5,6-trichloro- > 6- > 5- > 4- > 3-monochloro-2-pyridinols.

The effect of substituents on the lactam-lactim tautomeric equilibrium in 2-pyridones is discussed in reference 390.

In 6-chloro-2-hydroxypyridine–acetic acid mixtures, the spectral measurements in CCl$_4$ provided evidence for lactim-acid and lactam-acid cyclic heterodimer formation (**182** and **183**) and for a preferential association with the lactam tautomers **182**. King and coworkers (389) concluded that carboxylic acids should catalyze bifunctionally the tautomeric interconversion, thereby lowering the lifetime of the lactim forms; this may be an interesting result for molecular biology, since double hydrogen-bond formation is one of the interactions involved in DNA replication. For example, 2-dimethylamino-6-hydroxy-9-methylpurine was found to associate with butyric acid and these nucleic base–carboxylic acid interactions are expected to shift the tautomeric equilibria in favor of lactam forms, the correct ones for Watson-Crick base pairing. These results agree well with other papers either on 2-pyridone itself (391), where the lactam tautomers are favored by dimerization, or on specific association of 2-pyridones with water molecules (392).

182 **183**

Spectroscopic studies of isotopically substituted pyridones indicated that 4-pyridone is much more associated via hydrogen bonding than 2-pyridone (393). The IR data concerning the influence of isotopic substitution on the spectra of pyridones indicated that D- and ^{15}N-substitution have considerably less effect on the spectrum of 4-pyridone than in the case of 2-pyridone. An attempt to study the 4-pyridone-4-hydroxypyridine tautomerism via ^{15}N-substitution and ^{1}H-NMR spectroscopy was unsuccessful because of rapid proton exchange.

Infrared data on pyridones are given also in references 394 and 395. Charge-transfer complexes of pyridine and pyridone derivatives have been studied (396).

1.6.1.3.6. Other Derivatives. IR spectral studies on different complexes of pyridine and pyridine derivatives have been performed (397) not only to elucidate the structure of the complexes but also to establish the influence of metallic acceptors on the bands of the free ligands. For example, methylpyridine complexes of Pt(II) and Pd(II) were studied by Pfeffer and coworkers (397) and aluminum halide complexes with pyridine-N-oxides by Brown and coworkers (398).

The intensities of the characteristic modes in pyridines and in several N-substituted pyridinium derivatives and adducts depend on the nature of the substituents and their effect on the charge distribution in the ring. The position of the bands in the 1400–1600 cm^{-1} range and the apparent extinction coefficients in 4-substituted pyridines bearing various types of substituents were correlated with the variation in the value of the charge disturbance in the ring (399). In 2- or 3-substituted pyridines, because of lowered symmetry, the treatment is difficult. However, Katritzky and Topsom (400) accounted for the intensity variations by the degree of conjugation between the ring and the substituent. Recently, they used the intensities of the ring-breathing modes near 1600 cm^{-1} for pyridine and

N-substituted derivatives to obtain effective σ_R^0 values for the ring nitrogen or N-substituted nitrogen treated as a pseudosubstituent. The σ_R^0 obtained for the ring nitrogen corresponds to a moderately strong electron-acceptor substituent ($\sigma_R^0 = 0.24$), whereas the N-oxide nitrogen appears as a "donor" ($\sigma_R^0 = -0.21$). Thus, the use of infrared intensities appears to be particularly valuable for assessing the extent of intramolecular interactions in conjugated systems. For the alkyl derivatives of six-membered nitrogen heterocycles and their N-oxides, σ_R^0 values were calculated (401) from the intensities of ν_{CH} (2925 or 2912 cm^{-1}) bands with the correlations:

and
$$A^{1/2} = -12.4\sigma_R^0 + 34.5$$
$$\epsilon = -18.3\sigma_R^0 + 31.6$$

The σ_R^0 values calculated from intensity measurements in CCl$_4$ are equal with those derived from ^{1}H-NMR.

In pyridine N-nitroimines, the resonance interaction between the nitroimino group and the pyridine ring was investigated (402).

1.6.1.4. Pyrylium Cation

The unsubstituted pyrylium cation is planar and belongs to symmetry group C_{2v}. The assignments of the bands reported by Balaban and coworkers (403) are given in Table 1.41. This table also includes assignments for heteroanalogs with heteroatoms possessing progressively varying electronegativity: benzene **184**, pyridine **185**, pyridinium **186**, pyridine-acceptor (metallic) complexes **187**, and pyrylium **188**, thus obtaining a sequence with increasing electronegativity of the heteroatom. It may be seen from Table 1.41 that the frequency of deformations δ_{C-C}, γ_{C-C}, and δ_{C-H} increases and that the frequency of deformation γ_{C-H} decreases in the sequence **184–188**. The splitting $\Delta\nu 8a-8b$, $\Delta\nu 19a-19b$, and $\Delta\nu 20a-20b$ also increases monotonically in this sequence owing to the differences in symmetry, mass, and electronegativity of the heteroatoms.

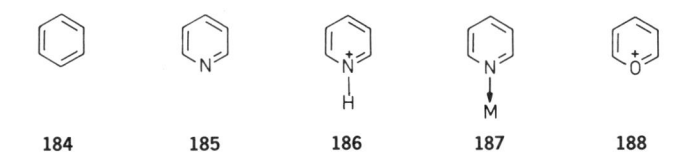

184   **185**   **186**   **187**   **188**

The variation of band position and intensity depending on substitution has been studied (405–407). Various charge-transfer complexes of pyrylium derivatives have been studied by S. Bădilescu and Balaban (408).

1.6.1.5. Phosphabenzene and Arsabenzene

Related to pyridine, the two heterocyclic compounds phosphabenzene (*Chemical Abstracts* name: phosphorin) **188A** and arsabenzene **188B** (*Chem. Abstr.* name:

Table 1.41. Vibrational Assignments in Pyrylium Cation[a]

Symmetry Class	Vibration No.	Benzene	Pyridine	Pyridinium	Pyridine complex	Pyrylium	Assignment
A_1	1	992	992	1010	1005–1010	995	ν_{C-C}
	2	3062	3054			3062	ν_{C-H}
	6a	606	605		662–641	582	δ_{C-C}
	8a	1596	1583	1638	1597–1610	1620	ν_{C-C}
	9a	1178	1218	1194	1215–1221	1216	δ_{C-H}
	12	1010	1030		1035–1042	1021	δ_{C-C}
	13	3060	3036	3060		3062	ν_{C-H}
	18a	1037	1068	1030	1069–1082	1039	δ_{C-H}
	19a	1485	1482	1484	1480–1490	1474	ν_{C-C}
	20a	3080	3054			3016 (3053)[b]	ν_{C-H}
B_2	3	1326	1288	1326		1216	δ_{C-H}
	6b	606	652			652	δ_{C-C}
	7b	3047	3036	3045		3016 (3053)[b]	ν_{C-H}
	8b	1596	1572	1608	1568–1575	1557	ν_{C-C}
	14	1309	1375		1379–1390	1387 (1349)	ν_{C-C}
	15	1147	1148	1161	1152–1165	1170	δ_{C-H}
	18b	1037	1085	1050		1060 (1120)	δ_{C-H}
	19b	1485	1439	1535	1437–1449	1412 (1472)	ν_{C-C}
	20b	3080	3083			3124	ν_{C-H}

B_1	4	703	749	738	750–769	779	γ_{C-C}
	5	985	942	980	930–950	958	γ_{C-H}
	10b	849	886	855		850	γ_{C-H}
						(755)	
	11	671	700	671	680–713	652	γ_{C-H}
	16b	405	405		410–426	400	γ_{C-C}

[a] Only IR-active fundamentals (in cm^{-1}).
[b] Bands enclosed in brackets are reassigned values after the study of three deuterated pyrylium cations (404).

78 Infrared Spectra of Heterocyclic Compounds

arsenin) are planar aromatic systems. Their IR spectra (at about 100 K as polycrystalline solids) were recently reported (408a), and assignments were made by analogy with pyridine and with the help of Raman spectra by studying the corresponding band polarization (at room temperature in the liquid), knowing that A_1 modes are polarized. The results are presented in Table 1.42.

188a   **188b**

The increased mass of the heteroatom affects strongly out-of-plane ring modes (which decrease by an average of 95 cm^{-1} between pyridine and phosphabenzene and by an average of 22 cm^{-1} between phospha- and arsabenzene) and in-plane ring modes (which decrease by an average of 90 cm^{-1} between pyridine and phosphabenzene and 33 cm^{-1} between phospha- and arsabenzene). There is little change in the CH stretching frequencies in the series pyridine-phosphabenzene-arsabenzene. More substantial structural changes (decreased force constant of the carbon-heteroatom X bond, increased C–X bond length, decreased C–X–C bond angle) occur on going from pyridine to phosphabenzene than from the latter to arsabenzene, in agreement with the above average changes in wavenumbers.

1.6.1.6. Quinolines

Far-infrared spectra of quinoline **189** and quinoline-*N*-oxide **190** were analyzed in order to establish correlations between the far-infrared features and intra- or intermolecular vibrations (409, 410). A similarity was found between ring skeletal vibrations of **189** and **190**, indicating that the addition of the oxygen atom leaves these modes almost unperturbed.

189   **190**

The ring skeletal modes are 400, 360, 280, 200, and 180 cm^{-1} for **189** and 380, 360, 270, 220, and 180 cm^{-1} for **190**. The β_{NO} for **190** was assigned in the range 320–340 cm^{-1}, and around 160 cm^{-1} an out-of-phase bending mode (γ_{NO}) was found; neither of these have a counterpart in the spectra of quinolines. A similarity between the far-infrared spectra of phenol **190** and those of its complexes was evidenced.

2-Quinolones **191** were reported (411) to absorb in the range 1641–1667 cm^{-1}, while 4-quinolones **192** absorb (412, 413) between 1620 and 1647 cm^{-1}.

191   **192**

Table 1.42. Assignments for the Infrared-Allowed Fundamental Vibrations of Pyridine, Phosphabenzene, and Arsabenzene (cm^{-1})

C_{2v} Species	Wilson Number	Pyridine[a]	Phosphabenzene[b]	Arsabenzene[b]	Description[c]
A_1	2	3054	3074	3065	C–H stretch
	20a	3054	3036	3027	C–H stretch
	13	3036, 3057	3014	3000	C–H stretch
	8a	1583	1563	1546	Ring stretch
	19a	1482	1396(?)	1377	Ring stretch
	9a	1218	1216(?)	1186	C–H bend
	18a	1068	994	977	C–H bend
	12	1030	910	890	Trig. ring
	1	992	761	674	Sym. breath
	6a	605	457	356	Ring deformation
B_1	5	942	910(?)	885(?)	C–H bend
	10b	886, 1007	820	803	C–H bend
	4	749	607	587	Ring deformation
	11	700	700	691	C–H bend
	16b	405	295	263	Ring deformation
B_2	20b	3080	?	?	C–H stretch
	7b	3036	?	?	C–H stretch
	8b	1572	1514	1510	Ring stretch
	19b	1439	1400	1403	Ring stretch
	14	1375, 1355	1283	1270	Ring stretch
	3	1288, 1227	1183	1190	C–H bend
	15	1148	1094	1089	C–H bend
	18b	1048, 1069	1025(?)	1021(?)	C–H bend
	6b	652	533	497	Ring deformation

[a] Assignment according to reference 352; the second value after the comma represents assignments after reference (356a) when it differs from the former.
[b] Assignments after reference 408a.
[c] The choice of axes for the C_{2v} species follows that recommended by the Joint Commission for Spectroscopy (408b) and results in an interchange of the B_1 and B_2 species relative to the earlier convention. Wilson's numbering of bands (408c) is conserved in Tables 1.37, 1.41, and 1.42.

80 Infrared Spectra of Heterocyclic Compounds

2-(Trifluormethyl)-4-quinolone and 4-(trifluormethyl)-2-quinolone were studied (414). The IR spectra of the two compounds showed strong absorptions at 1620 and 1610 cm^{-1}, respectively, outside the range established for 4-quinolones. IR spectra of substituted 4-quinolones with the general formula **193** (R = C$_5$H$_{11}$, C$_6$H$_{13}$) were studied, and carbonyl absorption at 1635 cm^{-1} has been reported (415).

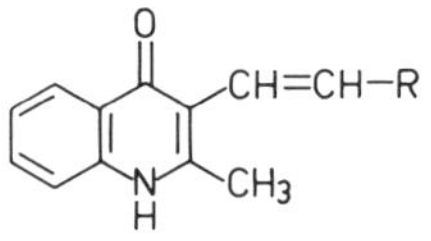

193

The infrared spectrum of quinolizin-4-one **194** and derivatives has been studied by Thyagarajan (416). The sensitivity of the amide carbonyl band to the presence of various substituents in the molecule has been evidenced. 2-Phenyl-4H-quinolizin-4-one **195** exhibits carbonyl absorption (417) at 1659 cm^{-1} and $\nu_{C=C}$ at 1630 cm^{-1}, whereas in 2-phenyl-4H-quinolizin-4-thione **196** the $\nu_{C=S}$ band is at 1110 cm^{-1}.

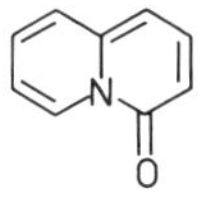

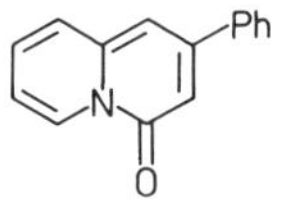

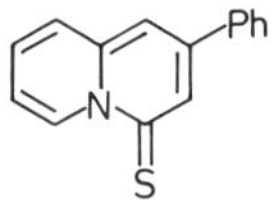

194   **195**   **196**

N-Acyldihydroquinaldonitriles **197** and N-acyldihydroisoquinaldonitrile **198** (Reissert compounds) do not present absorption, owing to the presence of a cyano group (418).

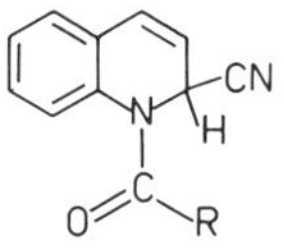

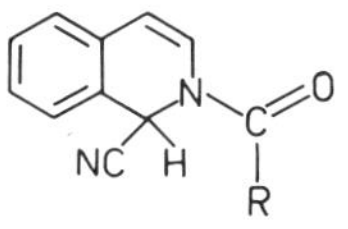

197   **198**

Infrared data for thienopyridines **199, 200** have also been reported (419, 420).

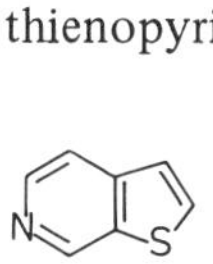

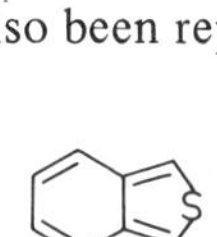

199   **200**

Azaindoles (pyrrolopyridines) were studied by Willette (421), and the principal bands for the four parent azaindoles **201–204** were listed and grouped into regions according to the scheme used by Katritzky and Ambler (1). Characteristic ring-stretching modes (combination of pyrrole and pyridine $\nu_{C=C}$ and $\nu_{C=N}$ modes) have been reported (421).

201   202

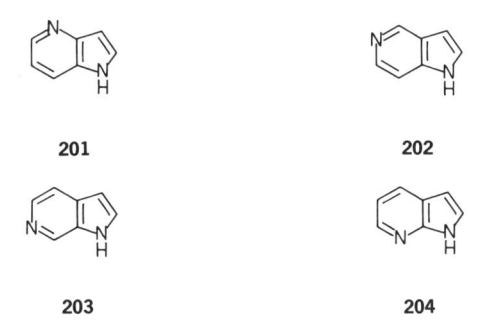

203   204

1.6.1.7. Carbolines

Kuchkova and coworkers (422–424) have studied and summarized IR data on carbolines **205–207**. They found in tetrahydro-β-carbolines **205** two ν_{NH} absorptions, at 3440 cm^{-1} (indole) and 3300 cm^{-1} (423), whereas the dihydro- and β-carboline derivatives exhibit (422, 424) only the ν_{NH} band in the spectral range 3250–3370 cm^{-1}. In all these compounds the $\nu_{C=N}$ band corresponding to rings was reported to be at 1635–1640 cm^{-1}.

205   206   207

1.6.1.8. Naphthyridines and Other Fused Systems with One Nitrogen Heteroatom per Ring

Attempts to establish correlations between the C–H out-of-plane deformation frequencies and the substitution patterns in the naphthyridine series were unsuccessful. Ikekawa (425) has studied IR spectra of twelve 2,7-, 1,6-, and 1,7-naphthyridines in the range 700–900 cm^{-1}, and Paudler and Kress (426, 427) studied the IR spectra of 1,6-naphthyridines. These studies have led to the assignments of absorption bands at 860–900 cm^{-1} for an isolated hydrogen, 830–850 cm^{-1} for two adjacent hydrogens, and 760–820 cm^{-1} for three adjacent hydrogens. However, Armarego and coworkers (428) concluded from a study of the IR spectra of all 1,X-diazanaphthalenes (X = 2 to 8) that no correlation between band position and substitution pattern exists in these spectra, since they show a wide variation from one isomer to another.

Czuba (429) compared the spectrum of 1,5-naphthyridine with that of a 5-substituted quinoline. Woźniak and Roszkiewitz (430) showed that with few exceptions the substituted 1,8-napthhyridines exhibit a ring-bending (skeletal) vibration at 690–740 cm^{-1}, three adjacent hydrogens give absorptions at 750–795 and 810–885 cm^{-1}, two adjacent hydrogens at 785–855 cm^{-1}, and one isolated hydrogen at 795–810 and 885–920 cm^{-1}. Woźniak (431) found the following characteristic absorptions for substituted 1,7-naphthyridines: ring bending at 680–740 cm^{-1}, three adjacent hydrogens at 750–795 cm^{-1} and 805–830 cm^{-1}, two adjacent hydrogens at 780–855 cm^{-1}, and one isolated hydrogen at 855–920 cm^{-1}. He concluded that the IR frequencies in the range 700–900 cm^{-1} are less appropriate for diagnostic use in 1,8-naphthyridines than in 1,7-naphthyridines. The broad overlap of the bands and the presence of uncharacteristic peaks in individual substituted 1,8-naphthyridines does not permit an unambiguous structural assignment of unknown 1,8-naphthyridines on the basis of IR spectra alone.

The keto-enol tautomerism in hydroxynaphthyridines has been studied (432).

Hawes and Gorecki (433) reported the following data for pyrimidonaphthyridines: 4-oxo-3,4-dihydropyrimido[4,5-b]-1,6-naphthyridine **208**: 3040 cm^{-1} (N–H), 2920–2600 cm^{-1} (N–H bonded), 1705 cm^{-1} (C=O); 4-oxo-3,4,5,10-tetrahydropyrimido-[4,5-b]-1,6-naphthyridine **209**: 3425 and 3125 cm^{-1} (NH), 1660 cm^{-1} (C=O).

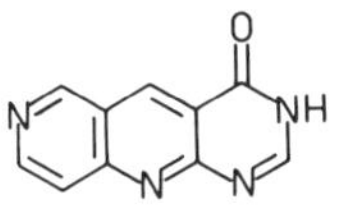

208 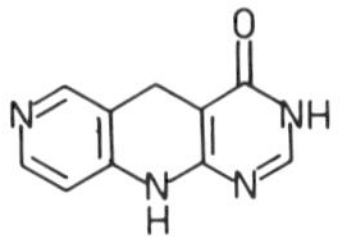   **209**

For pyrido[2,3-h]-2,6-naphthyridin-5-one **210, 211** the tautomeric equilibrium has been studied (434), and ν_{NH} at 3500 cm^{-1}, ν_{OH} at 2900 cm^{-1}, and $\nu_{C=O}$ at 1670 cm^{-1} have been reported.

210   **211**

IR spectra were given and discussed in detail for derivatives of 4-azafluoren-9-one **212** (435). For fused quinoline derivatives the data reported are: ν_{NH} 3410 and $\nu_{C=O}$ 1660 cm^{-1} for **213** (436); 1630, 1610, and 1580 cm^{-1} for **214** (437); and $\nu_{C=O}$ 1712 cm^{-1} and $\nu_{C=N}$ 1620 cm^{-1} for **215** (438).

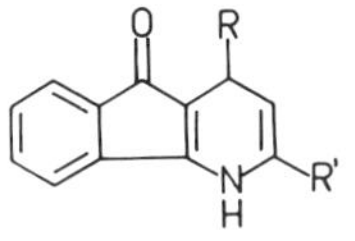

212 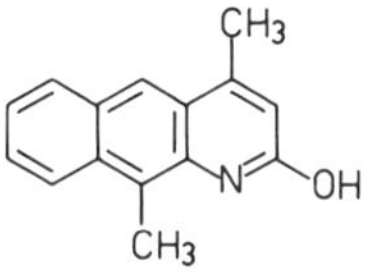   **213** 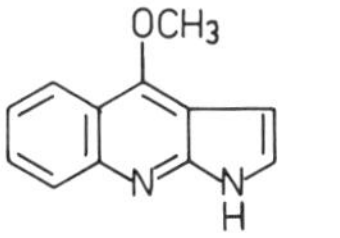   **214** 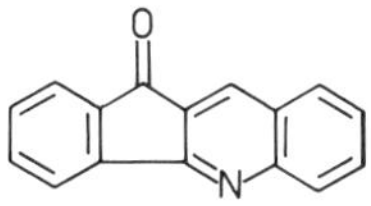   **215**

216 **217**

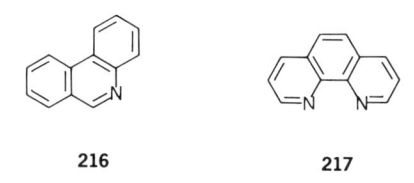

Infrared data (439) have been reported for phenanthridines **216** concerning the substitution pattern. The characteristic bands (in carbon disulfide) are 746 cm^{-1} (four hydrogens) and 888 cm^{-1} (one hydrogen). Data were also reported for 6-aminophenanthridine (440) and phenanthridones (441). IR spectra of 1,10-phenanthroline **217** have been studied (442, 443). A complete assignment of the vibrations in the Raman and IR spectra of ten isomeric phenenthrolines has been carried out (444).

1.6.2. Diazines

1.6.2.1. Pyridazines

Infrared spectra of pyridazine **218** and polypyridazines have been reported (445).

218

Joris and Schleyer (446) have determined the strength of the hydrogen bond between methanol and pyridazine on the basis of IR spectra.

1.6.2.2. Pyrimidines

Pyrimidine **219** is a planar molecule having C_{2v} symmetry. Its 24 normal vibrations are grouped into four species: nine a_1, two a_2, five b_1, and eight b_2. Under the selection rule of the point group C_{2v}, the a_1, b_1, and b_2 species are infrared-active.

219

The vibrational spectra of diazines have been investigated by Lord and coworkers (447). Ito and coworkers (448) suggested the first detailed assignment for **219**. A vibrational analysis of the electronic absorption bands of **219** and pyrimidine-d_4 (vapor) revised the assignment of the totally symmetric modes (449–451). Other modes were revised later by Sbrana and coworkers (452) on the basis of the crystal spectra of pyrimidine and pyrimidine-d_4 in polarized light, and by Foglizzo and Novak (453), who obtained the crystal spectra at liquid nitrogen temperature and the solution spectrum in nonpolar solvents. The IR and Raman spectra of pyrimidine complexes of iridium(III), with the general formula $K_{3-n}[IrL_nCl_{6-n}]$, where L

Table 1.43. Vibrational Assignment in Pyrimidine (cm^{-1})

C_{2v} Species	Vibration Number	Liquid (452)	Complex (454)	Description
A_1	20a	3084	3084	ν_{CH}
	2	3040	3050	ν_{CH}
	13	3001	3000	ν_{CH}
	8a	1560	1560	Ring
	19a	1395	1396	Ring
	9a	1134	1136	δ_{CH}
	12	1067	1066	Ring
	1	987	990	Ring
	6a	675	678	Ring
B_1	7b	3089	3039^a	ν_{CH}
	8b	1545	1569	Ring
	19b	1463	1465	Ring
	14	1355	1224	Ring
	3	1222	1355	δ_{CH}
	15	1155	1156	δ_{CH}
	18b	1075	1067^a	δ_{CH}
	6b	620	622	Ring
B_2	5	993	980	γ_{CH}
	11	809	810	γ_{CH}
	10b	718	955^a	γ_{CH}
	4	705	720^a	Ring
	16b	344	346	Ring
A_2	17a	870	927^a	γ_{CH}
	16a	394	398	Ring

^{a} Bands for which new frequencies were proposed on the basis of the proposed method.

denotes the pyrimidine ligand and $n = 1, 2, 3$, have been investigated, and a consistent vibrational assignment was deduced for pyrimidine itself (454). The method is based on the lowering of the symmetry on coordination involving only one nitrogen atom. The complex belongs to the C_s symmetry group in which all normal vibrations are active both in IR and Raman spectra. This complexation method has also been applied to solve other vibrational assignment problems, and has been very useful in deciding the correct description. The results of revised assignments in light of the new data are given in Table 1.43 together with Sbrana's liquid-phase spectral data (452).

Liquid-phase Raman spectra and liquid- and vapor-phase IR spectra were reported (455) for pyrimidines (pyrimidine-2-d, pyrimidine-2,5-d_2, pyrimidine-4,6-d_2, pyrimidine-2,4,6-d_3, and pyrimidine-d_4). The assignments given for the vibrations involved were consistent with the frequency sum rule and product rule criteria.

Table 1.44. Out-of-Plane Vibrations of the Pyrimidine Ring in Halogenated Derivatives (cm^{-1})

Substituent	Vibration 4	16a	16b
None	704	397	344
2-Chloro-	725	404	480
2,4-Dichloro-	763	435	621
	764	437	
2,6-Dichloro-	753	394	
2,6-Dibromo-	752	434	
2,4,6-Trichloro-	757	483	615
	755	490	
Tetrachloro-	755	543	639
2,4,6-Trifluoro-	769	512	794

The in-plane vibrations ν_{20a}, ν_{9a}, ν_{6a}, and ν_{18b} of **219** are sensitive to the substitution by chlorine in the 2-position (456). Bands observed at 1144, 754, and 434 cm^{-1} were assigned as ν_{20a}, ν_{9a}, and ν_{6a} (in 2-chloropyrimidine), and ν_{18b} (C–Cl bending) was estimated to be around 270 cm^{-1}. The modes of vibration for the remaining frequencies and their assignments remain the same as in pyrimidine.

Reactions of 2-chloropyrimidines with different nucleophiles enabled the isolation of a variety of 2-substituted pyrimidines exhibiting pharmaceutical activity (457) and strong absorption bands near 1600 cm^{-1}. In the spectra of 1-substituted-5-aryl- and 4,5-diaryl-2(1H)pyrimidones, the carbonyl group absorbs near 1660 cm^{-1} (in chloroform) (458). The coupling of the ν_{C-X} vibrations with the vibration of the cycle has been studied for halopyrimidines and the peculiarities of these couplings have been determined (459, 460). The out-of-plane vibrations of the pyrimidine ring (4, 16a, and 16b) are given in Table 1.44.

Some pyrimidine ring vibrations involving carbon-halogen motions for 2,6-dichloro- and 2,6-dibromo-pyrimidines **220A–220G** are given (460).

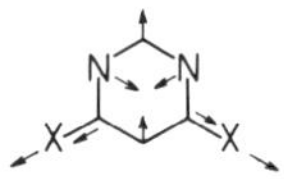

δ_{ring} ν_5 **220a**   X = Cl   1180 cm^{-1}

   X = Br   1144 cm^{-1}

   1153 cm^{-1}

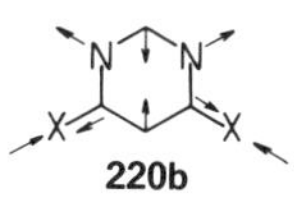

δ_{ring} ν_8 **220b**   X = Cl   730 cm^{-1}

   X = Br   710 cm^{-1}

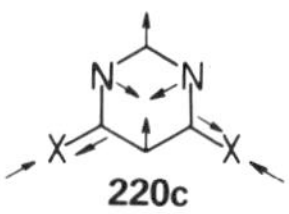

$\nu_{C\text{-hal}}$ ν_9 **220c**   X = Cl   411 cm^{-1}

   X = Br   294 cm^{-1}

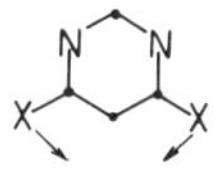

$\delta_{C\text{-hal}}$ ν_{10} **220d**   X = Cl   160 cm^{-1}

   X = Br   150 cm^{-1}

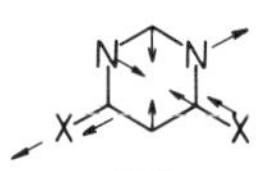

220e

δ_{ring} ν_{24} X = Cl 810 cm^{-1}
           X = Br 795 cm^{-1}

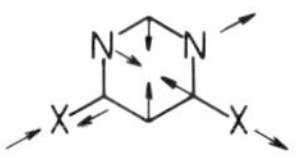

220f

δ_{C-hal} ν_{25} X = Cl 449 cm^{-1}
              X = Br 360 cm^{-1}

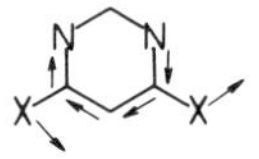

220g

δ_{C-hal} ν_{26} X = Cl 422 cm^{-1}
              X = Br 325 cm^{-1}

Spectroscopic data are discussed for mono-, di-, and trimethylpyrimidines in reference 461 and for 2-alkylthiopyrimidine **221** in reference 462. It was shown that only the 6a, 16b, and 4 vibrations of the pyrimidinic ring are not influenced by substitution in the 2-position. For a vibrational treatment of pyrimidinic derivatives with formula **221** (R = CH$_3$, C$_2$H$_5$, etc), the C_s symmetry group seems to be more correct than C_{2v}, indicating that the SR group does not rotate freely.

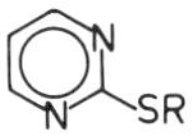

221

2- and 4-Amino-, -methylamino-, and -dimethylaminodichloro-5-methylthio-pyrimidines and their deuterated derivatives were studied by Gauthier and Lebas (463). Substituent-sensitive ring vibrations, ring out-of-plane vibrations, and amino vibrations were discussed. Spectroscopic data of the latter vibrations allowed estimation of the self-association of these compounds in solution and in the solid state.

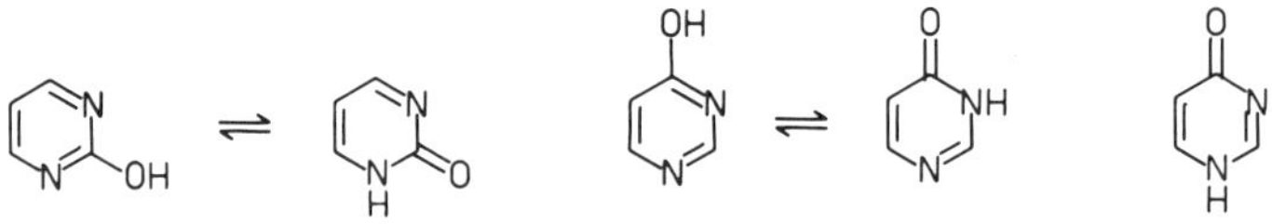

222 ⇌ **223**   **224** ⇌ **225**   **226**

Hydroxypyrimidines exhibit keto and enol forms. Tautomeric equilibria in 2- and 4-monooxopyrimidines **222–226** and some of their methyl derivatives were investigated in the gas phase, in low-temperature solid matrices, and in solution (464). Comparable populations of the keto and enol forms were found in all the 4-oxopyrimidines in the gas phase and in 4-oxo-6-methylpyrimidines in low-temperature matrices (the tautomeric equilibrium constant is about 1). The tautomerism in 2-ethoxy-4-pyrimidone **227–229** was evidenced both by its UV spectrum and by the stretching vibration of the NH bond (465). In aqueous solution both *o-* and *p-*quinonoid **227–229** forms are equally represented, and in chloroform the *ortho* form **227** predominates.

Aromatic Six-Membered Rings 87

Table 1.45. Assignments in the Spectral Range 1500–1800 cm^{-1} (in DMSO) for 4,6-Dihydroxypyrimidines and Derivatives

Compound	$\nu_{C=C}$	$\nu_{C=N}$	$\nu_{C=O}$
4,6-Dimethoxypyrimidine	1595	1555	–
1-Methyl-5-methoxy-6-pyrimidone	1605	1550	1673
5,5-Diethyl-2-phenyl-4,6-pyrimidinedione			1728, 1682
5-Methyl-4,6-pyrimidinedione			1660
5-Phenyl-4,6-pyrimidinedione			1658
5-Chloro-4,6-pyrimidinedione			1673
5-Phenoxy-4,6-pyrimidinedione			1672
5-Bromo-4,6-pyrimidinedione			1673
5-Nitro-4,6-pyrimidinedione			1640

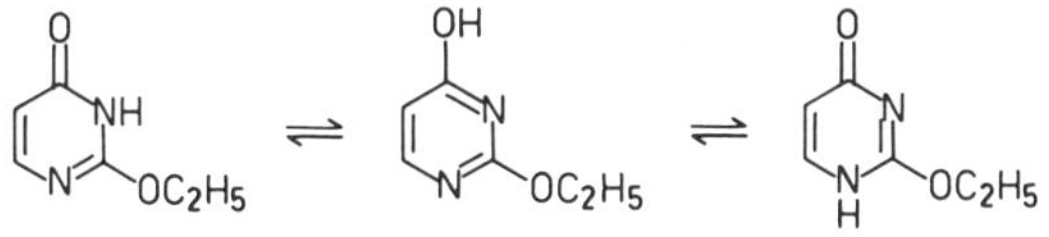

227   228   229

The structure of 4,6-dihydroxypyrimidine derivatives was studied (466), and assignments were given for the bands in the spectral range 1500–1800 cm^{-1}. Several model compounds with fixed enol and keto structures were studied; IR data are summarized in Table 1.45.

For 4,6-pyrimidinediones, only a single carbonyl band was observed, and a dipolar structure **230** has been suggested. The frequency of the carbonyl band depends on the nature of the substituent R' in position 5.

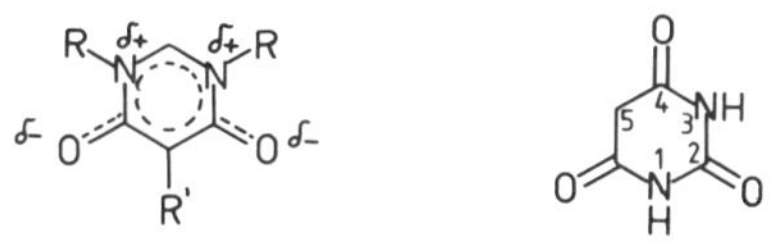

230   231

The IR and Raman spectra of barbituric acid **231** and derivatives, particularly the pharmaceutically important 5,5-disubstituted compounds (barbiturates, widely used in biological and medical studies), have been extensively studied. The first attempt to correlate the pharmaceutical activity of the barbiturates with their IR spectra was made by Price and coworkers (467) and Kyogoku and coworkers (468), who showed by IR spectroscopy that barbiturates selectively form hydrogen-bonded complexes with molecules containing adenine. Matrix isolation vibrational spectroscopy has enabled vibrational assignments for **231** and its 1-methyl- and 1,3-dimethyl-substituted derivatives (469, 470). These studies have been extended to the pharmaceutically important 5,5-disubstituted derivatives such as barbital (471).

88 Infrared Spectra of Heterocyclic Compounds

The interpretation of IR spectra of barbiturates may be of obvious value for understanding the binding of drugs to compounds such as adenine derivatives. Infrared spectra of these compounds were recorded in nitrogen and argon matrices or in an argon matrix doped with 1% nitrogen. The apparatus used for matrix deposition of barbituric acid derivatives and the experimental details have been described (469). The data enabled a vibrational assignment of the barbituric acid spectra for both the monomeric form and the solid. The data suggested that the compounds form dimers with a cyclical structure through hydrogen bonding; these dimers are further associated by hydrogen bonds between the NH and carbonyl groups 232.

232

Craven and coworkers (472) have reported crystal structures of three forms of solid barbital: barbital I, with a structure identical to that of solid barbituric acid (molecules linked through the carbonyls in the 2 and 4 positions), barbital II, with the molecules linked through the carbonyls in the 4 and 6 positions, and barbital IV, with the molecules linked by two hydrogen bonds to the carbonyl in the 4 position. Comparison of the spectra of barbital and barbituric acid (Table 1.46) showed that the vibrations of the pyrimidine ring occur with similar frequencies in the two compounds. Barnes and coworkers (469–471) concluded that several of the bands listed as characteristic of barbiturates by Mesley (473) in the infrared spectrum and by Willis and coworkers (474) in the Raman spectrum, actually originate in the substituents rather than in the pyrimidine ring. Table 1.46 summarizes only the characteristic bands of the NH and C=O groups, which are sensitive to intermolecular interactions and are thus particularly useful in studying complexes of the barbiturates.

Structural investigations of 5-nitrobarbituric acid 233 and its monosubstituted salts have been performed (475).

Sigma anionic complexes in the pyrimidine series were recently reported (476). The structure of sigma complexes of 5-nitropyrimidine and its methoxy derivatives with the acetonate ion (in the presence of KOH) 234 was studied by IR and ^{1}H-NMR and found to be similar to that of Meisenheimer complexes.

233   **234** K^+ R = R' = H, MeO

Table 1.46. Comparison of IR Spectra of Barbituric Acid and Barbital in Argon Matrixes (Monomers)[a]

Barbituric Acid	Barbital	Assignment
3437 s	3430 s	NH stretch
3431 sh	–	NH stretch
1764 s	1776 s	4,6-C=O sym. stretch
1754 vs	1757 vs	4,6-C=O asym. stretch
1433 m	1426 s	CN stretch
1418 s	1405 m	CN stretch
1322 s	1305 sh	NH i.p. bend
1319 s	1301 s	NH i.p. bend
1222 s	1225 m	CN stretch
1192 w	1167 w	CN stretch
768 w	758 w	C=O o.p. bend
754 ms	749 s	C=O o.p. bend
674 s	669 s	NH o.p. bend
670 s	662 sh	NH o.p. bend
–	654 w	C=O i.p. bend
–	607 w	C=O o.p. bend

[a] Data in cm^{-1}; i.p. = in-plane; o.p. = out-of-plane.

1.6.2.3. Pyrazines

The pyrazine molecule **235** has D_{2h} symmetry. In 24 normal vibrations, 10 are IR-active (four B_{1u}, two B_{2u}, and four B_{3u}), 12 are Raman active (five A_g, one B_{1g}, four B_{2g}, and two B_{3g}) and two are inactive. The vibrational assignment for pyrazine has been much discussed (477–482). A correct assignment for the pyrazine molecule was performed (483) by means of the complexation method. The coordinated **235** has C_{2v} symmetry. The complexes with iridium chloride are of the general formula $[IrL_3Cl_{6-n}]X_{3-n}$, where X = Na, K, and L = pyrazine. The most studied complex was $K_2[IrLCl_5]$, which is water-soluble, and thus it was possible to record its solution Raman spectra.

235

The vibrational assignment for **235** based on the complexation method is given in Table 1.47. It is interesting to note the agreement between data provided by different experimental methods. New assignments were made only for a few *u* modes (20*b*, 13, 16*a*, and 17*a*).

The influence of hydrogen-bond formation on pyrazine vibrational frequencies was studied (484), and the shift to higher frequencies was seen in hydrogen-donor solvents.

Table 1.47. Vibrational Assignment (cm^{-1}) for Pyrazine

D_{2h} Species	Vibration Number	Liquid (482)	Complex (483)	Description
A_g	2	3052	3055	ν_{CH}
	8a	1581	1580	Ring
	9a	1246	1233	δ_{CH}
	1	1016	1016	Ring
	6a	602	602	Ring
A_u	17a	997	960[a]	γ_{CH}
	16a	423	350[a]	Ring
B_{1g}	10a	930	927	γ_{CH}
	13	3082	3012[a]	ν_{CH}
	19a	1491	1483	Ring
	18a	1126	1130	δ_{CH}
	12	1022	1018	Ring
	7b	3034	3040	ν_{CH}
B_{2g}	8b	1522	1525	Ring
	3	1353	1346	δ_{CH}
	6b	699	704	Ring
B_{2u}	11	796	785	γ_{CH}
	16b	412	418	Ring
B_{3g}	5	974	983	γ_{CH}
	4	756	756	Ring
B_{3u}	20b	3011	3063[a]	ν_{CH}
	19b	1414	1411	Ring
	14	1149	1149	Ring
	15	1068	1063	δ_{CH}

[a] New assignments from reference 483.

Spectral data on some pyridazinols and pyridazindione have also been reported (485, 486).

The tautomeric enamine-imine equilibrium in aminothiazines **236** has been investigated (487).

236

Table 1.48. Infrared Data on Fused Pyridazines

Compound	IR Bands, cm^{-1}	Ref.
8-Amino-7-chloro-s-triazolo[4,3-b]pyridazine **237**	3340, 3150 ν_{NH_2} 1667, 1578 $\nu_{C=N}$	488
6-Carbethoxypyrido[2,3-c]pyridazin-5-8H-one-2-oxide **238**	1706 $\nu_{C=O}$ 1290 ν_{N-O}	489
Pyrrolo[2,3-d]pyridazine-5-oxide **239**	1620 $\nu_{C=N}$	490
2H-1-benzothieno[2,3-d]pyridazinethione-1 **240**	3120 ν_{NH} 1520 1440 1220 $\nu_{C=S}$	491
Pyrrolo[1,2-b]pyridazine **241**	2950, 1640, 1590	492
Pyrrolo[1,2-b]pyridazine **242**	1661, 1546, 1290, 1110, 800	493

1.6.2.4. Fused Diazines

1.6.2.4.1. Fused Pyridazines. Infrared data (488–493) on a number of fused pyridazine systems **237–242** are summarized in Table 1.48.

Chemical properties and tautomeric behavior of 1,4-dihydroxypyridazino[4,5-d]pyridazine **243** ⇌ **244** have been studied (494).

92 Infrared Spectra of Heterocyclic Compounds

243 ⇌ **244**

1.6.2.4.2. Fused Pyrimidines. The IR spectra of the four parent pyridopyrimidines **245** have been studied by Armarego and coworkers (495). The strong bands in the 1000–650 cm⁻¹ region have been assigned to the CH out-of-plane bending vibrations, and the bands observed in the range 2000–1750 cm⁻¹ to overtone and combination bands. On the basis of the low ν_{NH} (3389 cm⁻¹) and high $\nu_{C=O}$ (1745 cm⁻¹), Mason (496) suggested a quasi *o*-quinonoid form for pyrido[3,2-*d*]pyrimidin-4(3*H*)-one **246A**. In pyridopyrimidin-4(3*H*)-ones **246A–D** and -diones **247A–D** the oxo structures are the predominant forms in the solid phase.

245a   **245b**   **245c**   **245d**

246a   **246b**   **246c**   **246d**

247a   **247b**   **247c**   **247d**

Yale and coworkers (497–501) and others (502–508) have prepared and characterized by IR spectra several pyridopyrimidine-4-ones and -2,4-diones **248–262** (Table 1.49).

In 2,3-disubstituted pyrido[1,2-*a*]pyrimidin-4-ones, $\nu_{C=O}$ appears as a strong band at 1655 cm⁻¹ (509). In 2,3-cycloalkylene-4*H*-pyrido[1,2-*a*]pyrimidin-4-ones with the general formula **263** (*n* = 1–4), $\nu_{C=O}$ appears in the region 1665–1685 cm⁻¹, whereas in related compounds such as **264** $\nu_{C=O}$ is shifted to 1600–1610 cm⁻¹ (510).

A quantum chemical interpretation of the IR spectroscopic data of pharmacologically active homopyrimidazole derivatives has been reported (511).

Aromatic Six-Membered Rings 93

Table 1.49. Infrared Data on Fused Pyrimidines

Compound	IR Bands, cm^{-1}	Ref.
248	CCl$_4$: 1698–1702, 1650, 1415	497
249	KBr: 1700, 1600, 1510	498
250	1690, 1640, 1580, 1535, 1470	499
251	CDCl$_3$: 1650, 1635, 1575, 1540, 1470, 1430	500
252	1695, 1680, 1570, 1530, 1480, 1450, 1410	501
253	1700, 1630, 1580, 1540	501
254	1710, 1760	502
255	3440, 1725, 1680	502

94 Infrared Spectra of Heterocyclic Compounds

Table 1.49. (continued)

Compound	IR Bands, cm^{-1}	Ref.
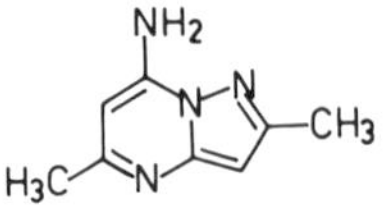 **256**	1647 (C=N)	503
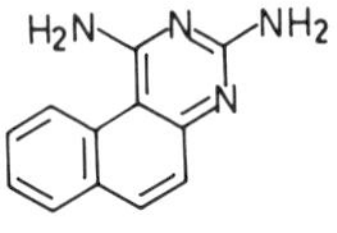 **257**	1640, 1564, 1515, 1490	504
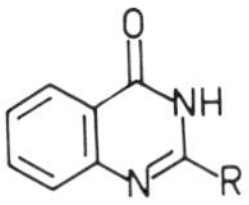 **258**	3450 (NH), 1620 (C=O), 1595 (NH bend)	505
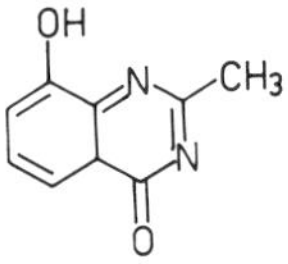 **259**	3300–2340 (OH), 1660 (C=O), 1640 (C=N)	499
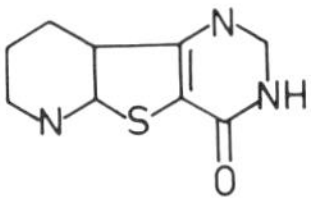 **260**	1681 (C=O), 3448 (NH)	506
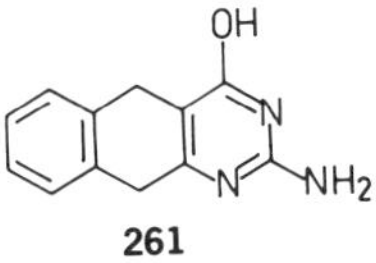 **261**	3350 (OH), 3220 (NH), 1670 (C=O), 1615 (C=C)	507
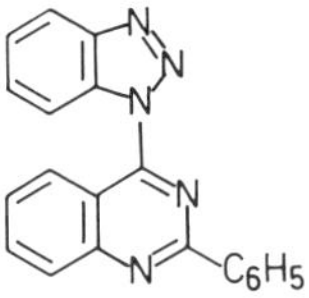 **262**	1613, 1595, 1565, 1548	508

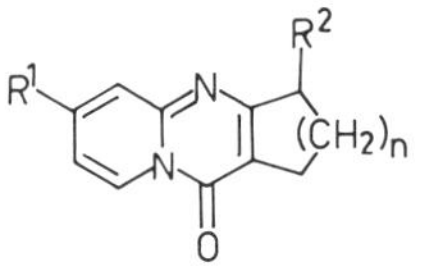

263

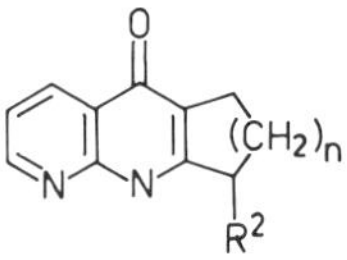

264

1.6.2.4.3. Pyrazine-N-oxides and Fused Pyrazines.

Infrared data on 2,5-diphenylpyrazine and 2,5-diphenylpyrazin-*N*-oxide have been reported by Gnichtel and coworkers (512). A strong band at 1256 cm^{-1} in 2,5-diphenylpyrazin-*N*-oxide was assigned to the N–O stretching vibration. Fused perhydrodiazines with the general formula **265** have been studied by Gnichtel and Hirte (513), and a strong carbonyl band has been shown in the 1730–1740 cm^{-1} range. The O–C–N stretching vibration has been established to be at 1100 cm^{-1}.

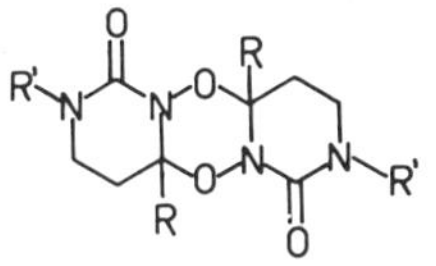

265

The absence of absorption bands in the 3700–3500 and 2600–2550 cm^{-1} regions in the IR spectrum of 4,6-dioxo-2-thio-1,2,3,4-tetrahydro-6*H*-indeno[2,1-*g*]pteridine **266 ⇌ 267**, and the presence of fairly intense bands at 3300 and 1700 cm^{-1} corresponding to the NH and amide carbonyl groups, respectively, led to the inference that this compound exists as tautomer **266** rather than **267** (514).

266   **267**

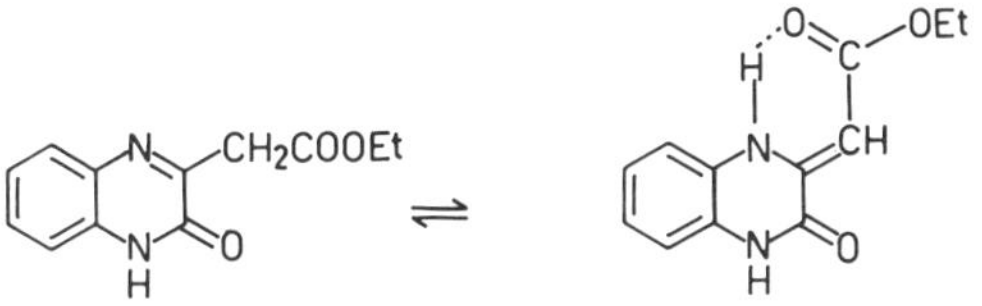

268   **269**

The side chain–ring tautomerism in six-membered nitrogen heterocycles **268, 269** has been studied (515). The 1,2-dihydro-2-oxo-3-ethoxycarbonylmethylquinoxaline **268**–1,2,3,4,-tetrahydro-2-oxo-3-ethoxycarbonylmethylenequinoxaline **269** equilibrium was shifted toward the left with an increase in temperature.

1.6.2.4.4. Quinoxaline Derivatives.

A spectrochemical study of quinoxaline-2,3-dithiol (a metal chelating agent) and related compounds was performed (516), and the IR spectra were compared with those of quinoxaline and its 2,3-substituted derivatives.

The thiol-thione **270–272** tautomerism was studied, and a weak absorption band was found between 3125 and 3150 cm^{-1} that is absent for other compounds in which no proton is available to give NH tautomeric forms. In 2,3-dihydroxyquinoxaline, a very intense band at 1674 cm^{-1} with a shoulder at 1700 cm^{-1} was assigned to ν_{CO} or ν_{NCO} modes and indicated that in this compound the keto form is prevalent or unique.

270 ⇌ **271** ⇌ **272**

1.6.2.4.5. Polycyclic Heteroaromatic Compounds. An infrared study of ring sulfoxides and their adducts with mercury and cadmium halides has been performed (517) in order to establish their conformation. X-ray structural analyses of various sulfoxides represented by the general formula **273** showed a folded structure **274** with a boat conformation of the central ring (X can be O, S, SO, SO$_2$, NH, or CO, while Y is SO). The S=O bond can assume two nonequivalent positions in all the sulfoxides: pseudo-equatorial **275** and pseudo-axial **276**.

Infrared studies (518) have shown that analogously to the hydroxyl or halogen substituents in cyclohexane, the equatorial S=O bond in the thianthrene system shows its stretching frequency at a higher energy than that of the axial S=O. Furthermore, on going from CCl$_4$ to CHCl$_3$ the axial isomer is displaced to lower wavenumbers more than the equatorial isomer. It was evidenced, by adduct formation, that the shift of the sulfoxide stretching frequency is different for the two conformers and this method may be useful to differentiate the axial and equatorial sulfoxide bands. The sulfoxide bands observed in the solid state spectra of various sulfoxide rings (Table 1.50) indicate a bathochromic shift on adduct formation, which is indicative of coordination through the oxygen atom (519). On the basis of the observed ν_{SO} frequencies it was concluded that the thioxanthone and phenoxathiin sulfoxides (lower $\nu_{C=O}$ frequencies) are axial, while the thianthrene mono- and trisulfoxides are equatorial. Additional evidence to distinguish between axial and equatorial sulfoxide groups was obtained by examining the sulfoxide deformation frequencies in the 300–500 cm^{-1} region.

All phenoxazines **277** exhibit NH frequency at 3330–3413 cm^{-1}, ring-stretching modes between 1449 and 1637 cm^{-1}, and γ_{CH} (four adjacent hydrogen atoms) between 735 and 765 cm^{-1} as in *ortho*-substituted benzene rings (520, 521).

Infrared data have been reported recently (522) for a new type of triazaphenothiazine heterocycle and 1,4,6-benzo[*b*]triazaphenothiazine **278**.

Aromatic Six-Membered Rings 97

Table 1.50. $\nu_{S=O}$ Frequencies in Sulfoxide Adducts (519)

Compound	$\nu_{S=O}$ (cm^{-1}) Ligand	HgCl$_2$ Adduct	Δ
Phenoxathiin sulfoxide	1038	983	−55
Thioxanthone sulfoxide	1025	997	−28
Thianthrene monosulfoxide	1077	1055	−22
Thianthrene disulfoxide "*cis*"	1088	1069	−19
Thianthrene disulfoxide "*trans*"	1039	982	−57
Thianthrene thioxide	1075	1054	−21
Phenothiazine sulfoxide	975	930	−45

Infrared data (523–528) concerning azaphenoxathiins **279–282** are summarized in Table 1.51.

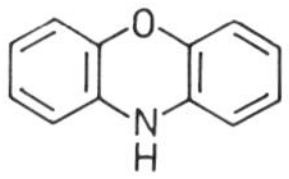

277

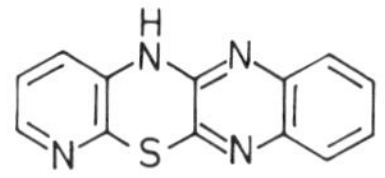

278

1.6.3. Vibrational Spectra of Nucleic Acid Constituents

The vibrational spectra of various nucleic acid constituents have been examined by several investigators, but the high complexity and low symmetry of these molecules makes detailed band assignments or normal coordinate calculations very difficult. Therefore, the first steps in this attempt were qualitative group-frequency assignments for some characteristic absorption bands and Raman lines. Later, it was possible to prepare deuterated analogs that facilitated the determination of force constants and thus permitted a better understanding of the vibrational motions giving rise to observed absorption bands. Recently, a great effort has been placed on measuring and identifying the low-frequency IR vibrations of these molecules. These vibrations of nucleic acids may have an important role in driving conformational changes and might therefore yield useful information on similar structures. This information will permit normal mode calculations, which will model the low-frequency force constants and nonbonded interactions in large biological molecules such as DNA and proteins.

The nucleic acid constituents exhibit some common features but also a number of peculiarities, and this is the reason for the individual treatment of pyrimidines and purines adopted here.

1.6.3.1. Pyrimidines

1.6.3.1.1. Cytosine. Cytosine is one of the four principal bases in native DNA. Beginning with the pioneering work of Blout and Fields (529) in the early

98   Infrared Spectra of Heterocyclic Compounds

Table 1.51.   Infrared Data on Azaphenoxathiins

Compound	Characteristic Bands in KBr (cm^{-1})	Ref.
1-Azaphenoxathiin 279	3400, 1480, 1440, 1420, 1280, 1225, 805, 755	523
7-Chloro-1-azaphenoxathiin	2900, 1600, 1560, 1470, 1410, 1280, 1210, 1195, 1100, 920, 850, 795	523
9-Chloro-1-azaphenoxathiin	3400, 2925, 1525, 1455, 1430, 1418, 1280, 1205, 1095, 915, 790, 765	524
7-Nitro-1-azaphenoxathiin	1600, 1510, 1420, 1345, 1280, 1225, 870, 815, 790	525
2-Azaphenoxathiin	1460, 1440, 1400, 1270, 1270, 1250, 1200, 880, 810	526
1,9-Diazaphenoxathiin 280	1590, 1455, 1420, 1280, 1220, 1100, 795	527
1,7-Diazaphenoxythiin 281	1590, 1560, 1540, 1480, 1440, 1420, 1280, 1210, 1080, 900, 870, 790	527
Benzo[1″,2″:5,6:5″,4″:5′,6′]bis[1,4]oxathiino[3,2-b:b′]dipyridine 282	3380, 2900, 2830, 1610, 1550, 1500, 1480, 1440, 1410, 1280, 1200, 1080	528

1950s, the vibrational spectra of cytosine **283** have been examined by several investigators (530, 531).

Susi and coworkers (532), by combining laser Raman and IR techniques, assigned the in-plane fundamentals of polycrystalline **283** on the basis of deuteration shift. A normal coordinate calculation was performed with a valence force field; frequencies and potential energy distribution were calculated. The in-plane fundamentals observed in IR spectra of polycrystalline **283** are given in Table 1.52, which also includes the same fundamentals for IR spectra of uracil **284** (533) and thymine **285** (534).

Table 1.52. Observed and Calculated In-Plane Fundamentals of Nucleic Acid Constituents (cm^{-1})

Uracil (534) Obs.	Calc.	Cytosine (532) Obs.	Calc.	Thymine (534) Obs.	Calc.	Assignment and Designation[a]
		3380 s	3356			ν_{NH_2} o.p.
		—	3240			ν_{NH_2} i.p.
3160 m	3134	3169 s	3190			ν_{NH} i.p.
3160 m	3132				3129	ν_{NH} o.p.
3100 s	3100	—	3121	3063	3126	ν_{CH} i.p.
3080 s	3092	—	3114			ν_{CH} o.p.
				2993	2990	ν_{CH_3} (asym)
				2930	2934	ν_{CH_3} (sym)
		1703 w	1689			δ_{NH_2}
1716 s	1684	1662 vs	1650	1735	1701	$\nu_{C_2=O}$ U(I)
1675 s	1653			1677	1674	$\nu_{C_4=O}\nu_{C=C}$ i.p. U(II)
—	1605			—	1606	$\nu_{C_4=O}\nu_{C=C}$ o.p. U(III)
		1615	1616			$\nu_{C=C}$
1508 m	1504	1505 m	1496	1495	1502	$\delta_{NH(1)}$
		1538 m	1530			δ_{NH}, ring
1453 s	1465	1465 s	1450	1483	1483	ν-ring U(IV)
				1447	1461	δ_{CH}, δ_{CN}
1417 s	1424			1406	1430	δ_{CH_3} (asym)
1390 m	1397	1364 m	1344	1366	1357	$\delta_{NH(3)}$
				1383	1378	δ_{CH} i.p. ring
1238 s	1248			1245	1250	δ_{CH_3} (sym)
1217 w	1211	1277 m	1295			ν-ring, δ_{CH} U(V)
		1236 m	1222	1203	1214	δ_{CH} o.p.
						ν-ring C–Me
						ν-ring
		1100 w	1126	1152	1153	ν-ring U(VI)
						ν-ring, NH$_2$

Table 1.52. (continued)

Uracil (534)		Cytosine (532)		Thymine (534)		Assignment and Designation[a]
Obs.	Calc.	Obs.	Calc.	Obs.	Calc.	
1003 m	1028	1010 w	1036	1028	1018	ν-ring U(VII)
993 m	980	994 w	1007			ν-ring U(VIII)
		966 w	955			ν-ring, NH$_2$
				984	984	r_{CH_3}
				815	815	ν-ring U(IX)
781 w	801	793 m	790			$\nu\delta$-ring
585 m	576	600 m	592	—		δ-ring U(X)
565 m	564	549 m	560	617	631	$\delta_{C=O}$ i.p. U(XI)
550 s	536	533	533	560	569	δ-ring U(XII)
—	384	—	386	—	392	$\delta_{C=O}$ o.p. U(XIII)
				321	319	δ_{C-Me}

[a] i.p., in-plane; o.p., out-of-plane; asym, asymmetrical; sym, symmetrical.

283   284   285

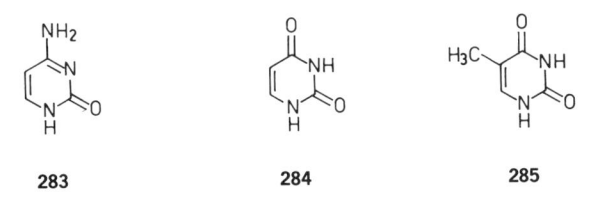

Skeletal vibrations appear to be characteristic for the pyrimidine bases in general. The mixed ring-stretching modes characteristic for the pyrimidine bases are in agreement with X-ray data concerning the lengths of the CN bonds: the C–NH$_2$ bond has been described as having about 30% double-bond character, and the ring CN "double bond" about 60% single-bond character. Recently, by selective substitution methods, Beetz and Ascarelli (535) provided detailed assignments of particular molecular vibrations. The carbonyl bending modes of 283 have been identified through the substitution of sulfur for oxygen at the C-2 position. Several ring vibrations, wagging and torsional modes of the amino group, and N$_1$–H out-of-plane bending modes have been identified by deuteration studies. Fermi resonance involving the carbonyl modes was evidenced in 283, as was the sulfur out-of-plane bending mode. Wagging and torsional modes of the amino group were evidenced in 2-thiocytosine. The frequencies observed in far-infrared spectra of crystalline powders of 283 are given in Table 1.53 together with those of other pyrimidine and purine bases obtained by the same method. Far-infrared spectra were obtained by Fourier transform of four to eight interferograms, averaging the resultant spectra.

Low-frequency vibrational Raman and IR spectra of crystalline cytosine monohydrate were measured (536) in order to explain the special role of water molecules in changing the effective pyrimidine-pyrimidine interaction in DNA. The results reflect an effective strengthening of pyrimidine-pyrimidine force, which, in the opinion of Kugel and coworkers (536) could be a contributing factor in hydration-induced conformational changes.

In the establishment of the biologically active conformations of transfer-RNA and ribosomal RNA and their specific recognition by various proteins, specifically placed methyl groups on pyridine and purine bases play an important role. Hydrogen bonding and restricted rotation studies demonstrated that the replacement of the amino group by methylamino in adenine and cytosine hinders normal base pairing of the free bases and destabilizes the double-helix structure of the respective polynucleotides (537).

In order to better elucidate the effect of *ortho*-methyl substitution on the conformation of the amino group in cytosine and in related heteroaromatic systems, IR spectra of amino and dimethylamino derivatives with and without an *ortho*-methyl group in 4- and 5-substituted pyrimidine and 4-substituted pyridine and in the respectively cytosine were recorded in the region of skeletal ring vibrations (538).

Katritzky and coworkers (539) showed that in *ortho*-disubstituted benzenes, steric interaction between an electron-donating substituent of pseudo C_{2v}

Table 1.53. The Far-Infrared Frequencies of Uracil, Cytosine, Thymine, and Xanthine (535)

Uracil	Cytosine	Thymine	Xanthine
853	1009		
827	990		
805	965		
792 β-ring	907 w		
762	862 sh	575 in-plane bend	355
740	819 NH w	545 in-plane O_7, O_8	325 out-of-plane bending
728	791 ring breath.		232
			206
616 r, δ-ring	780	433 out-of-plane	
587 β in-plane	760	521 O wagging	
567 β in-plane	744	485 ring bending	
548 δ-ring	726	407	
536 δ-ring	697		
515 sh	641	206	
437 sh, w	625	320 CH_3	
410 sh	600	285	
397 sh	582 δ-ring		
194 out-of-plane ring	565		241 libration mode
167 w	548 NH_2 w		
152 w	530 (114 + 422)		233 out-of-plane
129	513 (94 + 422)		211 out-of-plane
112	497 w		
88 w	485 broad NH_2		
76 w	481 w		

468
442
422 =, w
404
391 sh
232 out-of-plane ring
197
157
140
114
94
92

62 w

symmetry such as $-NR_2$ and another of cylindrical symmetry may cause a loss of resonance conjugation of the former with the ring by its twisting out of the molecular plane, thus decreasing the integrated intensity of the ν_8 bands. This observation provided a basis for examining the steric effects of *ortho*-methyl substitution on mesomeric coupling of the $-NR_2$ group with pyrimidine and pyridine moieties and thus indirectly on its conformation. Integrated intensities of ring vibration ν_8 at about 1600 cm^{-1} sensitive to the presence of electron-donating substituents were used to elucidate the steric effects of *ortho*-methyl on the mesomeric interaction between the $N(CH_3)_2$ group and the ring (538). Solution spectra of pairs with and without an *ortho*-methyl group of amino and dimethylamino derivatives of pyrimidine (4- and 5-substituted), pyridine (4-substituted), and methylcytosine were recorded, and the integrated intensity of bands in the region 1500–1700 cm^{-1} were measured. This intensity increases in all four series of compounds in the order: aminobenzenes < 5-aminopyrimidines < 4-aminopyridines < 4-aminopyrimidines, in agreement with Katritzky's data. The presence of an *ortho*-methyl group, in view of its electron-donating power being much smaller than that of $N(CH_3)_2$, is expected to have a small contribution to the intensity of the ν_8 band, and hence any intensity changes can be attributed to steric hindrance. The data point to a progressive twist of the dimethylamino group in hindered derivatives in the order pyrimidine-5 < pyridine-4 < pyrimidine-4 and are in agreement with the essential planarity of sterically crowded derivatives of cytosine shown by X-ray diffraction study.

1.6.3.1.2. Thymine. Normal coordinate calculations have been carried out for the planar modes of thymine, and their in-plane fundamentals were assigned (534). The observed and calculated fundamentals and their assignments are given in Table 1.53.

The crystal structure and lattice dynamics of the 1-methylthymine single crystal were studied (540). The four molecules in the unit cell are linked by double N–H···O hydrogen bonds at a distance of 2.83 Å forming dimers **286**. The hydrogen bonds occurring between pairs of molecules give rise to additional low-frequency vibration modes termed "internal vibrations." A set of force constants for hydrogen-bond vibrations was obtained and the internal vibrations were assigned. Some of the low frequencies of thymine, assigned by the selective substitution method, are given in Table 1.53.

286

1.6.3.1.3. Uracil. The vibrational assignments of in-plane fundamentals in crystalline uracil **284** were carried out by Susi and Ard (533). Four molecules (of C_s point symmetry) of **284** are arranged into two hydrogen-bonded dimers.

The molecular a' and a'' modes are split into A_g, B_u, A_u, and B_g crystal modes, and only the ungerade modes are IR-active. The frequencies were assigned on the basis of observed bands in polycrystalline uracil, N,N'-dideuterouracil, C,C'-dideuterouracil, and perdeuterouracil. The observed and calculated in-plane fundamentals of uracil and their assignment are given in Table 1.53.

1.6.3.2. Purines

The IR spectrum of unsubstituted purine **287A** has been studied by Willits and coworkers (541). They compared the spectra of sublimed films of a number of purines with the spectra of related pyrimidine and quinazoline structures and on this basis assigned two strong bands in the 1620–1550 cm^{-1} region to C=C and C=N vibrations of the purine ring and a broad band in the 3500–2500 cm^{-1} range (unsubstituted purine) to very strong intermolecular hydrogen bonding involving the hydrogen atoms attached to the 7(9)-nitrogen. From the comparative spectra of purine and its monodeuterated derivatives in specific positions, it was possible to make firm assignments of N–H and N–D stretching vibrations and tentative assignments of C–H deformation modes (542). Infrared difference spectra in KBr of equal concentrations (0.5 wt %) of purine and monodeuterated derivatives were compared in the relevant spectral regions and the isotopic effects were illustrated. A partial assignment of N–H and C–H bands resulting from Connolly's work (542) is given in Table 1.54.

The IR spectra of alcohol photoadducts of purine evidenced the addition of the purine ring at the pyrimidine moiety.

Assignments of IR bands of guanine **287B** have been proposed (543) by using specific ^{15}N-substitution together with specific D-substitution. For earlier work on guanine derivatives, see references 544–547.

Table 1.54. Assignment of NH and CH Bands in Purine 287A and Its Monodeuterated Derivatives (cm^{-1})

Purine	7(9)-Deutero-Purine	6-Deutero-Purine	8-Deutero-Purine	Assignment
2650	2100			ν_{NH}, ν_{ND}
1427	1247			$\nu_{=C-N-H}$ and $\nu_{=C-N-D}$ o.p. torsional
862	638			($\nu_H/\nu_D = 1.36$)
1100		1170		$\delta_{CH}(6)$
910		950		
800		870		$\delta_{CH}(6)$
1270		750	1170	$\delta_{CH}(8)$
1210			945	
910			860	
			760	

It is interesting to note the ^{15}N-substitution effect on vibrations involving various atomic motions in the guanine molecules. The isotopic shift is only 1–4% for vibrations involving hydrogen motions (ν_{NH}, $\nu_{sym\text{-}NH_2}$, $\nu_{asym\text{-}NH_2}$, δ_{HNH}, etc.) and doubles when the molecule is deuterated. For vibrations involving nitrogen atom motions (ν_{C-N}, $\nu_{C-(NH)}$, δ_{CNC}, etc.), the ^{15}N shifts decrease in deuterated **287B**. From analyses of ^{15}N and D shifts, pure ring vibrations have been assigned and the pyrimidinic, imidazolic, or purinic characters have been reported. Some bands of guanine spectra were compared with those assigned to guanine residues in DNA spectra. On analyzing the ^{15}N and D shifts, the ring character was established. By comparison with the frequency in deuterated guanine, the first band in Table 1.55 was assigned to a ring vibration with a pronounced pyrimidinic character, and the second one, which is coupled to the OH group of the molecule, has imidazolic character.

In the general research on the structure of biopolymers, vibrational analysis of DNA and RNA and their constituents may contribute essentially to the understanding of these systems.

Problems concerning the application of IR spectroscopy to structural studies of nucleic acids were detailed in the review of Tsuboi (548). Tsuboi's review showed the extent to which IR spectroscopy has been and will be useful for elucidating the molecular structure of nucleic acids and outlined the advantages of this method over X-ray and UV spectroscopy. General spectral features have been tabulated, and different techniques and experimental methods used in the vibrational assignments have been described (deuteration, ^{15}N- and ^{18}O-substitution, IR dichroism measurements of oriented DNA and RNA films, sensitivity to pH changes of aqueous solutions, etc.). Some peculiarities of IR spectra of deoxyribonucleic acids, deoxyribonucleoproteins, double-helix RNA, and DNA-RNA hybrids have been described, and various directions of applications of these data in constructing models of the nucleic acid structure have been detailed. Hydrogen bonding in DNA and related compounds has been investigated by Parker and Khare (549) and by Zundel and coworkers (550).

Table 1.55. Ring Vibrations of Guanine Residue and Guanine 287B (543)

Guanine Residue (545)	Guanine-d_0	Guanine-d_4	Guanine-d_5
1577[a] (− 17)		1567 (− 9, − 4)	1576 (− 10, − 1)
1481 ± 9	1468 (− 4, − 8)	1466 (− 6, − 6)	1420 (− 2, − 10)
1364 ± 4	1360 (− 6, − 4)	1342 (− 5, − 12)	1313 (− 1, − 11)
672 ± 8	651 (− 5, − 3)	640 (− 6, − 1)	636 (− 3, − 1)

[a] Shift for ^{15}N-enriched and deuterated product.

The sensitivity of the ring vibrations of the bases to temperature and pH changes was interpreted by Sukhorukov and coworkers (551). Other reviews on the vibrational spectroscopy of these compounds are included in references 552 and 553.

1.6.3.3. Hydrogen-Bonding Interactions of Nucleic Acid Constituents

The molecular basis for the transfer of information in biological systems is believed to be the specificity of hydrogen bonding between the constituent purines and pyrimidines of the nucleic acids. These interactions reinforce the geometrical specificity seen in the two-stranded nucleic acids. For example, with dilute solutions of 9-ethyladenine and 1-cyclohexyluracil in deuterochloroform, absorption bands in the IR spectrum demonstrate hydrogen bonding of the adenine and uracil derivatives with themselves (553a, 553b).

Significantly, the affinity of adenine **287C** for uracil (or thymine) derivatives and of guanine **287B** for cytosine derivatives is highly selective. Thus the hydrogen-bonding interactions of these molecules show a type of electronic complementarity that matches exactly the geometric complementarity that plays an important role in our understanding of the structure of the double-stranded helical nucleic acid molecules.

Estimates have been made of the pairwise association constants of the various bases in chloroform solutions (553c, 553d). It was found that some substituents enhance the hydrogen-bonding affinity of the adenine-uracil association, while others markedly decrease it. The selectivity found in chloroform corresponds to that seen in naturally occurring polynucleotides in aqueous solution.

1.6.4. Triazines

1.6.4.1. s-Triazines

The IR and Raman spectra of *sym*-triazine **288** and its derivatives have been much discussed (554–559), but most of these vibrational spectra have not been fully interpreted. The vibrational assignment for **288** of D_{3h} symmetry [$3A'$ (R) + $2A'$ (inact) + $5E'$ (R, IR) + $2A''$ (IR) + $2E''$ (R)] is given in Table 1.56.

288

Because of the high symmetry, the ring-stretching bands in **288** and its trisubstituted derivatives appear as a pair of doubly degenerate modes near 1560 and 1410 cm^{-1}, but in asymmetrical triazines the degeneracy is removed. The ring-stretching bands are sufficiently characteristic to enable the ring to be identified. Some authors (560) have used the method of local symmetry to interpret the spectra of uniformly substituted triazines, considering the vibration of the ring itself and the peripheral substituent groups. The infrared and Raman spectra of

Table 1.56. Vibrational Assignments of s-Triazines (554)

Species	cm^{-1}	Assignment
A_1'	3042	ν_{CH}
E'	3056	ν_{CH}
A_2'	1617	ν_{ring}
E'	1556	ν_{ring}
E'	1410	ν_{ring}
A_2'	1251	β_{CH}
E'	1174	β_{CH}
A_1'	1132	Breathing
A_1'	992	Breathing
E''	1031	γ_{CH}
A_2''	925	γ_{CH}
E'	675	β-ring
A_2''	737	γ-ring
E''	340	γ-ring

2,4,6-tris(trichloromethyl)-1,3,5-triazine were interpreted on the basis of characteristic group frequencies and temperature dependence of the solid state spectrum. For this molecule the overall molecular symmetry is less than D_{3h} (C_{3h}, C_{3v}) and depends on the location of the C–Cl bonds in the molecule; the skeletal vibrations are very sensitive to the bonding arrangement of the CCl$_3$ group. The method used consisted in trying to fit the observed skeletal frequencies to the D_{3h} model, and, on the basis of correlations from D_{3h} to C_{3v} and C_{3h}, to help in deciding the actual symmetry. The bands were assigned only tentatively, because no reliable data were available for a large number of structurally related molecules. Analytical applications of IR spectroscopy for s-triazines were reported (561–563).

There are few IR data reported for 1,2,3-triazines **289** and no vibrational analysis has yet been performed. IR spectra of **289** are complicated and have been used only to identify various functional groups in individual compounds (563, 564).

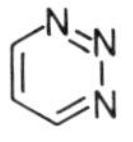

289

Although a separate section (Sec. 1.8.2) will discuss the vibrational spectra of borazines, it is appropriate to mention here (for comparison with s-triazine) that the vibrational spectra of N-trimethylborazines **290** on the basis of C_{3h} symmetry have been calculated and interpreted (565). Most vibrations were found to be delocalized and could not be attributed to vibrations of separate bonds.

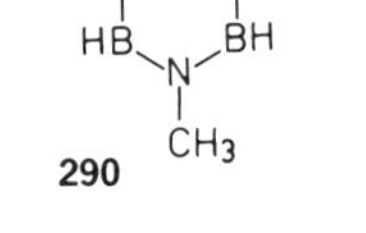

290

Aromatic Six-Membered Rings 109

In the IR spectra of B-(pentadeuteriophenyl)borazines a distinct coupling was observed (566) between the highest N—B ring stretching mode (ca. 1400 cm^{-1}) and the C—C stetching mode.

1.6.4.2. 1,2,4-Triazines

Slouka and coworkers (567–570) have prepared a great number of complex 1,2,4-triazine derivatives. The most characteristic IR bands for 1,2,4-triazine derivatives **291–294** are summarized in Table 1.57.

The tautomerism of **293** was also studied (569) by selective substitution of ^{14}N by ^{15}N. Substitution of ^{14}N by ^{15}N in position 1 or 2 lowered the frequencies

Table 1.57. Infrared Data on 1,2,4-Triazine Derivatives

Compound	IR Data	Ref.
4-Oxo-7-methyl-1,4-dihydropyrazolo[3,2-c]-1,2,4-triazine **291**	1711 (C=O) 1621 (C=N) 1608 (C=C)	567
1,2,3,4-Tetrahydropyrimido[4′,5′:5,6]-1,2,4-triazino[4,3-b]indazole-2,4-dione **292**	1570 1700 1602 1743 1633 3200 1680	568
2-Aryl-2,3-dihydro-9H-1,2,4-triazino[6,5-b]indole-3-ones **293**	1650 (C=N) 1252 (C—N) 1154 (C—N) 1122 (N—N) 893 skelet.	569
2-Aryl-2,3,4,6-tetrahydro-1,2,4-benzotriazine-3,6-dione **294**	1640 (C=O quinoid) 1712 (C=O)	570

110   Infrared Spectra of Heterocyclic Compounds

of some bands, and these changes enabled the following assignments: 1650 cm^{-1} ($\nu_{C=N}$); 1252 cm^{-1} (ν_{C-N}); 1154 cm^{-1} (ν_{C-N}); 1122 cm^{-1} (ν_{N-N}); 893 cm^{-1} (skeletal vibration of triazine ring). The band $\nu_{C=N}$ at 1650 cm^{-1} excludes the existence of 1H tautomeric forms. The 2H-3-hydroxy tautomeric form was also excluded on the basis of IR spectra of solutions of the compound in pyridine ($\nu_{C=O}$ of the CO group at 1668 and 3420 cm^{-1} for ν_{NH}). Slouka and coworkers (569) concluded that the most probable structures are the 9H or 4H tautomeric forms.

Poly(triazines) exhibit the same IR bands as the corresponding monomer molecules (571). Characteristic IR bands were reported for poly(1,2,4-triazines): 1490, 1445, 1385–1390 (C=C, C=N, N=N), and 1020 cm^{-1}; poly(1,3,5-triazines): 1500–1520, 1415, and 1360–1375 cm^{-1}, and a very weak band at 2220 cm^{-1} (C=N) in agreement with the earlier IR data on triazines (572).

1.6.5. Tetrazines

Spectral data were reported (572, 573) for s-tetrazine **295**, but detailed spectral assignments were possible only when Raman spectrometers equipped with laser sources became available, because **295** is a colored compound.

295

Table 1.58.   Vibrational Assignment of s-Tetrazine (cm^{-1}) (575)

Symmetry Species	Band Notation	s-Tetrazine Obs.	s-Tetrazine Calc.	Assignment
$A_g(A_1)$	2, 7a	3088.5	3090.0	CH stretch
	8a, 9a	1416.5	1425.5	Ring
	1	1015.0	1013.5	Ring
	6a	734.0	735.0	Ring
$B_{3g}(B_2)$	8b	1522.5	1519.0	Ring
	3, 9b	1302.0	1305.0	CH bend
	6b	649.0	646.5	Ring
$B_{2g}(B_1)$	5, 10b	1015.0		CH bend
	4	798.5		Ring
$A_u(A_2)$	16a	335.0		Ring
$B_{3u}(B_1)$	11, 17b	928.5		CH bend
	16b	235.5		Ring
$B_{1u}(A_1)$	13, 20a	3085.5	3088.5	CH stretch
	18, 19a	1203.5	1204.5	Ring
	12	1093.0	1089.5	Ring
$B_{2u}(B_2)$	19b	1447.5	1454.0	CH bend
	14	1109.0	1110.5	Ring
	15, 18b	893.0	897.5	Ring

Franks and coworkers (574) obtained the Raman spectrum and they also investigated the low-frequency IR spectrum. To select the best assignments for the in-plane modes, the low-temperature IR spectra of *sym*-tetrazine-d_0, -d_1, and -d_2 have been investigated (572). The observed and calculated frequency assignments for **295** (D_{2h}) are given in Table 1.58. The average difference $\nu_{calc} - \nu_{obs.}$ was found to be 0.28%. A normal coordinate analysis of **295** in a modified Urey-Bradley force field was performed (576).

1.7. SEVEN-MEMBERED RINGS

1,3-Dihydro-2*H*-azepin-2-one and its methyl derivatives **296–298** have been characterized by IR spectroscopy (577–579). In *N*-tosyl-benzazepinones such as **299** and **300**, the carbonyl bands appear at 1675 (**299**) and at 1715 and 1684 cm^{-1} (**300**), respectively (580).

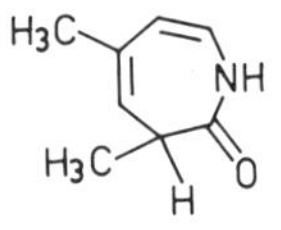

296 Nujol 3180 (ν_{NH})
1650 (ν_{CO})

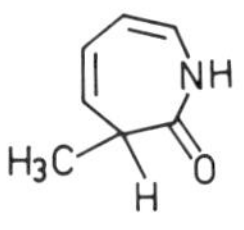

297 1695 (ν_{CO})

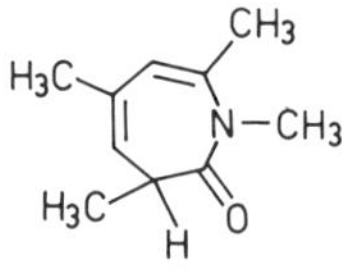

298 1670 (ν_{CO})

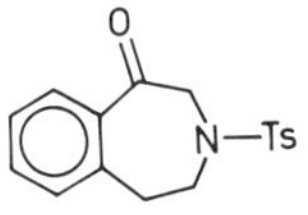

299

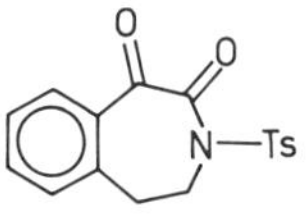

300

In azepinoindoles such as 5-oxo-3,4,5,6-tetrahydro-1*H*-azepino[4,3,2-*cd*]indole **301**, $\nu_{C=O}$ appears at 1645 cm^{-1} (581), whereas in 4-(indole-3-yl)-1-methylhexahydroazepin-2-one **302**, carbonyl absorption at 1714 cm^{-1} was evidenced (582). Other fused azepines such as **303** [obtained by Alper and Keung (583) by the Beckmann rearrangement of α-santonin-oxime] exhibit strong absorptions as indicated.

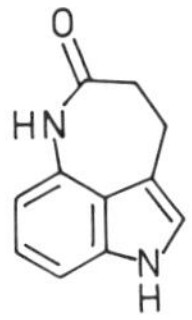

301

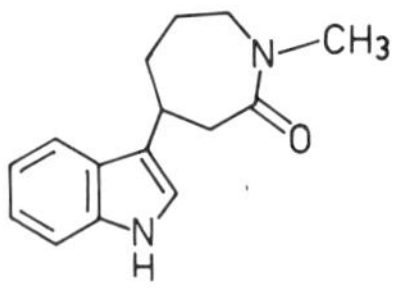

302

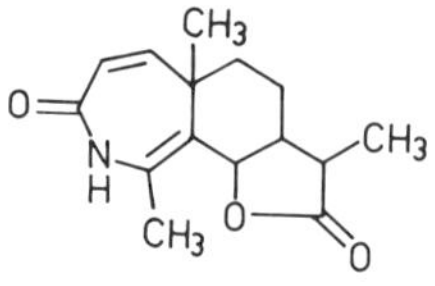

303
ν_{NH} 3520
$\nu_{C=O}$ (lactonic) 1750
$\nu_{C=O}$ (lactamic) 1660
$\nu_{C=C}$ 1560

The characteristic bands of dihydrodiazepinium salts are a band near 3500 cm^{-1} (ν_{NH}) and three strong bands at 1610–1650, 1510–1575, and 1305–1335 cm^{-1} (584). The ν_{NH} stretching vibration in the spectra of dihydrodiazepine bases appears at 3150–3190 cm^{-1}.

1,5-Benzodiazepines show no peaks due to NH groups, a fact that indicates the existence of the dianil form **304** rather than the isomeric conjugated forms **305** or **306** (585).

Chiral 1,4-benzodiazepinones **307** have been prepared (586), and carbonyl absorption has been reported in the region 1670–1690 cm^{-1}.

1,2,3,4-Tetrahydro-5H-1,4-benzodiazepin-5-one **308** (R = C$_6$H$_5$–CH$_2$) exhibits a strong $\nu_{C=O}$ band at 1652 cm^{-1} (587).

Fused diazepines such as **308** diimidazodiazepines **309** are characterized by strong bands at 1660 cm^{-1} ($\nu_{C=N}$) and ring vibrations at 1282 and 1175 cm^{-1} (588).

Pyrazolo- and pyridodiazepines have also been prepared (589, 590). For example, **310** exhibits a strong $\nu_{C=O}$ band at 1682 cm^{-1}.

Benzoxazepinones such as N-ethyl-4,5-dihydro-2,4-benzoxazepin-3(1H)-one **311** were prepared by Pifferi and coworkers (591), and the $\nu_{C=O}$ band appears at 1650 cm^{-1}. Pyrimidobenzoxazepin-5(11H)-one **312** absorbs at 1612 cm^{-1} ($\nu_{C=O}$) (592). 1-Benzooxepinediones **313** in KBr absorb at 1714 cm^{-1} ($\nu_{C=O}$) (593).

Table 1.59. Characteristic Bands for Some 3-Benzyl-7-oxo-4,5,6,7-tetrahydro-1,2,6-oxadiazepines 315 (cm^{-1}) (594)

R	$\nu_{C=O}$	$\nu_{C=N}$	ν_{C-O}	ν_{N-O}
C$_6$H$_5$	1690	1610	1234	921
p-C$_6$H$_4$CH$_3$	1694	1621	1238	924
p-C$_6$H$_4$CCH$_3$	1681	1606	1226	918
p-C$_6$H$_4$NO$_2$	1694	1603	1233	915
m-C$_6$H$_4$NO$_2$	1695	1618	1236	915

Recently, IR spectroscopy has been used to elucidate the stereochemistry of seven-membered heterocyclic oxepanes. In 3-alkyl-2,4-benzodioxepanes 314, a mixture of chair and twist conformers has been indicated. In oxadiazepinones 315, characteristic IR bands have been reported by Gnichtel and Hirte (594).

314   R = H, Me, Et   315   316

Characteristic bands for 3-benzyl-7-oxo-4,5,6,7-tetrahydro-1,2,6-oxadiazepines are summarized in Table 1.59.

Infrared data on seven-membered heterocyclic compounds with oxygen 316 are given in Table 1.60.

Characteristic bands for oxepin ring systems containing two rings are summarized in Table 1.61, and those for dioxepines and trioxepines in Table 1.62.

Table 1.60. Characteristic IR Absorptions for Oxepines 316 (cm^{-1})

Compound	IR Data	Ref.
2,7-Dimethyloxepine	1660 $\nu_{CH=CH-O-}$	595
2,3-Dihydrooxepine	1653 1621 $\nu_{CH=CH}$	596
4,5-Dihydrooxepine	3055 ν_{CH}(olefin) 2940 ν_{CH}(CH$_2$) 2860 1650 $\nu_{C=C}$ 1448 ν_{CH_2} (allylic enol ether)	597
2,5-Dihydrooxepine	1655 (vinyl ether) 1270 1105 ν_{C-O}	598
2,3,4,5-Tetrahydrooxepine	1655 1650 1641 $\nu_{C=C}$	599
2,3,4,7-Tetrahydrooxepine	1655 $\nu_{C=C}$	600
2,3,6,7-Tetrahydrooxepine	1660 $\nu_{C=C}$	596

Table 1.61. Oxepine Ring Systems Containing Two Rings

Compound	IR Data	Ref.
7,9,9-Trimethylperhydro-2-oxabicyclo [5.2.0]nonane-3-one **317**	1755 (C=O)	601
3,6-Dimethylperhydrofuro[3,4-d]oxepine **318**	1088 ν_{C-O} 1057	601
1-Benzoxepine **319**	1645 $\nu_{(C=C)}$ 1597	601
2,3,4,5-Tetrahydro-5,5-dimethyl-1-benzoxepine **319a**	1286 $\nu_{(C-O)}$ 1226	602

Seven-membered rings with two heteroatoms are summarized in Table 1.63. Compound **330** with one heteroatom in the seven-membered ring is included for comparison; it possesses two heteroatoms in the five-membered ring.

1.8. METALLO-ORGANIC AND ELEMENTO-ORGANIC CHELATE HETEROCYCLES

1.8.1. Metal Chelates

This section discusses the infrared spectra of a class of compounds that are seldom included among heterocyclic systems, although they should be: chelate complexes of bidentate ligands with metals or nonmetals.

In order to place the topic in its proper perspective, one should start by recalling Calvin's hypothesis (609, 610) that metal chelate complexes of 1,3-diketones

Table 1.62. **Infrared Data on Benzo-Fused Dioxepines, Trioxepines, and Thiepienes**

Compound	IR Data (Nujol) (cm^{-1})
2,3-Dihydro-5H-1,4-benzodioxepin-2,5-dione **320**	1795, 1730, 1613, 1586, 1320, 1223, 1204, 797, 785, 765, 702, 685
6,6'-Spirobidibenzo[d,f]-1,3-dioxepine **321**	1450, 1200, 1160, 1120, 1070, 1040, 760, 720
2,3-Dihydro-1-benzothiepin-1-oxide **322**	1030 $\nu_{S=O}$
1-Benzothiepin-3,5(2H,4H)-dione **323**	1732 $\nu_{C=O}$ 1673
Dibenzo[b,e]thiepin-11(6H)-one **324**	1651 $\nu_{C=O}$
Dibenzo[b,f]thiepin-10(11H)-one **325**	3600 ν_{OH} 1675 $\nu_{C=O}$

116 Infrared Spectra of Heterocyclic Compounds

Table 1.63. Characteristic Absorptions of Seven-Membered Rings with Two Heteroatoms (cm^{-1})

Compound	IR Data	Ref.
3H-(1,5)benzodiazepine **326**	1664 $\nu_{C=N}$	603
4-Phenyl-1H-1,5-benzodiazepin-2(3H)one **327**	3215 ν_{NH}	604
Dibenzo[b,f-1,4]thiazepin-11(10H)one-5,5-dioxide **328**	3100 ν_{NH} 1640 $\nu_{C=O}$ 1160 1310 ν_{SO_2} 3090 ν_{NH} 1638 $\nu_{C=O}$ 1575 ν_{NH}	605
Dibenzoxazepine **329**	1615 ν_{NH} 1550 1335 1130	606
3-Benzyl-7,8-methylenedioxy-1,3,4,5-tetrahydro-2H-3-benzazepine-2-one **330**	3000 ν_{NH} 2915 1650 1505	607

could possess aromatic character since they possess a sextet of electrons, as seen in formulas **331** and **332**. The metal can have either tetrahedral or octahedral coordination, since it is surrounded by four or six oxygen atoms, that is, there are two or three chelating diketone units. The fact that under certain conditions such metal chelates were shown to undergo nitration, halogenation, acylation, chloromethylation, or Vilsmeier formylation and other substitutions in position 3 was considered to support Calvin's hypothesis. The amine obtained by reduction of the nitro derivative can be diazotized to a stable diazonium salt and then coupled with β-naphthol or heated in aqueous solution to yield the HO compound, or converted into the tetrafluoroborate and hence into the fluoro derivative (611, 612). Another support was provided by the considerable deshielding experienced by the proton in position 3 in the ^{1}H-NMR spectra of the silicon chelate Si(acac)$_3^+$ analogously to the deshielding of benzene protons (613).

$$\underset{331}{\begin{array}{c}H_3C\diagdown\diagup CH_3\\ {}^+O\diagdown\diagup O\\ M^-\end{array}} \longleftrightarrow \underset{332}{\begin{array}{c}H_3C\diagdown\diagup CH_3\\ O\diagdown\diagup O^+\\ M^-\end{array}}$$

By extension one could argue that the similarly stable tropolene chelates **333**, **334** were aromatic like azulene. This hypothesis was proved to be erroneous, since the delocalization is not cyclic (it does not include the metal atom) but is limited to the diketone or tropolone systems (614–617). The ^{1}H-NMR spectra of the analogous boron chelates [M = BR$_2$ (618), where the boron atom has sp^3 hybridization interrupting the ring current] are quite similar to those of metallic chelates.

$$\underset{333}{} \longleftrightarrow \underset{334}{}$$

^{1}H-NMR spectra of 3-mesityl- and 3-(9-anthryl)-2,4-pentanedione, which indicates the absence of any shielding of protons above the chelate ring, also demonstrate the absence of aromaticity in this ring (619, 620).

We shall discuss here the IR spectra of metal chelates, because their assignments have been the subject of long controversy.

Copper(II)bis(acetylacetonate) presents four distinct bands in the 1600–1200 cm^{-1} range. The bands at 1578 and 1527 cm^{-1} have been assigned in various ways by different authors. Empirically, Collman (621) and West (622) assigned the highest frequency band to the C=O and the next band to the C=C stretching vibrations. Nakamoto (623, 624) concluded from a normal coordinate analysis, on the contrary, that the band at 1578 cm^{-1} has 75% C=C and 25% C=O character, while the band at 1527 cm^{-1} has reversed proportions for the C=C and C=O characters. Mecke and Funk (625) considered that all four vibrations are so extensively coupled that no distinct assignment can be made. This confusing situation prevailed until Musso and coworkers (620, 626) settled the assignments by labeling the chelates with ^{13}C, ^{18}O, and ^{2}H in several combinations indicated on formulas **335–342**.

335   336   337   338

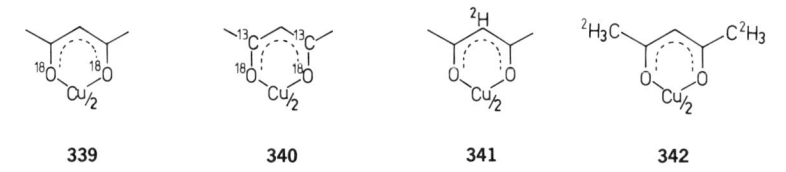

118 Infrared Spectra of Heterocyclic Compounds

339, **340**, **341**, **342**

The results of isotopic substitutions are depicted in the formulas **343–346** by circles with radii proportional to the long-wave shift caused when the corresponding atoms are replaced by heavier isotopes ^{13}C and ^{18}O:

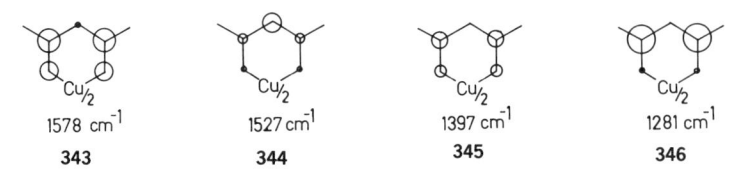

1578 cm^{-1}	1527 cm^{-1}	1397 cm^{-1}	1281 cm^{-1}
343	**344**	**345**	**346**

Clearly, the band at 1578 cm^{-1} is a C=O vibration (the symmetric one, as indicated by spectra of molecules oriented in stretched polyethylene films with polarized IR light). The band at 1397 cm^{-1} is antisymmetric, and the band at 1527 cm^{-1} is from antisymmetric C–C–C stretching (with C–C–H contribution). Two bands are almost insensitive to ^{13}C and ^{18}O labeling: the shoulder at 1430 cm^{-1} is the antisymmetric CH$_3$-deformation vibration, and the band at 1354 cm^{-1} (appearing with very low intensity in solution, more strongly in KBr pellets) is the symmetric CH$_3$-deformation vibration. The symmetrical C–C–C vibration at 1281 cm^{-1} undergoes an interesting triplet splitting on replacement of the 2- and 4-carbon atoms of the 1,3-pentanedionato ligand by 40–60%-enriched ^{13}C: the lowest frequency component of the triplet is due to the unlabeled compound, the middle one to the singly labeled compound, and the highest-frequency component to the doubly labeled compound. Such triplet splittings are also observed on ^{18}O substitution for the 685-cm^{-1} band and on 2,4-^{13}C substitution for the 652-cm^{-1} band. This triplet pattern indicates complete delocalization (equalization within ⩽ 0.01 Å) of the bonds in Cu(acac)$_2$; otherwise a quartet splitting would have resulted, as one observes in copper(II) bis(acetoacetate), where the bonds cannot be equivalent (620, 626).

With M = Cu, Mn, Zn, Ni, Cr, Al, Co, Zr, Fe, Rh, Ce, etc., the IR spectra are very similar. Beryllium bis(acetylacetonate) has practically the same bands as the preceding metal complexes above 1050 cm^{-1}, but marked differences appear below this limit owing to the low O–Be distance and to the low mass of beryllium. Assignments of metal-oxygen bands were confirmed by other authors with ^{18}O-labeling (627), by improved normal coordinate analysis (628), and by substitution with heavier metal isotopes for Fe, Cr, Cu, and Pd (629).

Interestingly, results are different with other metals, again as shown by IR spectra: mercury gives an insoluble acetylacetonate with the C–Hg bond involving the 3-carbon atom **347**, whereas Pt, Au, or Sb afford both oxygen- and carbon-metal bonded compounds e.g. **348** (631, 632). With CoBr$_2$, SnCl$_4$, TiCl$_4$, and

Metallo-organic and Elemento-organic Chelate Heterocycles 119

Table 1.64. Cu–O Stretching Vibrations and ν_{N-H} in Copper Acetoacetanilides 350 (636)

R	ν_{Cu-O}	ν_{N-H}
2-Br	410	3350
2-Cl	415	3250
3-NO$_2$	480	3340
p-bis	500	3280

ZnCl$_4$ ligands, the presence of bands at 1700–1720 cm^{-1} in the IR spectra indicates 1,3-diketo complexes **349** via the two oxygens. Some of these can be readily converted into normal chelates (633).

347   **348**   **349**

Spectra of copper(II) acetoacetarylamide complexes were investigated by Patel and coworkers (634, 635) in the 1000–3500 cm^{-1} region. Bădilescu and Bădilescu (636) have studied the far-IR spectra of copper complexes of several acetoacetanilides **350** using various polyethylene matrix isolation techniques. The compounds have been isolated in a matrix resulting from pyrolysis of polyethylene in a special infrared pyrolysis device. The ν_{NH} and ν_{Cu-O} bands observed in the far-IR spectra are listed in Table 1.64. The ν_{Cu-O} bands are shifted toward higher frequencies according to the electron-releasing capacity as determined by the Hammett parameters. The lower ν_{Cu-O} values in o-substituted compounds may be attributed to steric effects.

350

1.8.2. Boron and Silicon Heterocycles

Initially IR data for Si(acac)$_3^+$ had an incorrect assignment of C=C and C=O bands (637) following Nakamoto's first force field calculations (623, 624). Schott reversed this assignment and extended it to the newly prepared RSi(acac)$_2$Hal compounds (which are less stable) (638) after Musso and coworkers by experiment

(620, 626) and Nakamoto by improved calculations (628) showed the longest wavelength band in the 1500–1600 cm^{-1} range to be due to C=O stretching vibrations.

Infrared data were helpful in assigning chelated structures to the neutral boron heterocycles **351–355** (with R and R' being either aryl groups or benzodioxa fragments as in **354**). The complete delocalization in the 6- (639), 5- (640, 641), or 7-membered (642) chelate rings of **351, 354; 352, 355; 353**, respectively, is similar to that encountered in the transition metal chelates (643). Definitive proof for this complete delocalization was also obtained by Balaban and coworkers on the basis of dipole moments (644, 645), equivalence in ^{1}H and ^{13}C–NMR spectra (639, 646), and splittings observed for the diastereotropic groups in the chiral compounds **354** and **355** (647).

Unlike the boron chelates discussed above, the analogous crystalline arsenic chelate was considered on the basis of IR and other spectral evidence (648) to have a noncoordinated structure. However, the molecule **356 ⇌ 356A** undergoes a rapid automerization, which makes the tropolonic carbon atoms pairwise equivalent in the ^{13}C–NMR spectra even at low temperatures ($\Delta G^{\ddagger} < 6.5$ kcal/mol) (646); the structure of the arsenic compound has been established by X-ray analysis, a method with a much shorter time scale than the NMR methods.

Coming back to boron-oxygen heterocycles, cationic systems will be mentioned briefly. The bis(1,3-diketonato)boronium salt **357** has complete delocalization, exactly like the metal chelates and the neutral boron 1,3-diketonates discussed above, and hence its IR spectrum has a similar assignment (649). By contrast,

1,3,2-dioxaborinium perchlorates **358** (650, 651) have aromatic character because the boron atom is tricoordinated and has sp^2 hybridization, unlike the tetracoordinated boron atoms with sp^3 hybridization from all preceding systems. The relationship between systems **357**, **358**, and **359** is shown in the diagram. Infrared assignments for these three systems have been presented by Balaban and coworkers (651).

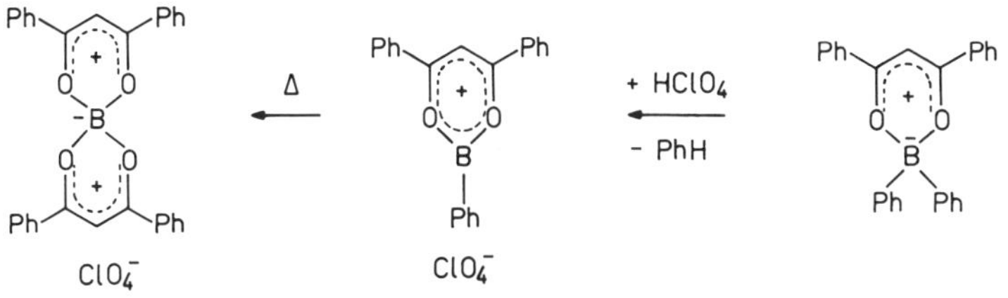

To continue with the boron heterocycles, a few data on infrared spectra of the "inorganic benzene" (heterocyclic borazines) follow.

β-Trichloro-*N*-trimethylborazine, easily obtained from methylamine and boron trichloride, has a strong band at 1392 cm^{-1}, which was assigned to the B–N stretching vibration (652). Kubo and coworkers (653), in agreement with others (654), assigned a band at 1087 cm^{-1} to the *N*-methyl rocking vibration, and a band in the 915–980 cm^{-1} range to the B–Cl stretching vibration; this assignment is confirmed by an IR study of (RB–NCD$_3$)$_3$, where R is Cl or NH$_2$ (655). The assignment was contested (656), but was confirmed by Gutmann and coworkers (657).

β-Trichloroborazine has an IR spectrum that could be assigned completely by means of the natural ^{10}B-^{11}B isotopic composition: the strong B–N ring vibration at 1442 cm^{-1} has a shoulder at 1452 cm^{-1} due to the ^{10}B isotopic content of 19%. The effect of substituents on the same strong band in symmetrically substituted borazines may be seen from Table 1.65.

Table 1.65. The Effect of Substituents on the B–N Ring Vibration of Borazines

β-Substituent	*N*-Substituent	$\nu_{\text{B-N}}$ (cm^{-1})
Me	Ph	1376
Ph	Ph	1368
Cl	Ph	1378
n-Bu	Ph	1382
Et	Ph	1387
Cl	Me	1392
Et	Et	1417
Cl	*n*-Bu	1418
H	D	1436
Cl	Et	1442
Cl	H	1442

Lappert (659) has drawn attention to a medium-intensity doublet around 720 cm^{-1} in the IR spectra of β-aminoborazines and β-alkoxyborazines as a good diagnostic for these structures. Extensive data on IR bands of borazines are given in tabular form in Steinberg's book (660). Gerrard's book (661) correlates the electron-donating ability of the β-substituent with the decrease of the B–N stretching frequency in (PhN–BR)$_3$: R = H, 1401 cm^{-1}, Me, 1380 cm^{-1}; Cl, 1376 cm^{-1}; Et$_2$N, 1285 cm^{-1}. For a calculation and interpretation of the IR spectra of N-trimethylborazine, see reference 565 and Section 1.6.4.1.

Boroxines are less aromatic than borazines, but the increased frequencies found by Lappert (662) for B–O stretching modes in trialkoxyborazines (methoxy, 1486 cm^{-1}) relative to the acyclic systems (around 1350 cm^{-1}) suggests higher B–O bond order.

ACKNOWLEDGMENT

Thanks are expressed to Dr. Cornelia Uncuta for help in the preparation of the manuscript.

REFERENCES

1. A. R. Katritzky and A. P. Ambler, "Infrared Spectra," in A. R. Katritzky, Ed., *Physical Methods in Heterocyclic Chemistry*, Vol. 2, Academic Press, New York, 1963, p. 161; A. R. Katritzky and P. J. Taylor, "Infrared Spectrocopy of Heterocycles," *ibid.*, Vol. 4, 1971, p. 265.
2. H. W. Thompson and W. T. Cave, *Trans. Faraday Soc.*, **47**, 951 (1951).
3. R. C. Lord and B. Nolin, *J. Chem. Phys.*, **24**, 656 (1956).
4. W. J. Potts, *Spectrochim. Acta*, **21**, 211 (1965).
5. J. Le Brumant and D. Maillard, *Compt. rend.*, **264**, 1107 (1967).
6. J. Le Brumant, *Compt. rend.*, **268**, 486 (1969).
7. K. Venkateswarlu and G. Thyagarian, *Proc. Indian Acad. Sci.*, **A52**, 101 (1960).
8. J. H. Wray, *Proc. Phys. Soc.*, Ser. 2, **1**, 485 (1968).
9. K. Venkateswarlu and P. A. Joseph, *J. Mol. Struct.*, **6**, 145 (1970).
10. N. W. Cant and W. J. Armstead, *Spectrochim. Acta*, **31A**, 839 (1975).
11. R. Mathis-Noel, R. Martino, A. Seches, and A. Lattes, *Compt. rend.*, **266B**, 926 (1968).
12. F. Imberlin, A. Lopez, R. Martino, A. Lalles, and R. Mathis, *Spectrochim. Acta*, **35A**, 1033 (1979).
13. R. Barlet, P. Barlet, H. Handel, and J. L. Pierre, *Spectrochim. Acta*, **30A**, 1471 (1974).
14. S. F. Bush, M. M. Allen, and W. C. Harris, *Spectrochim. Acta*, **31A**, 1509 (1975).
15. P. Rademacher and W. Lüttke, *Angew. Chem. Int. Ed.*, **9**, 245 (1970).
16. A. P. Schaap, *Tetrahedron Lett.*, **1971**, 1757.
17. P. A. Burns and C. S. Foote, *J. Amer. Chem. Soc.*, **96**, 4339 (1974).
18. W. Adam and J. C. Liu, *J. Amer. Chem. Soc.*, **94**, 2894 (1972).
19. J. M. Eyster and E. W. Prohofsky, *Spectrochim. Acta*, **30A**, 2041 (1974).
20. G. C. Engerholm, D. O. Harris, A. C. Luntz, and W. D. Gwinn, *J. Chem. Phys.*, **50**, 2446 (1969).

References

21. A. Palm and E. R. Bissel, *Spectrochim. Acta*, 16, 459 (1960).
22. W. H. Green and A. B. Harvey, *Spectrochim. Acta*, 25A, 723 (1969).
23. J. E. Katon and W. R. Fearheller, *Spectrochim. Acta*, 21, 199 (1965).
24. G. Fini and B. Fortunato, *Spectrochim. Acta*, 32A, 423 (1976).
25. G. Fini, P. Mirone, and B. Fortunato, *J. Chem. Soc., Faraday Trans. II*, 69, 1243 (1973).
26. G. Fini and P. Mirone, *J. Chem. Soc., Faraday Trans. II*, 70, 1776 (1974).
27. H. W. Thompson and R. B. Temple, *Trans. Faraday Soc.*, 41, 27 (1945).
28. B. Bak, S. Brodersen, and L. Hansen, *Acta Chem. Scand.*, 9, 749 (1955).
29. N. K. Sidorov and L. P. Kalashnikova, *Opt. Spektrosk.*, 24, 469 (1968); *Opt. Spectrosc.*, 24, 247 (1968).
30. G. Waddington, J. W. Knowlton, D. W. Scott, et al., *J. Amer. Chem. Soc.*, 71, 797 (1949).
31. P. J. Derrick, L. Asbrink, Q. Edqvist, and E. Lindholm, *Spectrochim. Acta*, 27A, 2525 (1971).
32. R. C. Lord and F. A. Miller, *J. Chem. Phys.*, 10, 328 (1942).
33. P. Mirone, *Gazzetta*, 86, 165 (1956).
34. J. Morcillo and J. M. Orza, *Anales real soc. españ. fis. quim.*, 56B, 231 (1960).
35. J. M. Lebas and M.-L. Josien, *Bull. Soc. Chim. France*, 1957, 251.
36. A. R. Katritzky and J. M. Lagowski, *J. Chem. Soc.*, 1958, 4155.
37. L. W. Pickett, *J. Chem. Phys.*, 10, 660 (1942).
38. J. Lecompte, *Bull. Soc. Chim. France*, 1946, 415.
39. J. Loisel and V. Lorenzelli, *Spectrochim. Acta*, 23A, 2903 (1967).
40. J. Garach and J. Lecompte, *Bull. Soc. Chim. France*, 1946, 423.
41. A. Hidalgo, *J. Phys. Radium*, 16, 366 (1955).
42. A. R. Katritzky and A. J. Boulton, *J. Chem. Soc.*, 1959, 3500.
43. E. N. Bolotina and L. M. Sverdlov, *Opt. Spektrosk. Suppl. 3, Mol. Spectrosc.*, II, 150 (1967); *Opt. Spectrosc. Suppl. Mol. Spectrosc.*, II, 75 (1968).
44. D. W. Scott, *J. Mol. Spectrosc.*, 31, 451 (1969).
45. V. T. Aleksanyan et al., *Opt. Spectrosc., Suppl. 3, Mol. Spectrosc.*, II, 85 (1968).
46. V. T. Aleksanyan, Ya. M. Kimelfeld, N. N. Magdesieva, and Yu. K. Yurev, *Opt. Spektrosk.*, 22, 216 (1967); *Opt. Spectrosc.*, 22, 116 (1967).
47. E. G. Treshchova, D. Ekhardt, and Yu. K. Yurev, *Zh. Fiz. Khim.*, 38, 295 (1964).
48. M. Rico, J. M. Orza, and J. Morcillo, *Spectrochim. Acta*, 21A, 689 (1965).
49. P. Tuomikoski, *J. Phys. Radium*, 15, 318 (1954).
50. P. Mirone and G. F. Fabri, *Gazzetta*, 86, 1079 (1956).
51. P. Mirone and A. M. Drusiani, *Atti Accad. Naz. Lincei, Rend. Classe Sci. Fis. Mat. Nat.*, 16, 69 (1954).
52. P. Chiorboli and P. Manaresi, *Gazzetta*, 84, 269 (1954).
53. A. B. Dempster, R. L. Morehouse, and H. Uslu, *Spectrochim. Acta*, 31A, 1775 (1975).
54. L. Asbrink, E. Lindholm, and O. Edqvist, *Chem. Phys. Lett.*, 5, 609 (1970).
55. K. Volka, J. Banki, K. Schmidt, I. Tibor, and Z. Ksandr, *Sci. Papers Prague Inst. Chem. Technol.*, H14, 169 (1979).
56. G. Palianti, R. Cataliotti, A. Poletti, F. Fringuelli, A. Taticchi, and M. G. Giorgini, *Spectrochim. Acta*, 32A, 1089 (1976).
57. C. G. Andrieu, D. Debruyne, and Y. Mollier, *Compt. rend.*, 280, 977 (1975).
58. F. Fringuelli, G. Marino, and A. Taticchi, *Advan. Heterocycl. Chem.*, 21, 119 (1977).
59. J. W. Sowell and C. De Witt Blanton, Jr., *J. Heterocycl. Chem.*, 10, 287 (1973).

60. P. G. Schultz and P. B. Dervan, *J. Amer. Chem. Soc.*, **102**, 878 (1980).
61. A. Lautié, M. F. Lautié, A. Gruger, and S. A. Fakhri, *Spectrochim. Acta*, **36A**, 85 (1980).
62. C. O. Bender and R. Bonnett, *J. Chem. Soc., C*, **1968**, 3036.
63. J. C. Emmett and W. Lwowski, *Tetrahedron*, **22**, 1011 (1966).
64. C. O. Bender and R. Bonnett, *Chem. Commun.*, **1966**, 198.
65. D. G. O'Sullivan and P. W. Sadler, *J. Chem. Soc.*, **1959**, 876.
66. D. G. O'Sullivan, *J. Chem. Soc.*, **1960**, 3278.
67. D. G. O'Sullivan and P. W. Sadler, *J. Org. Chem.*, **22**, 283 (1957).
68. R. A. Heacock and B. D. Scott, *Canad. J. Chem.*, **38**, 508 (1960).
69. C. C. Bond and M. Hooper, *J. Chem. Soc., Ser. C*, **1969**, 2453.
70. C. N. Robinson and R. C. Lewis, *J. Heterocycl. Chem.*, **10**, 395 (1973).
71. J. A. Faniran, *Spectrochim. Acta*, **32A**, 1159 (1976).
72. J. A. Faniran, *Spectrochim. Acta*, **34A**, 379 (1978).
73. J. H. S. Green, *Spectrochim. Acta*, **27A**, 2015 (1971).
74. A. Rogstad, *Spectrochim. Acta*, **31A**, 1749 (1975).
75. A. Arcoria, E. Maccarone, G. Musumarra, and G. Romao, *Spectrochim. Acta*, **29A**, 161 (1973).
76. A. Arcoria, E. Maccarone, and G. A. Tomaselli, *Spectrochim. Acta*, **29A**, 1601 (1973).
77. R. T. Hawkins, *J. Heterocycl. Chem.*, **11**, 291 (1974).
78. L. W. Deady, R. A. Shanks, and R. S. Thopsom, *Tetrahedron Lett.*, **1973**, 1881.
79. S. Fisichella, V. Librando, and E. Maccarone, *Spectrochim. Acta*, **32A**, 501 (1976).
80. G. Alberghina, A. Arcoria, S. Fisichella, and G. Scarlata, *Spectrochim. Acta*, **28A**, 2063 (1972).
81. J. Nakanishi, J. Umemura, and T. Takenaka, *Spectrochim. Acta*, **36A**, 109 (1980).
82. J. Nakanishi and T. Takenaka, *Bull. Chem. Soc. Japan*, **50**, 36 (1977).
83. Ya. M. Kimel'feld, M. A. Muskaleva, G. N. Zhizhin, V. P. Litvinov, S. A. Ozolin, and Ya. L. Goldfarb, *Opt. Spektrosk.*, **28**, 1112 (1970).
84. Y. Cozien and P. Saumagne, *Compt. rend. Ser. B*, **276**, 365 (1973).
85. G. Mille, G. Davidovics, and J. Chouteau, *J. Chim. Phys. Phys. Phys.-Chim. Biol.*, **69**, 1662 (1972).
86. A. Bigotto and V. Galasso, *Spectrochim. Acta*, **34A**, 923 (1978).
87. S. E. Wiberley and R. D. Gonzales, *Appl. Spectrosc.*, **15**, 174 (1961).
88. R. Zahradnik and K. Bocek, *Spectrochim. Acta*, **18**, 564 (1962).
89. D. Cagniant, P. Faller, and P. Cagniant, *Bull. Soc. Chim. France*, **1961**, 2410.
90. D. Cagniant, P. Faller, and P. Cagniant, *Bull. Soc. Chim. France*, **1964**, 1525.
91. D. Cagniant, P. Faller, and P. Cagniant, *Bull. Soc. Chim. France*, **1964**, 2423.
92. K. Gewald and G. Neumann, *Chem. Ber.*, **101**, 1933 (1968).
93. G. W. Stacy and T. E. Wollner, *J. Org. Chem.*, **32**, 3028 (1967).
94. T. P. C. Mulholland, R. I. W. Honeywood, H. D. Preston, and D. T. Rosevear, *J. Chem. Soc.*, **1965**, 4939.
95. A. Bree and R. Zwarich, *Spectrochim. Acta*, **27A**, 599 (1971).
96. E. Campaigne and R. B. Rogers, *J. Heterocycl. Chem.*, **10**, 297 (1973).
97. T. R. Bosin, R. P. Maickel, A. Dinner, and A. Snell, *J. Heterocycl. Chem.*, **9**, 1265 (1972).
98. P. Faller, *Bull. Soc. Chim. France*, **1969**, 941.
99. H. Hofmann and G. Salbeck, *Angew. Chem. Int. Ed.*, **8**, 456 (1969).

100. G. E. Heeres and H. Wynberg, *Tetrahedron*, **28**, 5237 (1972).
101. J. M. Angellini, A. R. Katritzky, R. G. Pinzelli, and R. D. Topsom, *Tetrahedron*, **28**, 2037 (1972).
102. M. Sénéchal and P. Saumagne, *J. Chim. Phys.*, **69**, 1246 (1972).
103. K. Volka, P. Adámek, J. Bánki, I. Stibor, and Z. Ksandr, private communication (see ref. 104).
104. K. Volka, P. Adámek, I. Stibor, and Z. Ksandr, *Spectrochim. Acta*, **32A**, 397 (1976).
105. M. Bambagiotti, E. Castellucci, and G. Sbrana, *Spectrochim. Acta*, **30A**, 1413 (1974).
106. P. Adámek, K. Volka, I. Stibor, and Z. Ksandr, *Sci. Papers Prague Inst. Chem. Technol. Anal. Chem.*, **H12**, 193 (1977).
107. K. Volká, I. Stibor, and Z. Ksandr, *Sci. Papers Prague Inst. Chem. Technol. Anal. Chem.*, **H13**, 157 (1978).
108. B. Pete, A. Kiss, J. Mink, K. Gál, and J. Bánki, *Spectrochim. Acta*, **36A**, 633 (1980).
109. C. Carrio, L. Ballester, J. Fernandez, and M. Sanfeliz, *Cienc. Fis.*, **3**, 23 (1971); *Chem. Abstr.*, **78**, 1510 (1973).
109a. V. N. Sheinker, A. D. Garnovskii, and O. A. Osipov, *Usp. Khim.*, **50**, 632 (1981).
109b. A. T. Balaban, M. D. Gheorghiu, and C. Draghici, *Israel J. Chem.*, **20**, 168 (1980).
110. F. Fringuelli and A. Taticchi, *J. Heterocycl. Chem.*, **10**, 89 (1973).
111. A. Perjessy, R. Frimm, and P. Hrnciar, *Collect. Czech. Commun.*, **37**, 3302 (1972).
112. T. Woldbaek, P. Klaboe, and C. J. Nielsen, *J. Mol. Struct.*, **27**, 283 (1975).
113. L. Le Gall, J. Lauransan, and P. Saumagne, *Canad. J. Spectrosc.*, **20**, 136 (1975).
114. M. G. Giorgini, B. Fortunato, and P. Mirone, *Atti Soc. Nat. Mat. Moldena*, **106**, 89 (1975).
115. A. J. Barnes, L. Le Gall, C. Madec, and J. Lauransan, *J. Mol. Struct.*, **38**, 109 (1977).
116. H. El Khadem and L. A. Kemler, *J. Heterocycl. Chem.*, **9**, 1413 (1972).
117. G. A. Zalesskaya, *Opt. Spektrosk. Akad. Nauk SSSR, Otd. Fiz. Mat. Nauk*, **3**, 121 (1967); *Chem. Abstr.*, **68**, 109 688 (1968).
118. N. N. Borisevich and N. N. Khovratovich, *Zh. Prikl. Spektrosk.*, **7**, 538 (1967).
119. A. Bigotto and V. Galasso, *Spectrochim. Acta*, **35A**, 725 (1979).
120. J. M. Markgraf, C. I. Heller, and N. L. Avery, *J. Org. Chem.*, **35**, 1588 (1970).
121. Y. Hase, *Spectrochim. Acta*, **36A**, 213 (1980).
122. Y. Hase, *Spectrochim. Acta*, **36A**, 413 (1980).
123. R. G. Fowler, L. R. Caswell, and L. I. Sue, *J. Heterocycl. Chem.*, **10**, 407 (1973).
124. A. Bigotto and V. Galasso, *Spectrochim. Acta*, **36A**, 95 (1980).
125. M. Renson and R. Collienne, *Bull. Soc. Chim. Belg.*, **73**, 491 (1964).
126. R. Huisgen, H. Brade, H. Walz, and I. Glogger, *Chem. Ber.*, **90**, 1437 (1957).
127. R. Mecke and R. Mecke, *Chem. Ber.*, **89**, 343 (1956).
128. H. E. Hallam and C. M. Jones, *J. Mol. Struct.*, **1**, 413, 425 (1967-68).
129. H. E. Hallam and C. M. Jones, *J. Chem. Soc. A*, **1969**, 1033.
130. H. K. Hall and R. Zbinden, *J. Amer. Chem. Soc.*, **80**, 6428 (1958).
131. C. Y. S. Chen and Ch. A. Swenson, *J. Phys. Chem.*, **73**, 2999 (1969).
132. S. E. Krikorian, T. A. Andrea, and M. Mahpour, *Spectrochim. Acta*, **32A**, 1447 (1976).
133. J. C. Sheehan and E. J. Corey, *J. Amer. Chem. Soc.*, **73**, 4756 (1951).
134. M. Lang, K. Prasad, W. Holick, J. Gosteli, I. Ernest, and R. B. Woodward, *J. Amer. Chem. Soc.*, **101**, 6296 (1979).
135. H. R. Pfaendler, J. Gosteli, and R. B. Woodward, *J. Amer. Chem. Soc.*, **101**, 6306 (1979).

136. I. Ernest, J. Gosteli, and R. B. Woodward, *J. Amer. Chem. Soc.*, 101, 6301 (1979).
137. H. R. Pfaendler, J. Gosteli, and R. B. Woodward, *J. Amer. Chem. Soc.*, 102, 2039 (1980).
138. G. S. Sidhu, G. Thyagarajan, and U. T. Bhalerao, *J. Chem. Soc.*, 1966, 969.
139. G. Thyagarajan, U. T. Bhalerao, S. Naseem, and V. S. Subramanian, *Indian J. Chem.*, 6, 625 (1968).
140. U. T. Bhalerao and G. Thyagarajan, *Indian J. Chem.*, 6, 176 (1968).
141. K. Schenker, *Helv. Chim. Acta*, 51, 413 (1968).
142. M. E. Derieg and C. H. Sternbach, *J. Heterocycl. Chem.*, 3, 237 (1966).
143. A. A. Santilli and T. S. Osdene, *J. Org. Chem.*, 31, 4268 (1966).
144. L. Bauer and R. E. Hewitson, *J. Org. Chem.*, 27, 3982 (1962).
145. A. P. Soll and F. Troxler, *Helv. Chim. Acta*, 51, 1864 (1968).
146. S. Minami, M. Tomita, H. Takamatsu, and S. Uyeo, *Chem. Pharm. Bull.*, 13, 1084 (1964).
147. D. Misiti, V. Rimatori, and F. Gatta, *J. Heterocycl. Chem.*, 10, 689 (1973).
148. I. I. Bădilescu, Ph.D thesis, 1973; *Rev. Roumaine Chim.*, 20, 1077 (1975).
149. J. M. Muchowski, *Canad. J. Chem.*, 48, 1946 (1970).
150. M. Uscovic, G. Grethe, J. Iacobelli, and W. Werner, *J. Org. Chem.*, 30, 3111 (1965).
151. J. Krapcho, E. R. Spitzmiller, and C. F. Turk, *J. Med. Chem.*, 6, 544 (1963).
152. D. C. Lankin, M. R. Scalise, J. C. Schmidt, and H. Zimmer, *J. Heterocycl. Chem.*, 11, 631 (1974).
153. Y. K. Kim, G. A. Grindahl, J. R. Greenwald, and O. R. Pierce, *J. Heterocycl. Chem.*, 11, 563 (1974).
154. K. Praefcke, M. M. Sidky, and F. H. Osman, *J. Heterocycl. Chem.*, 11, 845 (1974).
155. H. E. Zaugg, J. E. Leonard, R. W. Denet, and D. L. Arendsen, *J. Heterocycl. Chem.*, 11, 797 (1974).
156. F. Cristiani, F. A. Devillanova, and G. Verani, *J. Chem. Soc., Perkin II*, 1977, 324.
157. M. Guiliano, G. Davidovics, J. Chouteau, J. L. Larice, and J. P. Roggero, *J. Mol. Struct.*, 25, 329 (1975).
158. F. A. Devillanova, G. Verani, K. R. Gayathri Devi, and D. N. Sathyanarayana, *Spectrochim. Acta*, 36A, 199 (1980).
159. F. A. Devillanova, K. R. Gayathri Devi, and D. N. Sathyanarayana, *Spectrochim. Acta*, 35A, 1083 (1979).
160. F. A. Devillanova and G. Verani, *Spectrochim. Acta*, 36A, 371 (1980).
161. R. A. Pethrick and A. D. Wilson, *Spectrochim. Acta*, 30A, 1073 (1974).
162. H. N. Jallo and F. N. Al-Azawi, *Spectrochim. Acta*, 34A, 819 (1978).
163. R. A. Pethrick, E. Wyn-Jones, P. Chamblin, and R. F. M. White, *J. Chem. Soc. Ser. A*, 1969, 1852.
164. B. Fortunato and G. Fini, *Spectrochim. Acta*, 31A, 1233 (1975).
165. P. Mirone, B. Fortunato, and P. Canziani, *J. Mol. Struct.*, 5, 283 (1970).
166. A. C. Fabretti and G. Peyrone, *Spectrochim. Acta*, 34A, 667 (1978).
167. V. Knoppova, K. Antos, L. Drobnica, and P. Kristian, *Chem. Zvesti*, 26, 527 (1972).
168. N. Al. Kassab, M. H. Elnagdy, and N. A. Messeha, *J. Prakt. Chem.*, 314, 799 (1972).
169. P. Sohár, J. Nyitrai, K. Zauer, and K. Lempert, *Acta Chim. Hung.*, 58, 31 (1968).
170. J. T. Edward and I. Lantos, *J. Heterocycl. Chem.*, 9, 363 (1972).
171. A. R. Katritzky and J. M. Lagowski, *Adv. Heterocycl. Chem.*, 1, 311 (1963).
172. K. Lempert, J. Nyitrai, P. Sohár, and K. Zauer, *Tetrahedron Lett.*, 1964, 2679.

References

173. E. Schipper and E. Chinnery, *J. Org. Chem.*, **26**, 4480 (1961).
174. R. Jacquier, J. M. Lacombe, and G. Maury, *Bull. Soc. Chim. France*, **1971**, 1040.
175. H. Gnichtel, *Chem. Ber.*, **103**, 2411 (1970).
176. H. Gnichtel, R. Walentowski, and K. E. Schuster, *Chem. Ber.*, **105**, 1701 (1972).
177. H. Gnichtel, S. Exner, H. Bierbüsse, and M. Alterdinger, *Chem. Ber.*, **104**, 1512 (1971).
178. A. A. Santili, B. R. Hofmann, and Dong H. Khim, *J. Heterocycl. Chem.*, **11**, 879 (1974).
179. G. A. Newman and P. J. S. Pauwels, *Tetrahedron*, **25**, 4605 (1969).
180. W. A. Seth Paul and P. J. A. Demoen, *Bull. Soc. Chim. Belg.*, **75**, 524 (1966).
181. L. J. Bellamy, B. R. Connelly, A. R. Philpotts, and R. L. Williams, *Z. Electrochem.*, **64**, 563 (1960).
182. C. Fayat and A. Foucaud, *Compt. rend.*, **261**, 4018 (1965).
183. J. Bellanato, C. Avendano, P. Ballesteros, and M. Martinez, *Spectrochim. Acta*, **35A**, 807 (1979).
184. H. E. Zaugg, J. E. Leonard, and D. L. Arendsen, *J. Heterocycl. Chem.*, **11**, 833 (1974).
185. V. E. Marquez, T. Hirata, L. M. Twanmoh, H. B. Wood, Jr., and J. S. Driscol, *J. Heterocycl. Chem.*, **9**, 1145 (1972).
186. P. Sohár, *Acta Chim. Hung.*, **57**, 425 (1968).
187. P. J. F. Griffiths, G. D. Morgan, and B. Ellis, *Spectrochim. Acta*, **28**, 1899 (1972).
188. I. Ya. Kvitko, S.V. Bolkhovets, and A. M. Kokmina, *Khim. Geterotsikl. Soedin.*, **1972**, 1491.
189. V. E. Kononenko, B. E. Zhitar, and S. N. Baranos, *Zh. Org. Khim.*, **9**, 61 (1973).
190. K. A. V'yunov, A. I. Ginak and E. G. Sochilin, *Zh. Prikl. Spectrosk.*, **25**, 865 (1976).
191. P. J. Taylor, *Spectrochim. Acta*, **26A**, 153 (1970).
192. D. G. O'Sullivan, *Spectrochim. Acta*, **16**, 762 (1960).
193. J. R. Marvel and A. P. Lemberger, *J. Amer. Pharm. Assoc., Sci. Ed.*, **49**, 417 (1960).
194. I. I. Grandberg, A. N. Kost, and D. V. Sibiryakova, *Zh. Obshch. Khim.*, **30**, 2920 (1960).
195. I. I. Grandberg, A. N. Kost, and N. N. Zheltikova, *Zh. Obshch. Khim.*, **30**, 2931 (1960).
196. I. I. Grandberg, *Zh. Obshch. Khim.*, **31**, 2307 (1961).
197. P. Mirone, *Ann. Chim. (Rome)*, **46**, 39 (1956).
198. W. Ried and F. J. Königstein, *Liebigs. Ann. Chem.*, **625**, 53 (1959).
199. H. Wolff and H. Müller, *Spectrochim. Acta*, **32A**, 581 (1976).
200. H. Wolff and E. Wolff, *Spectrochim. Acta*, **27A**, 2109 (1971).
201. M. Majoube, *J. Mol. Struct.*, **61**, 129 (1980).
202. V. Lorenzelli and G. Randi, *Atti. Acad. Nazl. Lincei Rend. Classe Sci. Fis. Mat. Nat.*, **36**, 646 (1964); *Chem. Abstr.*, **62**, 7612 (1965).
203. S. T. King, *J. Phys. Chem.*, **74**, 2133 (1970).
204. A.-M. Bellocq and Ch. Garrigou-Lagrange, *Spectrochim. Acta*, **27A**, 1091 (1971).
205. A. Trani and E. Bellasio, *J. Heterocycl. Chem.*, **11**, 257 (1974).
206. T. K. Sevast'yanova and L. B. Volodarskij, *Izv. Akad. Nauk SSSR, Ser. Khim.*, **1972**, 2339.
207. P. Melloni, D. Fusar-Bassini, E. Dradi, and C. Confaconieri, *J. Heterocycl. Chem.*, **11**, 731 (1974).
208. D. M. Rackham, *Appl. Spectrosc.*, **33**, 561 (1979).
209. N. D. Cordes and J. L. Malter, *Spectrochim. Acta*, **24A**, 257 (1968).
210. J. K. Wilmshurst and H. J. Bernstein, *Canad. J. Chem.*, **35**, 1183 (1957).
211. D. Heinert and A. E. Martell, *J. Amer. Chem. Soc.*, **81**, 3933 (1959).

212. J. Elguero, R. Jacquier, and S. Mignonac-Mondon, *J. Heterocycl. Chem.*, **10**, 411 (1973).
213. N. K. Schel'tsyn, A. Ya. Kaminskii, T. P. Shapirovskaya, I. L. Vaisman, V. Fandrainov, and S. S. Gitis, *Khim. Geterotsikl. Soedin.*, **1973**, 115.
214. B. A. Tertov, A. V. Koblik, and Yu. Kolodazhny, *Tetrahedron Lett.*, **42**, 4445 (1968).
215. L. S. Efros and G. N. Kulbitskii, *Zh. Obshch. Khim.*, **38**, 981 (1968).
216. P. Sohár and L. Toldy, *Acta Chim. Hung.*, **75**, 99 (1973).
217. D. Bârcă-Gălăţeanu, F. Gagiu, and D. Iordache, *Rev. Roum. Phys.*, **14**, 231 (1969).
218. P. Söhár and L. Toldi, *Magy. Kém. Foly.*, **78**, 535 (1972).
219. D. J. Adam and M. Wharmby, *Tetrahedron Lett.*, **1969**, 3063.
220. S. Califano, F. Piacenti, and G. Sbrana, *Spectrochim. Acta*, **20**, 339 (1964).
221. G. Mille, J. Metzger, C. Pouchan, and M. Chaillet, *Spectrochim. Acta*, **31A**, 1115 (1975).
222. I. Gaile and E. Yu. Gudrinietze, *Izv. Akad. Nauk Latv. SSSR, Ser. Khim.*, **1967**, 54.
223. K. M. Ziemelis and E. Yu. Gudrinietze, *Izv. Akad. Nauk Latv. SSSR, Ser. Khim.*, **1967**, 445.
224. I. Gaile, E. Gudrinietze and G. Banag, *Izv. Akad. Nauk Latv. SSSR, Ser. Khim.*, **1962**, 523.
225. E. Borello, A. Zecchina, and A. Appiano, *Spectrochim. Acta*, **22**, 977 (1966).
226. G. Sbrana, E. Castellucci, and M. Ginanneschi, *Spectrochim. Acta*, **23A**, 751 (1967).
227. E. Borello, A. Zecchina, and A. Appiano, *Spectrochim. Acta*, **23A**, 1335 (1967).
228. P. Bassignana, C. Cogrossi, and M. Gandino, *Spectrochim. Acta*, **19**, 1885 (1963).
229. V. L. Koenig, F. N. Hayes, B. S. Rogers, and J. D. Perrings, *U.S. At. Energy Comm. AECU*, 2778 (1953).
230. D. Festel and Y. Mollier, *Tetrahedron Lett.*, **1970**, 1259.
231. R. Pinel and Y. Mollier, *Bull. Soc. Chim. France*, **1972**, 1385.
232. D. Festal, J. Tison, N. Kimson, R. Pinel, and Y. Mollier, *Bull. Soc. Chim. France*, **1973**, 3339.
233. H. Gotthardt and B. Christl, *Tetrahedron Lett.*, **1968**, 4743.
234. H. Gotthardt, M. C. Weisshuhn, and B. Christl, *Chem. Ber.*, **109**, 740 (1976).
235. E. Borello and A. Zecchina, *Ann. Chim. (Rome)*, **52**, 1302 (1962).
236. O. L. Stiefvater, H. Jones, and J. Sheridan, *Spectrochim. Acta*, **26A**, 825 (1970).
237. H. G. O. Becker, K. Heimburger, and M. J. Timpe, *J. Prakt. Chem.*, **313**, 795 (1971).
238. H. J. Timpe, *Z. Chem.*, **11**, 340 (1971).
239. A. P. Grekov and M. S. Solov'eva, *Zh. Obshch. Khim.*, **30**, 1644 (1960).
240. F. Maggio, G. Werber, and G. Lombardo, *Ann. Chim. (Rome)*, **50**, 491 (1960).
241. J. Barrans, *Compt. rend.*, **249**, 1096 (1959).
242. M. O. Losinskii, P. S. Pelkis, and S. N. Sanova, *Zh. Obshch. Khim.*, **33**, 2231 (1963).
243. W. R. Sherman, *J. Org. Chem.*, **26**, 88 (1961).
244. J. W. Anthonsen, D. H. Christensen, J. T. Nielsen, and O. F. Nielsen, *Spectrochim. Acta*, **32A**, 975 (1976).
245. D. H. Christensen, P. W. Jensen, J. T. Nielsen, and O. F. Nielsen, *Spectrochim. Acta*, **29A**, 1393 (1973).
246. E. Borello, A. Zecchina, and E. Guglichni-Notti, *Gazzetta*, **96** 852 (1966).
247. G. Sbrana, M. Ginanneschi, and M. P. Marzocchi, *Spectrochim. Acta*, **23A**, 1757 (1967).
248. D. H. Christensen and T. Stroyer-Hansen, *Spectrochim. Acta*, **26A**, 2057 (1970).
249. D. H. Christensen, *Acta Chem. Scand.*, **26**, 923 (1972).
250. T. S. Hansen, *Acta Chem. Scand.*, **A28**, 811 (1974).

251. G. Sbrana and M. Ginanneschi, *Spectrochim. Acta*, **22**, 517 (1966).
252. S. Brodersen and A. Langseth, *J. Mol. Spectrosc.*, **3**, 114 (1959).
253. E. Saegebart and A. P. Cox, *J. Chem. Phys.*, **43**, 166 (1965).
254. D. R. Jones, C. H. Wang, D. H. Christensen, and O. F. Nielsen, *J. Chem. Phys.*, **64**, 4475 (1976).
255. B. Soptrajanov and G. E. Ewing, *Spectrochim. Acta*, **22**, 1417 (1966).
256. N. E. Boyer, G. M. Czerniak, H. S. Gutowsky, and H. R. Snyder, *J. Amer. Chem. Soc.*, **77**, 4238 (1955).
257. J. H. Boyer, D. I. McCane, W. J. McCarville, and A. T. Tweedie, *J. Amer. Chem. Soc.*, **75**, 5298 (1953).
258. J. Goerdeler and O. Tegtmeyer, *Angew. Chem.*, **67**, 302 (1955).
259. K. S. Surgsh and C. N. R. Rao, *J. Indian Chem. Soc.*, **37**, 581 (1960).
260. H. J. Emeléus, A. Haas, and N. Sheppard, *J. Chem. Soc.*, **1963**, 3165.
261. G. D. Thorn, *Canad. J. Chem.*, **38**, 2349 (1960).
262. A. Zecchina, G. E. Andreoletti, and P. Sampietro, *Spectrochim. Acta*, **23A**, 2647 (1967).
263. K. Bast, M. Christl, R. Huisgen, and W. Mack, *Chem. Ber.*, **105**, 2825 (1972).
264. T. N'Gando, M. Pondo, C. Halavaud, and J. Barraws, *Compt. rend.*, **C274**, 2026 (1972).
265. M. Kuhn and R. Mecke, *Z. Anal. Chem.*, **181**, 487 (1961).
266. C. Christophersen and A. Holm, *Acta Chem. Scand.*, **25**, 2015 (1971).
267. G. Bianchi, A. J. Boulton, I. J. Fletcher, and A. R. Katritzky, *J. Chem. Soc. Ser. B*, **1971**, 2355.
268. D. Vedal, O. H. Ellestad, P. Klaboe, and G. Hagen, *Spectrochim. Acta*, **31A**, 339 (1975).
269. D. Vedal, O. H. Ellestad, P. Klaboe, and G. Hagen, *Spectrochim. Acta*, **31A**, 355 (1975).
270. D. Vedal, O. H. Ellestad, and P. Klaboe, *Spectrochim. Acta*, **32A**, 877 (1976).
271. R. Borsdorf, R. Broddack, and H. H. Schwartz, *Z. Chem.*, **8**, 378 (1968).
272. R. A. Jones, A. R. Katritzky, A. C. Richard, and R. J. Wyatt, *J. Chem. Soc., Ser. B*, **1970**, 127.
273. R. W. Baldock and A. R. Katritzky, *J. Chem. Soc., B*, **1968**, 1470.
274. R. W. Baldock and A. R. Katritzky, *Tetrahedron Lett.*, **1968**, 1159.
275. P. J. Kreuger and J. Jan, *Canad. J. Chem.*, **48**, 3229, 3236 (1970).
276. F. Moll, *Tetrahedron Lett.*, **1968**, 5201.
277. O. S. Anisimova, N. D. Chizhnikova, and L. G. Yudin, *Vest. Moskv. Univ. Ser. 2*, **23**, 155 (1968).
278. E. Brown, R. Dihal, and P. F. Casals, *Tetrahedron*, **28**, 5607 (1972).
279. L. A. Kotorlenko, V. S. Aleksandrova, and S. A. Samoilenko, *Zh. Prikl. Spektrosk.*, **31**, 851 (1979).
280. F. Cornea, C. Cercasov, and M. Ciureanu, *Spectrochim. Acta*, **36A**, 775 (1980).
281. M. P. Paradisi and A. Romeo, *J. Chem. Soc. Perkin Trans. I*, **1977**, 596.
282. G. P. Zecchini and M. P. Paradisi, *J. Heterocycl. Chem.*, **16**, 1589 (1979).
283. R. R. Wittekind and S. Lazarus, *J. Heterocycl. Chem.*, **10**, 217 (1973).
284. C. Schiele, *Z. Naturforsch.*, **24**, 976 (1969).
285. F. E. Malherbe and H. J. Bernstein, *J. Amer. Chem. Soc.*, **74**, 4408 (1952).
286. H. H. Kirchner, *Z. Phys. Chem.*, **29**, 166 (1961).
287. P. Tarte and P. A. Laurent, *Bull. Soc. Chim. Belg.*, **70**, 43 (1961).
288. T. Miyazawa, K. Fukushima, and Y. Ideguchi, *J. Chem. Phys.*, **37**, 2764 (1964).

289. R. G. Snyder and G. Zerbi, *Spectrochim. Acta*, **23A**, 391 (1967).
290. J. P. Marsault, Ph.D. thesis, Faculté des Sciences de Paris, 1968.
291. O. H. Ellestad, P. Klaboe, and G. Hagen, *Spectrochim. Acta*, **27A**, 1025 (1971).
292. J. Caillod, O. Saur, and J. C. Lavalley, *Spectrochim. Acta*, **36A**, 185 (1980).
293. Hung Chen and P. A. Giguère, *Spectrochim. Acta*, **29A**, 1611 (1973).
294. A. B. Dempster and H. Uslu, *Spectrochim. Acta*, **34A**, 71 (1978).
295. O. H. Ellestad, P. Klaboe, and G. Hagen, *Spectrochim. Acta*, **29A**, 1247 (1973).
296. O. H. Ellestad, P. Klaboe, and G. Hagen, *Spectrochim. Acta*, **28A**, 137 (1972).
297. P. Klaboe, *Spectrochim. Acta*, **25A**, 1437 (1969).
298. O. H. Ellestad, P. Klaboe, G. Hagen, and T. Stroyer-Hansen, *Spectrochim. Acta*, **28A**, 149 (1972).
299. T. Hirokawa, T. Kimura, K. Ohno, and H. Murate, *Spectrochim. Acta*, **36A**, 329 (1980).
300. G. Marcotrigeano and G. C. Pellacani, *Spectrochim. Acta*, **31A**, 1865 (1975).
301. J. Skolimowski, R. Skowronski, and M. Simalty, *Tetrahedron Lett.*, **1979**, 4833.
302. P. Klaboe and Z. Smith, *Spectrochim. Acta*, **34A**, 489 (1978).
303. V. V. Zamkova, A. E. Lyuts, V. I. Pesterev, O. V. Agashkin, L. A. Ignatova, and B. V. Unkovskii, *Dokl. Akad. Nauk SSSR*, **1980**, 250, 1144.
304. W. Ried and W. Ochs, *Synthesis*, **1972**, 311.
305. D. Bardard, J. M. Fabian, and H. P. Koch, *J. Chem. Soc.*, **1949**, 2442.
306. G. Pifferi and R. Moneuzzi, *J. Heterocycl. Chem.*, **9**, 1445 (1972).
307. J. G. Lombardino, *J. Heterocycl. Chem.*, **9**, 315 (1972).
308. M. Asai and K. Noda, *Spectrochim. Acta*, **34A**, 695 (1978).
309. L. Cazaux, J. D. Bastide, G. Chassaing, and P. Maroni, *Spectrochim. Acta*, **35A**, 15 (1979).
310. N. D. Heindel and C. C. Ho Ko, *J. Heterocycl. Chem.*, **11**, 1109 (1974).
311. H. Gnichtel, *Chem. Ber.*, **103**, 3442 (1970).
312. H. Gnichtel, S. Exner, H. Bierbüsse, and M. Alterdinger, *Chem. Ber.*, **104**, 1512 (1971).
313. H. Gnichtel and S. Thiele, *Chem. Ber.*, **104**, 1507 (1971).
314. B. Weinstein and H. H. Chang, *J. Heterocycl. Chem.*, **11**, 99 (1974).
315. J. E. Barnes, J. A. W. Dalziel, and S. D. Ross, *Spectrochim. Acta*, **27A**, 1247 (1971).
316. J. E. Barnes, J. A. W. Dalziel, and S. D. Ross, *Spectrochim. Acta*, **27A**, 1675 (1971).
317. K. Olsson, *Arkiv. Kemi*, **26**, 38 (1967).
318. A. Fredge and K. Olsson, *Arkiv. Kemi*, **9**, 163 (1955).
319. S. Hoffman, M. Herrmann, and E. Muehet, *Z. Chem.*, **8**, 417 (1968).
320. D. J. Daigle, A. B. Pepperman, and S. L. Vail, *J. Heterocycl. Chem.*, **11**, 407 (1974).
321. D. J. Daigle, A. B. Pepperman, and G. Boudreaux, *J. Heterocycl. Chem.*, **11**, 1085 (1974).
322. J. R. McDivitt and G. L. Humphrey, *Spectrochim. Acta*, **30A**, 1021 (1974).
323. P. Bruesch and H. H. Günthard, *Spectrochim. Acta*, **22**, 877 (1966).
323a. E. Grech, Z. Malarski, and L. Sobczyk, *Croat. Chem. Acta*, **55**, 47 (1982).
324. H.-P. Koopmann and P. Rademacher, *Spectrochim. Acta*, **32A**, 157 (1975).
325. A. R. Katritzky and R. A. Jones, *Spectrochim. Acta*, **17**, 64 (1961).
326. H. N. Al-Jallo and F. W. Al-Azawi, *J. Heterocycl. Chem.*, **11**, 1101 (1974).
327. I. El-Sayed El-Kholy, M. M. Mishrikey, and R. F. Atmeh, *J. Heterocycl. Chem.*, **11**, 487 (1974).
328. I. Lalezari, A. Ghanbarpour, F. Ghapgharan, M. Niazi, and R. Jafari-Namin, *J. Heterocycl. Chem.*, **11**, 468 (1974).

References

329. A. Ruwet and M. Renson, *Bull. Soc. Chim. Belg.*, **75**, 260 (1966).
330. C. Venturello and G. P. Chiusoli, *J. Heterocycl. Chem.*, **9**, 1155 (1972).
331. R. Mecke, R. Mecke, and A. Lüttringhaus, *Z. Naturforsch.*, **B10**, 367 (1955).
332. A. Ruwet and M. Renson, *Bull. Soc. Chim. Belg.*, **79**, 89 (1970).
333. J. W. Thompson Jr., G. E. Leroi, and A. I. Popov, *Spectrochim. Acta*, **31A**, 1553 (1975).
334. M. Asai, K. Noda, and A. Sado, *Spectrochim. Acta*, **30A**, 1147 (1974).
335. R. Granger, H. Orzelesi, and Y. Robbe, *Trav. Soc. Pharm. Montpellier*, **28**, 65 (1968); *Chem. Abstr.*, **69**, 105686 (1968).
336. B. W. Dominy and R. G. Cawton, *J. Org. Chem.*, **34**, 2013 (1969).
337. H. Sterk and E. Ziegler, *Monatsh. Chem.*, **100**, 739 (1969).
338. G. Winters, V. Aresi, and G. Nathanson, *J. Heterocycl. Chem.*, **11**, 997 (1974).
339. R. T. Courts, K. W. Hindermarsh, S. J. Powell, and J. L. Pound, *Canad. J. Pharm. Sci.*, **3**, 49 (1968).
340. L. Czuchajowski and M. Eckstein, *Rocz. Chem.*, **41**, 907 (1967).
341. G. Casini, F. Claudi, M. Grifantini, and S. Martelli, *J. Heterocycl. Chem.*, **11**, 381 (1974).
342. B. L. Shaw and T. H. Simpson, *J. Chem. Soc.*, **1955**, 655.
343. J. W. Looker and H. H. Hanneman, *J. Org. Chem.*, **27**, 381 (1962).
344. S. Balakhrisna, J. D. Ramanathan, T. R. Seshadri, and B. Venkataramani, *Proc. Roy. Soc.*, **268A**, 10 (1962).
345. C. I. Jose, P. S. Phadke, and A. V. Rama Rao, *Spectrochim. Acta*, **30A**, 1199 (1974).
346. D. R. Glusker and H. W. Thompson, *J. Chem. Soc.*, **1955**, 471.
347. K. Görlitzer, *Arch. Pharm.*, **307**, 523 (1974).
348. K. Görlitzer, *Arch. Pharm.*, **308**, 81 (1975).
349. K. Görlitzer, *Arch. Pharm.*, **308**, 272 (1975).
350. K. Görlitzer, *Arch. Pharm.*, **309**, 18 (1976).
351. K. Görlitzer, *Arch. Pharm.*, **309**, 356 (1976).
352. J. K. Wilmhurst and H. J. Berstein, *Canad. J. Chem.*, **35**, 1183 (1957).
353. D. A. Long and E. L. Thomas, *Trans. Faraday Soc.*, **59**, 783 (1963).
354. D. A. Long and W. O. George, *Spectrochim. Acta*, **19**, 1777 (1963).
355. M. A. Kovner, Yu. S. Korostelov, and V. I. Berezen, *Opt. Spectrosc.*, **10**, 233 (1961).
356. L. Corrsin, B. J. Fax, and R. C. Lord, *J. Chem. Phys.*, **21**, 1170 (1953).
356a. H. D. Stidham and D. P. DiLella, *J. Raman Spectrosc.*, **9**, 90 (1980).
356b. E. Wachsmann and E. W. Schmid, *Z. Phys. Chem.*, **27**, 145 (1961).
356c. J. H. S. Green, W. Kynaston, and H. M. Paisley, *Spectrochim. Acta*, **19A**, 549 (1963).
357. F. A. Andersen, B. Bak, S. Brodersen, and J. Rastrup-Anderson, *J. Chem. Phys.*, **23**, 1047 (1955).
358. J. S. Strukl and J. L. Walter, *Spectrochim. Acta*, **27A**, 209 (1971).
359. A. T. Hutton and D. A. Thornton, *Spectrochim. Acta*, **34A**, 645 (1978).
360. P. E. Rutherford and D. A. Thornton, *Spectrochim. Acta*, **35A**, 711 (1979).
361. G. A. Foulds, G. C. Percy, and D. A. Thornton, *Spectrochim. Acta*, **34A**, 1231 (1978).
362. J. E. Katon and N. T. McDevil, *J. Mol. Spectrosc.*, **14**, 308 (1964).
363. K. Krebs, S. Sandroni, and G. Zerbi, *J. Chem. Phys.*, **40**, 3502 (1964).
364. D. H. Whiffen, *Spectrochim. Acta*, **7**, 253 (1955).
365. J. S. Strukl and J. L. Walter, *Spectrochim. Acta*, **27A**, 223 (1971).
366. J. E. A. Otterstedt and R. Pater, *J. Heterocycl. Chem.*, **9**, 229 (1972).

367. Y. Tamura, Y. Miki, and M. Ideda, *J. Heterocycl. Chem.*, **10**, 447 (1973).
368. I. S. Ahuja and R. Sing, *Spectrochim. Acta*, **30A**, 2055 (1974).
369. R. Grzeskowiak, C. Whatley, and M. Goldstein, *Spectrochim. Acta*, **31A**, 1577 (1975).
370. R. A. Abramovich, J. Cambell, E. E. Knaus, and A. Silhankova, *J. Heterocycl. Chem.*, **9**, 1367 (1972).
371. J. H. S. Green, W. Kynaston, and H. M. Paisley, *Spectrochim. Acta*, **19**, 549 (1963).
372. G. Varsanyi, T. Farago, and S. Holly, *Acta Chim. Hung.*, **43**, 205 (1965).
373. D. A. Long and R. T. Bailey, *Trans. Faraday Soc.*, **59**, 599 (1963).
374. D. A. Long and D. S. Steele, *Spectrochim. Acta*, **19**, 1797 (1963).
375. R. T. Bayley and D. Steele, *Spectrochim. Acta*, **23A**, 2997 (1967).
376. S. Lui, S. Suzuki, and J. A. Ladd, *Spectrochim. Acta*, **34**, 583 (1978).
377. J. H. S. Green and D. J. Harrison, *Spectrochim. Acta*, **29A**, 293 (1973).
378. G. Forrest, G. Vilcins, and J. O. Lephardt, *Spectrochim. Acta*, **32A**, 511 (1976).
379. J. Barassin, G. Quequiner, and H. Lumbroso, *Bull. Soc. Chim. France*, **4**, 707 (1967).
380. G. Karabatsos and F. M. Vane, *J. Amer. Chem. Soc.*, **85**, 3886 (1963).
381. F. A. Miller, W. G. Fateley, and R. E. Witowski, *Spectrochim. Acta*, **23A**, 891 (1967).
382. V. V. Zverev, *Vopr. Stereokhim.*, **1**, 83 (1971).
383. P. Carmova, *Spectrochim. Acta*, **36A**, 705 (1980).
384. K. R. Wursthorn and E. H. Sund, *J. Heterocycl. Chem.*, **9**, 25 (1972).
385. A. T. Balaban and M. Gheorghiu, *Rev. Roumaine Chim.*, **23**, 1065 (1978).
386. R. A. Kydd, *Spectrochim. Acta*, **35A**, 409 (1979).
387. K. Sasaki and K. Aida, *J. Inorg. Nucl. Chem.*, **42**, 13 (1980).
388. J. Guillerez, O. Bensaude, and J. E. Dubois, *J. Chem. Soc., Perkin Trans. II*, **1980**, 620.
389. S. S. T. King, W. L. Dilling, and N. B. Tefertiller, *Tetrahedron*, **28**, 5869 (1972).
390. E. M. Peresleni, M. Ya. Uritskaya, V. A. Loginova, and Yu. N. Sheinker, *Dokl. Akad. Nauk SSSR*, **183**, 1102 (1968).
391. P. Beaks, *Acc. Chem. Res.*, **10**, 186 (1977).
392. O. Bensaude, M. Chevrier, and J. E. Dubois, *J. Amer. Chem. Soc.*, **101**, 2423 (1979).
393. R. A. Coburn and G. O. Dudek, *J. Phys. Chem.*, **72**, 3681 (1968).
394. M. Hamer and E. Patrick Lira, *J. Heterocycl. Chem.*, **9**, 215 (1972).
395. I. El-Sayed El-Kholy, M. M. Mishrikey, and R. F. Atmeh, *J. Heterocycl. Chem.*, **10**, 665 (1973).
396. J. Koziol and P. Tomasik, *Bull. Acad. Pol. Sci., Ser. Chim.*, **27** 335 (1979).
397. M. Pfeffer, P. Braunstein, and J. Dehand, *Spectrochim. Acta*, **30A**, 331 (1974).
398. D. H. Brown, D. T. Stewart, and D. E. H. Jones, *Spectrochim. Acta*, **29A**, 213 (1973).
399. A. R. Katritzky, *J. Chem. Soc.*, **1958**, 4162.
400. A. R. Katritzky and R. D. Topsom, *Chem. Rev.*, **77**, 639 (1977).
401. I. F. Tupitsyn, N. N. Zatsepina, N. S. Kolodina, and A. A. Kane, *React. Sposobnost Org. Soied.*, **5**, 931 (1968); *Chem. Abstr.*, **71**, 49041 (1969).
402. A. R. Katritzky, C. R. Palmer, F. J. Swinbourne, T. T. Tidwell, and R. D. Topsom, *J. Amer. Chem. Soc.*, **91**, 636 (1969).
403. A. T. Balaban, G. D. Mateescu, and M. Elian, *Tetrahedron*, **18**, 1083 (1962).
404. I. I. Stănoiu, M. Paraschiv, E. Romas, and A. T. Balaban, *Spectrochim. Acta*, **28A**, 1001 (1972).
405. D. Cook, *Canad. J. Chem.*, **39**, 2009 (1961).
406. J. K. Wilmhurst and H. J. Bernstein, *Canad. J. Chem.*, **35**, 1183 (1957).
407. N. N. Greenwood and K. Wade, *J. Chem. Soc.*, **1960**, 1130.

408. S. Bădilescu and A. T. Balaban, *Spectrochim. Acta*, **32A**, 1311 (1976).
408a. A. J. Ashe III, G. L. Jones, and F. A. Miller, *J. Mol. Struct.*, **78**, 169 (1982).
408b. Joint Commission for Spectroscopy, *J. Chem. Phys.*, **21**, 1170 (1953).
408c. E. B. Wilson, Jr., *Phys. Rev.*, **45**, 706 (1934).
409. S. Ghersetti, S. Giorgianni, and G. Punta, *Spectrosc. Lett.*, **6**, 505 (1973).
410. S. Ghersetti, S. Giorgianni, M. Minari, and G. Punta, *Spectrochim. Acta*, **31A**, 445 (1975).
411. H. Rapaport and K. G. Holden, *J. Amer. Chem. Soc.*, **82**, 4395 (1960).
412. J. R. Price and J. B. Willis, *Austr. J. Chem.*, **12**, 589 (1959).
413. M. F. Gudson and N. J. McCorkindale, *J. Chem. Soc.*, **1957**, 2177.
414. G. S. Bajwa and M. M. Joullie, *J. Heterocycl. Chem.*, **9**, 1403 (1972).
415. A. C. Casey, R. Neubeck, and S. Reynolds, *J. Heterocycl. Chem.*, **9**, 415 (1972).
416. B. S. Thyagarajan, *Adv. Heterocycl. Chem.*, **5**, 313 (1965).
417. H. N. Al-Jallo and F. W. Al-Azawi, *J. Heterocycl. Chem.*, **10**, 139 (1973).
418. W. E. McEwen and R. L. Cobb, *Chem. Rev.*, **55**, 511 (1955).
419. M. L. Dressler and M. M. Joullié, *J. Heterocycl. Chem.*, **7**, 1257 (1970).
420. L. H. Klemm, W. O. Johnson, and D. V. White, *J. Heterocycl. Chem.*, **9**, 843 (1972).
421. R. E. Willette, *Adv. Heterocycl. Chem.*, **9**, 84 (1968).
422. K. I. Kuchkova, E. P. Styngach, F. S. Rivilis, N. M. Frolova, and A. A. Semenov, *Khim. Geterotsikl. Soedin.*, **1973**, 386.
423. E. P. Styngach, K. I. Kuchkova, T. M. Efremova, and A. A. Semenov, *Khim. Geterotsikl. Soedin.*, **1973**, 1523.
424. K. I. Kuchkova, A. A. Semenov, and I. V. Terentieva, *Geterotsikl. Soedin.*, **1970**, 197.
425. N. Ikekawa, *Chem. Pharm. Bull. (Tokyo)*, **6**, 404 (1958).
426. W. W. Paudler and T. J. Kress, *J. Org. Chem.*, **31**, 3055 (1966).
427. W. W. Paudler and T. J. Kress, *Adv. Heterocycl. Chem.*, **11**, 129 (1970).
428. W. L. F. Armarego, G. B. Barlin, and E. Spinner, *Spectrochim. Acta*, **22**, 117 (1966).
429. W. Czube, *Bull. Acad. Polon. Sci. Chim.*, **11**, 423 (1963).
430. M. Woźniak and W. Roszkiewitz, *Zesz. Nauk. Uniw. Jagiellon., Prace Chem.*, **24**, 31 (1979).
431. M. Woźniak, *Zeszyty Naukowe Uniw. Jagiellon., Prace Chem.*, **23**, 55 (1978).
432. D. N. Bailey, D. M. Hercules, and T. D. Eck, *Anal. Chem.*, **39**, 877 (1967).
433. E. M. Hawes and D. K. J. Gorecki, *J. Heterocycl. Chem.*, **11**, 151 (1974).
434. R. L. Williams and M. G. El Fayoumy, *J. Heterocycl. Chem.*, **9**, 1021 (1972).
435. E. Ozola and G. Vanags, *Khim. Geterotsikl. Soedin.*, **1969**, 103.
436. J. G. Cannon, J. L. Born, and R. W. Krunnfusz, *J. Heterocycl. Chem.*, **9**, 959 (1972).
437. T. Tanaka, T. Iwakuma, M. Wagatsuma, and I. Iijima, *J. Heterocycl. Chem.*, **9**, 1355 (1972).
438. D. C. Lankin and H. Zimmer, *J. Heterocycl. Chem.*, **9**, 1133 (1972).
439. G. Coppens and J. Nasielski, *Bull. Soc. Chim. Belg.*, **70**, 136 (1961).
440. S. F. Mason, *J. Chem. Soc.*, **1959**, 1281.
441. D. H. Hey, J. A. Leonard, C. W. Rees, and A. R. Todd, *J. Chem. Soc. Ser. C*, **1967**, 1513.
442. A. A. Schilt and R. C. Taylor, *J. Inorg. Nucl. Chem.*, **9**, 211 (1959).
443. S. S. Singh, *Z. Naturforsch.*, **A24**, 2015 (1969).
444. H. H. Perkampus and W. Rother, *Spectrochim. Acta*, **30A**, 597 (1974).
445. I. Gabe, E. Mantaluta, and G. Neamtu, *Rev. Roumaine Chim.*, **14**, 1163 (1969).

446. L. Joris and P. R. Schleyer, *Tetrahedron*, **24**, 599 (1968).
447. R. C. Lord, A. L. Marston, and F. A. Miller, *Spectrochim. Acta*, **9**, 113 (1957).
448. M. Ito, R. Shimada, T. Kuraishi, and W. M. Zushima, *J. Chem. Phys.*, **25**, 597 (1956).
449. J. D. Simmons and K. K. Innes, *J. Mol. Spectrosc.*, **13**, 435 (1964).
450. K. K. Innes, J. A. Merritt, W. C. Tincher, and S. G. Tilford, *Nature*, **187**, 500 (1960).
451. J. E. Parkin and K. K. Innes, *J. Mol. Spectrosc.*, **15**, 407 (1965).
452. G. Sbrana, G. Adembri, and S. Califano, *Spectrochim. Acta*, **22**, 1831 (1966).
453. R. Foglizzo and A. Novak, *J. Chim. Phys.*, **64**, 1484 (1967).
454. L. Bokobza-Sebagh and K. Zarembowitch, *Spectrochim. Acta*, **32A**, 797 (1976).
455. F. Milani-Nejad and H. D. Stidham, *Spectrochim. Acta*, **31A**, 1433 (1975).
456. Y. Anantharama Sarma, *Spectrochim. Acta*, **30A**, 1801 (1974).
457. G. M. Coppola, G. E. Hardtmann, and B. S. Huegi, *J. Heterocycl. Chem.*, **17**, 1479 (1980).
458. G. M. Coppola, J. D. Fraser, G. E. Hardtmann, B. S. Huegi, and F. G. Kathawala, *J. Heterocycl. Chem.*, **16**, 545 (1979).
459. H. Gauthier and J. M. Lebas, *Spectrochim. Acta*, **35A**, 787 (1979).
460. E. Allenstein, P. Kiemle, E. Schlipf, and W. Podszun, *Spectrochim. Acta*, **34A**, 423 (1978).
461. E. Allenstein, W. Podszun, P. Kiemle, H. J. Mauk, E. Schlipf, and J. Weidlein, *Spectrochim. Acta*, **32A**, 777 (1976).
462. G. Mille, M. Guiliano, J. Kister, J. Chouteau, and J. Metzger, *Spectrochim. Acta*, **36A**, 713 (1980).
463. H. Gauthier and J. M. Lebas, *Spectrochim. Acta*, **36A**, 571 (1980).
464. M. J. Nowak, K. Szczepaniak, A. Barski, and D. Shugar, *J. Mol. Struct.*, **62**, 47 (1980).
465. J. Pitha, *J. Org. Chem.*, **35**, 903 (1970).
466. Yu. N. Boyarchuk, M. V. Vol'kenstein, G. M. Kheifets, and N. V. Khromov-Borisov, in S. Hiller Ed., *Khim. Geterotsikl. Soedin.*, Vol. 1, *Azotsoderzhashchie Geterotsikly*, Izd. Zinatne, Riga, USSR, 1967, p. 338; *Chem. Abstr.*, **71**, 80501 (1969).
467. W. C. Price, J. E. S. Bradley, R. D. B. Fraser, and J. P. Quilliam, *J. Pharm. Pharmacol.*, **6**, 522 (1954).
468. Y. Kyogoku, R. C. Lord, and A. Rich, *Nature (London)*, **218**, 69 (1968).
469. A. J. Barnes, M. A. Stuckey, W. J. Orville-Thomas, L. Le Gall, and J. Lauransan, *J. Mol. Struct.*, **56**, 1 (1979).
470. A. J. Barnes, L. Le Gall, and J. Lauransan, *J. Mol. Struct.*, **56**, 15 (1979).
471. A. J. Barnes, L. Le Gall, and J. Lauransan, *J. Mol. Struct.*, **56**, 29 (1979).
472. B. M. Craven, C. Cusatis, G. L. Gartland, and E. A. Vizzini, *J. Mol. Struct.*, **16**, 331 (1973).
473. R. J. Mesley, *Spectrochim. Acta*, **26A**, 1427 (1970).
474. J. N. Willis, R. B. Cook, and R. Jankow, *Anal. Chem.*, **44**, 1228 (1972).
475. F. Mihai and R. Nuțu, *Rev. Roumaine Chim.*, **13**, 39 (1968).
476. V. M. Cherkasov, G. Ya. Remennikov, A. A. Kisilenko, and E. A. Romanenko, *Khim. Geterotsikl. Soedin.*, **1980**, 239.
477. M. Ito, R. Shimada, T. Kuraishi, and W. Misushima, *J. Chem. Phys.*, **25**, 597 (1956).
478. R. C. Lord, A. Marston, and F. A. Miller, *Spectrochim. Acta*, **9**, 113 (1957).
479. H. Shindo, *Chem. Pharm. Bull.*, **8**, 33 (1960).
480. J. D. Simmons, K. K. Innes, and G. M. Begun, *J. Mol. Spectrosc.*, **14**, 190 (1964).
481. S. Califano, G. Adembri, and G. Sbrana, *Spectrochim. Acta*, **20**, 385 (1964).
482. G. Sbrana, V. Schettino, and R. Righini, *J. Chem. Phys.*, **59**, 2441 (1973).

References

483. J. Zarembowitch and L. Bokobza-Sebagh, *Spectrochim. Acta*, **32A**, 605 (1976).
484. H. Takahashi, K. Mamola, and E. K. Plyler, *J. Mol. Spectrosc.*, **21**, 217 (1966).
485. S. Kovac and V. Konecny, *Coll. Czech. Chem. Commun.*, **45**, 127 (1980).
486. J. M. Chalmers, M. E. A. Cudby, D. Smith, and P. J. Taylor, *Spectrochim. Acta*, **31A**, 1547 (1975).
487. E. Winterfeld and J. M. Nelka, *Chem. Ber.*, **100**, 3671 (1967).
488. D. K. Chesney and R. N. Castle, *J. Heterocycl. Chem.*, **11**, 167 (1974).
489. P. Kregar-Cadez, A. Pollak, B. Stanovnik, and M. Tisler, *J. Heterocycl. Chem.*, **9**, 351 (1972).
490. P. Dan Cook and R. N. Castle, *J. Heterocycl. Chem.*, **10**, 551 (1973).
491. M. Robba, G. Dore, and G. Bonhomme, *J. Heterocycl. Chem.*, **10**, 579 (1973).
492. W. Flitsch and U. Kramer, *Justus Liebigs Ann. Chem.*, **735**, 35 (1970).
493. M. Fraser, *J. Org. Chem.*, **36**, 3087 (1971).
494. G. Ademri, F. Desio, R. Nesi, and M. Scottom, *J. Chem. Soc., Ser. C*, **1968**, 2857.
495. W. L. F. Armarego, G. B. Barlin, and E. Spinner, *Spectrochim. Acta*, **22**, 117 (1964).
496. S. F. Mason, *J. Chem. Soc.*, **1957**, 4874.
497. H. Yale, B. Toeplitz, J. Z. Gougoutas, and M. Puar, *J. Heterocycl. Chem.*, **10**, 123 (1973).
498. H. Yale, *J. Heterocycl. Chem.*, **15**, 1047 (1978).
499. H. Yale and J. H. Sheehan, *J. Heterocycl. Chem.*, **10**, 143 (1973).
500. H. Yale and E. R. Smitzmiller, *J. Heterocycl. Chem.*, **13**, 797 (1976).
501. H. Yale and E. R. Smitzmiller, *J. Heterocycl. Chem.*, **14**, 637 (1977).
502. Kou-Yi Tsern and L. Bauer, *J. Heterocycl. Chem.*, **9**, 1433 (1972).
503. E. Alcade, J. De Mendoza, J. M. Garcia-Marquina, and C. Almera, *J. Heterocycl. Chem.*, **11**, 423 (1974).
504. A. Rosowsky, K. K. N. Chen, M. E. Madez, N. Papathasoioulus, and E. Modest, *J. Heterocycl. Chem.*, **9**, 275 (1972).
505. S. Palazzo and L. I. Giannola, *J. Heterocycl. Chem.*, **10**, 675 (1973).
506. S. W. Schneller and F. W. Clough, *J. Heterocycl. Chem.*, **11**, 975 (1974).
507. A. Rosowsky, E. P. Burrows, P. C. Huang, and E. J. Modest, *J. Heterocycl. Chem.*, **9**, 1239 (1972).
508. A. J. Hubert, *J. Heterocycl. Chem.*, **11**, 737 (1974).
509. F. Fülöp, I. Hermecz, Z. Mészáros, G. Dombi, and G. Bernáth, *J. Heterocycl. Chem.*, **16**, 457 (1979).
510. G. Bernáth, F. Fülöp, I. Hermecz, Z. Mészáros, and G. Tóth, *J. Heterocycl. Chem.*, **16**, 137 (1979).
511. G. Naray-Szabo, E. Dudar, and G. Horvath, *Acta Chim. Hung.*, **74**, 281 (1972).
512. H. Gnichtel, W. Griebenow, and W. Löve, *Chem. Ber.*, **105**, 1865 (1972).
513. H. Gnichtel and K. Hirte, *Chem. Ber.*, **108**, 3387 (1975).
514. G. Saint-Ruf, *J. Heterocycl. Chem.*, **11**, 13 (1974).
515. H. Sterk and T. Kappe, *Monatsh. Chem.*, **100**, 1274 (1969).
516. G. Peyronel, A. Pignedoli, and W. Malavasi, *Spectrochim. Acta*, **32A**, 1015 (1976).
517. S. Vazquez and J. Castrillon, *Spectrochim. Acta*, **30A**, 2021 (1974).
518. T. Cairns, G. Eglinton, and D. T. Gibson, *Spectrochim. Acta*, **20**, 159 (1964).
519. F. A. Cotton, R. Francis, and W. D. Horrocks, *J. Phys. Chem.*, **64**, 1534 (1960).
520. G. W. K. Cavill, P. S. Clezy, and F. B. Whitfield, *Tetrahedron*, **12**, 139 (1961).
521. H. Musso, *Chem. Ber.*, **92**, 2873 (1959).

522. C. O. Okafor, *J. Heterocycl. Chem.*, **17**, 149 (1980).
523. G. E. Martin, J. C. Turley, and L. Williams, *J. Heterocycl. Chem.*, **14**, 1249 (1977).
524. G. E. Martin and J. C. Turley, *J. Heterocycl. Chem.*, **15**, 609 (1978).
525. M. L. Steenberg and J. C. Buckley, *J. Heterocycl. Chem.*, **14**, 1067 (1977).
526. S. R. Caldwell and G. E. Martin, *J. Heterocycl. Chem.*, **17**, 989 (1980).
527. S. R. Caldwell, J. C. Turley, and G. E. Martin, *J. Heterocycl. Chem.*, **17**, 1153 (1980).
528. R. F. Miller, *J. Heterocycl. Chem.*, **15**, 101 (1978).
529. E. R. Blout and M. Fields, *J. Amer. Chem. Soc.*, **72**, 479 (1950).
530. L. N. Short and H. W. Thompson, *J. Chem. Soc.*, **1952**, 168.
531. C. L. Angell, *J. Chem. Soc.*, **1961**, 504.
532. H. Susi, J. S. Ard, and J. M. Purcell, *Spectrochim. Acta*, **29A**, 725 (1973).
533. H. Susi and J. S. Ard, *Spectrochim. Acta*, **27A**, 1549 (1971).
534. H. Susi and J. S. Ard, *Spectrochim. Acta*, **30A**, 1843 (1974).
535. F. P. Beetz, Jr., and G. Ascarelli, *Spectrochim. Acta*, **36A**, 299 (1980).
536. G. E. Kugel, X. Gerbaux, C. Carabatos, P. Martel, and B. M. Powell, *Spectrochim. Acta*, **35A**, 1155 (1979).
537. J. D. Engel and P. H. Von Hippel, *Biochemistry*, **13**, 4143 (1974).
538. E. Litońska, Z. Proba, I. Kuakowska, and K. L. Wierzchowski, *Acta Biochim. Polon.*, **26**, 39 (1979).
539. A. R. Katritzky, M. V. Sinnott, T. T. Tidwell, and R. D. Topson, *J. Amer. Chem. Soc.*, **91**, 628 (1969).
540. D. Kirin, L. Colombo, and K. Furic, *Spectrochim. Acta*, **31A**, 1721 (1975).
541. C. H. Willits, J. C. Decius, K. L. Dille, and E. Christensen, *J. Amer. Chem. Soc.*, **72**, 2569 (1955).
542. J. S. Connolly, *J. Heterocycl. Chem.*, **9**, 379 (1972).
543. J. M. Delabar and M. Mayoube, *Spectrochim. Acta*, **34A**, 129 (1978).
544. C. L. Angell, *J. Chem. Soc.*, **1961**, 504.
545. R. L. Lord and G. J. Thomas, *Spectrochim. Acta*, **23A**, 2551 (1967).
546. T. Shimanouchi, M. Tsuboi, and Y. Kyogoku, *Adv. Chem. Phys.*, **1964**, 434.
547. Y. Kyoguku, S. Higuchi, and T. Tsuboi, *Spectrochim. Acta*, **23A**, 969 (1967).
548. M. Tsuboi, *Appl. Spectrosc. Rev.*, **3**, 45 (1969).
549. G. P. Khare, *J. Mol. Spectrosc.*, **41**, 195 (1972).
550. G. Zundel, W. D. Lubos, and K. Kölkenbeck, *Canad. J. Chem.*, **49**, 3795 (1971).
551. B. I. Sukhoruv, L. A. Kozlova, and M. A. Mazo, *Zh. Fiz. Khim.*, **46**, 548 (1972).
552. H. Fritsche, *Z. Chem.*, **12**, 1 (1972).
553. G. P. Zhizhina and E. C. Oleinik, *Usp. Khim.*, **41**, 474 (1972).
553a. R. M. Hamlin, Jr., R. C. Lord, and A. Rich, *Science*, **148**, 1734 (1965).
553b. Y. Kyogoku, R. C. Lord, and A. Rich, *J. Amer. Chem. Soc.*, **89**, 496 (1967).
553c. Y. Kyogoku, R. C. Lord, and A. Rich, *Biochim. Biophys. Acta*, **179**, 10 (1969).
553d. Y. Kyogoku, R. C. Lord, and A. Rich, *Science*, **154**, 518 (1966).
554. J. E. Lancaster, R. F. Stamm, and N. B. Colthup, *Spectrochim. Acta*, **17**, 155 (1961).
555. J. Goubeau, E. L. Jahn, A. Kreutzberger, and C. Grundman, *J. Phys. Chem.*, **58**, 1078 (1954).
556. W. M. Padgett and W. F. Hamner, *J. Amer. Chem. Soc.*, **80**, 803 (1958).
557. R. D. Spencer, *Spectrochim. Acta*, **21**, 1553 (1965).
558. J. W. Dawson, J. B. Hynes, K. Niedenzu, and W. Sawodny, *Spectrochim. Acta*, **23A**, 1211 (1967).

References 137

559. H. F. Shurvell, J. A. Faniran, and H. Dodsworth, *Spectrochim. Acta*, **23A**, 1313 (1967).
560. A. O. Diallo, *Spectrochim. Acta*, **32A**, 1665 (1976).
561. S. Bădilescu, L. Tincu, and M. Dejan, *Rev. Chim. (Bucharest)*, **24**, 473 (1973).
562. V. Popescu, S. Bădilescu, E. Drăguşin, and E. Matei, *Rev. Chim. (Bucharest)*, in press.
563. M. S. Gibson, *J. Chem. Soc.*, **1963**, 3539.
564. A. C. Mair and M. F. G. Stevens, *J. Chem. Soc., C*, **1971**, 2317.
565. L. Ya. Rikhter and L. M. Sverdlov, *Zh. Prikl. Spectrosk.*, **17**, 491 (1972).
566. A. Meller and G. Beer, *Z. Anorg. Allg. Chem.*, **460**, 169 (1980).
567. J. Slouka, J. Kubata, and V. Bekárek, *Acta Univ. Palackianae Olomucensis*, **49**, 219 (1976).
568. J. Slouka, D. Buczkowská, and V. Bekárek, *Coll. Czech. Chem. Commun.*, **41**, 3090 (1976).
569. J. Slouka, V. Bekárek, and V. Stemberk, *Coll. Czech. Chem. Commun.*, **43**, 960 (1978).
570. J. Slouka, M. Sramkova, and V. Bekárek, *Coll. Czech. Chem. Commun.*, **42**, 3449 (1977).
571. B. Wahl and D. Wöhrle, *Makromol. Chem.*, **176**, 849 (1975).
572. H. Paul, S. Chatterjee, and G. Hilgetag, *Chem. Ber.*, **101**, 3696 (1968).
573. J. H. Kiefer, Ph. D. thesis, Cornell Univ., 1961.
574. L. A. Franks, A. J. Merer, and K. K. Innes, *J. Mol. Spectrosc.*, **26**, 458 (1968).
575. W. D. Sigworth and E. L. Pace, *Spectrochim. Acta*, **27A**, 747 (1971).
576. Y. Kawaguchi and R. H. Mann, *Spectrochim. Acta*, **31A**, 979 (1975).
577. L. A. Paquette and W. L. Farley, *J. Amer. Chem. Soc.*, **89**, 3595 (1967).
578. L. A. Paquette, *J. Amer. Chem. Soc.*, **85**, 3288 (1963).
579. L. A. Paquette, *J. Org. Chem.*, **28**, 3590 (1963).
580. M. A. Rehman and G. R. Proctor, *J. Chem. Soc., Ser. C*, **1967**, 58.
581. J. B. Hester, Jr., *J. Org. Chem.*, **32**, 4095 (1967).
582. J. B. Hester, Jr., *J. Org. Chem.*, **32**, 4098 (1967).
583. H. Alper and E. C. H. Keung, *J. Heterocycl. Chem.*, **10**, 637 (1973).
584. C. Barnett, D. Lloyd, and D. R. Marshall, *J. Chem. Soc., Ser. B*, **1968**, 1536.
585. W. Ruske and E. Hüfner, *J. Prakt. Chem.*, **18**, 156 (1962).
586. V. Sunjic, F. Kajfez, I. Stromer, N. Blazevic, and D. Kolbah, *J. Heterocycl. Chem.*, **10**, 591 (1973).
587. A. A. Santilli and T. S. Osdene, *J. Org. Chem.*, **31**, 4268 (1966).
588. P. Melloni, D. Fusar-Bassini, E. Dradi, and C. Confalonieri, *J. Heterocycl. Chem.*, **11**, 731 (1974).
589. H. A. De Wald, *J. Heterocycl. Chem.*, **11**, 1061 (1974).
590. A. Nawojski, *Rocz. Chem.*, **42**, 1641 (1968).
591. G. Pifferi, L. Fontanella, E. Occelli, and R. Monguzzi, *J. Heterocycl. Chem.*, **9**, 1209 (1972).
592. D. H. Kim, A. A. Santilli, and R. A. Feber, *J. Heterocycl. Chem.*, **9**, 1347 (1972).
593. J. H. Tyman, şi R. Pickles, *Tetrahedron Lett.*, **1966**, 4993.
594. H. Gnichtel and K. Hirte, *Chem. Ber.*, **108**, 3387 (1975).
595. E. Vogel, R. Schubart, and W. A. Boell, *Angew. Chem.*, **76**, 535 (1964).
596. J. Meinwald, D. W. Dicker, and N. Danieli, *J. Amer. Chem. Soc.*, **82**, 4087 (1960).
597. R. A. Braun, *J. Org. Chem.*, **28**, 1383 (1963).
598. S. J. Rhoads and R. D. Cockroff, *J. Amer. Chem. Soc.*, **91**, 2815 (1969).
599. D. R. Larkin, *J. Org. Chem.*, **30**, 335 (1965).

600. S. Olsen and R. Bredock, *Chem. Ber.*, **91**, 1589 (1958).
601. F. Sondenheimer and A. Shani, *J. Amer. Chem. Soc.*, **86**, 3168 (1964).
602. H. Hart, J. L. Corbin, C. R. Wagner, and C. Y. Wu, *J. Amer. Chem. Soc.*, **85**, 3269 (1963).
603. P. C. Unangst and P. L. Southwick, *J. Heterocycl. Chem.*, **10**, 399 (1973).
604. E. Ajello, O. Migliara, L. Ceraulo, and S. Petruso, *J. Heterocycl. Chem.*, **11**, 339 (1974).
605. S. Palazzo, L. O. Giannola, and S. Caronna, *J. Heterocycl. Chem.*, **11**, 839 (1974).
606. R. B. Petigara and H. L. Yale, *J. Heterocycl. Chem.*, **9**, 1275 (1972).
607. B. Pecherer, R. C. Sunbury, and A. Brossi, *J. Heterocycl. Chem.*, **9**, 611 (1972).
608. M. Calvin and K. W. Wilson, *J. Amer. Chem. Soc.*, **67**, 2003 (1945).
609. R. E. Martell and M. Calvin, *Chemistry of Metal Chelate Compounds*, Prentice-Hall, Englewood Cliffs, NJ, 1952.
610. M. Calvin and K. W. Wilson, *J. Amer. Chem. Soc.*, **67**, 2003 (1945).
611. J. P. Collman, *Angew. Chem.*, **77**, 154 (1965) and previous papers.
612. J. P. Collman, R. L. Marshall, and W. L. Young, *Chem. Ind.*, **1962**, 1380.
613. R. E., Hester, *Chem. Ind.*, **1963**, 1397.
614. R. H. Holm and F. A. Cotton, *J. Amer. Chem. Soc.*, **80**, 5658 (1958).
615. J. A. S. Smith and E. J. Wilkins, *J. Chem. Soc. Ser., A*, **1966**, 1749.
616. R. C. Fay and N. Serpone, *J. Amer. Chem. Soc.*, **90**, 5701 (1968).
617. J. P. Fackler, *Progr. Inorg. Chem.*, **7**, 374 (1966).
618. A. Trestianu, A. Niculescu-Majewska, I. Bally, A. Barabas, and A. T. Balaban, *Tetrahedron*, **24**, 2499 (1968).
619. M. Kuhn and H. Musso, *Angew. Chem. Intern. Ed.*, **8**, 147 (1969).
620. B. Bock, K. Flatan, H. Junge, M. Kuhr, and H. Musso, *Angew. Chem. Intern. Ed.*, **10**, 225 (1971).
621. A. F. Holtzclaw and J. P. Collman, *J. Amer. Chem. Soc.*, **79**, 3318 (1957).
622. R. West and R. Riley, *J. Inorg. Nucl. Chem.*, **5**, 295 (1958).
623. K. Nakamoto and A. E. Martell, *J. Chem. Phys.*, **32**, 388 (1960).
624. K. Nakamoto, P. J. McCarthy, A. Ruby, and A. E. Martell, *J. Amer. Chem. Soc.*, **83**, 1066, 1272 (1961).
625. R. Mecke and E. Funk, *Z. Elektrochem. Ber. Bunsenges. Phys. Chem.*, **60**, 1124 (1956).
626. H. Junge and H. Musso, *Spectrochim. Acta*, **24A**, 1219 (1968).
627. S. Pinchas, B. L. Silver, and I. Laulicht, *J. Chem. Phys.*, **46**, 1506 (1967).
628. T. Behnke and K. Nakamoto, *Inorg. Chem.*, **6**, 433 (1967).
629. K. Nakamoto, C. Udovich, and J. Takemoto, *J. Amer. Chem. Soc.*, **92**, 3973 (1970).
630. J. W. Macklin, *Spectrochim. Acta*, **32A**, 1459 (1976).
631. G. T. Behnke and T. Nakamoto, *Inorg. Chem.*, **7**, 330 (1968).
632. D. Gibson, B. F. G. Johnson, and J. Lewis, *J. Chem. Soc., Ser. A*, **1970**, 367.
633. A. L. Allred and D. W. Thompson, *Inorg. Chem.*, **7**, 1196 (1968).
634. M. N. Patel, C. B. Patel, and R. P. Patel, *J. Indian Chem. Soc.*, **51**, 512 (1974).
635. S. K. Shah, M. R. Patel, and B. N. Mankat, *Indian J. Chem.*, **8**, 607 (1970).
636. I. I. Bădilescu and S. Bădilescu, Abstracts of Papers, *3rd Conf. Phys. Chem. (General and Applied)*, Bucharest, 4–7 Sept. 1972, p. 56.
637. G. Schott and K. Golz, *Z. Anorg. Allg. Chem.*, **383**, 314 (1971).
638. G. Schott and K. Golz, *Z. Anorg. Allg. Chem.*, **399**, 7 (1973).
639. A. T. Balaban, C. N. Renţea, M. Mocanu-Paraschiv, and E. Romaş, *Rev. Roumaine Chim.*, **10**, 849 (1965).

640. A. T. Balaban, G. Mihai, R. Antonescu, and P. T. Frangopol, *Tetrahedron*, **16**, 68 (1961).
641. A. T. Balaban, *Rev. Roumaine Chim.*, **10**, 879 (1965).
642. I. Bally, A. Arsene, M. Paraschiv, E. Romaş, and A. T. Balaban, *Rev. Roumaine Chim.*, **11**, 1409 (1966).
643. A. T. Balaban, in "La nature et les propriétés des liaisons de coordination," *Coll. Intern.*, No. 191, CNRS, Paris, 1970, p. 233.
644. A. T. Balaban, I. Bally, R. J. Bishop, C. N. Renţea, and L. E. Sutton, *J. Chem. Soc.*, **1964**, 2382.
645. A. T. Balaban, I. Bally, V. I. Minkin, and A. I. Usachev, *Tetrahedron*, **33**, 3265 (1977).
646. V. I. Minkin, L. Polekhnovich, V. P. Metlushenko, N. G. Furmanova, I. Bally, and A. T. Balaban, *Tetrahedron*, **37**, Suppl. 1, 421 (1981).
647. A. T. Balaban, C. Uncuţa, M. Elian, and F. Chiraleu (in press).
648. A. Arsene, M. Paraschiv, and A. T. Balaban, *Rev. Roumaine Chim.*, **15**, 247 (1970).
649. A. Barabas, E. Isfan, M. Roman, M. Paraschiv, E. Romas, and A. T. Balaban, *Tetrahedron*, **24**, 1133 (1968).
650. A. T. Balaban, A. Barabas, and A. Arsene, *Tetrahedron Lett.*, **1964**, 2721.
651. A. T. Balaban, A. Arsene, I. Bally, A. Barabas, M. Paraschiv, M. Roman, and E. Romas, *Rev. Roumaine Chim.*, **15**, 635 (1970).
652. H. J. Becher and S. Frick, *Z. Anorg. Allg. Chem.*, **295**, 83 (1958).
653. H. Watanabe, M. Narisada, T. Nakagawa, and M. Kubo, *Spectrochim. Acta*, **16**, 78 (1960).
654. H. Watanabe, Y. Kuroda, and M. Kubo, *Spectrochim. Acta*, **17**, 454 (1961).
655. A. Meller and R. Schlegel, *Monatsh. Chem.*, **95**, 382 (1964).
656. J. M. Butcher, W. Gerrard, E. F. Mooney, R. A. Rothenbury, and H. A. Willis, *Spectrochim. Acta*, **18** 1487 (1962).
657. V. Gutman, A. Meller, and R. Schlegel, *Monatsh. Chem.*, **94**, 1071 (1963).
658. K. Niedenzu and J. W. Dawson, *Boron-Nitrogen Compounds*, Springer, Berlin, 1965, p. 114.
659. M. F. Lappert, *Proc. Chem. Soc.*, **1959**, 59.
660. H. Steinberg and R. J. Brotherton, *Organoboron Chemistry*, Vol. 2, Wiley-Interscience, New York, 1966, pp. 354–370.
661. W. Gerard, *The Organic Chemistry of Boron*, Academic Press, London, 1961, pp. 223–232.
662. M. F. Lappert, *J. Chem. Soc.*, **1958**, 2790, 3256.

2 NUCLEAR MAGNETIC RESONANCE SPECTROSCOPY

M. D. FENN

Medical Chemistry Group
The John Curtin School of Medical Research
Australian National University
Canberra City, Australia

2.1.	Introduction		142
	2.1.1.	Historical Development, 142	
	2.1.2.	Theory of Nuclear Magnetic Resonance, 143	
2.2.	NMR Spectrometers		148
	2.2.1.	The Magnetic Field, 150	
		2.2.1.1. Permanent magnets, 150	
		2.2.1.2. Electromagnets, 151	
		2.2.1.3. Superconducting Magnets, 152	
	2.2.2.	Receiver System, 154	
		2.2.2.1. Sweep Spectrometers, 154	
		2.2.2.2. Pulse Spectrometers, 156	
2.3.	Noise		158
	2.3.1.	Random Noise, 158	
	2.3.2.	Coherent Noise, 164	
		2.3.2.1. Acoustic Ringing, 165	
		2.3.2.2. Coil Resonance, 168	
2.4.	Multinuclear Magnetic Resonance		169
	2.4.1.	Other Isotopes of Hydrogen, 171	
		2.4.1.1. Deuterium, 171	
		2.4.1.2. Tritium, 171	
	2.4.2.	Nitrogen Nuclear Magnetic Resonance, 172	
		2.4.2.1. ^{14}N Spectroscopy, 173	
		2.4.2.2. ^{15}N Spectroscopy, 183	
	2.4.3.	Oxygen Spectroscopy, 203	
	2.4.4.	Sulfur Spectroscopy, 209	
2.5.	Reference Solvents		210
	2.5.1.	Recommended Reference Solvents, 215	
		2.5.1.1. ^{1}H and ^{13}C, 215	
		2.5.1.2. ^{15}N and ^{14}N, 215	
		2.5.1.3. ^{17}O, 217	
		2.5.1.4. ^{33}S, 217	
2.6.	Sample Temperature Measurement		217

2.7. Reduction of Noise   220
    2.7.1. Reduction of Baseline Roll, 220
    2.7.2. Decoupling Methods, 223

References   225

2.1. INTRODUCTION

2.1.1. Historical Development

Nuclear magnetic resonance spectroscopy is a relatively new technique compared to ultraviolet and infrared spectroscopy. Today, one must consider NMR spectroscopy to be equally as important as the two older branches of spectroscopy. Indeed in many aspects NMR is superior to UV and IR spectroscopy, but against this one must balance the higher cost of instrumentation.

The magnetic properties of nuclei were first postulated by Pauli (1) to explain certain hyperfine structural features of atomic spectra, and the phenomenon of nuclear magnetic resonance has long been known in molecular beams. This has been exploited effectively to give much useful information concerning nuclear properties (2). However, the first true experiments in nuclear magnetic resonance were made in 1945. In that year two independent groups of physicists, one at Stanford (3) and the other at Harvard (4), observed proton resonances. The Bloch equations (5) describing the phenomenon of NMR are now well known. Initially, the technique was used exclusively by physicists and applied to the measurement of such nuclear properties as magnetogyric ratios, nuclear spin relaxations, and internuclear distances in solids. Evidence accumulated from these studies show that for a particular nuclear species in a given magnetic field the magnetogyric ratio "depended" on the chemical environment of the nucleus. The term "chemical shift" was used to describe this apparent variation in magnetogyric ratio (6–8). Of course, the magnetogyric ratio is now known to be constant, but the frequency of resonance depends on the chemical environment of the nucleus. The term "chemical shift," however, has become established as the term for describing this variation of frequency. It is this effect which has become the basis of NMR spectroscopy.

The macroscopic experiments of the physicists merely indicated the presence of the phenomenon known as chemical shift, and it was not until 1951 that Arnold and his coworkers (9) reported the "resolution" of the spectrum of ethanol into three peaks (CH_3, CH_2, and OH). The first commercial NMR spectrometer was made available by Varian Associates in 1953. Since then the field has expanded enormously and is not confined to chemistry alone. NMR is now used in such diverse fields as medicine and the oil industry. For the years covered by this chapter, 1971–1981, there are 30,000 references to NMR in the literature, of which some 12,000 are for heterocyclic compounds. To review such a large number of references is an almost impossible task; hence this chapter will be concerned with techniques and nuclei other than ^1H and ^{13}C.

2.1.2. Theory Of Nuclear Magnetic Resonance

Although the literature contains several detailed mathematical treatments of the theory of nuclear magnetic resonance, it is intended here to give only a brief outline of the theory and refer interested readers to the literature (5, 10–13).

The nuclei of certain isotopes possess an intrinsic spin, that is, they are associated with angular momentum. One of the basic principles of quantum mechanics is that the total angular momentum of any isolated particle cannot have any arbitrary magnitude but may assume only certain discrete values. The angular momentum is said to be quantized, and its magnitude P is give by

$$P = \hbar[I(I+1)]^{1/2} \quad (1)$$

where I is known as the *nuclear spin quantum number* and may have any positive integer or half-integer value. As angular momentum is a vector quantity, direction must be specified also, and this is achieved by the use of a second quantum number M_I, where I may have the values

$$M_I = I, I-1, I-2, \ldots, -I$$

Thus for a given I there are $2I+1$ possible values of M_I. Any motion of a charged body has an associated field. On the macroscopic scale an electrical current, which is due to the motion of electrons, produces a magnetic field. This also applies to the nuclear scale, and any nucleus possessing angular momentum has a magnetic moment μ. The magnetic moment is related to angular momentum by

$$\mu = \gamma P \quad (2)$$

where γ is known as the *magnetogyric ratio* and is a constant for a particular isotope. Combing equations (1) and (2) gives

$$\mu = \gamma \hbar [I(I+1)]^{1/2} \quad (3)$$

Since I is a quantum number, μ must be quantized. In the absence of a magnetic field the energy of an isolated nucleus is independent of the quantum number M_I. That is, orientations of the nuclei are random. However, when the nucleus is placed in a magnetic field $\mathbf{B_0}$, the nucleus attempts to align itself with the field as a result of torque applied by $\mathbf{B_0}$ to the magnetic moment. Since the magnetic moment is a vector quantity, the direction of torque is always changing, and as a result the direction of μ rotates around $\mathbf{B_0}$ tracing out a cone (Fig. 2.1). Such motion is referred to as *Larmor precession*. Like gyroscopes in a gravitational field, the spin axis undergoes precession about the field direction. The frequency of the Larmor precession is designated ω_0 in radians per second. If one tries to force the nuclear moments to become aligned by increasing the magnitude B_0, the nuclear moments only process faster. The Larmor precession frequency is given by

$$\omega_0 = \mu B_0 \quad (4)$$

and the half-angle θ of the cone of precession is given by

$$\cos\theta = M_I [I(I+1)]^{-1/2} \quad (5)$$

144   Nuclear Magnetic Resonance Spectroscopy

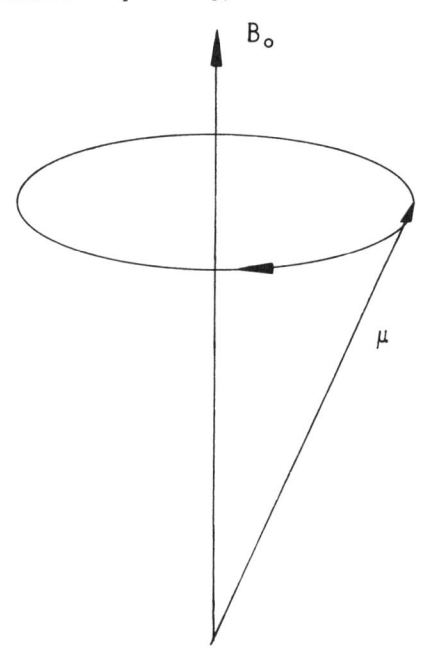

Fig. 2.1.   Precession of a nuclear magnetic moment about a magnetic field B_0.

Thus for a nucleus of spin I there will be $2I + 1$ precessional cones. For the simple case of $I = \frac{1}{2}$, θ is calculated to be $54°44'$, and there will be two precessional cones, one aligned toward the field and one aligned against the field. This is illustrated in Figure 2.2. If there is an additional magnetic field B_1 perpendicular to B_0, then this field will also exert torque on the magnetic moment, tending to change the angle θ. If B_1 is fixed in direction, then as μ precesses it will try to increase and decrease θ. However, if B_1 is not fixed but rotates at the same frequency as μ, then constant torque will be exerted on μ and the precessing magnetic moment and rotating magnetic field are said to be in resonance, resulting in the absorption of energy and a change in the angle θ (Fig. 2.3). The energy, of course, is derived from the rotating field B_1 which is supplied by radiofrequency electromagnetic radiation and the required frequency given by equation (4). It is the detection of this absorbed energy that forms the basis of a practical NMR experiment. It is obvious that the relative distribution of the number of nuclei in the various orientations available is important. If all nuclei were in the orientation aligned against the field, that is, the high-energy situation, no absorption of radiofrequency energy could occur and an NMR experiment would not be possible. The viability of an NMR experiment depends on a difference in population levels being established. The difference in population levels may be estimated from the Boltzmann distribution (9, 14). For example, for nuclei where $I = \frac{1}{2}$ in a noninteracting situation, the population difference $P_+ - P_-$ is given approximately by

$$P_+ - P_- = \frac{\mu \hbar B_0}{2kT} \qquad (6)$$

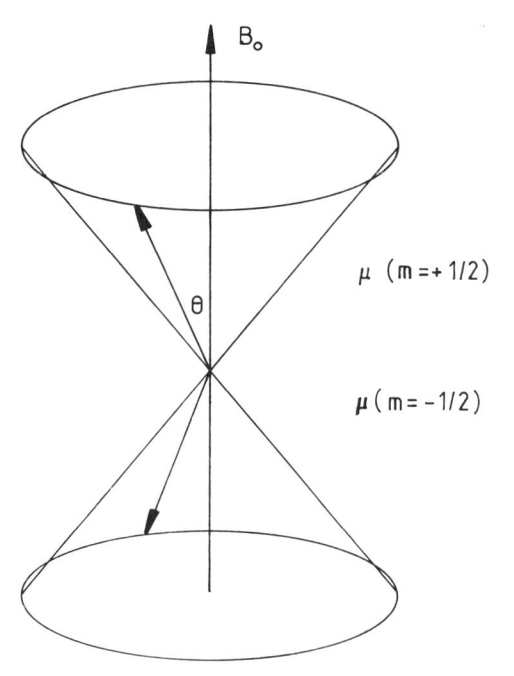

Fig. 2.2. Precessional cones for a nucleus of spin $\frac{1}{2}$ in a magnetic field. In this case θ is equal to 54°44′.

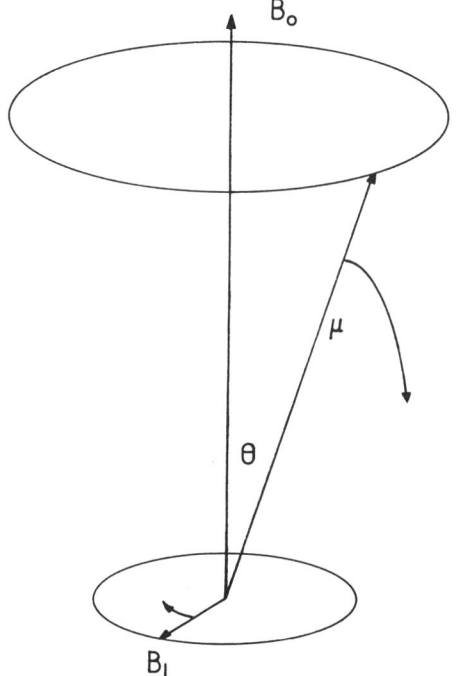

Fig. 2.3. Result of an applied rotating magnetic field B_1 on a precessing nuclear moment. When resonance occurs angle θ is altered.

Note that the population difference is related to the absolute temperature T, and the higher the temperature the smaller the difference in levels. Thus, a smaller amount of radiofrequency energy will be absorbed, and this is reflected in a decreasing signal-to-noise ratio with increasing temperature in the NMR experiment. It must also be obvious that once energy has been absorbed and a nucleus has been placed into the high energy state there must be some mechanism for the nucleus to lose the energy absorbed or the population levels will become equal, that is, $P_+ - P_- = 0$, and the NMR experiment would cease. This mechanism is known as *relaxation*.

The required relaxation may occur as each nucleus is not totally isolated from other nuclei in the immediate vicinity. These other nuclei may be from the rest of the assembly of molecules or even from the solvent and are commonly referred to as the *lattice*. A detailed discussion of relaxation may be found in references 12–14, but it is sufficient here to mention that two types of relaxation are recognizable, the spin-spin relaxation T_2 and spin-lattice relaxation T_1, and that rapid spin-lattice relaxation may be induced by interaction with unpaired electrons in paramagnetic substances or electric fields within the molecule. The last factor is of considerable importance in nuclear magnetic resonance. For nuclei with $I = \frac{1}{2}$ the nuclear charge has a spherical distribution, but when $I \neq \frac{1}{2}$ this charge distribution becomes asymmetric and an electric field gradient may be present. Such nuclei are said to possess electric quadrupole moments, and the moment may be positive (charge distribution corresponding to a prolate spheriod) or negative (charge distribution corresponding to an oblate spheroid). As some of the most common nuclei found in heterocyclic compounds, for example, ^{17}O and ^{14}N, possess an electric quadrupole moment, this is an important observation. Unless they are in a pseudosymmetrical environment, for example, the NH_4^+ ion, such nuclei will relax very quickly. Relaxations of the order of milliseconds are common for quadrupolar nuclei. This may be compared with relaxations of the order of seconds for nuclei of $I = \frac{1}{2}$ such as ^1H and ^{13}C. The presence of fast relaxation means that data may be acquired quickly by fast pulsing.

Finally, a factor of some importance in NMR spectroscopy is the nuclear Overhauser effect (NOE) (15–17), which is defined as the change in intensity of one nuclear spin when the resonance of another is irradiated. The magnitude of the effect depends on the balance between the various relaxation mechanisms available. Under normal circumstances the NOE lies between 1 (no effect) and a maximum given by

$$\text{NOE}_{max} = 1 + \frac{\gamma_s}{2\gamma_i} \qquad (7)$$

where γ_s is the magnetogyric ratio of the nuclei irradiated and γ_i is the magnetogyric ratio of the nuclei being observed. Thus, for the very common situation of irradiating proton nuclei while observing ^{13}C nuclei, the maximum NOE is close to 3. As γ_i in this case is positive, the NOE must at all times be positive. However, some nuclei such as ^{15}N have negative magnetogyric ratios, and it is possible under certain conditions for $\gamma_s/2\gamma_i$ to equal -1, in which case no spectrum will be obtained. In this situation special techniques must be used to suppress the NOE effect. Nuclear Overhauser effects are rarely observed for quadrupolar nuclei, because quadrupolar interactions provide the dominant relaxation mechanism and effectively suppress the NOE.

From the above discussion it might be supposed that at any particular frequency all nuclei of a given species would resonate at the same value of B_0. If this were strictly true then NMR would be of little use to chemists. Fortunately, the precise frequency of resonance for a given nucleus depends on its molecular environment. This difference has its origins in the electron density cloud that surrounds every nucleus. When a molecule is placed in a magnetic field, orbital currents are induced in the electron clouds, and these in turn give rise to small additional magnetic fields that are opposed to the main field B_0. Hence, a slightly higher value of B_0 is required to achieve resonance. This can be expressed by the equation

$$B_{local} = B_0(1 - \sigma) \tag{8}$$

where B_{local} is the actual field experienced by the nucleus and σ is a dimensionless quantity known as the screening constant or shielding coefficient. σ is independent of magnetic field but highly dependent on molecular structure. This chapter is not considered an appropriate place to include a detailed theoretical discussion of shielding, and interested readers are referred to the literature (12–14, 18). It is sufficient to point out that nuclear screening may arise from several sources and it is customary to separate the shielding arising from the electrons associated with the nucleus into several terms. The first term σ_D^L, known as *local diamagnetic shielding*, relates to the circulation of electrons around the nucleus and is a function of electron density. Any distortion of the electron density cloud around the nucleus by chemical bonding is accounted for by the term σ_P^L, the *local paramagnetic shielding*. This term is particularly important for nuclei such as ^{14}N but much less important for the hydrogen molecule (13). In addition to σ_D^L and σ_P^L, significant contributions may arise from circulation of electrons around neighboring atoms. Similarly there may be long-range diamagnetic σ_D^{LR} or long-range paramagnetic σ_P^{LR} shielding. Finally, one must include the effect of solvent, as all high-resolution NMR spectra are obtained as solutions in suitable solvents. The shielding of a nucleus can be expressed as a sum of five terms:

$$\sigma = \sigma_D^L + \sigma_P^L + \sigma_D^{LR} + \sigma_P^{LR} + \sigma_s \tag{9}$$

where σ_s is the solvent shielding contribution. σ_s is only the direct shielding contribution by the solvent, and the solvent may influence the magnitudes of the other four terms. Broadly speaking, nuclear shielding may be thought of as arising from

1. Variations in electron density due to the proximity of electronegative groups such as O, halogens, and NO_2
2. Special shielding effects produced by structures that allow circulation of electrons in preferred directions within the molecule, for example, benzene rings and cyano groups

2.2. NMR SPECTROMETERS

Nuclear magnetic resonance spectra may be obtained by either of two methods. The original and most obvious method of obtaining a spectrum is to keep the magnetic field constant and sweep the exciting frequency through the appropriate range. The frequency must be swept linearly and the value of nuclear magnetism plotted against frequency. In practice, it is difficult to obtain a linear sweep of frequency, and it is more usual to keep the frequency constant and change the magnetic field by means of a set of dc coils. The peaks of the plot occur at the resonant frequency of the corresponding nuclei, and the line widths are characterized by the relaxation time T_2. If the peak width at half-height in hertz is defined as δv, then it is easily shown that (14)

$$\delta v = \frac{1}{\pi T_2} \tag{10}$$

Thus, in theory one can measure T_2 from linewidths obtained by the sweep method.

The advantage of the sweep method is its experimental simplicity and consequent low cost. It is, however, a very inefficient method, as all resonances are detected sequentially at a slow rate. Typically, a sweep of 1 kHz for protons would require 4 min. It does have an advantage in that the area under the resonance peaks is directly proportional to the number of protons giving rise to the resonance.

The most efficient method of recording an NMR spectrum is to record the responses of all the nuclei simultaneously. This objective can be achieved by means of a short high-power radiofrequency pulse to excite all the nuclei simultaneously. The range of frequencies excited is determined by the width of the excitation pulse. The shorter the pulse, the wider the frequency range excited. If a pulse of width t s is placed in the middle of the spectrum, then all nuclei in the range $(2t)^{-1}$ Hz will be excited; that is, a 20-μs pulse will excite nuclei over the range ±12 kHz from the center of the pulse frequency. This dependence of bandwidth on pulse length is merely another way of expressing the uncertainty principle. After the termination of the pulse, the nuclei emit the energy absorbed; they relax. (What goes up must come down!) The relaxation toward equilibrium is by an exponential process that is characterized by the spin-spin relaxation T_2. During this time the energy released induces voltage in the receiver coil. This exponentially decaying signal is referred to as a *free induction decay* (FID) and is the time domain spectrum of the sample. Fourier transform of the FID yields the normal frequency domain spectrum. This is illustrated in Figure 2.4.

Pulse widths are sometimes described in terms of an angle α. The origin of this terminology arises from a vector description of the NMR experiment. The nuclear magnetism may be represented by a static vector parallel to the magnetic field. During excitation by radiofrequency energy the magnetization vector rotates through an angle α, which is given by (19)

$$\alpha = \gamma B_1 t \tag{11}$$

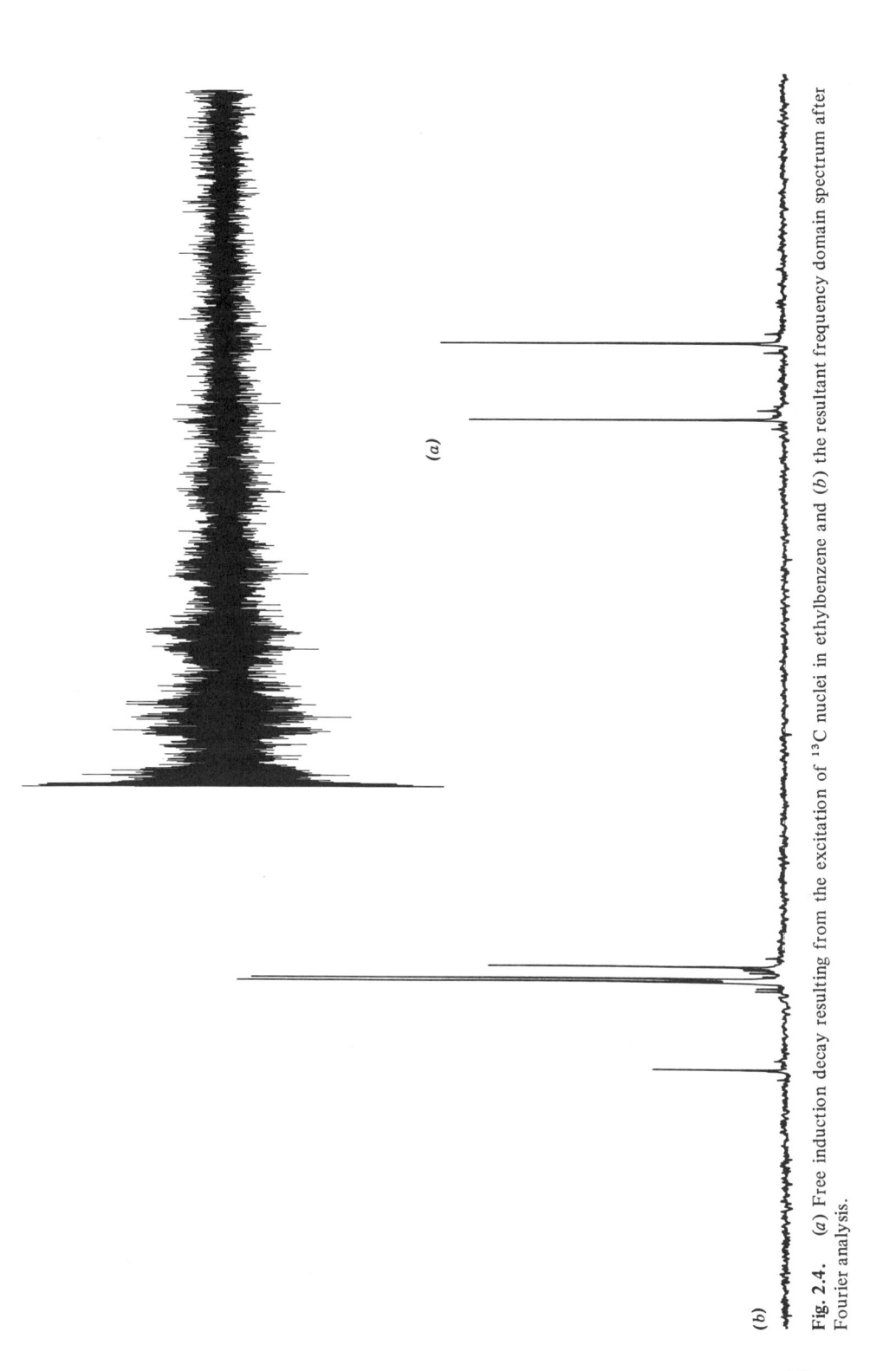

Fig. 2.4. (a) Free induction decay resulting from the excitation of ^{13}C nuclei in ethylbenzene and (b) the resultant frequency domain spectrum after Fourier analysis.

The magnitude of the signal induced in the receiver coil is proportional to sin α and is a maximum for $\alpha = 90$ or $270°$ and a minimum for $\alpha = 0$ and $180°$. As the angle α is proportional not only to the radiofrequency field B_1 but also to the pulse width t, t is often described by the angle it induces in the magnetization vector. α is an important NMR parameter, as it controls not only the signal magnitude but also the frequency range.

The advantage of pulse excitation is its efficiency. As all nuclei are excited simultaneously, the Fellgett or multichannel advantage is gained (20). This is typically 10–20 for protons and considerably more for nuclei of large spectral widths such as ^{13}C. Time is also gained, 4 s being the normal time for acquiring a single-pulse spectrum. One disadvantage of pulse excitation methods is that the area under a resonance peak is not always proportional to the number of nuclei giving rise to the resonance. The reason for this is not hard to see. The FID depends on the relaxation time, and if that time is long then not all the FID signal is acquired and consequently when a Fourier transform is performed the signal height will be less than expected. This problem can be expected for groups such as ^{13}C=O and ^{13}CR$_4$(R $\neq$ H) and it has been observed in this laboratory for heterocyclic amine protons. The sweep method does not suffer this disadvantage.

Whichever of the two methods is used to acquire spectra, any apparatus must consist of the following essentials:

1. A highly stable homogeneous magnetic field
2. A stable source of radiofrequency power
3. Some method of detection and display

2.2.1. The Magnetic Field

2.2.1.1. Permanent Magnets

The first NMR spectrometers constructed employed permanent magnets, but today the use of such magnets is confined to those spectrometers used for routine analysis, the so-called "bench-top" spectrometers. The advantages of a permanent magnet are:

1. *Low power consumption.* The only power required for the magnet is that needed to operate the thermostat system maintaining the magnet at a constant temperature. This low power consumption enables the use of a standby battery system to maintain readiness in the event of a power supply failure.
2. *Low cost.* The cost of a permanent magnet spectrometer would be approximately one-third that of an equivalent spectrometer employing an electromagnet.
3. *Small size.* With modern ceramic magnet technology, a 1.6-Tesla magnet, operating at 60 MHz for protons, would occupy a 30-cm cube.

However, these magnets have serious disadvantages:

1. The field of a permanent magnet is very sensitive to temperature fluctuations. It must be carefully thermostatted to better than 0.1 degree and it is necessary to preheat samples to magnet temperature to avoid misleading values of chemical shifts.
2. It is sensitive to the movement of ferromagnetic bodies such as metal chairs, elevators, or service equipment in the vicinity of the magnet, and great care must be exercised in siting of such a spectrometer.
3. There is only a narrow pole gap. The main disadvantage is the need for a very narrow gap between the poles of the magnet to enable high field strengths to be obtained. It is unusual to find a permanent magnet with a gap sufficiently large to enable sample tubes of greater than 5 mm diameter to be used.
4. Only sweep methods can be used. Because of the high rates of change of radiofrequency fields and consequent heating effects, pulse methods cannot be used with a permanent magnet.

Today most research spectrometers employ electromagnets.

2.2.1.2. *Electromagnets*

Advantages of an electromagnet are:

1. *Small effect of temperature upon field strength.* Samples do not require preheating to magnet temperature.
2. *Low sensitivity to ferromagnetic materials.* There is little if any interference from ferromagnetic bodies, and siting is not as critical as for a permanent magnet.
3. *Large pole gap.* Large pole gaps are available at high field strengths, enabling the use of sample tubes of up to 15 mm diameter.
4. *Pulse methods may be used.*

Disadvantages of electromagnets are not as serious as for permanent magnets:

1. *High consumption of power.* An electromagnet will typically require 5 kVA of power and must have a power source separate from any other instrument, often requiring the installation of a separate main power supply. Shared power supplies invariably result in magnet instability.
2. *Generation of heat.* A considerable amount of heat is generated by an electromagnet and must be removed by some form of water circulation, which may be of the order of 20 liters/min. The large quantities of heat generated are not needed to sustain the magnetic field. Indeed, once the magnetic field has been established no additional energy is required to

sustain it. The energy is dissipated solely in overcoming the resistance of the copper in the magnet windings. If the windings had no resistance, then once a field was established no extra energy would be required to maintain the field. This is a property of superconducting magnets. Although such magnets invariably incorporate safety devices to guard against overheating or cooling water failure, abrupt termination of water circulation around the magnet may lead to the development of "hot spots" in the magnet and subsequent damage to its windings.

All electromagnets today consist of a copper coil wound around a core of iron or an iron-based alloy. The function of the copper coil is to supply a field called the *excitation field* to the iron core. Each atom of iron or other ferromagnetic material has a small magnetic moment, which under the influence of the excitation field is induced to align with the field. The proportion of the atoms that are aligned depends on the strength of the applied field and hence on the excitation current. The alignment of the atomic moments results in a field that is greater than the field attributed to the coils alone. It is obvious that there is a practical limit to the field strength that can be obtained. Once all the magnetic moments are aligned, no further increase in field strength can occur regardless of any increase in excitation current. For iron, this point occurs at a field strength of slightly over 2 T. This corresponds to a proton frequency of 90 MHz. Slightly higher yields may be obtained by using cores of rare earth iron alloys. However, because of the larger cost involved, spectrometers operating at frequencies greater than 90 MHz that use electromagnets are uncommon.

2.2.1.3. *Superconducting Magnets*

It will be seen from Section 2.1 that numerous advantages can be obtained by increasing the field strength of the magnet beyond the 2-T limit of electromagnets. Increased sensitivity, greater signal-to-noise ratio, and simplified coupling are some of the advantages.

Fields higher than those obtained by electromagnets may be achieved by the use of superconducting materials. Such superconducting magnets are now commercially available for fields up to 11 T, that is, 500 MHz for the proton frequency.

Superconductivity, the complete loss of electrical resistance, has been known for over 70 years (20–22), although only recently has the technology become available for fabricating superconducting magnets with facility.

Pure metals such as mercury, tin, and lead exhibit superconductivity at 4.2 K, the boiling point of liquid helium. However, on increasing the current these superconductors known as type I quickly revert to the resistive state at a distinct and sharply defined critical field strength (i.e., the superconductivity is quenched). For pure metals this critical field strength is usually less than 0.1 T. Since the 1960s a number of alloy materials known as type II superconductors have been discovered that satisfy the criteria for high field, high current operation. Only two, however, are in general use in superconducting magnet technology: the niobium-tin

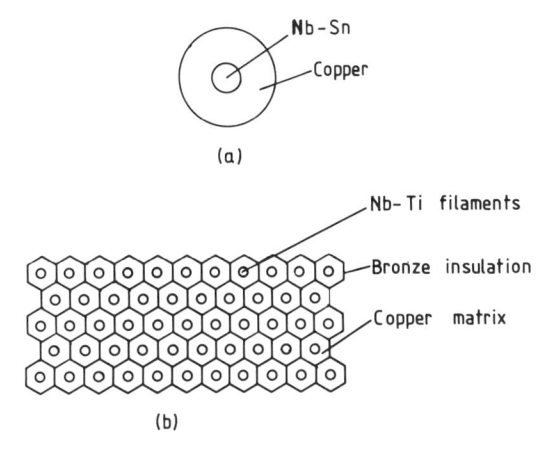

Fig. 2.5. Superconducting magnet winding configurations: (a) single filament and (b) modern multiple strand winding.

compound Nb$_3$Sn and the niobium-titanium compound Nb$_{40}$Ti$_{60}$. Both materials have extreme type II properties. The critical field for Nb$_3$Sn is over 25 T, while that for Nb$_{40}$Ti$_{60}$ is 11 T (20).

Even with the availability of suitable materials, the building of a superconducting magnet is not a straightforward undertaking. If some small region of the winding exceeds the critical current field density and is thereby quenched, such a region will be an isolated link of high resistance and will immediately be subject to strong resistive heating. The heat would, of course, quench adjacent regions and the resistive region would grow, with catastrophic results to the magnet as the stored electrical energy was released as heat. A first step in avoiding such a failure of the magnet was to clad the superconducting material with a low-resistance normal conductor such as copper and circulate liquid helium directly over the surface of the windings (Fig. 2.5a). By these means all the heat envolved by a small resistive region can be drawn off and the growth of a normal zone curtailed.

Further problems arise during excitation of the magnet; the sudden movement of flux lines introduces a spike of heat that can raise the temperature enough to quench the superconductor. As the thermal energy released by a flux surge is proportional to the diameter of the conducting wire, it is possible to reduce the effect of flux surge instability by replacing a single conductor with a composite of small superconducting filaments embedded in a copper matrix. For niobium-tin alloys the filament diameter is of the order of 10 μm.

Embedding the filaments in a matrix of copper allows eddy currents to arise in the copper matrix, which results in magnet instability. These eddy currents may be minimized by twisting and transposing the filaments and incorporating a resistive barrier in the copper. The barrier can be formed of a high-resistance copper alloy such as bronze (Fig. 2.5b).

Because of the brittleness of superconducting materials, fabrication is difficult and it is difficult or impossible to join the superconducting wires. Generally, a filament of niobium metal is embedded in a medium of bronze, the composite wire drawn to the required size, the conductor made by twisting the transposing filaments, and the magnet armature wound. Subsequent heat treatment of the whole magnet causes tin from the bronze alloy to migrate into the niobium, where it reacts to form Nb_3Sn and the windings become superconducting. As can be imagined, there is a high failure rate in magnet production, and undoubtedly this is one of the causes of the high cost of superconducting magnets.

Without doubt, the future of NMR spectrometers lies with superconducting magnets. Although the theoretical maximum field that may be obtained with present alloys is of the order of 50 T, in practice a field of 17.5 T has been obtained (20). Fields are limited chiefly by technical difficulties in the fabrication of finely divided conductors and in joining such conductors together.

All superconducting materials discovered so far require liquid helium as coolant. Commercial quantities of helium are obtained from a limited number of natural gas fields in the south of the United States and Indonesia. This supply is expected to be exhausted by the end of the century, necessitating extraction of helium from the atmosphere by liquefaction processes at a greatly increased cost (23). The use of liquid hydrogen (bp 20.3 K) as a coolant would offer important advantages over helium, including cost, availability, and thermodynamic efficiency. Unfortunately, as yet no superconductor is known that has the required critical current and field to operate in liquid hydrogen. A continuing search for high temperature superconductors is clearly justified.

2.2.2. Receiver System

The receiver system is one of the most important components of an NMR spectrometer, as the ultimate signal-to-noise ratio that can be obtained depends on careful design of the detection system (see Sec. 2.3).

2.2.2.1. Sweep Spectrometers

Two types of detection are common:

1. The single-coil arrangement whereby energy from the transmitter is applied to a sample coil, which forms one arm of a radiofrequency bridge (24) (Fig. 2.6a). The radiofrequency bridge functions in very much the same way as the well-known Wheatstone bridge. It is balanced in such a manner that for no absorption of energy by the sample no voltage appears across the conjugate arms of the bridge. As the radiofrequency or the field is swept through the spectrum, no change occurs until a resonant frequency is reached. At this point energy is absorbed, the permeability of the sample coil alters, and the radiofrequency bridge becomes unbalanced. This results in the appearance of a voltage across the conjugate arms of the bridge, which is amplified, detected, and recorded on a chart.

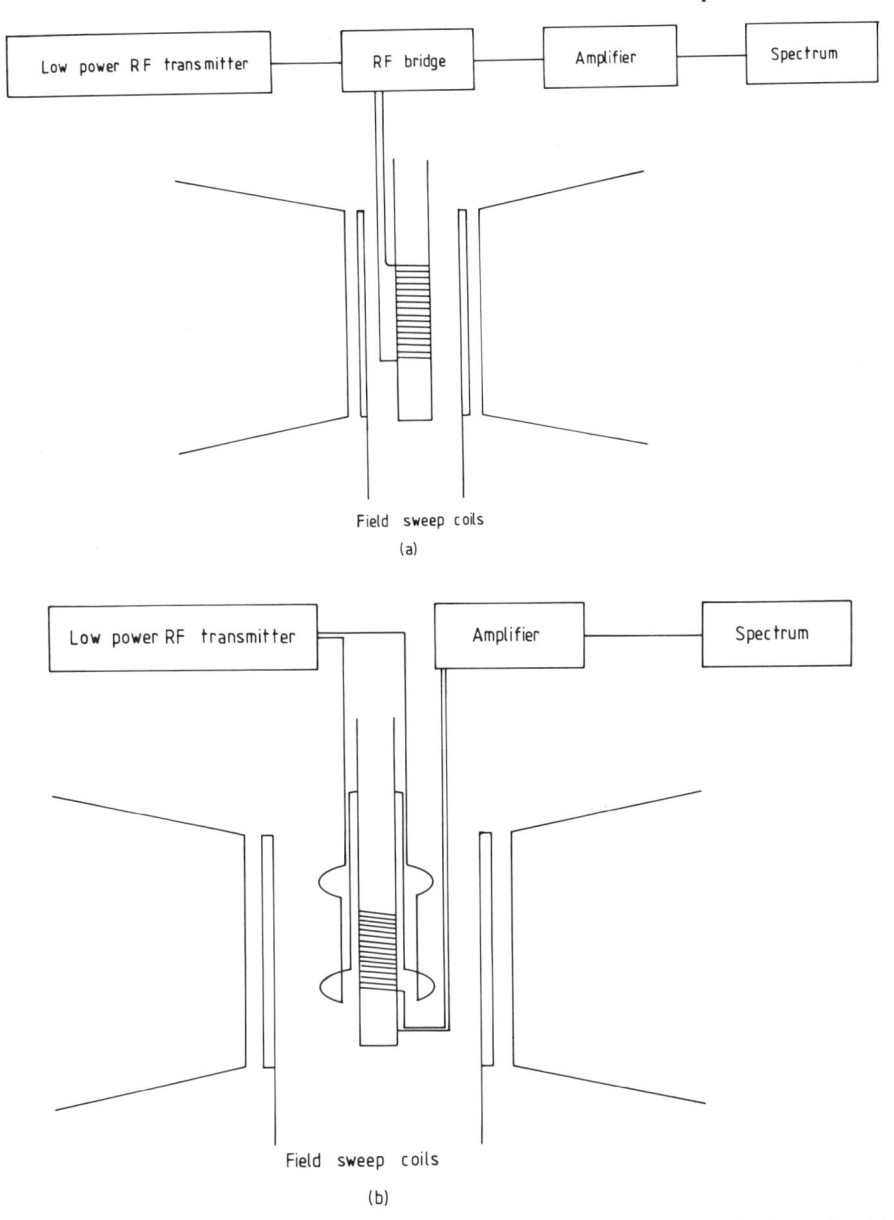

Fig. 2.6. Continuous wave spectrometer. Diagrammatic representation of circuit for (*a*) single-coil configuration and (*b*) crossed coil configuration.

2. The crossed coil or nuclear induction method in which use is made of two coils arranged with their axes othogonal to each other and to the axis of the magnetic field (25) (Fig. 2.6*b*). In theory, when two radiofrequency coils are orthogonal to each other, coupling between coils is zero; that is, a radiofrequency source in one coil will not induce a voltage in the other.

Thus if a continuous radiofrequency energy is applied to the transmitter coils, no voltage should be induced in the receiver coil until resonance occurs. At resonance the nuclei absorb energy and rotate, inducing a voltage into the receiver coil, which is detected, amplified, and recorded. In practice, some leakage does exist between the transmitter and receiver coils and must be balanced out by means of mechanical metal "paddles." The main disadvantages of paddles are their limited range of action, a pronounced temperature dependence, and restricted access for adjustment. Several solutions to this problem have been suggested, all of which employ auxiliary circuits; some of the circuits provide inductive or capacitive compensation for balancing of the leakage (26–29), while one uses a radiofrequency bridge to compensate for voltages induced in the receiver coil (30).

2.2.2.2. *Pulse Spectrometers*

The situation here is slightly different. Figure 2.7 illustrates a typical pulse spectrometer probe arrangement. A high-power pulse transmitter is required to deliver a large amount of power to the transmitter coil, yet the preamplifier must be protected from any high voltage generated by the transmitter, or else expensive low-noise transistors in the preamplifier will be destroyed. Although many, varied methods of protection have been employed (31–33, 57–62), only the most common will be discussed here.

The most prevalent method of protection involves the use of crossed diodes. Conduction through diodes does not take place unless there is sufficient voltage present across the diode to overcome the potential barrier that exists at the semiconductor junction. This is usually of the order of 0.6 V. This property may be used to advantage in spectrometer probe circuits. On operation of the transmitter, the high voltage present causes the pair of crossed diodes to conduct. Since the sample coil forms part of a series tuned circuit, there will be a low-impedance path to ground and most power will flow through this path and hence through the sample coil. Any high voltage that accumulates at the preamplifier input will cause the second pair of crossed diodes to conduct and short out the voltage to ground,

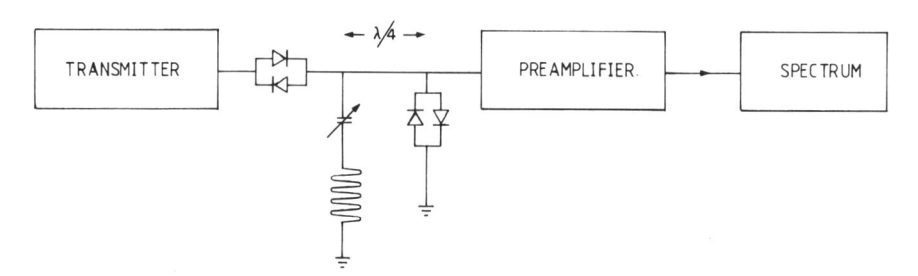

Fig. 2.7. Diagrammatic representation of a circuit commonly used for pulse spectrometers.

ensuring no destructive voltage to the preamplifier. Further protection may be added by ensuring that the length of coaxial cable between the probe and the input to the preamplifier is equal to one-quarter of the wavelength of the transmitter frequency. For reasons that will not be entered into here, any wave traveling along a coaxial conductor that is terminated in a short circuit will be totally reflected at the conductor terminal when the length of the conductor is equal to one-quarter of the wavelength of the wave. Thus no power from the transmitter can proceed past the end of the quarter-wavelength line, and the preamplifier is effectively protected. The disadvantage of using quarter-wavelength cables is the length, particularly at low frequencies (2.4 m at 20 MHz), and the need to change the length of cable with change in frequency.

After the termination of the transmitter pulse, the induced emf in the sample coil is insufficient to overcome the potential barrier in the diodes, and both sets of crossed diodes revert to a nonconductive state. This effectively isolates the transmitter from the probe circuit, and there is a direct path from the sample coil to the preamplifier input.

One point often neglected by designers is the need to match the output impedance of the transmitter to the impedance of the sample coil and that of the sample coil to the input impedance of the preamplifier. It is important that impedances be matched correctly; otherwise power delivered to the coil will be less than optimum and the induced emf in the sample coil will be reduced. Mismatching usually means longer 90° pulses and lower signal-to-noise ratios due to excessive power losses. A typical output impedance for the transmitter would be 50 Ω, although 75 and 300 Ω are common and one would expect the sample coil to have an impedance of 1 or 2 Ω. Such conditions present no design problems, and an impedance-matching circuit may be constructed using design criteria available in most electronics manuals.

Matching of the coil impedance to the input impedance of the preamplifier presents a different situation. The difficulty lies in the fact that the input impedance of an amplifier for maximum signal amplification is not the same as that required for maximum signal-to-noise ratio (34, 35). The two impedances may differ by several orders of magnitude! Not only is the impedance for optimum signal-to-noise difficult to calculate, but it can be shown to be dependent on frequency. Here, one first encounters the design limitations of wideband amplifiers. Multinuclear NMR spectrometers often employ a wideband amplifier so that a wide range of nuclei may be investigated without the operator having to change probe circuits. Unfortunately, impedance-matching circuits for such wideband amplifiers must, of necessity, be fixed, and thus the amplifier will be matched correctly for one frequency only. At all other frequencies mismatching will occur, with a consequent loss in signal-to-noise. Some manufacturers incorporate switching mechanisms that allow different matching circuits to be used for different frequencies. Even so, the above observations still apply, and one must expect some loss in signal-to-noise ratio when employing wideband amplifiers. The effects of wideband amplifiers and the implications of bandwidth on signal-to-noise ratios are discussed more fully in the next section.

2.3. NOISE

The most obvious problem that confronts the spectroscopist is signal-to-noise ratio. At some time or other, all spectroscopists will have wished for an improved signal-to-noise ratio. A major goal of electronics engineers is the design of a virtually noiseless amplifier. Noise can never be completely eliminated, but the amount of noise introduced by the various components of an NMR spectrometer may be reduced by the correct design of preamplifiers and use of various techniques. Such an enhancement of signal-to-noise ratios is important when nuclei of low sensitivity such as ^{15}N and ^{17}O are under investigation. This section is an abridged discussion on the theory of noise, which the author hopes will correct some common misconceptions concerning the nature of noise. Interested readers requiring more detail are referred to the literature (34, 35).

There are two types of noise that are important in NMR spectroscopy, random noise and a type of noise peculiar to NMR spectrometers known as coherent noise.

2.3.1. Random Noise

In the broadest sense, *noise* can be defined as any unwanted disturbance that obscures or interferes with a desired signal. Noise is a totally random signal consisting of frequency components that are random in both amplitude and phase. As a random quantity, the instantaneous value of noise amplitude cannot be predicted, although the long-term value (the root mean square value) can be measured. Hence, in discussing noise one uses the term E_t, the root mean square voltage, and i_t, the root mean square current. A second symbol commonly used for noise voltages is the mean square noise voltage $\overline{e_t^2}$. Hence E_t is equivalent to $\sqrt{\overline{e_t^2}}$ and the two symbols are commonly interchanged.

Noise has an unusual property that is very different from what would be expected from conventional voltage and current theory. When sinusoidal voltage sources of the same frequency and amplitude are connected in series, then the resultant voltage has twice the common voltage if they are in phase and zero voltage if the phases differ by 180°. When independent noise generators are connected in series, the separate sources neither help nor hinder each other, and the resultant mean square voltage is the sum of the individual mean square voltages:

$$E_{1+2}^2 = E_1^2 + E_2^2 \tag{12}$$

Thus noise voltages are always additive, and once a source of noise has been introduced into a system it cannot be removed by any means whatsoever.

There are three main types of noise mechanisms, which are referred to as *thermal noise*, *shot noise*, and *flicker noise*. In NMR spectroscopy, one is concerned primarily with the first two types.

Thermal noise is caused by the random thermal motion of charge carriers such as electrons in a conductor. This motion is similar to the Brownian motion of particles. In every conductor at temperatures above absolute zero, the electrons are in random motion and the amplitude of this motion is dependent on temperature.

It has been shown that for a resistance R at absolute temperature T, the root mean square noise voltage E_t is given by (34)

$$E_t = \sqrt{4kTR\Delta f} \qquad (13)$$

where k is Boltzmann's constant (1.38×10^{-23} J/deg) and Δf is the bandwidth of the measuring system, in hertz. Hence thermal noise is present in every electrical connector that is at a temperature above absolute zero.

Thus at room temperature for a resistance of 1 Ω, a typical value for the resistance of the sample coil in most NMR spectrometers and a bandwidth of 10 kHz, the noise voltage obtained from equation (13) will be 13 nV. If the temperature of the resistance is lowered to $-100°$C, the noise voltage decreases to 10 nV. However, if the bandwidth is limited to 1 kHz, the noise voltage obtained will be 4 nV. Although noise voltage is reduced by lowering the temperature, it is evident from the above values that a far more substantial reduction in noise is obtained by limiting the bandwidth of the system. One consequence of equation (13) is that the thermal noise associated with a narrowband tuned amplifier, that is, an amplifier used for a single nucleus only, will always be smaller than that associated with a wideband multinuclear amplifier. This must be so because E_t is proportional to $\sqrt{\Delta f}$.

Typically, the induced emf in the sample of the spectrometer will be between 1 nV and 1 μV (36), depending on the nucleus being observed and its concentration in the sample tube; that is, it will be of the same order as the thermal noise voltage. Equation (13) is very important, as it provides the limit that must be kept in mind when undertaking NMR spectroscopy measurements. Note that neither inductive nor capacitive reactance appear in the equation. This does not, however, imply that practical inductors or capacitors are noiseless, for it should be pointed out that the resistance used in equation (13) is not simply the dc resistance of the conductor as one would measure with an ohmmeter but should be termed more properly the real part of a complex impedance:

$$Z = R + jX \qquad (14)$$

For practical inductors and capacitors, R is very small, but nevertheless it is real and in the case of inductors is due to eddy current losses, while for a capacitor R may be due to dielectric losses.

Discrete resistive components used in a circuit also contribute noise to the overall spectrum. Apart from fundamental thermal noise generated as explained at the beginning of this section, there is often excess noise introduced into the system by the passage of current through the resistor. Of the four commonly used types of resistor – carbon composition, carbon film, metal film, and wirewound – the first two introduce substantial excess noise into the circuit. This excess noise is due to the method of manufacture of carbon resistors. Such resistors consist of carbon granules mixed with a suitable binder and compressed together. Unfortunately, this technique results in an uneven conductivity across the resistor, and when current flows microarcs tend to occur between the carbon granules, causing excess noise. Wirewound resistors do not suffer from this disadvantage but must be used with caution. In the manufacture of wirewound resistors, resistive wire is wound on a tubular insulator to give the desired value of resistance. This method of construction, while forming a resistor of little or no excessive noise, does constitute an inductance, and the impedance of the resistor is represented by equation (14). jX may be calculated from equation (15), and although at low frequencies jX may be small enough to be ignored, at frequencies commonly used in NMR spectroscopy it may be substantial enough to become the major component of equation (14).

$$jX = 2\pi fL \qquad (15)$$

Metal film resistors generally do not suffer from any of the disadvantages listed above and are the best choice for amplifier construction in NMR spectroscopy. Even so, one must be careful in selecting metal film resistors, as some are sputtered onto a former in a helical configuration so as to increase the value of resistance, while others are sputtered unevenly. Although they have a lower excess noise figure than carbon resistors, these types of resistors are not the most suitable for low-noise electronic design. Helically sputtered resistors suffer the same disadvantage as wirewound resistors. Termination of lead wires onto the resistor body may also have considerable bearing on the amount of excess noise present in a resistor. Ultimately, one must rely on manufacturers' specifications of noise figures in order to choose the most suitable type of metal film resistors.

In general, in order to minimize thermal noise voltage, both temperature and bandwidth must be limited, metal film resistors must be employed, and any reactive components used must be of high quality.

The second type of noise of concern in NMR spectrometers is known as *shot noise*. Shot noise is associated with the flow of current across a potential barrier, such as exists at every *pn* junction in semiconductor devices such as bipolar transistors and diodes. No such junction exists in a simple conductor, and therefore no shot noise is present. It can be shown that the rms value of shot noise current I_{sh} is given by (35)

$$I_{sh} = \sqrt{2qI_{dc}\Delta f} \qquad (16)$$

where q is the electronic charge ($= 1.59 \times 10^{-19}$ C), I_{dc} is the direct current in amperes crossing the potential barrier, and Δf is the bandwidth of the system in hertz. Notice that the temperature of the device does not appear in equation (16). Like thermal noise, shot noise is proportional to the square root of the bandwidth, and the comments made in regard to thermal noise also apply here. That is, limiting the bandwidth of the system reduces shot noise substantially. Shot noise may also be minimized by limiting the current passing across the potential barrier. In practice, this amounts to employing transistors with high gain and very low (of the order of microampere) collector currents, the so-called super β transistors. Field effect transistors (FETS) do not have a junction; they are inherently low-noise devices and are useful in the construction of low-noise amplifiers (37). In this laboratory, metal oxide semiconductor field effect transistors (MOSFETs) have been show to be highly effective in the construction of low-noise amplifiers.

It is apparent from the above that for a given temperature there is a minimum value of noise voltage that cannot be reduced. As the spectroscopist is concerned primarily with improving signal-to-noise ratio, if it is difficult to reduce the absolute value of the noise signal one must look at procedures that increase signal voltage while retaining a smaller increase in noise voltage. Taking advantage of equation (12) leads to the most well-known method for obtaining an improvement in signal-to-noise ratios by averaging spectra. On the addition of spectra, both the signal voltage and the noise voltage increase. Fortunately for spectroscopists, signal voltage increases at a faster rate than noise voltage, and hence the signal-to-noise ratio increases. An estimate of expected increase in signal-to-noise ratio may be obtained by considering the following. Let V_1 and E_1 be the signal voltage and rms noise voltage, respectively, of the first spectrum and V_n and E_n the signal voltage and rms noise voltage of the nth spectrum. As the signal voltage is of a sinusoidal nature and in phase, the total signal votage for n spectra will be $\Sigma_1^n V_i = nV_1$. However, from equation (12), only mean square noise voltages are additive, and hence the total noise voltage after the accumulation of n spectra will be $\Sigma_1^n E_i^2 = nE_1^2$ and the root mean square noise voltage will be $E_{1-n} = \sqrt{n}E_1$ and the signal-to-noise ratio for n accumulations becomes $(V_1/E_1)\sqrt{n}$. That is, an improvement of $\sqrt{n}$ has been obtained. Note that both noise and signal voltages increase as the number of spectra accumulated increase and that averaging spectra *does not* decrease the absolute value of noise voltage, it is simply that the signal voltage increases at a faster rate than the increase in noise voltage. The above discussion assumes that all signal voltages from n accumulations are in phase. In practice, a slight phase change occurs from one spectrum to another and the total accumulated signal voltage will always be less than nV_1; consequently the improvement in signal-to-noise ratio obtained will always be slightly less than $\sqrt{n}$.

To consider other methods of improving signal-to-noise, one must examine expressions for the amplitude of the signal voltage induced in the sample coil. Abragam (38) developed an expression for the signal-to-noise ratio in the NMR experiment that is more conveniently written as (39)

$$\psi_{\text{rms}} = K\eta M_0 \left(\frac{\mu_0 Q \omega_0 V_c}{4FkT_c\Delta f}\right)^{1/2} \tag{17}$$

where K is a numerical factor approximately equal to unity and dependent on the sample coil geometry

η is the "filling factor," a measure of the coil volume V_c occupied by the sample

M_0 is the nuclear magnetization, which is proportional to field strength, $M_0 = N\gamma^2\hbar^2 I(I+1)B_0/3kT_s$

μ_0 is the permeability of free space

Q is the quality factor of the sample coil

ω_0 is the Larmor angular frequency

F is the noise figure of the amplifier

T_c is the temperature of the probe

Δf is the bandwidth, in hertz

Using equation (17) Hill and Richards [39] showed that at resonance and under nonsaturating conditions the expression for signal-to-noise becomes

$$\psi_{rms} = \frac{K\eta Q^{1/2} \omega_0^{3/2} V_c^{1/2} \chi_0 T_2 B_1}{8kT\Delta fF} \qquad (18)$$

where T_2 is the spin-spin relaxation time

χ_0 is the magnetic susceptibility, which is given by $(\mu_0/4\pi)(M_0/B_0)$

B_1 is the magnitude of the radiofrequency field

Both equation (17) and equation (18) show that the signal-to-noise ratio is dependent on a wide range of factors. ψ_{rms} is dependent on the volume of the coil, the filling factor and the Q of the coil, and the concentration and temperature of the sample. Optimizing any of these quantities will increase the signal-to-noise ratio, and the careful probe designer makes full use of equations (17) and (18). It is also predicted that increasing the bandwidth of the amplifier or raising the temperature of the sample will decrease the signal-to-noise ratio. An important conclusion illustrated by equation (18) is that the signal-to-noise ratio of an NMR experiment is proportional to the Larmor frequency raised to the three-halves power. That is to say, the higher the magnetic field, the greater must be the signal-to-noise ratio obtained. A simple example will suffice to show the magnitude of the expected increase in the signal-to-noise ratio. If the Larmor frequency were increased from 90 MHz to 270 MHz (for ^{1}H nuclei), then an improvement in signal-to-noise ratio of 5.2 would be predicted from equation (18). This has been a major driving force behind the development of high-field NMR spectrometers employing superconducting magnets. In actual practice, the improvement obtained in signal-to-noise ratio by increasing the Larmor frequency from a 90 MHz electromagnet spectrometer to a 270-MHz superconducting magnet spectrometer is usually less than 2. In fact, experience in this laboratory has shown that between a 90-MHz electromagnet and a 200-MHz superconducting magnet there is little if any improvement in the signal-to-noise ratio. This was usually considered to be due to inadequate electronic design of the preamplifier and associated components. Recent developments in the design of preamplifiers (40) have indicated that this is not the case, and the reason for the failure of equation (18) to predict the signal-to-noise ratio obtained experimentally must be sought elsewhere.

If one considers the geometry of both an electromagnet and a superconducting magnet (Fig. 2.8), it is apparent that the two are very different. For an electromagnet the magnetic field is parallel to the pole gaps of the magnet. The necessary condition that the radiofrequency field be at right angles to the magnetic field and still allow convenient placement of a sample is easily achieved by the use of a solenoid coil (Fig. 2.8a). The situation for a superconducting magnet is different. Here, because of manufacturing limitations, the superconducting coil is wound in solenoid form, and egress to the center of the magnetic field can be achieved only along the magnetic axis (Fig. 2.8b). To achieve the necessary requirement of a radiofrequency field at right angles to the field and still allow easy access to the center of the magnetic field, saddle coils must be employed as the sample coil.

Equations (17) and (18) are known to relate to solenoids only and cannot be applied to spectrometers employing saddle coils. Hoult and Richards (41) have reexamined the problem and derived the following equation from first principles:

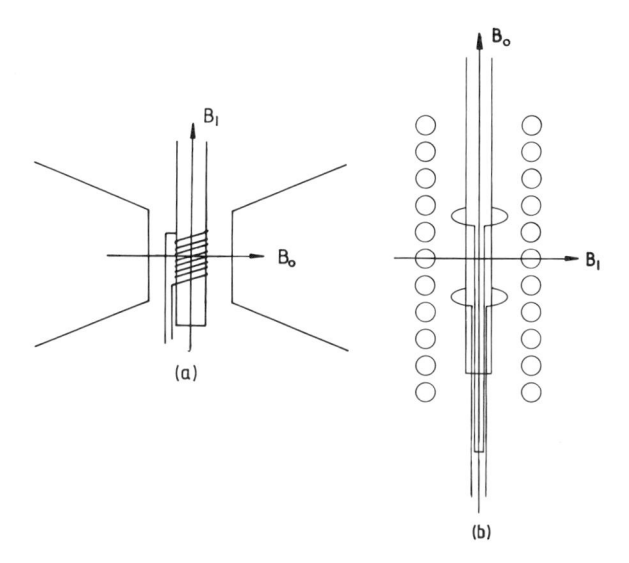

Fig. 2.8. Orientation of B_0, B_1, and receiver coil for (a) permanent magnets and electromagnets and (b) superconducting magnets.

$$\psi_{rms} = \frac{K(B_1)_{xy} V_s N \gamma \hbar^2 I(I+1)}{7.12 k T_s} \left(\frac{P}{F k T_c l \zeta \Delta f} \right)^{1/2} \frac{\omega_0^{7/4}}{[\mu \mu_0 \rho(T_c)]^{1/4}} \quad (19)$$

where $(B_1)_{xy}$ is the effective field over the sample volume V_s produced by unit current flowing in the sample coil
- P is the perimeter of the conductor used to wind the coil
- l is the length of the conductor
- ρ is the resistivity of the conductor
- T_c is the temperature of the sample coil
- ζ is a proximity factor that is known from experience

The signal-to-noise ratio is predicted to be proportional to the Larmor frequency to the power, $1\frac{3}{4}$ a dependence that has been postulated by Soutif and Gabillard (42). Of major interest is the factor $(B_1)_{xy}$, which depends primarily on coil geometry. $(B_1)_{xy}$ is generally much smaller for saddle coils than for solenoid coils, and Hoult has shown both theoretically and experimentally that this results in a factor of approximately 3 in the final signal-to-noise ratio obtained (41). Thus, because of the need to use saddle coils, superconducting magnets show considerably lower signal-to-noise ratios than would be expected from taking into account the three halves dependence on Larmor frequency of equation (18).

The limitations imposed by the use of saddle coils are not easily overcome. The significant loss in signal-to-noise ratio may be avoided by using a solenoid coil design (43). However, such an arrangement, although successful, requires complete removal of the probe in order to change samples, a time-consuming, and in all probability undesirable, operation. At very high frequencies (> 200 MHz) it is possible to use cavity resonators (44), which again would avoid the problems of saddle coils. Another solution would be to design the superconducting magnet with geometry similar to that of an electromagnet, which would allow the use of solenoid coils. Such magnets are technically difficult to fabricate and are unavailable as yet. As superconducting magnets of very high field are becoming common, the author feels that the problem of saddle coils is an important one that should be given a great deal of attention.

Equations (17), (18), and (19) contain the factor F, which is the amount of noise introduced into the system by the amplifier in the NMR spectrometer receiver circuit. It is of interest to note that the amount of noise at each stage of the amplifier contributes to the overall noise figure of the amplifier. That is to say, if a transistor has an associated noise voltage F_a, does all or only part of F_a appear in the total noise figure for the amplifier? If the amplifier consists of two stages, with noise figures F_1 and F_2 and gains G_1 and G_2 associated with the first and second stage, respectively, then it can be shown that the total noise of the amplifier expressed in terms of F_1, F_2, G_1, and G_2 will be given by (35)

$$F = F_1 + \frac{F_2 - 1}{G_1} \qquad (20)$$

For an amplifier of three stages with noise factors F_1, F_2, and F_3 and gains G_1, G_2, and G_3, the expression becomes (45, 46)

$$F = F_1 + \frac{F_2 - 1}{G_1} + \frac{F_3 - 1}{G_1 G_2} \qquad (21)$$

Equation (21) indicates quite clearly that if the gains of the first two stages are very high, then the noise introduced by subsequent stages will be very small. As a consequence, careful noise design is required for the first two stages, and the design of further stages for low noise operation is not critical. One other point must be mentioned in relation to the design of low-noise amplifiers. It is often forgotten that the power supply for the amplifier is part of the low-noise design. Very often an excellent low-noise amplifier design has been ruined by using a high-noise power supply. Low-noise power supplies are essential for low-noise amplifiers.

2.3.2. Coherent Noise

This is a phenomenon that is generally considered to pertain to solid state NMR spectroscopy, low frequencies, and samples where very fast relaxation times are likely to be encountered. However, the author has observed this phenomenon at frequencies as high as 20 MHz. This unfortunately implies that observation of low-frequency nuclei such as ^{14}N, ^{15}N, and ^{17}O may be complicated by the occurrence of coherent noise that renders interpretation of spectra difficult. As the name implies, coherent noise is not random in nature; it manifests itself as "baseline roll" in the spectra (Fig. 2.9a) rather than the furry or grasslike appearance of random noise. Baseline role may be so severe in some cases as to make it impossible to distinguish between a broad resonance such as would be obtained in

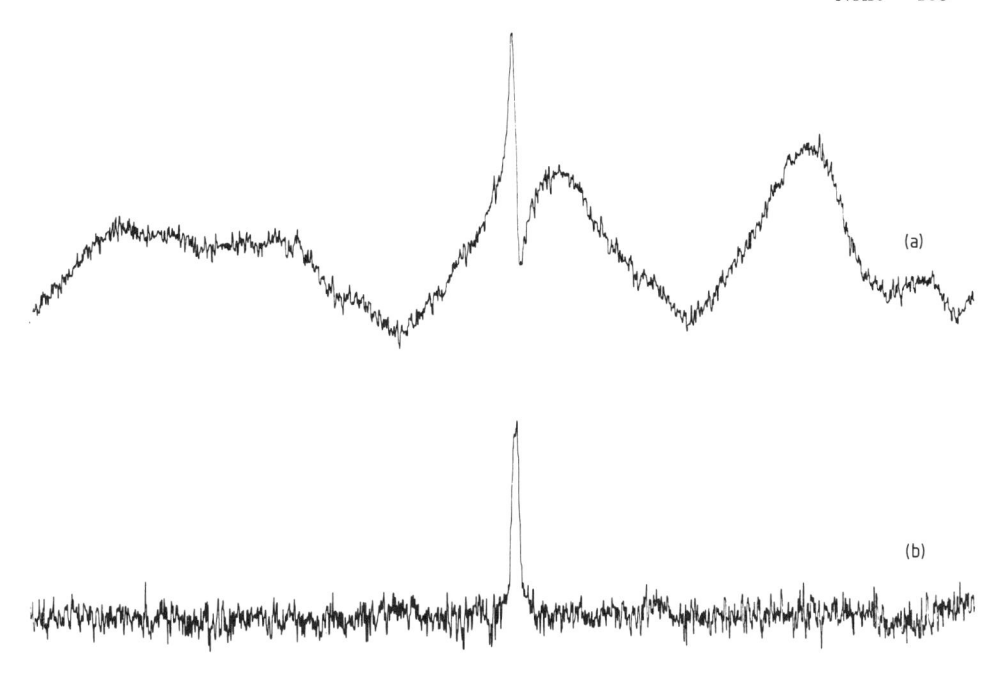

Fig. 2.9. ^{14}N spectra of a 2% v/v solution of pyrrol in acetone illustrating (a) substantial baseline roll and (b) baseline roll eliminated by methods described in Section 2.7.

^{14}N spectroscopy and the baseline itself. It is therefore considered appropriate to devote some space in this chapter to the causes of coherent noise and the methods of eliminating or at least minimizing baseline roll. The most common cause of baseline roll, particularly at higher frequencies, is maladjustment of the Golay coils in the spectrometer, which results in a nonuniform magnetic field across the sample coil with subsequent variations in baseline height. The remedy in this case is very simple. Readjust the Golay coils according to the manufacturer's directions. Other causes of coherent noise, acoustic ringing of the probe body and self-oscillation of the receiver coil, are not so easily described, nor are they easily overcome.

2.3.2.1. Acoustic Ringing

When a radiofrequency field of angular velocity ω interacts with a metal of density m and resistivity ρ then an eddy current will be induced within the skin depth δ of the metal. The classical skin depth δ of the material is given by $2\rho/(4\pi \times 10^{-7}\omega)^{1/2}$. Where a static magnetic field $\mathbf{B_0}$ orthogonal to the radiofrequency field exists, both electrons and the lattice of the metal will experience a Lorentzian force. This is the driving mechanism for the generation of an acoustic wave of velocity v_s that propagates through the metal along the same axis as the static magnetic field. A magnetic field orthogonal to a radiofrequency field describes precisely experimental conditions required for NMR spectroscopy, and it is certain that an acoustic wave will be generated in the probe of an NMR spectrometer. When the skin depth is very much smaller than the wavelength of the acoustic wave, then the amplitude $|\mu|$ of the induced acoustic wave will be given by (47, 48)

$$|\mu| = \frac{B_0 B_1}{(4\pi m v_s \omega)(1 + \beta^2)} \quad (22)$$

where $\beta^2 = 2.5 \times 10^{13}(\rho^2/v_s^4)(\omega/2\pi)$.

The inverse process also occurs. When the acoustic waves reach an interface in the presence of the magnetic field, an electromagnetic wave is emitted from the material. Any nearby radiofrequency coil will detect this signal and therefore, in the NMR experiment at least, introduce a spurious noise signal into the spectrum acquisition circuit. The amplitude of this signal may be obtained from the equation (47–49)

$$E = \frac{k B_0^2}{m v_s (1 + \beta^2)} \quad (23)$$

where k is a constant.

Thus, in theory, a radiofrequency coil near a metal object in a static magnetic field can be used both to generate acoustic waves in the metal and to detect the echoes as they are reflected from the surfaces of the object. Indeed, this technique has been proposed as a method for the nondestructive testing of materials (50, 51). At each reflection energy is lost and the acoustic signal slowly decays to zero. The time required for decay may be of the order of millisecond. However, under the multiple pulsing conditions that frequently occur in Fourier transform NMR spectroscopy, the energy that has been acquired by the acoustic wave will be "topped up" and an acoustic standing wave set up. When such a situation occurs, the acoustic pulses are reflected back and forth with the probe, emitting radiation each time a surface is reached. Thus, whenever the pulse returns to the surface at which it was generated — the receiver coil — a voltage will be induced, and consequent Fourier analysis from the time domain to the frequency domain will result in a rolling baseline.

Equations (22) and (23) do not allow the frequency of the acoustic wave to be determined. The resonant frequency will depend on the geometry of the probe (55), and a parallel can be drawn with the resonance frequency of sound waves in organ pipes.

It has been shown that the acoustic resonance frequency f_m is given by (48)

$$f_m = \frac{M v_s}{2D} \quad (24)$$

where M is an integer and D depends on the geometry of the body. The presence of M in equation (24) indicates that large-amplitude harmonics exist, which may cause more trouble than the fundamental resonant frequency. For example, the FT spectrometer in this laboratory shows a fundamental harmonic frequency at approximately 1.5 MHz, yet large spurious peaks are obtained when observing a nucleus at 6 MHz, the fourth harmonic!

Acoustic ringing may be suppressed by several methods. Examination of equation (23) indicates that the amplitude of the acoustic wave may be minimized

Table 2.1. Acoustic Parameters of Metals and Alloys at Room Temperature (19, 20)

Metal or Alloy	m 10^3kg/m^3	v_s m/s	β^2 at 5 MHz	$mv_s(1+\beta^2)$ $\times 10^{-3}$
Stainless steel	8.0	3100	3.61	114.33
Lead	11.4	690	118	93.60
Platinum	21.4	1730	0.783	66.01
Tungsten	19.3	2640	0.042	53.09
Titanium	4.5	3125	1.156	30.32
Gold	19.7	1200	0.173	27.73
Copper	8.9	2270	0.007	20.34
Brass	8.5	2110	0.121	20.11
Zinc	7.1	2440	0.061	18.38
Silver	10.4	1610	0.024	17.14
Tin	7.3	1670	0.097	13.37
Aluminum	2.7	3040	0.005	8.25

by selecting a probe material that maximizes the acoustic number $mv_s(1+\beta^2)$. Practical considerations such as machinability, rigidity, economic cost, and a factor often forgotten by designers, thermal conductivity, must be taken into account as well. Table 2.1 lists acoustic parameters m, v_s, β^2, and $mv_s(1+\beta^2)$ for a series of alloys and pure metals. Surprisingly, aluminum, which for practical considerations has been used extensively for probes, is one of the worst materials. Tin is not much better, while gold is marginally better than copper, and silver is slightly worse. Stainless steel appears to be the most suitable. Unfortunately, stainless steel is difficult to machine and has a low heat conductivity, and many stainless steels have magnetic properties. Tungsten, titanium, and zinc may be impractical materials for similar reasons. Platinum, with an acoustic parameter of 66, is suitable but economically unviable. Lead is very attractive on acoustic considerations, but its lack of rigidity precludes its use. One is left with brass or copper as a suitable probe material. Of the two, copper is preferred because of its higher thermal conductivity.

Buess and Petersen (48) have been able to reduce acoustic ringing by coating the metal surfaces of the probe body with an acoustically lossy material such as silicon rubber. Speight and coworkers (54) reported success in eliminating acoustic ringing by inserting a grounded copper shield between the receiver coil and the conducting surface of the probe. In this laboratory, some success in reducing acoustic ringing has been obtained by winding the receiving coil with multistranded copper wire. A computer-assisted method has been developed also, and details of this technique are given in a later section.

Another approach that is likely to prove fruitful is being evaluated in this laboratory. As indicated at the beginning of this section, an acoustic wave is caused by the induction of eddy currents within the skin depth of the probe body. The skin depth is related to the resistivity of the metal (36) and for most metals will be certainly less than 100 μm. There exists the possibility of fabricating the probe out of one metal and then coating it either by electrolytic deposition or by hot dipping (to the skin depth) with a second metal that is more acoustically desirable. An obvious choice would be copper for the probe body and hot dipping with lead. The technique of hot dipping or "tinning" copper wire with either lead or solder is a well-established practice in electrical engineering. Solder would not be suitable for acoustical damping purposes, as the presence of tin with its low acoustic number in the solder would degrade the performance of the probe. Electrolytic deposition of platinum on the probe body, while expensive, would achieve the desired effect. The common practice of gold or silver plating probes achieves little, In fact, silver plating would degrade the performance of the probe and is not recommended.

One must not forget that there is a possibility of an acoustic standing wave being induced within the receiver coil itself (48, 56). When this situation arises no acceptable solution exists. One may diminish the amplitude of the induced acoustic wave either by reducing the size of the wire used to wind the receiver coil (47) or by using a more suitable material such as tungsten. In either case, the resistivity of the coil is increased, and as a consequence the Q of the coil is decreased, with a subsequent loss in signal-to-noise.

2.3.2.2. Coil Resonance

In every spectrometer the reciever coil forms part of a tuned circuit, and unfortunately this presents problems when fast relaxing nuclei such as ^{17}O are investigated. When a tuned circuit is excited at resonance the energy stored in the tuned circuit oscillates between the capacitive components and the inductive components. If the circuit components were ideal this oscillation would continue indefinitely, but because of such unavoidable effects as flux loss from the inductive components and dielectric loss from the capacitive components the oscillations will decrease to zero after a finite time. The greater the Q of the tuned circuit − the more ideal it becomes − the longer the oscillations will persist. Indeed, measurement of decay of oscillation in a tuned circuit has been used to determine Q for that circuit (36).

A problem arises where the free induction decays of nuclei are of the same order of magnitude in intensity and persistence as that of the ringing of the probe circuit, and resonances may be completely obscured by the baseline roll caused by coil oscillation. Nuclei with strong absorption and long relaxation times do not present a problem. Equations (17) and (18) indicate quite clearly that the signal-to-noise ratio is proportional to $Q^{1/2}$. Hence one is faced with a dilemma. Increase Q to increase the signal-to-noise ratio, and coil ringing becomes a problem; decrease Q to decrease coil ringing, and the signal-to-noise ratio suffers.

Several schemes have been proposed to reduce the ringing problem and have met with various degrees of success. One approach is to reduce the Q of the coil immediately after the main pulse by means of an electronic switching circuit, and then after ringing has decreased to a satisfactory level remove the damping and restore Q. The problem with this approach is that activation of the switch results in renewed coil ringing. Many experimenters simply lower the Q of the coil and accept a loss in signal-to-noise ratio as inevitable (57, 58). Lowe and Engelsberg (59) used a lumped parameter delay line, whereas Hoult (60) has described a more sophisticated technique that employs a short phase-inverted pulse immediately after the main transmitter pulse to reduce drastically the ring downtime for the receiver coil. The approach of Engle (61, 62) merits mention here. A ferrite toroid switch is employed to dampen oscillations. The author has investigated the use of ferrite toroid switches and observed that isolations of greater than 60 dB are easily obtained and no coil ringing is apparent when the toroid is switched on or off. Unfortunately, ferrite alters its properties when placed near a magnetic field, and the technique of Engle appears to have limitations in this respect. In this laboratory the computer-assisted technique developed to counter acoustic ringing has also proved useful in controlling coil ringing.

2.4. MULTINUCLEAR MAGNETIC RESONANCE

Nearly all the elements in the periodic table have at least one isotope that is potentially suitable for NMR studies (63). Improvements in NMR spectroscopy in recent years, particularly the advent of Fourier transform pulse techniques and the remarkable improvement in low-noise electronics, have led to a wide range of nuclei becoming readily accessible for study. Chemical shifts, coupling constants, and relaxation mechanisms are closely related to the electronic environments of the nuclei concerned. A study of the NMR spectra of several different nuclei in a given molecule can lead to an understanding of the electronic distribution in that molecule and an elucidation of its structure. Indeed, nuclei occupying positions in the skeleton of the molecule such as carbon, nitrogen, oxygen, and sulfur, may provide more useful information than those occupying peripheral sites such as hydrogen. Pulse spectrometers with multinuclear capabilities are now very common, and this has meant that the heterocyclic chemist has access to a wide range of nuclei for study. In particular, the heterocyclic chemist may obtain spectra of ^{14}N, ^{15}N, ^{17}O, and ^{33}S nuclei as well as the more usual ^{1}H and ^{13}C. One must not forget that both deuterium and tritium also possess nuclear spins and may be used for spectral studies (64–67). The spin properties of these nuclei reflect the ease (or the difficulty) with which spectra may be obtained and are listed in Table 2.2.

In view of the large number of references in the literature to NMR spectra of heterocyclic compounds (over 12,000 for the period 1971–1981), it is considered inappropriate, indeed impossible, for this chapter to deal *in toto* with all nuclei of interest to heterocyclic chemists. Accordingly, ^{1}H and ^{13}C spectra have virtually been ignored in view of the large number of comprehensive texts and atlases of chemical shifts for ^{1}H and ^{13}C that are currently available, and readers are referred to these very comprehensive references (68–73). The author's efforts have been concentrated on the lesser known but still very useful nuclei ^{14}N, ^{15}N, ^{17}O, and ^{33}S, for which no recent reviews for heterocyclic compounds are available.

Table 2.2. Spin Properties of Nuclei

Isotope	Spin	Natural Abundance (at %)	Magnetic Moment μ/μ_N	Magnetogyric Ratio $\gamma/10^7$ (rad T^{-1} s^{-1})	Quadrupole Moment $Q/10^{-28}$ M^2	NMR Frequency (MHz)	Relative Sensitivity to protons at Natural Abundance
^{1}H	$\frac{1}{2}$	99.985	4.8371	26.7510	—	100.00	1.00
^{2}H	1	0.015	1.2125	4.1064	2.73×10^{-3}	15.35	1.45×10^{-6}
^{3}H	$\frac{1}{2}$	$<1 \times 10^{-16}$	5.1594	28.5335	—	106.66	—
^{13}C	$\frac{1}{2}$	1.108	1.2162	6.7263	—	25.14	1.76×10^{-4}
^{14}N	1	99.63	0.5706	1.9324	1.6×10^{-2}	7.22	1.00×10^{-3}
^{15}N	$\frac{1}{2}$	0.37	−0.4901	−2.7107	—	10.14	3.85×10^{-6}
^{17}O	$\frac{5}{2}$	0.037	−2.2398	−3.6266	-2.6×10^{-2}	13.56	1.08×10^{-5}
^{33}S	$\frac{3}{2}$	0.76	0.8296	2.0517	-6.4×10^{-2}	7.67	1.71×10^{-5}

2.4.1. Other Isotopes Of Hydrogen

2.4.1.1. *Deuterium*

The deuterium nucleus with its low magnetogyric ratio and spin number of 1 has low sensitivity and very poor spectral dispersion. ^{2}H-NMR chemical shifts expressed in parts per million on the δ scale are virtually identical to those of ^{1}H, or ^{3}H for that matter. However, because of its much lower magnetogyric ratio, chemical shifts expressed in hertz are compressed by a factor equal to the ratio of the magnetogyric constants, that is, $26.7510/4.1064 = 6.5$. This means that resonances that may be 1 Hz apart in the ^{1}H spectrum will be 0.15 Hz apart in the ^{2}H spectrum, a figure that is very close to the maximum resolution of most spectrometers. Because of the quadrupolar spin number $I = 1$, shortened relaxation times (generally less than 1 s) and broadened ^{2}H spectral lines occur. The width of ^{2}H resonances may be expected to be on the order of 1 Hz. Hence any resonances closer than 2 Hz cannot be expected to be resolved. To achieve spectral dispersion of ^{2}H comparable to that accepted for ^{1}H at 100 MHz, at least a sixfold increase in the applied magnetic field would be required – a superconducting magnet capable of achieving a field of 12 T. A further consequence of the low magnetogyric constant is that ^{2}H coupling constants to other nuclei compared to ^{1}H coupling constants are also reduced by a factor equal to the ratio of the magnetogyric constants. Hence one would expect coupling constants of 0–3 Hz for proton-deuterium and 18–30 Hz for carbon-deuterium (64). Coupling constants are, of course, field indpendent, and no enhancement can be expected from using spectrometers with higher fields. The main use of ^{2}H spectra will probably continue to be in the assignment of complex proton spectra. The author has, however, had success in using ^{2}H in the assignment of ^{13}C spectra by making use of the deuterium isotope shift. When hydrogen is replaced by deuterium, a slight change is induced in the electron density of nearby nuclei, which causes a small shift in the ^{13}C resonance of the adjacent carbon nucleus. The effect decreases with distance, and hence comparison of ^{13}C-^{1}H spectra with ^{13}C-^{2}H spectra readily identifies resonances close to the site of deuteration. Thus ^{13}C spectra of several triazaindenes have been assigned using this technique (74). The method has proved useful in assigning ^{13}C spectra of carbohydrates where both ^{13}C-^{1}H and ^{13}C-^{2}H spectra are obtained simultaneously (75, 76). The technique will undoubtedly prove useful in assignment of ^{13}C and ^{15}N resonances in heterocycles providing an easily exchangeable proton is available close to the site of the ambiguous nucleus.

2.4.1.2. *Tritium*

Tritium is an interesting isotope. It has a magnetogyric ratio and sensitivity higher than ^{1}H and a nuclear spin $I = \frac{1}{2}$. Chemical shifts expressed in parts per million on the δ scale will be virtually identical to ^{1}H shifts but with slightly better dispersion. Relaxation times are similar to those expected for protons, and consequently narrow linewidths can be expected. Unfortunately, tritium is radioactive and classed as a soft β emitter with a half-life of 12.3 years. Tritium enriched samples may be handled safely if they are stored in sealed NMR tubes, as the β-radiation emitted will not penetrate the glass walls. The level of enrichment required (approximately 1 mCi per site) implies considerable gas evolution, partly from the disruption of chemical bonds in the compound and partly from the disintegration of tritium into helium-3. The possibility of tube explosions or breakages cannot be disregarded, and any laboratory contemplating ^{3}H-NMR spectroscopy must be equipped to deal rapidly with any radioactive spillage. The maximum permissible dose for ^{3}H is $5 \times 10^{-6} \mu$Ci per cubic centimeter of air for 40 hr exposure per week (77), a figure very much lower than the isotopic enrichment required for ^{3}H-NMR. In spite of the difficulties, some work on exchange processes, catalytic hydrogenation, and biochemical transformations has been published (64). Undoubtedly, there are many applications to which ^{3}H-NMR can be applied, for example, hydrogen bonding, specific coupling constants, and chemical shifts, but because of the safety hazards involved ^{3}H-NMR is likely to remain the preserve of a very few specially equipped NMR laboratories.

2.4.2. Nitrogen Nuclear Magnetic Resonance

Nitrogen would be the nucleus of greatest interest to heterocyclic chemists. The element possesses two isotopes the magnetic moments: ^{14}N, which is 99.6% abundant, has a relative sensitivity of 1×10^{-3} (i.e., approximately 20 times better than ^{13}C), but with a spin number $I = 1$ has a moderate quadrupole moment. ^{15}N, with a natural abundance of 0.37%, has a very low sensitivity of 3.85×10^{-6} (approximately one-hundredth that of ^{13}C and one-third that of ^{17}O) but the advantage of possessing a spin number $I = \frac{1}{2}$. ^{15}N also has a negative magnetogyric ratio, which presents problems.

Chemical shifts for the two isotopes are very similar and have a range of approximately 1000 ppm although for a given class of compounds the range is normally only 50 ppm. The chemical shifts of one isotope may be used to assign chemical shifts of the other. The chemical shifts for ^{14}N and ^{15}N are so similar that many of the early reviews do not distinguish between ^{14}N and ^{15}N and report both shifts together. However, the author prefers to separate ^{14}N shifts from ^{15}N shifts due to the uncertainty associated with the measurement of ^{14}N chemical shifts. The chemical shifts of both nuclei are strongly influenced by temperature, solvent, concentration, and pH (78–81). Whenever possible, sample temperature and concentration should be reported to allow comparison of data. Techniques for accurate measurement of sample temperature are outlined in Section 2.6. Choice of a suitable reference compound is a serious problem, as any internal reference is quite likely to be influenced by solvent and interactions with the sample. External references are recommended for both ^{14}N and ^{15}N, and correction must be made for differences in magnetic susceptibilities. The accepted reference for both ^{14}N and ^{15}N is external neat nitromethane. The problem of references is considered in more detail in a later section.

2.4.2.1. ^{14}N Spectroscopy

The high natural abundance and relatively high sensitivity of this nucleus ensured that initial attempts to obtain nitrogen NMR spectra were made with it. It possesses a moderately strong quadrupole moment. Thus ^{14}N relaxation is dominated by quadrupolar relaxation except in cases of very high symmetry such as NH$_4^+$. This leads to relaxation times of the order of 10 ms. Such fast relaxation times lead to large linewidths. Linewidths are generally of the order of 100 Hz, but in the highly asymmetrical environments that arise in heterocyclic compounds this may be of the order of 1 kHz or more with a consequent overlap of resonances. In a few cases resonances are so broad as to be indistinguishable from baseline noise. The low resonance frequency of ^{14}N makes the observation of this nucleus susceptible to probe ringing and consequent baseline roll, which may obscure resonances. The fast relaxation rates observed for this nucleus imply that data may be acquired quickly by fast pulse repetition rates. Pulse repetition rates between 50 and 100 ms are commonly employed in this laboratory. Tables 2.3–2.5 list the available ^{14}N chemical shifts for a wide variety of heterocyclic compounds. All chemical shifts are from external heat nitromethane, and wherever possible concentrations are listed.

Table 2.3. ^{14}N Shifts of Heterocycles with One Heteroatom[a]

Compound	Shift	Solvent Conditions	Ref.
Acridine N-oxide	− 109 ± 2	Acetone	82
	− 110	Acetone	83
5-NO$_2$ furan	− 28.3	30% in DMSO	84
2-acetyl	− 29.7	30% in DMSO	84
2-acetoxy	− 28.6	30% in DMSO	84
2-cyano	− 31.1 (NO$_2$)	30% in DMSO	84
	−[b] (CN)		
2-formyl	− 28.8	30% in DMSO	84
2-iodo	− 27.3	30% in DMSO	84
2-methyl	− 29.6	30% in DMSO	84
2-trifluormethyl	− 31.0	30% in DMSO	84
Indole	− 251 ± 5	Dioxan (satd)	85
	− 246 ± 5	Methanol (satd)	85
1-methyl	− 250 ± 3	Neat	86
	− 248 ± 5	Neat	85
2-Methylisoindole	− 218 ± 2	Ether (satd)	86
Indolizine	− 191.2 ± 0.2	Ether (satd)	87
Carbazole	− 260 ± 5	Acetone (satd)	85
9-methyl	− 260 ± 4	Acetone (satd)	85
Isoquinoline N-oxide	− 90	Acetone	83
	− 89 ± 1	Acetone	82
	+ 93 ± 1	Chloroform	82
	+ 110 ± 2	Methanol	82
Pyridine	− 62.01 ± 0.14	Neat	87
	− 64.01 ± 0.30	0.3 M in acetone	87
	− 59.20 ± 0.31	0.3 M in ether	87
	− 58.01 ± 0.37	0.3 M in carbon-tetrachloride	87
	− 84.38 ± 0.59	0.5 M in water	87

Table 2.3. (continued)

Compound	Shift	Solvent Conditions	Ref.
2-amino	−116 ± 3 (N$_1$) −310 ± 3 (NH$_2$)	Acetone 1:3 v/v	88
3-amino	−66 ± 3 (N$_1$) −334 ± 3 (NH$_2$)	Acetone 1:3 v/v	88
4-amino	−106 ± 3 (N$_1$) −323 ± 3 (NH$_2$)	Acetone 1:3 v/v	88
2-chloro-3-cyano-4-ethylthio	−185.7	20% w/v DMSO	89
2-chloro-3-cyano-4-methoxy	−185.4	20% w/v DMSO	89
2-dimethylamino	−109 ± 3 (N$_1$) −319 ± 3 (NMe$_2$)	Acetone 1:3 v/v	88
3-dimethylamino	−64 ± 3 (N$_1$) −342 ± 3 (NMe$_2$)	Acetone 1:3 v/v	88
4-dimethylamino	−102 ± 3 (N$_1$) −329 ± 3 (NMe$_2$)	Acetone 1:3 v/v	88
2-hydroxy (2-pyridone)	−209 ± 2 −212 ± 2	Acetone 1:3 v/v Methanol 1:3 v/v	90 90
3-hydroxy (3-pyridone)	−67 ± 4 −71 ± 4	Acetone 1:3 v/v Methanol 1:3 v/v	90 90
4-hydroxy (4-pyridone)	−222 ± 4 −227 ± 3	Acetone satd Methanol 1:3 v/v	90 90
2-mercapto	−187 ± 1 −186 ± 1	Acetone 1:3 v/v Methanol 1:3 v/v	88 88
3-mercapto	−74 ± 1 −88 ± 5 −72 ± 5	Acetone 1:3 v/v Methanol 1:3 v/v Acetone-DMSO 1:4 v/v	88 88 88
4-mercapto	−225 ± 2 −218 ± 1 −222 ± 1	Acetone 1:3 v/v Methanol 1:3 v/v Acetone-DMSO 1:4 v/v	88 88 88
2-methoxy	−109 ± 3 −111 ± 2 −119 ± 2	Neat Acetone 1:3 v/v Methanol 1:3 v/v	90 90 90
3-methoxy	−60 ± 3 −64 ± 3 −67 ± 4	Neat Acetone 1:3 v/v Methanol 1:3 v/v	90 90 90
4-methoxy	−89 ± 4 −91 ± 3 −95 ± 4	Neat Acetone 1:3 v/v Methanol 1:3 v/v	90 90 90
1-*N*-methyl-2-imino	−242 ± 3 (N$_1$) −192 ± 3 (NH) −228 ± 2 (N$_1$) −^{b} (NH)	Acetone 1:3 v/v Acetone 1:3 v/v	88 86
1-*N*-methyl-4-imino	−260 ± 3 (N$_1$) −168 ± 3 (NH)	Acetone 1:3 v/v	88
1-*N*-methyl-2-methylimino	−237 ± 3 (N$_1$) −195 ± 3 (NHMe)	Acetone 1:3 v/v	88
1-methyl-2-mercapto	−189 ± 2 −188 ± 1	Acetone 1:3 v/v Methanol 1:3 v/v	86 88
1-methyl-4-mercapto	−225 ± 2 −222 ± 1	Acetone-DMSO 1:4 v/v Methanol 1:3 v/v	88 88

Table 2.3. (continued)

Compound	Shift	Solvent Conditions	Ref.
1-methyl-2-oxo	−215 ± 3	Neat	90
	−216 ± 2	Acetone 1:3 v/v	90
	−214 ± 1	Methanol 1:3 v/v	90
1-methyl-4-oxo	−248 ± 2	Acetone 1:3 v/v	90
	−240 ± 3	Methanol 1:3 v/v	90
2-methylamino	−110 ± 3 (N$_1$)	Acetone 1:3 v/v	88
	−308 ± 3 (NHMe)		
3-methylamino	−65 ± 3 (N$_1$)	Acetone 1:3 v/v	88
	−336 ± 3 (NHMe)		
4-methylamino	−105 ± 3 (N$_1$)	Acetone 1:3 v/v	88
	−318 ± 3 (NHMe)		
2-methylthio	−79 ± 3	Acetone 1:3 v/v	88
	−88 ± 4	Methanol 1:3 v/v	88
3-methylthio	−64 ± 1	Acetone 1:3 v/v	88
	−82 ± 5	Methanol 1:3 v/v	88
	−64 ± 3	Acetone-DMSO 1:4 v/v	88
4-methylthio	−77 ± 2	Acetone 1:3 v/v	88
	−86 ± 4	Methanol 1:3 v/v	88
	−82 ± 4	Acetone-DMSO 1:4 v/v	88
2-oxo-3-acetoxy-4-hydroxy	−176.2	20% w/v DMSO	89
2-oxo-3-amido-4-hydroxy	−173.6	20% w/v DMSO	89
2-oxo-3-carboxy-4-hydroxy	−171.0	20% w/v DMSO	89
Pyridinium ion (Cl$^-$)	−178.96 ± 0.09	0.5 M in 10 M HCl	87
1-(4′-bromophenyl)	−166.1	1 M in water	91
1-(4′-chlorophenyl)	−164.0	1 M in water	91
1-(4′-iodophenyl)	−164.5	1 M in water	91
1-(4′-methoxyphenyl)	−164.1	1 M in water	91
1-(4′-methylphenyl)	−162.9	1 M in water	91
1-phenyl	−162.6	1 M in water	91
Pyridinium (Br$^-$)-1,3-diethyl	−163.7	1.5 M in methanol	92
1,4-diethyl	−169.3	1.5 M in methanol	92
1-(3′-nitrophenyl)	−168.2 (N$_1$)	1 M in water	91
	−^{b} (NO$_2$)		
1-(4′-nitrophenyl)	−166.3 (N$_1$)	1 M in water	91
	−^{b} (NO$_2$)		
N-Methyl-3-oxypyridylbetaine	−181 ± 1	Neat	90
Pyridine N-oxide	−86	Acetone	93
	−96	Methanol	93
	−89	Chloroform	93
	−99	Water	93
2-acetyl	−104 ± 2	Acetone	93
3-acetyl	−88	Acetone	93
4-acetyl	−77	Acetone	93
2-amino	−140 ± 3 (N$_1$)	Methanol	93
	−306 ± 20 (NH$_2$)		
3-amino	−102 ± 2 (N$_1$)	Methanol	93
	−330 ± 20 (NH$_2$)		
4-amino	−139 ± 10 (N$_1$)	Methanol	93
	−290 ± 20 (NH$_2$)		
2-bromo	−87	Acetone	93
3-bromo	−80	Acetone	93
4-bromo	−85	Acetone	93

Table 2.3. (continued)

Compound	Shift	Solvent Conditions	Ref.
2-carboxy	−103	Acetone	93
	−106 ± 3	Water	93
3-carboxy	−82 ± 6	Water	93
4-carboxy	−79 ± 8	Water	93
2-chloro	−89	Acetone	93
3-chloro	−81	Acetone	93
4-chloro	−85	Acetone	93
2-cyano	−89 (N_1)	Methanol	93
	−[b] (CN)		
3-cyano	−92 (N_1)	Methanol	93
	−[b] (CN)		
4-cyano	−90 (N_1)	Water	93
	−[b] (CN)		
2-dimethylamino	−106 (N_1)	Acetone	93
	−[b] (NMe_2)		
	−122 ± 2 (N_1)	Methanol	93
	−[b] (NMe_2)		
4-dimethylamino	−152 ± 8 (N_1)	Methanol	93
	−[b] (NMe_2)		
	−135 ± 5 (N_1)	Chloroform	93
	−[b] (NMe_2)		
2-formyl	−81	Acetone	93
3-formyl	−82	Acetone	93
4-formyl	−73	Acetone	93
2-hydroxy	−158 ± 10	Chloroform	93
	−162 ± 10	Water	93
3-hydroxy	−98 ± 4	Methanol	93
	−101 ± 4	Water	93
4-hydroxy	−144 ± 5	Methanol	93
	−136 ± 8	Water	93
2-hydroxymethyl	−107	Methanol	93
3-hydroxymethyl	−102 ± 2	Methanol	93
4-hydroxymethyl	−99 ± 3	Methanol	93
2-methoxy	−146 ± 10	Acetone	93
3-methoxy	−91	Acetone	93
4-methoxy	−103	Acetone	93
2-methyl	−86	Acetone	93
	−99 ± 2	Methanol	93
3-methyl	−83	Acetone	93
	−97 ± 2	Methanol	93
4-methyl	−88	Acetone	93
	−107 ± 2	Methanol	93
2-nitro	−99 (N_1)	Acetone	93
	−98 (NO_2)		
3-nitro	−82 (N_1)	Acetone	93
	−83 (NO_2)		
	−92 ± 2 (N_1)	Methanol	93
	−72 (NO_2)		
	−82 ± 2 (N_1)	Chloroform	93
	−[b] (NO_2)		
4-nitro	−71 (N_1)	Acetone	93
	−[b] (NO_2)		

Table 2.3. (continued)

Compound	Shift	Solvent Conditions	Ref.
	−79 (N$_1$)	Methanol	93
	−^{b} (NO$_2$)		
	−73 (N$_1$)	Chloroform	93
	−^{b} (NO$_2$)		
2-(2-pyridyl)	−98 ± 2 (N$_1$)	Water	93
	−^{b} (N$'_1$)		
3-(3-pyridyl)	−94 ± 2 (N$_1$)	Water	93
	−^{b} (N$'_1$)		
4-(4-pyridyl)	−98 ± 5 (N$_1$)	Water	93
	−^{b} (N$'_1$)		
2-Aminopyrone			
3-acetoxy	−291.7	20% w/v in DMSO	89
2-acetyl	−229.7	20% w/v in DMSO	89
3-carboxy	−282.7	20% w/v in DMSO	89
Pyrrole	−231.4 ± 0.4	Neat	94
	−229.6 ± 0.4	0.15 M in acetone	94
	−229 ± 2	ca. 40% in acetone	95
	−222.3 ± 0.4	0.15 M in DMSO	94
	−236.4 ± 0.4	0.10 M in carbontetrachloride	94
	−224 ± 1	0.55 M in water	94
1-methyl	−230 ± 1	0.15 M in acetone	94
	−230 ± 1	ca. 40% in acetone	95
	−234 ± 2	0.10 M in carbontetrachloride	94
1-methyl-2-nitro	−228 ± 5 (N$_1$)	ca. 40% in acetone	95
	−22.5 ± 1 (NO$_2$)		
1-methyl-2,4-donitro	−225 ± 5 (N$_1$)	ca. 40% in acetone	95
	−^{b} (NO$_2$)		
2-nitro	−227 ± 5 (N$_1$)	ca. 40% in acetone	95
	−22.5 ± 1 (NO$_2$)		
2,4-dinitro	−227 ± 5 (N$_1$)	ca. 40% in acetone	95
	−^{b} (NO$_2$)		
2,5-dinitro	−236 ± 5	ca. 40% in acetone	95
	−25 ± 1 (NO$_2$)		
Quinoline N-oxide	−94 ± 1	Acetone	82
	−95 ± 1	Chloroform	82
	−105 ± 1	Methanol	82

^{a} Shifts are from external nitromethane in ppm.
^{b} Shifts not reported.

178 Nuclear Magnetic Resonance Spectroscopy

Table 2.4. ^{14}N Spectra of Heterocycles with two Heteroatoms[a]

Compound	Shift	Solvent Conditions	Ref.
Benzimidazole	−185 ± 5	Acetone (satd)	85
	−192 ± 5	Methanol (satd)	85
	−185 ± 2	0.15 M in acetone	94
	−237 ± 4	0.15 M in DMSO	94
1-methyl	−231 ± 1 (N$_1$)	0.15 M in acetone	94
	−134 ± 1 (N$_3$)		
	−237 ± 6 (N$_1$)	0.10 M in carbon tetrachloride	94
	−127 ± 6 (N$_3$)		
	−131 ± 5 (N$_1$)	Acetone (satd)	86
	−229 ± 5 (N$_3$)		
Benzoxazole	−140 ± 3	Neat	85
Benzothiazole	−60 ± 3	Neat	85
	−61 ± 1	0.15 M in acetone	94
2-methyl	−63 ± 5	Neat	85
Benzo[d]isothiazole	−76 ± 3	Ether (satd)	86
Benzo[c]isothiazole	−121 ± 2	Neat	86
Benzo[d]isoxazole	−27 ± 1	Neat	86
Benzo[c]isoxazole	−8 ± 1	Neat	86
Benzopyrazole	−75 ± 8	Acetone (satd)	85
	−197 ± 5		
1-methyl	−62 ± 3 (N)	Acetone (satd)	86
	−201 ± 2 (N$_1$)		
2-methyl	−85 ± 5 (N)	Acetone (satd)	85
	−162 ± 2 (N$_2$)		
Cinnoline 1-oxide	−58 ± 1 (NO)	Acetone	82
	−[b] (N)		
Cinnoline 2-oxide	−52 ± 1 (NO)	Acetone	82
	−[b] (N)		
Cinnoline N,N'-dioxide	−69 ± 5	Acetone:DMSO 1:1	82
Imidazole	−171 ± 5	Dioxane (satd)	85
	−171 ± 3	Methanol 1:1	85
	−167 ± 3	ca. 40% in DMSO	95
	−171 ± 1	0.15 M in acetone	94
	−177 ± 2	0.25 M in acetone	94
	−168 ± 1	0.15 M in DMSO	94
	−176 ± 1	0.55 M in water	94
1-methyl	−123 ± 1 (N$_3$)	0.15 M in acetone	94
	−217 ± 1 (N$_1$)		
	−218 ± 1 (N$_1$)	0.10 M in carbon tetrachloride	94
	−109 ± 3 (N$_3$)		
	−210 ± 1 (N$_1$)	0.55 M in water	94
	−123 ± 2 (N$_3$)	Neat	85
	−221 ± 1 (N$_1$)		
	−116 ± 3 (N$_3$)	Carbon tetrachloride 1:5	85
	−218 ± 1 (N$_1$)		
1-methyl-4-nitro	−208 ± 5 (N$_1$N$_3$)	ca. 40% in DMSO	95
	−17 ± 2 (NO$_2$)		
2-methyl-4(5)-nitro	−203 ± 5 (N$_1$N$_3$)	ca. 40% in DMSO	95
	−16 ± 2 (NO$_2$)		

Table 2.4. (continued)

Compound	Shift	Solvent Conditions	Ref.
2-methyl-4,5-dinitro	−160 ± 3 (N$_1$N$_3$)	ca. 40% in acetone	95
	−27 ± 1 (NO$_2$)		
4(5)-nitro	−202 ± 5 (N$_1$N$_3$)	ca. 40% in DMSO	95
	−16 ± 2 (NO$_2$)		
4,5-dinitro	−158 ± 3 (N$_1$N$_3$)	ca. 40% in acetone	95
	−28 ± 1 (NO$_2$)		
	−28 ± 2 (NO$_2$)	ca. 40% in DMSO	95
	−^{b} (N$_1$N$_3$)		
Oxazole	−124 ± 1	Carbon tetrachloride	85
	−125 ± 1	Carbon tetrachloride: methanol 1:1:1	85
Isooxazole	+2 ± 1	Neat	85
	+4 ± 2	DMF 1:1	85
	−6 ± 2	Methanol 1:1	85
Phenazine	−90 ± 2 (NO)	Acetone:DMSO 1:4	82
	−^{b} (N)		
	−94 ± 3 (NO)	Dibromomethylene	82
	−^{b} (N)		
N,N'-dioxide	−103 ± 5	Dibromomethylene	82
	−106 ± 5	Hexachloroacetone	82
Phthalazine N-oxide	−66 ± 1 (NO)	Acetone	82
	−^{b} (N)		
	−75 ± 1 (NO)	Methanol	82
	−^{b} (N)		
Pyrimidine	−84 ± 1	0.15 M in acetone	94
	−92 ± 2	0.5 M in water	94
	−86 ± 4	ca. 5% in acetone	96
2-acetoxy	−73 ± 4	ca. 5% in acetone	96
2-amino	−135 ± 4	ca. 5% in acetone	96
2-bromo	−86 ± 4	ca. 5% in acetone	96
2-chloro	−92 ± 4	ca. 5% in acetone	96
2-cyano	−67 ± 4 (N)	ca. 5% in acetone	96
	−^{b} (CN)		
2-fluoro	−133 ± 4	ca. 5% in acetone	96
2-iodo	−69 ± 4	ca. 5% in acetone	96
2-methoxy	−127 ± 4	ca. 5% in acetone	96
2-methoxysulfinyl	−101 ± 4	ca. 5% in acetone	96
2-methylthio	−101 ± 4	ca. 5% in acetone	96
N-oxide	−99 ± 10 (N)	Acetone	82
	−90 ± 1 (NO)		
Pyrazine	−44 ± 1	0.15 M in acetone	94
N-oxide	−77 ± 5 (N)	Acetone	82
	−67 ± 1 (NO)		
	−66 ± 1 (NO)	Chloroform	82
	−^{b} (N)		
	−71 ± 1 (NO)	Methanol	82
	−^{b} (N)		
N,N'-dioxide	−98 ± 5	Acetone:DMSO 1:2	82
	−92 ± 5	Water	82
Pyridazine N-oxide	−54 ± 1 (NO)	Acetone	82
	−^{b} (N)		

180 Nuclear Magnetic Resonance Spectroscopy

Table 2.4. (continued)

Compound	Shift	Solvent Conditions	Ref.
	−65 ± 5 (N)	Chloroform	82
	−53 ± 1 (NO)		
	−70 ± 5 (N)	Methanol	82
	−57 ± 1 (NO)		
Pyrazole	−135 ± 3	Dioxan 1:1	85
	−133 ± 3	Methanol 1:1	85
	−133 ± 1	3 M in chloroform	97
	−129 ± 2	0.15 M in acetone	94
	−134 ± 1	0.55 M in water	94
	−176 ± 1	0.15 M in DMSO	94
1,3-dimethyl	−182 ± 1 (N$_1$)	3 M in chloroform	97
	−78 ± 3 (N$_3$)		
1,5-dimethyl	−183 ± 1 (N$_1$)	3 M in chloroform	97
	−71 ± 3 (N$_3$)		
1-methyl	−174 ± 2 (N$_1$)	0.15 M in acetone	94
	−72 ± 2 (N$_2$)		
	−179 ± 3 (N$_1$)	0.10 M in carbon tetrachloride	94
	−68 ± 2 (N$_2$)		
	−178 ± 2 (N$_1$)	0.55 M in water	94
	−93 ± 3 (N$_2$)		
	−180.8 ± 0.3 (N$_1$)	3 M in chloroform	97
	−72.6 ± 0.8 (N$_2$)		
	−68 ± 2 (N$_2$)	Carbon tetrachloride 1:2	85
	−178 ± 2 (N$_1$)		
	−78 ± 3 (N$_2$)	Methanol:carbon tetrachloride 1:1:2	85
	−178 ± 2 (N$_1$)		
3-methyl	−134 ± 2	3 M in chloroform	97
Pyrazolo[1,5-a]pyridine (3-azaindolizine)	−79.2 ± 1.7 (N$_3$)	0.5 M in ether	87
	−145.4 ± 3 (N$_4$)		
Imidazolo[1,5-a]pyridine (2-azaindolizine)	−141.6 ± 4.2 (N$_2$)	0.5 M in ether	87
	−190.8 ± 0.2 (N$_4$)		
Imidazolo[1,2-a]pyridine	−135.5 ± 1 (N$_1$)	0.5 M in ether	87
	−181.6 ± 0.2 (N$_4$)		
1,2,3-Triazolo[1,5-a]pyridine (2,3-diazaindolizine)	−44.8 ± 3.2 (N$_2$)	0.5 M in ether	87
	+26.7 ± 3.9 (N$_3$)		
	−123.8 ± 0.2 (N$_4$)		
1,2,4-Triazolo[4,3-a]pyridine (1,2-diazaindolizine)	−64.9 ± 4.2 (N$_1$)	0.5 M in ether	87
	−122.0 ± 9 (N$_2$)		
	−189.4 ± 0.2 (N$_4$)		
1,2,4-Triazolo[1,5-a]pyridine (1,3-diazaindolizine)	−153.9 ± 1 (N$_1$)	0.5 M in ether	87
	−98.0 ± 2.2 (N$_3$)		
	−144.6 ± 0.3 (N$_4$)		
Tetrazolo[1,5-a]pyridine (1,2,3-triazaindolizine)	−65.7 ± 1.2 (N$_1$)	Ether:acetone 1:1 (satd)	87
	−74.4 ± 11 (N$_2$)		
	−13.9 ± 2.4 (N$_3$)		
	−133.1 ± 0.2 (N$_4$)		
Quinazoline 3-oxide	−91 ± 2 (N$_3$)	Acetone	82
	−[b] (N)		
	−103 ± 2 (N$_3$)	Methanol	82
	−[b] (N)		

Table 2.4. (continued)

Compound	Shift	Solvent Conditions	Ref.
Quinazoline N,N'-dioxide	−90 ± 5	Acetone:DMSO 1:1	82
	−92 ± 5	Acetone:DMSO 1:2	82
Quinoxaline N-oxide	−82 ± 3 (N)	Acetone	82
	−76 ± 1 (NO)		
	−80 ± 4 (N)	Chloroform	
	−76 ± 1 (NO)		
	−82 ± 1 (NO)	Methanol	82
	−[b] (N)		
Quinoxaline N,N'-dioxide	−102 ± 1	Dibromomethylene	82
	−105 ± 3	DMSO	82
	−108 ± 5	Chloroform	82
Thiazole	−56 ± 1	Neat	94
	−55 ± 1	0.15 M in acetone	94
	−53 ± 2	0.15 M in DMSO	94
	−62 ± 4	0.55 M in water	94
	−56 ± 2	Neat	85
	−68 ± 2	Methanol 1:1	85
Isothiazole	−80 ± 1	Neat	85
	−85 ± 2	Methanol	85

[a] Shifts are from external nitromethane in ppm.
[b] Shifts not reported.

Table 2.5. ^{14}N Spectra of Heterocycles with Three or More Heteroatoms[a]

Compound	Shift	Solvent Conditions	Ref.
Benzo-2,1,3-oxadiazole	+36 ± 2	Ether (satd)	86
	+32 ± 5	Methylene chloride	86
Benzo-1,2,3-thiadiazole	+62 ± 1 (N$_3$)	Ether (satd)	86
	+38 ± 2 (N$_2$)		
Benzo-2,1,3-thiadiazole	−50 ± 1	Ether (satd)	86
Benzotriazole	−81 ± 7	Dioxan (satd)	85
	−89 ± 7	Methanol (satd)	85
1-methyl	−40 ± 8 (N$_2$N$_3$)	Acetone (satd)	85
	−148 ± 5 (N$_1$)		
	−40 ± 5 (N$_3$)	Acetone (satd)	86
	−4 ± 3 (N$_2$)		
	−162 ± 2 (N$_1$)		
2-methyl	−50 ± 8 (N$_1$N$_3$)	Neat	85
	−118 ± 2 (N$_2$)		
	−62 ± 5 (N$_2$N$_3$)	Neat	86
	−119 ± 1 (N$_1$)		
Furoxan	−11 ± 3 (N$_5$)	Acetone 1:1 v/v	98
3,4-dimethyl	−25 ± 1 (N$_2$)		
Benzofuroxan	−5.3 (N$_3$)	Acetone (satd)	98
	−19.0 ± 0.4 (N$_1$)		
	−18.4 ± 2 (NO$_2$)		
Benzodifuroxan	+2 ± 3	Acetone (satd)	98
	−16 ± 5		
	−24 ± 1		
	−23 ± 0.2 (NO$_2$)		

Table 2.5. (continued)

Compound	Shift	Solvent Conditions	Ref.
1,2,3-Triazole	−60 ± 8 (N$_3$N)	Neat	85
	−132 ± 4 (NH)		
	−60 ± 8 (N,N)	Methanol 1:1	85
	−128 ± 6 (NH)		
1-methyl	−22 ± 1 (N$_2$N$_3$)	Neat	85
	−143 ± 1 (N$_1$)		
	−28 ± 1 (N$_2$N$_3$)	Methanol 1:1	85
	−144 ± 1 (NMe)		
2-methyl	−51 ± 1 (N$_1$N$_3$)	Neat	85
	−130 ± 1 (N$_2$)		
	−53 ± 2 (N$_1$N$_3$)	Methanol 1:1	85
	−132 ± 2 (N$_2$)		
1,2,4-Triazole	−135 ± 1	0.15 M in acetone	94
	−132 ± 3 (N$_2$N$_4$)	0.15 M in DMSO	94
	−174 ± 2 (N$_1$)		
	−142 ± 3	0.55 M in water	94
	−134 ± 2	Dioxane (satd)	85
	−136 ± 3	Methanol 1:1	85
1-methyl	−126 ± 3 (N$_2$N$_4$)	Neat	85
	−170 ± 2 (N$_1$)		
	−130 ± 5 (N$_2$N$_4$)	Methanol 1:1	85
	−170 ± 1 (N$_1$)		
	−152 ± 2 (N$_1$)	0.15 M in acetone	94
	−119 ± 2		
	−51 ± 3		
	−170 ± 2 (N$_1$)	0.10 M in carbon tetrachloride	94
	−113 ± 2		
	−42 ± 2		
	−160 ± 4 (N$_1$)	0.55 M in water	94
	−140 ± 2		
	−46 ± 3		
4-methyl	−80 ± 4 (N$_1$N$_2$)	Methanol (satd)	85
	−220 ± 2 (N$_4$)		
Pentazole p-N,N-dimethylaminophenyl	−73 ± 1 (N$_1$)	Methanol:dichloromethane 1:1	100
Tetrazole	−15 ± 3	Acetone (satd)	85
	−106 ± 2		
	−25 ± 5	Methanol (satd)	85
	−106 ± 4		
1-methyl	−17 ± 5 (N,N,N)	Chloroform 1:1	85
	−150 ± 2 (N$_1$)		
2-methyl	−5 ± 6 (N$_3$N$_4$)	Carbon tetrachloride 1:1	85
	−44 ± 3 (N$_1$)		
	−101 ± 1 (NMe)		
	−10 ± 6 (N$_3$N$_4$)	Methanol 2:1	85
	−55 ± 4 (N$_1$)		
	−103 ± 2 (N$_2$Me)		
1,3,5-Triazine	−97 ± 1	0.15 M in acetone	94
	−104 ± 10	0.5 M in water	94
1,2,4-Triazine 1-N-oxide	−43 (NO)	Acetone	83
	−[b] (N)		

Table 2.5. (continued)

Compound	Shift	Solvent Conditions	Ref.
Benzo-1,2,4-triazine 1-N-oxide	−46 (NO) −[b] (N)	Acetone	83
Tetrazole-4,5-trimethylene	−79 ± 3 (N$_1$) −29 ± 4 (N$_2$N$_3$) −11 ± 4 (N$_2$N$_3$) −133 ± 1 (N$_4$)	Water (satd)	99

[a] Shifts are from external nitromethane in ppm.
[b] Shifts not reported.

2.4.2.2. ^{15}N Spectroscopy

In spite of the vast improvements in NMR technology, obtaining ^{15}N spectra still remains one of the most difficult of tasks. Three problem areas can be identified.

1. The very long relaxation times that are likely to be encountered in heterocyclic compounds. For instance, the T_1 relaxation times for indole, imidazole, pyrrole, pyrrolidine, pyridine, and quinoline are 3, 8, 40, 58, 85, and 200 s, respectively (129).
2. The very low sensitivity (Table 2.2), which cannot be overcome by rapid pulse repetition rates as in the case of the fast relaxing nuclei ^{17}O and ^{14}N.
3. The negative magnetogyric ratio, which implies that a negative nuclear Overhauser enhancement factor and the nuclear Overhauser effect may lie between −3.93 (maximum) and +1 (no NOE). Thus there is the distinct possibility that the NOE may be close to zero and the signal is then canceled.

It is not proposed to deal here in any great depth with the various methods that have been used to obtain ^{15}N spectra, and the reader is referred to a recent monograph for a more detailed account (130). Needless to say, reducing the relaxation time and suppressing the nuclear Overhauser effect would be an advantage in ^{15}N spectroscopy. Since the relaxation times of nitrogen nuclei may be very long, "relaxation agents" that are capable of inducing considerably shortened T_1 values without causing line broadening or displacement of chemical shifts are very useful in ^{15}N spectroscopy. The most widely used agent is the tris-acetylacetonate complex of chromium, Cr(acac)$_3$, and concentrations of the order of 0.05 M generally give good results. Fe(acac)$_3$ has also been used, but while it is more efficient it is a less stable compound. Chromium tris-dipivaloylmethane, Cr(dpm)$_3$, is also popular but is somewhat less efficient than Cr(acac)$_3$. The gadolinium complexes of acetylacetone and dipivaloylmethane induce selective decreases in T_1 values of nitrogen nuclei and may be useful in assignment problems. Cr(acac)$_3$ reduces the T_1 values of pyrrolidine, pyridine, and pyrrole to 2–3 s, while the dipivaloylmethane complex reduces the values to 4–5 s (129). It should be noted that the addition of

Table 2.6. ^{15}N Shifts of Heterocycles[a]

Compound	Shift	Solvent Conditions	Ref.
Benzofurazan	+36.3	2 M in Acetone	101
	+29.7	2 M in Trifluoroethanol	101
	+22.3	2 M in TFA	101
Benzofurazan N-oxide (benzofuroxan)	−4.5 (N$_3$)	Acetone 1:1	98, 101
	−18.0 (NO)		
4-NO$_2$	−4.7 (N$_3$)	Acetone 1:1	98
	−19.4 (NO)		
	−19.4 (4-NO$_2$)		
difuroxan	+2.4 (N)	Acetone 1:1	98
	−6.5 (N)		
	−18.8 (NO)		
	−22.4 (NO)		
	−22.4 (NO$_2$)		
trifuroxan	−1.5 (N)	Acetone 1:1	98
	−20.9 (NO)		
Benzothiadiazole	−50.5	2 M in DMSO	101
	−49.6	2 M in Acetone	101
	−60.6	2 M in Trifluoroethanol	101
	−72.3	2 M in TFA	101
Cinnoline	+44.6 (N$_3$)	In DMSO[c]	102
	+41.3 (N$_4$)		
Furan-5-nitro	−27.7	5% in DMSO	84
2-diacetoxymethyl	−28.7	5% in DMSO	84
2-semicarbazido	−28.3	5%iin DMSO	84
diacetate	+5.8 (trans)	5% in DMSO	84
	+7.6 (cis)		
Furazan-3,4-dimethyl	+24.8	20% v/v in Acetone	101
	+18.3	20% v/v in trifluoroethanol	101
	+5.8	20% v/v in TFA	101

Furazan-3,4-dimethyl N-oxide (furoxan-3,4-dimethyl)	−13.2 (N$_5$)	Acetone 1:1	98, 101
	−25.3 (NO)		
	−12.5 (N$_5$)	20% v/v in TFA	101
	−31.5 (NO)		
Imidazole	−177.2 (N$_1$N$_3$)	In water	103
	−208.2	In acid	103
4-acetoxy	−186.2 (N$_1$)	In water	103
	−161.8 (N$_3$)		
	−209.2 (N$_1$)	In acid	103
	−205.9 (N$_3$)		
1-methyl	−218.7 (N$_1$)	In methanol	104
	−134.0 (N$_3$)		
	−217.7 (N$_1$)	In water	103
	−134.7 (N$_3$)		
	−210.3 (N$_1$)	In acid	103
	−209.8 (N$_3$)		
4-methyl	−179.0 (N$_1$)	In water	103
	−170.6 (N$_3$)		
	−208.8 (N$_1$)	In acid	103
	−204.8 (N$_3$)		
Indole	−247.3	2 M in DMSO	105
	−249.0	DMSO (satd)	105
	−253.4	Chloroform (satd)	105
	−259.3	DMSO (satd)	105
3-acetoxy	−237.9	1.6 M in DMSO	105
3-acetyl	−238.4	DMSO (satd)	105
5-amino	−250.4 (N)	1 M in DMSO	105
	−^{b} (NH$_2$)		
	−251.2 (N)	DMSO (satd)	105
	−^{b} (NH$_2$)		
5-bromo	−245.6	1.3 M in DMSO	105

Table 2.6. (continued)

Compound	Shift	Solvent Conditions	Ref.
2-carboxy	−246.6	2 M in DMSO	105
5-carboxy	−243.7	2 M in DMSO	105
	−244.6	DMSO (satd)	105
3-carboxyethyl	−253.0	DMSO (satd)	105
3-carboxymethyl	−250.5	1 M in DMSO	105
	−251.3	DMSO (satd)	105
5-chloro	−256.0	Chloroform (satd)	105
	−245.8	1.4 M in DMSO	105
5-cyano	−242.5 (N)	DMSO (satd)	105
	[b] (CN)		
	−249.3 (N)	Chloroform (satd)	105
	[b] (CN)		
2,3-dimethyl	−247.1	1 M in DMSO	105
2,5-dimethyl	−244.5	1.8 M in DMSO	105
5-fluoro	−257.0	Chloroform (satd)	105
3-formyl	−234.1	2 M in DMSO	105
	−234.3	DMSO (satd)	105
5-hydroxy	−251.3	DMSO (satd)	105
5-methoxy	−248.6	1 M in DMSO	105
	−250.2	DMSO (satd)	105
6-methoxy	−256.7	Chloroform (satd)	105
7-methoxy	−255.7	Chloroform (satd)	105
	−259.6	Chloroform (satd)	105
2-methyl	−243.3	2 M in DMSO	105
	−244.7	DMSO (satd)	105
	−250.6	Chloroform (satd)	105
3-methyl	−261.6	2 M in DMSO	105
	−260.4	Chloroform (satd)	105

5-methyl	−248.1	0.5 M in DMSO	105
	−256.6	Chloroform (satd)	105
7-methyl	−247.9	1 M in DMSO	105
	−257.6	Chloroform (satd)	105
5-nitro	−240.4 (N)	1.5 M in DMSO	105
	−[b] (NO$_2$)		
Phthalozine	−10.3 (N$_2$N$_3$)	In DMSO[c]	102
3 N-oxide	−68.9 (N$_3$)	In DMSO[c]	102
	−53.2 (N$_2$)		
Purine	−128.6 (N$_1$)	1.25 M in water	106
	−113.4 (N$_3$)		
	−189.6 (N$_7$)		
	−185.8 (N$_9$)		
	−100.7 (N$_1$)	1 M in DMSO	107
	−119.8 (N$_3$)		
	−193.2 (N$_7$)		
	−193.2 (N$_9$)		
Pyrimidine	−84.8 (N$_1$N$_3$)	0.5 M in DMSO[c]	102
	−134.8 (N$_1$N$_3$)	0.5 M in TFA[c]	102
	−84.8 (N$_1$N$_3$)	0.5 M in DMSO[c]	108
N-oxide	−90.0 (N$_1$)	In DMSO[c]	102
	−80.3 (N$_3$)		
	−89.9 (N$_1$)	In chloroform[c]	102
	−79.5 (N$_3$)		
2-amino	−129.9 (N$_1$N$_3$)	0.5 M in DMSO[c]	102
	−297.9 (2-NH$_2$)		
	−178.8 (N$_1$N$_3$)	0.5 M in TFA[c]	102
	−294.3 (2-NH$_2$)		
	−126.3 (N$_1$)	4.5 M in DMSO	109
	−126.4 (N$_3$)		
	−293.0 (2-NH$_2$)		

Table 2.6. (continued)

Compound	Shift	Solvent Conditions	Ref.
2-amino 1-oxide	−134.0 (N₁) −130.1 (N₃) −304.8 (2-NH₂)	In DMSO^c	102
	−231.5 (N₁) −146.3 (N₃) −305.5 (2-NH₂)	In water^c	102
2-amino-4,6-dichloro	−141.5 (N₁N₃) −292.0 (NH₂)	0.5 M in DMSO^c	108
2-amino-4,6-dimethoxy	−180.2 (N₁N₃) −296.4 (NH₂)	0.5 M in DMSO^c	108
2-amino-4,6-dimethyl	−138.0 (N₁N₃) −300.4 (NH₂)	0.5 M in DMSO^c	108
2-amino-4-methyl	−138.2 (N₁) −130.3 (N₃) −299.5 (NH₂)	0.5 M in DMSO^c	108
4-amino-5-(3,4,5-trimethoxybenzyl)	−120.9 (N₁) −132.7 (N₃) −298.4 (NH₂)	0.5 M in DMSO^c	102
	−217.9 (N₁) −218.6 (N₃) −278.3 (NH₂)	0.5 M in TFA^c	102
2-chloro	−88.2 (N₁N₃)	0.5 M in DMSO^c	102
2-chloro-5-nitro	−91.3 (N₁N₃) −17.6 (NO₂)	0.5 M in DMSO^c	108
2,4-diamino	−160.1 (N₁) −169.2 (N₃) −295.4 (2-NH₂) −294.1 (4-NH₂)	4.5 M in DMSO	109

	— 164.5 (N$_1$)	
	— 173.4 (N$_3$)	0.5 M in DMSO^c	102
	— 301.6 (2-NH$_2$)	
	— 299.6 (4-NH$_2$)	
	— 168.2 (N$_1$)	
	— 175.4 (N$_3$)	4.5 M in water (pH 11)	109
	— 300.5 (2-NH$_2$)	
	— 296.1 (4-NH$_2$)	
	— 253.5 (N$_1$)	
	— 231.6 (N$_3$)	0.5 M in TFA^c	102
	— 296.4 (2-NH$_2$)	
	— 276.2 (4-NH$_2$)	
2,4-diamino 3-oxide	— 166.8 (N$_1$)	
	— 167.8 (N$_3$)	In DMSO^c	102
	— 306.2 (2-NH$_2$)	
	— 307.4 (4-NH$_2$)	
2,4-diamino N-oxide	— 162.4 (N$_1$)	
	— 163.4 (N$_3$)	4.5 M in DMSO	109
	— 301.4 (2-NH$_2$)	
	— 300.1 (4-NH$_2$)	
2,4-diamino-6-chloro	— 166.1 (N$_1$)	
	— 178.7 (N$_3$)	0.5 M in DMSO^c	102
	— 297.7 (2-NH$_2$)	
	— 296.6 (4-NH$_2$)	
	— 227.4 (N$_1$)	
	— 229.1 (N$_3$)	0.5 M in TFA^c	102
	— 297.9 (2-NH$_2$)	
	— 283.5 (4-NH$_2$)	
2,4-diamino-5-(4′-chlorophenyl)-6-ethyl	— 176.3 (N$_1$N$_3$)	
	— 300.9 (2-NH$_2$)	0.5 M in DMSO^c	102
	— 299.4 (4-NH$_2$)	

Table 2.6. (continued)

Compound	Shift	Solvent Conditions	Ref.
2,4-diaminocyclohexyl	−247.5 (N_1N_3) −294.5 (2-NH_2) −274.9 (4-NH_2)	0.5 M in TFA[c]	102
	−174.1 (N_1N_3) −305.1 (2-NH_2) −302.6 (4-NH_2)	0.5 M in DMSO[c]	102
	−247.1 (N_1) −245.1 (N_3) −296.4 (2-NH_2) −278.7 (4-NH_2)	0.5 M in TFA[c]	102
2,4-diamino-5-(methoxymethyl) 3-oxide	−164.8 (N_1) −169.2 (N_3) −308.1 (2-NH_2) −311.1 (4-NH_2)	In DMSO[c]	102
2,4-diamino-5-(3,4,5-trimethoxybenzyl)	−163.0 (N_1) −174.4 (N_3) −304.6 (2-NH_2) −302.3 (4-NH_2)	0.5 M in DMSO[c]	102
	−156.1 (N_1) −168.4 (N_3) −299.2 (2-NH_2) [b] (4-NH_2)	0.4 M in DMSO	109
2,4-diamino-5-(3′,4′,5′-trimethoxybenzyl) N-oxide	−160.5 (N_1) −164.2 (N_3) −304.6 (2-NH_2) −303.9 (4-NH_2)	4.5 M in DMSO	109
2,4-diamino-5-(3,4,5-trimethoxybenzyl) 3-oxide	−164.9 (N_1) −168.8 (N_3) −308.6 (2-NH_2) −309.5 (4-NH_2)	In DMSO[c]	102

4,5-diamino	—132.6 (N₁)	0.5 M in DMSO^c	102
	—133.5 (N₃)		
	—305.9 (4-NH₂)		
	—338.0 (5-NH₂)		
	—215.1 (N₁N₃)	0.5 M in TFA^c	102
	—282.3 (4-NH₂)		
	—335.8 (5-NH₂)		
4,6-diamino	—149.4 (N₁N₃)	0.5 M in DMSO^c	102
	—309.1 (4,6-NH₂)		
	—210.3 (N₁N₃)	0.5 M in TFA^c	102
	—289.0 (4,6-NH₂)		
4,6-diamino-5-(4'-chlorophenyl)	—151.8 (N₁N₃)	0.5 M in DMSO^c	102
	—304.5 (4,6-NH₂)		
	—209.9 (N₁N₃)	0.5 M in TFA^c	102
	—297.2 (4,6-NH₂)		
2,4-dichloro	—92.6 (N₁)	0.5 M in DMSO^c	108
	—94.5 (N₃)		
4,6-dichloro	—93.7 (N₁N₃)	0.5 M in DMSO^c	108
2-dimethylamino	—132.0 (N₁N₃)	0.5 M in DMSO^c	102
	—311.9 (NMe₂)		
2,5-dimethyl-4-amino	—121.3 (N₁)	0.5 M in DMSO^c	108
	—134.0 (N₃)		
	—299.6 (NH₂)		
2,6-dimethyl-4-amino	—121.0 (N₁)	0.5 M in DMSO^c	108
	—139.2 (N₃)		
	—299.0 (NH₂)		
2,4-dimethyl-5-ethoxycarbonyl	—94.5 (N₁)	0.5 M in DMSO^c	108
	—85.4 (N₃)		
2,4-dimethoxy	—150.0 (N₁)	0.5 M in DMSO^c	108
	—160.9 (N₃)		

Table 2.6. (continued)

Compound	Shift	Solvent Conditions	Ref.
2-ethyl-4-amino-5-cyano	−108.8 (N$_1$) −121.7 (N$_3$) −288.3 (NH$_2$) −134.7 (CN)	0.5 M in DMSO^c	108
1-methyl-2-aminoiodide	−229.2 (N$_1$) −115.7 (N$_3$) −283.3 (NH$_2$)	0.5 M in DMSO^c	102
1-methyl-2-imino	−246.9 (N$_1$) −98.6 (N$_3$) −194.4 (NH)	0.5 M in DMSO^c	102
2-methyl-4-amino-5-cyano	−108.9 (N$_1$) −120.2 (N$_3$) −288.3 (NH$_2$) −134.2 (CN)	0.5 M in DMSO^c	108
2-methyl-5-phenyl	−87.5 (N$_1$N$_3$)	0.5 M in DMSO^c	108
2-methylthio-4-amino-5-ethoxycarbonyl	−128.8 (N$_1$) −142.0 (N$_3$) −289.4 (NH$_2$)	0.5 M in DMSO^c	108
2-methylthio-4-chloro	−100.5 (N$_1$) −102.4 (N$_3$)	0.5 M in DMSO^c	108
4-methyl	−93.1 (N$_1$) −84.7 (N$_3$)	0.5 M in DMSO^c	108
5-methyl	−85.9 (N$_1$N$_3$)	0.5 M in DMSO^c	108
2-piperidyl-4,6-diamino	— −190.2 (N$_1$N$_3$) −308.2 (2-N) −306.2 (4,6-NH$_2$)	0.5 M in DMSO^c	102
	−266.1 (N$_1$N$_3$) −295.8 (2-N) −293.3 (4,6-NH$_2$)	0.5 M in TFA^c	102

2-phenyl-4-amino-5-(3,4,5-trimethoxybenzyl)	—127.8 (N$_1$) —141.3 (N$_3$) —297.8 (NH$_2$)	0.5 M in DMSO[c]	102
	—227.6 (N$_1$N$_3$) —281.2 (NH$_2$)	0.5 M in TFA[c]	102
4-phenyl	—89.2 (N$_1$) —94.5 (N$_3$)	0.5 M in DMSO[c]	108
2,4,5-triamino	—[b] (N$_1$N$_3$) —311.2 (2-NH$_2$) —316.3 (4-NH$_2$) —357.3 (5-NH$_2$)	0.5 M in DMSO[c]	102
2,4,5-triamino-1-hydrochloride	—218.8 (N$_1$N$_3$) —310.1 (2-NH$_2$) —309.0 (4-NH$_2$) —357.3 (5-NH$_2$)	0.5 M in DMSO[c]	102
2,4,5-triamino-6-benzylmethyl	—173.0 (N$_1$) —180.2 (N$_3$) —303.4 (2-NH$_2$) —311.5 (4-NH$_2$) —345.9 (5-NH$_2$)	0.5 M in DMSO[c]	102
2,4,5-triamino-6-benzylmethyl-1-hydrochloride	—260.8 (N$_1$N$_3$) —303.1 (2-NH$_2$) —279.4 (4-NH$_2$) —336.1 (5-NH$_2$)	0.5 M in DMSO[c]	102
	—253.3 (N$_1$) —226.3 (N$_3$) —299.0 (2-NH$_2$) —279.0 (4-NH$_2$)	0.5 M in TFA[c]	102
2,4,6-triamino	—189.5 (N$_1$N$_3$) —304.0 (2-NH$_2$) —306.0 (4,6-NH$_2$)	0.5 M in DMSO[c]	102

Table 2.6. (continued)

Compound	Shift	Solvent Conditions	Ref.
2,4,6-triamino-5-(3,4,5-trimethoxybenzyl)	−261.7 (N$_1$N$_3$) −291.3 (2,4,6-NH$_2$)	0.5 M in TFA^c	102
	−191.6 (N$_1$N$_3$) −305.5 (2-NH$_2$) −306.4 (4,6-NH$_2$)	0.5 M in DMSO^c	102
	−261.5 (N$_1$N$_3$) −297.5 (2-NH$_2$) −295.9 (4,6-NH$_2$)	0.5 M in TFA^c	102
4,5,6-triamino	−152.7 (N$_1$N$_3$) −309.8 (4,6-NH$_2$) −346.9 (5-NH$_2$)	0.5 M in DMSO^c	102
4,5,6-triamino-1-hydrochloride	−181.7 (N$_1$N$_3$) −301.5 (4,6-NH$_2$) −343.2 (5-NH$_2$)	0.5 M in DMSO^c	102
Pyrazine	−46.3	In DMSO^c	102
N-oxide	−75.7	0.5 M in DMSO^c	102
	−70.4 −75.2	0.5 M in TFA^c	102
	−69.1		
Pyridazine	+20.3 −55.1 (N$_1$) −33.6 (N$_2$)	In DMSO^c In DMSO^c	102 102
N-oxide	−54.7 (N$_1$) −32.8 (N$_2$)	In CHCl$_3$^{c}	102
Pyridine	−63.5	Neat^c	104
	−63.0	0.5 M in DMSO^c	108, 102
	−63.8	1 M in DMSO	110
	−90.2	1 M in trifluoroethanol	110

N-oxide	−179.0	1 M in TFA	110
	−182.5	0.5 M in TFA[c]	102
	−168.6	TFA 2:1	111
	−86.8	2 M in DMSO	110
	−87.5	In DMSO[c]	111
	−99.5	2 M in trifluoroethanol	110
	−135.7	2 M TFA	110
2-acetamido	−73.5 (N$_1$)	0.5 M in DMSO[c]	108
	−282.0 (NH$_2$)		
3-acetamido	−64.5 (N$_1$)	0.5 M in DMSO[c]	108
	−277.1 (NH$_2$)		
4-acetamido	−56.4 (N$_1$)	0.5 M in DMSO[c]	108
	−275.9 (NH$_2$)		
2-acetyl	−65.7	0.5 M in DMSO[c]	108
2-amino	−116.0 (N$_1$)	1 M in DMSO	107
	−307.8 (NH$_2$)		
	−113.8 (N$_1$)	0.5 M in DMSO[c]	102
	−307.3 (NH$_2$)		
	−226.0 (N$_1$)	0.5 M in TFA[c]	102
	−305.6 (NH$_2$)		
2-amino-4,6-dimethyl	−118.7 (N$_1$)	0.5 M in DMSO[c]	108
	−308.7 (NH$_2$)		
2-amino-4-methyl	−119.3 (N$_1$)	0.5 M in DMSO[c]	108
	−308.1 (NH$_2$)		
2-amino-6-methyl	−112.4 (N$_1$)	0.5 M in DMSO[c]	108
	−307.4 (NH$_2$)		
2-amino-5-nitro	−117.8 (N$_1$)	0.5 M in DMSO[c]	108
	−11.4 (NO$_2$)		
	−287.6 (NH$_2$)		
3-amino	−63.9 (N$_1$)	0.5 M in DMSO[c]	102
	−325.3 (NH$_2$)		

Table 2.6. (continued)

Compound	Shift	Solvent Conditions	Ref.
4-amino	−184.7 (N$_1$)	0.5 M in TFA^c	102
	−325.2 (NH$_2$)		
	−107.2 (N$_1$)	1 M in DMSO	107
	−312.8 (NH$_2$)		
	−103.7 (N$_1$)	0.5 M in DMSO^c	102
	−312.0 (NH$_2$)		
	−220.7 (N$_1$)	0.5 M in TFA^c	102
	−293.0 (NH$_2$)		
4-acetyl	−51.8	Neat	104
	−64.3	In methanol	104
3-acetoxy	−63.2	0.5 M in DMSO^c	108
2-benzyl	−64.3	0.5 M in DMSO^c	108
4-benzyl	−69.7	0.5 M in DMSO^c	108
2-bromo	−64.3	0.5 M in DMSO^c	108
3-bromo	−56.7	0.5 M in DMSO^c	108
4-bromo	−67.4	0.5 M in DMSO^c	108
2-carboxy	−65.4	0.5 M in DMSO^c	108
3-carboxy	−64.4	0.5 M in DMSO^c	108
4-carboxy	−52.0	0.5 M in DMSO^c	108
2-chloro	−72.0	0.5 M in DMSO^c	108
2-chloro-5-nitro	−70.6 (N$_1$)	0.5 M in DMSO^c	108
	−14.6 (NO$_2$)		
3-chloro	−57.1	0.5 M in DMSO^c	108
4-chloro	−67.5	0.5 M in DMSO^c	108
N-oxide	−106.2	1 M in trifluoroethanol	110
	−141.7	1 M in TFA	
2-cyano	−62.2 (N$_1$)	0.5 M in DMSO^c	108
	−126.2 (CN)		

2,3-diamino	— 114.5 (N$_1$)	0.5 M in DMSO^c	102
	— 313.5 (2-NH$_2$)		
	— 330.4 (3-NH$_2$)		
	— 219.4 (N$_1$)	0.5 M in TFA^c	102
	— 305.0 (2-NH$_2$)		
	— 337.0 (3-NH$_2$)		
2,6-diamino	— 149.0 (N$_1$)	0.5 M in DMSO^c	102
	— 309.1 (2-NH$_2$)		
	— 309.1 (6-NH$_2$)		
	— 239.3 (N$_1$)	0.5 M in TFA^c	102
	— 312.3 (2-NH$_2$)		
	— 312.3 (6-NH$_2$)		
3,4-diamino	— 99.3 (N$_1$)	0.5 M in DMSO^c	102
	— 336.9 (3-NH$_2$)		
	— 322.0 (4-NH$_2$)		
	— 216.9 (N$_1$)	0.5 M in TFA^c	102
	— 339.0 (3-NH$_2$)		
	— 290.4 (4-NH$_2$)		
2,6-dichloro	— 80.6	0.5 M in DMSO^c	108
2,3-dimethyl	— 62.3	Neat^c	111
	— 173.1	TFA 2:1	111
2,4-dimethyl	— 71.0	Neat^c	111
	— 179.0	TFA 2:1	111
2,5-dimethyl	— 62.7	Neat^c	111
	— 168.6	TFA 2:1	111
2,6-dimethyl	— 62.4	Neat^c	111
	— 171.7	TFA 2:1	111
N-oxide	— 92.9	2 M in DMSO	110
	— 107.8	2 M in trifluoroethanol	110
	— 143.1	2 M in TFA	110
3,4-dimethyl	— 68.8	Neat^c	111
	— 175.6	TFA 2:1	111

Table 2.6. (continued)

Compound	Shift	Solvent Conditions	Ref.
3,5-dimethyl	−61.7	Neat[c]	111
	−174.1	TFA 2:1	111
2,6-di-*tert*-butyl	−70.4	Neat[c]	111
2-ethyl	−64.0	Neat[c]	111
3-ethyl	−61.4	0.5 M in DMSO[c]	108
4-ethyl	−65.5	Neat[c]	111
3-ethoxycarbonyl	−61.5	0.5 M in DMSO[c]	108
2-ethylene	−71.0	0.5 M in DMSO[c]	108
4-ethylene	−65.1	0.5 M in DMSO[c]	108
2-formyl	−60.4	0.5 M in DMSO[c]	108
3-formyl	−63.3	0.5 M in DMSO[c]	108
4-formyl	−47.8	0.5 M in DMSO[c]	108
2-isopropyl	−67.3	Neat[c]	111
4-isopropyl	−64.8	Neat[c]	111
4-methoxy	−86.6	Neat[c]	104
	−103.3	In methanol[c]	104
4-methoxy N-oxide	−106.4	2 M in DMSO	110
	−126.2	2 M in trifluoroethanol	110
	−161.5	2 M in TFA	110
1-methyl iodide	−179.1	DMSO[c]	102
2-methyl	−62.6	Neat[c]	111
	−167.5	TFA 2:1	111
2-methyl N-oxide	−90.9	2 M in DMSO	110
	−89.6	In DMSO[c]	111
	−103.8	2 M in trifluoroethanol	110
	−141.8	2 M in TFA	110
	−102.7	2 M in water	111
3-methyl	−61.7	Neat[c]	111
	−178.4	TFA 2:1	111

198

3-methyl N-oxide	−86.9	2 M in DMSO	110
	−87.5	In DMSO[c]	111
	−105.2	2 M in trifluoroethanol	110
	−141.8	2 M in TFA	111
	−101.0	2 M in water	110
3-methyl-4-nitro N-oxide	−77.1 (N$_1$)	1 M in DMSO	110
	−13.8 (NO$_2$)		
	−90.4 (N$_1$)	1 M in trifluoroethanol	110
	−16.6 (NO$_2$)		
	−122.0 (N$_1$)	1 M in TFA	110
	−20.1 (NO$_2$)		
4-methyl	−70.2	Neat[c]	111
	−181.5	TFA 2:1	111
4-methyl N-oxide	−96.6	2 M in DMSO	110
	−96.3	In DMSO[c]	111
	−106.8	2 M in trifluoroethanol	110
	−146.0	2 M in TFA	110
	−110.2	2 M in water	111
2-(N-methylpyrrolidine)	−63.6 (N$_1$)	Neat	112
	−330.1		
3-(N-methylpyrrolidine)	−60.9 (N$_1$)	Neat	112
	−327.6		
4-(N-methylpyrrolidine)	−64.5 (N$_1$)	Neat	112
	−329.4		
4-nitro N-oxide	−74.4 (N$_1$)	1 M in DMSO	110
	−17.1 (NO$_2$)		
	−85.8 (N$_1$)	1 M in trifluoroethanol	110
	−20.2 (NO$_2$)		
	−112.9 (N$_1$)	1 M in TFA	110
	−24.2 (NO$_2$)		
2-phenyl	−71.2	0.5 M in DMSO[c]	108

Table 2.6. (continued)

Compound	Shift	Solvent Conditions	Ref.
4-phenyl	−67.7	0.5 M in DMSO[c]	108
4-phenyl N-oxide	−110.4	1 M in trifluoroethanol	110
	−145.9	1 M in TFA	110
2-tert-butyl	−64.7	Neat[c]	111
4-tert-butyl	−64.7	Neat[c]	111
2,4,6-tri-tert-butyl	−76.4	Benzene[c]	111
2,4,6-trimethyl	−68.0	Neat[c]	111
Pyrrole	−230.6	Neat	113
2-acetyl	−226.4	Acetone	114
2,4-dinitro	−18.3 (2-NO$_2$)	ca. 30% in water	95
	−25.7 (4-NO$_2$)		
	−b (N)		
2,5-dinitro	−25.5 (2,5-NO$_2$)	ca. 30% in water	95
	−b (N)		
2-formyl	−229.7	Acetone	114
1-methyl-2-carboxy	−226.1	Acetone	114
1-methyl-2-carboxy 4-NO$_2$	−223.0 (N)	Acetone	114
	−b (NO$_2$)		
1-methyl-2-carboxy 5-NO$_2$	−230.4 (N)	Acetone	114
	−b (NO$_2$)		
2-nitro	−23.1 (NO$_2$)	ca. 30% in water	95
	−b (N)		
3-nitro	−13.5 (NO$_2$)	ca. 30% in water	95
	−b (N)		
Quinazoline	−85.5 (N$_1$)	In DMSO[c]	102
	−96.9 (N$_3$)		
	−87.0 (N$_1$)	1 M in DMSO	107
	−98.4 (N$_3$)		

Compound	Shift	Solvent	Ref
2-oxide	−92.9 (N₁)	4.5 M in DMSO	109
	−81.8 (N₃)		
Quinaxoline	−89.5 (N₁N₃)	In DMSO	102
	−49.8 (N₁N₄)	In DMSO[c]	102
4-oxide	−76.8 (N₁)	In DMSO[c]	102
	−80.7 (N₄)		
Quinoline N-oxide	−101.5	1 M in DMSO	110
	−112.8	1 M in trifluoroethanol	110
	−150.3	1 M in TFA	110
Quinoline N-oxide-8-hydroxy	−111.0	1 M in DMSO	110
	−116.1	1 M in trifluoroethanol	110
	−144.3	1 M in TFA	110
1,3,5-Thiadiazine-bis-anilino	−147.7 (N₃)	DMSO	116
	−211.6 (N₅)		
	−277.5 (2-N′)		
	−280.9 (4-N′)		
1,3,5-Thiadiazine-4-anilino-2-methylanilino	−151.2 (N₃)	Chloroform	116
	−209.8 (N₅)		
	−283.3 (2-N′)		
	−291.0 (4-N′)		
1,3,5-Thiadiazine-2,3-dihydro-4-anilino-2-imino-3-phenyl	−230.3 (N₃)	DMSO	116
	−233.3 (N₅)		
	−183.8 (2-N′)		
	−288.1 (4-N′)		
1,3,5-Thiadiazine-2,3-dihydro-4-anilino-2-(N-methyl-thioformadino) imino	−218.1 (N₃)	DMSO	116
	−199.6 (N₅)		
	−151.3 (2-N′)		
	−251.4 (2-NMe)		
	−288.1 (4-N′)		
1,3,5-Thiadiazine-2,3-dihydro-4-anilino-2-oxo-3-phenyl	−216.9 (N₃)	DMSO	116
	−229.8 (N₅)		
	−287.3 (4-N′)		

Table 2.6. (continued)

Compound	Shift	Solvent Conditions	Ref.
1,3,5-Thiadiazinetetrahydro-4-anilino-3,5-dimethyl-2-phenylimino	−255.8 (N$_3$) −330.2 (N$_5$) −146.6 (2-N′) −186.7 (4-N′)	Chloroform	116
Uracil	−247.8 (N$_1$) −220.2 (N$_3$)	0.8 M in DMSO	106
	−246.4 (N$_1$) −218.9 (N$_3$)	In DMSO	115
1-methyl	−220.1 (N$_3$) −^{b} (N$_1$)	In DMSO	115
	−219.5 (N$_3$) −^{b} (N$_1$)	In water	115
3-methyl	−244.9 (N$_1$) −^{b} (N$_3$)	In water	115

^{a} Shifts in ppm from external nitromethane.
^{b} Shifts not reported.
^{c} 0.1 M Cr(acac)$_3$ added.

relaxation agents is likely to introduce chemical shift variations of up to 1.5 ppm, and if accurate changes in chemical shift values are required relaxation agents should be used with caution. These changes occur not only through intermolecular interaction with the agent but also through magnetic susceptibility variations, since an external reference must be used. Moreover, such reagents invariably act as quenchers for the nuclear Overhauser enhancement, and there is some risk of canceling the nitrogen signal altogether.

Suppression of the nuclear Overhauser effect is normally achieved by the use of gated decoupling techniques, and many modern spectrometers usually contain facilities for doing so. Unfortunately, such pulse sequences require long time delays between pulses and complete suppression of the effect is not always achieved. The effect may be ignored if fully coupled spectra are acquired, but in this case sensitivity is lost and a long time is required to obtain a spectrum.

Attention is now centered on methods of spin polarization transfer techniques in which the more favorable Boltzmann distribution of hydrogen may be transferred to the less favored distribution in nitrogen. Theoretically, the increase in sensitivity to be gained by this technique is very large. Such methods as selective population transfer (131), J-cross polarization (132), and the INEPT sequence (133) are now becoming well known. Disadvantages of spin polarization transfer methods are the need for complex pulse sequences and the requirement that the transmitter be capable of reproducing very stable pulses.

In spite of the difficulties encountered in ^{15}N spectroscopy, a great number of heterocyclic compounds have been examined and their chemical shifts are listed in Table 2.6. Chemical shifts are in parts per million from external nitromethane, and wherever possible concentrations and solvents are given.

2.4.3. Oxygen Spectroscopy

Of the naturally occurring isotopes of oxygen, only one, ^{17}O, has a magnetic moment. This ^{17}O isotope has a very low natural abundance (0.037%), low sensitivity (approximately one-tenth that of ^{13}C), a negative magnetogyric ratio, and a quadrupole moment of moderate strength. The nuclear spin number is high, $I = \frac{5}{2}$. The high spin number is a decided advantage as linewidth is inversely proportional to I (all other factors being equal) (117). Hence, one might expect narrow resonance lines for ^{17}O compared to ^{14}N with a nuclear spin number of $I = 1$ (all other factors being euqal). The high spin number also implies that quadrupolar relaxation mechanisms dominate and this very efficient method of relaxation leads to very short relaxation times. Relaxation times for ^{17}O nuclei are normally of the order of milliseconds, which allows for fast pulse repetition rates and thus tends to compensate somewhat for the low sensitivity of the nucleus. As the quadrupolar relaxation mechanism dominates, little if any nuclear Overhauser effect is present and the negative magnetogyric ratio is not a problem. Linewidths vary considerably, ranging from some few tens of hertz to over 1 kHz. The range of observed ^{17}O chemical shifts is approximately 800 ppm, although for a given class of compound this is somewhat less than 100 ppm. Comments similar to those made for nitrogen

Table 2.7. Variation in Chemical Shift and Linewidth of the ^{17}O Spectra of N-Methylacetamide with Solvent[a]

Solvent	δ (ppm)[b]	ν(Hz)
Nitromethane	313.5	132
Acetonitrile	316.7	136
Acetone	317.5	136
Tetrahydrofuran	319.9	148
Methanol	331.1	168
2 N DCl	260.4	192
Water	282.9	208
Dioxane (30°C)	315.2	280
(50°C)	307.6	200
Diethylcarbonate	316.7	296
Benzene	307.0	368
Trifluoroethanol	378.1	480
Dimethylsulfoxide	308.6	520
Chloroform	301.2	572
Chloroform plus formamide (satd)	308.6	415

[a] 20% w/v concentration at 30°C except for dioxane.
[b] ppm from external neat H$_2$O.

also apply to oxygen, although in the case of ^{17}O both linewidth and chemical shift show a marked dependence on temperature, solvent, concentration, and pH (118, 119). As data on the behavior of heterocycles are not available, the effects of solvent and concentration on ^{17}O chemical shifts and linewidths are illustrated for some simple amides in a variety of solvents by Tables 2.7 and 2.8 and Figures 2.10 and 2.11. The effect of solvent choice (Table 2.7) is quite remarkable. A fourfold range in linewidth and a 30% variation in chemical shift is possible at a given concentration of 20% w/v. The effect appears to be quite general for amides (Table 2.8), and there is no reason to believe that a similar situation does not exist with heterocyclic ^{17}O resonances. The "best" solvents appear to be acetone, acetonitrile, nitromethane, and tetrahydrofuran, while the common NMR solvents

Table 2.8. ^{17}O Chemical Shifts and Linewidths of Some Simple Amides in Acetone and Chloroform[a]

	Chloroform		Acetone	
Compound	δ (ppm)[b]	ν(Hz)	δ (ppm)[b]	ν(Hz)
Formamide	325.5	115	320.7	40
N-Methylformamide	306.2	244	317.5	64
N,N'-Dimethylformamide	310.2	160	323.9	48
Acetamide	322.9	116	328.7	104
N-Methylacetamide	301.2	572	317.5	136
N,N'-Dimethylacetamide	331.1	160	340.8	64
Tetramethylurea	[c]		279.7	160

[a] 20% w/v concentration at 30°C.
[b] ppm from external neat H$_2$O.
[c] Not observed even after several million pulses.

Multinuclear Magnetic Resonance 205

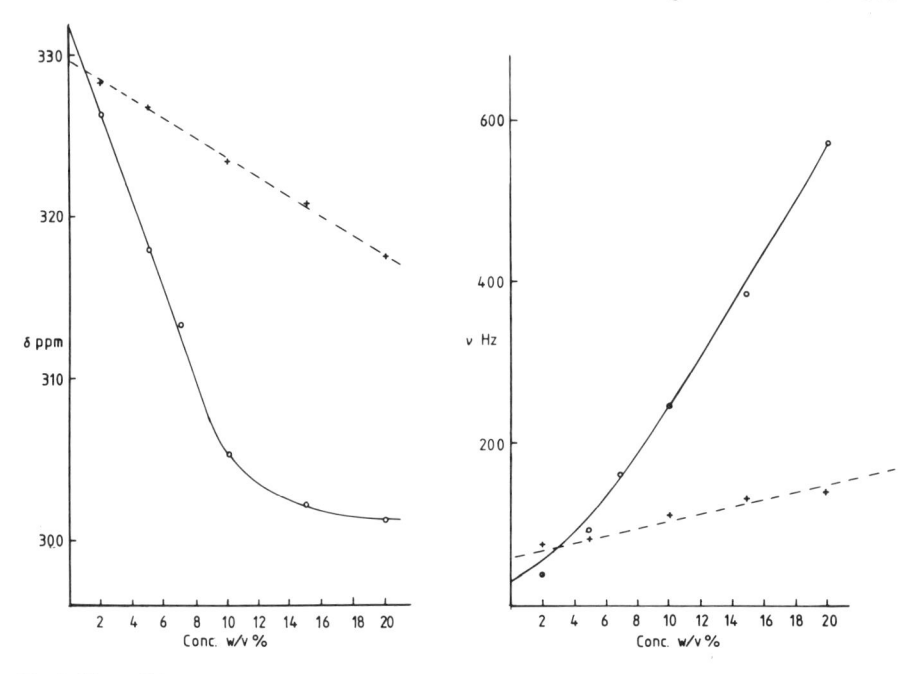

Fig. 2.10. ^{17}O NMR of N-methylacetamide in CdCl$_3$ (o) and acetone (+) showing variation of chemical shift and linewidth with concentration.

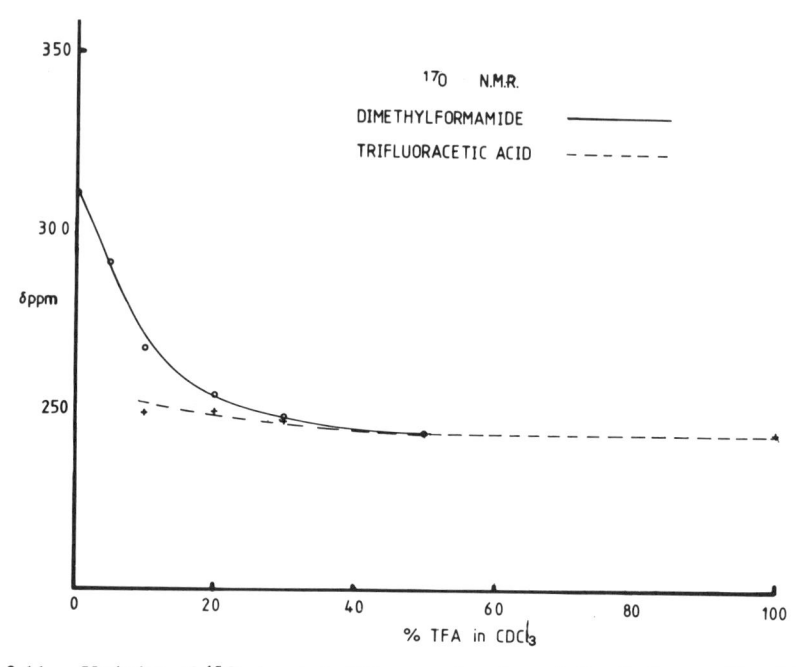

Fig. 2.11. Variation of ^{17}O chemical shift on protonation of dimethylformamide with trifluoroacetic acid.

Table 2.9. ^{17}O Shifts of Oxygen Heterocyclic and Some Related Compounds[a]

Compound	Shift	Solvent Conditions	Ref.
1,3-Dioxolane	+28	Neat	120
	+34	Chloroform 1:1	121
	+34.8	Neat	122
	+33	Neat	123
	+35.5	Neat	124
2,2-dimethyl	+64.2	Neat	122
	+61.0	Neat	124
	+61	Neat	123
2-methyl	+53	Neat	123, 124
	+56.0	Neat	122
4-methyl	+61.9	Neat	122
	+36.9		
1,3-Dioxane	+37	Chloroform 1:1	121
	+38	Neat	121
	+35.3	Neat	122
2,2-dimethyl	+51.8	Neat	122
cis-2,4-dimethyl	+48.8	Neat	122
	+78.3		
trans-2,4-dimethyl	+47.4	Neat	122
	+63.6		
4,4-dimethyl	+30.9	Neat	122
	+69.3		
cis-4,6-dimethyl	+58.0	Neat	122
trans-4,6-dimethyl	+52.4	Neat	122
5,5-dimethyl	+31.1	Neat	122
2-ethyl	+48.8	Neat	122
2-isopropyl	+46.2	Neat	122
2-methyl	+52.5	Neat	122
4-methyl	+32.6	Neat	122
	+61.7		

5-methyl	+ 35.5	Neat	122
cis-5-methyl-2-isopropyl	+ 34.9	Neat	122
trans-5-methyl-2-isopropyl	+ 46.7	Neat	122
2,2,4,6-pentamethyl	73.6	Neat	122
	82.8		
2-tert-butyl	+ 42.5	Neat	122
2,2,5,5-tetramethyl	+ 47.7	Neat	122
2,2,4-trimethyl	+ 45.5	Neat	122
	+ 76.5		
2,5,5-trimethyl	+ 47.7	Neat	122
r-2,cis-4,cis-6-trimethyl	76.4	Neat	122
r-2,trans-4,trans-6-trimethyl	+ 63.3	Neat	122
1,4-Dioxane	− 1.7	Neat	122
	− 6	Neat	123
	0	Chloroform 1:1	121
Furan	+ 228.5	Neat	124
	+ 228	Neat	123
	240	Neat	121
	+ 241	Neat	125
2-acetyl	+ 240.0	Neat	124
dibenzyl	+ 158.0	Neat	124, 123
1-formyl	+ 237	Neat	125
	+ 234.5	Neat	124
2-formylmethyl	+ 228	Neat	123
2,5-dihydro	237	Neat	121
	− 7.0	Neat	124
2,5-dimethoxy	26.5	Neat	124
tetrahydro	+ 16.2	Neat	122
	+ 18.0	Neat; chloroform 1:1	121
2-methyl	+ 43.9	Neat	122
3-methyl	+ 15.5	Neat	122
2,5-dimethyl	+ 66.7	Neat	122
	+ 74.5		

Table 2.9. (continued)

Compound	Shift	Solvent Conditions	Ref.
Furazan, dimethyl	+460		125
Furoxan, dimethyl	+350		125
4-Methyl-sydnone	+387 (O$_2$)	5 M in benzene	126
	+232 (O$_1$)		
Morpholine	+2.6	Neat	122
2,6-dimethyl	+49.4	Neat	122
Pyridine-3-formyl	+613	Neat	120
1,4-Pyran-dibenzyl	+100	Neat	123, 124
Pyran-tetrahydro	+8.8	Neat	122
	10	Chloroform 1:1	121
Pyran-2,3-dihydro	+53.5	Neat	124
2-methyl	+33.6	Neat	122
3-methyl	+10.3	Neat	122
4-methyl	+7.7	Neat	122
Pyrrole-2-nitro	+520 ± 10	Ethyl alcohol (satd)	95
	+405 ± 10	Water	95
Pyrrole-1-methyl-2-nitro	+536 ± 10	Ethyl alcohol (satd)	95
Pyrolidine-2-oxo-1-methyl (N-methyl-2-pyrolidone)	+286	5 M in chloroform	126
1,4-Oxathiane	+5.9	Neat	122
Oxathianone	+365 (CO)	Neat	127
	+159 (O)		
Oxepane	14	Neat	121
Oxetane	−12	Neat	121
1,3,5-Trioxane	+65	Chloroform	121
Uridine	+248 ± 7	Water	128
isopropylidine	+255 ± 9	Water	128
	+263 ± 4	Acetonitrile	128
3-methyl	+253 ± 9	Water	128
Vinylene carbonate	+189 (O)	Neat	120
	+228 (CO)		

[a] Shifts are in ppm from external neat water.

chloroform and dimethylsulfoxide are very poor indeed. No less disturbing are the large variations in chemical shifts experienced on dilution (Fig. 2.10). The ^{17}O resonance for N-methylacetamide shows a downfield movement of nearly 30 ppm on dilution in chloroform from 20% w/v to 2% w/v. Linewidths narrow considerably on dilution. In the same solvent the linewidth for N-methylacetamide is 35 Hz at 2% w/v but close to 600 Hz at 20% w/v. One must point out that line narrowing effectively improves the signal-to-noise ratio, and hence the spectrum for 2% N-methylacetamide in chloroform was obtained in a much shorter time than that at 20% w/v. It may be that the reported problems in obtaining ^{17}O spectra are due in many cases to workers attempting to obtain spectra on neat solutions or in very concentrated solutions using an unsuitable solvent. Note that protonation (Fig. 2.11) shows an upfield shift in ^{17}O resonances, indicating greater shielding, which is the reverse of what one might expect from ^{1}H-NMR.

In view of the comments made above and the fact that most ^{17}O chemical shifts for heterocyclic compounds have been obtained on neat solutions or very concentrated solutions, the chemical shifts listed in Table 2.9 must be treated with caution.

2.4.4. Sulfur Spectroscopy

Sulfur has one isotope, ^{33}S, that possesses a nuclear magnetic moment. In spite of ^{33}S having a natural abundance some 30 times that of ^{17}O and a quadrupole moment of the same order, it remains one of the least known of the observable nuclei. In fact, remarkably little work has been done and the chemical shifts for only a few compounds are known. The main problem appears to be the extraordinarily large linewidths that occur. Linewidths may be as large as 5 kHz (which corresponds to 1100 ppm), and this poses a problem as the chemical shift range for sulfur appears to be of the order of only 500 ppm. From what few data are available, it appears that highly symmetrical sulfur environments lead to very narrow lines, while asymmetry causes excessive broadening. No investigation of the dependence of linewidth on concentration has been reported. The few heterocycles that have been examined by ^{33}S-NMR spectroscopy are listed in Table 2.10.

Table 2.10. ^{33}S Chemical Shifts of Heterocyclic Compounds

Compound	Shift	Solvent Conditions	Ref.
Thiophene	−115.2	90% in carbon disulfide	1
2-bromo	−201.8	Neat	
2-methyl	−157.8	Neat	1
3-methyl	−138.8	Neat	1
tetrahydro	−424.8	Neat	1
Tetramethylenesulfone	+42 ± 1.5	DMSO	2
3-methyl	+37 ± 1.5	DMSO	2
3-amino	+33 ± 1.5	DMSO	2
3-hydroxy	+36 ± 1.5	DMSO	2
2,4-dimethyl	+37 ± 1.5	DMSO	2
3,4-dehydro	+32 ± 1.5	DMSO	2

2.5. REFERENCE SOLVENTS

The choice of a suitable reference system is one of the most vexatious problems contronting the NMR spectroscopist. With the advent of modern frequency-measuring equipment, it is possible, of course, simply to quote the absolute frequency of the absorption line and the strength of the magnetic field at which it was measured. Indeed, for nuclei where the resonances occupy a wide frequency range, for example ^{195}Pt, this method has been found to be acceptable (63). However, one must remember that in spite of the high precision with which a frequency may be measured, the absolute accuracy of that value depends ultimately on the accuracy of the calibration of the frequency-measuring equipment and the stability of that calibration with time. For, say, the measurement of proton resonances accurate to 0.1 Hz at a field of 2.35 T, the frequency equipment would need to have an absolute accuracy of one part in 10^9 and be stable with time. Such conditions are not easy to achieve.

These problems may be overcome by using an arbitrary reference substance dissolved in the sample and referring all chemical shift displacements in resonance to this internal reference. Ideally, the reference substance in the solution under investigation should have the following properties:

1. It must be chemically inert, interacting with neither the solvent nor the sample.
2. It must be soluble in a wide variety of solvents so as to facilitate comparison of spectra.
3. It must be magnetically isotropic.
4. It should give a single sharp resonance well removed from the resonances of the sample.
5. It should be volatile in order to facilitate the recovery of samples.

Such conditions are in practice impossible to achieve, and one is left with two alternatives: (1) to choose a substance that approximates the above conditions or (2) to use an external reference, a system where the reference compound is physically separated from the sample but still in the magnetic field.

An external reference is advantageous in eliminating problems arising from intermolecular interactions or chemical reaction with the solvent or compound under investigation. Solubility problems are also avoided. There is, however, a serious difficulty arising from the difference in bulk magnetic susceptibility between sample and reference. In all molecules with completely paired electrons, the motion of the electrons in a magnetic field is such as to make the compound diamagnetic. The magnetism per unit volume, $\mathbf{M}$, induced by a field $\mathbf{B_0}$ in the sample is given by

$$\mathbf{M} = K\mathbf{B_0} \qquad (25)$$

where K is the volume magnetic susceptibility and is negative, that is, repellent to the field for all diamagnetic materials. The volume magnetic susceptibility for the solution K_s will be a weighted average of the values for solvent and sample. But the volume magnetic susceptibility of the reference K_r will not be the same as that for the solution. As a result molecules in the solution will experience a slightly different field to those in the reference, and a correction has to be made to the observed chemical shift due to the differences in volume magnetic susceptibilities. This correction takes the form

$$\delta_{int} = \delta_{obs} + f(K_r - K_s) \qquad (26)$$

where δ_{int} is the intrinsic chemical shift, δ_{obs} is the observed chemical shift, and f is a form factor that depends not only on the geometry of the sample but also the orientation of the sample geometry to the magnetic field. If the sample is spherical, then $f = 0$ and no correction for magnetic susceptibilities is required (134). Thus in theory spherical reference cells may be used to avoid susceptibility errors. However, in practice such cells are generally inconvenient and imperfections in the glass wall may introduce spurious results (135).

For the more usual coaxial tube arrangement the shape factor is different depending on whether the magnetic field is collinear with the tube as in superconducting magnets or orthogonal to the tube as in electromagnets and permanent magnets. The correction factor, it should be noted, is independent of nuclei and magnetic field strength. For superconducting magnets (135),

$$\delta_{int} = \delta_{obs} + \frac{4\pi}{3}(K_r - K_s) \qquad (27)$$

and for electromagnets and permanent magnets (136),

$$\delta_{int} = \delta_{obs} - \frac{2\pi}{3}(K_r - K_s) \qquad (28)$$

where volume magnetic susceptibilities are given in cgs units. In principle, volume magnetic susceptibilities may be obtained by measurement of chemical shifts at high field with a superconducting magnet and at low field using an electromagnet. Substitution of the observed shifts into equations (27) and (28) will enable an unknown volume magnetic susceptibility to be calculated (137).

Caution should be exercised when using tables of volume magnetic susceptibilities as some tables give values in SI units, not cgs units. The conversion factor between cgs units and SI units is $K_{SI} = 4\pi K_{cgs}$. Volume magnetic susceptibilities for a range of solvents commonly used in NMR spectroscopy are given in Table 2.11. Note that the magnetic susceptibilities quoted are for protic solvents, not deuterated solvents and a small discrepancy in $K_r - K_s$ will exist. This discrepancy is unlikely to be greater than 15% and some correction may be made by use of Pascal's constants (138).

As an example of the magnitude of shifts induced by external references, consider the common NMR solvents chloroform and nitromethane. If volume magnetic susceptibilities for these solvents are substituted into equations (27) and (28), then the correction will be -0.73 ppm for an electromagnet and $+1.46$ ppm for a superconducting magnet. These corrections are not negligible, and it should be noted that working with a high field superconducting magnet introduces corrections that are not only twice as large as those for electromagnets but also opposite in

Table 2.11. Physical Properties of Some Common NMR Solvents

Solvent	MP	BP	Dielectric Constant	Magnetic Susceptibility $\times 10^6$	Chemical Shift ^{1}H	Chemical Shift ^{13}C
Acetic Acid	16.7	117.9	6.1	.551	2.1	21.1
						177.3
Acetone	−94.7	56.3	20.7	.460	2.2	30.2
						205.1
Acetonitrile	−44	81.6	37.5	.534	2.0	0.3
						117.2
Benzene	5.5	80.1	2.3	.699	7.4	128.7
Carbon disulfide	−111.6	46.2	2.6	.532	—	192.8
Carbon tetrachloride	−23	76.7	2.2	.691	—	96.7
Chloroform	−63.5	61.1	4.8	.740	7.3	77.7
Cyclohexane	6.6	80.7	2.0	.627	1.4	27.8
Cyclopentane	−93.8	49.3	2.0	.629	1.5	26.5
1,2-Dichloroethane	−35.7	83.5	10.4		3.7	51.7
Dichloromethane	−95.1	39.8	8.9	.733	5.3	54.2
N,N-Dimethylformamide	−60.4	153.0	36.7		2.9	31.36
					8.0	161.7
Dimethylsulfoxide	18.5	189.0	46.7		2.6	43.5
Dioxane	11.8	101.3	2.2	.606	3.7	67.8
Ethanol	−114.1	78.3	24.5	.575	1.2	17.9
					3.7	57.3
Ethyl acetate	−83.9	77.1	6.0	.554	1.2	14.3
						60.1
					4.1	170.4
Formamide	2.6	210.5	109	.551	7.2	165.1
					8.1	

Hexachloroacetone	−30	203			123.7
					126.4
Hexamethylphosphoramide[a]	7.2	233	30.0	2.4	36.6
				2.6	
Methanol	−97.7	64.7	32.7	.530	3.5	49.3
Methyl chloride	−97.7	−24.1	12.6			25.1
Nitromethane	−28.5	101.2	35.9	.391	4.3	57.3
Pyridine	−41.6	115.3	12.4	.611	7.1–8.8	124–150
Tetrachloroethane	−22.3	121.2	2.3	.802		120.4
Tetrahydrofuran	−66	66.0	7.6		1.8	26.7
					3.7	68.6
Trifluoroacetic acid	−15	72	39.5		11.61	
Toluene	−94.9	110.6	2.4	.618	2.3	21.3
					7.2	125–138
Water (H$_2$O)	0	100	78.5	.719	≃4.8	

[a] Potent carcinogen.

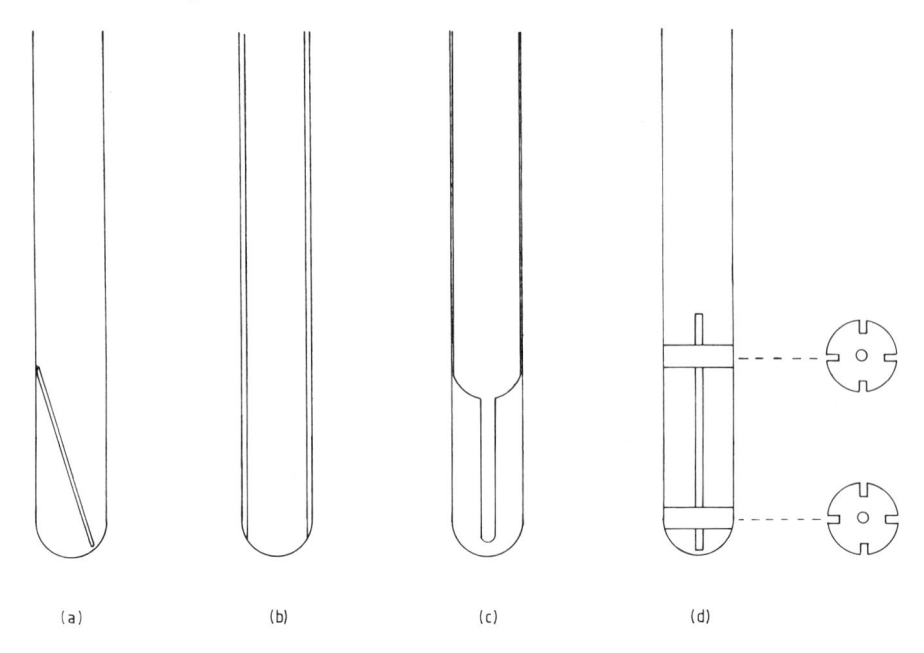

Fig. 2.12. Methods for insertion of an external reference. (*a*) Simple method using a sealed capillary; (*b*) coaxial tube arrangement; (*c*) commercially available insert; (*d*) arrangement using machined spacers and capillary tube.

sign. This may account for some of the discrepancies in chemical shifts that occur in the literature.

Unfortunately, the shape factor is very critical and doubt exists as to the accuracy of some volume magnetic susceptibility values. Also, it appears difficult to obtain reproducible results when the cells are changed (139). Caution must therefore be exercised when interpreting spectra obtained using an external reference.

Various tube configurations suitable for the use of external references are illustrated in Figure 2.12.

The simplest method, Figure 2.12*a*, is to drop a small sealed capillary containing the reference into the NMR tube, This method is satisfactory for rough approximations, but it may be found that spinning sidebands are intolerably large and, as a rigid geometry cannot be maintained, equations (27) and (28) cannot be applied.

In method (*b*) two NMR tubes, one of a slightly smaller diameter than the other, are sealed at the base and a thin film of reference is held in the outer cavity. Although this arrangement ensures a constant geometry for the cell, it is fragile and difficult to clean.

In the arrangement shown in Figure 2.12(*c*) the reference capillary is attached to the bottom of a close-fitting insert. This method is very popular and is available commercially (140). In order to avoid flexing of the capillary and excessive fragility, the glass walls of the design are thick. The insert occupies a large amount of space in the NMR tube, thus reducing signal-to-noise ratios by a significant amount.

Method (*d*) is used extensively in this laboratory. Two snugly fitting spacers are machined out of Teflon or a similarly inert material and drilled centrally to hold a small capillary rigidly. This method avoids the disadvantages of method (*c*) by making use of a small (1 mm) capillary, and the Teflon spacers keep the geometry rigid. The assembly is easily inserted and removed by use of a hooked rod.

2.5.1. Recommended Reference Solvents

2.5.1.1. ^1H and ^{13}C

The accepted reference for both ^1H and ^{13}C-NMR spectroscopy is tetramethylsilane (TMS) (141), which has stood the test of time remarkably well. This material has several advantages: the 12 magnetically equivalent protons in the molecule give rise to a single intense sharp peak for both ^1H and ^{13}C that is well removed from resonances normally found for organic molecules, thus avoiding overlapping of signals; it has a high degree of magnetic isotropy in most solvents; it is relatively inert chemically and miscible with or soluble in a wide range of solvents; and it has high volatility, enabling easy recovery of sample.

The high volatility of TMS is sometimes a problem, and it is generally advisable to store TMS in a refrigerator. The volatility may also be troublesome for work at high temperatures, for its signal may become weak due to evaporation into the vapor space of the tube (14). Some concentration and temperature dependence have been reported for TMS, but this appears to be minor and may be ignored except in high precision work (142). The main disadvantage of TMS is its very low solubility in aqueous solutions. For aqueous solutions the accepted reference is the sodium salt of 3-(trimethylsilyl) propane sulfonic acid (TPS) (143). TPS has several disadvantages. It gives rise to absorptions in a region of the spectrum that is frequently of interest, appears to be sensitive to the presence of aromatic compounds (144), and is not entirely above suspicion of undergoing shifts due to molecular association and changing ionic atmospheres (4). TPS also has a distinct disadvantage in that it cannot be removed easily from the sample and in this laboratory it is customary to use a more volatile secondary reference for aqueous solutions. Preferred secondary references for aqueous solutions and also high temperature where TMS would evaporate are dioxane ($\delta = 3.68$) and acetonitrile ($\delta = 2.0$) (145).

The absorption signal of water itself is also a useful secondary reference in the pH range 2–12. Outside this range the HDO signal is strongly dependent on pH and is also temperature dependent.

2.5.1.2. ^{15}N and ^{14}N

As solvent effects on nitrogen chemical shifts are very pronounced, internal references are not recommended. External references are used exclusively in ^{15}N and ^{14}N NMR spectroscopy. Although complete agreement on a suitable choice for reference compounds has not been reached, it is generally accepted that neat

Table 2.12. References for Nitrogen Chemical Shifts

Reference Solvent	Concentration and solvent	Conversion factor	Ref.
CH_3NO_2	Neat	0	79
	0.3 M in benzene	− 4.4	79
	0.3 M in DMSO	+ 2.0	79
	0.3 M in $CHCl_3$	− 3.8	79
HNO_3	1 M in D_2O	− 4.4	79
	9 M in H_2O	− 14.4	80
	10 M in H_2O	− 18.2	79
	15 M in H_2O	− 31.3	79
NH_3	Liquid	− 381.9	79
$NaNO_3$	1 M in D_2O	− 4.2	130
	2 M in 1 M HNO_3	− 1.5	130
	2 M in 2 M HNO_3	− 3.1	130
	7.9 M in H_2O	− 3.7	79
$NaNO_2$	0.3 M in H_2O	+ 227.6	79
KNO_3	0.3 M in H_2O	− 3.5	79
KNO_2	H_2O[a]	+ 237.1	146
$NH_4\underline{N}O_3$	5 M in 2 M HNO_3	− 4.6	79
	5 M in 2 M HCl	− 5.2	79
	12.3 M in H_2O	− 4.0	79
$\underline{N}H_4NO_3$	1 M HNO_3[a]	− 358.6	80
	2 M HNO_3		147
	5 M in 2 M HCl	− 358.0	79
	12.3 M in H_2O	− 359.5	79
	15 M in 2 M HCl	− 357.4	148
NH_4Cl	5.6 M in H_2O	− 352.9	79
	5.6 M in 1 M HCl	− 355.3	149
	5 M in 2 M HCl	− 352.5	79
	2.9 M in 1 M HCl	− 351.8	150
	2 M in 1 M HCl	− 355.3	80
$N(Me)_4Cl$	6.03 M in H_2O	− 336.7	79
	12 M in H_2O	− 336.7	151
$N(Et_4)Cl$	11 M in H_2O	− 316	152
	4.6 M in H_2O	− 315.8	79
$N(Me)_4I$	0.3 M in H_2O	− 337.3	79

[a] Concentration not given.

nitromethane should be used in the interim as the primary reference for both ^{15}N and ^{14}N NMR spectroscopy (79, 80). Due to the low sensitivity of the ^{15}N nucleus it is advisable to use enriched material for the reference compound. Here using an external reference is an advantage, as it will not be necessary to recover expensive labeled material from samples as would occur in using an internal reference. The necessity for using ^{15}N-enriched compounds for references has led to some problems in ^{15}N-NMR spectroscopy. A wide range of compounds have been used over the years as reference compounds. As ^{15}N chemical shifts are more often than not dependent on solvent, pH, and concentration, direct comparison of shifts is difficult. Undoubtedly, choice of reference has been influenced by the availability of ^{15}N-enriched compounds. For example, in this laboratory $2 M$ NH$_4$Cl in $1 M$ DCl is used as a reference simply because ^{15}N-enriched ammonium chloride is available. Table 2.12 lists compounds that have been used as references with conversion factors to the nitromethane scale. The effects of solvent and concentration on nitrogen shifts are very apparent from the table. For instance, the concentration dependence of HNO$_3$ spreads over 25 ppm, while that of NH$_4$Cl is somewhat better at 6 ppm. The author cannot emphasize too strongly the effects of solvent and concentration on nitrogen chemcial shifts and the need for care when measuring nitrogen chemical shifts.

2.5.1.3. ^{17}O

Remarks made in the preceding section on ^{15}N and ^{14}N reference solvents with respect to concentration and solvent dependence also apply to ^{17}O NMR spectroscopy. Internal references are not recommended, and external references should be used. There is no accepted primay reference for ^{17}O chemical shifts. Neat water or neat acetone is recommended. ^{17}O-enriched water is available commercially, and acetone is easily enriched by exchange, enabling small quantities of reference to give good signals. Nitromethane has also been used as a reference (153). The conversion between the chemical shift scales based on nitromethane or acetone and the scale based on water are as follows:

$$\delta_{H_2O} = \delta_{CH_3NO_2} + 605 \text{ ppm}$$
$$\delta_{H_2O} = \delta_{acetone} + 569 \text{ ppm}$$

2.5.1.4. ^{33}S

Remarkably little work has been done on ^{33}S-NMR spectroscopy, so it is no surprise that a suitable reference has not been recommended. Neat carbon disulfide has been used as a reference, and cesium sulfate has been proposed as a suitable reference material in view of its small temperature and concentration dependence (154). In this laboratory $4 M$ NH$_4$SO$_4$ in D$_2$O is used. This gives a very sharp resonance compared to neat carbon disulfide, and the high concentration enables a spectrum to be obtained in a short time. It is recommended that workers intending to obtain ^{33}S spectra use an external reference of $4 M$ ammonium sulfate. The conversion between the carbon disulfide scale and the ammonium sulfate scale is (155)

$$\delta_{(NH_4)_2SO_4} = \delta_{CS_2} - 335 \text{ ppm}$$

2.6. SAMPLE TEMPERATURE MEASUREMENT

The accurate determination of sample temperature may be of great importance in NMR spectroscopy. Where solutions of high ionic strength are being investigated or high decoupling power is used, the sample temperature may be increased to a substantial degree above that of the probe. Such increases in sample temperature may cause problems where the sample is temperature sensitive. Variable-temperature NMR spectroscopy is used widely for measurements of rates and activation parameters of chemical reactions and rotational barriers. The accuracy of the method is often limited by the lack of precision in measuring sample temperature. Knowledge of an exact sample temperature is also significant when investigating nitrogen and oxygen nuclear magnetic resonance where chemical shifts are strongly temperature dependent. Thus accurate sample temperature measurement is of considerable importance in NMR spectroscopy.

The simplest method is to place a mercury NMR thermometer in the sample. However, this method is very inaccurate, as radiofrequency heating effects invariably result in a reading several degrees higher than that of the sample (156).

Van Geet (157) has proposed the insertion of a thermocouple or thermistor in the sample. In practice, it is difficult to obtain accurate results with this method. The sample cannot be spun, and the large temperature gradients that occur in the sample make placement of the sensing element critical. Further complications arise from the fact that the insertion of the sensing element results in broadening of spectral lines (158).

A more reliable method is the use of melting point reference compounds, which are placed in capillaries coaxial with the sample tube (159). No spectrum of the reference compound is obtained when the compound is solid, but on liquefaction a spectrum is easily obtained. This method is very accurate but suffers the disadvantage of allowing calibration at predetermined fixed points only.

The most popular method is the use of a secondary standard, that is, the application of thermometric shift solutions that allow temperature measurements to be made simultaneously with spectral acquisition. The thermometric shift solution is first calibrated using one of the primary methods outlined above, and a curve of temperature against shift is drawn. The solution is then placed in a small capillary and inserted in the sample tube by one of the methods described by the use of the external references.

For ^{1}H-NMR measurements, two thermometric solutions are recommended, one for temperatures in the range -100 to $+60°$C and one for the range 40–140°C. For the lower temperature range, methanol to which a trace of concentration HCl (0.03% v/v) has been added is recommended (160, 161). The addition of concentrated acid is necessary so as to cause complete collapse of the multiplet structure, and sharp resonance lines are obtained at all temperatures. The difference in chemical shifts between the methanol hydroxyl and methyl protons is temperature dependent. For the higher range, ethylene glycol containing a trace of concentrated acid (0.03% w/v conc. HCl) is recommended (160–162). For ^{13}C-NMR measurements, several thermometric probes have been proposed. Vidrine and Peterson (163) have proposed a 1:3 v/v mixture of methyl iodide and tetramethylsilane for the range -60 to $+20°$C and a 1:5 v/v mixture of diiodomethane and cyclooctane for the range 20–100°C. Both mixtures show temperature-dependent shifts suitable for temperature calibration. Led and Petersen have used a 1:1

Table 2.13. Thermometric Shift Solutions for Temperature Calibration

Nucleus	Temp. Range (°C)	Solution	Shift	Ref.
^{1}H	−100 to +60	Methanol−0.03% v/v conc. HCl	$T = 156.2 - 62.26\Delta\delta - 13.85208(\Delta\delta)^2$	160
	+40 to +140	Ethylene glycol−0.03% v/v conc. HCl	$T = 193.5 - 101.42\Delta\delta$	160
	−70 to +40	$(CH_3)_2CO/Yb(fod)_3$	$T = 1264\Delta\delta^{-1} - 745\Delta\delta^{-2} - 184.4$	165
	−130 to +30	$C_6H_5CH_3-C_6D_5CD_3-CHFCl_2$ 2:1:2	Not given	166
^{13}C	−60 to +20	CH_3I−TMS 3:1 v/v	$T^{-1} = 0.0161165 - 0.000570057\Delta\delta$	163
	+20 to +100	CH_2I_2−cyclooctane 5:1 v/v	$T^{-1} = 0.0220027 - 0.000223362\Delta\delta$	163
	−90 to +90	$(CD_3)_2CO-CCl_4$ 1:1 v/v	$T = 5529.3 - 50.7342\Delta\delta$	164
	−70 to +40	$(CH_3)_2CO/Yb(fod)_3$	$T = 4505\Delta\delta^{-1} + 4640\Delta\delta^{-2} - 202.7$	165
^{19}F	−130 to +30	$C_6H_5CH_3-C_6D_5CD_3-CHFCl_2$ 2:1:2	Not given	166
^{31}P	−130 to +30	$C_6H_5CH_3-C_6D_5CD_3-CHFCl_2$ 2:1:2	Not given	166
	−90 to +90	0.1 M OP$(C_6H_5)_3$−0.1 M P$(C_6H_5)_3$ toluene-d_8	$T = 937.244 - 30.929\Delta\delta$	167

mixture of deuteroacetone and carbon tetrachloride (164). However, the temperature dependence of the ^{13}C=O shift is small, and this system would be useful only at high fields. Schneider and coworkers (165) made use of the known temperature dependence of lanthanide-induced pseudo-contact shifts to develop a chemical shift thermometer consisting of the complex of deuteroacetone and ytterbium(III) (fod)$_3$ shift reagent. The authors reported that a mixture of 736 mg Yb (fod)$_3$, 783 mg (CD$_3$)$_2$CO, 46 mg (CH$_3$)$_2$CO, 2.646 g CS$_2$, 5.74 g CFCl$_3$, and 1.23 g TMS gave a shift of 15 ppm for ^{13}C=O and 5 ppm for ^{1}H over 100°C. Another thermometric solution that has been proposed consists of perdeuterotoluene, hexafluorobenzene, and fluorodichloromethane in the ratio 2:1:2 (166). This system may also be employed for ^{1}H and ^{19}F nuclear magnetic resonance.

Mention should also be made of an indirect method developed by Smolenaers and Beattie (156). In this method ^{1}H spectra of a capillary of acidified methanol are acquired at various times after the completion of ^{13}C spectrum acquisition, and the methanol chemical shift is extrapolated back to zero time. The accuracy of the method is purported to be ±1°C.

No thermometric solutions have been proposed for either ^{15}N or ^{17}O NMR measurements. However, the indirect method of Smolenaers and Beattie may prove useful for these nuclei.

Finally, a thermometric solution of P(C$_6$H$_5$)$_3$ and OP(C$_6$H$_5$)$_3$ in deuterotoluene has been proposed for ^{31}P measurements (167) that gives a satisfactory change in chemical shift per degree (1.3 Hz at 40 MHz).

Table 2.13 summarizes the various methods with shifts in ppm and temperatures in °C.

2.7. REDUCTION OF NOISE

The four nuclei ^{14}N, ^{15}N, ^{17}O, and ^{33}S that have been examined in this chapter are all of the very low sensitivity group, and it is of interest to detail two methods that are used in this laboratory to improve signal-to-noise ratios. The first deals with a method of eliminating coherent noise and has application to fast-relaxing nuclei such as ^{17}O and ^{14}N. For nuclei with long free induction decays, coherent noise and the consequent baseline role are not a problem, and the second method based on decoupling techniques is of use for low-sensitivity nonquadrupolar nuclei such as ^{15}N.

2.7.1. Reduction of Baseline Roll

As explained in Section 2.3, difficulty is encountered when free induction decays are of the same order in intensity and magnitude as that obtained from coherent noise. Coherent noise or coil ringing is a combination of two factors, oscillation of energy within the tuned circuit that forms part of the receiver system of the spectrometer and the occurrence of an acoustic standing wave within the probe. In a well-designed probe, acoustic ringing may be small but still be of sufficient intensity

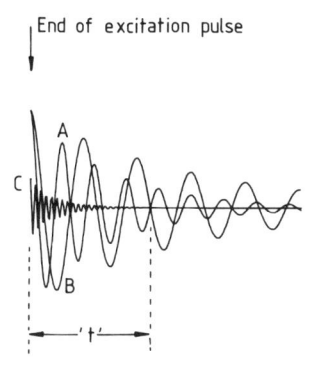

Fig. 2.13. Representation of fast-relaxing free induction decays and probe ringing immediately after cessation of excitation pulse. *A* and *B* represent free induction decays and *C* decaying oscillation due to probe ringing.

to modulate the electronic ringing of the receiver coil and will undoubtedly cause some variation in phase and amplitude of the composite ringing decay pattern. The problem of ringing is illustrated schematically in Figure 2.13, where *A* and *B* represent two fast-decaying free induction decays and *C* represents coil ringing of a similar intensity and decay. If a time to frequency transform using Fourier analysis were applied to Figure 2.13, the ringing pattern would manifest itself as the phenemenon known as baseline roll.

Where the expected resonances are sharp, a situation where the free induction decays are several orders of magnitude larger than that of coil ringing, no trouble is experienced in distinguishing real peaks from baseline roll. However, when free induction decays are of the same order as that from coil ringing, as for broad resonances, baseline roll may completely submerge the resonances and render measurement of chemcial shifts impossible. A common procedure to overcome baseline role is to program the microprocessor controlling the spectrometer transmitter and data acquisition system so as to leave a time t between the termination of the excitation pulse and the beginning of acquisition of data in such a manner that coil ringing may decay to zero intensity before acquisition of data commences (120). Similar results may be obtained by using a trapezoidal window during Fourier transform analysis, setting the trapezoidal window to zero until coil ringing has ceased. These two methods have several disadvantages. It is evident from Figure 2.13 that advantage cannot be taken of the full amplitude of the free induction decay and signal-to-noise ratio or the experiment will suffer. Furthermore, if the two free induction decays decay at different rates as illustrated in Figure 2.13, the phase of one relative to the other will vary with time. At the particular "dead time" t chosen, the free induction decays are 180 degrees out of phase and a spectrum similar to that illustrated in Figure 2.14*a* would be obtained. A further complication is for very fast free induction decays whose lifetimes are smaller than t. In this case no free induction decays would be obtained on acquisition of data.

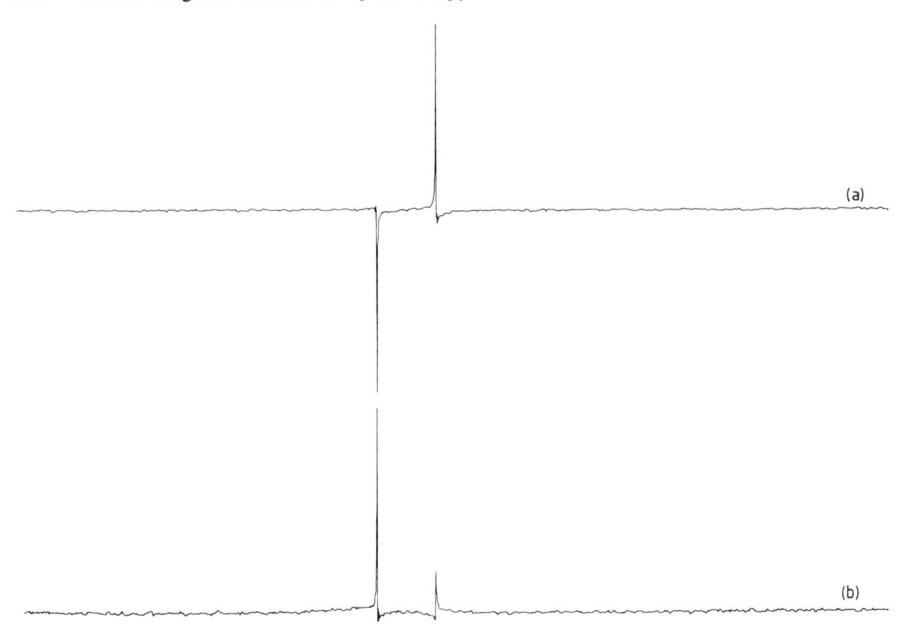

Fig. 2.14. ^{49,47}Ti spectra of titanium tetrachloride. (a) Spectrum obtained using deadtime method; (b) Spectrum obtained using method described in text. Note that phasing problems occur in (a) but not (b).

The following method has been developed in this laboratory to overcome the above disadvantages. If a 90-degree pulse is applied with no delay, the acquired data will contain all free induction decays plus coil ringing. But if a 180-degree pulse is applied, then only coil ringing is obtained. Subtraction of the 180-degree pulse from the 90-degree pulse should result in free induction decays without the complication of coil ringing. Unfortunately, if a 90 degree minus 180 degree pulse sequence is used, no substantial reduction in baseline roll is evident. Recent evidence (168) suggests that a slight change in phase and amplitude of the composite ringing pattern is responsible. Thus one needs to subtract a block of time-averaged 180-degree free induction decays from a block of 90-degree free induction decays in order to overcome the change in phase and amplitude. The results from such an experiment are illustrated in Figure 2.15. Figure 2.15a is the ^{17}O spectrum of a time-averaged block of one hundred 90-degree pulse free induction decays with no delay for 60% acetone in benzene. Baseline roll is apparent. Figure 2.15b shows the effect of subtracting a block of one hundred 180-degree pulse free induction decays. Baseline roll has been reduced substantially. Note that the difference in signal-to-noise ratios is in the ratio of $1:\sqrt{2}$ as predicted from Section 2.3, equation (12). No decrease in signal-to-noise was evident between the method outlined above and the method of Canet and coworkers (120).

Figure 2.14b illustrates the second advantage of this method. Both resonances are now in phase and it is very evident from comparison of Figures 2.14a and b that the free induction decay of one resonance is very much shorter than the other.

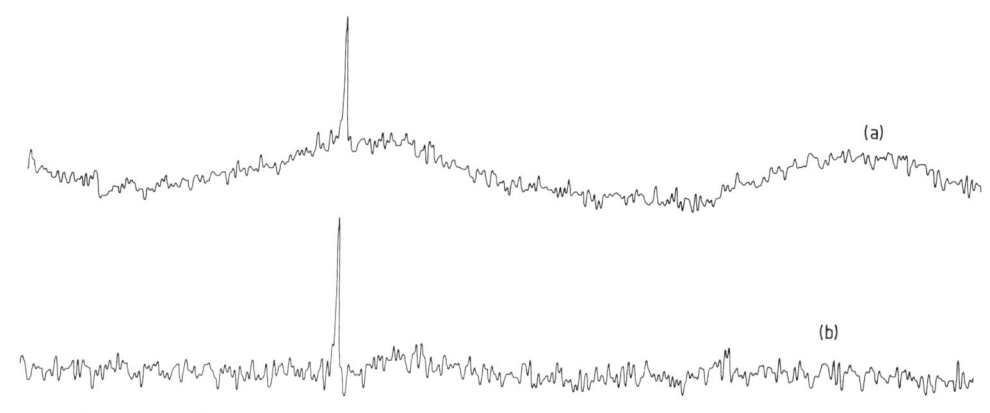

Fig. 2.15. ^{17}O spectra of acetone. (*a*) Spectrum obtained using normal conditions; (*b*) spectrum with baseline roll eliminated using method described in the text.

In fact, this method was developed to obtain free induction decays of ^{17}O and $^{49,47}Ti$ which were shorter than the ringdown time of the spectrometer in this laboratory and has proven very useful in obtaining otherwise inaccessible resonances.

2.7.2. Decoupling Methods

The use of spin decoupling to improve signal-to-noise ratio in ^{13}C spectra is well known. It is obvious that the signal-to-noise ratio must improve by decoupling protons from ^{13}C nuclei. Not only is an improvement in sensitivity gained from the nuclear Overhauser effect but the simple physical fact that a multiplet pattern is reduced to a single peak also increases the signal-to-noise ratio. For example, a $^{13}C-H$ resonance exists as a doublet under normal experimental conditions. However, on decoupling the proton from the ^{13}C nucleus this doublet becomes a singlet; in effect the signal-to-noise ratio has been doubled.

Ideally, each separate proton resonance should be irradiated with a continuous wave radiofrequency. In practice this is not possible without a large number of separate frequency generators and power amplifiers. Methods have been developed to overcome this difficulty. The most widely used method in commercial spectrometers was developed by Ernst (169). In this method the proton frequency is modulated with pseudo-random noise of up to 5-kHz bandwidth. This random noise modulation technique does have a disadvantage. Random noise decoupling is essentially a stochastic process. Noise by its very nature is random in phase, frequency, and amplitude (see Sec. 2.3). Thus no proton frequency is subjected to continuous irradiation. The resulting "decoupled" spectra are in essence a time-averaged mixture of the fully coupled and the fully decoupled spectra. Such a method leads to residual couplings being present, with consequent broadening of the spectra lines and lowering of the signal-to-noise ratio. Fortunately for spectroscopists, the time-averaged spectra may be moved toward the fully decoupled spectra by increasing the power of the decoupling transmitter. But these residual couplings can never be removed completely. Satisfying the requirements of a

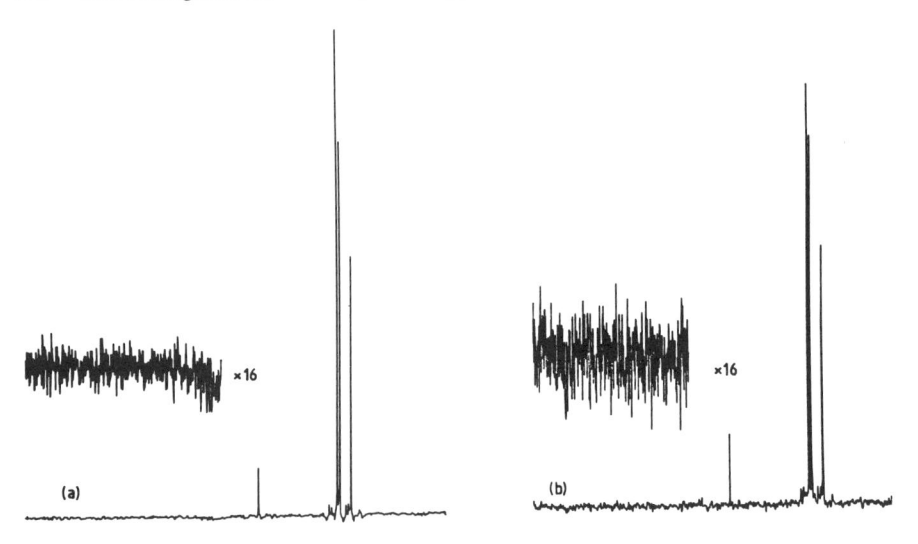

Fig. 2.16. ^{13}C spectra of the aromatic region of 90% ethylbenzene in aceton, single pulse, 30° angle. (*a*) Double modulation decoupling; (*b*) random noise decoupling. The improvement in signal-to-noise ratio afforded by double modulation decoupling is more than 100%.

sufficiently strong decoupling field requires a radiofrequency power amplifier capable of supplying several tens of watts of power to the decoupler coil in the probe. Obviously, the limitation here is the ability to remove heat generated by the decoupler from the probe and the sample with a reasonable flow of cooling air. Such a technique is not always successful, and an increase of several degrees in sampled temperature can be expected. Experience in this laboratory has shown that air used for cooling the decoupler coils must be scrupulously dry and oil free. Wet air results in extensive dielectric heating, and it is not unknown for samples to boil when "wet" air has been used for cooling purposes. Devices are available for the complete removal of both water and oil from compressed air supplies, and even where the compressed air supply is reputed to be water- and oil-free the use of these devices is strongly recommended (170).

An alternative approach to decoupling was proposed by Grutzner and Santini (171). In their approach the decoupling frequency is modulated with a 100-Hz square wave using a passive modulator. Residual coupling was reduced to such an extent that an improvement in signal-to-noise ratio by a factor of 2 was reported. Further work along these lines by Dykstra (172) employed a double modulation technique. In the case the proton frequency was modulated by 128 and 32 Hz. An improvement in signal-to-noise ratio of a factor of 3 was reported not only for ^{13}C spectra but also ^{31}P spectra. Dykstra also reported that power required for complete decoupling was 25% of that required for complete decoupling with random noise modulation decoupling. No increase in sample temperature was apparent when the double modulation technique was used. These factors indicate the effectiveness of double modulation decoupling. The double modulation technique has been found to be highly effective for ^{13}C, ^{31}P, and ^{15}N under continuous decoupling conditions (Fig. 2.16) but its effectiveness for gated as opposed to decoupling irradiation seqeunces has not yet been evaluated (173). Dykstra's circuit is simple to build, and the choice of active components is not critical. It is critical to use passive modulators, as the effectiveness of the technique depends on the production of a large number of harmonic frequencies, which would not be available by the use of active modulators. Installation is simple, being immediately before the decoupler power amplifier. Because of the large improvement in signal-to-noise ratio gained by this method, particularly for the ^{15}N nucleus, this technique can be expected to have major applications for heterocyclic NMR spectroscopy.

REFERENCES

1. W. Pauli, *Naturwiss.*, 12, 741 (1924).
2. I. Rabi, S. Millman, P. Kusch, and J. P. Zacharias, *Phys. Rev.*, 55, 526 (1939). J. B. M. Kellogg, I. Rabi, N. F. Ramsey, and J. P. Zacharias, *Phys. Rev.* 56, 728 (1939).
3. F. Bloch, W. W. Hansen, and M. E. Packard, *Phys. Rev.*, 69, 127 (1946).
4. E. M. Purcell, H. C. Torrey, and R. V. Pound, *Phys. Rev.*, 69, 37 (1946).
5. F. Bloch, *Phys. Rev.*, 70, 460 (1946).
6. W. D. Knight, *Phys. Rev.*, 76, 1259 (1949).
7. W. G. Proctor and F. C. Yu, *Phys. Rev.*, 77, 717 (1950).
8. W. C. Dickinson, *Phys. Rev.*, 77, 736 (1950).
9. J. T. Arnold, S. S. Dharmatti, and M. E. Packard, *J. Chem. Phys.*, 19, 507 (1951).
10. F. Bloch, W. W. Hansen, and M. E. Packard, *Phys. Rev.*, 70, 474 (1946).
11. G. E. Pake, *Amer. J. Phys.*, 18, 438, 473 (1950).
12. A. Abragam, *Principles of Nuclear Magnetism*, Clarendon Press, Oxford, 1961.
13. E. D. Becker, *High Resolution NMR*, Academic Press, London, 1969.
14. F. A. Bovey, *Nuclear Magnetic Resonance Spectroscopy*, Academic Press, New York, 1969.
15. J. K. Saunders and J. W. Easton, *Determination of Organic Structure by Physical Methods*, Vol. 6, F. C. Nachod and J. J. Zuckerman, Eds., Academic Press, New York, (1976).
16. J. H. Noggle and R. E. Schirmer, *The Nuclear Overhauser Effect*, Academic Press, New York, 1971.
17. J. R. Lyerla and D. M. Grant, in *MTP International Review of Science* (Phys. Chem. Ser. 1, Vol. 4), C. A. McDowell, Ed., Chemical Society, London, 1972, Chapter 5.
18. J. W. Emsley, J. Feeney, and L. H. Sutcliff, *High Resolution Nuclear Magnetic Resonance Spectroscopy*, Pergamon Press, Oxford, 1965.
19. R. M. Lynden-Bell and R. K. Harris, *Nuclear Magnetic Resonance Spectroscopy*, Nelson, London, 1969.
20. *International Conference on the Science of Superconductivity Rev. Mod. Phys.*, Vol. 36, D. Douglass and R. W. Schmitt, Eds., 1964, p. 1.
21. F. A. Nelson and H. E. Weaver, *Science*, 146, 223 (1964).
22. R. C. Ferguson and W. D. Phillips, *Science*, 157, 257 (1967).
23. T. H. Geballe and J. K. Hulm, *Sci. Amer.* 243(11), 112 (1980).
24. N. Bloembergen, E. M. Purcell, and R. V. Pound, *Phys. Rev.*, 73, 679 (1948).
25. F. Bloch, W. W. Hansen, and M. Packard, *Phys. Rev.*, 70, 474 (1946).

26. M. A. Packard, *Rev. Sci. Instr.*, **19**, 435 (1948).
27. R. J. Blume, *Rev. Sci. Instr.*, **33**, 1472 (1962).
28. D. H. Hughes and K. Reed, *Rev. Sci. Instr.*, **41**, 293 (1970).
29. L. E. Drain, *Rev. Sci. Instr.*, **43**, 1648 (1972).
30. R. R. Lembo and V. J. Kowalewski, *J. Phys. E*, **8**, 632 (1975).
31. S. Kan, P. Gonord, C. Duret, J. Salset, and C. Vibet, *Rev. Sci. Instr.*, **44**, 1725 (1973).
32. D. I. Hoult and R. E. Richards, *J. Magn. Reson.*, **22**, 561 (1976).
33. M. E. Stoll, *Rev. Sci. Instr.*, **52**, 391 (1981).
34. F. H. N. Robinson, *Noise and Fluctuations in Electronic Devices and Circuits*, Clarendon Press, Oxford, 1974.
35. C. D. Motchenbacher and F. C. Fitchen, *Low Noise Electronic Design*, John Wiley, New York, 1973.
36. D. I. Hoult, *Progr. N.M.R. Spectrosc.*, **12**, 41 (1978).
37. "Field Effect Transistors," Publications Dept, Philips Ind., Eindhoven, 1972.
38. A. Abragam, *The Principles of Nuclear Magnetism*, Oxford University Press, London, 1961.
39. H. D. W. Hill and R. E. Richards, *J. Sci. Instr.*, **1**, 977 (1968).
40. D. I. Hoult and R. E. Richards, *Electron Lett.*, **11**, 596 (1975).
41. D. I. Hoult and R. E. Richards, *J. Magn. Reson.*, **24**, 71 (1976).
42. M. Soutif and R. Gabillard, in P. Grivet, Ed., *La Résonance Paramagnétique Nucléaire*, Centre National de La Recherche Scientifique, Paris, 1955.
43. J. T. Bailey, R. C. Rosanske, and G. C. Levy, *Rev. Sci. Instr.*, **52**, 548 (1981).
44. W. N. Hardy and L. A. Whitehead, *Rev. Sci. Instr.*, **52**, 213 (1981).
45. H. F. Friis, *Proc. IRE*, **32**, 419 (1944).
46. G. H. Krauss, *Ham Radio*, 50 (1979).
47. E. Fukushima and S. B. W. Roeder, *J. Magn. Reson.*, **33**, 199 (1979).
48. M. L. Buess and G. L. Petersen, *Rev. Sci. Instr.*, **49**, 1151 (1978).
49. M. R. Gaerttner, W. D. Wallace, and B. M. Maxfield, *Phys. Rev.*, **184**, 702 (1969).
50. W. D. Wallace, *Int. J. Nondestructive Test.*, **2**, 309 (1971).
51. R. B. Thompson, *IEEE Trans. Sonics*, **SU-20**, 340 (1973).
52. *Handbook of Chemistry and Physics*, 57th ed., Chemical Rubber Co., Cleveland, Ohio, 1976.
53. T. Lyman, Ed., *Metals Handbook*, Amer. Soc. for Metals, Metals Park, Ohio, 1961.
54. P. A. Speight, K. R. Jeffrey, and J. A. Courtney, *J. Phys. E*, **7**, 801 (1974).
55. S. I. Aksenov, B. P. Vikin, and K. V. Vladimirskii, *Sov. Phys. JETP*, **1**, 609 (1955).
56. W. G. Clark, *Rev. Sci. Instr.*, **35**, 316 (1964).
57. I. J. Lowe and C. E. Tarr, *J. Sci. Instr.*, **1**, 320 (1968).
58. W. G. Clark and J. A. McNeil, *Rev. Sci. Instr.*, **44**, 844 (1973).
59. I. J. Lowe and M. Engelsberg, *Rev. Sci. Instr.*, **45**, 631 (1974).
60. D. I. Hoult, *Rev. Sci. Instr.*, **50**, 193 (1979).
61. J. L. Engle, *Rev. Sci. Instr.*, **49**, 1356 (1978).
62. J. L. Engle, *J. Magn. Reson.*, **37**, 547 (1980).
63. R. K. Harris and B. E. Mann, *NMR and the Periodic Table*, Academic Press, London, 1978.
64. J. A. Elvidge, *Isotopes: Essential Chemistry and Applications*, Spec. Pub. Vol. 35, The Chemical Society, London, 1980, p. 123.

65. J. A. Elvidge, J. R. Jones, R. B. Mane, and J. M. A. Al-Rawi, *J. Chem. Soc. Perkin II,* **1979**, 386.
66. J. P. Bloxsidge, J. A. Elvidge, J. R. Jones, R. B. Mane, and M. Saljoughian, *Org. Magn. Reson.,* **12**, 574 (1979).
67. H. H. Mantsch, H. Saito, and I. C. P. Smith, *Progr. NMR Spectrosc.* **11**, 211 (1977).
68. "NMR Spectral Catalogue," Varian Associates, Palo Alto, 1963.
69. "Selected ^{13}C Nuclear Magnetic Resonance Spectral Data," Texas A & M University, College Station, Texas.
70. "Selected ^{1}H Nuclear Magnetic Resonance Spectral Data," Texas A & M University, College Station, Texas.
71. *Aldrich Library of NMR Spectra*, Aldrich Chemical Co., Milwaukee, U.S.A.
72. ^{13}C *NMR Spectral Data*, Ed. Weinheim, Verlag Chemie, Berlin, 1978.
73. T. J. Batterham, *NMR Spectra of Simple Heterocycles*, John Wiley, New York 1973.
74. G. B. Barlin and M. D. Fenn, *Aust. J. Chem.,* **34**, 1341 (1981).
75. P. E. Pfeffer, K. M. Valentine, and F. W. Parrish, *J. Amer. Chem. Soc.* **101**, 1265 (1979).
76. P. J. Archbald, M. D. Fenn, and A. B. Roy, *Carbohydr. Res.,* **93**, 177 (1981).
77. G. D. Mair, Ed., *Hazards in the Chemical Laboratory*, The Royal Institute of Chemistry, London, 1972.
78. M. Alei, A. E. Florin, W. M. Litchman, and J. F. O'Brien, *Phys. Chem.,* **75**, 932 (1971).
79. M. Witanowski, L. Stefaniak, S. Szymanski, and H. Januszewski, *J. Magn. Reson.,* **28**, 217 (1977).
80. P. R. Srinivasan and R. L. Lichter, *J. Magn. Reson.,* **28**, 227 (1977).
81. M. I. Burgar, T. E. St. Amour, and D. Fiat, *J. Phys. Chem.,* **85**, 502 (1981).
82. L. Stefaniak, *Spectrochim. Acta,* **32A**, 345 (1976).
83. M. Witanowski, L. Stefaniak, B. Kamienski, and G. A. Webb, *Org. Magn. Reson.,* **14**, 305 (1980).
84. E. E. Liepin'sh, R. M. Zolotoyabko, Y. P. Stradyn, M. A. Trushule, and K. K. Venter, *Khim. Geterotsikl, Soedin,* 741 (1980).
85. M. Witanowski, L. Stefaniak, H. Januszewski, Z. Grabowski, and G. A. Webb, *Tetrahedron,* **28**, 637 (1972).
86. L. Stefaniak, *Org. Magn. Reson.,* **11**, 385 (1978).
87. M. Witanowski, L. Stefaniak, S. Szymanski, Z. Grabowski, and G. A. Webb, *J. Magn. Reson.,* **21**, 185 (1976).
88. L. Stefaniak, *Org. Magn. Reson.,* **12**, 379 (1979).
89. F. Cavagna and H. Pietsch, *Org. Magn. Reson.,* **11**, 204 (1978).
90. L. Stefaniak, *Tetrahedron,* **32**, 1065 (1976).
91. A. Lyčka, *Coll. Czech. Chem. Commun.,* **45**, 2766 (1980).
92. D. Ghesquière and C. Chachaty, *Org. Magn. Reson.,* **9**, 392 (1977).
93. L. Stefaniak and A. Grabowska, *Bull. Acad. Polon. Sci. Ser. Sci. Chem.,* **22**, 267 (1974).
94. H. Saitô, Y. Tanaka, and S. Nagata, *J. Amer. Chem. Soc.,* **95**, 324 (1973).
95. E. Lippmaa, M. Mägi, S. S. Novikov, L. I. Khmelnitskii, A. S. Prikhodo, O. V. Lebedov, and L. V. Epishina, *Org. Magn. Reson.,* **4**, 153 (1972).
96. W. McFarlane and C. J. Turner, *Bull. Soc. Chim. Belg.,* **87**, 271 (1978).
97. M. Witanowski, L. Stefaniak, H. Januszewski, and J. Elguero, *J. Chim. Phys.,* **1973**, 697.
98. M. Witanowski, L. Stefaniak, S. Biernat, and G. A. Webb, *Org. Magn. Reson.,* **14**, 356 (1980).

99. E. B. Baker and A. I. Popov, *J. Phys. Chem.*, **76**, 2403 (1972).
100. M. Witanowski, L. Stefaniak, H. Januszewski, K. Bahadur, and G. A. Webb, *J. Cryst. Mol. Struct.*, **5**, 137 (1975).
101. I. Yavari, R. E. Botto, and J. D. Roberts, *J. Org. Chem.*, **43**, 2542 (1978).
102. W. Stadeli, W. von Philipsborn, A. Wick, and I. Kompis, *Helv. Chim. Acta.* **63**, 504 (1980).
103. W. W. Bachovchin and J. D. Roberts, *J. Amer. Chem. Soc.*, **100**, 8041 (1978).
104. D. O. Duthaler and J. D. Roberts, *J. Amer. Chem. Soc.*, **100**, 4969 (1978).
105. E. Rosenberg, K. L. Williamson, and J. D. Roberts, *Org. Magn. Reson.*, **8**, 117 (1976).
106. G. E. Hawkes, E. W. Randall, and W. E. Hull, *J. Chem. Soc. Perkin II*, **1977**, 1268.
107. V. Markowski, G. R. Sullivan, and J. D. Roberts, *J. Amer. Chem. Soc.*, **99**, 714 (1977).
108. W. Städeli and W. von Philipsborn, *Org. Magn. Reson.*, **15**, 106 (1981).
109. W. B. Cowden and P. Waring, *Aust. J. Chem.*, **34**, 1539 (1981).
110. I. Yavari and J. D. Roberts, *Org. Magn. Reson.*, **12**, 87 (1979).
111. A. J. Digioia, G. T. Furst, L. Psola, and R. L. Lichter, *J. Phys. Chem.*, **82**, 1644 (1978).
112. J. F. Whidby, W. B. Edwards, and T. P. Pitner, *J. Org. Chem.*, **44**, 794 (1979).
113. M. D. Fenn, unpublished data.
114. M. M. King, H. J. C. Yeh, and G. O. Dudek, *Org. Magn. Reson.*, **8**, 208 (1976).
115. R. L. Lipnick and J. D. Fissekis, *J. Labelled Comp. Radiopharm.*, **17**, 247 (1980).
116. A. R. Butler, C. Glidewell, I. Hussain, and P. R. Maw, *J. Chem. Res. (S)*, **1980**, 114.
117. J. W. Akitt, *Ann. Rep. NMR Spectrosc.*, **5A**, 465 (1972).
118. M. I. Burgur, T. E. St. Amour, and D. Fiat, *J. Phys. Chem.*, **85**, 502 (1981).
119. B. Valentine, T. E. St. Amour, R. Walter, and D. Fiat, *J. Magn. Reson.*, **38**, 413 (1980).
120. D. Canet, C. Goulon-Ginet, and J. P. Marchal, *J. Magn. Reson.*, **22**, 541 (1976).
121. T. Sugawara, Y. Kawada, M. Katoh, and H. Iwamura, *Bull. Chem. Soc. Japan*, **52**, 3391 (1979).
122. E. L. Eliel, K. M. Pietrusiewicz, and L. Jewell, *Tetrahedron Lett.*, **20**, 3649 (1979).
123. J. P. Kintzinger, *Oxygen NMR*, Springer-Verlag, Berlin, 1981.
124. J. P. Kintzinger, C. Delseth, and T. T. Nguyen, *Tetrahedron*, **36**, 3431 (1980).
125. H. A. Christ, P. Diehl, H. R. Schneider, and H. Dahn, *Helv. Chim. Acta.* **44**, 865 (1961).
126. J. P. Kintzinger and T. T. Nguyen, *Org. Magn. Reson.*, **13**, 464 (1980).
127. W. Nakanishi, T. Jo, K. Miura, Y. Ikeda, T. Sugawara, Y. Kuwada, and H. Iwamura, *Chem. Lett.* 387 (1981).
128. H. M. Swartz, M. Maccoss, and S. S. Danyluk, *Tetrahedron Lett.* **21**, 3837 (1980).
129. G. C. Levy, J. J. Dechter, and J. Kowalewski, *J. Amer. Chem. Soc.*, **100**, 2308 (1978).
130. G. J. Martin, M. L. Martin, and J. P. Gouesnard, ^{15}N *NMR Spectroscopy*, Springer-Verlag, Berlin, 1981.
131. H. J. Jakobsen and W. S. Brey, *J. Amer. Chem. Soc.*, **101**, 774 (1979).
132. R. D. Bertrand, W. B. Moniz, A. N. Garroway, and G. L. Chingas, *J. Amer. Chem. Soc.*, **100**, 5227 (1978).
133. G. A. Morris and R. Freeman, *J. Amer. Chem. Soc.*, **101**, 760 (1979).
134. J. K. Beconsall, *Mol. Phys.*, **15**, 129 (1968).
135. D. J. Frost and G. E. Hall, *Mol. Phys.*, **10**, 191 (1966).
136. W. C. Dickinson, *Phys. Rev.*, **81**, 717 (1951).
137. J. Homer and P. M. Whitney, *J. Chem. Soc. Chem. Commun.*, 153 (1972).

138. F. E. Mabbs and D. J. Machin, *Magnetism and Transition Metal Complexes*, Chapman and Hall, London, 1973.
139. A. A. Bothner-by and R. E. Glick, *J. Chem. Phys.*, **26**, 1647 (1957).
140. Wilmad Glass Company, Inc., Buena, New York.
141. G. V. D. Tiers, *J. Phys. Chem.*, **62**, 1151 (1958).
142. A. K. Jameson and C. J. Jameson, *J. Amer. Chem. Soc.*, **95**, 8559 (1973).
143. G. V. D. Tiers and R. I. Coon, *J. Org. Chem.*, **26**, 2097 (1961).
144. E. S. Hand and T. Cohen, *J. Amer. Chem. Soc.*, **87**, 133 (1965).
145. R. A. Y. Jones, A. R. Katritzky, J. N. Murrell, and N. Sheppard, *J. Chem. Soc.*, 2576 (1962).
146. E. Fanghänel, R. Radeglia, B. Hauptmann, B. Tyszkiewicz, and M. Tyszkiewicz, *J. Prakt. Chem.*, **320**, 618 (1978).
147. J. M. Briggs and E. W. Randall, *Mol. Phys.*, **26**, 699 (1973).
148. J. M. Briggs, L. F. Farnell, and E. W. Randall, *J. Chem. Soc. Chem. Commun.*, 680 (1971).
149. E. Lipmaa, *Org. Magn. Reson.*, **5**, 429 (1973).
150. M. P. Sibi and R. L. Lichter, *J. Org. Chem.*, **42**, 2999 (1977).
151. J. P. Warren and J. D. Roberts, *J. Phys. Chem.*, **78**, 2507 (1974).
152. J. Chatt, M. E. Fakley, R. L. Richards, J. Mason, and I. A. Stenhouse, *J. Chem. Res (S)*, **1979**, 44.
153. J. P. Kintzinger, *NMR, Basic Principles and Progress*, Springer-Verlag, Berlin, 1981.
154. O. Lutz, A. Nolle, and A. Schwenk, *Z. Naturforsch.*, **28A**, 1370 (1973).
155. M. D. Fenn, results obtained in this laboratory at 30°C.
156. P. J. Smolenaers and J. K. Beattie, *JEOL News*, **16A**, 51 (1980).
157. A. L. van Geet, *Anal. Chem.*, **40**, 2227 (1968).
158. R. Duerst and A. Merbach. *Rev. Sci. Instr.*, **36**, 1896 (1965).
159. M. L. Kaplan, F. A. Bovey, and H. N. Cheng, *Anal. Chem.*, **47**, 1703 (1975).
160. A. L. van Geet, *Anal. Chem.*, **42**, 679 (1970).
161. D. S. Railford, C. L. Fisk, and E. D. Becker, *Anal. Chem.*, **51**, 2050 (1979).
162. R. C. Neuman and V. Jones, *J. Amer. Chem. Soc.*, **90**, 1970 (1968).
163. D. W. Vidrine and P. E. Peterson, *Anal. Chem.*, **48**, 1301 (1976).
164. J. J. Led and S. B. Petersen, *J. Magn. Reson.*, **32**, 1 (1978).
165. H. J. Schneider, W. Freitag, and M. Schommer, *J. Magn. Reson.*, **18**, 393 (1975).
166. J. Bornais and S. Brownstein, *J. Magn. Reson.*, **29**, 207 (1978).
167. F. L. Dickert and S. W. Hellman, *Anal. Chem.*, **52**, 996 (1980).
168. R. K. Harris, personal communication.
169. R. R. Ernst, *J. Chem. Phys.*, **45**, 3845 (1966).
170. Puregas Equipment Corp., Copiague, New York.
171. J. B. Grutzner and R. E. Santini, *J. Magn. Reson.*, **19**, 173 (1975).
172. R. W. Dykstra, *J. Magn. Reson.*, **46**, 503 (1982).
173. M. D. Fenn, work currently in progress in this laboratory.

3 ULTRAVIOLET PHOTOELECTRON SPECTROSCOPY OF HETEROCYCLIC COMPOUNDS

C. N. R. RAO AND P. K. BASU

Solid State and Structural Chemistry Unit
Indian Institute of Science
Bangalore, India

3.1.	Introduction	231
3.2.	Saturated Heterocyclic Compounds	233
	3.2.1. Three-Membered Rings, 238	
	3.2.2. Four-Membered Rings, 241	
	3.2.3. Five-Membered Rings, 241	
	3.2.4. Six-Membered Rings, 246	
3.3.	Heteroaromatic Compounds	248
	3.3.1. Five-Membered Compounds, 248	
	3.3.2. Six-Membered Compounds, 255	
	3.3.2.1. Pyridine Derivatives, 257	
	3.3.2.2. Pyridine Analogs, 261	
	3.3.2.3. Azabenzenes, 261	
3.4.	Fused-Ring Heterocyclic Compounds	263
3.5.	Lone-Pair Interactions in Heterocyclic Compounds	270
Acknowledgment		273
References		274

3.1. INTRODUCTION

Electronic absorption spectroscopy of heterocyclic compounds in the UV and visible regions has been widely discussed in the literature (1, 2). An important development in recent years that has provided valuable information on the electron states of organic molecules is photoelectron spectroscopy. Molecular photoelectron spectroscopy involves irradiation of free molecules (in the vapor phase)

Table 3.1. Ionization Energies of 3-Membered Saturated Heterocyclic Compounds[a,b]

Compound	Ionization Energy			Ref.
Oxirane	10.56 (n_O; b_1)	11.85 (σ_S; a_1)	13.73 (σ_A; b_2)	14.16 (π_{CH_2}; a_2)	8–11
	16.52 (n; a_1)	17.20 (π_{CH_2}; b_1)			
2-Methyl oxirane	10.26 (b_1)	11.23 (a_1)	12.88 (b_2)	13.33 (a_2)	8
2-Ethyl oxirane	10.15 (b_1)	11.08 (a_1)	12.26 (b_2)	12.7 (a_2)	8
2,2-Dimethyloxirane	10.00 (b_1)	10.75 (a_1)	12.45 (b_2)	12.94 (a_2)	8
2,3-Dimethyloxirane	9.98 (b_1)	10.73 (a_1)	12.42 (b_2)	12.87 (a_2)	8
2-Bromomethylene oxirane	10.46 (b_1)	10.70 (n_{Br})	10.99 (n_{Br})	11.50 (a_1)	8
	13.34 (b_2)	13.72 (σ)	14.65 (a_2)		
2-Chloromethylene oxirane	10.60 (b_1)	11.28 (n_{Cl})	11.59 (a_1)	13.61 (b_2)	8
	14.98 (a_2)				
2-Fluoromethylene oxirane	10.78 (b_1)	11.75 (a_1)	13.02 (n_F)	14.17	8
	15.06	16.29			
Azirane	9.83 (n_N; a')	11.79 (σ_A; a'')	12.16 (σ_S; a')	13.45 ($\pi^A_{CH_2}$; a'')	11, 12
	15.69 (σ; a')	17.19 ($\pi^S_{CH_2}$; a')			
1-Methylazirane	9.57				12
1,1-Dimethylazirane	9.29	11.07			12
1,1-Dimethyl-N-methylazirane	8.68	10.58			12
Thiirane	9.03 (n_S; b_1)	11.37 (σ_A; b_2)	11.93 (σ_S; a_1)	13.51 (π_{CH_2}; a_2)	9, 11
	15.33 (n; a_1)	16.58 (π_{CH_2}; b_1)			
Thiirane sulfoxide	10.20	11.57	11.98	12.03	13
	13.92	14.62			
1,2-Dimethylthiirane sulfoxide	9.82 (π_{SO_2}; b_1)	11.10 (σ_{SO_2}, σ_{SO_2}, π_{SO_2}; a_1, b_1, a_2)	11.65		14
1-Methylthiirene sulfoxide	10.40 (π_{CC})	10.63 (π_{SO_2}; b_2)	11.88 (σ_{SO_2}; a_1)	12.17 (σ_{SO_2}; b_1)	14
	12.43 (π_{SO_2}; a_2)				
1,2-Dimethylthiirene sulfoxide	9.89 (π_{CC})	10.14 (π_{SO_2}; b_2)	11.57 (σ_{SO_2}; a_1)	11.76 (π_{SO_2}; b_1)	14
	11.94 (π_{SO_2}; a_2)				

[a] All values in this table as well as subsequent tables are in eV.
[b] Values in parentheses are adiabatic values.

with photons whose energy $h\nu$ is much greater than that of the first ionization energy (IE) of the molecule. By determining the kinetic energy (KE) of the electrons ejected, one can obtain the IEs of the various orbitals as given by the equation

$$h\nu = \text{IE} + \text{KE} + E_{\text{vib}} \tag{1}$$

The IEs to a first approximation are directly related to the orbital binding (IE = $-$ BE) energies (Koopmans' theorem).

Depending on the energy of the photon, one can obtain energies of the valence or core electron levels of molecules. Thus, UV radiation [He(I), 21.2 eV; He(II), 40.8 eV] is employed to obtain information on the valence levels of molecules while X-rays ($\geqslant 10^3$ eV) are employed to obtain information of the core electronic levels [e.g., C (1s), N (1s), etc.]. Ultraviolet photoelectron spectroscopy (UVPS) has been employed effectively not only to characterize molecules but also to understand their electronic states. One generally makes use of energy levels (ordering scheme) obtained from molecular orbital calculations in assigning various bands in UV photoelectron spectra. Analysis of the vibrational structure of UVPS bands provides structural information on the molecular ions and is also valuable in making band assignments.

Ultraviolet photoelectron spectroscopy of organic molecules was first reviewed by Turner and coworkers (3) and later by Heilbronner and Maier (4) and Rao and coworkers (5). UV photoelectron spectroscopy of heterocyclic compounds was briefly reviewed by Heilbronner and coworkers (6) in 1974. Considerable information has since become available in the literature, and we shall discuss the important findings in this article. We shall present much of the data in the form of correlation charts and tables. We shall not discuss the theory and instrumentation of UV photoelectron spectroscopy since these aspects have been adequately covered elsewhere in the literature (3, 7).

3.2. SATURATED HETEROCYCLIC COMPOUNDS

The most important feature in the UVPS of saturated heterocyclic compounds is the band due to the ionization of the lone-pair orbitals of the heteroatoms. Interactions between the lone-pair (n) orbitals on different atoms manifest themselves in UVPS. Other than the bands due to lone-pair ionization, we find those due to π ionization if double bonds are also present in the molecules. In compounds containing carbonyl or thiocarbonyl or similar functional groups, we see bands due to the ionization of the lone-pair as well as the π orbitals. Bands due to ionization of σ orbitals appear at much higher energies than those due to n or π orbitals and are generally not as useful for characterization or structure elucidation.

In Figures 3.1 and 3.2, we have shown the spectra of a few typical saturated heterocyclic compounds; the lone-pair ionization bands appearing at the lowest ionization energies can be clearly seen in the spectra. The important UVPS data on 3- and 5-membered ring heterocyclic compounds are presented in Tables 3.1 and 3.2, respectively.

Table 3.2. Ionization Energies of Five-Membered Heterocyclic Compounds

Compound	Ionization Energy			Ref.
Tetrahydrofuran	9.65 (n_π)	11.4 (C—O—C; a_1)	13.0 (C—O—C; b_2)	15
2,5-Dihydrofuran	9.14 (n_O)	10.59 (π)	11.4 (σ)	16
Tetrahydropyrrole	8.77 (n_N)	11.49		12, 17, 18
N-Methyltetrahydropyrrole	8.41 (n_N)	11.16		12, 17
2,5-Dihydropyrrole	8.61 (n_N)	9.77 (π)		17
N-Methyl-2,5-dihydropyrrole	8.21 (n_N)	9.66 (π)		17
Tetrahydrothiophene[a]	8.42 (n_π; b_1)	10.9 (C—S—C; a_1)	11.9 (C—S—C; b_2)	15
2,5-Dihydrothiophene	8.54 (n_π)	9.86 (π)	11.24 (n_σ)	20
2,5-Dihydrothiophene-1,1-dioxide	10.44 ($\pi_{SO_2}^+ - \bar\sigma_{SC}$)	10.66 (n_O^+)	11.25 ($\bar\pi_{SO_2} - \bar\sigma_{CH_2}$)	21
	11.63 (n_O^-)	11.99 ($\bar\sigma_{SC} + \pi_{SO_2}^+$)		21
Tetrahydroselenophene	8.14 (n_π; b_1)	10.5 (C—Se—C; a_1)	11.4 (C—Se—C; b_2)	15
Tetrahydrotellurophene	7.73 (n_π; b_1)	10.0 (C—Te—C; a_1)	10.7 (C—Te—C; b_2)	15
P-n-Butyltetrahydrophosphole	8.25 (n)	10.6 (P—C)	10.7 (P—C)	22
P-Phenyltetrahydrophosphole	8.35 (n)			22
α-Butyrolactone	10.26 (10.06) (n_O)	10.93 (π_{OCO})	12.44 (π_{CO})	18, 23
α-Crotonolactone	10.70			24
α-Butyrolactam	9.53 (n_{CO})	9.76 (n_N)	11.91 (π_{CO})	18
β-Butyrolactam	8.83 (n_N)	9.53 (n_{CO})	12.24 (π_{CO})	18
Maleic anhydride	11.1 (σ; $9b_2$)	12.0 (σ, π; $12a_1, 3b_1$)	12.5 ($2b_1/3b_1$)	25
	14.5 (σ, π; $11a_1, 1a_2/2a_2$)		15.3 (σ; $8b_2$)	
	~16.5 (σ, σ, π; $10a_1, 7b_2, 1b_1$)	17.4 (σ; $9a_1$)	19.4 (σ; $8a_1$)	
	20.4 (σ; $6b_2$)	22.3 (σ; $5b_2$)	26.3 (σ; $7a_1$)	
Succinic anhydride	10.84 (n_{CO}^-)	11.62 (n_{CO}^+)	12.12 (n_O)	18, 25, 26
	13.16 (π_{CO}^+)	13.83 (π_{CO}^-)		
	~15.3 (σ, σ, π; $8b_2, 10a_1, 1a_2$)	16.4 (σ; $7b_2$)	17.0 (σ; $9a_1$)	
	17.7 (π; $1b_1$)	19.0 (σ; $8a_1$)	20.6 (σ; $6b_2$)	
	22.9 (σ; $5b_2$)			

Compound			Ref.
Succinimide	10.01 (n_{CO}^-) 14.18 (π_{CO}^-)	10.88 (n_{CO}^+, n_N)	12.4 (π_{CO}^+)	18, 26
N-Chlorosuccinimide	10.29 (n_{CO}^-) 12.74 (π_{CO}^-)	11.12 (n_N/n_{CO}^+)	12.33 (n_{Cl})	18, 26
N-Bromosuccinimide	10.12 (n_{CO}^-) 11.52 (n_{Br})	10.99 (n_N/n_{CO}^+) 12.15 (π_{CO})	11.41 (n_{Br})	18, 26
1,3-Dioxalane	10.1 (a)	10.65 (b)		27, 28
2,2-Dimethyl-1,3-dioxalane	9.71	10.20		27
1,3-Dithiolane	8.75 (a)	9.05 (b)		28
Ethylene carbonate	10.89 (10.70) (n_O; $5b_2$)	11.38 (π; $2a_2$)	11.45 (π; $3b_1$)	23
Vinylene carbonate	9.87 (9.68) (π; $3b_1$)	11.73 (11.66) (n_O; $5b_2$)	12.95 (12.95) ($4b_2$)	23
Propylene carbonate	10.71 (10.52) (n_O; b_2)	11.17 (π; a_2)	11.27 (π; b_1)	23
Ethylene dithiocarbonate	9.5 (n_{σ_O}; $n_{\sigma_{SS}}^-$) 11.85 (n_{σ_O}, $n_{\sigma_{SS}}^-$)	9.88 ($n_{\pi_{SS}}^-$) 12.64 ($n_{\sigma_{SS}}^+$)	10.08 ($n_{\pi_{SS}}^+$, π_{CO}) 13.5 ($n_{\pi_{SS}}^+$, π_{CO})	29
2-Oxazolidinone	10.21 (n_{CO}) 12.82 (π_{CO})	10.71 (n_N)	11.07 (n_O)	18
5-Methyl-2-oxazolidinone	9.99 (n_{CO}) 12.55 (π_{CO})	10.64 (n_N)	10.86 (n_O)	18
5,5-Dimethyl-2-oxazolidinone	9.88 (n_{CO}) 12.32 (π_{CO})	10.45 (n_N)	10.68 (n_O)	18
4-Methyl-2-oxazolidinone	9.95 (n_{CO}) 12.45 (π_{CO})	10.63 (n_N)	10.92 (n_O)	18
4,4-Dimethyl-2-oxazolidinone	9.80 (n_{CO}) 12.13 (π_{CO})	10.31 (n_N)	10.81 (n_O)	18
4,5-Dimethyl-2-oxazolidinone	9.84 (n_{CO}) 12.10 (π_{CO})	10.48 (n_N)	10.72 (n_O)	18
N-Chlorooxazolidinone	9.68 (n_{CO}) 11.90 (n_{Cl})	10.79 (n_N) 12.36 (π_{CO})	11.05 (n_O)	18
N-Bromooxazolidinone	9.45 (n_{CO}) 11.08 (n_{Br})	10.67 (n_N) 11.89 (π_{CO})	10.91 (n_O)	18

(*Continued on next page*)

Table 3.2. (continued)

Compound	Ionization Energy			Ref.
Ethylene trithiocarbonate	8.4 ($n_{\sigma_S^+}$)	8.87 (π_{CS}; $n_{\pi_{CSS}}^+$)	9.42 ($n_{\pi_{SS}}^-$)	29
	11.42 ($n_{\sigma_{SS}}^-$)	12.05 (σ_{CS}, $n_{\sigma_{SS}}^+$)	12.6 (π_{CS}, $n_{\pi_{CSS}}^+$)	
	12.85 (σ_{CS}, $n_{\sigma_{SS}}^+$)			
1,3-Dithiole-2-thione	8.33 (n, σ; $5b_2$)	8.56 (π; $3b_1$)	10.60 (π; $1a_2$)	30
	10.90 (π; $2b_1$)	12.00 (σ; $4b_2$)	12.37 (σ; $7a_1$)	
4,5-Dimethyl-1,3-dithiole-2-thione	7.96 (σ; b_2)	8.29 (π; b_1)	10.15 (π; a_2)	30
	10.33 (π; b_1)	11.50 (σ; b_2)		
3-Selena-1,3-dithiole-2-thione	8.28 (σ; b_2)	8.52 (σ; b_1)	10.17 (π; a_2)	30
	10.75 (π; b_1)	11.72 (σ; b_2)	12.12 (σ; a_1)	
1,3-Diselena-1,3-dithiole-2-thione	8.24 (σ; b_2)	8.47 (π; b_1)	9.80 (π; a_2)	30
	10.60 (π; b_1)	11.37 (σ; b_2)	11.89 (σ; a_1)	
1,3-Dithiole-2-selenone	7.81 (σ; b_2)	8.06 (π; b_1)	10.54 (π; a_2, b_1)	30
	11.66 (σ; b_2)	11.98 (σ; a_1)		
1-Selena-1,3-dithiole-2-selenone	7.83 (σ; b_2)	8.08 (π; b_1)	10.01 (π; a_2)	30
	10.40 (π; b_1)	11.51 (σ; b_2)	11.77 (σ; a_1)	
1,2,3-Triselena-1,2-dithiole-2-thione	7.85 (σ; b_2)	8.08 (π; b_1)	9.71 (π; a_2)	30
	10.18 (π; b_1)	11.26 (σ; b_2)	11.53 (σ; a_1)	
4,5-Dimethyl-1,2,3-triselena-1,3-dithiole-2-thione	7.68 (σ; b_2)	7.90 (π; b_1)	9.46 (π; a_2)	30
	9.73 (π; b_1)	10.90 (σ; b_2)	11.17 (π; a_1)	
N-Methylimidazoline-2-thione	7.41 (π)	7.90 (n)	10.09 (π)	31
	10.68 (π)	12.4 (σ)		
N,N-Dimethylimidazoline-2-thione	7.27 (π; b_1)	7.78 (n; b_2)	9.87 (π; b_1)	31, 32
	10.30 (π; a_2)	12.23 (σ; a_1)		
Imidazolidine-2-thione	8.15 (n; π)	9.46 (π)	12.6 (σ)	31
N,N-Dimethylimidazolidine-2-thione	7.95 (n)	7.99 (π)	8.62 (n)	31
	11.77 (σ)			
Thiazolidine-2-thione	8.25 (n)	8.48 (π)	9.47 (π)	31
	11.97 (σ)			

N-Methylthiazolidine-2-thione	8.05 (n)	8.23 (π)	31
	11.71 (σ)		
4,4-Dimethylthiazolidine-2-thione	8.18 (n)	8.36 (π)	31
	11.47 (σ)		
Thiazoline-2-thione	7.74 (π)	8.12 (n)	31
	10.82 (π)	12.84 (σ)	
N-Methylthiazoline-2-thione	7.68 (π)	8.02 (n)	31, 32
	10.53 (π)	12.18 (σ)	
4,N-Dimethylthiazoline-2-thione	7.55 (π)	7.98 (n)	31
	10.28 (π)	12.00 (σ)	
4,5,N-Trimethylthiazoline-2-thione	7.45 (π)	7.94 (n)	31
	10.00 (π)	11.94 (σ)	
4,5-Dimethylthiazoline-2-thione	7.56 (π)	8.04 (n)	31
	10.44 (π)	12.16 (σ)	
Oxazolidine-2-thione	8.37 (n)	8.70 (π)	31
	12.81 (σ)	13.28 (σ)	
4,5-Dimethyloxazoline-2-thione	7.74 (π)	8.22 (n)	31
	11.71 (π)	12.58 (σ)	
4,5,N-Trimethyloxazoline-2-thione	7.54 (π)	8.03 (n)	31
	11.15 (π)	12.33 (σ)	

[a] See ref. 19 for data on thiophene and its derivatives.

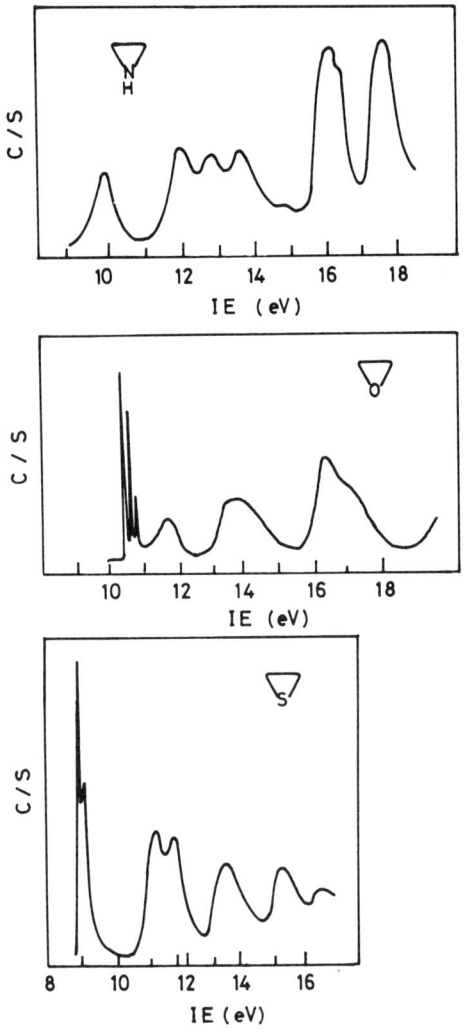

Fig. 3.1. UV photoelectron spectra of azirane, oxirane, and thiirane. [After Basch and coworkers (10) and Aue and coworkers (11).]

3.2.1. Three-Membered Rings

Ultraviolet photoelectron spectra of oxirane and substituted oxiranes have been studied by many workers (8–11, 33). In all these compounds, the first IE is assigned to the lone-pair orbital of the oxygen atom. The assignments of the second and third bands of oxirane to a_1 and b_2 orbitals made by Basch and coworkers (10) have been confirmed by McAlduff and Houk (8) by analysis of the UVPS bands of a series of substituted oxiranes. Spectra of thiirane and its dirivatives have been compared with those of oxirane for the purpose of band assignments (8). Alkyl

Saturated Heterocyclic Compounds 239

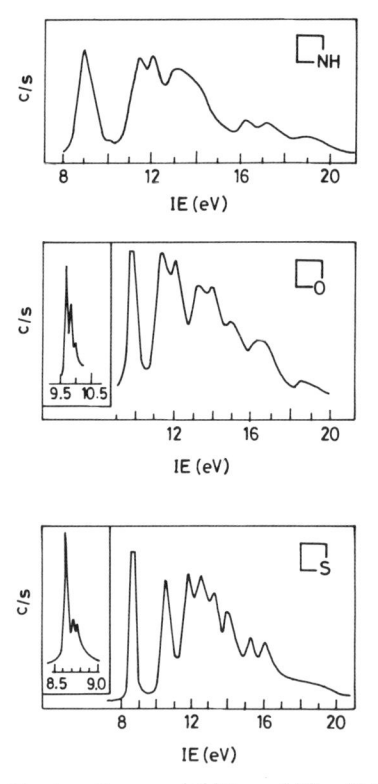

Fig. 3.2. UVPS of azetane, oxetane, and thietane. [After Mollere and Houk (35).]

substitution in the ring decreases the IEs, but this does not obviously relate to the type of orbital involved. Effects of the alkyl and other groups of the four lowest IEs of oxirane are linearly related to the electronegativities of groups. Changes in the observed IEs agree with those predicted by Koopmans' theorem using *ab initio* (STO-3G) MO calculations. Using the He(I)/He(II) intensity ratio of the UVPS bands of oxirane and its –CH$_2$– and –S– analogs, Schweig and Thiel (9) have reassigned the two bands in the region 15–18 eV. These assignments are, however, opposite to the sequence of the respective *ab initio* eigenvalues. Aue and coworkers (11) have measured proton affinites and UVPS bands of oxirane, aziridine, phosphirane, and thiirane. Based on *ab initio* (STO-4-31G) calculations and structural correlations within these molecules, these workers propose band assignments that are in essential agreement with those of earlier workers.

Yashikawa and coworkers (12) have studied the UVPS of aziridine and its 2-methyl and 2,2-dimethyl derivatives along with other cyclic amines. The lone-pair IEs are linearly related to their *s*-character and also to their basicities as determined by ion cyclotron resonance spectroscopy in the vapor phase. Basch and coworkers (10) have found a vibrational spacing of 700 cm^{-1} in the first band of aziridine (9.8 eV). The decrease in the first IE due to methyl substitution at the 2-position of aziridine seems to be additive as in the oxirane system.

240 Ultraviolet Photoelectron Spectroscopy of Heterocyclic Compounds

Ultraviolet photoelectron spectra of thiirane have been studied by several workers (8, 9, 11, 34). Ionization energies of thiiranes have been assigned on the basis of both semiempirical and *ab initio* MO calculations employing different basis sets (8, 11, 34). The first IE around 9.05 eV is undoubtedly due to the sulfur lone-pair. Using the relative band intensities in the He(I) and He(II) photoelectron spectra of thiirane, Schweig and Thiel (9) assigned the last two bands in the region 15–18 eV to $3a_1(\sigma)$ and $1b_1(\sigma)$ orbitals.

Ultraviolet photoelectron spectra of phosphirane have been examined by Aue and coworkers (11). A correlation chart for the UVPS spectra of the three-membered N, O, and S heterocyclic compounds and of cyclopropane is shown in Figure 3.3.

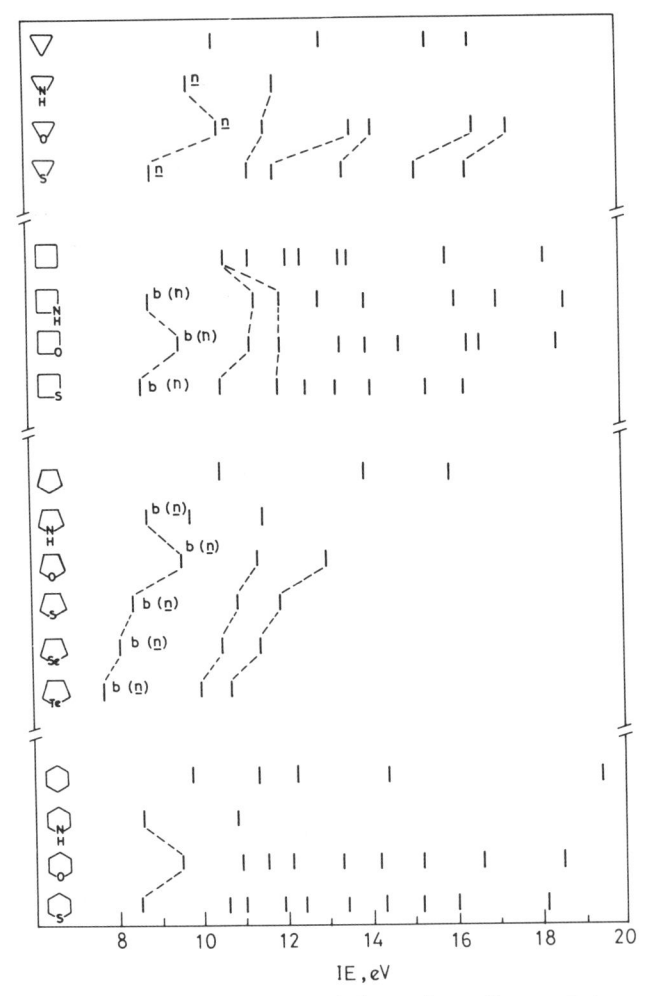

Fig. 3.3. Correlation chart for the UVPS of three-, four-, five-, and six-membered rings. [After Rao and coworkers (5).]

Ultraviolet photoelectron spectra of thiirane dioxide (13), 1,2-dimethyl thiirane dioxide (14), and thiirene dioxide (14) have been reported. The first bands in thiirane dioxide and its methyl derivative are assigned to a π_{SO_2} orbital of b_2 symmetry. Müller and coworkers (14) have shown that in 1-methylthiirene dioxide as well as 1,2-dimethylthiirene dioxide there is strong hyperconjugative interaction between the π_{CC} molecular orbital and the σ_{SO_2} orbital. There is moderate mixing between the π_{CC} orbital and the vacant $\sigma_{SO_2}^*$ orbital, which is a nearly pure d orbital of sulfur. There is stabilization of the π_{SO_2} orbital in thiirene dioxide compared to the π_{SO_2} orbital in thiirane dioxide.

3.2.2. Four-Membered Rings

Photoelectron spectra of heterocyclobutanes have been reported (12, 35). The first band in these compounds arises from the ionization of the lone-pair orbital (Fig. 3.3). The presence of the heteroatom perturbs both the energies and shapes of the MOs of cyclobutane and also lifts the degeneracy of the highest occupied MO (of e_u symmetry), which is a pair of Walsh σ orbitals (36). The UVPS bands of these derivatives have been assigned by comparison with minimal basis set STO-3G *ab initio* calculations and by correlations with the spectra of acyclic analogs and of three-membered ring heterocycles (35). The shifts in orbital energies of both the three-membered and four-membered systems have been explained by Mollere and Houk (35) employing two models: (1) considering orbital energy changes resulting from cyclization and ring size and (2) orbital energy changes resulting from interactions due to mixing of orbitals of heteroatom and alkyl fragments. The second and third bands of oxetane, azetidine, and thietane assigned to $a_1(\sigma)$ and $b_2(\sigma)$ orbitals, respectively, can be correlated to the e_u orbital of cyclobutane.

3.2.3. Five-Membered Rings

Photoelectron spectral bands of tetrahydrofuran and its analogs have been reported (15, 16, 37). Bain and coworkers (16) have assigned the first IE of tetrahydrofuran to the lone-pair orbital of the oxygen atom. Pignataro and Distefano (15) have investigated the UVPS bands of tetrahydrofuran on the basis of local C_{2v} symmetry of the CH_2-O-CH_2 fragment. Mixing of the nonbonding electrons of the heteroatom with the σ_π system seems apparent from the shape of the band and also from the comparable values of IEs of H_2O molecule. Bain and coworkers (16) suggest a strong n-π mixing of orbitals and a weak n-σ and π-σ mixing in 2,5-dihydrofuran. Based on UVPS data and MINDO/2 calculations, Schmidt and Schweig (37) have proposed that n-π mixing is hyperconjugative rather than transannular in 2,5-dihydrofuran. Bloch and coworkers (38) have reported UVPS bands of 2,5-dihydrofuran as well as its 3,4-dimethyl derivative and have compared experimental IEs with those obtained from *ab initio* STO-3G MO calculations; the study confirms the suggestion of Schmidt and Schweig that the transannular interaction is less important.

The aza analog of tetrahydrofuran shows the first UVPS band, which is due mainly to the lone-pair on nitrogen at 8.77 eV (12, 17). When the NH group is replaced by an NMe group, the first IE decreases by 0.4 eV. In amines and enamines of pyrrolidine, lone-pair IEs of nitrogen are related to their gas-phase basicites (39) and the pK_a values vary inversely with the lone-pair IEs. The relative basicity varies in the order: secondary amines < tertiary amines < enamines. In β,γ-unsaturated five-membered cyclic amines, the nitrogen lone-pair is perturbed and destabilized due to interaction with the nonadjacent $\pi_{C=C}$ MO. Bis-homoallyl conjugation favors through-space n-π interaction with destabilization of the lone-pair orbital. When the NH group is replaced by an NMe group, the first IE decreases by ~ 0.4 eV in these systems as well; the second IE due to the $\pi_{C=C}$ MO decreases by only about 0.1 eV.

The phosphorus analog of tetrahydrofuran with a P–n-C$_4$H$_9$ group shows three bands in the UVPS at 8.25 (n_P), 10.6 (σ_{P-C}) and 10.7 (σ_{P-C}) eV (22). When the n-C$_4$H$_9$ group is replaced by a phenyl group, the lone-pair orbital is destabilized and the first UVPS band occurs at 8.35 eV (22).

Schmidt and Schweig (20) have studied the UVPS of 2,3,4,5-tetrahydrothiophene and 2,5-dihydrothiophene and have shown that in the latter, the $\pi_{C=C}$ part interacts hyperconjugatively with the lone-pair of sulfur, transannular interaction being absent. The first band in both these compounds is due to the ionization of the lone-pair orbital. The sulfur lone-pair is stabilized in 2,5-dihydrothiophene, unlike in the oxygen and nitrogen analogs, where the lone-pair is destabilized (20). Analyzing the shape of the first UVPS band of 2,3,4,5-tetrahydrothiophene and by comparing the IE with that of H$_2$S, Pignataro and Destefano (15) proposed that there is mixing of the nonbonding electrons of the heteroatom with the σ_π system. The photoelectron spectrum of 2.5-dihydrothiophene-1,1-dioxide has been reported by Solouki and coworkers (21). Employing a procedure similar to that with tetrahydrofuran and tetrahydrothiophene, Pignataro and Distefano (15) interpret the UVPS bands of tetrahydroselenophene and tetrahydrotellurophene. The correlation diagram of saturated five-membered heterocyclic compounds is shown in Figure 3.3.

In the UVPS of α-butyrolacetone and α-crotonolactone (24), the band due to the lone-pair orbital of the carbonyl group is found at 10.26 and 10.70 eV, respectively. In α-butyrolactone, the second band at 10.94 eV is assigned to the lone-pair orbital of the ring oxygen, and the band at 12.44 eV to the $\pi_{C=O}$ MO (18, 23). When the oxygen atom in the ring is replaced by an NH group, the lone-pair IE from the carbonyl group decreases to 9.53 eV and the lone-pair IE due to the nitrogen atom occurs at 9.76 eV; the π_{CO} IE also decreases to 11.91 eV (18). If the carbonyl group is in the β position relative to the ring N(>NCH$_3$ group), the first IE due to the carbonyl lone pair remains at 9.53 eV, but the second IE due to lone-pair ionization of the nitrogen decreases to 8.83 eV. This behavior shows that there is little through-space interaction between the nitrogen lone-pair and the carbonyl π system. The aza analog of α-butyrolactone shows a large through-space interaction leading to a stabilization of the nitrogen lone-pair and destabilization of the π system.

Ultraviolet photoelectron spectra of maleic anhydride, succinic anhydride, and succinimide, as well as of substituted succinimides, have been reported (18, 24–26). The first band in maleic anhydride is reported at 11.45 eV by Bain and Frost (24); Almemark and coworkers (25) report this band at 11.1 eV. A study of the UVPS bands of succinic anhydride and succinimide as well as its N-chloro and N-bromo derivatives suggests that succinic anhydride and succinimide are skeletally planar systems. The two lone-pair orbitals therefore split, due to interaction between the two carbonyl groups, giving two IEs (18, 26); the π_{CO} orbitals are also split. The magnitude of n-orbital splittings in the anhydride and the imide are about 0.72 eV and 0.87 eV, respectively. In N-chloro and N-bromo succinimides there is no such interaction, since they are substantially nonplanar in the gas phase (18, 26).

UVPS of 1,3-dioxalane and 1,3-dithiolane have been reported (27, 28, 40). Baker and coworkers (28) have shown that in both dioxalane and dithiolane, the sulfur $3p$ AOs interact by a through-bond rather than through-space mechanism. CNDO/2 calculations on these two systems based on a planar geometry gave splittings of lone-pair orbitals that agree well with the experimental results. In 1,4,6,9-tetrathia-spiro[4.4]nonane, there is evidence for through-space spiro interaction of sulfur $3p$ AOs giving rise to four MOs, one of a_2 symmetry, one of b_2 symmetry, and two (a degenerate pair) of e symmetry (28); Kobayashi and coworkers (40), however, propose through-bond interaction based on MO calculations with variation in dihedral angles.

UVPS bands of vinylene carbonate, ethylene carbonate, and propylene carbonate have been assigned on the basis of the observed vibrational structures and MO calculations (23, 24). Jinno and coworkers (23) have assigned the first two bands in these compounds to the oxygen lone-pair and the nonbonding π orbitals. Wittel and coworkers (41) have studied UVPS bands of vinylene and ethylene carbonates as well as thiocarbonates and also of methylene thiocarbonate. They assign the low-energy bands (which are well separated in the unsaturated compounds) to the lone-pair and π-type ionizations.

Guimon and Pfister-Guillonzo (29) find that in both ethylene trithiocarbonate and ethylene dithiocarbonate the first band is due to the lone-pair orbital of the exocyclic heteroatom. The second band in the trithiocarbonate has a contribution from π_{CS}, while the third band in the dithiocarbonate has a contribution from $\pi_{C=O}$. In both these compounds there is through-bond mixing of the lone-pair and π orbitals. Electronic structures of 1,3-dithiole-2-thione and its selenium analogs have been studied by UVPS and polarized electronic spectra by Spanget-Larsen and coworkers (30). Based on comparisons of the UVPS bands of these compounds with the results of MO calculations, the first five bands have been assigned to $5b_2(\sigma)$, $3b_1(\pi)$, $1a_2(\pi)$, $2b_1(\pi)$, $4b_2(\sigma)$, and $7a_1(\sigma)$ orbitals, respectively. Bands in UV polarized absorption spectra were assigned to π-π^* and n-π^* transitions. Guimon and coworkers (31, 32) have assigned UVPS bands of imidazoline-2-thione, imidazolidine-2-thione, thiazoline-2-thione, thiazolidine-2-thione, oxazoline-2-thione, and oxazolidine-2-thione by studying the methyl derivatives of these compounds. The assignements were confirmed by CNDO calculations. In imidazoline-, thiazoline-, and oxazoline-2-thiones, the first band is due to a $\pi(C=S)$ orbital, and

244 Ultraviolet Photoelectron Spectroscopy of Heterocyclic Compounds

all the bands show strong vibrational structure; the second band is assigned to the lone-pair of sulfur in C=S. That there is a through-band coupling in these systems becomes evident from the vibrational structure appearing in the second band due to lone-pair ionization. In imidazolidine-, thiazolidine-, and oxazolidine-2-thiones,

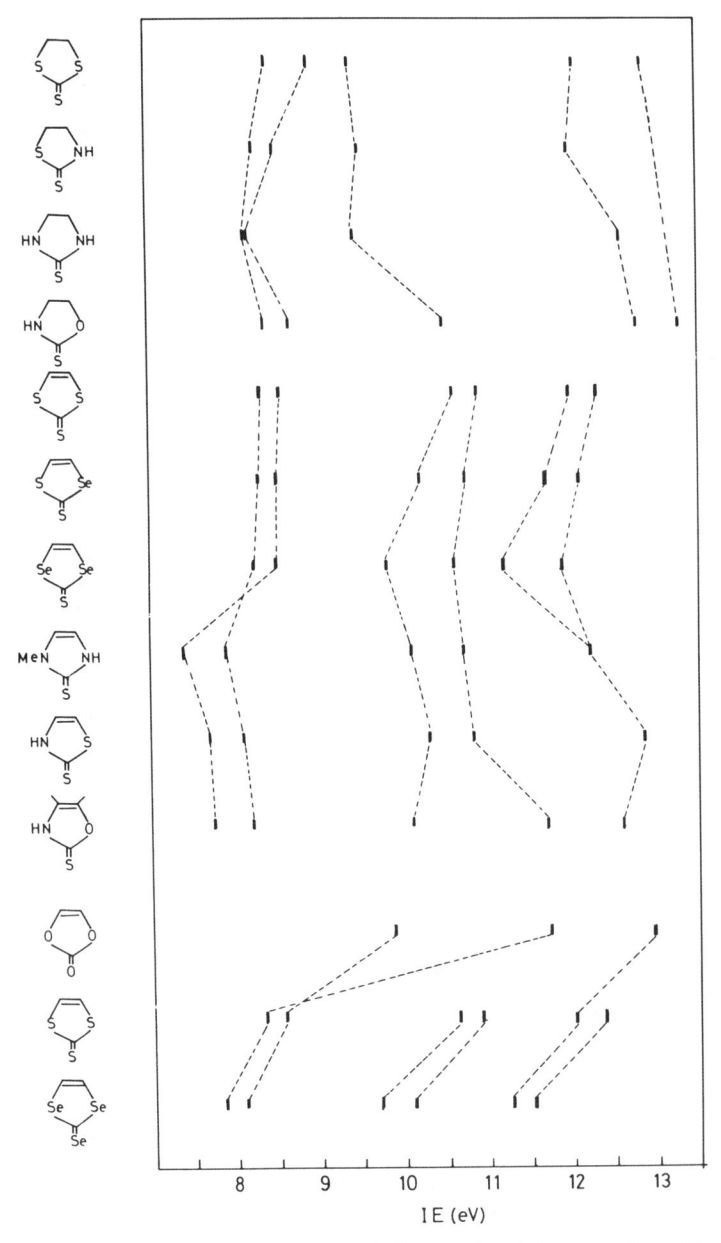

Fig. 3.4. Correlation chart for the UVPS of five-membered heterocyclics with exocyclic C=S.

there is a reversal of bands, the first one being due to a lone-pair of sulfur from the C=S group and the second to the π orbital of C=S. The second band in all these systems shows about the same vibration spacing (700–800 cm^{-1}). In the chemically related compound where the C=S group is replaced by a C=C(CN)$_2$ group, the first UVPS band at 8.31 eV is due to the π orbital of b_2 symmetry followed by a band due to the lone-pair orbital of ring nitrogen (42). Figure 3.4 is a correlation chart for the UVPS bands of five-membered heterocyclic comppounds containing exocyclic thiocarbonyl groups.

Guimon and coworkers (43) have studied UV photoelectron spectra of 1,3,4-triazoline-2-thione, -2-one, and 1,3,4-thiadiazoline-2-thione derivatives. Based on these data, the influence of exocyclic groups on the nature and energies of the MOs have been evaluated; furthermore, the nucleophilic reactivity of these and other heterocyclic thiocarbonyl compounds have been related to the lone-pair IEs of the thiocarbonyl sulfur. UVPS spectra of pyrazolinethione and 1,2,3-triazolinethione and their α-C—Me derivatives have also been studied by Guimon and coworkers (44). Using perturbation theory it has been shown that the C—Me groups stabilize the MOs in these compounds with a conjugated thiocarbonyl system. The first bands at 7.55 and 7.97 eV of pyrazolinethione and 1,2,3-triazolinethione, respectively, have contributions from both the π and n_s orbitals. The UVPS bands of thiadiazole have been discussed in the light of CNDO-CI calculations (45). Solouki and coworkers (46) have studied thiadiazole-type cumulene molecules by UVPS and assigned bands by comparison with the data on chemically related compounds as well as with results from CNDO calculations.

Cetinkaya and coworkers (47) have studied the UVPS bands of imidazolidine dimer and have assigned the bands to π levels of various symmetries based on Hückel calculations. UVP spectra of tetrathiafulvalene and its derivatives have been studied by Gleiter and coworkers (48, 49), Barlinsky and coworkers (50), and Schweig and coworkers (51). Gleiter and coworkers (48, 49) have interpreted the UVPS bands with the aid of semiempirical MO calculations. The first six bands of tetrathiafulvalene are due to ionizations of $4b_{1u}(\pi)$, $3b_{3g}(\pi)$, $2a_u(\pi)$, $2b_{2g}(\pi)$, $9b_{1g}(\sigma, n)$, and $3b_{1u}(\pi)$ orbitals. Barlinsky and coworkers (50) have, however, assigned the bands to $3b_{1u}(\pi)$, $2b_{2g}(\pi)$, $2b_{1u}(\pi)$, $2a_u(\pi)$, $b_{3g}(\pi)$, and $1b_{2g}(\pi)$ orbitals. Electronic absorption spectra of tetrathiafulvalene and its tetramethyl-, tetramethyltetraselena-, and tetrakis (trifluoromethyl) derivatives have been reported (48, 49). The first bands in all these cases are due to π-π^* transitions. Schweig and coworkers (51) have studied the UVPS bands of tetrathiafulvalene as well as its diselena and tetraselena analogs. They have correlated the IEs with those obtained from CNDO/S calculations. They have also shown that the bands at 9.63, 10.03, and 11.01 eV of tetrathiafulvalene are possibly due to orbitals localized on the heteroatoms, since they are strongly affected by varying the heteroatom. Tetrahydrotetrathiafulvalene, a sulfur analog of imidazolidine dimer, shows UVPS bands due to orbitals whose ordering is the same as that of tetrathiafulvalene (49). Barlinsky and coworkers (50), however, propose a different ordering of orbitals for this compound.

3.2.4. Six-Membered Rings

Ultraviolet photoelectron spectral bands of tetrahydropyran, tetrahydrothiopyran, and piperidine are correlated with those of cyclohexane in Figure 3.3. The first IE of tetrahydropyran at 9.48 eV is due to the equatorial type of lone-pair orbital (n_{eq}) of a' symmetry as assigned to Kobayashi and Nagakura (52), while the second band at 10.90 eV is due to the axial lone-pair orbital (n_{ax}) of a' symmetry. The corresponding IEs of tetrahydrothiopyran are 8.39 and 10.55 eV (53). The first band of the sulfur analog shows vibrational structure (spacing of 800 cm^{-1}), and there appears to be no evidence for $3d$ orbital participation of sulfur. Piperidine shows a band due to the lone-pair of nitrogen at 8.64 eV; the corresponding NCH$_3$ derivative shows the first band at 8.29 eV (39).

Photoelectron spectra of N-chloro and N-bromo derivative of piperidine have been measured and compared with the spectrum of piperidine (54). IEs associated with the nitrogen and halogen (X) lone-pair orbitals indicate that there is only a small n_N-n_X interaction. The halogen character is localized in both the N-halopiperidines, with most IEs showing simple inductive shifts resulting from halogenation. Bodor and coworkers (55) have studied the UVPS of 2,2,6,6-tetramethyl- and 2,2,6,6-tetramethyl-4-one derivatives of piperidine and N-chloropiperidine. They have defined the extent of the interaction between the nitrogen lone-pair and the carbonyl group or the chloro group. In these compounds, the first band is due to the nitrogen lone-pair. The IE corresponding to the onset of the σ band increases from piperidine to piperidine-4-one. Amines and enamines drived from piperidine have been studied by UVPS by Colonna and coworkers (39). Relating the first UVPS bands (due to the nitrogen lone-pair) to gas-phase basicities, these authors have shown that the relative order of basicity is secondary amines < tertiary amines < enamines.

Spectra of 1,3-dioxane and 1,3-dithiane as well as the corresponding 1,4 derivatives have been studied (27, 28, 40, 52, 56). The first two bands of these compounds are assigned to lone-pair ionizations. In the case of 1,3-dioxane, Kobayashi and Nagakura (52) assigned the first IE to the asymmetric combination (n_{eq}^-) of the two n_{eq} orbitals of the oxygen atoms. The ordering of the orbitals is $n_{eq}^-, n_{ax}^+, \sigma, n_{ax}^-, \sigma, n_{eq}^+$. For 1,4-dioxane, the ordering of the molecular orbitals is n_{eq}^{anti} (antiparallel combination of the two oxygen atoms), $n_{ax}^{anti}, \sigma, n_{ax}^{para}$ (parallel combination of the two n_{ax} orbitals of the oxygen atoms), $\sigma, \sigma, n_{eq}^{para}$. These interactions have been interpreted in terms of through-space and through-bond mechanisms. In both dioxane and dithiane systems, the interaction seems to be predominantly through-bond in the 1,4 compounds and the interaction is greater in dioxane; in 1,3 systems, the interaction is predominantly through-space and is greater in the sulfur compound. In 4-thia-1,4-dioxane, the lone-pair of oxygen and sulfur interact by through-bond mechanism and the two bands are at 8.67 and 10.00 eV. In 1,3 and 1,4-dioxane, the interaction strength is larger for 1,4-dioxane (higher value of splitting), while for the sulfur analogs the values are comparable. UVPS of amines and enamines of morpholine have been reported, and the lone-pair IE due to nitrogren has been correlated with gas-phase basicities (39).

In 1,3,5-trioxane as well as 1,3,5-trithiane, lone-pair orbitals on the heteroatoms interact both by through-space and through-bond mechanisms. The Δ(IE) between the first two bands of the oxygen-containing molecule is about 0.35 eV, while for the sulfur analog the value is around 0.51 eV. Furthermore, there is some evidence for Jahn-Teller effect operating in 1,3,5-trioxane (27). UVPS bands of 1,5,7,11-tetrathiaspiro[5.5]undecane have been reported by Kobayashi and coworkers (40) and have been assigned based on MO calculations. The calculated IEs agree with experimental IEs when through-bond interaction is taken into account, and the splitting is 0.55 eV.

The first UVPS band of 2,3-dihydropyran appears at 8.56 eV (53, 57). The band exhibits strong vibrational structure with a spacing of about 1350 cm^{-1}. Based on *ab initio* calculations using the STO-3G basis set, all the bands of 2,3-dihydropyran have been assigned by Bloch and coworkers (57). The first band arises from a π orbital. In 4,4-dimethyl-2,3-dihydrothiopyran, the first band due to the π orbital appears at 8.06 eV (53). Unsaturated ethers and thioethers exhibit a π-π^* band in UV spectra on which the s,p,d-type Rydberg series converging to the first (π-type) IE is superimposed. There is no evidence for the participation of the sulfur 3d orbital in the low-lying excited states of unsaturated thioethers, unlike in the case of saturated thioethers. In the saturated thioether (tetrahydrothiopyran), the UV spectrum is interpreted in terms of at least five closely spaced Rydberg series of the s, p, or d type; they converge to the first IE (53). 2,3-Dihydro-6-methylpyran shows a band at 8.40 eV due to the ionization from a π orbital (58). Bloch and coworkers (57) have reported UVPS bands of 1,4-dioxin, 2,3-dihydro-1,4-dioxin, and pyran and have assigned the bands based on *ab initio* (STO-3G) calculations. The first bands in pyran and 2,3-dihydro-1,4-dioxin show vibrational structure with spacings of 1570 ± 50, 530 ± 50 cm^{-1} and 1560 ± 50, 470 ± 50 cm^{-1}.

Lauer and coworkers (59) have reported the UV photoelectron spectrum of dithiatricyclotetradecadiene and have show it to have an undistorted [4]annulene system in the singlet ground state. Worley and coworkers (60) have reported the UVPS of three piperizine-2,5-diones and their N,N-dichloro derivatives. Based on the oxygen and nitrogen lone-pair IEs of these molecules as well as MINDO/3 calculations on the nonchlorinated derivatives, band assignments have been proposed and the predominant conformations of the molecules predicted. Solouki and coworkers (46) have assigned the UVPS bands of six-membered thiadiazole radical cation states based on comparison between chemically related compounds and by CNDO calculations. Betteridge and coworkers (42) have reported the UVPS bands of piperimidine with a cyano-ethylenic group attached at the 2-position. Bischof and coworkers (61) have studied the UVPS bands of tetrazene and have assigned the first four bands to $b(\pi)$, $b(n_-)$, $a(\pi)$, and $a(n_+)$ orbitals. In Figure 3.5 we have presented a correlation diagram of unsubstituted six-membered saturated heterocyclic compounds containing more than one oxygen or sulfur atom.

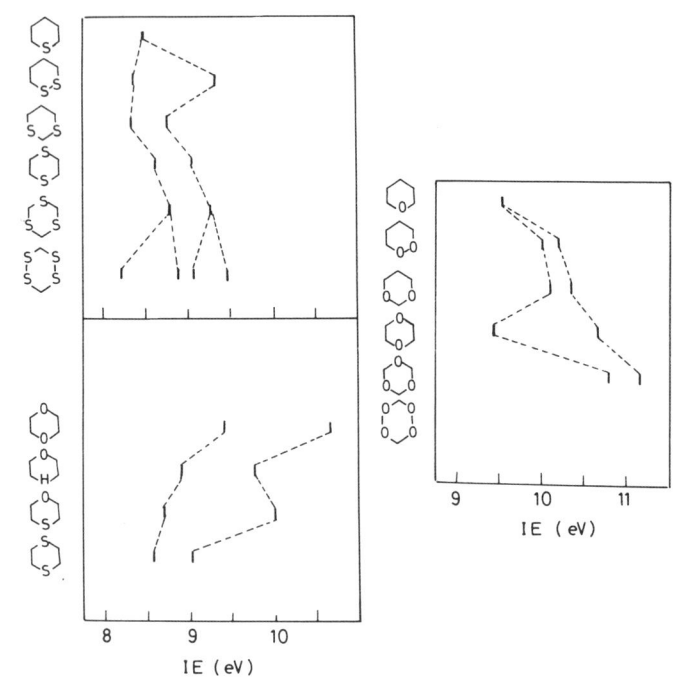

Fig. 3.5. Correlation chart for the UVPS of six-membered heterocyclics with one or more X (X = O, S).

3.3. HETEROAROMATIC COMPOUNDS

Heteroaromatic compounds show characteristic bands in UVPS due to the π orbitals of the aromatic system in addition to the bands due to the lone-pair (n) orbitals of the heteroatoms. It may often be difficult to distinguish the π and n bands, owing to their close proximity. In this section we shall discuss the spectra of five- and six-membered heteroaromatic compounds.

3.3.1. Five-Membered Compounds

Figure 3.6 shows typical spectra of three heteroaromatic five-membered compounds, and Figure 3.7 presents a correlation chart for the UVPS bands of these systems. Table 3.3 lists data on some important systems.

UVPS of furan, pyrrole, selenophene, and tellurophene have been investigated by several workers (62–69). Derrick and coworkers (62, 63) have tried to assign the bands of furan on the basis of the UV absorption spectrum, vibrational analysis, and extended Hückel calculations. De Alti and coworkers (71) have calculated the photoelectron and electronic absorption spectra of furan by the MS–SCF–Xα method. The first band of furan at 8.88 eV arises from the $1a_2$ orbital (of π_{CC} nature), while the second band at 10.31 eV is due to the $2b_1$ orbital,

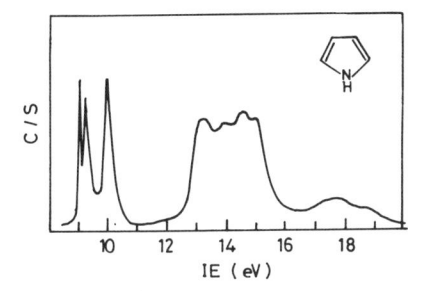

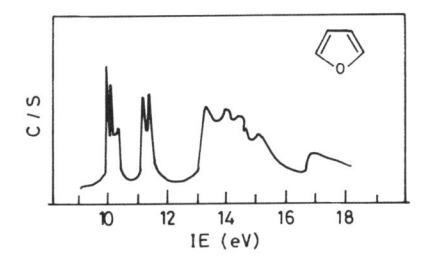

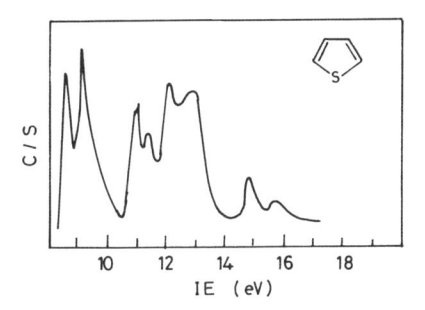

Fig. 3.6. UV photoelectron spectra of pyrrole, furan, and thiopene. [After Baker and co-workers (67).]

which has a major contribution from the lone-pair orbital of the ring heteroatom. Based on the relative intensities of the first bands in furan, pyrrole, and thiophene in He(I) and He(II) spectra, Schweig and coworkers (64, 70) have proposed the same assignments for the first bands of these three molecules.

The spectra of five-membered heteroaromatics are compared in Figure 3.7. Based on vibrational analysis of the UVPS bands of thiophene and its 2- and 3-bromo derivatives, Rabalais and coworkers (65) have shown that bromine-sensitive ring-stretching and ring-breathing modes are excited together in many of the bands, indicating possible participation of the Br atomic orbitals along with the ring system. Vibrational analysis and MO calculations have also shown the presence of two types of lone-pair orbitals, the more bonding out-of-plane a''

Table 3.3. Ionization Energies of Five-Membered Heteroaromatic Compounds

Compound	Ionization Energies				Ref.
Furan	8.89 (8.88) (π_{CC}; $1a_2$)	10.31 (10.30) (n, π_{CC}; $2b_1$)		12.87 (12.68)	63, 69, 70
	13.70	14.35	15.02	17.4 ($7a_1$)	
2-Methylfuran	8.37 (π_3)	10.13 (π_2)			69
2.Carboxylfuran	9.16 (π_3)	10.72 (π_2)			69
2-N,N-Dimethylamidefuran	8.86 (π_3)	9.65	10.41 (π_2)		69
2-Acetoxyfuran	9.00 (π_3)	10.56 (π_2)	11.1		69
2.Nitrofuran	9.75 (π_3)	~ 10.9	11.13 (π_2)	~ 11.50	69
2-Cyanofuran	9.47 (π_3)	10.99 (π_2)			69
2-Methylthiofuran	8.58 (π_3)	10.32 (π_2)			69
Pyrrole	8.22 (8.22) ($1a_2$)	9.22 (9.22) ($2b_1$)	12.70	12.95	63, 67, 70
	13.6	14.4	14.95	17.8	
N-Methylpyrrole	7.95 (7.95)	8.80 (8.80)	12.67 (11.92)	13.8	67
	17.1				
P-n-Butylphosphole	8.45 (n; a_2)	10.6 (P–C; b_1)			22
3,4-Dimethyl-P-n-butylphosphole	8.15 (n; a_2)	10.20 (P–C; b_1)			22
3,4-Dimethylphosphole	8.05 (n; a_2)	9.95 (P–C; b_1)			22
3.4.P-Trimethylphosphole	8.25 (n; a_2)	10.35 (P–C; b_1)			22
Thiophene	8.87 (π; $1a_2$)	9.49 (π; $2b_1$)	11.86 (11.46)	12.4	19, 63, 67–69
	13.1	14.2	16.4	17.5	
2-Methylthiophene	8.43 (π_3)	9.23 (π_2)	11.8	12.4	67, 69
	13.0	13.7	16.1		
3-Methylthiophene	8.54 (8.40)	9.11 (9.11)	11.94	12.5	67
	13.1	13.8	15.9	17.2	
2,5-Dimethylthiophene	8.09 (π_{CC}; $1a_2$)	8.93 (n; $2b_1$)			64
2-Chlorothiophene	8.87 (8.70)	9.62 (9.62)	11.55 (11.55) (n_{Cl})		67
	12.02 (n_{Cl})	12.36	13.88	14.37	
	16.4	17.15			
2,5-Dichlorothiophene	8.80 (8.60)	9.78 (9.71)	11.58 (n_{Cl})	11.9 (n_{Cl})	67
	12.8	13.1	14.1	14.4	
	16.7	17.9			

2-Bromothiophene	8.82 (π_3)	9.58 (π_2)	10.74 (10.74) (n_{Br})	67, 69
	11.28 (n_{Br})	12.18	13.25 13.85	
	16.4			
3-Bromothiophene	9.00 (8.90)	9.51 (9.51)	10.67 (10.67) (n_{Br})	67
	11.47 (n_{Br})	12.3	13.43 14.1	
	14.4	16.4	17.75	
2-Iodothiophene	8.55 (8.55)	9.47 (9.47)	10.00 (10.00) (n_I)	67
	10.65 (n_I)	12.17	12.59 13.5	
	16.0			
2-Carboxylthiophene	9.14 (π_3)	9.73 (π_2)	10.61 11.67	69
2-Acetoxythiophene	8.98 (π_3)	9.61 (π_2)	10.37 11.03	69
2-Nitrothiophene	9.73 (π_3)	10.21 (π_2)	~ 10.80 11.40	69
2-Chloromethylthiophene	8.89 (π_3)	9.49 (π_2)		69
Furan-2-aldehyde	9.37 (π_3)	9.87 (π_2)		69
2-Methylthiofuran	8.63 (π_3)	9.37 (π_2)		69
2-N,N-Dimethylamidethiophene	8.84 (π_3)	9.40 (π_2)		69
Selenophene	8.92 (π; a_2)	9.18 (π; $2b_1$)	9.63	69
2-Methylselenophene	8.40 (π_3)	8.96 (π_2)		69
2-Chloroselenophene	8.83 (π_3)	9.34 (π_2)	11.34 (n_{Cl}) 11.70 (n_{Cl})	69
2-Carboxylselenophene	9.19 (π_3)	9.45sh (π_2)	10.58 11.81	69
2-Acetoxyselenophene	9.05 (π_3)	9.26sh (π_2)	10.33 11.00	69
2-Nitroselenophene	9.64 (π_3)	9.88sh (π_2)	10.65 11.22	69
2-N,N-Dimethylamideselenophene	8.85 (π_3)	9.10 (π_2)	9.63	69
Tellurophene	8.40 (π_3)	8.81 (π_2)	10.8	67, 69
2-Methyltellurophene	8.20 (π_3)	8.43sh (π_2)		69
2-Chlorotellurophene	8.68 (π_3)	8.89 (π_2)	10.86 (n_{Cl}) 11.24 (n_{Cl})	69
2-Bromotellurophene	8.59 (π_3)	8.84 (π_2)	10.43 (n_{Br}) 10.96 (n_{Br})	69
2-Iodotellurophene	8.34 (π_3)	8.52 (π_2)	9.71 (n_I) 10.40 (n_I)	69
2-Carboxyltellurophene	8.62 (π_3)	9.15 (π_2)	9.71 10.40	69
2-Acetoxytellurophene	8.51 (π_3)	9.00 (π_2)	10.21 11.00	69
2-N,N-Dimethylamidetellurophene	8.39 (π_3)	8.89 (π_2)	9.54	69

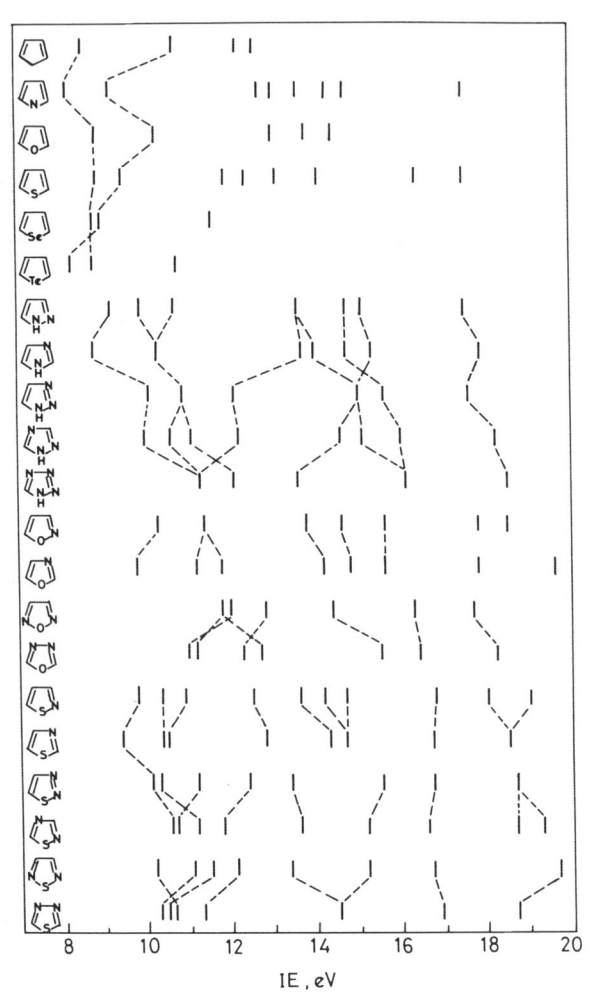

Fig. 3.7. Correlation diagram of ionization energies of five-membered heteroaromatic compounds. [After Rao and coworkers (5).]

orbital and the in-plane a' orbital. Intensification of the bands due to π IEs in bromo derivatives with respect to σ IEs facilitates assignment of the bands due to the π orbitals (65, 66).

Baker and coworkers (67) have studied UVPS bands of halogen-substituted and methyl-substitued thiophene and other five-membered heteroaromatic compounds. These workers have interpreted the shifts in IEs of related molecules qualitatively in terms of the substituent electronegativity. From studies of a series of α-substituted five-membered heteroaromatic compounds, Schäfer and coworkers (68) and Fringuelli and coworkers (69) have shown that there is reversal in the sequence of the two highest occupied molecular orbitals from furan $(1a_2/2b_1)$, thiophene, through selenophene and tellurophene $(2b_1/1a_2)$. De Alti and Decleva (72) have calculated IEs of furan, thiophene, selenophene, and tellurophene by the MS-SCF-Xα method, and their assignments are similar to those of Schäfer and Coworkers (68) and Fringuelli and coworkers (69).

2,5-Di-*tert*-butylthiophene-1,1-dioxide shows an extraordinarily strong through-bond conjugation through the sulfone group, which has been verified by Müller and coworkers (73) by UVPS. Schäfer and coworkers (22) have studied the electronic structure of a series of phospholes by UVPS as well as theoretical methods. The theoretical study shows that aromatic stabilization of these compounds, despite their pyramidal structures, is gained from $n\pi^*$ conjugative and $P-C/\pi^*$ hyperconjugative interactions between two subunits *cis*-butadiene and P–R, where R is an alkyl group. Because of the strong electron-accepting power of the phosphorus d orbitals, there is no appreciable π-charge transfer from the P–R unit to the π part. The UVPS results confirm the strong $P-C/\pi$ hyperconjugation but weak n/π conjugation. Because of this mixing, the n-orbital energy does not provide a suitable means of deducing the aromatic nature of phospholes.

UVPS bands due to the nitrogen lone-pair and the two highest occupied orbitals of imidazole and 1-substituted imidazoles have been examined by Ramsay (74) in the light of the results from CNDO/2 calculations. It has been found that the n and π_2 IEs are resolvable only in 1-acetyl and 1-trifluoroacetylimidazoles. Methyl-substituted imidazoles have been studied by Klasinc and coworkers (75), who also discuss the UVPS bands of imidazole on the basis of MO calculations. These workers have shown that methyl substitution in imidazole, apart from a nearly constant destabilization of the low-energy UVPS bands, causes little change in the band shapes, indicating the absence of two tautomers of imidazole. Ramsay (74) has shown a linear relationship between the pK_a values of substituted imidazoles and the CNDO lone-pair orbital energies; the relation is limited to those substituents which leave the n-orbital electron density at the nitrogen unchanged. Using this relation and a perturbation MO method, he proposes that substituent interaction is mainly through inductive and field effects, with the exclusion of through-bond resonance interaction between the substituent and the nitrogen lone-pair orbital. Kajfez and coworkers (76) have reported the UVPS bands of methylnitroimidazole and have discussed the substituent effects on the electronic structure of 5-nitroimidazole.

UVPS bands of oxazole and isoxazole have been assigned on the basis of *ab initio* calculations by Palmer and coworkers (77). The first band of oxazole at 9.83 eV (a'') is about 0.4 eV lower than the first band of isoxazole at 10.26 eV (a''), possibly due to a better through-bond interaction in oxazole. Palmer and coworkers (78), Bernardi and coworkers (79), and Salmona and coworkers (80) have studied UVPS of thiazole, isothiazole, and their derivatives. Palmer and coworkers (78) propose that the S(3d) orbital polarizes the MOs. Including these orbitals, they get better agreement with the observed dipole moments. Bernardi and coworkers (79) have assigned UVPS bands of thiazole by making use of *ab initio* calculations; the spectra of halo derivatives of thiazole are assigned on the basis of perturbation calculations. Salmona and coworkers (80) have performed *ab initio* (STO-3G), CNDO/S, and EHT calculations to compare the electronic structures of thiazole, isothiazole, and their derivatives, and find that in the case of isothiazole there is an inversion of the two highest occupied MOs between STO-3G and CNDO/S calculations. The first three UVPS bands of both thiazole and isothiazole are assigned by all three groups of workers to π_3, π_2, and n_N orbitals. Alkyl-substituted derivatives of thiazole and isothiazole show UVPS bands due to π_3, π_2, and n_N orbitals that vary linearly with σ^* constants of Taft (Fig. 3.8). Such correlations with the Taft σ^* constants have been proposed (5) for several related series of compounds.

Ultraviolet photoelectron spectral bands of 1,3.4-oxadiazole and 1,2,5-oxadiazole have been analyzed by correlating the bands with related oxazoles and also based on *ab initio* calculations (77). In both the oxadiazoles the first band appears at a much higher energy than in oxazole or isoxazole. Cariati and coworkers (81) have studied eleven 3-substituted-5-amino-1,2,4-oxadiazoles by UVPS, IR, Raman, and ^{1}H- and ^{13}C-NMR methods. All the compounds are found in the aminic form; the presence of a carbonyl group in the 3-position effects the electron distribution in the ring, causing a decrease in the C—N bond order.

Guimon and coworkers (82) have studied the UVPS of chloro- and bromo-1,2,4-triazoles using both He(I) and He(II) radiations and also by STO-3G calculations. They show that the preferred tautomers of the title compounds in the vapor phase are 3-chloro- and 5-bromo-1,2,4-1*H*-triazole. Palmer and coworkers (78) have studied the UVPS of isomeric thiadiazoles. By comparing the UVPS bands of 1,2,5-, 1,3,4-, 1,2,3-, and 1,2,4-thiadiazoles with those predicted by *ab initio* calculations as well as with the observed UVPS bands of thiazole and

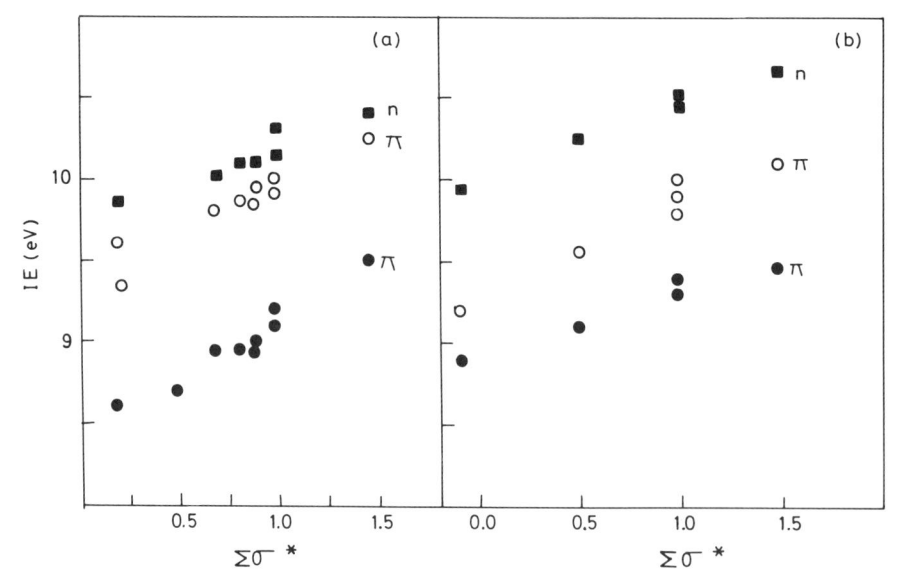

Fig. 3.8. Plot of lone-pair IE of nitrogen and two π IEs versus Taft σ^* constants of substituents. (*a*) thiazoles, (*b*) isothiazoles.

isothiazole, they have shown that the S(3d) atomic orbital mainly polarizes the MOs in these compounds. Inclusion of these S(3d) AOs gives better agreement with the observed dipole moments in both magnitude and direction. They have also shown that 1,3,4-thiadiazole is less aromatic.

Flitsch and coworkers (83) have studied conformations of bipyrryls by UVPS. The splittings of the π_2 ionizations indicate that the most stable conformation of 1,1-bipyrryl and its methyl derivatives is one with the pyrole rings perpendicular to each other. In bipyrrole, the pyrrole rings are coplanar. Meunier and coworkers (84) and Meunier and Pfister-Guillouzo (85) have studied the conformations of 2,2'- and 3,3'-bithienyl and their bromo derivatives by UVPS. Based on the UVPS data, they propose the existence of twisted conformations, the degree of twist depending on the orientation of the substituents; the results agree with those from electron diffraction studies.

3.3.2. Six-Membered Compounds

Ultraviolet photoelectron spectra of six-membered heteroaromatic compounds have been widely studied. In Table 3.4 we have summarized the important data on the spectra of six-membered heteroaromatics, while in Figure 3.9 we present a correlation chart of energy levels in these compounds. In what follows, we shall discuss the characteristic UVPS bands due to the lone pair and the π orbitals. We shall first examine the spectra of pyridine and its derivatives.

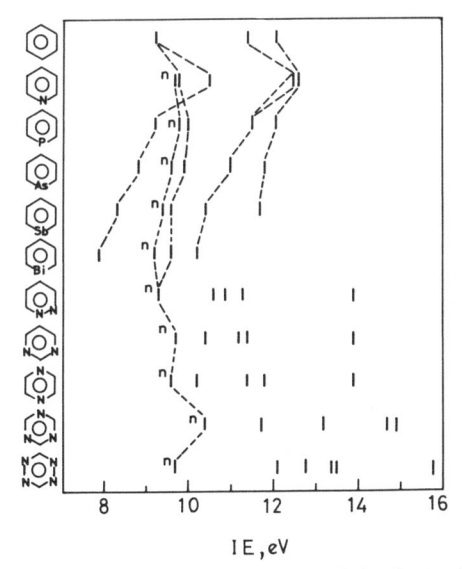

Fig. 3.9. Correlation chart for heteroaromatic compounds isoelectronic with benzene. [After Rao and coworkers (5).]

Table 3.4. Ionization Energies of Six-Membered Heteroaromatic Compounds

Compound	Ionization Energies			Ref.
Pyridine	9.67 (n_N; a_1)	9.80 (π_2; a_2)	10.50 (10.42) (π_3; b_1)	87–89
	12.45 (12.27) (σ; b_2)	12.6 (π; b_1)	13.1	
	13.8	14.5	15.6 (15.6)	
	17.1			
Phosphabenzene	9.2 (π; $3b_1$)	9.8 (π; $1a_2$)	10.0 (σ, n; $3a_1$)	89
	11.5 (σ; $8b_2$)	12.1 (π; $2b_1$)		
Arsabenzene	8.8 (π; $5b_1$)	9.6 (π; $2a_2$)	9.9 (σ,n; $17a_1$)	89
	11.0 (σ; $10b_2$)	11.8 (π; $4b_1$)		
Stibabenzene	8.3 (π; $7b_1$)	9.4 (π; $3a_2$)	9.6 (σ, n; $21a_1$)	89
	10.4 (σ; $12b_2$)	11.7 (π; $6b_1$)		
Bismabenzene	7.9 (π; $11b_1$)	9.2 (π; $5a_2$)	9.6 (σ; $27a_1$)	90
	10.2 (σ; $16b_2$)			
Pyridazine	9.31 (8.64) (σ, n_-; b_2)	10.61 (10.49) (π_S; a_2)	11.3 (π_A, σ, n_+; b_1, a_1)	91, 92
	13.9 (13.45) (π; b_1)			
Pyrimidine	9.73 (9.23) (σ, n_-; b_2)	10.41 (10.41) (π_S; b_1)	11.23 (11.10) (σ, n_+; a_1)	91–93
	11.39 (π_A; a_2)	13.9 (13.51) (π; b_1)		
Pyrazine	9.63 (9.29) (σ, n_+; a_g)	10.18 (10.18) (π_A; b_{2g})	11.35 (11.00) (σ, n_-; b_{2u})	91, 93
	11.77 (11.66) (π_S; b_{1g})	13.9 (π; b_{3u})		
s-Triazine	10.41 (10.1) (σ)	11.71 (11.71) (π_2)	12.2 (π_3)	87
	13.25 (13.25) (σ)	14.65 (14.65) (π_1)	14.85	
	17.94 (17.7)			
1,2,4-Triazine	9.61 (n)	11.30 (π)	11.82 (n)	94
	12.4 (n)	12.43 (π)	~15 (π_σ)	
s-Tetrazine	9.72 (9.14) (σ, n_{AA}; b_{3g})	12.05 (11.75) (π_S, n_{SA}; b_{1g}, b_{2u})		91
	12.78 (n_{AS}, σ; b_{1u})	13.36 (n_{SS}, σ; a_g)	13.5 (π_A; b_{2g})	
	15.84 (15.53) (π; b_{3u})			

3.3.2.1. Pyridine Derivatives

The UV photoelectron spectrum of pyridine has been investigated by many workers. There were discrepancies in the assignment of the UVPS bands of pyridine in the earlier years, but recent theoretical and experimental investigations have afforded correct assignments. In Figure 10, we have compared the UV photoelectron spectrum of pyridine with the spectra of di-, tri-, and tetraaza derivatives of benzene.

Baker and Turner (86) reported the first three bands of pyridine as 9.2(π), 9.5(π), and 10.5(n_N) eV. Spanget-Larsen (95) applied the modified iterative extended Hückel method to the UVPS of pyridine. Guided by *ab initio* Hartree-Fock calculations, the MOs have been divided into four categories. Correlation between the calculated orbital energies and the experimental IEs was then established separately for each group of orbitals. The resulting assignment, which is in general agreement with Gleiter and coworkers (93), shows that the first UVPS band around 9.6 eV is due mainly to the nitrogen lone pair; the bands at 9.73 and 10.50 eV are due to two π orbitals. Using the perfluoro effect, Brundle and coworkers (87) made the following assignments: 9.67 eV (n_N), 9.80 eV (π), and 10.50 eV (π). Based on He(I) and He(II) intensity ratios of a series of substituted pyridines, Daamen and Oskam (88) have shown that the first IE of pyridine is due to n_N (9.67 eV; a_1). The next two IEs (at 9.79 and 10.51 eV) are due to π orbitals of symmetry a_2 and b_1, respectively. Because of its lower symmetry (C_{2v}) compared to benzene, pyridine shows more complex vibrational structure. The first band shows a vibrational spacing of 560 cm^{-1}, while the band at 12.5 eV shows two spacings of 560 and 1650 cm^{-1}.

Table 3.5 presents some useful data on the UVPS of pyridine derivatives. Daamen and Oskam (88) have assigned the UVPS bands of 2-methylpyridine using He(I)/He(II) intensity ratios of bands originating from the same type of orbitals. These bands are at 9.25 (π_3), 9.39 (n_N), and 10.29 (π_2) eV. They have also assigned the bands of 2-fluoropyridine where the sequence is same as in the 2-methyl derivative and the IEs are at 9.85 (π_3), 10.45 (n_N), and 10.85 (π_2) eV. In pentafluoropyridine, the first band at 10.27 eV is assigned to the nitrogen lone pair (87); vibrational structure (1450 cm^{-1}) is associated with this band. The band at 11.37 (π_2) eV also shows vibrational structure with a spacing of 1530 cm^{-1}. The other bands of pentafluoropyridine are at 12.08 (π_3), 13.62, 14.38, 15.45, 16.27, 17.6, 18.6, 20.5, 22.2, and 21.7 eV. The band at 13.62 eV shows vibrational spacings of 1330 and 650 cm^{-1}, while the band at 14.38 eV shows vibrational spacings of 1370 and 980 cm^{-1}. The band at 15.45 eV shows vibrational structure with a spacing of 1200 cm^{-1}. A classification of the UVPS bands of chloropyridines has been made on the basis of perturbations of the states of the six-membered ring by Cl and N (96). An atom-additive model correlates the shifts of the π and n states of pyridine by Cl, but the correlation is not as good as for the shifts by F.

Cook and coworkers (97) and Guimon and coworkers (98) have investigated different hydroxy- and alkoxypyridines as well as mercapto- and alkylmercaptopyridines. In hydroxypyridines, the first band (π) shows an increase in energy from

Table 3.5. Ionization Energies of Pyridine and Substituted Pyridines

Compound	Ionization Energies			Ref.
Pyridine	9.67 (n_N; a_1)	9.80 (π_2; a_2)	10.50 (10.42) (π_3; b_1)	12.45 (12.27) (σ; b_2)	87–89
	12.6 (π; b_1)	13.1	13.8	14.5	
	15.6 (15.6)	17.1			
2-Methylpyridine	9.25 (π_3)	9.39 (n_N)	10.29 (π_2)		88
2-Fluoropyridine	9.85 (π_3)	10.45 (n_N)	10.85 (π_2)		88
2-Hydroxypyridine	9.11 (π)	10.08 (n_N)	10.48 (π)	12.22 (π/n_O)	97
2-Methoxypyridine	8.82 (π)	9.82 (n_N)	10.20 (π)	11.45 (π/n_O)	97
2-Mercaptopyridine	8.79 (π)	9.86 (n_N)	10.47 (π)	11.02 (π)	97
	12.15 (n_S)				
2-Methylthiopyridine	8.24 (π)	9.56 (n_N)	10.25 (π)	10.62 (π)	97
	11.47 (n_S)				
2-Aminopyridine	8.34 (π)	9.57 (n_N)	10.15 (π)	11.16 (π)	99
2-Cyanopyridine	10.12 (π)	10.42 (n_N)	11.10 (π)		88, 99
3-Hydroxypyridine	9.15 (π)	9.71 (n_N)	10.37 (π)	12.56 (π/n_O)	97
3-Mercaptopyridine	8.89 (π)	9.78 (n_N)	10.25 (π)	11.32 (π)	97
3-Aminopyridine	8.44	9.47 (n_N)	10.09	11.59	99
3-Cyanopyridine	10.10	10.30 (n_N)	11.01	12.33	99

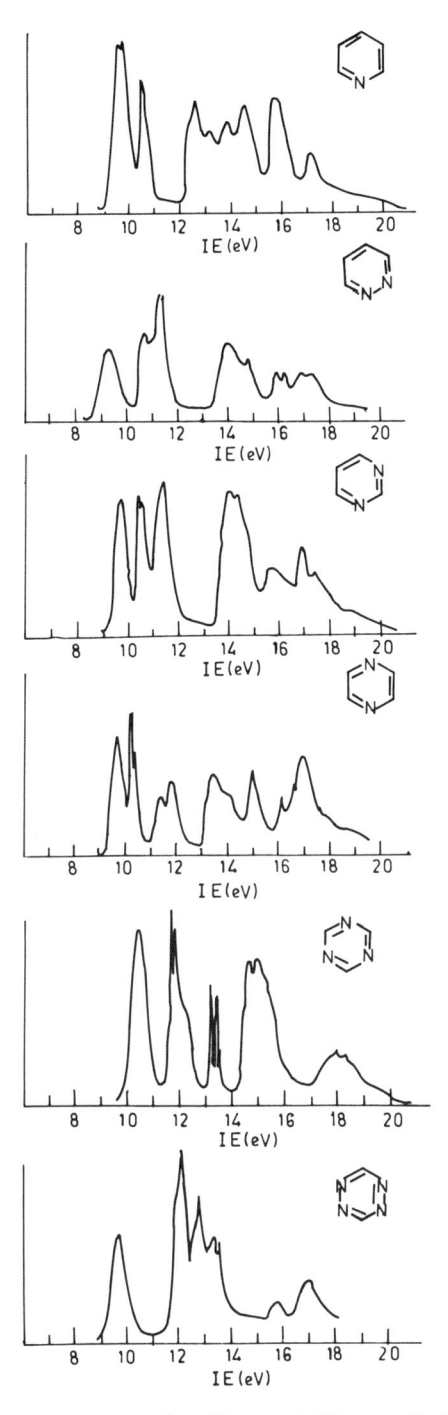

Fig. 3.10. UV photoelectron spectra of pyridine, pyrimidine, pyridazine, pyrazine, *s*-triazine, and *s*-tetrazine. [After Gleiter and coworkers (91).]

the 2-hydroxy to the 4-hydroxy derivative, but the lone-pair IEs of nitrogen and oxygen (second and fourth bands) do not show any order. The third band, due to a π level, shows a decreasing trend in the order 2-hydroxy > 3-hydroxy > 4-hydroxy. The fourth band of the 2-hydroxy derivative shows a vibrational structure with a spacing of 800 cm^{-1}, while the first band of the 3-hydroxy derivative shows a vibrational spacing of 800 cm^{-1}. Alkoxy derivatives show all the IEs lower than the corresponding hydroxyridines.

Mercaptopyridines have lower IEs than hydroxy derivatives, as expected. The first band, due to a π orbital, shows an increase in energy from 2-mercapto to 4-mercaptopyridine, just as in the hydroxy derivatives, but the nitrogen lone pair is unaffected. The third band (π) again shows a decreasing trend as seen earlier in the case of hydroxy derivatives. The lone pair due to sulfur comes at rather high energy at 12.15 eV in 2-mercaptopyridine; the nitrogen lone pair comes at a comparatively lower energy and is unaffected by substitution. Alkylmercaptopyridines show a decrease in all the IEs compared to the mercapto derivatives, but the different trends with the position of the substituent remain the same. In 2-hydroxy- and 2-mercaptopyridine, the predominant tautomers in the vapor phase are the hydroxy and mercapto conformers, although a significant percentage of the carbonyl and the thiocarbonyl tautomers exist. The 3- and 4-hydroxy- and 3- and 4-mercaptopyridines exist as hydroxy and mercapto derivatives with less than 5% of the carbonyl or thiocarbonyl tautomers (pyridones and thiopyridones). Table 3.5 includes data on hydroxy- and mercaptopyridines.

Kobayashi and Nagakura (99) and Daamen and Oskam (88) have assigned the UVPS bands of 2-aminopyridine. The first band at 8.34 eV is due to a π orbital, while the second band at 9.57 eV is due to the lone-pair orbital of the ring nitrogen. The third and fourth bands at 10.15 and 11.16 eV, respectively, are assigned to the π orbitals. The first IE of 2-, 3-, and 4-substituted aminopyridines varies in the order 2- > 3- > 4-. The lone-pair IE as well as the third IE show a decreasing trend in the order 2- < 3- < 4-. The fourth band again shows an increasing trend like the first band.

Cyanapyridines exhibit some new features. The first band assigned to a π orbital in 2- and 3-cyanopyridines shows the same IE, while the first band of 4-cyanopyridine is assigned to the lone-pair orbital of the ring nitrogen; there is an increase in IE of the first π orbital of 4-cyanopyridine (99). The lone-pair IEs of 3- and 4-cyanopyridines appear at the same positions. Also, the first band assigned to a π orbital decreases slightly from 2- to 4-cyanopyridine.

Dougherty and coworkers (100) have reported the UVPS bands of nicotinic acid, nicotinic acid methyl ester, nicotinamide, and *N,N*-diethylnicotinamide. They have investigated the validity of Pullman *k*-index approach to IEs and have also generated an experimental scale for electron-donating ability of groups.

UVPS of *cis*- and *trans*-1-phenyl-2-(4-pyridyl)cyclopropanes have been reported and compared with charge-transfer and protonation data (101). The low-IE regions of the spectra of the *cis* and *trans* isomers are assigned by analogy with those of related molecules. Gas-phase IE values for both the isomers are the same. This probably means that there is no significant through-space interaction between the MOs of the 2-*cis* aryl groups.

3.3.2.2. Pyridine Analogs

Helium(I) photoelectron spectra of phosphabenzene, arsabenzene, stibabenzene, and bismabenzene have been reported (89, 90), and in Figure 3.9 we have presented a correlation diagram to illustrate how the spectra of these compounds are related. Spectra of these compounds bear a linear relation to the IEs of the free heteroatoms; the slope of this correlation corresponds to the electron density of the π orbital at the heteroatom. The linear relation obtained for the first $\pi(b_1)$-type orbital follows the equation

$$I_v(\pi(b_1)) = 5.22 + 0.363 I(X) \qquad (2)$$

where X = N, P, As, Sb, Bi. The second b_1-type π orbital also follows a linear relationship with $I(X)$:

$$I_v(\pi(b_1)) = 10.25 + 0.163 I(X) \qquad (3)$$

Ionization energies calculated for benzene using this equation are quite close to the experimental values. The relative intensities of the first three bands of phosphabenzene and arsabenzene in He(I) and He(II) spectra (102) support the orbital sequences proposed earlier on the basis of the theoretical calculations of Von Niessen and coworkers (103). Inversions of the orbitals of a_2 and b_1 type from nitrogen to phosphorus appear to be due to the electron-accepting nature of nitrogen, which stabilizes $\pi(b_1)$ compared to $\pi(a_2)$; phosphorus appears to be essentially electron-donating in nature. Graph-theoretical resonance energies of benzene and heterobenzenes have been calculated using experimental π IEs, nonsubjective linear regression techniques, and graph theory. The calculated values are comparable to the experimental values (104).

3.3.2.3. Azabenzenes

Ultraviolet photoelectron spectra of pyridazine, pyrimidine, and pyrazine have been reported by many workers and the bands assigned by making use of semiempirical calculations (91, 93, 105–108). Using simple first-order perturbation theory and MINDO calculations, Dewar and Worley (105) assign the first band of pyridazine and pyrimidine to a nonbonding orbital and that of pyrazine to a π orbital. Spanget-Larsen (95) has assigned the UVPS bands by employing the modified iterative extended Hückel method and comparing the results obtained from *ab initio* Hartree-Fock calculations. Åsbrink and coworkers (109–111) have investigated the UVPS of pyridazine, pyrimidine, and pyrazine and have assigned the first and third IEs of pyrazine to the lone pair with bonding properties; the second and fourth IEs are assigned to π orbitals with nearly nonbonding properties.

Gleiter and coworkers (93) have discussed the UVPS bands of azabenzenes by comparison with the spectra of benzene and pyridine. As expected, there is nonbonded interaction in these diazines. In pyridazine, the predominant nature of interaction is through-space, while in pyrazine through-bond interaction is dominant since the through-space interaction becomes negligible beyond $\sim$ 2.5 Å

(112). Based on the arguments of Hoffmann and coworkers (112), the first IE of pyridazine and pyrazine can be assigned to the lone-pair IE of the n_- ($= n_1 - n_2$) type. In pyrimidine, the first IE is due to ionization from n_+ ($= n_1 + n_2$) orbital (93, 106). The values of the n-orbital splitting in pyridazine, pyrimidine, and pyrazine are 2.0, 1.5, and 1.7 eV, respectively. According to Gleiter and coworkers (93), contribution to n-orbital splitting from the through-space interaction should vary in the order pyridazine > pyrimidine > pyrazine, owing to the increasing distance between the lone-pair orbitals, through-bond interaction remaining approximately the same. This explanation seems erroneous, since there can be no competition between the two types of interactions (112). If we obtain the values of splittings using the adiabatic IEs of Gleiter and Coworkers (91), they fall in the order pyridazine (1.66 eV) < pyrazine (1.71 eV) < pyrimidine (1.87 eV). The higher value for pyrimidine may be understood in terms of both mechanisms operating to increase the interaction and hence the splitting. The perfluoro effect in all these diazenes clearly brings out the fact that there is considerable through-bond interaction. The lone-pair orbitals are the most perturbed by fluorine substitution (92), while the π orbitals are much less affected.

Haselbach and coworkers (108), while trying to relate the IE to the excitation energy E in the case of pyrazine and 2,6-dimethylpyrazine, showed that in these systems the energy gap ΔE between the two lowest lying $^1n, \pi^*$ states is significantly smaller than that between the corresponding ^2n states (ΔIE) of the parent radical cations. The relation $\Delta E < \Delta$IE in these systems has been discussed and related to the different shapes of the lone-pair MOs involved in the excitation and the ionization processes. Heilbronner and coworkers (6) have shown clearly the necessary corrections to be applied before correlating these two energies.

Vibrational analysis of the UVPS bands of pyrazine (106) seems to confirm the assignment of the first band to a π orbital and that of the fourth band to a lone-pair orbital. The first band shows vibrational structure with three different species; ν_{6a}-type ring-breathing mode (75 meV), ν_1-type ring-breathing mode (132 meV), and H-bend (170 meV). The fourth band shows vibrational spacings of 71 meV (ν_{6a}-type ring-breathing mode), 117 meV (ν_1-type ring-breathing mode), and 191 eV (ν_{8a}-type ring-breathing mode). Muszkat and Schäublin (107) have assigned IEs of pyrazine by using the analysis of vibrational structure as well as the calculation of the potential energy distribution of the totally symmetric normal modes. In 2,6-dimethylpyrazine and 2,3,5,6-tetramethylpyrazine, all the IEs are lowered relative to pyrazine due to the inductive effect of the methyl groups. In the case of 2,3,5,6-tetramethylpyrazine, there is a reversal in the sequence of the two highest occupied MOs (113). This leads to a lower value of the splitting (~ 1.5 eV) of the lone-pair orbital, while the splitting due to π orbitals increases to ~ 2.12 eV, the value for pyrazine being ~ 1.59 eV.

Fridh and coworkers (114) have interpreted the electronic structure of s-triazine by means of UVPS as well as by Rydberg transitions (observed in an electron-impact energy-loss spectrum). In a small energy range, Rydberg transitions in the UV spectrum were also used. The lowest IE corresponds to ionization of a lone-pair electron with bonding properties. The next to IEs are due to π electrons that appear to be nearly nonbonding. Brundle and coworkers (87) have investigated the UVPS bands of s-triazine as well as its perfluoro analog. They have shown that the first and fourth bands of s-triazine arise from the lone-pair combinations of the nitrogen atomic orbitals. The first band is of e' symmetry and the fourth band is of a' symmetry. The second and third bands, which arise from π_3 and π_2 MOs, respectively, show Jahn-Teller splitting. In the perfluoro analog, the π_1 band (12.0 eV) precedes those of lone pairs, while in s-triazine the π_1 band appears at 14.65 eV.

Many of the UVPS bands of s-triazine show strong vibrational structure. The first band shows a vibrational spacing of 970 cm^{-1} due to the symmetric breathing mode (v'_2), while the second and third bands show vibrational spacings of 1050 and 1190 cm^{-1}, respectively both due to the symmetric ring-breathing mode (v'_2). The band at 17.94 eV shows a vibrational structure of 2900 cm^{-1} (v'_1). The perfluoro analog shows vibrational structure with spacings of 1370 cm^{-1} in the second band and 1000 cm^{-1} in the third, fourth, and fifth bands due to symmetric v_{C-F} stretching.

Gleiter and coworkers (94) have investigated the UVPS of 1,2,4-triazine, its 3-monomethyl derivative and two dimethyl and one trimethyl derivatives, and have interpreted the spectra by comparison with the results of MO calculations performed by ZDO and energy-weighted maximum overlap as well as *ab initio* methods and empirical correlations. The lone-pair splitting decreases from 2.46 eV in the 3-methyl derivative to a nearly constant value in the other methyl derivatives as well as in the parent compound ($\Delta E = 2.2$ eV); π-orbital splitting ($\Delta E \sim 1.4$ eV) remains unchanged in all the methyl derivatives, the splitting in the parent compound being only 1.1 eV.

Ultraviolet photoelectron spectra of s-tetrazine and 3,6-dimethyltetrazine have been reported (91, 115). The lowest IE corresponds to the ionization of a lone-pair orbital with bonding properties. No other orbitals can be classified as lone-pair orbitals. IEs of s-tetrazine and other azines have been compared using the EHMO method with special parameterization. Gleiter and coworkers (91) assign the first four IEs based on an HMO-type model to the π orbitals and linear combinations of lone-pair orbitals and considering both through-space and through-bond interactions.

3.4. FUSED-RING HETEROCYCLIC COMPOUNDS

The ultraviolet photoelectron spectrum of 3,6-dehydrooxepin has been reported by Müller and coworkers (116). Based on MO calculations, the first two and the last bands (at 8.05, 8.95, and 11.40 eV) are assigned to π-type orbitals, while the third band (11.85 eV) is assigned to a σ-type orbital.

Coustale and coworkers (117) and Gleiter and coworkers (118) have studied the UVPS of thieno[2,3-b]thiophene. Coustale and coworkers (117) assign the first five bands at 8.45, 8.72, 10.23, 11.43, and 12.10 eV to b_1, a_2, b_1, a_1, and b_2, a_2-type orbitals, respectively. Gleiter and coworkers (118) report the first four bands originating from π-type orbitals. Coustale and coworkers (117) have also reported the

UVPS bands of monohalo and monomethyl derivatives of thieno[2,3-*b*]thiophene as well as its di-, tri-, and tetrabromo derivatives. Methyl substitution at the β position changes the ordering of the first two orbitals and increases the IE due to the first b_1-type orbital (compared to the IE of the same orbital in the α-substituted compound); ordering of the other three orbitals remains the same. Bromine substitution at the α position decreases the first three IEs by about 0.1 eV, and the last two bands of the parent compound merge to give only one band. Substitution at the β position, like methyl substitution, changes the ordering of the first two orbitals, while the fourth and fifth bands of the parent compound merge to give a single band. The lone-pair splitting of the bromine lone pair is different in the α- and β-bromo derivatives, the magnitudes being 0.22 and 0.46 eV, respectively. Iodine substitution at the α position changes the ordering of the first two bands of the parent compound although the β-substituted derivative gives the same ordering of levels as the parent compound. In the β-substituted derivative, the first three IEs decrease by 0.2 eV, but the last two bands do not merge as in the case of the α-substituted derivative. The lone-pair splitting of iodine is different for the α-(0.62 eV) and β-substituted (0.80 eV) derivatives. The di-, tri-, and tetrabromo derivatives show the same ordering of orbitals except for the β,β'-derivative, where there is a reversal of the first two orbitals. Substitution at the β or the β' position causes a greater stabilization of the first orbital compared to substitution at the α or α' positions; the effect is opposite in the case of the second and third orbitals. In the spectra of all these compounds the last two bands of the parent compound merge to give a single band.

Gleiter and coworkers (118) have studied the UVPS of seleno and pyrrolo analogs of thieno[2,3-*b*]thiophene and thieno[3,2-*b*]thiophene and have found that replacement of S by Se causes a minor perturbation while the replacement of S or Se by the NH group changes the π-electronic system appreciably. This change is governed by two factors: a stabilizing effect due to an increase in the resonance integral of the bonds involving the heteroatom and a strong destabilizing inductive effect due to the polar N—H bond. Müller and coworkers (119) have calculated the ground-state multiplicity, aromaticity, ordering, and energies of electronically excited and ionic states of thieno[3,4-*c*]thiophene and have shown that the compound is unstable owing to its particular HOMO structure and energy. UVPS of tetraphenyl thieno[3,4-*c*]thiophene reported by the same workers (119) indeed shows a low first IE (6.19 eV). Based on localized orbital interaction schemes, these workers have attempted to rationalize the strong charge transfer from sulfur to the C-skeleton as well as its high 1,3 reactivity.

Gleiter and coworkers (120) have studied the UVPS of 6*a*-thiathiophthene and several of its methyl and phenyl derivatives. Based on several theoretical models they assign the first three bands at 8.11, 8.27, and 9.58 eV to π_1, n, and π_2 orbitals, respectively. Methyl substitution at the 2-position lowers the first IE by 0.3 eV. Further substitution at the 5-position lowers the first IE by only 0.1 eV. The second IE due to lone-pair ionization, however, decreases by 0.2 eV with every methyl substitution. The third band, due to the π_2 orbital, also decreases by 0.25 eV for each methyl substitution. Methyl substitution at the 3- and 4-positions lowers the first IE to a greater extent (by 0.48 eV) compared to 2,5-substitution (by 0.38 eV). The lone-pair IEs in the 2,5- and 3,4-dimethyl derivatives remain the same and the third IE differs only marginally. Phenyl substitution decreases the first and second IEs to an extent comparable to the 3,4-dimethyl derivative, and the third IE seems to be due to the π orbital from the benzene ring. Palmer and Findlay (118, 121) have calculated the ground-state wave functions of 6a-thiathiophthene and its 1-oxa-, 1,6-dioxa-, and 1-aza analogs and have shown by bond energy analysis that these molecules have low resonance energies; these workers point out that in 6a-thiathiophthene there are two lone-pair orbitals at the sulfur center and there is no appreciable $3d$ orbital participation in bonding.

Lin and coworkers (122) have studied the electronic structure and gas-phase tautomerism of hypoxanthine and guanine as well as their methyl derivatives by UVPS. They have carried out semiempirical and *ab initio* calculations and have shown that the lone-pair MO of the 9-methyl derivatives of hypoxanthine and guanine have similar structures; the first four π IEs of 9-methylhypoxanthine and first five π IEs of 9-methylguanine are found to be in the energy region 8.0–13.5 eV. In isolated environments, both hypoxanthine and guanine are stable in the N(1)H–N(7)H tautomeric form. Ajo and coworkers (123) have studied UVP spectra of xanthine, theophylline, theobromine, and caffeine and have assigned the bands based on CNDO calculations and comparison with related molecules.

The UV photoelectron spectrum of thiobenzopropiolactone was studied by Schulz and Schweig (124), who have assigned the bands with the aid of MINDO and configuration interaction calculations.

Behan and coworkers (36) have studied the UVPS of 2,3-dihydrobenzofuran and have assigned the first four bands at 8.20, 8.90, 10.90, and 11.34 eV to π, π, n, and π orbitals, respectively. Maier and Turner (125) have reported the UVPS of indoline and have assigned the first three bands to π orbitals. Galasso and coworkers (126) and Distefano and coworkers (127) have reported UVPS of the heterocyclic analogs of 1,2- and 1,3-indanediones and have assigned the bands with the aid of *ab initio* calculations; they find the effect of the bridgehead heteroatom to be a balance between that of a σ acceptor and a π donor. Comparing the UVPS bands of quinolinimide with those of *N*-methylquinolinimide, phthalimide, and *N*-methylphthalimide, Distefano and coworkers (127) showed that quinolinimide prefers a hydroxyimide structure stabilized by OH–N hydrogen bonding; evidence from IR, NMR, and XPS also supports this finding.

The electronic structures of benzo-substituted thiocarbonyl compounds were studied by Guimon and coworkers (32) using UVPS, UV absorption spectroscopy, and CNDO/S calculations. All the UVPS bands of benzo-substituted derivatives of *N,N'*-dimethylimidazoline-2-thione, thiazoline-2-thione and its *N*-methyl derivative, and oxazoline-2-thione and its *N*-methyl derivative have been assigned by CNDO/S calculations and annelation effects on the structure of these compounds have been discussed. The spectrum of 1,3-dithiol-2-thione has been compared with the spectra of benzo-substituted 1,3-dithiol, ethyl trithiocarbonate, and 2,2-methylethyl-benzo-1,3-dithiol by Guimon and coworkers (128), who have analyzed the spectra in terms of orbital interactions to determine the effect of spatial configuration and of the thiocarbonyl group on the electronic structure.

UVPS of chroman has been examined and the bands appropriately assigned (36). The first two bands (at 8.13 and 8.82 eV) of chroman have been assigned to π orbitals and the third and fourth bands (at 10.27 and 10.63 eV) to n-type orbitals. The fifth band at 11.50 eV may be due to a π orbital. Maier and Turner (125) assigned the first three UVPS bands of 1,2,3,4-tetrahydroquinoline to π orbitals. Loutfy and coworkers (129) have reported UVPS, UV absorption, and electrochemical redox potential studies of a series of thiochromanone and thiochroman-4-one derivatives. The first IE increases markedly as the sulfur atom is oxidized and there is an addition of a double bond in the nonaromatic ring. They have also shown a correlation between the gas-phase IEs and solution electrochemical oxidation potentials.

UVPS of several heteroaromatics where a six-membered carbocyclic aromatic ring is fused to a five-membered heterocyclic aromatic ring have been reported in the literature. Figure 3.11 shows a correlation chart for the UVPS bands of some of these derivatives.

UVPS of benzofuran, indole, isoindole, N-methyl derivatives of indole and isoindole as well as benzothiophene and benzo[c]thiophene have been reported (130–133). Palmer and Kennedy (130) have assigned the UVPS bands of benzofuran, indole, and benzothiophene based on *ab initio* calculations using a GTO basis set. Dolby and coworkers (131) have assigned the UVPS bands of N-methylindole and N-methylisoindole using INDO and Hückel MO studies. Palmer and Kennedy (134, 135) have reported UVPS studies of aza derivatives of indole, N-methylisoindole, benzofuran, benzothiophene, and related compounds and have assigned the spectra on the basis of *ab initio* calculations. Variations in lone-pair levels and

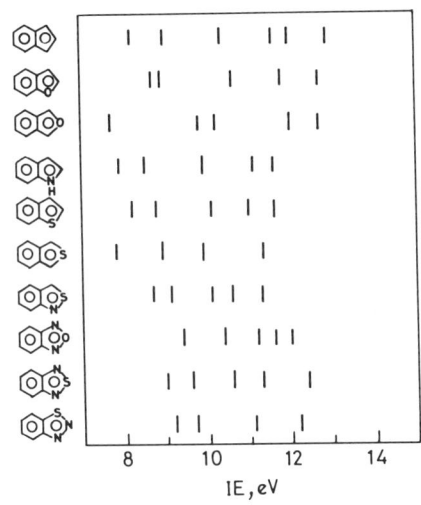

Fig. 3.11. Correlation diagram of ionization energies of heteroaromatic compounds with a five-membered ring fused to a benzene ring. [After Rao and coworkers (5).]

π levels between these compounds and the monocyclic compounds are discussed. These bycyclic compounds have lower π IEs than the corresponding monocyclic compounds. Substituent effects on the UVPS of benzothiazole have been studied by Salmona and coworkers (136) and the IEs assigned based on CNDO/2 and CNDO/S calculations. The CNDO/S method gives better results than the CNDO/2 method. Palmer and Kennedy (134, 135) as well as Clark and coworkers (133) have studied the UVPS of 1H-benzotriazole, 1-methylbenzotriazole, 1,2,3-benzothiadiazole, 2,1,3-benzothiadiazole, 2-methylbenzotriazole, 2,1,3-benzoxadiazole (and its 5-methoxy derivative), and 2,1,3-benzoselenadiazole and have assigned the bands based on *ab initio* as well as semiempirical PPP and EHT calculations. The first three IEs in these compounds are due to π orbitals, and the sulfur 3d orbital participation is very small. Schulz and Schweig (137) have, however, assigned the third band of 1,2,3-benzoxadiazole to the nitrogen lone-pair.

Lin and coworkers (138) have examined gas-phase tautomerism of purine and adenine by studying the UVPS of these compounds as well as their 7- and 9-methyl derivatives. They have performed HAM/3 (hydrogenic atoms in molecules) MO calculations to assign the UVPS bands. The results show that the lone-pair orbitals in purine and adenine are similar in spatial distribution and have similar IEs. The presence of a planar exocyclic amino group in adenine changes the π structure of this molecule significantly from that of purine. The 9-methyl derivatives show UVPS bands similar to those of the parent compounds, while the bands of the 7-methyl derivatives are considerably different, suggesting that the N(9)–H tautomers of these parent compounds are more stable than N(7)–H tautomers in the free state.

UVPS of quinoline (105, 139-141), isoquinoline (105, 117, 140, 141), and 2-phosphanaphthalene (117) as well as their perfluoro and phenyl derivatives have been reported. In all these compounds, the first two bands are due to ionization from the π orbitals while the third band is due to lone-pair ionization; the fourth band is also due to π ionization. Though the UVPS bands of quinoline and isoquinoline are similar, the perfluoro derivatives show remarkable differences in the magnitude of IEs. All the first four bands change drastically in the perfluoro derivatives. The IEs of π bands in the perfluoro derivative of quinoline increase by 0.9–1.0 eV, while those of perfluoroisoquinoline increase by 0.8–1.2 eV, the maximum shift in both these cases being in the case of the π_2 orbital. The lone-pair IE, however, changes by about 2.2 eV in both the perfluoro derivatives. Such an increase in π as well as lone-pair IEs suggests that there is an inductive or combined inductive-conjugative interaction of these compounds. These models were proposed by Van den Ham and Van der Meer (141) based on Hückel calculations. Schäfer and coworkers have studied the UVPS of 2-phosphanaphthalene and assigned the bands with the aid of generalizations pertaining to the energies of π and n_P ionizations as well as CNDO/S calculations. The large FWHM of the n_P band indicates that the lone-pair has bonding character arising from orbital mixing with other orbitals of the molecule. This delocalization is almost as large as the delocalization of lone-pair in pyramidal phosphorus atom. Schäfer and coworkers (139) have also shown that there is a large destabilization of the P π_1 orbital, when phosphorus is pentavalent (compared to trivalent phosphanaphthalene) by studying the UVPS of 2-phenylnaphthalene, 2-phenylquinoline, 2-phenyl-1-phosphanaphthalene, and 1,1-dibenzyl-2-phenyl-1-phosphanaphthalene.

UVPS of 1,2 and 2,3-diazanaphthalenes have been reported (105, 139, 142, 143). The first IEs in these compounds are due to lone-pair orbitals arising from the asymmetric combination of the lone-pair orbitals (n_-) on the two nitrogen atoms. The second, third, and fourth UVPS bands of 1,2-diazanaphthalene are assigned to π orbitals, while the fifth band is assigned to a symmetrical combination of lone-pair orbitals (n_+) on the nitrogen atoms. In the case of the 2,3-diaza derivative, the fourth band is due to the lone pair while the fifth band is due to the π orbital. The lone-pair splitting in 1,2- and 2,3-diazanaphthalenes are 2.15 and 1.90 eV, respectively. Van den Ham and Van der Meer (142) have carried out Hückel, PPP-SCF, extended Hückel, and CNDO/2 calculations and have shown that IEs calculated by the CNDO/2 method were unsatisfactory. In the UVPS of the hexafluoro derivatives there is a reversal of the first two levels (143). In perfluoro-1,2-diazanaphthalene the lone-pair splitting increases to 2.6 eV, while in perfluoro-2,3-diazanaphthalene it increases to 2.2 eV. This is because in the perfluoro-2,3-diaza derivative, there is a reversal of energy levels of the n and π_3 orbitals. Fluorine substitution increases the energy of all the levels. In the fluoro derivative of 1,2-diazanaphthalene all the π IEs increase by 0.5–0.7 eV, while in the 2,3 derivative the π_1 and π_2 IEs increase by 0.7–0.8 eV and the π_3 IE increases by 0.4 eV. These large changes in π IEs as well as lone-pair IEs show that there is significant σ contribution to the MOs of these levels that should be investigated.

UVPS of 1,3-, 1,4-, 1,5-, 1,6-, 1,7-, 1,8-, 2,6-, and 2,7-diazanaphthalenes as well as many of the perfluoro derivatives have been reported (105, 140, 142–144). Based on various kinds of semiempirical calculations, Van den Ham and Van der Meer (142) have shown that the first IE in 1,5- and 1,8-diazanaphthalenes is due to an n_--type of orbital, whereas the first IE in 1,3-, 1,6-, 1,7-, 2,6-, and 2,7-diazanaphthalenes is due to a π orbital. The lone-pair splittings in 1,3- and 1,4-diazanaphthalenes are 1.2 and 1.6 eV, respectively. This is perhaps due to the smaller through-space interaction in the 1,3 derivative and large through-bond interaction in the 1,4 derivative. Fluorination at different positions of the 1,3 derivative increases all the IEs. Fluorination at the 2- position causes merger of the n_- and π_2 levels as well as the n_+ and π_3 levels, and the lone-pair splitting is 1.6 eV (144). Lone-pair splittings in the 4-fluoro and 2,4-difluoro derivatives are 1.3 and 1.45 eV, respectively. In the hexafluoro derivative, the splitting is 1.4 eV. This indicates that there is an upper limit to the lone-pair splitting; the limit is apparently reached in the difluoro derivative. 2,3-Difluoro, 5,6,7,8-tetrafluoro, and hexafluoro derivatives of 1,4-diazanaphthalene show lone-pair splittings of 0.9, 1.50, and 1.0 eV, respectively. This clearly shows that the effect of fluorination on the lone-pair splittings is not additive. Thus the effect in 2,3-difluoro and 5,6,7,8-tetrafluoro derivatives seems to be in the opposite direction, leading to a lower lone-pair splitting in hexafluoro derivatives.

UVPS of 1,5-, 1,6-, 1,7-, and 1,8-diazanaphthalenes show lone-pair splittings of 1.2, 0.4, 0.7, and 0.9 eV, respectively (142). The three π IEs in these compounds remain quite similar, the maximum variation for any series being about 0.3 eV. In 1,8-diazanaphthalene, there is only one band due to π_1 and π_2 at 9.40 eV. Fluorination at the 2,7 positions separates π_1 and π_2 and lowers the lone-pair splitting (to 0.75 eV), although all the IEs increase. Also, there is a reversal of the first two bands of the parent compound, causing the lone-pair band to appear at a higher energy than the bands due to π_1 and π_2 orbitals. The ordering of the π_3 and lone-pair orbitals, however, remain unchanged. Fluorination at all six positions leads to a further decrease in the lone-pair splitting to 0.55 eV although all the IEs increase; the first IE is this case is due to π_1 and π_2. UVPS of 2,6- and 2,7-diazanaphthalenes show similar lone-pair splittings of 0.6 and 0.75 eV, respectively (142, 144). Fluorination of 2,7-diazanaphthalene does not change the lone-pair splitting significantly, the splitting being 0.5 eV in 1,3,6,8-tetrafluoro-2,7-diazanaphthalene and 0.7 eV in the hexafluoro derivative.

Sandman and coworkers (145) have studied the UVPS bands of naphthalene-1,8-disulfide and have assigned the bands based on perturbation MO and CNDO/2 calculations. These workers have shown that the highest MO is determined by π interactions between the sulfur lone pairs and naphthalene π levels. The lone-pair splitting for coplanar S—S lone-pair interaction was found to be 2.1 eV. Meunier and Pfister-Guillouzo (146) have studied the conformations of dihydro epines by UVPS. The UPVS bands of dithieno[c,e]dihydroazepin, oxepin, and thiepin have been reported and analyzed by these workers; it appears that these molecules have a twisted conformation in the vapor phase.

Eweg and coworkers (147) have reported UVPS [He(I) and He(II)] of alloxazines and isoalloxazines. The bands were interpreted using various methyl-substituted derivatives and by comparison with the results of CNDO/S calculations and photoionization cross sections derived from it. Analysis of the experimental and theoretical results reveal a planar moelcular conformation to be the most probable one, although the bent conformation was established in solution and in the solid state. Bending appears to be due to environmental effects on the molecule. UVPS of acridan, carbazole and 10,11-dihydro-1,H-dibenz[b,f]azepin were examined and analyzed by CNDO calculations by Haink and coworkers (148). Bigelow and Caesar (149) have examined the effect of hydrogen bonding and N-alkylation on the electronic structure of carbazole based on UVPS, UV absorption spectrocopy, and semiempirical CNDO/S-CI studies on carbazole and its N-methyl and N-ethyl derivatives. They have shown that the primary interaction responsible for the observed shifts is the selective delocalization of the π orbitals largely localized on the N atoms.

Hush and coworkers (150) and Jongsma and coworkers (151) have reported the UVPS of azaphenanthrenes and azaanthracenes. Using a two-parameter first-order perturbation expression, Hush and coworkers have correlated the π IEs of azaphenanthrenes and azaanthracenes with those of the corresponding parent compounds. For monoaza compounds, the lone-pair IE is almost constant at 9.3 ± 0.1 eV. This constancy is consistent with experimental evidence. Schäfer and coworkers (152) have studied the UVPS of heteroatom-substituted analogs of 9-phenylanthracene, the heteroatoms being N, P, and As. They have shown that in these heteroanalogs of anthracene, the π MO with nonzero coefficients on the heteroatom was higher in energy than the π MO with its node on the heteroatom. Jongsma and coworkers (151) have reported UVPS studies of 10-methyl-9-phosphaanthracene as well as the 10-phenyl derivative. They have shown that steric effects are mainly responsible for stabilizing the 9-phosphaanthracene system and that the inductive or congugative effects of the substitutent have minor importance. Hush and coworkers (150) have also shown that lone-pair interactions can lead to the stabilization of either the symmetric or antisymmetric orbital combination. The lone-pair splitting can be large (2.3–2.5 eV) as in 4,5- and 9,10-diazaphenanthrene or negligibly small as in 1,8-diazaphenanthrene. Based on semiempirical SCF calculations for both π and σ levels, the energy level ordering has been explained.

Haink and Huber (153), Domelsmith and coworkers (154), and Basu and Rao (155) have reported UV photoelectron spectra of phenothiazine and related molecules. Haink and Huber (153) studied the spectra of 5,10-diemthyl-5,10-dihydrophenazine, 5-methyl-10-phenyl-5,10-dihydrophenazine, phenoxazine, and phenothiazine and assigned the lowest IEs on the basis of CNDO calculations. Basu and Rao (155) studied the spectra of phenothiazine and chlorpromazine and assigned the bands based on CNDO calculations. They have also shown that the difference in the first and second IEs as well as the first and third IEs correspond quite well with the UV absorption bands of the radical cations. Domelsmith and coworkers (154) have studied the UVPS of phenothiazine, N-methyl phenothiazine, promazine, chlorpromazine, thioridiazine, and trifluoroperazine and have assigned the bands on the basis of qualitative models, correlations with IEs of similar molecules, substituent effects, and CNDO/2 calculations. N-Alkylation seems to lower the first and higher π IEs in phenothiazine derivatives. Discrepancies between gas-phase IEs and solvent oxidation potentials and charge-transfer studies are attributed to differential solvation effects rather than conformational effects. Ricci and coworkers (156) have studied the oxidation rates, carbon-13 NMR, and UVPS of 10,10-dimethylphenothiasilin, -germin, and -stanin derivatives and of 9,9-dimethylthioxanthen. The trends in the IEs of these compounds seem to be governed by the variation in the conjugation between n_S and the benzene rings. As the dihedral angle between the two benzene rings decreases, the sulfur atom goes out of plane, causing a decrease in the conjugation; the first IE therefore shows a greater lone-pair character.

3.5. LONE-PAIR INTERACTIONS IN HETEROCYCLIC COMPOUNDS

Ultraviolet photoelectron spectroscopy provides a direct means of investigating interactions between lone-pair electrons, through-space as well as through-bond, in organic molecules. This subject has been reviewed by Heilbronner (4, 7) and Rao and coworkers (5). While discussing the spectra of diaza and similar heterocyclic molecules, we mentioned the splitting of lone-pair bands due to orbitals arising from

Lone-Pair Interactions in Heterocyclic Compounds 271

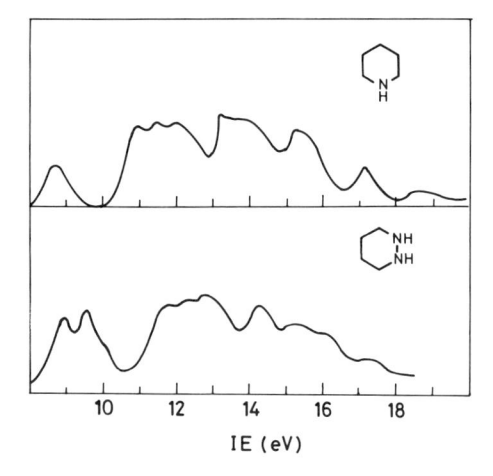

Fig. 3.12. UVPS of piperidine and cyclohexahydrazine. [After Gan and Peel (54) and Rademacher and Koopmann (159).]

the antisymmetric (n_-) and symmetric (n_+) combinations of the lone-pair orbitals. The splitting depends crucially on the dihedral angle. In Figure 3.12 we show typical spectra to illustrate the splittings.

Haselbach and coworkers (157) have examined the lone-pair electronic structure, conformation, and oxidation behavior of diaziridines by UVPS and have shown that the nitrogen lone pairs in monocyclic diaziridine prefer the *trans* conformation and that the lone-pair interaction is comparable to that encountered in alkyl hydrazines. Haselbach and coworkers (158) have shown that in 3,3-dimethyldiazirine there is a delocalization of lone-pair electrons. The $b_2(n_-)$ lone-pair orbital is localized to only 56% on the nitrogen atoms, while the localization of the $a_1(n_+)$ orbital is about 43%. Oxidation of the cyclic hydrazo compounds to azo compounds cannot be accounted for by lone-pair interactions. Rademacher and Koopman (159) and Rademacher (160) have studied conformations of several cyclic N,N'-dimethylhydrazines and bicyclic hydrazines by UVPS and have analyzed the UVPS bands by correlating the dihedral angle between the two lone pairs with the (n_+, n_-) splitting, ΔE, of the lone-pair orbitals. The dihedral angles vary between 45 and 180°. Nelsen and coworkers (161, 162) have studied the conformations of hexahydropyridazines by UVPS and have shown that in the case of the 1,2-dimethyl derivative there is a mixture of conformations, with two equatorial methyl groups (*ee*) and another conformation most probably with axial and equatorial methyl groups (*ae*) in the ratio 2:1. Both the *cis*- and *trans*-1,2,3,6-tetramethyl derivatives have a small proportion of the *ee* conformation, while in the 1,2,3-trimethyl derivative the ratio of *ee* to other conformers is around 1:5. Nelsen and Buschek (163) have derived empirical parameters from the UVPS bands of five cyclic dialkylhydrazines, which are used to calculate average IEs for 12 cyclic tetraalkylhydrazines. These parameters seem to be fairly successful for cyclic hydrazines exhibiting lone-pair splittings of less than 1 eV but not so satisfactory

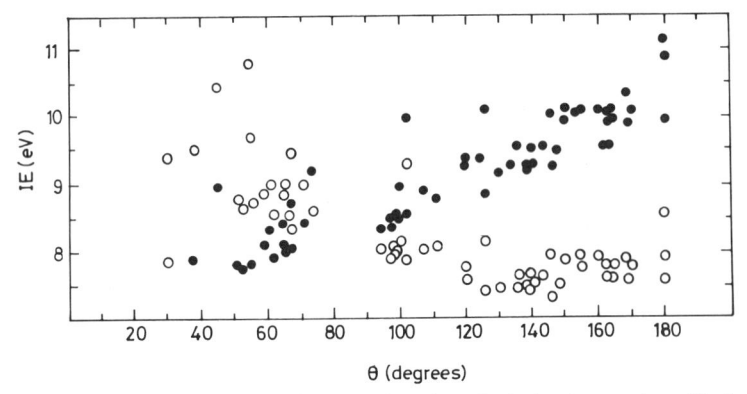

Fig. 3.13. Plots of n_+ and n_- ionization energies of cyclic hydrazines against dihedral angle. [After Rao and coworkers (5).]

for tetraalkylhydrazines with large lone-pair splittings. The lone-pair splittings in cyclic hydrazines are related to the dihedral angle in Fig. 3.13.

Nelsen and Buschek (164) have studied the UVPS of 28 1,2-cycloalkylhydrazines containing pyrazolidine, hexahydropyrazolidine, and 1,2,3,6-tetrahydropyrazolidine and have discussed the structures in terms of conformations. The direction and amount of torsion in 2,3-diazabicyclo[2.2.2]heptyl and pyrazolidine rings were obtained from the spectra; in some cases more than one conformation was found. Katritzky and coworkers [165] have shown by UVPS that 2-oxa-4a,8a-diazadecalin exists predominantly in the axial-axial (aa) form with a minor proportion of the equatorial-axial (ea) form; 3,4-dimethyl-1-thia (or N-methyl)-3,4-diazacyclohexene exists in the ae (ea) conformation.

Rademacher and coworkers (166) have studied the UVPS and conformations of hydrazobenzenes and related compounds. The conformations of these compounds were determined by employing an empirical relation between the splittings of the lone-pair band and the torsional angle (about the N–N group) obtained by MINDO/2 calculations. In all the hydrazobenzenes, the anti arrangement of the Ph group is the preferred conformation in the gas phase, unless geometric considerations force the molecule into another conformation. Rademacher and coworkers (167) have shown by UVPS that the tricyclic hexahydro-1,2,4,5-tetrazines exist in the tetraequatorial form, while the bicyclic compounds exist in the diequatorial diaxial forms as in the gas phase.

Domelsmith and coworkers (168) have carried out UVPS studies as well as semiempirical MINDO/3 and *ab initio* STO-3G calculations on ring-size effects on azo group IEs. Except for 3,3,4,4-tetramethyldiazetine, where the ordering of the n_-, n_+, and π IEs are $n_- < n_+ < \pi$, the ordering in the three-, five-, and six-membered rings is $n_- < \pi < n_+$. Boyd and coworkers (169) have interpreted the UVPS bands of 2,3-diazabicyclo[2.2.n]alk-2-enes, where n = 1, 2, 3, 4, based on CNDO/2 calculations, fine structure analysis of the second band, and comparison with the literature data; these workers assign the first three bands to n_-, π, and n_+ orbitals. In all four compounds the lone-pair splitting is greater than 2.8 eV. Boyd and coworkers (170) have also studied the UVPS of diazabasketene, diazadeltacyclene, and related compounds. In all these compounds, the sequence of the first three levels is $n_- < \pi < n_+$. The lone-pair splitting is very high in these systems, the magnitudes being 3.8 eV for diazabasketene and 4.0 eV for diazadeltacyclene.

The ultraviolet photoelectron spectra of 3,3,5,5-tetramethyl-1,2-dioxane and 3,3,6,6-tetramethyl-1,2-dioxane were studied by Batich and Adam (171), and the dihedral angles estimated to be 0 and 90°, respectively. Brown (172) has studied the UVPS of several peroxides and has shown that the effects of substituents can be separated from vicinal orbital interactions using the effect of similar substitution on the ether analogs as a guideline. The HOMO of peroxides is antibonding with respect to the O—O linkage. UVPS of symmetric bisallylic oxygen compounds like ascaridole as well as their saturated analogs have been studied by Brown (173). Rademacher and Elling (174) studied UVPS of peroxides and ozonides and showed the splitting of the first two bands to be related to the O—O torsional angle θ by the equation

$$E = 2.08 \cos \theta + 0.15 \quad (4)$$

UVPS of tetroxanes and ozonides have been interpreted in terms of gas-phase conformations.

Caughlin and coworkers (175) have studied the prostaglandin endoperoxide nucleus and related bicyclic peroxides by UVPS. They have shown that with the exception of the highly strained bicyclo[2.2.1]peroxide, the first IE of bicyclic peroxides increases with the increasing C—O—O—C dihedral angle θ. UVPS of the homologous series of monocyclic peroxides have also been reported by them. The lone-pair splitting ΔE is linearly related to θ, with a slope of about -0.025 eV/deg and an intercept of 2.24 eV.

A study of the electronic structure of 2-(dialkylamino)-1,3-dimethyl-1,3,2-diazaphospholanes and related molecules by UVPS has shown that in the parent compound the two N lone-pair basis orbitals are oriented such that their interaction with the P lone-pair orbital is π type (176); the exocyclic N lone-pair orbital is orthogonal to the other three, and the HOMO of the parent compound has partial P lone-pair character. Howalla and coworkers (177) have studied the UVPS of bicyclic phosphoranes and related phospholanes and have also carried out MO calculations. In the vapor phase, bicyclic phosphorane has been found to exist in the ring-opened tautomeric form.

Comparison of the angular dependence of the UVPS of dialkyl disulfides, lipoic acid, 1,2,4-trithiolane, and 3,3,5,5-tetramethyl-1,2,4-trithiolane indicate that 1,2,4-trithiolane exists in half-chair conformation in the vapor phase (178, 179). Guimon and coworkers (180) have shown by UVPS studies that the preferential conformation for 3,3,6,6-tetramethyl-S-tetrathiane is twist while it is chair for the 3,3:6,6-bis (tetramethylene) as well as 3,3:6,6-bis (pentamethylene) derivatives.

ACKNOWLEDGMENT

The authors thank the Department of Science and Technology, Government of India, and the Indian National Science Academy for support of this work.

REFERENCES

1. W. L. F. Armarago, in A. R. Katritzky, Ed., *Physical Methods in Heterocyclic Chemistry*, Vol. 3, Academic Press, New York, 1971, p. 67.
2. C. N. R. Rao, *Ultraviolet and Visible Spectroscopy*, 3rd. ed., Butterworths, London, 1975.
3. D. W. Turner, C. Baker, A. D. Baker, and C. R. Brundle, *Molecular Photoelectron Spectroscopy*, Wiley-Interscience, New York, 1970.
4. E. Heilbronner and J. P. Maier, in A. D. Baker and C. R. Brundle, Eds., *Electron Spectroscopy: Theory, Techniques and Applications*, Vol. 1, Academic Press, London, 1977, p. 205.
5. C. N. R. Rao, P. K. Basu, and M. S. Hegde, *Appl. Spectrosc. Rev.*, 15, 1 (1979).
6. E. Heilbronner, J. P. Maier, and E. Haselbach. in A. R. Katritzky, Ed., *Physical Methods in Heterocyclic Chemistry*, Vol. 6, Academic Press, New York, 1974, p. 1.
7. A. D. Baker and C. R. Brundle, Eds., *Electron Spectroscopy: Theory, Techniques and Applications*, Vol. 1, Academic Press, London, 1977.
8. E. J. McAlduff and K. N. Houk, *Canad. J. Chem.*, 55, 318 (1977).
9. A. Schweig and W. Thiel, *Chem. Phys. Lett.*, 21, 541 (1973).
10. H. Basch, M. B. Robin, N. A. Kuebler, C. Baker, and D. W. Turner, *J. Chem. Phys.*, 51, 52 (1969).
11. D. H. Aue, H. M. Webb, W. R. Davidson, M. Vidal, M. T. Bowers, H. Goldwhite, L. E. Vertal, J. E. Douglas, P. A. Kollman, and G. L. Kenyon, *J. Amer. Chem. Soc.*, 102, 5151 (1980).
12. K. Yashikawa, M. Hashimoto, and I. Morishima, *J. Amer. Chem. Soc.*, 96, 288 (1974).
13. B. Solouki, H. Bock, and R. Appel, *Chem. Ber.*, 108, 897 (1975).
14. C. Müller, A. Schweig, and H. Ver Meer, *J. Amer. Chem. Soc.*, 97, 982 (1975).
15. S. Pignataro and G. Distefano, *Chem. Phys. Lett.*, 26, 356 (1974).
16. A. D. Bain, J. C. Bünzli, D. C. Frost, and L. Weiler, *J. Amer. Chem. Soc.*, 95, 291 (1973).
17. I. Morishima, K. Yoshikawa, M. Hashimoto, and K. Bekki, *J. Amer. Chem. Soc.*, 97, 4283 (1975).
18. S. H. Gerson, S. D. Worley, N. Bodor, J. J. Kaminski, and T. W. Flechner, *J. Electron Spectrosc. Relat. Phenom.*, 13, 421 (1978).
19. A. D. Baker and D. Betteridge, *Photoelectron Spectroscopy*, Pergamon Press, Oxford, 1972.
20. H. Schmidt and A. Schweig, *Tetrahedron Lett.*, 1973, 1437.
21. B. Solouki, H. Bock, and R. Appel, *Angew. Chem. Int. Ed.*, 11, 927 (1972).
22. W. Schäfer, A. Schweig, and F. Mathey, *J. Amer. Chem. Soc.*, 98, 407 (1976).
23. M. Jinno, J. Watanabe, Yu. Yokoyama, and S. Ikeda, *Bull. Chem. Soc. Japan*, 50, 547 (1977).
24. A. D. Bain and D. C. Frost, *Canad, J. Chem.*, 51, 1245 (1973).
25. M. Almemark, J. E. Bäckvall, C. Moberg, B. Akermark, L. Åsbrink, and B. Roos, *Tetrahedron*, 30, 2503 (1974).
26. S. D. Worley, S. H. Gerson, N. Bodor, J. J. Kaminski, and T. W. Flechtner, *J. Chem. Phys.*, 68, 1313 (1978).
27. D. A. Sweigart and D. W. Turner, *J. Amer. Chem. Soc.*, 94, 5599 (1972).
28. A. D. Baker, M. A. Brisk, T. J. Venanzi, Y. S. Kwon, S. Sadka, and D. C. Lidta, *Tetrahedron Lett.*, 1976, 3415.

References

29. C. Guimon and G. Pfister-Guillouzo, *J. Electron Spectrosc. Relat. Phenom.*, **7**, 191 (1975).
30. J. Spanget-Larsen, R. Gleiter, M. Kobayashi, E. M. Engler, P. Shu, and D. O. Cowan, *J. Amer. Chem. Soc.*, **99**, 2855 (1977).
31. C. Guimon, G. Pfister-Guillouzo, M. Arbelot, and M. Chanon, *Tetrahedron*, **30**, 3831 (1974).
32. C. Guimon, M. Arbelot, and G. Pfister-Guillouzo, *Spectrochim. Acta*, **31A**, 985 (1975).
33. G. Levy and P. De Loth, *Compt. Rend.*, **279C**, 331 (1974).
34. D. C. Frost, F. G. Herring, A. Katrib, and C. A. McDowell, *Chem. Phys. Lett.*, **20**, 401 (1973).
35. P. D. Mollere and K. N. Houk, *J. Amer. Chem. Soc.*, **99**, 3226 (1977).
36. J. M. Behan, F. M. Dean, and R. A. W. Johnstone, *Tetrahedron*, **32**, 167 (1976).
37. H. Schmidt and A. Schweig, *Chem. Ber.*, **107**, 725 (1974).
38. M. Bloch, E. Heilbronner, T. B. Jones, and J. L. Ropoll, *Heterocycles*, **11**, 443 (1978).
39. F. P. Colonna, G. Distefano, S. Pignataro, G. Pitacco, and E. Valentin, *J. Chem. Soc., Faraday Trans. 2*, **71**, 1572 (1975).
40. M. Kobayashi, R. Gleiter, D. L. Coffen, H. Bock, W. Schulz, and U. Stein, *Tetrahedron*, **33**, 433 (1977).
41. K. Wittel, E. E. Astrup, H. Bock, G. Graeffe, and H. Julsen, *Z. Naturforsch.*, **30B**, 862 (1975).
42. D. Betteridge, L. Hendriksen, J. Sandstorm, I. Wennerbeck, and M. A. Williams, *Acta Chem. Scand.*, **31A**, 14 (1977).
43. C. Guimon, G. Pfister-Guillouzo, and M. Arbelot, *Tetrahedron*, **31**, 2769 (1975).
44. C. Guimon, G. Pfister-Guillouzo, and M. Begtrup, *J. Amer. Chem. Soc.*, **100**, 1275 (1978).
45. J. Kroner, W. Strack, F. Holsbar, and W. Kosbahn, *Z. Naturforsch.*, **28B**, 188 (1973).
46. B. Solouki, H. Bock, and O. Glemser, *Z. Naturforsch.*, **33B**, 284 (1978).
47. B. Cetinkaya, G. H. King, S. S. Krishnamurthy, M. F. Lappert, and J. B. Pedley, *J. Chem. Soc., Chem. Commun.*, **1971**, 1370.
48. R. Gleiter, E. Schmidt, D. O. Cowan, and J. P. Ferraris, *J. Electron Spectrosc. Relat. Phenom.*, **2**, 207 (1973).
49. R. Gleiter, M. Kobayashi, J. Spanget-Larsen, J. P. Ferraris, A. N. Bloch, K. Bechgaard, and D. O. Cowan, *Ber. Bunsenges. Phys. Chem.*, **79**, 1218 (1975).
50. J. A. Barlinsky, J. F. Carolan, and L. Weiler, *Canad. J. Chem.*, **52**, 3373 (1974).
51. A. Schweig, N. Thon, and E. M. Engler, *J. Electron Spectrosc. Relat. Phenom.*, **12**, 335 (1977).
52. T. Kobayashi and S. Nagakura, *Bull. Chem. Soc. Japan*, **46**, 1558 (1973).
53. A. A. Planckaert, J. Doucet, and C. Sandorfy, *J. Chem. Phys.*, **60**, 4846 (1974).
54. T. Gan and J. B. Peel, *Aust. J. Chem.*, **32**, 475 (1979).
55. N. Bodor, J. J. Kaminski, S. D. Worley, R. J. Cotton, T. H. Lee, and J. W. Rabalais, *J. Pharm. Sci.*, **63**, 1387 (1974).
56. H. Bock and G. Wagner, *Angew. Chem. Int. Edn.*, **11**, 150 (1972).
57. M. Bloch, F. Brogli, E. Heilbronner, T. B. Jones, H. Prinzbach, and O. Schweikert, *Helv. Chim. Acta*, **61**, 1388 (1978).
58. C. Batich, E. Heilbronner, C. B. Quinn, and J. R. Wiseman, *Helv. Chim. Acta*, **59**, 512 (1976).
59. G. Lauer, C. Müller, K-W. Schulte, A. Schweig, and A. Krebs, *Angew. Chem.*, **86**, 597 (1974).

60. S. D. Worley, S. H. Gerson, N. Bodor, and J. J. Kaminski, *Chem. Phys. Lett.*, **60**, 104 (1978).
61. P. Bischof, R. Gleiter, R. Dach, D. Enders, and D. Seebach, *Tetrahedron*, **31**, 1415 (1975).
62. P. J. Derrick, L. Åsbrink, O. Edquist, B. O. Jonsson, and E. Lindholm, *Int. J. Mass Spectrom. Ion Phys.*, **6**, 161 (1971).
63. P. J. Derrick, L. Åsbrink, O. Edquist, and E. Lindholm, *Spectrochim. Acta*, **27A**, 2525 (1971).
64. A. Schweig and W. Thiel, *Mol. Phys.*, **27**, 265 (1974).
65. J. W. Rabalais, L. O. Werme, T. Bergmerk, L. Karlsson, and K. Siegbahn, *Int. J. Mass Spectrom. Ion Phys.*, **9**, 185 (1972).
66. T. Bergmerk, J. W. Rabalais, L. O. Werme, L. Karlsson, and K. Siegbahn, in D. A. Shirley, Ed., *Proceedings International Conference 1971*, North-Holland, Amsterdam, 1972, p. 413.
67. A. D. Baker, D. Betteridge, N. R. Kemp, and R. E. Kirby, *Anal. Chem.*, **42**, 1064 (1970).
68. W. Schäfer, A. Schweig, S. Gronowitz, A. Taticchi, and F. Fringuelli, *J. Chem. Soc., Chem. Commun.*, **1973**, 541.
69. F. Fringuelli, G. Marino, A. Taticchi, G. Destefano, F. P. Colonna, and S. Pignataro, *J. Chem. Soc., Perkin Trans. 2*, **1976**, 276.
70. P. Dechant, A. Schweig, and W. Thiel, *Angew. Chem.*, **85**, 358 (1973).
71. G. De Alti, P. Decleva, and A. Sgamelotti, *Gazzetta*, **110**, 49 (1980).
72. G. De Alti and P. Decleva, *Chem. Phys. Lett.*, **77**, 413 (1981).
73. C. Müller, A. Schweig, and W. L. Mock, *J. Amer. Chem. Soc.*, **96**, 280 (1974).
74. B. G. Ramsey, *J. Org. Chem.*, **44**, 2093 (1979).
75. L. Klasinc, B. Ruscic, F. Kajfez, and V. Sunjic, *Int. J. Quantum Chem., Quantum Biol. Symp.*, **5**, 367 (1978).
76. F. Kajfez, L. Klasinc, and V. Sunjic, *J. Heterocycl. Chem.*, **16**, 529 (1979).
77. M. H. Palmer, R. H. Findlay, and R. G. Egdell, *J. Mol. Struct.*, **40**, 191 (1977).
78. M. H. Palmer, R. H. Findlay, J. N. A. Ridyard, A. Barrie, and P. Swift, *J. Mol. Struct.*, **39**, 189 (1977).
79. F. Bernardi, L. Forlani, P. E. Todesco, F. P. Colonna, and G. Distefano, *J. Electron Spectrosc. Relat. Phenom.*, **9**, 217 (1976).
80. G. Salmona, R. Faure, E. J. Vincent, C. Guimon, and G. Pfister-Guillouzo, *J. Mol. Struct.*, **48**, 205 (1978).
81. F. Cariati, C. Cauletti, M. L. Ganadu, M. N. Riancastelli, and A. Sgamellotti, *Spectrochim. Acta*, **36A**, 1029 (1980).
82. C. Guimon, G. Pfister-Guillouzo, A. Bernardini, and P. Viallefont, *Tetrahedron*, **36**, 107 (1980).
83. W. Flitsch, H. Peeters, W. Schulten, and P. Rademacher, *Tetrahedron*, **34**, 2301 (1978).
84. P. Meunier, M. Coustale, C. Guimon, and G. Pfister-Guillouzo, *J. Mol. Struct.*, **36**, 233 (1977).
85. P. Meunier and G. Pfister-Guillouzo, *Canad. J. Chem.*, **55**, 3901 (1977).
86. A. D. Baker and D. W. Turner, *Phil. Trans. Roy. Soc. London*, **268A**, 131 (1970).
87. C. R. Brundle, M. B. Robin, and N. A. Kuebler, *J. Amer. Chem. Soc.*, **94**, 1466 (1972).
88. H. Daamen and A. Oskam, *Inorg. Chim. Acta*, **27**, 209 (1978).
89. C. Batich, E. Heilbronner, V. Hornung, A. J. Ashe III, D. T. Clark, U. T. Cobley, D. Kilcast, and I. Scanlon, *J. Amer. Chem. Soc.*, **95**, 928 (1973).

References 277

90. J. Bastide, E. Heilbronner, J. P. Maier, and A. J. Ashe III, *Tetrahedron Lett.*, **1976**, 411.
91. R. Gleiter, E. Heilbronner, and V. Hornung, *Helv. Chim. Acta*, **55**, 255 (1972).
92. R. J. Suffolk, *J. Electron Spectrosc. Relat. Phenom.*, **3**, 53 (1974).
93. R. Gleiter, E. Heilbronner, and V. Hornung, *Angew. Chem. Int. Ed.*, **9**, 901 (1970).
94. R. Gleiter, M. Kobayashi, H. Neunhoeffer, and J. Spanget-Larsen, *Chem. Phys. Lett.*, **46**, 231 (1977).
95. J. Spanget-Larsen, *J. Electron Spectrosc. Relat. Phenom.*, **2**, 33 (1973).
96. J. N. Murrell and R. J. Suffolk, *J. Electron Spectrosc. Relat. Phenom.*, **1**, 471 (1973).
97. M. J. Cook, S. El-Abbady, A. R. Katritzky, C. Guimon, and G. Pfister-Guillouzo, *J. Chem. Soc., Perkin Trans. 2*, **1977**, 1652.
98. C. Guimon, G. Garrabe, and G. Pfister-Guillouzo, *Tetrahedron Lett.*, **1979**, 2585.
99. T. Kobayashi and S. Nagakura, *J. Electron Spectrosc. Relat. Phenom.*, **4**, 207 (1974).
100. D. Dougherty, E. S. Younathan, R. Voll. S. Abdulnur, and S. P. McGlynn, *J. Electron Spectrosc. Relat. Phenom.*, **13**, 379 (1978).
101. G. Distefano, A. Modelli, and V. Mancini, *Z. Naturforsch.*, **34A**, 245 (1979).
102. A. J. Ashe III, F. Burger, M. Y. El-Sheik, E. Heilbronner, J. P. Maier, and J. F. Muller, *Helv. Chim. Acta*, **59**, 1944 (1976).
103. W. Von Niessen, G. H. Dickersen, and L. S. Cederbaum, *Chem. Phys.*, **10**, 345 (1975).
104. W. C. Herndon, *Tetrahedron Lett.*, **1979**, 3283.
105. M. J. S. Dewar and S. D. Worley, *J. Chem. Phys.*, **51**, 263 (1969).
106. L. Åsbrink, E. Lindholm, and O. Edqvist, *Chem. Phys. Lett.*, **5**, 609 (1970).
107. K. A. Muszkat and J. Schäublin, *Chem. Phys. Lett.*, **13**, 301 (1972).
108. E. Haselbach, Z. Lanyiova, and M. Rossi, *Helv. Chim. Acta*, **56**, 2889 (1973).
109. L. Åsbrink, C. Fridh, B. O. Jonsson, and E. Lindholm, *Int. J. Mass Spectrom. Ion Phys.*, **8**, 215 (1972).
110. L. Åsbrink, C. Fridh, B. O. Jonsson, and E. Lindholm, *Int. J. Mass Spectrom. Ion Phys.*, **8**, 229 (1972).
111. C. Fridh, L. Åsbrink, B. O. Jonsson, and E. Lindholm, *Int. J. Mass Spectrom. Ion Phys.*, **8**, 101 (1972).
112. R. Hoffmann, A. Inamura, and W. J. Hehre, *J. Amer. Chem. Soc.*, **90**, 1499 (1968).
113. P. Bischof, R. Gleiter, and P. Hofmann, *J. Chem. Soc., Chem. Commun.*, **1974**, 767.
114. C. Fridh, L. Åsbrink, B. O. Jonsson, and E. Lindholm, *Int. J. Mass Spectrom. Ion Phys.*, **8**, 85 (1972).
115. C. Fridh, L. Åsbrink, B. O. Jonsson, and E. Lindholm, *Int. J. Mass Spectrom. Ion Phys.*, **9**, 485 (1972).
116. C. Müller, A. Schweig, W. Thiel, W. Grahn, R. G. Bergman, and K. P. C. Vollhardt, *J. Amer. Chem. Soc.*, **101**, 5579 (1979).
117. M. Coustale, C. Guimon, J. Arriau, and G. Pfister-Guillouzo, *J. Heterocycl. Chem.*, **13**, 231 (1976).
118. R. Gleiter, M. Kobayashi, J. Spanget-Larsen, S. Gronowitz, A. Konar, and M. Fernier, *J. Org. Chem.*, **42**, 2230 (1977).
119. C. Müller, A. Schweig, M. P. Cava, and M. V. Lakshmikanthan, *J. Amer. Chem. Soc.*, **98**, 7187 (1976).
120. R. Gleiter, V. Hornung, B. Lindberg, S. Högberg, and N. Lozach, *Chem. Phys. Lett.*, **11**, 401 (1971).
121. M. H. Palmer and R. H. Findlay, *J. Chem. Soc., Perkin Trans. 2*, **1974**, 1885.
122. J. Lin, C. Yu, S. Peng, I. Akiyama, J. Li, L. K. Lee, and P. R. Le Breton, *J. Phys. Chem.*, **84**, 1006 (1980).

123. D. Ajo, I. Frogala, G. Gronozzi, and E. Tondello, *Spectrochim. Acta*, **34A**, 1235 (1978).
124. R. Schulz and A. Schweig, *Tetrahedron Lett.*, **1979**, 59.
125. J. P. Maier and D. W. Turner, *J. Chem. Soc., Faraday Trans. 2*, **69**, 521 (1973).
126. V. Galasso, F. P. Colonna, and G. Distefano, *J. Electron Spectrosc. Relat. Phenom.*, **10**, 227 (1977).
127. G. Distefano, D. Jones, F. P. Colonna, A. Bigotto, V. Galasso, G. C. Pappolardo, and G. Scarlata, *J. Chem. Soc., Perkin Trans. 2*, **1978**, 441.
128. C. Guimon, G. Pfister-Guillouzo, and M. Arbelot, *J. Mol. Struct.*, **30**, 339 (1976).
129. R. O. Loutfy, I. W. J. Still, M. Thompson, and T. S. Leong, *Canad. J. Chem.*, **57**, 638 (1979).
130. M. H. Palmer and S. M. F. Kennedy, *J. Chem. Soc., Perkin Trans. 2*, **1974**, 1893.
131. L. J. Dolby, G. Hanson, and T. Koenig, *J. Org. Chem.*, **41**, 3537 (1976).
132. W. Rettig and J. Wirz, *Helv. Chim. Acta*, **59**, 1054 (1976).
133. P. A. Clark, R. Gleiter, and E. Heilbronner, *Tetrahedron*, **29**, 3085 (1973).
134. M. H. Palmer and S. M. F. Kennedy, *J. Mol. Struct.*, **43**, 33 (1978).
135. M. H. Palmer and S. M. F. Kennedy, *J. Mol. Struct.*, **43**, 203 (1978).
136. G. Salmona, R. Faure, and E. J. Vincent, *Compt. Rend.*, **280C**, 605 (1975).
137. R. Schulz and A. Schweig, *Angew, Chem.*, **91**, 737 (1979).
138. J. Lin, C. Yu, S. Peng, I. Akiyama, K. Li, L. K. Lee, and P. R. LeBreton, *J. Amer. Chem. Soc.*, **102**, 4627 (1980).
139. W. Schäfer, A. Schweig, G. Märkl, and K-H. Heier, *Tetrahedron Lett.*, **1973**, 3743.
140. F. Brogli, E. Heilbronner, and T. Kobayashi, *Helv. Chim. Acta*, **55**, 274 (1972).
141. D. M. W. Van den Ham and D. Van der Meer, *Chem. Phys. Lett.*, **15**, 549 (1972).
142. D. M. W. Van den Ham and D. Van der Meer, *Chem. Phys. Lett.*, **12**, 447 (1972).
143. D. M. W. Van den Ham and D. Van der Meer, *J. Electron Spectrosc. Relat. Phenom.*, **2**, 247 (1973).
144. D. M. W. Van den Ham, M. Beerlage, D. Van der Meer, and D. Feil, *J. Electron Spectrosc. Relat. Phenom.*, **7**, 33 (1975).
145. D. J. Sandman, G. P. Caeser, P. Nielsen, A. J. Epstein, and T. J. Holmes, *J. Amer. Chem. Soc.*, **100**, 202 (1978).
146. P. Meunier and G. Pfister-Guillouzo, *Canad. J. Chem.*, **55**, 2867 (1977).
147. J. K. Eweg, F. Müller, H. Van Dam, A. Terpstra, and A. Oskam, *J. Amer. Chem. Soc.*, **102**, 51 (1980).
148. H. J. Haink, J. E. Adams, and J. R. Huber, *Ber. Bunsenges. Phys. Chem.*, **78**, 436 (1974).
149. R. W. Bigelow and G. P. Caeser, *J. Phys. Chem.*, **83**, 1790 (1979).
150. N. S. Hush, A. S. Cheung, and P. R. Hilton, *J. Electron Spectrosc. Relat. Phenom.*, **7**, 385 (1975).
151. C. Jongsma, H. Vermeer, F. Bichelhaupt, W. Schäfer, and A. Schweig, *Tetrahedron*, **31**, 2931 (1975).
152. W. Schäfer, A. Schweig, F. Bickelhaupt, and H. Vermeer, *Angew, Chem. Int. Ed.*, **11**, 924 (1972).
153. H. J. Haink and J. R. Huber, *Chem. Ber.*, **108**, 1118 (1975).
154. L. N. Domelsmith, L. L. Munchausen, and K. N. Houk, *J. Amer. Chem. Soc.*, **99**, 6506 (1977).
155. P. K. Basu and C. N. R. Rao, *Spectrochim. Acta*, **34A**, 845 (1978).
156. A. Ricci, D. Pietropaolo, G. Distefano, D. Macciantelli, and F. P. Colonna, *J. Chem. Soc., Perkin Trans. 2*, **1977**, 689.

157. E. Haselbach, E. Heilbronner, A. Mannschreck, and W. Seitz, *Angew. Chem. Int. Ed.,* **9**, 902 (1970).
158. E. Haselbach, A. Mannschreck, and W. Seitz, *Helv. Chim. Acta,* **56**, 1614 (1973).
159. P. Rademacher and H. Koopmann, *Chem. Ber.,* **108**, 1557 (1975).
160. P. Rademacher, *Tetrahedron Lett.,* **1974**, 83.
161. S. F. Nelsen and J. M. Buschek, *J. Amer. Chem. Soc.,* **95**, 2011 (1973).
162. S. F. Nelsen, J. M. Buschek, and P. J. Hintz, *J. Amer. Chem. Soc.,* **95**, 2013 (1973).
163. S. F. Nelsen and J. M. Buschek, *J. Amer. Chem. Soc.,* **96**, 6982 (1974).
164. S. F. Nelsen and J. M. Buschek, *J. Amer. Chem. Soc.,* **96**, 6987 (1974).
165. A. R. Katritzky, V. J. Baker, F. M. S. Brito-Palma, R. C. Patel, G. Pfister-Guillouzo, and C. Guimon, *J. Chem. Soc., Perkin Trans. 2,* **1980**, 91.
166. P. Rademacher, V. M. Boss, M. Wildemann, and H. Weger, *Chem. Ber.,* **110**, 1939 (1977).
167. P. Rademacher, H. Breier, and R. Poppek, *Chem. Ber.,* **112**, 853 (1979).
168. L. N. Domelsmith, K. N. Houk, J. W. Timberlake, and S. Szilagyi, *Chem. Phys. Lett.,* **48**, 471 (1977).
169. R. J. Boyd, J. C. Bünzli, J. P. Snyder, and M. L. Heymen, *J. Amer. Chem. Soc.,* **95**, 6478 (1973).
170. R. J. Boyd, J. C. Bünzli, and J. P. Snyder, *J. Amer. Chem. Soc.,* **98**, 2398 (1976).
171. C. Batich and W. Adam, *Tetrahedron Lett.,* **1974**, 1467.
172. R. S. Brown, *Canad. J. Chem.,* **53**, 3439 (1975).
173. R. S. Brown, *Canad. J. Chem.,* **54**, 805 (1976).
174. P. Rademacher and W. Elling, *Liebigs Ann. Chem.,* **1979**, 1473.
175. D. J. Caughlin, R. S. Brown, and R. G. Salomon, *J. Amer. Chem. Soc.,* **101**, 1533 (1979).
176. S. D. Worley, J. H. Harjis, L. Chang, G. A. Mattson, and W. B. Jennings, *Inorg. Chem.,* **18**, 3581 (1979).
177. D. Howalla, M. Sanchez, D. Gonbeau, and G. Pfister-Guillouzo, *Nouv. J. Chim.,* **3**, 507 (1979).
178. M. F. Guimon, C. Guimon, and G. Pfister-Guillouzo, *Tetrahedron Lett.,* **1975**, 441.
179. M. F. Guimon, C. Guimon, F. Metras, and G. Pfister-Guillouzo, *Canad. J. Chem.,* **54**, 146 (1976).
180. M. F. Guimon, C. Guimon, F. Metras, and G. Pfister-Guillouzo, *J. Amer. Chem. Soc.,* **98**, 2078 (1976).

4 DIAMAGNETISM OF HETEROCYCLIC COMPOUNDS

E. A. BOUDREAUX

Department of Chemistry
University of New Orleans
New Orleans
Louisiana

R. R. GUPTA

Department of Chemistry
University of Rajasthan
Jaipur
India

4.1.	Introduction	282
4.2.	Measurement of Diamagnetic Susceptibility of Heterocyclic Compounds	282
	4.2.1. Gouy Method, 282	
	4.2.1.1. Experimental Technique, 284	
	4.2.1.2. Calibration of the Gouy Tube, 285	
	4.2.1.3. Instrumentation, 286	
	4.2.2. Faraday Method, 287	
	4.2.2.1. Torsion Head Arrangement, 287	
	4.2.2.2. Optical Arrangement, 288	
	4.2.3. Quincke Method, 289	
	4.2.4. Nuclear Magnetic Resonance Method, 291	
4.3.	Theoretical Calculation of Diamagnetic Susceptibility of Heterocyclic Compounds	292
	4.3.1. Pacault Method, 293	
	4.3.2. Semiempirical Methods, 293	
	4.3.2.1. Hameka Method, 293	
	4.3.2.2. Hartree–Fock–Roothaan Coupled Method, 298	
4.4.	Results and Discussions	298
	4.4.1. Heterocyclic Compounds Containing Nitrogen, 298	
	4.4.2. Heterocyclic Compounds Containing Oxygen, 300	
	4.4.3. Heterocyclic Compounds Containing Sulfur, 301	
	4.4.4. Heterocyclic Compounds Containing Nitrogen and Sulfur, 301	
	4.4.5. Heterocyclic Compounds Containing Nitrogen and Oxygen, 302	
	4.4.6. Heterocyclic Compounds in Solution, 302	
Appendix		309

Acknowledgments 309

References 309

4.1. INTRODUCTION

In the present age of advances in instrumentation and quantum mechanical approximation, studies on diamagnetism are providing information on chemical structures that complement results attained by other approaches, such as infrared and nuclear magnetic resonance spectroscopy. Diamagnetism is gaining much importance in view of its relationship to the phenomenon of chemical shift and the solution of numerous structural problems in chemistry. Recently it has been used in structural studies of organic derivatives (1–10). In the present review an account of diamagnetism of heterocyclic compounds is presented.

Diamagnetic susceptibilities and related quantities are expressed in cgs units $[\times (-10^{-6})]$ throughout this chapter.

4.2. MEASUREMENT OF DIAMAGNETIC SUSCEPTIBILITY OF HETEROCYCLIC COMPOUNDS

A number of methods have been reported for measuring magnetic susceptibility, but because of space limitation only those most commonly used for heterocyclic compounds will be discussed here.

4.2.1. Gouy Method

Perhaps the most widely and conveniently employed method for measuring magnetic susceptibilities was developed in 1889 by Gouy (11). This method has been discussed in detail by Selwood (12, 26), Bhatnagar and Mathur (13), Goyal (14), Bates (15), Nyholm (16), Figgis and Lewis (17), Earnshaw (18–21), Mulay (22–24), and Muller (25), among others. In the Gouy method the sample must be in the form of a rod of uniform cross section. To achieve this requirement a container (a glass or quartz tube known as the Gouy tube) of known cylindrical dimensions is filled with the sample. When the sample is a solid powder, the tube is filled very carefully to ensure uniformity of packing in order to satisfy the condition of uniform cross section. The Gouy tube is suspended vertically from the arm of a sensitive balance $(v \pm 10^{-5}$ g$)$ in such a manner that its lower end is located in the center of the magnetic field between the magnetic poles and its upper end is in a region well outside the pole gape where the strength of the magnetic field is negligible. The force dF acting on an element of sample dV, dm along the field gradient dH/dx in a homogeneous field of H gauss at the point dV has been worked out by Bates (15) and is given by equation (1):

$$dF = H\kappa\, dV \frac{dH}{dX} \quad (1)$$

When the ends of the tube are in the regions of the field H and H_0 ($H \gg H_0$), integration of equation (1) over the range of field gradients gives expression (2) for the force acting on the Gouy tube,

$$F = \frac{A}{2} \kappa (H^2 - H_0^2) \qquad (2)$$

where A is the area of cross section of the tube and κ is the volume susceptibility of the substance contained in the tube (-10^{-6} emu/cm^3).

The Gouy tube itself comprises a form of hollow specimen and consequently develops a force that is always present and has to be subtracted from the observed force to obtain the net force on the specimen sample. This force (δg, where δ is the apparent change in the weight of the empty Gouy tube in the field) is negative, as the Gouy tube is usually made of glass. The susceptibility is generally measured in air, which itself posssesses an appreciable magnetic susceptibility, and therefore the amount of air displaced by the sample must also be accounted for.

On applying these two corrections, equation (2) is reduced to

$$F = \frac{A}{2} (\kappa - \kappa_a)(H^2 - H_0^2) + \delta g \qquad (3)$$

where κ_a is the volume susceptibility of air.

Since the Gouy tube is hung with its lower end at the center of the pole gap and its upper end out of the influence of the field, H maintains a region of large and fairly constant field while H_0 is very nearly zero. Substituting $H_0 \approx 0$ into equation (3) yields

$$F = \frac{A}{2H^2} (\kappa - \kappa_a) + \delta g \qquad (4)$$

When the tube is suspended from the balance, the force exerted by the field on the substance causes an apparent change dW in the weight of the substance, and hence $F = g dW$ (where g is the gravitation constant). On setting $\kappa = \chi_s d_s = \chi_s (m/V)$ in equation (4), equation (5) is obtained

$$g\, dW = \frac{AH^2}{2} \left(\chi_s \frac{m}{V} - \kappa_1 \right) + \delta g \qquad (5)$$

or

$$\chi_s = \frac{1}{m} \left[V\kappa_1 + \frac{2Vg}{AH^2}(dW - \delta) \right] \qquad (6)$$

where
- χ_s = mass or specific susceptibility of the sample
- d_s = density of the sample
- $(m/V)_s$ = the mass-to-volume ratio of the sample
- $V\kappa_a$ = α (a constant) obtained by multiplying the volume susceptibility of air and volume of the sample in the tube
  = 0.029×10^{-6} times the volume of the Gouy tube up to the mark to which the substance is filled, in cgs units
- $2Vg/AH^2$ = β (another constant) because A and V are constants for the Gouy tube, H is kept constant for a set of measurements, and g is also constant for a fixed location

Thus equation (6) is expressed as

$$\chi_s = \frac{\alpha + \beta(dW - \delta)}{m_s} \tag{7}$$

where dW and δ are in milligrams and m_s in grams. β is known as the tube calibration constant.

Having determined the values of α and β for the Gouy tube under a set of given conditions, χ_s may be calculated.

4.2.1.1. Experimental Technique

The experimental procedure is illustrated by tracing the steps of an actual experiment.

The empty Gouy tube is weighed and then filled up to the mark with distilled water. It is again weighed, and the weights are recorded.

Weight of empty Gouy tube = 6.12710 ± 0.00002 g
Weight of Gouy tube and water = 7.82250 ± 0.00002 g
Weight of distilled water = $(7.82250 - 6.12710)$
= 1.69540 ± 0.00004 g
Volume of water in the tube = 1.6954 ml

In order to get the effective volume of water in the Gouy tube filled up to the mark, the correction for the meniscus is applied. Since the Gouy tube is of small cross-sectional diameter (e.g., 3–5 mm), it may be assumed with sufficient accuracy that the volume of the meniscus is $\frac{1}{3}\pi r^3$, where r is the radius of water in the tube (i.e., the inner radius of the tube).

The volume V of the cylinder is given by the equation

$$V = \pi r^2 l \tag{8}$$

where l is the length of the Gouy tube filled up to the mark. If $l = 7.4$, then with $V = 1.6954$,

$$V = 1.6954 = \tfrac{22}{7} r^2 \times 7.4 \tag{8a}$$

$$r^3 = \left(\frac{1.6954 \times 7}{7.4 \times 22}\right)^{3/2} \tag{9}$$

$$\text{Meniscus volume} = \frac{\pi r^3}{3} = \frac{1}{3} \times \frac{22}{7} \left(\frac{1.6954 \times 7}{7.4 \times 22}\right)^{3/2}$$

$$= 0.0206 \text{ ml} \tag{10}$$

The effective volume of water in the tube filled up to the mark is therefore $1.6954 - 0.0206 = 1.6748$ ml, and

$$\alpha = 0.029 \times 10^{-6} \times 1.6748$$

$$= 0.048569 \times 10^{-6}, \quad \text{or} \quad 0.049 \times 10^{-6} \text{ mg} \tag{11}$$

The Gouy tube is cleaned and dried and weighed in the magnetic field. If the weight of the empty Gouy tube in the magnetic field is, say, 6.12190 g, the difference in weight is 6.12190 g − 6.12710 g = 0.0052 g = 5.2 mg.

4.2.1.2. Calibration of the Gouy Tube

The calibration constant β of the tube is determined by measuring the susceptibility χ_s of a standard substance. χ_s of this substance, α, δ, dW, and m are substituted into equation (7), and β is calculated. A number of substances such as water (14), benzene (27–29), platinum (30), naphthalene (31), benzoic acid (32), ferrous ammonium sulfate (17), tris-(ethylenediamine)nickel(II) thiosulfate (35), copper sulfate (17), mercury tetrathiocyanato cobalt (33), aqueous nickel chloride solution (34), and others, have been recommended for calibration purposes.

The calibration of the Gouy tube will be illustrated with aqueous nickel chloride solution.

The χ_{NiCl_2} solution is obtained from equation (12), for which $\chi_{H_2O} = -0.720 \times 10^{-6}$ cgs.

$$\chi_{NiCl_2} = \left[\frac{10{,}030}{T} Y - 0.720(1-Y)\right] \times 10^{-6} \text{ (cgs)} \quad (12)$$

where T is the absolute temperature at which the measurement is made and Y is the specific concentration of nickel chloride expressed in g/ml. For this purpose a solution of $NiCl_2$ containing 26–30% nickel chloride is prepared and the $NiCl_2$ content (g/ml) is estimated gravimetrically. Let the weight of nickel chloride per milliliter of solution be 0.026598 g. Since the measurement of χ_s for $NiCl_2$ is made at room temperature (say 28°C), T is equal to 273.2 + 28 = 301.2, and χ_{NiCl_2} for this solution is given by equation (12).

$$\chi_{NiCl_2} = \frac{10{,}030(0.26598)}{301.2} - 0.72(1.0 - 0.26598)$$

$$\chi_{NiCl_2} = 8.3575 \times 10^{-6} \text{ (cgs)} \quad (13)$$

The Gouy tube is filled up to the mark with nickel chloride solution and is weighed in both the absence and presence of the magnetic field. The recorded weights are:

Weight of Gouy tube filled to the mark with $NiCl_2$ = 8.07590 g

Weight of Gouy tube filled to the mark with $NiCl_2$ in magnetic field = 8.11025 g

$$dW = 8.11025 - 8.07790 = 0.03435 \text{ g} = 34.35 \text{ mg}$$

$$m = 8.07590 - 6.12710 = 1.94880 \text{ g}$$

Substituting these values into equation (12), β is calculated as

$$8.3575 \times 10^{-6} = \frac{0.0486 \times 10^{-6}}{1.94880} + \beta[34.25 - (-5.20)]$$

$$\beta = 0.4130 \times 10^{-6} \text{ (cgs)}$$

286 Diamagnetism of Heterocyclic Compounds

Hence upon obtaining the values of δ, dW, and m, for the sample under examination and for the given Gouy tube, χ_s for any substance may be calculated.

4.2.1.3. Instrumentation

The instrument used in the Gouy method is considered to comprise three parts: (1) A balance, (2) the Gouy tube, and (3) the magnet.

4.2.1.3.1. Balance.
A semimicro balance with a precision of ± 0.02 or ± 0.05 mg is employed to measure the apparent change in weight of the specimen in the field, and thus to determine the force experienced by the specimen in the magnetic field. The Gouy tube is hung from one of the arms of the balance in such a manner that the closed end of the tube remains in the center of the pole gap (H), and the other end is out of the influence of the magnetic field (H_0). The Gouy tube and suspension are surrounded by an external glass tube or plastic casing to exclude drafts.

4.2.1.3.2. Gouy Tube.
The Gouy tube is generally made of Pyrex glass or quartz and is 6–10 cm in length and 3–9 mm in diameter. It is used as a container. However, Gouy tubes of varying dimensions, either single- or double-ended, may be employed depending on the nature of the measurement.

Three types of Gouy tubes reported in the literature are shown in Figure 4.1. The best suspensions are thin quartz fibers.

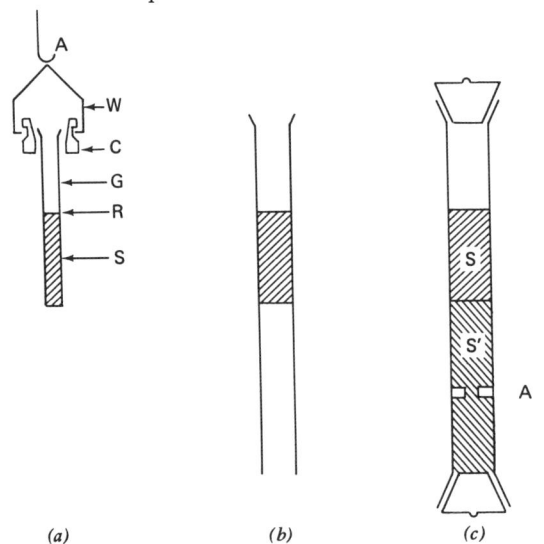

Fig. 4.1. Gouy tubes of three types. (*a*) Simple Gouy tube: *A*, suspension from balance; *W*, wire loop; *C*, collar; *G*, Gouy tube; *R*, reference mark; *S*, specimen. (*b*) Double-ended Gouy tube to eliminate δ. (*c*) Double-ended Gouy tube to eliminate δ and solvent correction: *S*, solution; *S'*, solvent; *A*, air bubble to permit expansion and contraction of solvent. Reproduced with permission from *Modern Coordination Chemistry*, J. Lewis and R. G. Wilkins, Eds., Wiley-Interscience, New York, 1960.

4.2.1.3.3. Magnet.

A magnet capable of producing a homogeneous field of 5000–12,000 G is normally employed in making magnetic susceptibility measurements. Of course, the field should be constant at the center of the pole gap. For measurements at room temperature, a pole gap of 1.5–2.5 cm is convenient and sufficient, and can be achieved if the minimum pole diameter is 3.5 cm. For carrying out measurements at other temperatures a pole gap of more than 6 cm is usually required to accommodate the Dewar flask, etc., but this requires a pole diameter of 8–10 cm. In the Gouy method both permanent magnets and electromagnets have been employed. The pole gap between the two magnets can be varied by means of a mechanical arrangement. Permanent magnets can produce fields up to about 8000 G, but the field strength for a given pole gap connot be conveniently changed. However, such limitations are not associated with electromagnets. Electromagnets can provide homogeneous fields up to about 15,000 G, but current stability is always a major concern. Field strength at a given pole gap is conveniently altered by varying the current energizing the magnet. The field strength of the electromagnet for a given pole gap can be calibrated from a plot of the field current. The main drawback of electromagnets is that a current-stabilized regulated power supply unit is required.

4.2.2. Faraday Method

In this method a small amount (a few milligrams) of the sample (say of volume dV and mass m) is placed in a nonuniform magnetic field. The force displacement experienced by the sample in the magnetic field is given by equations (14a–c).

$$dF = \kappa dV \frac{H dH}{dx} \quad (14a)$$

$$dF = \frac{\kappa d V d}{d} \frac{H dH}{dx} \quad (14b)$$

$$dF = \chi_s dm \frac{H dH}{dx} \quad (14c)$$

and is thus directly related to mass susceptibility of equation (14c), and as the volume of the substance is small, $H dH/dx$ may be considered constant. In such a situation dF is directly proportional to χ_s and thus forms the basis of the Faraday method (36). In the Faraday method a torsion-head arrangement (37–41), an optical (mirror, lamp, and scale) arrangement (12, 19, 42), or a Cahn electrobalance may be used.

4.2.2.1. Torsion Head Arrangement

The sample is suspended by a quartz torsion fiber of a torsion balance (which consists of a torsion head and a quartz beam) in the pole gap and is free to move horizontally. When the field is switched on, the sample is displaced from the zero position. The torsion head is twisted (by adjusting the weights) until the beam

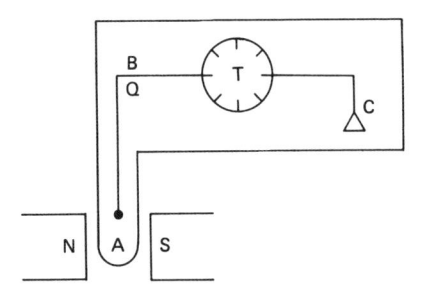

Fig. 4.2. A simple apparatus for the Faraday method (with torsion head). Reproduced with permission from *Modern Coordination Chemistry*, J. Lewis and R. G. Wilkins, Eds., Wiley-Interscience, New York, 1960.

returns to the original position. This twist of the torsion head is a measure of the force required to just balance the magnetic force at the zero position. (See Figure 4.2.)

From measurements of both a standard substance of known susceptibility and a sample under investigation, χ_s of the sample can be calculated from equation (15).

4.2.2.2. Optical Arrangement

To measure the force experienced by a sample in a magnetic field, an optical system consisting of a mirror, lamp, and scale may also be used. The sample (generally placed in a fused quartz bucket of about 1 mm internal diameter) is suspended by a quartz fiber from a phosphor bronze ring fitted with the optical system (Fig. 4.3). On switching on the field, the sample is displaced and the

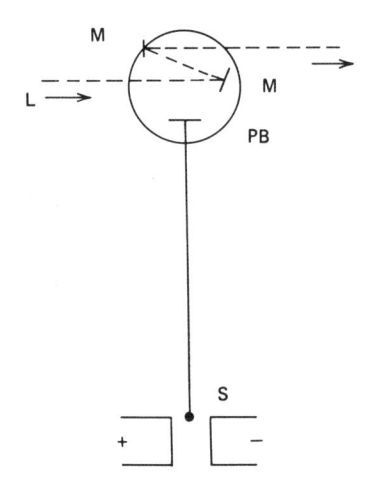

Fig. 4.3. Apparatus for Faraday method with optical arrangement. *M*, mirror; *L*, beam of light; PB, phosphor bronze ring; *S*, sample.

displacement is magnified several hundred times in this arrangement. The balance, suspension, and sample are enclosed. Let θ_b, θ_r, and θ_s be the respective angular deflections in the field for the empty bucket, the bucket containing a standard reference substance, and the bucket containing the test sample whose magnetic susceptibility is to be determined. It can be shown that the deflection is proportional to the force experienced by the sample in the field:

$$dF = \chi_s dm \frac{H\,dH}{dx} \tag{15}$$

and upon substituting $H\,dH/dx = C_1$ (a constant),

$$\frac{dF}{C_1} = \chi_s dm$$

Again, substituting $dF = C_2 \theta$, where C_2 is a constant,

$$\frac{C_2 \theta}{C_1} = \chi_s dm$$

(C_2/C_1 is again a constant, say C.) Hence

$$C\theta = \chi_s dm \tag{16}$$

If χ_{sr} is the susceptibility of the reference substance, equation (16) is reduced to

$$C(\theta_r - \theta_b) = \chi_{sr}(dm_r - dm_b) \tag{17}$$

Similarly,

$$C(\theta_s - \theta_b) = \chi_s(dm_s - dm_b) \tag{18}$$

where dm_b = mass of the empty bucket
dm_r = mass of the bucket containing the reference substance
dm_s = mass of the bucket containing the test sample

Dividing equation (18) by equation (17) yields

$$(\theta_s - \theta_b) = \frac{\chi_s}{\chi_{sr}} \frac{dm_s - dm_b}{dm_r - dm_b}$$

or

$$\chi_s = \chi_{sr} \frac{\theta_s - \theta_b}{\theta_r - \theta_b} \frac{dm_r - dm_b}{dm_s - dm_b} \tag{19}$$

Thus by noting the successive deflections for the empty bucket, the bucket containing the standard reference substance, and the bucket containing the test sample, and substituting the values of θ_s, θ_b, and θ_r into equation (19), χ_s can be calculated.

4.2.3. Quincke Method

The Quincke (43) method is most suitable for the determination of magnetic susceptibility of liquids. It is based on the same principle as the Gouy method, except that the force on a liquid sample is measured in terms of the hydrostatic

290 Diamagnetism of Heterocyclic Compounds

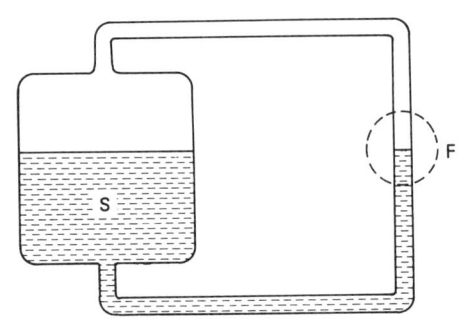

Fig. 4.4. Apparatus for the Quinck method. *S*, sample in reservoir; *F*, capillary tube in magnetic field.

pressure developed when the capillary tube containing the liquid is placed in a strong uniform magnetic field (25,000 G). In the presence of a magnetic field the liquid meniscus may either rise or fall depending on whether the liquid is paramagnetic or diamagnetic.

The force due to the applied magnetic field, represented by equations (14), can equally be represented by equation (20) if $H \gg H_0$.

$$F = \frac{A}{2} \kappa H^2 \qquad (20)$$

In this method a reservoir of much larger diameter than the capillary tube is used (Fig. 4.4), and if the susceptibility of the vapor above the meniscus is negligible, the force F can be equated to the force developed by hydrostatic pressure, i.e., $\Delta h \times (dAg)$, where Δh is the change in the height of meniscus, d is the density of liquid, A is the area of cross section, and g is the acceleration due to gravity. Hence the susceptibility is related to Δh by the equation

$$F = A \Delta h d g = \frac{A}{2} \kappa H^2 \qquad (21)$$

or

$$\frac{\kappa}{d} = \chi_s = \frac{2 \Delta h g}{H^2}$$

where the magnetic field strength must be on the order of 15–20 kG. The magnetic susceptibility of a liquid is calculated in this method by comparing the value of Δh (say, Δh_r) for a reference liquid of known susceptibility (χ_r). The latter is obtained by the equation

$$\chi_r = \frac{2 \Delta h_r g}{H^2} \qquad (22)$$

From equations (21) and (22), an equation is obtained that gives the value of χ_s in terms of Δh, Δh_r, and χ_r:

$$\chi_s = \frac{\Delta h}{\Delta h_r} \chi_r \qquad (23)$$

Using equation (23), χ_s is calculated from Δh, which is measured (within 10^{-3} cm) by a traveling microscope, and sometimes sensitivity is improved by employing the null method. The meniscus is returned to its original position by adjusting the pressure of the vapor above it.

4.2.4. Nuclear Magnetic Resonance Method

An NMR method based on the chemical shift in NMR spectra has been reported by Evans (44). Evans' method can be applied only to measure magnetic susceptibilities of paramagnetic substances and will not be discussed here. Frei and Bernstein (45) have also reported an NMR technique based on chemical shifts in NMR spectra to measure diamagnetic susceptibility. This method has been considered as accurate as classical methods, while other methods (46–50) have been reported less accurate for the determination of magnetic susceptibilities of diamagnetic liquids. Bernstein's method is based on the difference in chemical shifts arising from the difference in shape factors for a sphere and a cylinder. Bernstein and coworkers inserted spherically and cylindrically shaped reference tubes [each containing distilled water as a reference material in a conventional spinning sample tube (Fig. 4.5) of 4 mm outer diameter] and obtained two sharp signals (Fig. 4.6) corresponding to the two different reference tubes. The separation of signals has been found to be linearly dependent on the volume susceptibility of the sample in the sample tube. The susceptibility κ has been calculated from the equation.

$$\delta_{cyl}(\text{ref}) - \delta_{sph}(\text{ref}) = (g_{cyl} - g_{sph})[\kappa(\text{ref}) - \kappa(\text{sample})] \qquad (24)$$

where δ is the chemical shift and $\delta_{cyl}(\text{ref}) - \delta_{sph}(\text{ref})$ is obtained from the NMR spectra, g is the geometrical constant, which for ideal geometry $(g_{cyl} - g_{sph}) = 2\pi/3 = 2.0952$. Bernstein and Frei (45), have found this factor to be 2.058 from

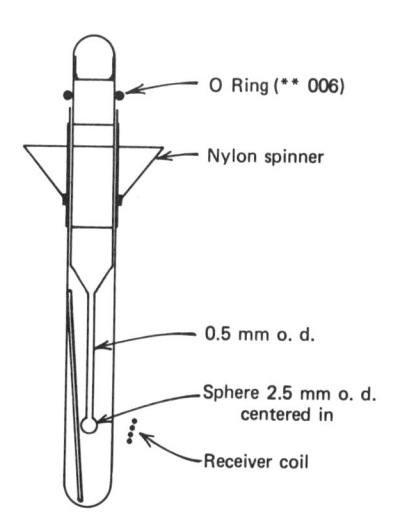

Fig. 4.5. Sample tube containing cylindrical and spherical reference tubes. Reproduced with permission from *J. Chem. Phys.*, 37, 1891 (1962).

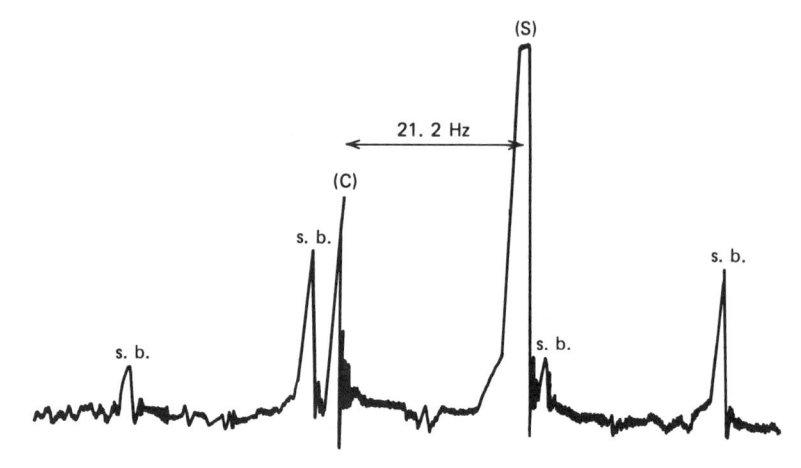

Fig. 4.6. NMR spectrum at 60 MHz of reference substance (H$_2$O) in cylindrical (c) and spherical (s) tubes immersed in tetramethylsilane. Sweep rate 0.64 Hz, audio modulation 24.7 Hz. Reproduced with permission from *J. Chem. Phys.*, **37**, 1891 (1962).

a slope of least-squares treatment for 15 compounds. By substituting values of different constants and factors, κ (sample is calculated from equation (24) and is converted into χ_s by dividing κ (volume susceptibility) by the density of the sample.

4.3. THEORETICAL CALCULATION OF DIAMAGNETIC SUSCEPTIBILITY OF HETEROCYCLIC COMPOUNDS

Two methods have been developed to theoretically calculate diamagnetic susceptibilities of heterocyclic compounds; the empirical Pacault and semiempirical methods.

Table 4.1. Atomic Susceptibility Data

Atom	Atomic Susceptibility Contribution
C	6.00
H	2.93
N	5.55
S	15.00
O (in alcohols and ethers)	4.60
O (in aldehydes and ketones)	−1.72
O (in amides)	1.60
O$_2$ (in acids)	7.95
Cl	16.65
Br	24.75

Table 4.2. Contribution of Different Rings to Molecular Susceptibility of Heterocyclic Compounds

Ring	Value
Benzene	1.4
Naphthalene	5.3
Triazine	1.4
Pyrimidine	−6.5
Pyrazine	−9.0
Furan	2.5
Pyrrole	3.5
Pyrazole	−8.0
Imidazole	−9.0
Thiophene	7.0
Thiophene moiety	3.5
Thiazole	3.0
Thiazole moiety	1.0
Isoxazole	−1.0
Cyclohexane	−3.0
Phenothiazine	−7.0
Pyridine	−0.5
Contribution on C_3^α	1.3

4.3.1. Pacault Method

Pacault (51) has developed a method for calculating diamagnetic susceptibilities of heterocyclic compounds using Pascal atomic susceptibility data (52–63) and accounting for the magnetic susceptibilities for heterocyclic rings. Using empirically derived magnetic susceptibilities for various heterocyclic rings, he has calculated molecular diamagnetic susceptibilities of heterocyclics. Atomic susceptibility data, susceptibility contributions of rings used in Pacault's calculations, and the calculated susceptibilities, are summarized in Tables 4.1, 4.2, and 4.3, respectively.

4.3.2. Semiempirical Methods

4.3.2.1. Hameka Method

Recently Hameka and coworkers (64) have developed a semiempirical method to calculate diamagnetic susceptibilities of heterocyclics containing nitrogen (pyridine and pyrrole series). In this method χ_M is represented by the equation

$$\chi_M = \chi_\epsilon + \chi_\pi - \chi_{\epsilon\pi} \qquad (25)$$

where χ_ϵ is the susceptibility of σ electrons, χ_π is the susceptibility contribution of

Table 4.3. Diamagnetic Susceptibilities of Heterocyclic Compounds Calculated by Pacault's Method

Compounds	Contribution of Atoms	Structural Increments	χ_M
Coumarin	74.46	8.10	82.56^a
4,7-Dimethyl coumarin	98.18	8.10	106.28^a
4,6-Dimethyl coumarin	98.18	8.10	106.28^a
4-Methyl-7,8-dihydrocoumarin	95.20	11.02	106.22^a
4-Methyl-7-hydroxycoumarin	90.32	10.00	100.32^a
3-Anisyl-4-methyl-7-hydroxycoumarin	155.1	11.6	166.7^a
3-Methyl chromone	86.32	8.10	94.42^a
Pyridine	50.20	0.50	49.7^a
Quinoline	80.06	6.56	86.62^a
Hydroxyquinoline	84.66	7.00	91.66^a
Tetrahydroquinoline	91.78	+.60	90.18^a
Acridine	109.92	2.90	122.82^a
1,10-Dimethyl-5,6-benzoacridine	163.50	19.60	183.10^a
1,3,4,10-Tetramethyl-5,6-benzoacridine	187.22	19.60	206.82^a
2-Amino-4-methyl pyrimidine	67.16	+7.45	59.71^a
2,4-Dimethyl-6-hydroxypyrimidine	75.14	+6.20	69.94^a
6-Methyl-4-hydroxy-2-thiopyrimidine	78.28	+5.20	73.08^a
2-Methyl-4-hydroxyquinazoline	93.14	1.40	94.54^a
2-Isobutyl-4-hydroxyquinazoline	128.72	1.50	130.22^a
2-Phenyl-4-hydroxyquinazoline	129.00	2.90	131.90^a
Pyrazine	46.82	+9.00	37.82^a
Diphenylmethyltriazine	150.74	4.20	154.94^a
Triethyltriazine tricarbonate	156.45	6.05	162.50^a
Furan	40.32	2.50	42.82^a
2,6-Dimethylfuran	64.04	2.50	66.54^a
Furfural	44.6	4.2	48.4^a
2-Phenylbenzofuran	117.90	11.60	128.50^a
Pyrrole	44.20	1.50	45.7^a
2-Methyl pyrrole	56.06	3.50	59.56^a
2,5-Dimethyl pyrrole	67.92	3.50	71.42^a
2,4-Dimethyl pyrrole	67.92	3.50	71.42^a

Compound			
2,3,5-Trimethyl pyrrole	79.78	3.50	83.28[a]
1,2-Dimethyl-5-ethyl pyrrole	91.74	3.50	95.24[a]
2,5-Dimethyl-3-ethyl pyrrole	91.74	3.50	95.24[a]
2,5-Dimethyl-3-propyl pyrrole	103.60	3.50	107.10[a]
Indole	74.16	9.20	83.36[a]
Carbazole	103.92	13.40	117.32[a]
Tetrahydrocarbazole	115.64	3.70	119.34[a]
Dimethyl pyrazole	64.54	+ 8.54	56.0[a]
Methyl diphenyl imidazole	148.12	+ 5.20	142.92[a]
Thiophene	50.72	6.80	57.52[a]
2-Methylthiophene	62.58	4.50	68.08[a]
2-Methyl-5-acetylthiophene	78.72	5.05	83.77[a]
5-Acetylthiophene	66.86	5.05	71.91[a]
Thiazole	47.34	3.00	50.34[a]
2-Methyl thiazole	59.20	1.00	60.20[a]
Dimethyl isoxazole	67.7	+ 11.83	55.87[a]
Phenothiazine	118.9	+ 4.20	114.70[a]
1,2,3,4-Dibenzophenothiazine	178.64	9.20	187.84[a]
1,3-Dinitrophenothiazine	129.92	1.60	128.32[b]
1,3-Dinitro-7-methylphenothiazine	141.68	+ 1.50	140.18[b]
1,3-Dinitro-7-methoxyphenothiazine	146.38	+ 0.30	146.08[b]
1,3-Dinitro-7-chlorophenothiazine	143.64	+ 0.30	143.34[b]
1,3-Dinitro-7-bromophenothiazine	151.74	+ 0.30	151.44[b]
1-Nitro-7-methylphenothiazine	136.28	+ 2.90	133.38[b]
1-Nitro-7-bromophenothiazine	146.24	1.60	144.64[b]
1-Nitro-7-chlorophenothiazine	159.96	+ 0.30	159.66[b]
1-Nitrophenothiazine	124.42	+ 2.90	121.52[b]
3-Nitro-4-chlorophenothiazine	138.14	+ 1.60	136.54[b]
3-Nitro-7-methylphenothiazine	136.88	+ 3.50	133.38[b]
2-Nitro-7-chlorophenothiazine	138.14	1.60	136.54[b]

[a] Reference 51.
[b] Reference 14.

π electrons, and $\chi_{\epsilon\pi}$ is the susceptibility contribution due to the interaction between π and σ electrons. The χ_π susceptibility contribution for a molecule is expressed by the equation

$$\chi_{\pi,M} = K_M \chi_{\pi,benz} \tag{26}$$

K_M is the London (65) parameter, the ratio of theoretical London values for the molecule M and for benzene. The ratio $K(\text{pyr})$ between the π-electron susceptibilities of pyridine and benzene has been determined to be 0.9886 from the molecular ground-state energy equations of benzene (27) and pyridine (28):

$$E_0(\text{benz}) = -8.0 + 4r^2 \tag{27}$$

$$E_0(\text{pyr}) = -8.314 + 3.9543 r^2 \tag{28}$$

where r^2 is the mean square radius of π electrons in the six-membered aromatic ring.

$\chi_\pi(\text{pyr})$ is expressed by the equation

$$\chi_\pi(\text{pyr}) = 0.9866 D \tag{29}$$

where D is the π-electron susceptibility contribution of benzene. π-Electron susceptibility contributions for a number of six-membered heterocyclic compounds have been determined analogously and are summarized in Table 4.4. By introducing a new parameter for $\chi_\epsilon - \chi_{\epsilon\pi}$, χ_M of pyridine is expressed by the equation

$$\chi_M(\text{pyr}) = A + P + 0.9866 D \tag{30}$$

where P is the parameter to differentiate the σ-electron susceptibility contribution of pyridine from benzene; that is, it represents a change in diamagnetic susceptibility on account of σ-electron contributions when a C–H group in the benzene ring is replaced by a nitrogen atom to convert benzene into pyridine. To express χ_M of a number of heterocyclics, other parameters B, M, N, U, and Y have been introduced. B represents the susceptibility contribution of a benzene ring attached to a heterocyclic moiety, and M the susceptibility contribution of a methyl group when it is present at the *para* or *meta* positions to nitrogen in the heterocyclic ring. $M + N$ is the susceptibility contribution of a methyl group attached *ortho* to the nitrogen atom. U and Y are the susceptibility contributions due to interactions between two nitrogen atoms in a heterocyclic ring at the *meta* and *para* positions, respectively, and their σ-electron susceptibility contributions are given by $A + 2P + U$ and $A + 2P + Y$, respectively. Values of A, B, and D are those reported for aromatics (66, 67), and the parameters P, M, N, U, and Y have been determined from the experimental molecular diamagnetic susceptibilities of heterocyclics and are summarized in Table 4.5.

For calculating molecular susceptibilities of the pyrrole series, two new parameters L and R have been introduced. L represents σ-electron susceptibility and interactions of the pyrrole ring, and R represents π-electron susceptibility. Thus the molecular diamagnetic susceptibility of pyrrole is $L + R$. The π-electron susceptibility contribution for other compounds in the pyrrole series has been determined in terms of the π-electron susceptibility contribution of pyrrole and is expressed by the equation

Table 4.4. Diamagnetic Susceptibilities of Heterocyclic Compounds Calculated by Hameka's Method

Compound	χ_M in Terms of Parameters	χ_M
Pyridine	$A + P + 0.9886D$	48.034
α-Picoline	$A + P + M + N + 0.9921D$	60.351
β-Picoline	$A + P + M + 0.9848D$	60.381
γ-Picoline	$A + P + M + 0.9911D$	60.452
2,4-Lutidine	$A + P + 2M + 2N + 0.9939D$	72.760
2,6-Lutidine	$A + P + 2M + 2N + 0.9949D$	72.661
2,4,6-Collidine	$A + P + 3M + 2N + 0.9959D$	85.062
Pyridazine	$A + 2P + Y + 0.9937D$	40.500
Pyrimidine	$A + 2P + U + 0.9654D$	42.177
s-Triazine	$A + 3P + 3U + 0.9878D$	38.207
Pyrazine	$A + 2P + 0.9902D$	41.085
Quinoline	$A + P + B + N + 2.1638D$	84.613
Isoquinoline	$A + P + B + 2.1677D$	84.769
2-Methylquinoline	$A + P + B + 2N + 2.1723D$	96.986
4-Methylquinoline	$A + P + B + M + N + 2.1691D$	97.062
6-Methylquinoline	$A + P + B + M + N + 2.1597D$	96.956
7-Methylquinoline	$A + P + B + M + N + 2.1657D$	97.024
8-Methylquinoline	$A + P + B + M + N + 2.1585D$	96.942
2,4-Dimethylquinoline	$A + P + B + 2M + 2N + 2.1762D$	109.421
9-Methyl-3,4-benzoacridine	$A + P + 3B + M + 2N + 4.3672D$	168.610
5,9-Dimethyl-1,2-benzoacridine	$A + P + 3B + 2M + 2N + 4.3534D$	180.845
5,7,9-Trimethyl-3,4-benzoacridine	$A + P + 3B + 3M + 2N + 4.3529D$	193.230
5,7,9-Trimethyl-1,2-benzoacridine	$A + P + 3B + 3M + 2N + 4.3425D$	193.112
5,7,8,9-Tetramethyl-3,4-benzoacridine	$A + P + 3B + 4M + 2N + 4.3572D$	205.667
5,7,8,9-Tetramethyl-1,2-benzoacridine	$A + P + 3B + 4M + 2N + 4.3498D$	205.584
1,2,5,6-Dibenzoacridine	$A + P + 4B + 2N + 5.4152D$	191.474
Pyrrole	$L + R$	47.4382
2-Methylpyrrole	$L + M + N + 0.7929R$	59.1010
2,4-Dimethylpyrrole	$L + 2M + N + 1.1245R$	72.4759
2,5-Dimethylpyrrole	$L + 2M + 2N + 1.0227R$	72.0616
2,3,5-Trimethylpyrrole	$L + 3M + 2N + 0.6824R$	83.4406
Indole	$L + B + N + 5.2056R$	83.2310
Carbazole	$L + 2B + 2N + 8.8382R$	117.3212

$$\chi_{\pi, M} = K_M, \quad \chi_{\pi(\text{pyrr})} = K_M, R \tag{31}$$

The values of L and R have been determined from the experimental molecular diamagnetic susceptibilities and are summarized in Table 4.5.

Calculated diamagnetic susceptibilities are summarized in Table 4.4.

Table 4.5. Values of the Parameters

Parameter	Value
A	43.8286
B	23.4107
D	11.3003
P	−6.9664
M	12.3899
N	−0.1120
Y	−0.6252
U	1.3719
L	44.4675
R	2.97077

The Hameka semiempirical method can be applied to predict molecular diamagnetic susceptibility with reasonable accuracy.

4.3.2.2. Hartree-Fock-Roothann Coupled Method

Boucekkine and Gayso (68) have reported the calculations of diamagnetic susceptibilities of 14 heterocycles of biochemical interest by the Hartree-Fock-Roothann coupled semiempirical method. The reported results are satisfactory but not as accurate as semiempirical methods.

4.4. RESULTS AND DISCUSSIONS

For purposes of discussion, the heterocyclic compounds have been divided into six classes according to their content: nitrogen, oxygen, sulfur, nitrogen and sulfur, and nitrogen and oxygen. The sixth series deals with diamagnetism in solutions.

4.4.1. Heterocyclic Compounds Containing Nitrogen

In this class of heterocyclic compounds pyrroles, pyridines, and diazines series have been studied. Bonino and Manzoni-Ansidei (69, 70) extensively studied the diamagnetic susceptibilities of pyrrole and its derivatives. They measured the diamagnetic susceptibilities of pyrrole and its monoalkyl, dialkyl, and trialkyl carbon-linked pyrroles and compared them with those calculated from Pascal's additivity rule considering double bonds as ethylenic double bonds. They observed poor agreement between the calculated and measured diamagnetic susceptibilities, but good agreement when the diamagnetic susceptibilities were calculated from the Pascal additivity rule, considering the benzene ring as one moiety and giving due consideration to the diamagnetic susceptibility contribution of the ring. Such observations led them to conclude that the double bonds in pyrrole and its derivatives behave differently than ethylenic double bonds and are delocalized, which accounts for their aromatic character. N-Alkyl pyrroles (70) behave similarly, and diamagnetic studies (69, 71) have shown that their double bonds are delocalized.

Pyridines, picolines, lutidines, and collidines have been studied extensively (72), and some useful information has been obtained regarding the structural derivations from diamagnetic measurements. Measured diamagnetic susceptibilities agree well with those calculated by the Gray and Cruickshank method (73), in which due consideration is made for depression in molecular diamagnetism due to ethylenic double bonds. It is interesting to note that the diamagnetic contribution of π bonds is much less if they are considered as ethylenic π bonds, and is much higher when π bonds are delocalized. Such magnetochemical studies are useful in studying delocalization of π electrons in six-membered nitrogen heterocyclics.

Manzoni-Ansidei and Ghe (74) have reported the diamagnetic susceptibilities of quinoline and its derivatives. They have observed large deviations between measured susceptibilities and those calculated following Pascal's additivity rule. Such large deviations between measured and calculated diamagnetic susceptibilities are ascribed to the delocalization of π bonds, which contribute significantly to molecular diamagnetism but are ignored in calculating magnetic susceptibilities with the Pascal addivity rule. The phenomenon of hyperconjugation in quinoline as well as in pyridine derivatives has been studied using diamagnetic susceptibility investigations. Hyperconjugation of the methyl groups to aromatic heterocyclic nuclei has a unique effect in lowering molecular diamagnetism.

Wilson (75) has used diamagnetic susceptibility data as evidence for the delocalization of lone pairs in diazines. He has measured and compared diamagnetic susceptibilities of pyrimidine, pyridazine, and pyrazine and has noted a decreasing trend (43.1, 40.5, 38.1, respectively). This has been excellently explained by the findings of Baldeschwieher and Randall (76) and Gil and Murrell (77) that the unshared pair of electrons on the nitrogen atom in pyridinelike molecules contribute substantially to the susceptibility of molecules because of the low-energy $n-\pi^*$ transition associated with electronic structural features and its influence on the magnitude of the Van Vleck paramagnetic contribution, which is inversely proportional to the energy of this electronic transition. Such findings have been reported by Clementi (78) for pyrazine, and by Hoffman and coworkers (79) for diazine-type systems, in which orbitals of the "lone-pair" type are in general delocalized throughout the π system. The orbitals in these species interact more or less strongly depending on the geometry, and such interaction results in raising the energy of the highest filled n orbital, which causes red shifts in the $n-\pi^*$ transitions in diazines, thus increasing the Van Vleck paramagnetic contribution. It has been further predicted (78) that the increase in the Van Vleck paramagnetic contribution is small in pyrimidine but substantial in pyridazine and pyrazine. This explains why diamagnetic susceptibilities of pyridazine and pyrazine are lower than that of pyrimidine. On the basis of these observations, Wilson (75) predicted that the susceptibilities of the three triazines would follow the order $1,3,5 > 1,2,3 > 1,2,4$.

Diamagnetic studies (80) have also been extended to 5,5'-disubstituted barbituric acids and thiobarbituric acids. Such studies suggest that 5,5'-disubstituted barbituric acids have structures between the neutral and ionic structures. Thiobarbituric acids possess large diamagnetic exaltations which suggest the existence of molecular π-electron orbitals.

4.4.2. Heterocyclic Compounds Containing Oxygen

In this class of heterocyclic compounds, the furan and coumarin series have been studied. Diamagnetic susceptibilities of furan and its derivatives (69, 70) have been measured and compared with those calculated theoretically from the Pascal additivity rule, considering double bonds as ethylenic double bonds. The poor agreement observed leads to the conclusion that double bonds in furans do not behave as ethylenic double bonds. When their diamagnetic susceptibilities are calculated from the Pascal additivity rule and are considered as the contribution for the entire ring, they agree well with measured values. Such observations show that delocalization of π bonds exists in furans as in pyrrole, which accounts for the aromatic character in furans.

Mathur (81, 82) has investigated the diamagnetism of coumarins and has analyzed the experimental data in the light of structural factors. He has calculated the molecular diamagnetic susceptibility of a methylene group attached directly to coumarin, by subtracting the diamagnetic susceptibility of coumarin from the diamagnetic susceptibility of methyl coumarin. Similarly, he calculated the molecular diamagnetic susceptibility of a methylene group attached to coumarin through the oxygen atom by subtracting the diamagnetic susceptibility of hydroxycoumarin from that of methoxycoumarin. The diamagnetic susceptibility of the methylene group has been reported to be 11.67 when it is attached to coumarin as CH_3 and 10.68 when it is attached as OCH_3 (see in Table 4.6). The diamagnetic susceptibility of the methylene group when attached to coumarin as CH_3 is higher than that when attached as OCH_3, by 0.99. Lowering in the diamagnetic susceptibility of the methylene group has been attributed to (1) the distortion of sp^3 tetrahedral orbitals and (2) its linking with the oxygen atom. A decrease of 0.99 in diamagnetic susceptibility of the methylene group when it is present as OCH_3 has led to the

Table 4.6. Diamagnetic Susceptibilities of Coumarins

Compound M_1	χ_{M_1}	Compound M_2	χ_{M_2}	χ_{CH_2}[a,b]
4,6-Dimethyl coumarin	106.20	Coumarin	83.0	11.60
7-Hydroxy-4-methyl coumarin	99.96	7-Hydroxy coumarin	88.22	11.74
7-Methoxy-4-methyl coumarin	110.50	7-Hydroxy-4-methyl coumarin	99.96	10.54
6-Methoxy-4-methyl coumarin	109.50	6-Hydroxy-4-methyl coumarin	98.69	10.81

[a] $\chi_{CH_2} = \chi_{M_1} - \chi_{M_2}$.
[b] Average $\chi_{CH_2} = 11.67$ and 10.68, respectively, when it is present as CH_2 and OCH_2.

conclusion that ionic character due to the difference in the electronegativities of carbon and oxygen in C—O is 22% in the coumarin series (82).

4.4.3. Heterocyclic Compounds Containing Sulfur

In this series thiophene and its methyl derivatives have been studied (69, 70). Their diamagnetic susceptibilities have been compared with those calculated from the Pascal additivity rule giving due consideration to the delocalization of π bonds. Good agreement between measured and calculated diamagnetic susceptibilities is due to the delocalization of the π-bond system in thiophene and its derivatives.

4.4.4. Heterocyclic Compounds Containing Nitrogen and Sulfur

In this class of heterocyclic compounds, phenothiazines and thiazoles have been studied. Goyal (14) has used diamagnetic susceptibility measurements in studying hydrogen bonding in phenothiazines. He has reported diamagnetic susceptibilities of substituted phenothiazines and the corresponding 1-nitrophenothiazines (summarized in Table 4.7). He has calculated their diamagnetism by the Pacault method (51) and compared the calculated values with experimental results. It was observed that experimental values of molecular susceptibility agree well with the corresponding theoretical values in the case of 1-unsubstituted phenothiazines, while the 1-nitrophenothiazines differ by 2.0–3.3%, as summarized in Table 4.7. The large deviation between experimental and calculated susceptibilities has been ascribed to the formation of intramolecular hydrogen bonding due to involvement of the nitro group in 1-nitrophenothiazines. The formation of a six-membered ring of high stability as a consequence of strong NH---O—N bonding, exalts molecular susceptibilities of 1-nitrophenothiazines **1**.

Table 4.7. Diamagnetic Susceptibilities of Phenothiazines

Compound	χ_M Experimental	χ_M Calculated	$\Delta\chi_M$ (%)
Phenothiazine	114.80	114.72	0.07
3-Nitrophenothiazine	121.97	121.52	0.37
3-Nitro-7-chlorophenothiazine	136.02	136.54	0.38
4-Nitro-7-methylphenothiazine	132.74	133.38	0.48
2-Nitro-7-chlorophenothiazine	135.59	136.54	0.70
1,3-Dinitrophenothiazine	130.98	128.32	2.0
1,3-Dinitro-7-methylphenothiazine	143.17	140.18	2.1
1,3-Dinitro-7-methoxyphenothiazine	149.52	146.08	2.3
1,3-Dinitro-7-chlorophenothiazine	146.85	143.34	2.4
1,3-Dinitro-7-bromophenothiazine	154.96	151.44	2.3
1-Nitro-7-methylphenothiazine	137.88	133.38	3.3
1-Nitro-7-bromophenothiazine	147.91	144.64	2.2
1-Nitro-3-chloro-7-bromophenothiazine	163.75	159.66	2.5

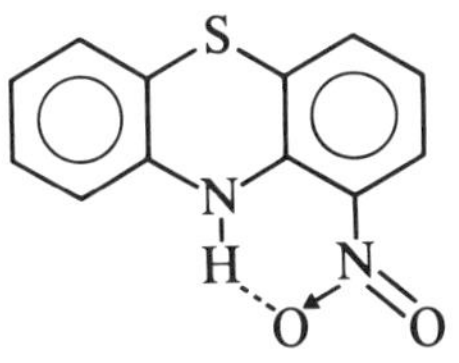

1

Goyal (14) has provided a thorough rationalization for the role of hydrogen bonding based upon IR spectra, mass fragmentation, and diamagnetic studies of 1-nitrophenothiazines. Hydrogen bonding in 1-nitrophenothiazines lowers the N–H stretching frequency from 3320–3345 cm^{-1} to 3280–3300 cm^{-1} and keeps the NO$_2$ group and H attached to nitrogen in the same plane, which creates suitable conditions for McLafferty rearrangement and increases diamagnetism.

Yoshida (83) has studied the diamagnetic behavior of 2-mercaptobenzothiazole, di-2-benzothiazolyl disulfide, and N-cyclohexylbenzothiazole sulfenemide. He has measured diamagnetic susceptibilities of these compounds and compared them with those calculated theoretically. Large diamagnetic exaltations were observed. Such large exaltations have shown the existence of molecular π-electron orbitals. Similar results have also been reported for thiazole and 2-methylthiazole (69).

4.4.5. Heterocyclic Compounds Containing Nitrogen and Oxygen

Diamagnetic studies have been extended to only one compound of this series, dimethyl oxadiazole. It has been observed (84) that its diamagnetic susceptibility agrees well with the theoretical value if calculated without considering it as a system of conjugated double bonds. Diamagnetic studies show that there is no conjugation of double bonds in dimethyl oxadiazole and that it possesses an open-chain aliphatic structure similar to diazo derivatives **2**, which contradicts a cyclic structure **3**.

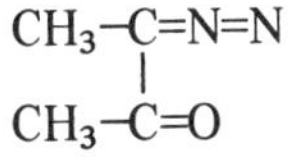

$$CH_3-C=N=N$$
$$|$$
$$CH_3-C=O$$

2

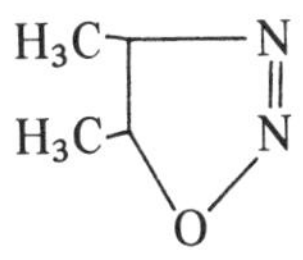

3

4.4.6. Heterocyclic Compounds In Solution

Diamagnetic studies have been used quite successfully in studying hydrogen bonding between heterocycles and solvents. Cini and coworkers (85, 86) selected

1,4-dioxane–water, tetrahydrofuran–water, and pyridine–water mixtures. They studied 1,4-dioxane–water mixtures over the full concentration range and found positive deviations from Wiedmann's additivity rule, with the highest deviation of 0.7% in the range of 0.60 mole fraction of dioxane (Table 4.8). It has been inferred that dioxane breaks the hydrogen-bonded water structure and forms an associated complex with it in a molecular ratio 1:2. Positive deviations have also been reported (85) for tetrahydrofuran–water and pyridine–water mixtures (Tables 4.9 and 4.10) in the full concentration range. In tetrahydrofuran–water mixtures, breaking of water structures by tetrahydrofuran prevails over their associations. A minimum derivation at 30% is reported, which corresponds to an association in the molecular ratio 2:1. It has been reported that in pyridine–water mixtures the combined breaking action of pyridine on water structure and breaking of water on pyridine (as pyridine is self-associated) prevails over the association of a water–pyridine complex, in molecular ratio 1:1. On the basis of these studies Cini and coworkers (85, 86) reported that hydrogen bonds are formed between heterocyclic compounds and other components in the mixture if deviations in molecular susceptibility with respect to Wiedmann's additivity rule are negative, and are broken if the deviations are positive.

Gopalakrishnan (87) has studied pyridine, α-picoline, and 2,4-lutidine in chloroform. He has observed large negative deviations for pyridine–chloroform, which were attributed to the formation of hydrogen bonds between pyridine and chloroform. In the case of α-picoline–chloroform and 2,4-lutidine–chloroform mixtures, small deviations have been observed but have not led to any significant conclusion.

Table 4.8. χ_M and $\Delta\chi_M$ for Water-1,4-Dioxane Mixtures

Mole fraction of Dioxane	20°C χ_M	20°C $\Delta\chi_M$	25°C χ_M	25°C $\Delta\chi_M$
0.0	12.971	0.0	12.979	0
0.06489	15.513	0.035	—	—
0.10098	16.933	0.060	16.944	0.063
0.14516	18.689	0.110	—	—
0.20589	21.070	0.144	21.085	0.149
0.24991	22.754	0.127	—	—
0.30310	24.802	0.120	24.796	0.103
0.34584	26.449	0.116	—	—
0.40255	28.643	0.119	28.669	0.133
0.50243	32.512	0.129	32.518	0.121
0.59410	35.630	0.091	—	—
0.60311	36.332	0.059	36.356	0.068
0.63629	—	—	—	—
0.70505	—	—	40.275	0.048
0.70612	40.287	0.034	—	—
0.80540	44.127	0.038	44.147	0.042
0.89670	47.648	0.031	47.679	0.045
1.0	51.608	0	51.626	0

Table 4.9. χ_M and $\Delta\chi_M$ of Water–Tetrahydrofuran

Mole Fraction of Tetrahydrofuran	χ_M	$\Delta\chi_M$
0	12.980	0
0.04101	14.559	−0.001
0.10245	16.976	0.048
0.15303	18.955	0.077
0.20032	20.792	0.092
0.25140	22.776	0.108
0.27465	23.682	0.094
0.30708	24.890	0.076
0.35257	26.666	0.098
0.35859	28.462	0.121
0.45037	30.463	0.126
0.50591	32.453	0.130
0.54905	34.283	0.143
0.59702	36.101	0.113
0.59753	36.120	0.112
0.69381	39.811	0.092
0.78934	43.458	0.058
0.89993	47.690	0.028
1	51.519	0

Table 4.10. χ_M and $\Delta\chi_M$ of Water–Pyridine Mixtures

Mole Fraction of Pyridine	χ_M	$\Delta\chi_M$
0	12.980	0
0.05190	14.831	0.028
0.10091	16.572	0.048
0.14690	18.171	0.066
0.24669	21.749	0.103
0.29348	23.412	0.122
0.34516	25.240	0.135
0.38998	26.811	0.131
0.43910	28.542	0.137
0.48717	30.235	0.141
0.53626	31.956	0.138
0.59014	33.856	0.144
0.63819	35.538	0.139
0.68536	37.196	0.140
0.74093	39.116	0.107
0.78813	40.760	0.093
0.83255	42.312	0.085
0.88867	44.264	0.065
1	48.109	0

Table 4.11. Diamagnetic Susceptibilities of Heterocyclic Compounds

Heterocyclic	Molecular formula	χ_M	Ref.
Acetylthiophene	C_6H_6OS	71.7	51
Acridine	$C_{13}H_9N$	123.3	87
1-Allyl pyrrole	C_7H_9N	73.80	68
Aminomethyl diethyl diazine	$C_9H_{15}N_3$	114.8	88
2-Aminothiazole	$C_3H_4N_2S$	56.0	51
Barbuturic acid (anhydrous)	$C_4H_4O_3N_2$	53.8	79
Barbuturic acid ($2H_2O$)	$C_4H_8O_5N_2$	78.6	79
Carbazole	$C_{12}H_9N$	117.40	51
2,4,6-Collidine	$C_8H_{11}N$	85.062, 83.22	96, 71
Coumarin	$C_9H_6O_2$	82.5	51
1,2,5,6-Dibenzoacridine	$C_{21}H_{13}N$	186.4	64
5,9-Dimethyl-1,2-benzoacridine	$C_{19}H_{15}N$	184.3	64
1,10-Dimethyl-5,6-benzoacridine	$C_{19}H_{15}N$	184.3	51
5,7-Dihydroxycoumarin	$C_{10}H_8O_4$	106.9	80
7,8-Dihydroxycoumarin	$C_{10}H_8O_4$	105.1	80
4,6-Dimethylcoumarin	$C_{11}H_{10}O_2$	106.2	80
4,7-Dimethylcoumarin	$C_{11}H_{10}O_2$	107.6	80
Dimethyl diethyl keto tetrahydrofurfuran	$C_{10}H_{18}O$	116.2	87
2,5-Dimethyl-1-ethylpyrrole	$C_8H_{11}N$	94.61	68
2,5-Dimethyl-3-ethylpyrrole	$C_8H_{11}N$	93.87	68
2,5-Dimethyl furan	C_6H_8O	66.37	68
Dimethyl furazan	$C_4H_6ON_2$	57.27	83
Dimethyl isoxazole	C_5H_7ON	59.7	51
Dimethyl keto tetrahydrofurfurane	$C_6H_{10}O_2$	68.5	87
3,4-Dimethyl oxadiazole	$C_4H_6ON_2$	57.17	83
2,5-Dimethyl-3-propyl pyrrole	$C_9H_{15}N$	106.07, 105.62	68, 69
3,5-Dimethyl pyrazole	$C_5H_8N_2$	56.2	89

(*Continued on next page*)

Table 4.11. (continued)

Heterocyclic	Molecular formula	χ_M	Ref.
2,4-Dimethyl pyridine	C_7H_9N	71.50	71
2,6-Dimethyl pyridine	C_7H_9N	71.72	71
2,4-Dimethyl pyrrole	C_6H_9N	69.64	68
2,5-Dimethyl pyrrole	C_6H_9N	71.92	68
2,4-Dimethylquinoline	$C_{11}H_{11}N$	108.54	73
1,3-Dinitro-7-bromophenothiazine	$C_{12}H_6BrN_3O_4S$	154.96	14
1,3-Dinitro-7-chlorophenothiazine	$C_{12}H_6ClN_3O_4S$	146.85	14
5,7-Dinitro-7-hydroxy-4-methyl coumarin	$C_{10}H_6O_7N_2$	111.9	80
1,3-Dinitro-7-methoxy phenothiazine	$C_{13}H_9N_3O_5S$	149.52	14
1,3-Dinitro-7-methyl phenothiazine	$C_{13}H_9N_3O_4S$	143.17	14
1,3-Dinitrophenothiazine	$C_{12}H_7N_3O_4S$	130.98	14
1,4-Dioxane	$C_4H_8O_2$	52.16	90
Diphenyl dihydrotetrazine	$C_{14}H_{12}N_4$	129.9	88
Flavanthrone	$C_{28}H_{11}N_2O_2$	241.0	68
Furan	C_4H_4O	43.09, 43.18	68, 70
Furfural	$C_5H_4O_2$	47.1	51
Guanosine	$C_{10}H_{13}O_5N_5$	149.1	91
7-Hydroxy coumarin	$C_9H_6O_3$	88.62	80
5-Hydroxy-4,7-dimethyl coumarin	$C_{11}H_{10}O_3$	113.5	80
6-Hydroxy-4-methyl coumarin	$C_{10}H_8O_3$	98.69	80
7-Hydroxy-4-methyl coumarin	$C_{10}H_8O_3$	99.96	80
7-Hydroxy-4-methyl-3-bromo coumarin	$C_{10}H_7O_3Br$	127.3	80
7-Hydroxy-6-nitro-4-methyl coumarin	$C_{10}H_7O_5N$	106.0	80
6-Hydroxy-5-nitro-4-methyl coumarin	$C_{10}H_7O_5N$	105.6	80
7-Hydroxy-8-nitro-4-methyl coumarin	$C_{10}H_7O_5N$	106.0	80
Indole	C_8H_7N	85.0	51
2,4-Lutidine	C_7H_9N	72.76, 71.50	96, 71
2,6-Lutidine	C_7H_9N	72.66, 71.72	96, 71

Compound	Formula		
Melamine	$C_3H_6N_6$	61.8	71
6-Methoxy-4-methyl coumarin	$C_{11}H_{10}O_3$	109.5	80
7-Methoxy-4-methyl coumarin	$C_{11}H_{10}O_3$	110.5	80
2-Mercaptobenzothiazole	$C_7H_5NS_2$	99.4	82
Methyldiphenylthiazine	$C_{16}H_{13}N_3$	155.1	87
9-Methyl-3,4-benzoacridine	$C_{18}H_{13}N$	161.1	64
2-Methylpyridine	C_6H_7N	60.3	71
3-Methylpyridine	C_6H_7N	59.8	71
4-Methylpyridine	C_6H_7N	59.8	71
1-Methylpyrrole	C_5H_7N	58.8	68
2-Methylpyrrole	C_5H_7N	61.10	68
2-Methylquinoline	$C_{10}H_9N$	99.86	73
4-Methylquinoline	$C_{10}H_9N$	94.71	73
6-Methylquinoline	$C_{10}H_9N$	97.43	73
7-Methylquinoline	$C_{10}H_9N$	97.86	73
8-Methylquinoline	$C_{10}H_9N$	96.57	73
2-Methylthiazole	C_4H_5NS	59.56	68
2-Methylthiophane	C_5H_6S	66.35	68
Morphine	$C_{17}H_{19}O_3N$	55.0	51
Nicotine	$C_{10}H_{14}N_2$	113.328	92
1-Nitro-7-bromophenothiazine	$C_{12}H_7BrN_2O_2S$	147.91	14
1-Nitro-7-bromo-3-chlorophenothiazine	$C_{12}H_6BrClN_2O_2S$	163.75	14
1-Nitro-7-methylphenothiazine	$C_{13}H_{10}NS$	137.88	14
Nitrosopiperidine	$C_5H_{10}ON_2$	63.4	87
Phenothiazine	$C_{12}H_9N_5$	114.8	51
2-Phenylbenzofuran	$C_{14}H_{10}O_2$	130.5	51
α-Picoline	C_6H_7N	61.22, 60.33	96, 71
β-Picoline	C_6H_7N	62.15, 59.81	96, 71
γ-Picoline	C_6H_7N	61.83, 59.84	96, 71
Piperazine	$C_4H_{10}N_2$	56.8	88
Piperidine	$C_5H_{11}N$	64.2	88

(*Continued on next page*)

Table 4.11. (continued)

Heterocyclic	Molecular formula	χ_M	Ref.
Pyramidone	$C_{13}H_{17}ON_3$	149.0	51
Pyrazine	$C_4H_4N_2$	37.6, 38.0	51, 74
Pyridazine	$C_4H_4N_2$	40.50	74
Pyridine	C_5H_5N	49.21, 48.4	93, 71
Pyrimidine	$C_4H_4N_2$	43.10	74
Pyrrole	C_4H_5N	47.6, 48.7	51, 68, 69, 92
		49.11, 46.029	
Pyrrolidine	C_4H_9N	54.8	51
Quinoline	C_9H_7N	86.0	87
Tetrahydroquinoline	$C_9H_{11}N$	89.0	51
5,7,8,9-Tetramethyl-3,4-benzoacridine	$C_{21}H_{19}N$	196.3	64
5,7,8,9-Tetramethyl-1,2-benzoacridine	$C_{21}H_{19}N$	209.2	64
1,3,4,10-Tetramethyl-5,6-benzoacridine	$C_{21}H_{19}N$	209.2	51
Tetraiodopyrrole	$C_4H_5I_4N$	188.9	51
Tetramethyl keto tetrahydrofurfurane	$C_8H_{14}O_2$	104.7	87
Thiocoumarin	C_9H_6OS	93.6	51
Thiazole	C_3H_3NS	50.55	68
s-Triazine	$C_3H_3N_3$	37.90	74
Quinolic acid	$C_7H_5O_4N$	72.3	94
Quinoline	C_9H_7N	86.62	73, 96
Isoquinoline	C_9H_7N	83.9	73
Thiobarbituric acid	$C_4H_7O_2N_2S$	72.9	79
Thiophene	C_4H_4S	57.38, 59.95	68, 70
Thymine	$C_5H_6O_2N_2$	57.1	91
Triethyl triazine tricarbonate	$C_{12}H_{15}O_6N_3$	164.1	87
5,7,9-Trimethyl-3-benzoacridine	$C_{20}H_{17}N$	184.3	64
5,7,9-Trimethyl-1,2-benzoacridine	$C_{20}H_{17}N$	183.7	64
2,3,5-Trimethylpyrrole	$C_7H_{11}N$	82.31	68
Tryptophan	$C_{11}H_{12}O_2N_2$	132.00	95
Xanthone	$C_{13}H_8O_2$	108.1	87

APPENDIX

The diamagnetic susceptibilities of a number of heterocyclic compounds are presented in Table 4.11.

ACKNOWLEDGMENTS

We gratefully acknowledge the granting of permission by the American Chemical Society, The American Institute of Physics, The Chemical Society (London), The Bulletin of the Chemical Society (Japan), and Nuovo Cimento (Italy) to reproduce diagrams and data from their respective journals. Our sincere thanks are also due to H. F. Hameka, R. Cini, L. N. Mulay, A. Pacault, H. J. Bernstein, R. Alan Earnshaw, and R. Gopalakrishnan for sending reprints.

REFERENCES

1. R. L. Mital and R. R. Gupta, *J. Amer. Chem. Soc.*, 91(17), 4664 (1964).
2. R. L. Mital and R. R. Gupta, *J. Chem. Phys.*, 54(7), 3230 (1971).
3. R. D. Goyal, R. R. Gupta, and R. L. Mital, *J. Phys. Chem.*, 76, 1579 (1972).
4. R. L. Mital, R. D. Goyal, and R. R. Gupta, *Inorg. Chem.*, 11, 1924 (1972).
5. R. L. Mital and R. R. Gupta, *Inorg. Chem. Acta Rev.*, 4, 97 (1970).
6. R. R. Gupta and R. L. Mital, *Bull. Soc. Chim. Belg.*, 76, 631 (1967).
7. R. R. Gupta and R. L. Mital, *Indian J. Chem.*, 4(8), 370 (1966).
8. R. L. Mital and R. R. Gupta, *Indian J. Pure Appl. Phys.*, 8(3), 197 (1970).
9. R. D. Goyal, R. R. Gupta, and R. L. Mital, *Indian J. Chem.*, 9, 696 (1971).
10. R. R. Gupta, *J. Phys. Chem.*, 88, 2047 (1976).
11. L. G. Gouy, *Compt. Rend.*, 109, 935 (1889).
12. P. W. Selwood, *Magnetochemistry*, 2nd ed., Interscience, New York, 1956.
13. S. S. Bhatnagar and K. N. Mathur, *Physical Principles and Applications of Magnetochemistry*, Macmillan, London, 1935.
14. R. D. Goyal, Ph.D. thesis, Rajasthan University, Jaipur, India, 1972.
15. L. F. Bates, *Modern Magnetism*, 4th ed., Cambridge University Press, London, 1961.
16. R. S. Nyholm, *Quart. Rev.*, 1953, 377.
17. B. N. Figgis and J. Lewis, "The Magnetochemistry of Complex Compounds," in J. Lewis and R. G. Wilkins, Eds., *Modern Coordination Chemistry*, Interscience, New York, 1960.
18. A. Earnshaw, *Lab. Pract.*, 10, 157 (1961).
19. A. Earnshaw, *Introduction to Magnetochemistry*, Academic Press, London, 1968.
20. A. Earnshaw, *Lab. Pract.*, 10, 294 (1961).
21. A. Earnshaw, *Lab. Pract.*, 10, 89 (1961).
22. L. N. Mulay and I. L. Mulay, *Anal. Chem.*, 36(5), 404R (1964).
23. L. N. Mulay, *Anal. Chem.*, 34(5), 343R (1962).
24. L. N. Mulay, *Magnetic Susceptibility*, Interscience, New York, 1966.
25. E. Muller, A. Rieker, K. Scheffler, and A. Moosmayer, *Angew. Chem. Int. Ed.*, 5(1), 6 (1966).

26. R. L. Mital, Ph.D. thesis. Rajasthan University, Jaipur, India (1956).
27. S. Sriraman and D. Shanmugasundaram, *Bull. Chem. Soc. (Japan)*, **36**(5), 547 (1963).
28. W. R. Angus and W. K. Hill, *Trans. Faraday Soc.*, **40**, 185 (1943).
29. S. Sriraman, R. Sabesan, and S. Srinivasan, *Bull. Chem. Soc. (Japan)*, **36**(9), 1080 (1963).
30. F. E. Hoare and J. C. Matthews, *Proc. Roy. Soc. (London)*, **A212**, 137 (1952).
31. R. R. Gupta, Ph.D. thesis, Rajasthan University, Jaipur, India, 1968.
32. E. A. Bourdeaux, H. B. Jonassen, and L. J. Theriot, *J. Am. Chem. Soc.*, **85**, 2039 (1963).
33. B. N. Figgis and R. S. Nyholm, *J. Chem. Soc.*, 4190 (1958).
34. H. Nettleton and S. Sudgen, *Proc. Roy. Soc. (London)*, **A173**, 313 (1939).
35. N. E. Curtis, *J. Chem. Soc.*, **1961**, 3147.
36. M. Faraday, *Exp. Res. (London)*, 7, 27, 497 (1855).
37. L. Sacconi and R. Cini, *J. Sci. Instr.*, **31**, 56 (1954).
38. M. C. Day, L. D. Hulett, and D. E. Willis, *Rev. Sci. Instr.*, **31**, 1142 (1960).
39. C. Cheneveau, *Phil. Mag.*, **20**, 357 (1910).
40. A. E. Oxley, *Phil. Trans. Roy. Soc. (London)*, **A215**, 79 (1915).
41. E. Wilson, *Proc. Roy. Soc. (London)*, **A96**, 429 (1920).
42. W. Sucksmith, *Phil. Mag.*, **8**, 158 (1929).
43. G. Quincke, *Ann. Physik*, **24**, 347 (1885).
44. D. F. Evans, *J. Chem. Soc.*, **1959**, 2003.
45. K. Frei and H. J. Bernstein, *J. Chem. Phys.*, **37**, 1891 (1962).
46. A. A. Bothner-by and R. E. Glick, *J. Chem. Phys.*, **26**, 1647 (1957).
47. G. Feher and W. D. Knight, *Rev. Sci. Instr.*, **26**, 293 (1955).
48. M. G. Morin, G. Paulett, and M. E. Hobbs, *J. Phys. Chem.*, **60**, 1594 (1956).
49. C. A. Reilly, H. H. McConnel, and R. G. Meisenheimer, *Phys. Rev.*, **60**, 1594 (1956).
50. J. R. Zimmerman and M. R. Foster, *J. Phys. Chem.*, **61**, 282 (1957).
51. A. Pacault, *Ann. Chim.*, **1**, 527 (1946).
52. P. Pascal, *Compt. Rend.*, **147**, 56, 242, 742 (1908).
53. P. Pascal, *Ann. Chim. Phys.*, **19**, 5 (1910).
54. P. Pascal, *Compt. Rend.*, **148**, 413 (1909).
55. P. Pascal, *Compt. Rend.*, **150**, 1167 (1910).
56. P. Pascal, *Compt. Rend.*, **152**, 862 (1911).
57. P. Pascal, *Ann. Chim. Phys.*, **25**, 289 (1912).
58. P. Pascal, *Compt. Rend.*, **152**, 1010 (1911).
59. P. Pascal, *Ann. Chim. Phys.*, **28**, 289 (1912).
60. P. Pascal, *Compt. Rend.*, **156**, 323 (1913).
61. P. Pascal, *Ann. Chim. Phys.*, **28**, 218 (1913).
62. P. Pascal, *Compt. Rend.*, **158**, 37 (1914).
63. P. Pascal, *Compt. Rend.*, **172**, 144 (1921).
64. L. H. Haley and H. F. Hameka, *J. Am. Chem. Soc.*, **96**, 2020 (1974).
65. F. London, *J. Phys. Radium*, **8**, 397 (1937).
66. P. S. O'Sullivan and H. F. Hemeka, *J. Am. Chem. Soc.*, **92**, 1821 (1970).
67. M. E. Stockham and H. F. Hameka, *J. Am. Chem. Soc.*, **94**, 4076 (1972).
68. A. Boucekkine and J. Gayso, *Compt. Rend.*, **288**, 307 (1979).
69. G. B. Bonino and R. Manzoni-Ansidei, *Chem. Ber.*, **76**, 553 (1943).

70. G. B. Bonino and R. Manzoni-Ansidei, *Rec. Sci.*, 7(1), No. 3-4, 1 (1936); *Chem. Abstr.*, **31**, 3352[09].
71. G. B. Bonino and R. Manzoni-Ansidei, *Rec. Sci.*, 7(1), No. 3-4, (1936); *Chem. Abstr.*, **31**, 3352[10].
72. C. M. French, *Trans. Faraday Soc.*, **47**, 1056 (1951).
73. W. Gray and H. Cruickshank, *Trans. Faraday Soc.*, **31**, 1491 (1935).
74. R. Manzoni-Ansidei and A. M. Ghe, *Boll. Soc. Fac. Chim. Ind. Univ. Bologna*, **5**, 8 (1944-47); *Chem. Abstr.*, **43**, 8767[h].
75. J. D. Wilson, *J. Chem. Phys.*, **53**, 467 (1970).
76. J. D. Baldeschwieher and E. W. Randall, *Proc. Chem. Soc.*, **1961**, 303.
77. N. M. S. Gil and J. N. Murrell, *Trans. Faraday Soc.*, **60**, 248 (1964).
78. E. Clementi, *J. Chem. Phys.*, **46**, 4737 (1967).
79. R. Hoffman, A. Imanura, and W. J. Hehre, *J. Am. Chem. Soc.*, **90**, 1499 (1968).
80. Y. Sato, *Bull. Chim. Res. Inst. Non-Aqueous Solutions, Tohoku Univ.*, 6(1), 1 (1956); *Chem. Abstr.*, **51**, 1783[1].
81. R. M. Mathur, *Trans. Faraday Soc.*, **54**, 1609 (1958).
82. R. M. Mathur, *Trans. Faraday Soc.*, **56**, 325 (1960).
83. K. Yoshida, *Sci. Rep. Res. Inst. Tohoku Univ. Ser. A*, **11**, 422 (1959); *Chem. Abstr.*, **54**, 8188[f].
84. R. Monzoni-Ansidei, *Bull. Sci. Fac. Chim. Ind. Univ. Bologna*, **5**, 70 (1944-1947); *Chem. Abstr.*, **43**, 8767[i].
85. R. Cini, G. Taddie, and M. Torrini, *Nuovo Cimento*, **408**, 432 (1965).
86. R. Cini and G. Taddie, *Nuovo Cimento*, **43B**, 354 (1966).
87. R. Gopalakrishnan, *Bull. Chem. Soc. (Japan)*, **43**, 1607 (1970).
88. International Critical Tables.
89. P. Pascal, *Compt. Rend.*, **180**, 1596 (1925).
90. P. Pascal, *Ann. Chim. Phys.*, **181**, 656 (1925).
91. K. Venkateswarlu and S. Sriraman, *Trans. Faraday Soc.*, **53**, 433 (1957).
92. D. L. Woernley, *J. Biol. Chem.*, **207**, 717 (1954).
93. C. Courty, *Bull. Soc. Chim.*, **5**, 3, 420 (1936).
94. B. K. Singh, O. N. Petri, M. Singh, and S. L. Agrawal, *Proc. India Acad. Sci.*, **29A**, 309 (1949).
95. N. Perakis and L. Capatos, *J. Phys. Rad.*, **7**, 391 (1936).
96. R. Perceau, *Compt. Rend.*, **236**, 76 (1953).
97. S. A. Zaveri and M. G. Datar, *Indian J. Chem.*, **3**, 11 (1965).

5 X-RAY STUDIES OF SMALL-RING NITROGEN HETEROCYCLES

L. M. TREFONAS

Department of Chemistry
University of Central Florida
Orlando, Florida

R. J. MAJESTE

Department of Chemistry
Southern University in New Orleans
New Orleans, Louisiana

5.1.	Introduction	314
5.2.	Status of X-Ray Diffraction Techniques	315
5.3.	Aziridines	316
	5.3.1. Aziridine and Simply Substituted Aziridines, 316	
	5.3.2. Metal-Aziridine Complexes, 319	
	5.3.3. Fused Aziridine Systems, 321	
	5.3.3.1. Fused 2,3-Aziridine Moieties, 321	
	5.3.3.2. Fused 1,2-Aziridine Moieties, 326	
5.4.	Aziridine Variants	327
	5.4.1. Azirines, 327	
	5.4.2. Diazirines, 329	
	5.4.3. Diaziridines, 330	
	5.4.4. Thiadiaziridines, 334	
	5.4.5. Oxaziridines, 335	
5.5.	Azetidines	338
	5.5.1. Simple Azetidines, 338	
	5.5.2. β-Lactams, 340	
	5.5.3. Fused β-Lactam Systems, 342	
5.6.	Azetidine Variants	347
References		348

5.1. INTRODUCTION

The theoretical models that explain the vast accumulation of knowledge of structural chemistry are the result of a great deal of practical and theoretical effort. Thousands of articles appear annually concerning the structures of carbon compounds alone. Chemists have become comfortable with various structural concepts regarding carbon in its compounds. One routinely refers to sp^3-hybridized carbon with bond angles around 109° and sp^2-hybridized carbon with bond angles near 120°. Saturated and unsaturated systems present no unusual conceptual difficulties.

Small-ring compounds, on the other hand, have a certain fascination for scientists interested in such structural features. Three-membered rings force one to consider bond angles near 60°, while four-membered rings lead to bond angles near 90°. If one forms polycyclic systems involving these rings, the molecule as a whole must often make adjustments to accommodate the rings. The structural chemist thus looks at dihedral angles (angles between various planar portions in the molecule); fusion angles (compromise angles at the juncture of two rings of disparate dimensions); and torsion angles, which measure the "twist" of certain portions of the molecule from values that have come to be accepted as standards.

Small-ring nitrogen compounds have held a personal fascination for the authors not only for all of the above reasons but also because certain three-membered nitrogen-containing rings (aziridines) have shown promise in the battle against cancer (1), while four-membered ring systems containing nitrogen (azetidines) have been of great practical interest as antibiotics (2).

For these reasons, a large number of reviews have been written, almost on an annual basis, on these compounds. These reviews can be general, covering a range of interests regarding these compounds (3), or more specific, dealing with just one aspect such as their synthesis or chemistry, nitrogen inversion, biochemical properties, theoretical treatments and so on (4).

This particular review will focus primarily on one aspect of these systems. Namely, we have gathered the detailed structural information that has been obtained on any compound that includes either a three-membered or four-membered ring with at least one nitrogen heteroatom. If these compounds contain a second heteroatom in the same small-ring moiety they are arbitrarily classified here as "variants" and dealt with in a different subsection.

A variety of techniques are used to obtain quantitative structural information. These are normally either diffraction (X-ray, electron, and neutron) or spectroscopic (microwave, near-infrared) techniques. X-ray diffraction is by far the most common technique, with over 4000 structures reported annually (as compared to 4 in 1940, 18 in 1950, 118 in 1960, and 800 in 1970). For that reason, a short discussion of the present status of X-ray diffraction techniques follows.

5.2. STATUS OF X-RAY DIFFRACTION TECHNIQUES

Diffraction techniques are currently used to study the structure of matter as it applies to various chemical, physical, and biological systems. In the broad range they deal with crystalline solids, powders, fibers, films, membranes, liquids, and gases. Of all the experimental techniques, X-ray diffraction is still the most used because of its high degree of accuracy and the broad range of systems that can be studied [it is still the only technique that permits the determination of the three-dimensional structures of large systems such as biological macromolecules (5)]. Besides the studies of these biological systems, X-ray techniques are widely used in the areas of chemical crystallography, diffraction physics, earth sciences, and materials research. X-ray crystallography is still a growing field. It is increasingly in the forefront of research, since it is necessary to provide precise structural information useful in and of itself, as well as for the interpretation of other spectral and physical parameters obtained from other techniques such as Mössbauer, magnetic, and optical studies, As these areas grow, so does the need for X-ray structural data. But X-ray crystallography is by no means a stagnating technique used only for support studies. Besides the increasing number of applications for structural data, the area itself is still expanding by continued improvements in basic theory, techniques, procedures, and instrumentation (6, 7), including extended X-ray absorption fine structure (EXAFS) measurements (8) and high resolution X-ray measurements of the electron density distribution (EDD) and molecular electrostatic potential (MEP) (9).

To look at the state of the art and its future, it is necessary to examine the new innovations and the areas of research involved with improving X-ray crystallography directly. The development of new programs and systems of programs such as TAILS (10) and XTAL (11) are making it possible for researchers with a weaker background in X-ray diffraction theory to use the techniques to solve crystal structures as part of their studies. There even exist companies (12) that will commercially make X-ray determinations on simple compounds for those who have no background or facilities to perform such analyses. This information combined with NMR, UV, IR, ESR, and mass spectroscopical data opens many new areas of research. It must be noted, however, that X-ray crystallography is still far from a simple analytical tool. While these services and simplifying programs remove the drudgery of data analysis and take care of the routine centrosymmetric structures, they work only when conditions are optimum. If the crystals are poor, extremely small or twinned, if they decompose with time or exposure to radiation, or if they are disordered or have more than one molecule per asymmetric unit, the problem becomes much more complex. For crystals with pseudosymmetry, ambiguities in space ground determination, disordering, or twinning, or for larger molecules such as proteins, only an experienced crystallographer in a well-equipped laboratory can be expected to solve the structure.

The most striking advances in the field lie in the area of new equipment and the development of the techniques needed to accompany them. Synchrotron radiation (13–18) is fulfilling its promise as an X-ray source. While it still does not offer a significant increase in intensity over the rotating anode, it does offer a powerful source of X-rays that produces a polarized beam with a high degree of natural vertical collimation. More important is the fact that this beam can be "tuned" to a variety of wavelengths. This allows researchers to select an optimum wavelength to maximize the f' and f'' components of the anomalous scattering factor for any atom in the available range. It also allows for multiple measurements, for example, at three different wavelengths, so that in the case of metalloproteins no heavy-atom derivatives are required.

Another growing area of study is the position-sensitive linear detector (19–22). These detectors allow the measurements of blocks of data at one time, thus reducing the total time required for data taking, the exposure time, for any one crystal. Stronger beams and multiple counters are only a few of the new techniques being looked at today. New ideas such as the energy-dispersive detector (23, 24), which measures 2θ as a function of energy, new oscillation techniques (25), and diffraction-profile analysis (26–28) are also being developed. And still to come in the future ... perhaps a practical X-ray laser (29).

5.3. AZIRIDINES

5.3.1. Aziridine and Simply Substituted Aziridines

A series of microwave studies over a 20-year period (30–35) coupled with an electron diffraction study (36) of aziridine have led to a very accurate structural picture of the molecule (Fig. 5.1).

The unique structural feature of this molecule is the rigidity of the three-membered ring under a variety of simple substitutions. Regardless of whether the substitution is only on the nitrogen (37–48) or is on one or more of the carbons (49–52), a "composite molecule" (Fig. 5.2) constructed from these results is essentially identical to the original aziridine in its molecular parameters.

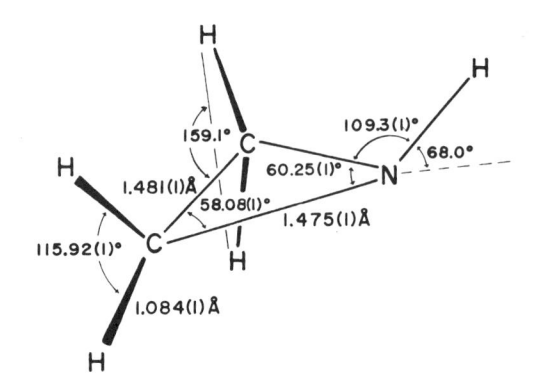

Fig. 5.1. Structural parameters for the "free" aziridine molecule.

Aziridines 317

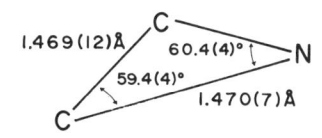

```
        C
       /  \
1.469(12)Å  60.4(4)°
   /    59.4(4)°   N
  C_____
         1.470(7)Å
```

Fig. 5.2. Averaged structural parameters for the aziridine ring.

The averages of all parameters (of those measured) agree within two e.s.d. with the original unsubstituted aziridine, although these studies represent a variety of substitutions. If one compares the molecular parameters obtained in each of these studies to the "composite aziridine" molecule shown in Figure 5.2, then the largest deviation is in the values obtained for the NCC angles in *meso*-1,4-diaziridinyl-2,3-butanediol (39) (Fig. 5.3), which are quoted at 58.9 and 59.1°, respectively. However, a pattern of hydrogen bonding creating a layerlike structure for this solid diol may be contributing to the difference.

The uniformity of these structures is evidenced further in the retention of angles between dihedral planes. For example, in the original aziridine the N–H bond forms an angle of 112° with the plane of the aziridine (CCN). In the addition compound, aziridine-borane (37) (Fig. 5.4), this angle is increased only to 115° even though the nitrogen goes from pseudo-trigonal to quarternary bonding and must, in fact, also accommodate the N–B bond.

A structural study of ethyleneimine quinone (40) at three different temperatures [110 ± 20, 240 ± 10, and 300 ± 5 K] further supports the relative inflexibility of this ring (Fig. 5.5). The crystal structures at the three temperatures are the same. The three-membered ring is approximately a regular triangle with N–C = 1.469, 1.457 Å, and C–C = 1.498 Å. Difference electron density maps show three residual peaks (increasing in height to $0.26e/Å^3$ at the lowest temperature), which always

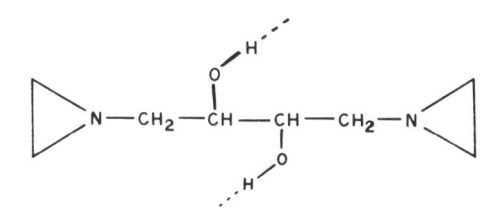

```
         H
         |
         O
         |
  ▷N—CH₂—CH—CH—CH₂—N◁
                |
                O
                |
                H
```

Fig. 5.3. *meso*-1,4-Diaziridinyl-2,3-butanediol.

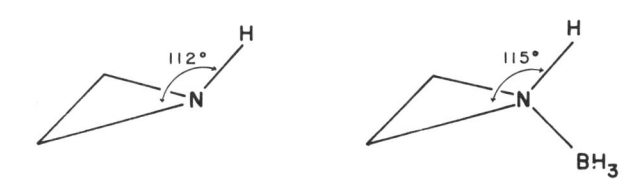

```
       H                         H
      /                         /
 112°/                    115°/
   ▷—N                    ▷—N
                                \
                                 BH₃
```

Fig. 5.4. The dihedral angle between the imine-hydrogen and the aziridine plane.

318   X-Ray Studies of Small-Ring Nitrogen Heterocycles

Fig. 5.5.   Ethyleneimine quinone.

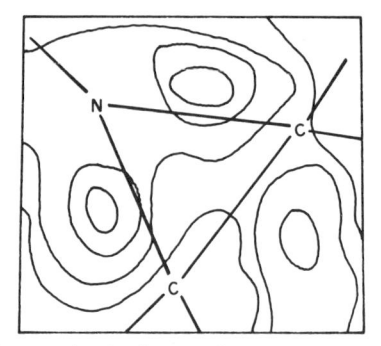

Fig. 5.6.   Electron density "bulging" along the aziridine ring bonds.

appear outside and approximately at the middle of each side of the triangle made by connecting the three atoms of the aziridine ring. These peaks are more than twice the standard deviations of the electron density and systematically increase in size with decreasing temperature, and so they can be attributed to the bonding electrons (Fig. 5.6).

The most unique of the simply substituted aziridines is the aziridinone (51). (Fig. 5.7). Unfortunately, the accuracy of the results has been dramatically reduced by the presence of crystal disorder and further complicated by two 1-adamantyl substituents in the ring. The bond lengths and angles are shown in Figure 5.7. The two adamantyl groups are *trans* relative to the ring, with the nitrogen atom pyramidal, lying 0.534 Å from the plane defined by its three substituents. The pyramidal nitrogen implies no amide resonance in the molecule, yet the shortened C–N bond strongly suggests double-bond character. The UV spectra of aziridinones

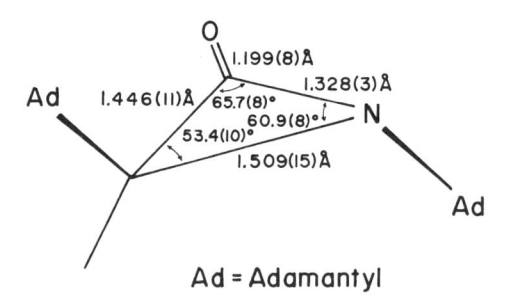

Fig. 5.7.   *t*-2,3-Diadamantyl aziridinone.

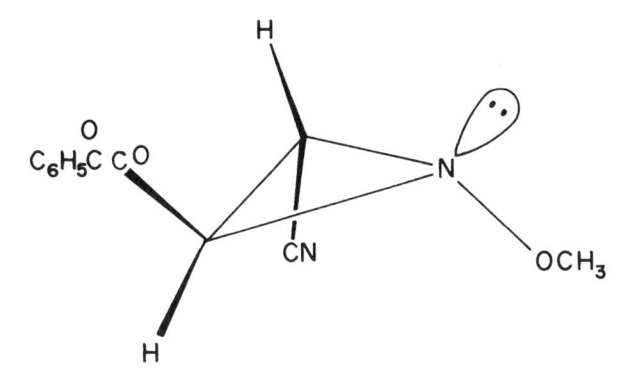

Fig. 5.8. Stable invertomer of aziridinone.

have been interpreted in terms of interactions of the lone-pair electrons on nitrogen with the carbonyl group (53. 54). Adsorption maxima at 210 (55) and 250 nm (53–55) have been reported in aziridinones.

3-Cyano-2-phenylglyoxyl-N-methoxyaziridine (35) (Fig. 5.8) was the first stable invertomer of aziridine to be unambiguously determined by X-ray diffraction techniques. The relative configuration of the ring C atoms of the aziridine had been previously established by NMR spectroscopy, but it was not possible to establish the nitrogen stereochemistry by spectroscopic techniques, particularly for aziridine with *trans* H atoms.

In summary, the aziridine ring itself forms a plane perpendicular to the individual planes formed by each carbon and its two substituents. Each of the latter CH$_2$ planes makes an angle of about 159° with the C–C bond in aziridine. Additionally, the substituent on the nitrogen does not lie in the plane of the aziridine but is tilted up from that plane by as much as 68°. These features, in addition to the actual values of the molecular parameters, remain relatively constant for simply substituted aziridines. The rigidity of structure causes the pyrimidal inversion rate of nitrogen to be very slow. Thus, the resulting configurations are amenable to NMR studies.

5.3.2. Metal-Aziridine Complexes

The earliest reported structure with the aziridine nitrogen bonded directly to a metal was the ethyleniminodimethylaluminum trimer (56) (Fig. 5.9a, b). Additionally, the structure is of interest because it represents the first such study of a molecule in which the aziridinium nitrogen acts as a bridging atom between two metals.

Another structure determination of a bridged aziridine trimer, the aziridine trimer, was subsequently reported (57). The structure of the gallium trimer was quite similar to the aluminum trimer, lacking only the metal methyls but otherwise having the same stereochemistry (Fig. 5.9a). Distances and angles in the two metal complexes are compared in Figures 5.9b and c.

The *trans*-diiodotetra(ethylenimine)rhodium(III)iodide (58) is the first structure of a complex containing aziridine as a coordinated ligand (Fig. 5.10a). The struc-

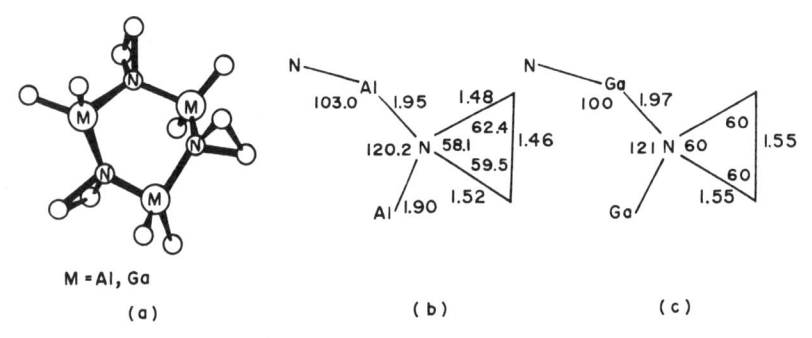

320   X-Ray Studies of Small-Ring Nitrogen Heterocycles

Fig. 5.9.   Metal-bridged aziridine trimers.

ture of an analogous compound, [Cu(Az)₄](NO₃)₂ (59) (Fig. 5.10b), has also been reported (59). In both, a pronounced asymmetry in the supposedly equivalent C–N aziridine distances, together with a very short aziridine C–C distance, are reported. C–N distances of 1.47 and 1.58 Å and a C–C distance of 1.36 Å are reported for the rhodium complex. In the copper complex the same distances are cited as C–N = 1.40, 1.53 Å and C–C = 1.40 Å. In no other aziridine structure studies is there such a disparity in the (equivalent?) C–N distances nor such short C–C distances cited.

The structures of both *cis*- and *trans*-dichloro-bisaziridine platinum(II) have

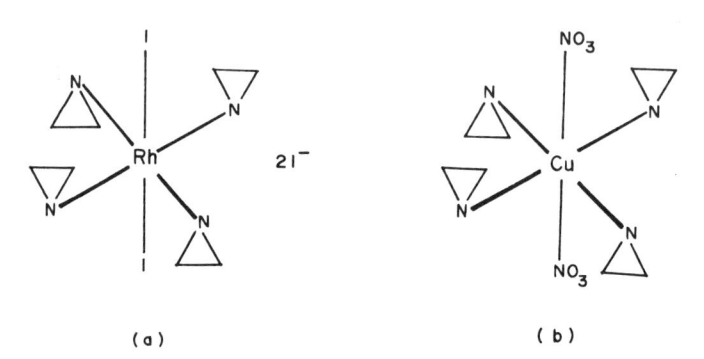

Fig. 5.10.   Aziridine as a coordinating ligand.

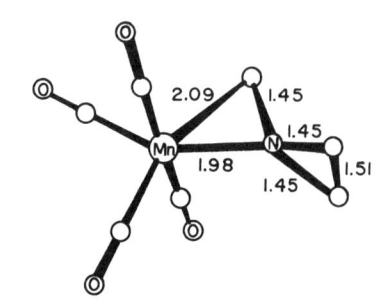

Fig. 5.11.   Spiro-fused "aziridinyl" rings.

been determined (60). There are no significant differences in the magnitudes of the molecular parameters for the aziridine ring in either isomer.

The structure of a rather unusual tetracarbonyl methyl aziridinyl manganese complex has been reported (61) (Fig. 5.11) in which the aziridinyl nitrogen attaches to the methylene carbon and back to the manganese, in essence creating a seond aziridinyl ring with manganese as an additional heteroatom in the ring.

5.3.3. Fused Aziridine Systems

5.3.3.1. Fused 2,3-Aziridine Moieties

The bycyclic fusion of an aziridine ring to a second (or multiple) ring system leads to a number of structural changes in the aziridine and in the ring(s) adjacent to the aziridine. If one focuses on those systems where only aziridine and a second ring are involved in a bicyclic fusion, these changes can be systematized and, conceivably, any trends that may exist can be uncovered and understood. One can then attempt to corroborate these trends by looking at similar types of bicyclic fusions in polycyclic systems containing an aziridine ring.

Fortunately, the structures of a number of simple bicyclic compounds are known. The structures of a series of compounds with *cis* fusions of an aziridine ringe to 5-, 6-, 7-, 8-, 10-, and 12-membered homocyclic rings have been determined (62–67) in addition to a *cis* fusion with a five-membered heterocyclic (oxygen) ring (68) and a *trans* fusion to a cyclododecane ring (69) (Fig. 5.12).

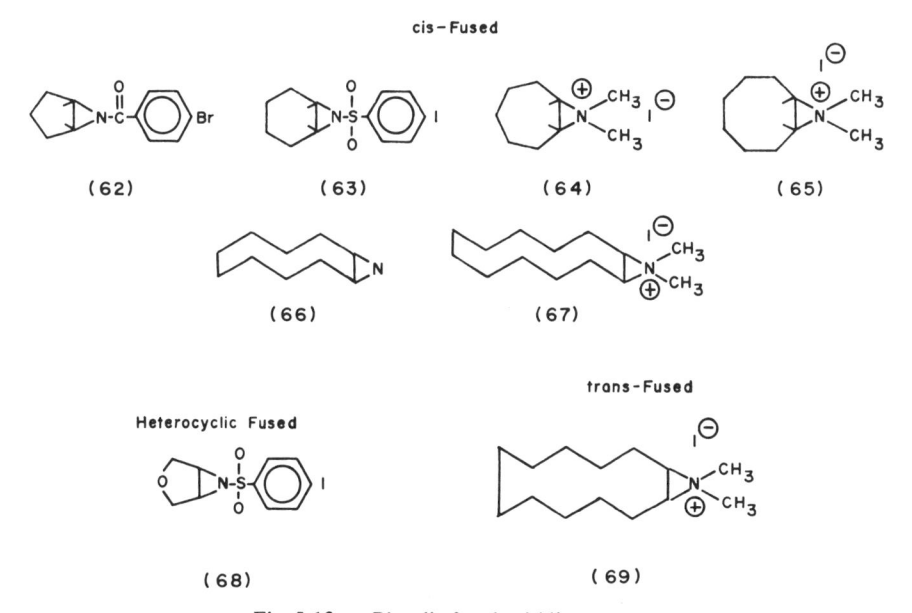

Fig. 5.12. Bicyclic fused aziridine systems.

The effects of this fusion on the aziridine ring are small and gradual. The most evident effects are a relative lengthening of both the C–C and C–N bonds. As the size of the ring increases from five to twelve atoms, the values of both bond distances increase from those found in free aziridine (34) (C–C 1.480 Å, C–N 1.488 Å), with both distances approaching a limiting value of 1.54 Å. This result is directly opposite in direction from that anticipated, since ring strain would be greatest when an aziridine is fused to a smaller ring. It is impossible, within the constraints set by the estimated standard deviation values, to draw any conclusion regarding angular trends within the aziridine ring in these cases except to assume that the angles are all equal to 60 ± 1°.

There are significant distortions in the larger ring as a result of the fusion to aziridine. One of the more obvious immediate localities for such distortion to occur is at the fusion points. In aziridine the dihedral angles between the HCH plane and the C–C bond is 159.4°. In single-ring systems, regardless of their size, the bond angles do not exceed 112°. If one defines the fusion angles as those angles in the larger ring at the points of fusion (Fig. 5.13), then these angles should define the manner in which each bicyclically fused molecule resolves this conflict between 159.4 and 112°. Obviously, the dihedral angle must be reduced drastically, while the bond angles in the second ring are simultaneously increased. The situation is further complicated by a number of other factors. First, if the second ring is sufficiently large it may be able to accommodate a large fusion angle more easily than in the case where the second ring is itself small and severe limitations would exist on the magnitude of the fusion angles. Second, there may be a difference in this angle depending on the bicyclic fusion of *cis* or *trans*. Third, the chemical nature of the second ring (e.g., homocyclic or heterocyclic) may affect this fusion angle.

If the second ring contains six or more atoms, the determining factor seems to be the type of fusion involved regardless of the resulting distortion to the ring system. Thus, in *cis*-fused systems the fusion angle is 123 ± 1°, whereas in the only *trans*-fused system studied (69) the fusion angles are 127 ± 1.5°.

In those cases where this larger ring system contains eight or more atoms, a rationale suggests itself (65, 67, 69). These systems have a high degree of flexibility, which is hindered only by close transannular interactions between hydrogens on carbon atoms at least three links apart in the ring. These transannular interactions were first observed in studies of homocyclic medium-sized rings by Dunitz and coworkers (70–72). They were reconfirmed in studies on bicyclic systems containing aziridine (65–67). The bicyclic fusion would eliminate at least

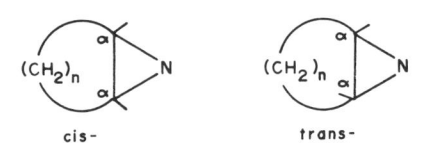

α = Fusion Angle

Fig. 5.13. Fusion angle in bicyclic fused aziridine systems.

two of these close-contact repulsions of transannular hydrogens, and this could compensate for the energy loss resulting from the expansion of these same angles from their preferred values. If one takes the carbon-carbon bending force constant as 0.8×10^{-11} erg radian^{-2}, the energy required to deform a bond angle is given (73) by $E = 0.175X^2$, where E is the deformation energy in kilocalories per mole and X is the deformation angle in degrees. In the *cis*-fused compounds this leads to a value of 0.18 kcal/mol per bond angle, which is a small enough value to account for its relative constancy in these compounds. In the *trans*-fused compound, this value would be 3.73 kcal/mol per bond angle, a significantly larger value, leading to an average deformation of 14.6° in each of the two fusion angles.

In all of these systems where this high degree of molecular flexibility can occur (subject only to the minimization of transannular interactions), the possibility exists of a molecule and its enantiomorphic mate being able to occupy almost exactly the same available space without significant crowding resulting. This is contrary to the more general case where, in the process of crystallization, it is usual for the crystal to distinguish between a molecule and its mirror image and consequently to crystallize in one enantiomorphic form. Hence these "flexible" molecules quite often crystallize with orientational disorder existing in the crystal. This is the situation which is, in fact, found for these bicyclic compounds where the larger ring has eight or more atoms and in the analogous monocyclic compounds where the ring system again contains eight or more carbons.

The cases where the second ring in this *cis* bicyclic fusion contains either six (63) or seven (64) atoms do not have this disordering and they do not have close transannular interactions, and yet they maintain *cis*-fusion angles of 123 ± 1°. The small energy cost of 0.28 kcal/mol per bond angle in the *cis*-fused compound may account for this. However, these compounds do pay a significant price in terms of the resultant changes in conformation of this second ring. It is difficult to rationalize the "correctness" of the resulting conformations because of the paucity of appropriate theoretical models and the necessary theoretical calculations. In the case of the aziridine fused to a cycloheptane ring (64), for example, early calculations (74, 75) based upon energetic considerations tacitly assumed transannular interactions and hence predicted a conformation different from the one that was, in fact, observed. In the case of an aziridine ring fused to a cyclohexane ring (63), although the cyclohexane ring is still able to maintain a chair conformation, there is an appreciable flattening of this ring system to accommodate these fusion angles (Fig. 5.14). The resultant structure is much more analogous to cyclohexene (76). These results lend credence to the theoretical conclusions (84) that at three-membered ring has π-bond character and that the fusion site has an effect analogous to that of the double bond in cyclohexene.

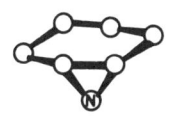

Fig. 5.14.   Bicyclic fusion of aziridine and hexane.

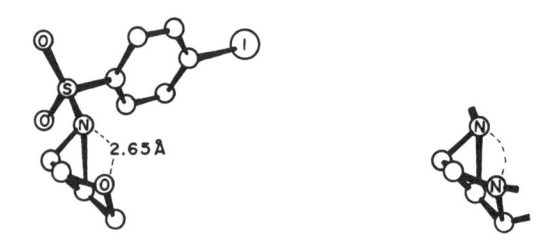

324 X-Ray Studies of Small-Ring Nitrogen Heterocycles

Fig. 5.15. Bicyclic fusion of aziridine to hetero- and homocyclic five-membered rings.

If the second ring, bicyclically fused to the aziridine, contains fewer than six atoms, even severe distortions in the conformation of that ring cannot cause these fusion angles to enlarge as before. In two cases, where the second ring contains five atoms [one homocyclic (62) and the other with oxygen as an additional heteroatom (68)], this fusion angle is 112°. In the other known case (78), *N*-brosylmitomycin A, the fusion angles are 107.6 and 111.4°. This deviation may be attributed at least in part to steric interference of the valence shell electrons with the free *p*-orbitals of the other nitrogen in the five-membered ring. Again, they are the largest angles in the second ring, although admittedly they are significantly less than the value of 123 ± 1° found for *cis*-fused compounds in which the second ring contains six (or more) atoms.

The five-membered rings fused to an aziridine provide an interesting comparison. For, if this larger ring is homocyclic (62), all nonbonding distances were greater than van der Waals distances. But where this larger ring is heterocyclic (68, 77), the heteroatom in the larger ring and the aziridine nitrogen approached each other much more closely than van der Waals separations would indicate, strongly suggesting an interaction in these cases (Fig, 5.15). Table 5.1 summarizes these trends for the aziridine bicyclic systems.

An interesting extension of these concepts developed on bicyclically fused systems occurs when one is dealing with polycyclic systems in which one of the rings is an aziridine. Azatetracyclo[5.3.1.0^{2,6}.0^{8,10}]undec-4-ene, which contains an aziridine ring fused simultaneously to five-, six- and nine-membered rings (79) (Fig. 5.16) is one such example. If one calculates the dihedral angle between the aziridine ring and the plane containing the four atoms of the tricyclo system starred in the diagram, the value of 112° is in agreement with previous fusion angles

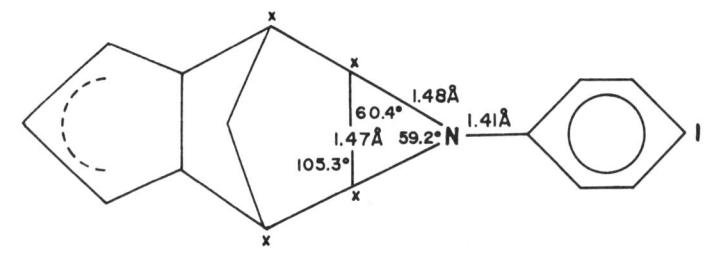

Fig. 5.16. Polycyclic fused aziridine (five-, six-, and nine-membered rings).

Table 5.1. Comparison of Parameters for Aziridine Bicyclic System

Journal reference	62	68	63	64	65	66	67	69
Atoms in large ring	5	5[a]	6	7	8	10	12	12
Type of fusion	cis	cis	cis	cis	cis	cis	cis	trans
Aziridine Parameters								
C—C	1.49	1.53	1.49	1.52	1.55	1.51 ± 0.01	1.53	1.54
C—N	1.49 ± 0.01	1.53 ± 0.03	1.52 ± 0.02	1.54 ± 0.03	1.52 ± 0.03	1.54 ± 0.01	1.55 ± 0.03	1.54 ± 0.02
∠NCC	60 ± 1°	61.1 ± 1°	60.7 ± 1°	61.1 ± 1°	59.5 ± 1°	60.2 ± 1°	60 ± 1°	60 ± 1°
∠CNC	60°	58.2°	58.6°	58°	61°	58.5 ± 1°	60°	60°
Fusion angles	112 ± 1.5°	112 ± 1.7°	122 ± 1°	123 ± 1°	123 ± 1°	121°	123.5 ± 1°	127 ± 1°
Other Ring Parameters								
C—C distance (Å) (excluding common bond)	1.57 ± 0.04	1.53 ± 0.04	1.49 ± 0.02	1.52 ± 0.03	1.55 ± 0.03	1.54 ± 0.03	1.53 ± 0.03	1.54 ± 0.0
∠CCC (excluding fusion angles)	106 ± 2°	—	118 ± 3°	114 ± 2°	112 ± 2°	113 ± 4°	111 ± 3°	115 ± 3°
Comments								
Close transannular contacts?	No	No	No	No	1.98–2.05 Å	1.80–2.13 Å	1.95–2.04 Å	2.05 Å
Form stable ionic methiodide?	No	No	No	Yes	Yes	Yes	Yes	Yes
Crystalline disorder?	No	No	No	No	Yes	Yes	Yes	Yes
Usual conformation of larger ring?	Boat[b] (bicyclic) system	Boat[b] (bicyclic) system	Chair[b] (almost planar)	No	No	No[b]	Yes	No

[a] Cyclopentane ring also contains oxygen as heteroatom.
[b] Distorted ring.

325

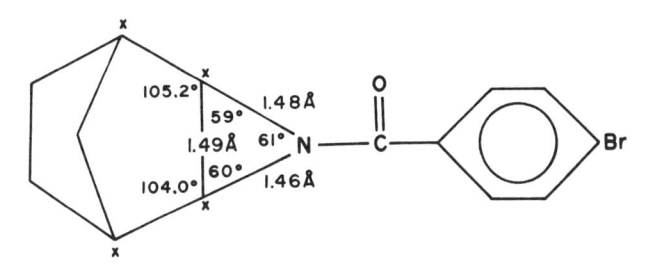

326 X-Ray Studies of Small-Ring Nitrogen Heterocycles

Fig. 5.17. Polycyclic fused aziridine (five- and six-membered rings).

for *cis*-fused bicyclic systems containing five-membered rings (62, 68). The C–C and C–N distances in the aziridine ring, on the other hand, correspond closely to the values in the free aziridine (34) and not those for bicyclic-fused systems.

The results found in the structure study of a second polycyclic system *N-p*-bromobenzoyl-*exo*-2,3-aziridinobicyclo[2.2.1]heptane (80) confirm that the previous agreement is not merely coincidental. In this latter study, an aziridine ring is fused simultaneously to five- and six-membered rings (Fig. 5.17), yet the dihedral angle between the aziridine ring and the plane defined by the four starred atoms in Figure 5.17 has a value of 111°, a number in conformity with the results in the previous structure determination. Again, the C–C and C–N distances in the aziridine ring and the angles correspond closely to values found for free aziridine (34) rather than those predicted by bicyclic-fused systems.

5.3.3.2. Fused 1,2-Aziridine Moieties

Two examples are in the literature where there is a bicyclic fusion between an aziridine ring and a second ring and one of the fusion points contains the aziridinium nitrogen.

In the 1,3-diazabicyclo[3.1.0]hexane (81) (Fig. 5.18), the dihedral angle of the aziridine and the adjacent four members of the fused five-member ring is 104.7°, a significant departure from the value of 112° found in other studies (64–69). The C–C, C–N distances are close to, but intermediate between, those studies and that for free aziridine. As in all bicyclic fusions involving aziridine and five-membered rings (62, 68, 78), the resultant six-membered ring is in the boat conformation.

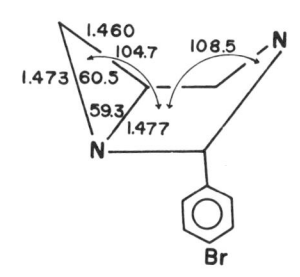

Fig. 5.18. 1,2-Aziridine fusion (to a five-membered ring).

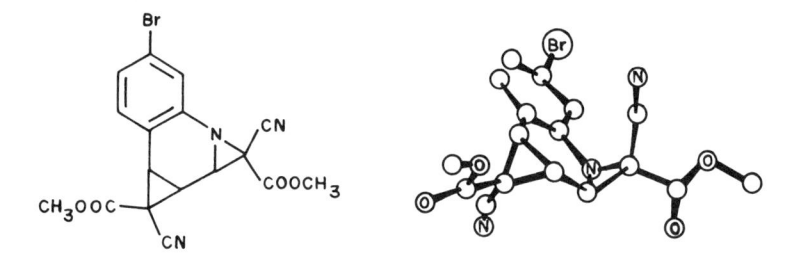

Fig. 5.19. 1,2-Aziridine fusion (to a six-membered ring).

In the other such bicyclic compound, the tetrahydroquinone aziridine (82), (Fig. 5.19), the conformation is given but no molecular parameters are quoted.

5.4. AZIRIDINE VARIANTS

5.4.1. Azirines

The experimental results obtained in the structural determination of a substituted azirine (83) give distances and angles that are in close agreement with theoretical calculations for an analogous unsubstituted amino-2-azirine (Fig. 20a, b). Interestingly enough, even though there are sizeable distortions introduced by the double bond (C–N = 1.279 Å and the resultant decrease in the azirine ∠CCN to 51.9°) the dimethylamino group has a H$_2$N-CC torsional angle of 151.2°, which is very similar to the H$_2$C-CC torsional angles of 159.4° found in the study of aziridine itself. One other point of interest concerns the C–N distances both in and outside the ring. The endocyclic C–N bond is longer than the C=N bond in an imine (1.279 Å instead of 1.24 Å), and the *exo*-cyclic bond is shorter than the C–N bond in amines (1.317 Å instead of 1.47 Å). This suggests an important contribution of the polar mesomeric form to the electronic structure (Fig. 5.21).

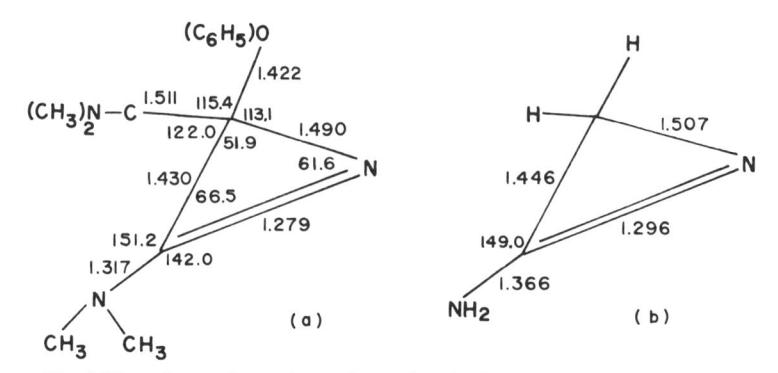

Fig. 5.20 Comparison of experimental and calculated azirine parameters.

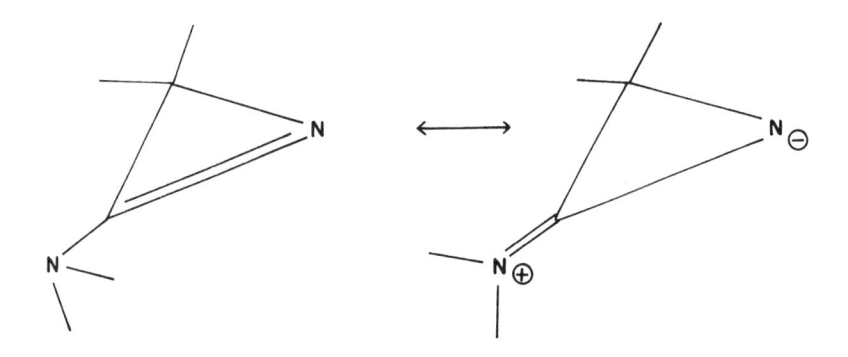

328   X-Ray Studies of Small-Ring Nitrogen Heterocycles

Fig. 5.21.   Mesomeric form of azirine.

The three-membered ring in the spiro-fused azirine (84) (Fig. 5.22) parallels the parameters obtained earlier (Fig. 5.20), since both have a short N=C distance (1.28 vs. 1.279 Å), a lone N–C distance (1.57 vs. 1.49 Å), and an intermediate C–O distance (1.46 vs. 1.43 Å). In a similar fashion the interior angles in the azirine ring follow the same pattern although, again, in this compound the actual numerical values are more extreme (CC=N, 70 vs. 66.5°; C=NC, 60 vs. 61.6°; NCC, 50 vs. 51.9°, respectively). The indanedione ring is at 91.8° to the azirine ring, ensuring maximum conjugation of these rings and allowing the electron-acceptor carbonyl groups to cause a pronounced shift of the electron density in the direction of the indanedione ring. The long C–N bond (1.57 Å) implies that conjugation of the three-membered ring with the indanedione is realized along the N–C–C "chain" in the azirine. Consequently, in the azirine ring the C–C bond becomes more polarized, and the polarization of the C=N bond decreases. This accounts for the facile cleavage of the C–C bond by nucleophiles and the relative inertness of the C=N

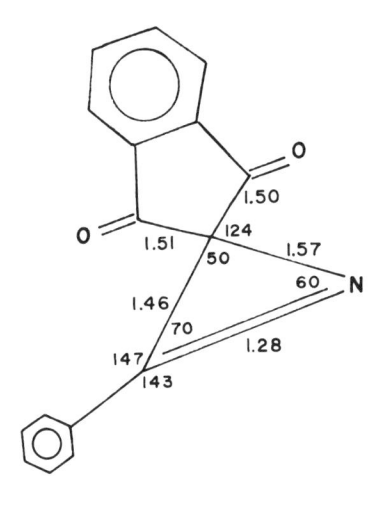

Fig. 5.22.   Spiro-fused azirine.

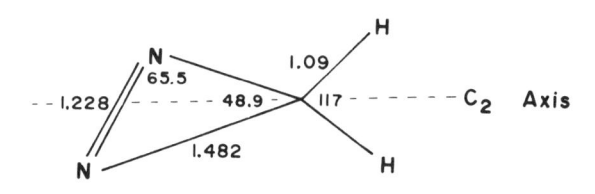

Fig. 5.23. Diazirine.

bond (85). At the same time the C–N bond is lengthened and easily cleaved when the compound is reduced by HI.

5.4.2. Diazirines

A microwave study of the rotational spectrum of the various isotopically substituted diazirines (86–89) led to the structure shown in Figure 5.23. Aside from the obvious higher symmetry (the C atom lies on the C_2 axis), the basic ring structure is comparable to the previous azirine. The NN double bond is shorter than the CN double bond (1.228 vs. 1.29 Å), and the CN single bonds are quite close to the single CN bond in azirine (1.482 vs. 1.507 Å).

The methyl-substitued diazirine microwave determinations of both the monomethyl (90) (Fig. 5.24a) and the dimethyl (91) (Fig. 5.24b) offer the closest comparisons of ring distortion due to substituent effects. As anticipated, the rigidity of this system is maintained, and the substituent distortion, based entirely on steric effects, is negligible. The only statistically significant difference between the three involves the angle between the substituents on the azirene carbon. This angle changes from 117° for the HCH angle (to 119.7° for the H_3C–C–CH_3 angle), which parallels the trend observed between formaldehyde (92) (HCH = 115.8°) and acetone (93) (H_3C–C–CH_3 = 117.2°). Intriguingly enough, the H–C–CH_3 angle of 122.3° (Fig. 5.24a) in the monomethyl diazirine (90) does not follow the aforementioned trend, but then neither does the same H–C–CH_3 angle in acetaldehyde (118.2°) (94) follow the trend in the acyclics.

It was anticipated that a severe test of the rigidity of the diazirene system would occur if one could make the structure with a strong electron-withdrawing substituent. The perfluorodiazirene was synthesized (95), its cyclic structure and C_{2v} symmetry initially deduced (96) through a detailed analysis of its infrared and Raman spectra, and its complete structure solved by electron diffraction

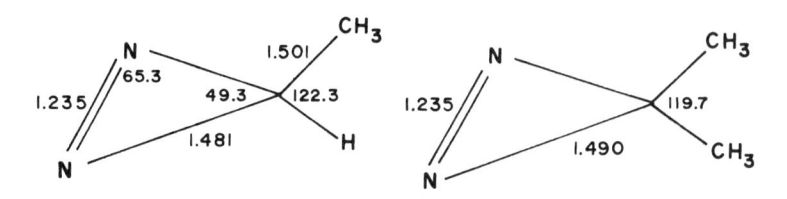

Fig. 5.24. Methyl-substituted diazirines.

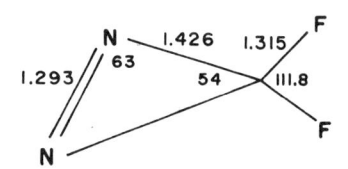

Fig. 5.25. Difluoro diazirine.

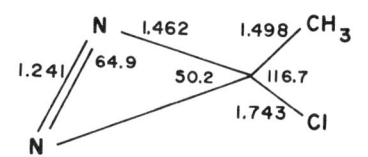

Fig. 5.26. Chloromethyl diazirine.

techniques (97) (Fig. 5.25). The magnitude of the differences in C—N and N—N bond lengths distances between this compound and the previous two diazirenes was greater than anticipated. The replacment of H by F onto a carbon atom generally leads to compression of the charge density about that carbon and hence to a shortening of adjacent bond lengths. It is also established that when carbon is involved in π bonding in fluorinated alkanes, the C—F distances are shorter than when all substituents are attached to the central atom by single bonds. A comprehensive series of actual examples (97) led to the conclusion that orbitals associated with the three-membered ring interact with the $p\pi$ orbitals associated with the C—F bond. The shorter C—N bond distance in this compound (1.426 Å) follows the established pattern observed in adjacent C—C distances for HFCCH$_2$ (1.337Å) (98), HFCCFH (1.324 Å) (99), and F$_2$CCH$_2$ (1.315 Å) (100, 101). The most puzzling aspect is the lengthening of the N—N bond from 1.228 Å in azirene to 1.293 Å in these compounds (Fig. 5.25).

In an attempt to corroborate this difference, the structure of a compound of intermediate strength, a monohalodiazirene (102) (Fig, 5.26), was determined, and this distance was compared. As anticipated, the distance for the N—N bond is 1.241 Å, which is intermediate between the aforementioned values of 1.228 and 1.293 Å.

Assumed Parameters
   CN = 1.462 Å
   CH = 1.09 Å
   ∠HCN = 108.0°
   ∠CCN = 110.0°

5.4.3. Diaziridines

The structure of bis(*p*-bromo-2,2-dimethylbenzyl)diaziridinone (103) (Fig. 5.27) was determined by X-ray diffraction techniques to be the transoid structure with

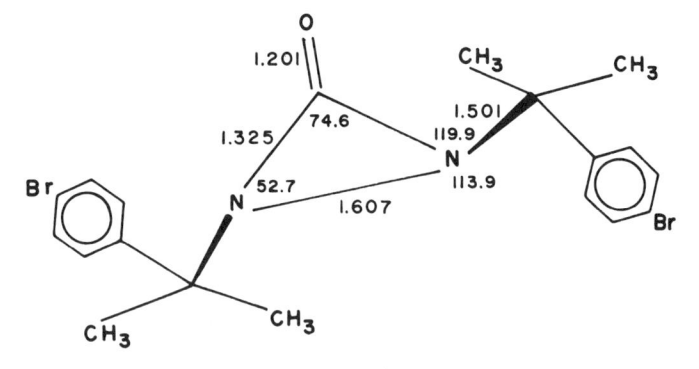

Fig. 5.27. Substituted diaziridinone.

the two substituents 56° above and below, respectively, the plane defined by the diaziridine ring (Fig. 5.28). The C-N distances of 1.325 Å are typical of planar amide systems, 1.33 ± 0.005 Å (104), and considerably shorter than the N-C sp^2 (1.48 Å) of 2,4,6-trimethylnitrobenzene (105) or N-C sp^3 average values of 1.48 Å (106). However, the C=O distance of 1.201 Å is the same within experimental error as the typical ketone C=O lengths found in cyclopropanone (107) and aziridinone (51). The geometry of this molecule and its IR carbonyl absorptions (1855-1880 cm^{-1}) as compared to those of the aziridinones (1837-1850 cm^{-1}) and cyclopropanones (1813-1840 cm^{-1}) (107), are not in accord with amide resonance stabilization in diaziridinones. The relative reactivity toward nucleophiles of diaziridinones and cyclopropanones (108) remains somewhat of a puzzle; the lower reactivity of diaziridinones may be associated, in part, with a steric effect resulting from the bulky side groups attached to the ring to stabilize the compound or with the larger internal carbonyl angle (74.6°), with repulsion between a nitrogen lone pair and an attacking nucleophile, and with amide resonance (reduced, but presumably not absent). In summary, this compound is unusual, even in comparison to other diaziridines, in having a very long N-N bond (1.607 Å) and very short C-N bonds (1.325 Å).

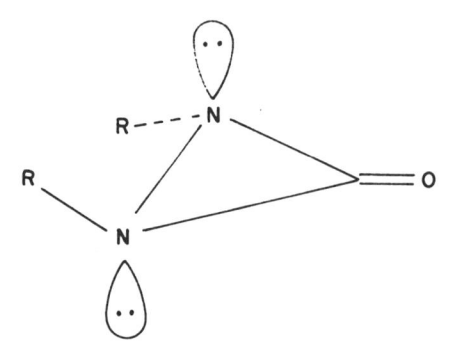

Fig. 5.28. Geometry of diaziridinone.

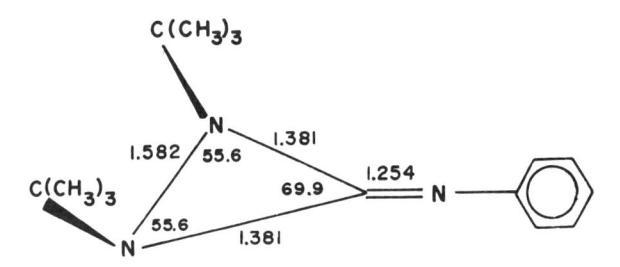

332 X-Ray Studies of Small-Ring Nitrogen Heterocycles

Fig. 5.29. Substituted "aniline," diaziridine.

Only one other diaziridine bears a close structural resemblance to the above diaziridinone. N-(1,2-Di-*tert*-butyl-diaziridin-3-yliden)-2,4,6-trimethyl aniline (109) (Fig. 5.29) has a very long N–N bond (1.582 Å), a short C–N bond (1.381 Å), and a large NCN angle (69.9°). Stereochemically, this molecule has a torsion angle of 58.6° between the diaziridine ring and the best plane through the aniline ring. Each of the *tert*-butyl groups is rotated (in opposite sense) so that the torsion angle between them ($N_1-C_{11} \cdots N_2-C_{16}$) is 126.8°, making each tilt at an angle of 53.4° with respect to the aziridine ring, a value comparable to the torsional angle of 56° for the substituents in aziridinone (103). In retrospect, the one common feature that distinguishes both of these molecules from the diaziridines to be discussed is that each has an external double bond on the unique carbon in the diaziridine ring. In the diaziridinone this is the CO (1.201 Å) and in aziridine-3-yliden it is the CN (1.254 Å) distance.

In those cases where the diaziridine has a "tetrahedral" carbon in the ring, the molecular parameters of the diaziridine are relatively constant and uniform even though a wide disparity exists between the substituents (and their mode of attachment to the diaziridine). In the 1-cyclohexyl-3-(*p*-bromophenyl) derivative (110) (Fig. 5.30), the C atom has hydrogen and *p*-bromophenyl as substituents; in 2,4,6-trimethyl-1,3-5-triazabicyclo[3.1.0]hexane (111) (Fig. 5.31), as in 3-methyldiaziridine (112) (Fig. 5.32), the C atom has a hydrogen and a methyl as substituents; whereas in (–)-1R,2S-1-(S)-α-phenylethylcarbamoyl-2-methyl-3,3-pentamethylenediazirdene (113) (Fig. 5.33) the C atom is spiro-fused to a six-membered ring.

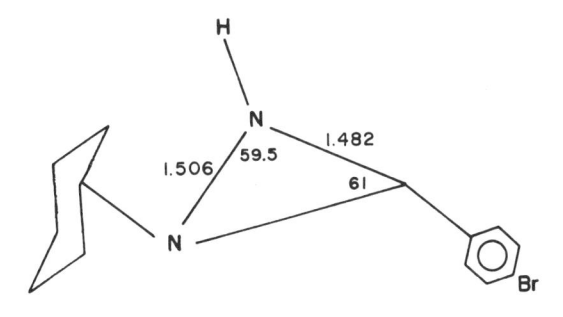

Fig. 5.30. C,N-Substituted diaziridine.

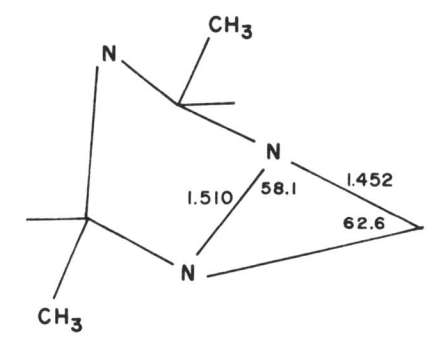

Fig. 5.31. *N,N*-Fused diaziridine.

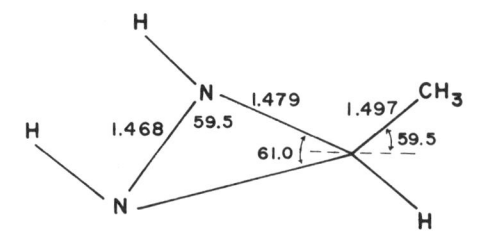

Fig. 5.32. *C*-Substituted diaziridene.

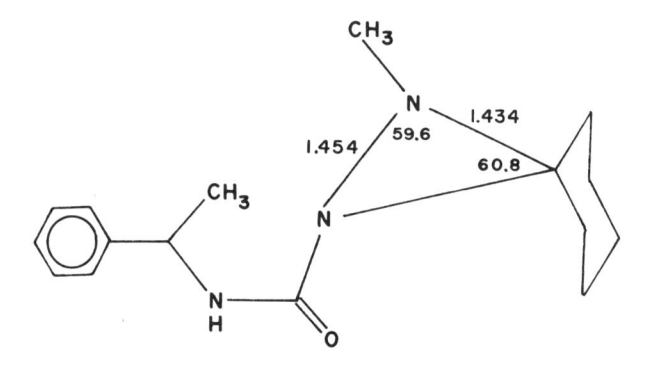

Fig. 5.33. *C,N,N*-Substituted diaziridine.

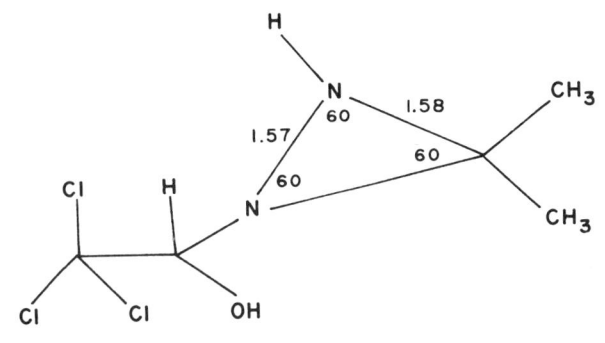

334   X-Ray Studies of Small-Ring Nitrogen Heterocycles

Fig. 5.34. "Equilateral" diaziridine.

1-α-Hydroxy-β-trichloroethyl-3,3-dimethyl diaziridine (114) (Fig. 5.34) has an angular conformation similar to the others but with what appears to be an equilateral triangle arrangement for the diaziridine. This difference may not be real, as the structure has not been refined below a value of $R = 0.15$.

5.4.4. Thiadiaziridines

Bis(1,1,3,3-tetramethylbutyl)thiadiaziridine-1,1-dioxide (115) (Fig. 5.35) is the only thiadiaziridine whose structure has been reported. The two octyl chains are *trans* to each other. The N–S distance (1.62 Å) is shorter than other analogous but acyclic N–S distances (116), whereas the N–N bond (1.67 Å) is longer than in any other diaziridine (1.607 Å or less). This value is bracketed by the values of 1.64–1.71 Å found for the admittedly lengthened (117, 118) N–N bond in N_2O_4. Such lengthening, in this compound, is probably the only way of maintaining the NSN angle (62° in this study) at a value near 60°. For example, if one assumed an N–N distance of 1.48 Å and still maintained the shortened N–S distances, the resultant NSN angle would be lowered to 52°, a value far below any such values found in diaziridines.

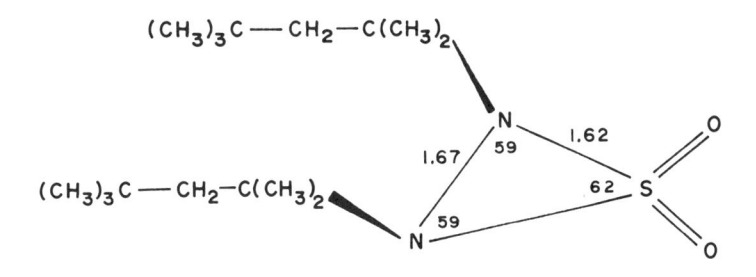

Fig. 5.35.   Substituted thiadiaziridine.

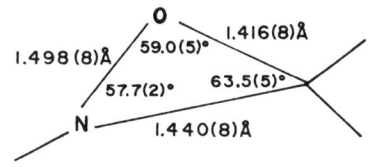

Fig. 5.36. "Averaged" oxaziridine.

5.4.5. Oxaziridines

A sufficient number of structural determinations of oxaziridines, substituted at both the tetrahedral C and the trigonal N, have been completed so that this ring system is well characterized. The most striking feature, again, is the rigidity of the ring. If one averages the analogous distances and the analogous angles for all the compounds, they are almost identical, with average deviations of 0.008 Å or less in bond distances and 0.5° or less in bond angles (Fig. 5.36).

The inversion of the pyramidal configuration of nitrogen in three-membered rings is fairly slow on the NMR time scale (119) and hence measurable. Changes in the values of the inversion barrier of nitrogen (120) as one goes to *N*-acyloxaziridines indicate an increase in the stability of the nitrogen. If there is sufficient stabilization, then the trivalent nitrogen is noninverting under normal circumstances and oxaziridines of the type shown in Figure 5.37 are isolable in optically active *cis* and *trans* forms due to the asymmetric carbon and nitrogen atoms. If one (or more) of these substituents is itself optically active, then a series of diastereomers can be prepared, isolated and characterized. The structures of the diastereomers can be determined readily by X-ray diffraction techniques and thus the absolute configuration can be found.

The product of the UV irradiation of *N*-methylated anti-2,6-dimethyl-4-bromobenzaldoxime was proven to be *trans*-2-methyl-3-(2,6-dimethyl-4-bromo) oxaziridine by a two-dimensional X-ray structure study (121). The corresponding chlorine derivative is isomorphous to the bromine derivative and so a more detailed analysis of the structure containing the chlorine has been reported (122)(Fig, 5.38).

The *p*-toluene sulfonyl derivatives of oxaziridines (132, 133) are of additional oxaziridines (123–125), again in *trans* isomers, carried out before the suggestion that two isomers were in fact possible and available synthetically.

Finally, two forms of α-isopropyl-3-(4-nitrophenyl)-oxaziridine were isolated.

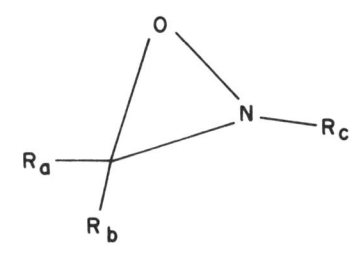

Fig. 5.37. Inversion substituents on oxaziridine.

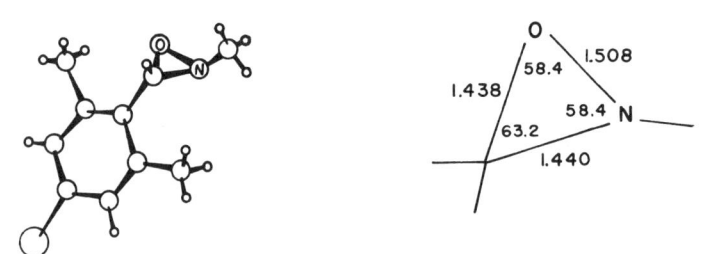

Fig. 5.38. Structure of isolated *trans*-substituted oxaziridine.

Spectral and chemical properties suggested that the higher melting isomer, which was present in the greater amount, was the *cis* isomer. A structural determination by X-ray diffraction techniques verified that this was the *cis* isomer (Fig. 5.39) (126). *Ab initio* SCF-LCAO-MO calculations (127) using "standard" bond lengths for unsubstituted oxaziridine predict a barrier to inversion of 32.4 kcal/mol. This barrier is consistent with the fact that *cis-trans* isomers of substituted oxaziridines are isolable.

Subsequently, oxaziridines were made by *m*-chloroperbenzoic acid oxidation of the Schiff bases resulting from the reaction of chiral $R(+)$-α-phenylethylamine and achiral aldehydes. All four nonracemic diastereoisomers were obtained and separated, with the *trans* presumably in excess. One of these in particular (the highest melting product, mp 128–129°) was 58.5% of the mixture, and a structural determination was carried out in this compound (128, 129) (Fig. 5.40).

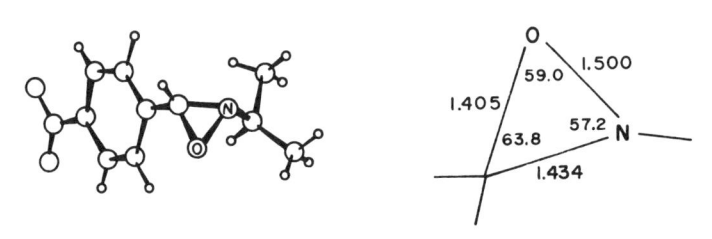

Fig. 5.39. *cis* Structure (both isomers isolated) of a substituted oxaziridine.

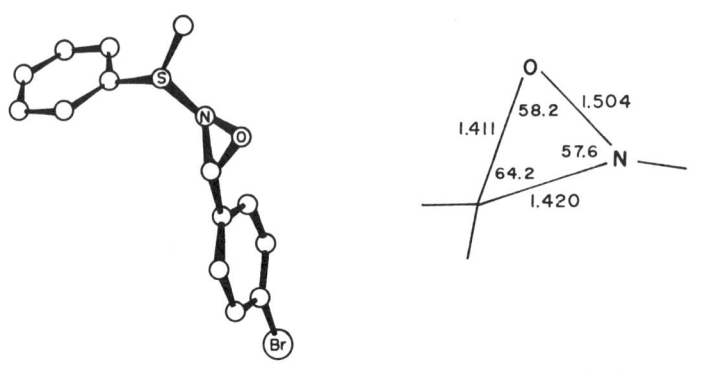

Fig. 5.40. RRS *trans* configuration of a substituted oxaziridine.

Aziridine Variants 337

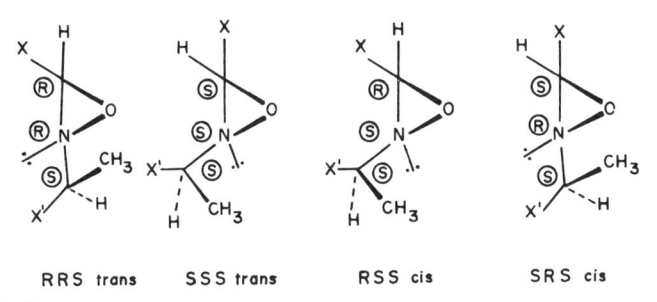

Fig. 5.41. Four possible nonracemic diastereomers of a substituted oxaziridine.

The structure was found to have the *RRS trans* configuration with the other *trans* structure having the *SSR* configuration (Fig. 5.41), and the two *cis* conformations were *RSS* and *SRS*.

Another structure in which the absolute configuration at the chiral nitrogen was determined was the optically active aziridine, 2S(−)-N-(R)-α-methyl benzyldiphenyloxaziridine (130) (Fig. 5.42).

The most definitive study of the absolute configuration at the chiral nitrogen was done on *both* epimers of 2-α-methylbenzyl-3,3-diphenyl oxaziridine (131) (Fig. 5.43).

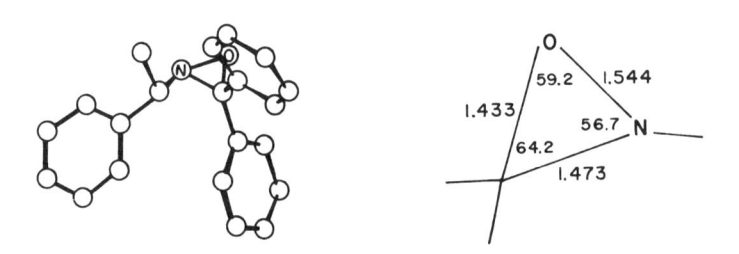

Fig. 5.42. Optically active oxaziridine.

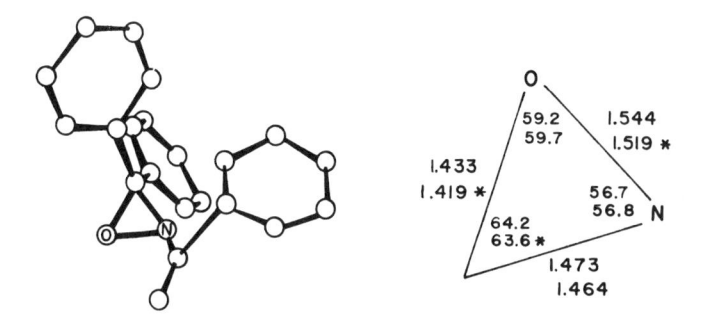

Fig. 5.43. Parameters for both epimers of an active oxaziridine.

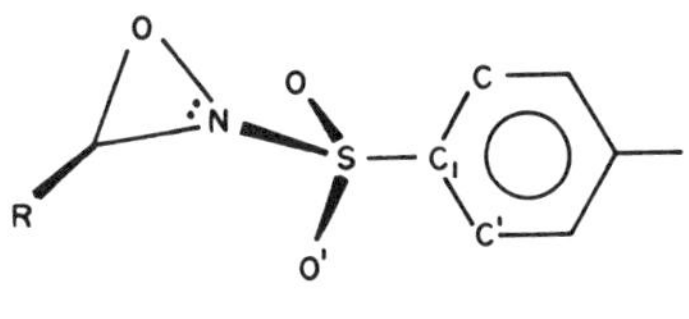

Fig. 5.44. *p*-Toluene sulfonyl derivatives of oxaziridine.

The corresponding bond distances and bond angles are close in the two molecules with the exception of the O—C and O—N distances and OCN angle (see * values in Fig. 5.43), which differ significantly owing to the different intramolecular environment of the oxygen.

The *p*-toluene sulfonyl derivatives of oxaziridines (132, 133) are of additional interest because of the conformation about the S—N bond. The three-membered heterocyclic ring makes it energetically unfavorable for the N atom to achieve a pseudoplanar conformation which would favor lone-pair *d*-orbital interactions, and so it is assumed that electron-repulsion interactions play a dominant role in the determination of conformation. In both cases (Fig. 5.44), the two sulfonyl O atoms and the N atom are symmetrically oriented relative to the toluene ring (torsion angles of $+29°$, $-19°$ for the OSC_1C and $O'SC_1C'$, respectively, and, analogously, $+89°$, $-88°$ for the NSC_1C' and NSC_1C, respectively). The N lone pair lies opposite the sulfonyl O atoms, not allowing lone pair–*d*-orbital interaction, and the N—S distances are 1.738(4) Å (132) and 1.728(4) Å (133). In *N,N,N'N'*-tetramethylsulfamide (134), *p*-methoxylbenzene-sulfon-*p*-anisidine (135), and *p*-methoxybenzene sulfon-*N*-isoproyl-*p*-anisidine (136), the lone pairs of the N atoms lie between the two sulfonyl O atoms. These three compounds exhibit a flattened geometry about the N and an average S—N bond of 1.628(4) Å, which is significantly shorter than the same distance in the oxaziridines.

As is evident from Figure 5.44, the substituents on the C and N atoms are *trans* to each other relative to the oxaziridine ring. Other molecular parameters for the oxaziridine ring (bond distances and bond angles) in these two compounds are in complete conformity with the average structure (Fig. 5.36).

5.5. AZETIDINES

5.5.1. Simple Azetidines

Although the decline in basicity as one goes from the aziridines to the azetidines makes the azetidines correspond more closely to the five-membered pyrrolidine ring systems (137), the strain in the four-membered azetidine ring is almost the same as that in the three-membered aziridines. Nevertheless, one cannot forget that the stability and overall geometry of these small ring systems is greatly affected by both the nature of the heteroatom and the substituents attached to the ring. For example, in variously substituted cyclobutanes studied by X-ray diffraction techniques, the four-membered ring is found to be either planar or significantly

Azetidines 339

nonplanar (138, 139) but not intermediate between these extremes. One can measure the planarity in terms of the dihedral angle (θ) as defined in Figure 5.45. For the planar structures, θ is obviously 180° but, interestingly enough, for the nonplanar structures θ averages 154 ± 3°, with no known examples of cyclobutane rings with this dihedral angle in the range of 162–179°. However, in four-membered rings containing a nitrogen atom, the azetidines, the results of X-ray structure studies indicate that substitutions lead to significant differences. For the moment, exclude fused systems (bicyclic and polycyclic) and consider only systems with exocyclic bonds and focus on the variously substituted simple azetidines. There are nine such compounds presently reported in the literature **(140–147)** as illustrated in Figure 5.46. The structures in Figure 5.46 have dihedral angles (θ) that range from 153 to 180°. Those rings that deviate the most from planarity have dihedral angles of 153° **(141a)**, 154° **(140)**, and 156° **(141b)**. These three structures have hydrogen-bonding patterns that involve both the OH group and the N atom in the ring. These bonding patterns produce additional strain that cause the rings to fold to an exaggerated angle. On the other hand, the one planar structure **(147)** is found to be packed in the crystal between the two phenyl rings, which are bent back above and below the four-membered ring. The repulsion of the hydrogens on the *gem*-dimethyl group and the phenyl electrons is apparently

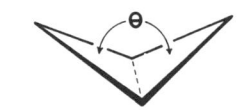

Fig. 5.45. Dihedral angle in azetidines.

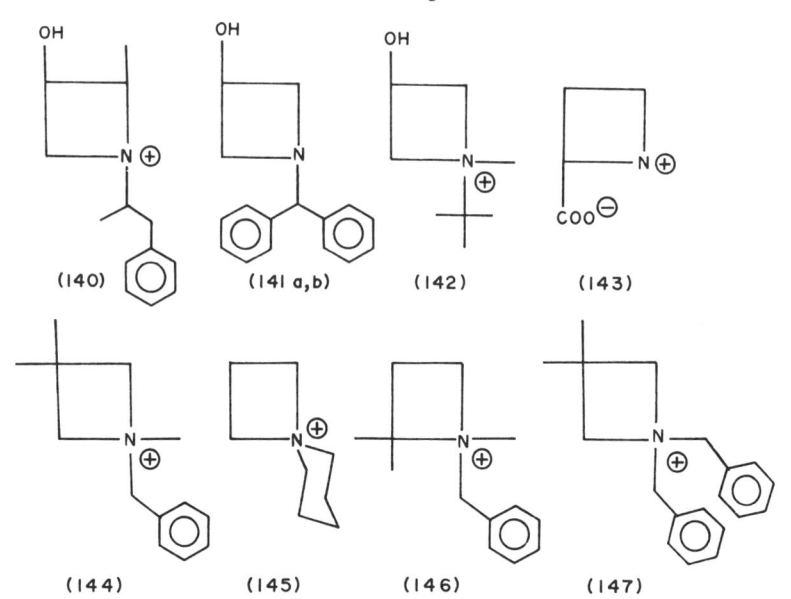

Fig. 5.46. Structures of simply substituted azetidine.

Table 5.2. Average Bond Distances and Angles for Unstrained Azetidine Ring

C–N	1.51 ± 0.03 Å
C–C	1.53 ± 0.02 Å
∠CNC	90 ± 1°
∠NCC	90 ± 2°
∠CCC	89 ± 2°
Dihedral angle	171 ± 3°

enough to flatten out the azetidine ring. As a consequence, the remaining known structures (142–146) can be considered more representative of an unstrained azetidine ring system.

Table 5.2, based on these representative compounds, lists "typical" values for bond distances and angles in simple azetidines. The table indicates that simple azetidines possess a symmetric structure with no significant differences in the various bond angles and only a slight tendency for the C–N distances to be shorter the the C–C distances. The ring is puckered so that the nitrogen bends out of the plane of the three carbon atoms with a 9° tilt. The degree of buckling is affected by the steric influence of the substituents, especially the cross ring or 1,3 interactions.

Electron diffraction studies of azetidine in the gas phase (148) give a somewhat different picture. The C–C distance is longer, 1.560(9) Å, than the average 1.53(2) Å found in the solid-state X-ray studies, and the C–N is shorter, 1.477(9) Å, than the average 1.51(3) Å. The CNC angle is only 88.0(8)° compared to the average 90(1)°, and the ring has an extreme buckle with a dihedral angle of 143(2)°. This is consistent with the differences found in X-ray solid-state and electron gas-phase studies of cyclobutane (149). In the latter case the dihedral angle for cyclobutane was found to be 145°.

5.5.2. β-Lactams

The addition of an exocyclic double bond to either the nitrogen or one of the carbons causes a change in the hybridization and with it a corresponding change in the preferred ring geometry. All of the unfused four-membered rings with exocyclic double bonds tend to be planar (150–155). The majority of the compounds in this category whose structures have been determined are of a special class of small-ring heterocycles known as β-lactams. While the β-lactams are probably best known from their fused ring analogs, several unfused structures have been reported (Fig. 5.47). Using these structures as representative of an unstrained β-lactam ring, the average parameters are listed in Table 5.3.

The β-lactam system shows several differences from simple azetidine rings. Besides the difference in the dihedral angle, 171° for simple azetidines and approximately 180° for the planar β-lactams, there are some shifts in the ring distances and angles. In azetidine all ring angles average between 89 and 90°. In β-lactam the

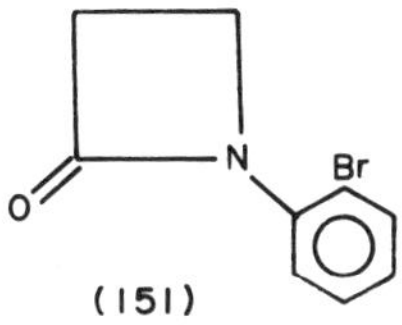

(151)

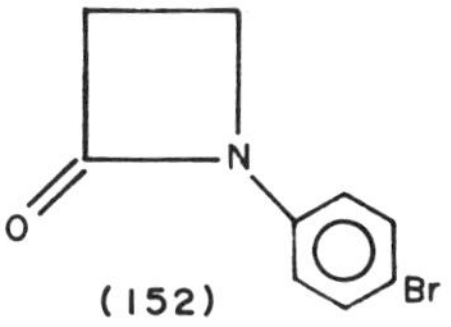

(152)

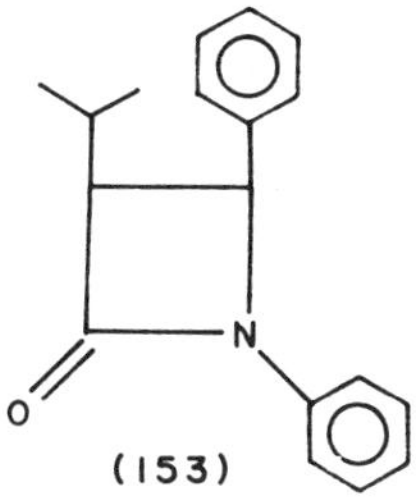

(153)

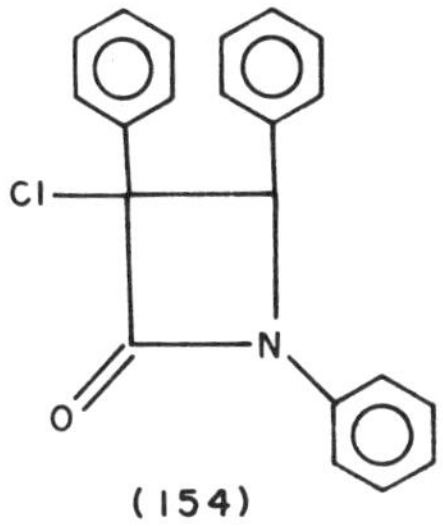

(154)

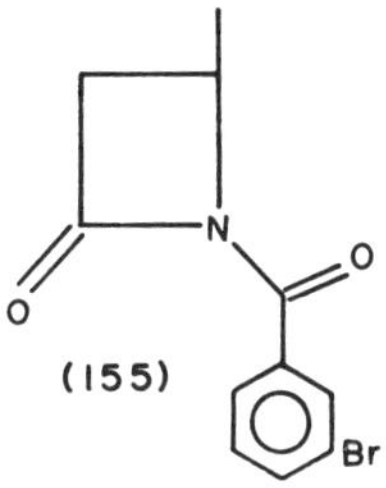

(155)

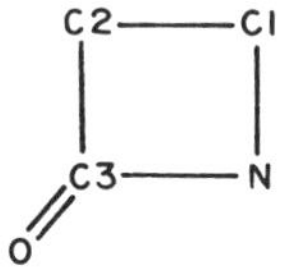

Fig. 5.47. β-Lactam "azetidines."

Table 5.3. Average Bond Distances and Angles for the "Standard" β-Lactam Ring

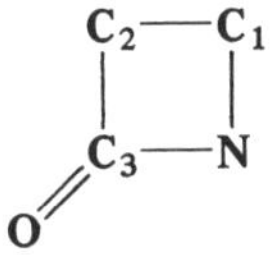

$N-C_1$	1.49 ± 1 Å
C_1-C_2	1.58 ± 2 Å
C_2-C_3	1.52 ± 1 Å
C_2-N	1.36 ± 1 Å
C_3-O	1.21 ± 1 Å
$\angle N-C_1-C_2$	$86 \pm 1^\circ$
$\angle C_1-C_2-C_3$	$86 \pm 1^\circ$
$\angle C_2-C_3-N$	$93 \pm 1^\circ$
$\angle C_3-N-C_1$	$95 \pm 1^\circ$
$\angle N-C_3-O$	$131 \pm 1^\circ$
$\angle C_2-C_3-O$	$136 \pm 1^\circ$

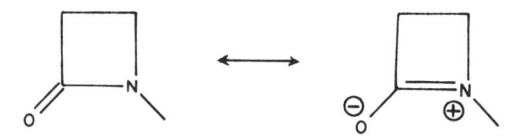

Fig. 5.48. Canonical forms of β-lactam ring.

angles around the nitrogen and the carbonyl carbon are consistently larger, 95 and 93°, respectively, while the other two ring angles are consistently smaller, both 86°. This is readily explained in terms of the change in hybridization from sp^3 to sp^2, and the difference is taken up by the remaining angles. The nitrogen-carbonyl carbon bond is significantly shorter in β-lactams than the nitrogen-carbon average in azetidines, 1.36–1.51 Å, whereas the other two distances are practically unaffected. These changes can be understood if one looks at the two canonical forms postulated for the unfused β-lactam ring shown in Figure 5.48. The proposed resonance structures account for the fact that the nitrogen-carbon distance in β-lactams is shorter than the azetidine average but not as short as the 1.30 Å found for a "pure" nitrogen-carbon double bond (156). The distance is, in fact, identical to the average amide distance, 1.36 Å, found in acyclic chains (152).

5.5.3. Fused β-Lactam Systems

In bicyclically fused systems, the fusion of a four-membered ring to a larger ring can cause strain and distortion. This is seen in structures where the normally puckered azetidine ring becomes planar (157) and the normally planar β-lactam ring becomes buckled (158–160). The amount of distortion is dependent on the size and nature of the fused ring as well as the substituents on both rings. The question of distortion becomes extremely important in the study of biologically active molecules containing small fused rings such as the penicillins and the cephalosporins (Fig. 5.49).

To understand the importance that structure plays in biological activity it is necessary to review the mechanism by which the penicillin and cephalosporin antibiotics inhibit the synthesis of bacterial cell walls (161–163). Normal bacterial cell wall growth depends on a many-step process to build up individualized strands

General Formula
for Penicillins

General Formula
for Cephalosporins

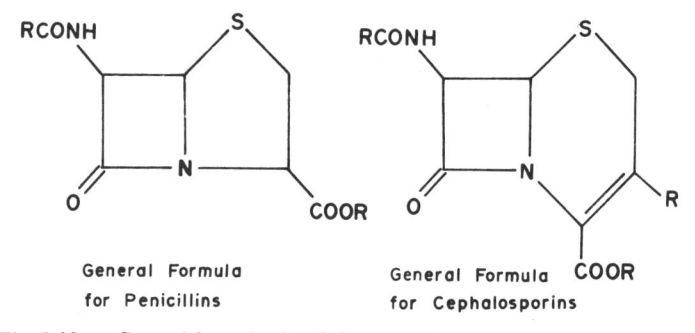

Fig. 5.49. General formulas for (*left*) penicillins and (*right*) cephalosporins.

of mucoglucosaccharides with pentapeptide side chains terminating in D-alanyl-D-alanine units. The interweaving of these strands to form a cell wall occurs in the peptidoglycan transpeptidase enzyme. This occurs by cleavage of a C-terminal D-alanine residue. This enzyme removes the D-alanine from the short pentapeptide chain which terminates with D-ala-D-ala and attaches the rest of the chain to an adjacent peptidoglycan strand by means of a free amino group. This process produces a three-dimensional crosslinking of the peptidoglycan strands. When penicillin or cephalosporin is introduced into the system the enzyme mistakes it for the proper substrate. In fact, the penicillin conformation fits the enzyme better than the D-ala-D-ala, which has to be distorted approximately 45° around the dihedral angle of the peptide bonds in order to react (164). Hence the rate of reaction is faster for penicillin. Therefore, even in small concentrations, penicillin can compete effectively with the substrate for the transpeptidase. The enzyme mistakes the antibiotic for its normal substrate and cleaves the amide bond in the β-lactam ring of the penicillin or cephalosporin. Since the rest of the molecule is still held by the carbon structure from the β-lactam it cannot break away and there is no longer a free end to continue the crosslinking. The chain is terminated. As more of the crosslinking is stopped, the cell wall cannot maintain its structural integrity and it ruptures, destroying the bacterial content.

Thus two things are necessary for an antibiotic to be able to function by this mechanism. First, the molecule must have the precise stereochemical nature to fool the enzyme. This means that the chemistry and conformation must be suitable to react with the complicated enzyme structure. Second, the net result of this reaction must be a product that halts the chain propagation.

The key to the mechanism that stops the chain-linking process involves the hydrolysis and breaking of the amide bond. Anything that increases the susceptibility of the carbonyl carbon to nucleophilic attack increases the effectiveness of the antibiotic. Looking back at Figure 5.42 one can see the adverse effect that resonance has on any potentially antibacterial properties of the free β-lactam. The carbon-nitrogen bond is shortened, the carbon-oxygen bond is lengthened, and the structure becomes planar and the C—N bond becomes more stable (or less susceptible to rupture by hydrolysis within the transpeptidase enzyme). In fact, for the nitrogen to be involved in π bonding the ring must remain planar. Fortunately, this is not so in many of the fused ring systems.

Detailed structural studies have been carried out on many penicillins (165–179) and cephalosporins (162, 180–189). A summary of some of the important parameters for both active and inactive compounds is given in Table 5.4. In the active penicillin molecules the nitrogen is bent an average of 0.40 ± 0.01 Å out of the plane of the other three carbons in the β-lactam ring. This decreases the ability of the nitrogen electrons to participate in π bonding, which would then lead to planarity and subsequent stability. Instead it conforms more to tetrahedral geometry, favoring a nucleophilic attack at the carbon-nitrogen bond. Spectral studies show that carbonyl stretching frequencies shift as one progresses from unfused β-lactams to penicillins, with the shift implying electron delocalization. The stretching frequency shifts from $1730\,\text{cm}^{-1}$ to $1780\,\text{cm}^{-1}$ as the planarity of

Table 5.4. Selected Structural Parameters for Penicillins and Cephalosporins[a]

Compounds[b]	C=O Stretch (cm^{-1})	Nitrogen Distance Above Plane (Å)	C=O Bond Distance (Å)	C–N Bond Distance (Å)
Penicillins (162)	1780–1770			
Active				
Amoxycillin (176)		0.38	1.200(8)	1.381(9)
Ampicillins (162, 171)		$\overline{0.38}$	$\overline{1.19}$ (1)	$\overline{1.37}$ (1)
Bacmecillin (175)		142	1.190(7)	1.382(7)
Benzylpenicillin (172)		0.38	1.206(6)	1.377(5)
Cloxacillins (169)		$\overline{0.40}$	$\overline{1.19}$ (1)	$\overline{1.39}$ (1)
Dicloxacillins (174)		0.41	1.20 (2)	1.40 (2)
Methicillin (170)		0.44	1.196(2)	1.389(2)
Penicillin G's (165, 178)		$\overline{0.40}$	$\overline{1.19}$ (1)	$\overline{1.37}$ (3)
Penicillin V (165)		0.40	1.21 (2)	1.45 (2)
Phenoxyacetamidopenam (179)		0.37	1.205(4)	1.390(4)
Piperacillin (177)		0.44	1.81 (9)	1.393(9)
Anhydropenicillins (165)	1810			
Inactive				
Anhydropenicillins (165, 167)		$\overline{0.41}$	$\overline{1.17}$ (1)	$\overline{1.42}$ (1)

344

Δ^3-Cephalosporins	1776–1764		
Active			
Cephaloridine (162)		0.24	1.214(8)
Cephaloglycine (162)		0.22	1.28 (5)
Cephapyrine (184)		0.22	1.21 (1)
Δ^3-Cephems (182, 183, 187)		$\overline{0.19}$	$\overline{1.19}$ (1)
Inactive			
Δ^3-Cephalothin (188)		0.13	1.21 (2)
Δ^3-Cephazolin (189)		0.07	1.22 (2)
Δ^2-Cephalosporins (162)	1760–1730		
Inactive			
Δ^2-Cephalosporins V (162)		0.7	1.223(7)
Δ^2-Cephem (181)		0.3	1.23 (1)
Unfused β-lactam (162)	1760–1730		
Inactive			
β-lactams (Table 5.2)		0	$\overline{1.21}$ (1)

	1.382(8)
	1.48 (5)
	1.38 (1)
	$\overline{1.39}$ (1)
	1.35 (2)
	1.36 (2)
	1.339(7)
	1.33 (1)
	$\overline{1.36}$ (1)

[a] An average is given for those structures for which two or more forms have been determined.
[b] Numbers in parentheses are literature data sources.

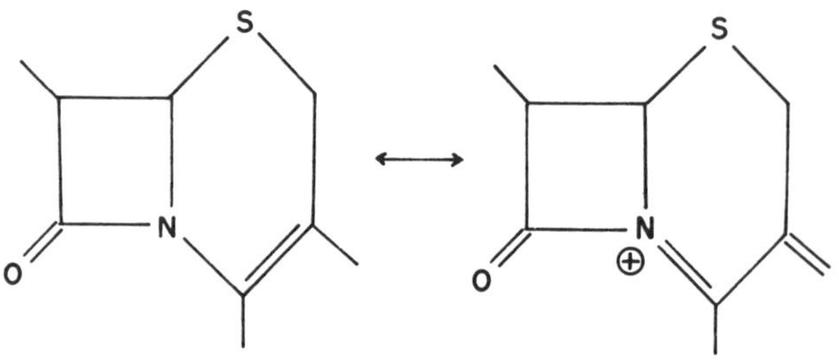

Fig. 5.50. Canonical forms for cephalosporins.

the nitrogen decreases. This is consistent with lengthening of the carbon-nitrogen bond from 1.36 to 1.39 Å and the shortening of the carbon-oxygen bond from 1.21–1.19 Å in comparison with the free β-lactam.

In the active cephalosporins the deviation from planarity of the nitrogen atom is 0.21 ± 0.01 Å less than in the penicillins. However, the other parameters compare more favorably. The carbonyl stretching still shows a loss of delocalization, up to 1776 cm^{-1}, an average carbon-nitrogen bond of 1.40 Å, and an average carbon-oxygen bond length of 1.21 Å. The reason that the cephalosporins do not require as much deviation from planarity to be active is that they have a second mechanism for isolating the carbonyl double bond. The six-membered ring can interact with the nitrogen electrons as shown in Figure 5.50.

Since the nitrogen electrons can be involved to some degree in enamine resonance, this cuts down on the amide resonance and helps to encourage the nucleophilic attack necessary to block the crosslinking process.

The only discrepancy seems to come with the anhydropenicillins. The carbonyl frequency is 1810 cm^{-1}, the average nitrogen out-of-plane distortion is 0.41 Å, the carbon-nitrogen distance is as long as 1.42 Å, and the carbon-oxygen as short as 1.17 Å. According to all the structural arguments they should show high antibacterial activity, but, on the contrary, they are biologically inactive. The reason for this lack of activity may be chemical since there is no carboxylate group on the position adjacent to the β-lactam of the anhydropenicillin. This serves as a good example to demonstrate that while the nonplanarity of the nitrogen in the β-lactam ring is necessary for biological activity it is not sufficient.

Still more recent studies have begun to explore the relationship of structure to biological activity in new classes of antibiotics such as those shown in Figure 5.51.

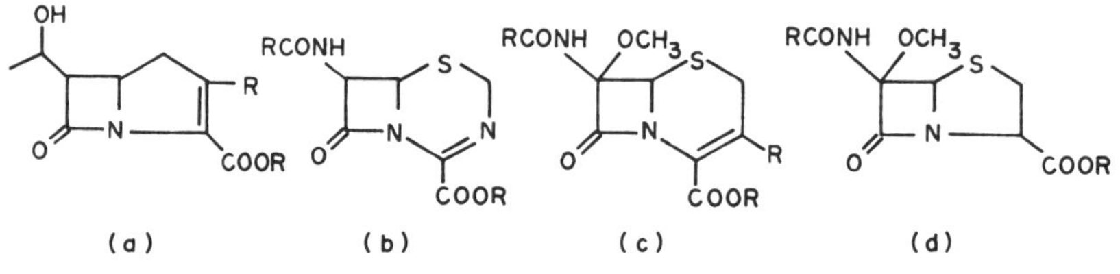

Fig. 5.51. Penicillin and cephalosporin analogs.

The carbapenem-type structures (Fig. 5.51a) such as thienamycin and its derivatives (190), have a double bond in the five-membered ring and no sulfur atom. They show highly desirable antibacterial activity (against Gram-positive bacteria as well as a wide range of Gram-negative bacteria). Structural studies have played a key role in developing a synthetic means for their preparation (191). Also currently under investigation are systems where other heteroatoms are added to the larger fused ring such as the 3-azacephalosporins (192) (Fig. 5.51b) and systems where different groups are added as substituents to the fused ring system as in the 7α-methoxycephalosporins (193) (Fig. 5.51c) and 6α-methoxypenicillins (193) (Fig. 5.51d). Each of these new systems offers a variety of derivatives, and each is active against a unique range of bacteria, Because of the selective nature of each and the stereochemical requirements. X-ray structure determinations are playing an important role in understanding the mechanism of activity and in the derivation of new synthetic techniques necessary for their preparation.

5.6. AZETIDINE VARIANTS

A number of four-membered heterocycles exist that have more than one heteroatom in the ring. Here we shall consider only those that have at least one nitrogen atom among the heteroatoms.

In all cases where the structure has been determined, there has also been an exocyclic double bond. Thus it is not surprising that all are planar or nearly so. The first group under consideration, the diazetidinium compounds (194–199), have their two nitrogen atoms in adjacent positions in the ring (Fig. 5.52). Structures **XIV** and **XV** have short N–N distances (1.415 and 1.40 Å, respectively), whereas the N–N distance in **XVI** is significantly longer at 1.450 Å. Thus, the double-bond character for the first two is more pronounced, although in both cases the nitrogen is quaternary. However, both compounds contain two exocyclic double bonds.

Finally, in a diazatetracyclo compound (197) (Fig. 5.53), there is a more localized N–N double bond in the four-membered ring. It is not unexpected to find a very short bond length (1.29 Å) for N–N in this compound.

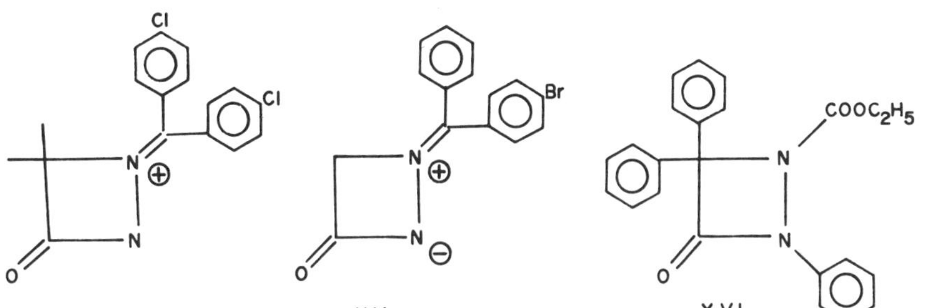

Fig. 5.52. Known substituted 1,2-diazetidines.

Fig. 5.53. Substituted 1,2-diazetirine.

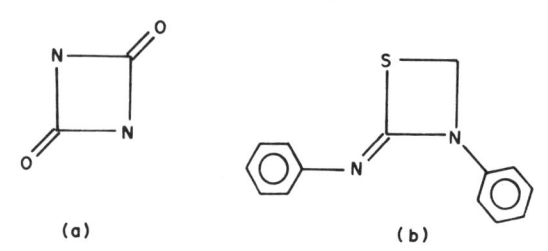

(a)　　　　　　　　　(b)

Fig. 5.54. Other heterosubstituted azetidines.

In the last category, the two heteroatoms in the ring are not adjacent to each other. As anticipated, the 1,3-diazetidinium compound (198) (Fig. 5.54a) is planar, but it is surprising that the 1,3-thiacetidinium compound (199) (Fig. 5.54b) is also very close to planar. This occurs despite the presence of the sulfur, which would result in the long S–N distances causing strain in the ring that could be alleviated readily by buckling the ring, and despite the fact that three of the four bond lengths in the ring indicate essentially single-bond characters. The strain is taken up by increasing the CNC angle to 100.4° and decreasing the CSC angle to 74.3°. This contrasts with the solution to ring strain in the three-membered thiadiaziridine structure (107), where the compensation for the long N–S distances in the ring occurs by having a very long N–N bond (1.67 Å) and maintaining ring angles very near the value of 60° ($\angle$NSN = 62°). It is apparently more critical to preserve the double-bond character (and the resultant planarity) in the four-membered rings, and hence the mechanism for alleviating the strain is different from that found in the three-membered ring.

REFERENCES

1. O. C. Dermer and G. E. Ham, *Ethylenimine and Other Aziridines,* Academic Press, New York, 1969, p. 592.
2. E. P. Abraham, *Biosynthesis and Enzymic Hydrolysis of Penicillins and Cephalosporins,* Univ. of Tokyo Press, 1974, pp. 1–86.
3. *Specialist Periodical Report,* Vol. 5, Royal Chemical Society, London, 1978, p. 44; and W. L. F. Armarego, *Stereochemistry of Heterocyclic Compounds,* Wiley-Interscience, New York, 1977, p. 12.
4. G. W. Griffin and A. Padwa, "Photochemistry of 3,4-Membered Heterocyclics," in O. Buchardt, Ed., *Photochemistry of Heterocyclic Compounds,* Wiley-Interscience, New York, 1976; D. S. C. Black and J. C. Doyle in A. R. Katritzky and A. J. Boulton, Eds., *Advances in Heterocyclic Chemistry,* Vol. 27, Academic Press, New York, 1980, p. 1.

References

5. D. C. Wiley, I. A. Wilson, and J. J. Skehel, *Nature,* **289**, 366 (1981).
6a. M. G. Vincent and H. D. Flack, *Acta Cryst.,* **A36**, 610 (1980).
6b. P. T. Beurskens and P. A. J. Patric, *Acta Cryst.,* **A37**, 180 (1981).
7a. G. R. Mitchell and R. Lovell, *Acta Cryst.,* **A37**, 189 (1981).
7b. S-H. Hong and S. Åsbrink, *J. Appl. Cryst,* **14**, 43 (1981).
8. J. C. Phillips, J. Bordas, A. M. Foote, M. H. J. Koch, and M. F. Moody, *Biochem.* **21**, 830 (1982).
9. E. D. Stevens, *J. Amer. Chem. Soc.* **103**, 5087 (1981).
10. K. D. Rouse and M. J. Cooper, *J. Appl. Cryst.,* **14**, 72 (1981).
11. S. R. Hall, J. M. Stewart, and R. J. Munn, *Acta Cryst.,* **A36**, 979 (1980).
12. Three recent advertisements from United States companies which have crossed my desk are from: Crystalytics Company, Lincoln, Nebraska; Latticeworks, Inc., Cranford, New Jersey; and Molecular Structure Corporation, College Station, Texas.
13. H. D. Bartunik, P. N. Clout, and B. Robrahn, *J. Appl. Cryst.,* **14**, 134 (1981).
14. J. C. Phillips and K. O. Hodgson, *Acta Cryst.,* **A36**, 856 (1980).
15. V. E. Dmitrienko and V. A. Belyakor, *Acta Cryst.,* **A36**, 1044 (1980).
16. H. Hart, M. Sauvage, and D. P. Siddon, *Acta Cryst.,* **A36**, 947 (1980).
17. D. H. Templeton, L. K. Templeton, J. C. Phillips, and K. O. Hodgson, *Acta Cryst.,* **A36**, 856 (1980).
18. D. H. Templeton and L. K. Templeton, *Acta Cryst.,* **A36**, 237 (1980).
19. S. G. Baru, G. I. Proviz, G. A. Savinov, V. A. Sidorov, I. G. Fel'dman, A. G. Khabakhpashev, and B. N. Shuvalov, *Sov. Phys. Cryst.,* **25**, 212 (1980).
20. T. D. Mokulskaya, S. V. Kuzev, G. E. Myshko, A. A. Khrenov, M. A. Mokulskii, Z. D. Dobrokhotova, A. Ya. Volodenkov, V. P. Rubanov, N. A. Ryanzina, B. I. Shitikov, S. E. Baru, A. G. Khabakhpashev, and V. A. Sidorov, *J. Appl. Cryst.,* **14**, 33 (1981).
21. R. Hamlin, C. Cork, A. Howard, C. Nielsen, W. Vernon, D. Matthews, and Ng. H. Xuong, *J. Appl. Cryst.,* **14**, 85 (1981).
22. R-J. Roe, J. C. Chang, M. Fishkis, and J. J. Curro, *J. Appl. Cryst.,* **14**, 139 (1981).
23. T. Sakamaki, S. Hosoya, and T. Fukamachi, *Acta Cryst.,* **A36**, 183 (1980).
24. T. Fukamachi, S. Hosoya, T. Fawamura, and M. Okunuki, *Acta Cryst.,* **A35**, 828 (1979).
25. K. Wilson and D. Yeates, *Acta Cryst.,* **A35**, 146 (1979).
26. M. Hecq, *J. Appl. Cryst.,* **14**, 60 (1981).
27. W. Clegg, *Acta Cryst.,* **A37**, 22 (1981).
28. M. Zocchi, *Acta Cryst.,* **A36**, 164 (1980).
29. W. A. Denne, *Acta Cryst.,* **A34**, 1028 (1978).
30. W. S. Wilcox, K. C. Brannock, W. DeMore, and J. H. Goldstein, *J. Chem. Phys.,* **21**, 563 (1953).
31. T. G. Turner, V. C. Fiora, W. M. Kendrick, and B. L. Hicks, *J. Chem. Phys.,* **21**, 564 (1953).
32. R. D. Johnson, R. J. Myers, and W. D. Gwinn, *J. Chem. Phys.,* **21**, 1425 (1953).
33. R. W. Mitchell, J. C. Burr, Jr., and J. A. Merritt, *Spectrochim. Acta,* **23A**, 195 (1967).
34. T. E. Turner, V. C. Fiora, and W. M. Kendrick, *J. Chem. Phys.,* **23**, 1966 (1955).
35. B. Bak and S. Skaarup, *J. Mol. Struct.,* **10**, 385 (1971).
36. M. Igarashi, *Bull. Chem. Soc. Japan,* **34**, 369 (1961).
37. H. Ringertz, *Acta Chem. Scand.,* **23**, 137 (1969).
38. C. Dickinson, J. R. Holden, E. G. Boonstra, and J. M. Stewart, *A. C. A. Abstr. Papers* (Summer), 74 (1971).

39. E. S. Gould and R. A. Pasternak, *J. Amer. Chem. Soc.*, **83**, 2658 (1961).
40. T. Ito and T. Sakurai, *Acta Cryst.*, **B29**, 1594 (1973).
41. R. P. Sibaeva, L. O. Atovmyan, and R. G. Kostanovskij, *Acta Cryst.*, **21**, A129 (1966).
42. R. P. Sibaeva, L. O. Atovmyan, and R. G. Kostanovskij, *Dokl. Akad. Nauk, SSR*, **175(3)**, 586 (1967).
43. T-Mko, L. Olansky, and J. W. Moncrief, *Acta Cryst.*, **B31**, 1875 (1975).
44. J. Iball, S. N. Scrimgeour, and B. C. Williams, *Acta Cryst.*, **B31**, 1121 (1975).
45. J. C. Barnes, J. Iball, and W. R. Smith, *Acta Cryst.*, **B33**, 848 (1977).
46. V. I. Andrianov, R. G. Kostyanovskii, R. P. Shibaeva, and L. O. Atovmyan, *Zh. Strukt. Khim.*, **8**, 100 (1967).
47. M. Bolognesii and G. Rossi, *Acta Cryst.*, **B33**, 122 (1977).
47a. E. Subramanian and J. Trotter, *J. Chem. Soc., A*, **1969**, 2309.
48. Y. Delugeard, M. Vaultier, and J. Meinnel, *Acta Cryst.*, **B31**, 2885 (1975).
49. H. M. Zacharis and L. M. Trefonas, *J. Heterocycl. Chem.*, **7**, 1301 (1970).
50. A. H-T. Wang, I. C. Paul, E. R. Talaty, and A. E. Dupuy, Jr., *Chem. Commun.*, **1972**, 43.
51. S. A. Giller, Y. Y. Bleidelis, A. A. Kemme, A. V. Eremeev, and V. A. Kholodnikov, *Zh. Strukt. Khim. (Engl. Transl.)*, **16**, 411 (1975).
52. J. C. Sheehan and M. Mehdi Nafissi-V, *J. Amer. Chem. Soc.*, **91**, 1176 (1969).
53. E. R. Talaty, A. E. Dupuy, Jr., and T. H. Golson, *Chem. Commun.*, **1969**, 49.
54. S. Sarel, B. A. Weissman, and Y. Stein, *Tetrahedron Lett.*, 373 (1971).
55. R. Grée and R. Carié, *Chem. Commun.* **1975**, 112.
56. J. L. Atwood and G. D. Stucky, *J. Amer. Chem. Soc.*, **92**, 285 (1970).
57a. W. Harrison, A. Storr, and J. Trotter, *Chem. Commun.*, 1101 (1971).
57b. W. Harrison, A. Storr, and J. Trotter, *J. Chem. Soc. Dalton*, **1972**, 1554.
58. R. Lussier, J. O. Edwards, and R. Eisenberg, *Inorg. Chim. Acta*, **3(3)**, 468 (1969).
59. W. B. Bang, Ph.D. thesis, Brown University, 1963.
60. J. C. Barnes, J. Iball, and T. J. R. Weakley, *Acta Cryst.*, **B31**, 1435 (1975).
61. E. W. Abel, R. J. Rowley, R. Mason, and K. M. Thomas, *Chem. Commun.*, 72 (1974).
62. H. M. Zacharis and L. M. Trefonas, *J. Heterocycl. Chem.* **5**, 343 (1968).
63. L. M. Trefonas and R. Majeste, *J. Heterocycl. Chem.*, **2**, 80 (1965).
64. L. M. Trefonas and R. Towns, *J. Heterocycl. Chem.* **1**, 19 (1965).
65. L. M. Trefonas and R. J. Majeste, *Tetrahedron*, **19**, 929 (1963).
66. H. Zacharis and L. M. Trefonas, *J. Heterocycl. Chem.* **7**, 755 (1970).
67. L. M. Trefonas, R. Towns, and R. Majeste, *J. Heterocycl. Chem.*, **4**, 511 (1967).
68. L. M. Trefonas and T. Sato, *J. Heterocycl. Chem.*, **3**, 404 (1966).
69. L. M. Trefonas and J. Couvillion, *J. Amer. Chem. Soc.*, **85**, 3184 (1963).
70. R. F. Bryan and J. D. Dunitz, *Helv. Chim. Acta*, **43**, 3 (1960).
71. J. D. Dunitz and H. M. Shearer, *Helv. Chim. Acta*, **43**, 18 (1960).
72. A good discussion of medium-sized rings and their conformations, strain minimization, and actual structures is given in the following review: *Perspectives in Structural Chemistry*, Vol. 2, J. D. Dunitz and J. A. Ibers, Eds., John Wiley, New York, 1968, pp. 1–70.
73. E. L. Eliel, *Stereochemistry of Carbon Compounds (Advanced Chemistry)*, McGraw-Hill, 1962, p. 252.
74. N. L. Allinger, *J. Amer. Chem. Soc.*, **81**, 5727 (1959).
75. J. B. Hendrickson, *J. Amer. Chem. Soc.*, **83**, 4537 (1961).

References   351

76. R. A. Pasternak, *Acta Cryst.*, **4**, 316 (1951).
77. K. B. Wilberg, *Physical Organic Chemistry,* John Wiley, New York, 1968, p. 123.
78. A. Tulinsky and J. H. van den Hende, *J. Amer. Chem. Soc.*, **89**, 2905 (1967).
79. J. N. Brown, R. L. R. Towns, and L. M. Trefonas, *J. Heterocycl. Chem.*, **7**, 1321 (1970).
80. E. M. Gopalakrishna, *Acta Cryst.*, **B28**, 2754 (1972).
81. S. A. Hiller, Y. Y. Bleidelis, A. A. Kemme, and A. V. Eremeyev, *Chem. Commun.*, **1975**, 130.
82. Y. Kobayashi, T. Kutsuma, Y. Hanzawa, Y. Iitala, and H. Nakamura, *Tetrahedron Lett.*, **1969**, 5337.
83. J. Galloy, J. P. Putzeys, G. Germain, J. P. Declerq, and M. van Meerssche, *Acta Cryst.*, **B30**, 2462 (1974).
84. A. F. Mishnev, Y. A. Bleidelis, and L. S. Geita, *Khim. Geterotsikl. Soedin.*, **9**, 1217 (1977), Engl. Transl. 1977.
85. L. S. Geita, I. E. Dalberga, A. K. Grinvalde, and I. S. Yankova, *Khim. Geterotsikl. Soedin.*, **1**, 65 (1976).
86. S. R. Paulsen, *Angew. Chem.*, **72**, 781 (1960).
87. E. Schmitz and R. Ohme, *Tetrahedron Lett.*, **1961**, 612.
88. W. H. Graham, *J. Amer.Chem. Soc.*, **84**, 1063 (1962).
89. L. Pierce and V. Dobyns, *J. Amer. Chem. Soc.*, **84**, 2651 (1962).
90. L. H. Sharpen, J. E. Wollrab, and D. P. Ames, *J. Chem. Phys.*, **50**, 2063 (1969).
91. J. E. Wollrab, L. Scharpen II, D. P. Ames, and J. A. Merritt, *J. Chem. Phys.*, **49**, 2405 (1968).
92. T. Oka, *J. Phys. Soc. Japan*, **15**, 2274 (1960).
93. R. Nelson and L. Pierce, *J. Mol. Spectrosc.*, **18**, 344 (1965).
94. R. W. Schendeman, *Dissert. Abstr.*, **18**, 1823 (1958).
95. R. A. Mitsch, *J. Heterocycl. Chem.*, **3**, 245 (1966).
96. C. W. Bjork, N. C. Craig, R. A. Mitsch, and J. Overend, *J. Amer. Chem. Soc.*, **87**, 1186 (1965).
97. J. L. Hencher and S. H. Bauer, *J. Amer. Chem. Soc.*, **89**, 5527 (1967).
98. B. Bak, D. Christensen, L. Nygaard, and J. R. Andersen, *Spectrochim. Acta*, **13**, 120 (1958).
99. V. W. Laurie and D. T. Pence, *J. Chem. Phys.*, **38**, 2693 (1963).
100. W. Edgelc, P. A. Kinsey, and J. W. Amy, *J. Amer. Chem. Soc.*, **79**, 2691 (1957).
101. V. W. Laurie, *J. Chem. Phys.*, **34**, 291 (1961).
102. J. E. Wollrab and L. H. Scharpen, *J. Chem. Phys.*, **51**, 1584 (1969).
103. P. E. McGann, J. W. Groves, F. D. Greene, G. M. Stack, R. J. Majeste, and L. M. Trefonas, *J. Org. Chem.*, **43**, 922 (1975).
104. A. H-J. Wang, I. C. Paul, E. R. Talaty, and A. E. Dupuy, Jr., *Chem. Commun.*, **1972**, 43.
105. J. Trotter, *Acta Cryst.*, **12**, 605 (1969).
106. L. E. Sutton, Ed. *Tables of Interatomic Distances (1956-1959) (Spec. Publ. #18)*, Chemical Society, London, 1965.
107. F. D. Greene, J. C. Stowell, and W. R. Bergmark, *J. Org. Chem.*, **34**, 2254 (1969).
108. J. M. Pochan, J. E. Baldwin, and W. H. Flygare, *J. Amer. Chem. Soc.*, **91**, 1896 (1969).
109. K. Peters and H. G. von Schnering, *Chem. Ber.*, **109**, 1384 (1976).
110. A. Nabeya, Y. Tamura, T. Kodama, and Y. Iwakura, *J. Org. Chem.*, **38**, 3758 (1973).
111. G. B. Ansell, A. T. Nielsen, D. W. Moore, R. L. Atkins, and C. D. Stanifer, *Acta Cryst.*, **B35**, 1505 (1979).

112. V. S. Mastryukov, O. V. Dorefeeva, and L. V. Vilkov, *Chem. Commun.*, **1974**, 397.
113. O. A. D'yachenko, L. O. Atovmyan, S. M. Aldoshin, A. E. Polyakeu, and R. C. Kostyanovskii, *Chem. Commun.*, **1976**, 50.
114. E. Hoehne, *J. Prakt. Chem.*, **312**, 862 (1970).
115. L. M. Trefonas and L. D. Cheung, *J. Amer. Chem. Soc.*, **95**, 636 (1973).
116. L. M. Trefonas and R. Majeste, *J. Heterocycl. Chem.*, **2**, 80 (1965).
117. J. S. Broadley and J. M. Robertson, *Nature*, **164**, 915 (1949).
118. B. S. Cartwright and J. M. Robertson, *Chem. Commun.*, **1966**, 82.
119. R. G. Kostyanovsky, K. S. Zakharov, M. Zaripova. and V. F. Rudtchenko, *Tetrahedron*, **48**, 4207 (1974).
120. D. R. Boyd, *J. Chem. Soc., Perkin Trans. II*, **1973**, 1575.
121. B. Jensen and B. Jerslev, *Acta Cryst.*, **21**, A118 (1966).
122. B. Jerslev, *Acta Cryst.*, **23**, 645 (1967).
123. D. R. Boyd, *Tetrahedron Lett.*, 4561 (1968).
124. F. Montenari, I. Moretti, and G. Torre, *Chem. Commun.*, **1968**, 1994.
125. L. Brehm, K. G. Jensen, and B. Jerslev, *Acta Chem. Scand.*, **20**, 915 (1966).
126. J. F. Cannon, J. Daly, J. V. Silverton, D. R. Boyd, and D. M. Jerina, *J. Chem. Soc. Perkin II*, **1972**, 1137.
127. J. M. Lehn, B. Munsch, P. Millie, and A. Veillard, *Theor. Chem. Acta*, **13**, 313 (1969).
128. M. Bogucka-Ledochowska, A. Konitz, A. Hempel, Z. Dauter, E. Borowski, C. Belzecki, and D. Mostowicz, *Tetrahedron Lett.*, **1976**, 1025.
129. M. Bogucka-Ledochowska, A. Konitz, A. Hempel, Z. Dauter, and E. Borowski, *Z. Krist.*, **149**, 49 (1979).
130. M. Bucciarelli, I. Moretti, G. Torre, G. D. Andretti, G. Bocelli, and P. Sgarabotto, *Chem. Commun.*, **1976**, 60.
131. A. Forni, G. Garuti, I. Moretti, G. Torre, G. D. Andretti, G. Bocelli, and P. Sgarabotto, *J. Chem. Soc., Perkin II*, **1978**, 401.
132. J. S. Chen and W. H. Watson, *Acta Cryst.*, **B34**, 2861 (1978).
133. M. Kimura and W. H. Watson, *Acta Cryst.*, **B36**, 234 (1979).
134. T. Jordan, H. W. Smith, L. L. Lohr, and W. N. Lipscomb, *J. Amer. Chem. Soc.*, **85**, 846 (1976).
135. S. Pokrywiecki, C. M. Weeks, and W. L. Duax, *Cryst. Struct. Commun.*, **2**, 63 (1973).
136. S. Pokrywiecki, C. M. Weeks, and W. L. Duax, *Cryst. Struct. Commun.*, **2**, 67 (1973).
137. S. Searles, M. Taures, F. Black, and L. A. Quaterman, *J. Amer. Chem. Soc.*, **78**, 4917 (1950).
138. F. A. Cotton and B. A. Frenz, *Tetrahderon*, **30**, 1587 (1974).
139. R. M. Moriarty, *Topics Stereochem.*, **8**, 271 (1974).
140. J. B. Wetherington and J. W. Moncrief, *Acta Cryst.*, **B30**, 534 (1974).
141. S. Ramakumar, K. Venkatesan, and S. T. Rao, *Acta Cryst.*, **B33**, 824 (1977).
142. E. L. McGandy, H. M. Berman, J. W. Burgner II, and R. L. VanEtten, *J. Amer. Chem. Soc.*, **91**, 6173 (1969).
143. H. M. Berman, E. L. McGandy, J. W. Burgner II, and R. L. VanEtten, *J. Amer. Chem. Soc.*, **91**, 6177 (1969).
144. R. L. Towns and L. M. Trefonas, *J. Amer. Chem. Soc.*, **93**, 1761 (1971).
145. H. M. Zacharis and L. M. Trefonas, *J. Amer. Chem. Soc.* **93**, 2935 (1971).
146. C. L. Moret and L. M. Trefonas, *J. Heterocycl. Chem.*, **5**, 549 (1968).
147. R. L. Snyder, E. L. McGandy, R. L. VanEtten, L. M. Trefonas, and R. L. Towns, *J. Amer. Chem. Soc.*, **91**, 6187 (1969).

References

148. O. V. Dorofeeva, V. S. Mastryukov, and L. V. Vilko, *Chem. Commun.* **1973**, 773.
149. A. Almenninger, O. Bastiansen, and P. N. Skancke, *Acta Chem. Scand.*, **15**, 711 (1961).
150. M. Cesari, L. D'Ilario, E. Giglio, and G. Perego, *Acta Cryst.*, **B31**, 49 (1975).
151. H. Fujiwara, R. L. Varley, and J. M. van der Veen, *J. Chem. Soc. Perkin II*, **1977**, 547.
152. G. Kartha and G. Ambady, *J. Chem. Soc. Perkin II*, **1973**, 2043.
153. R. Parthasarathy, *Acta Cryst.*, **B26**, 1283 (1970).
154. A. Colens, J. P. Declercq, G. Germain, J. P. Putzeys, and M. van Meerssche, *Cryst. Struct. Commun.*, **3**, 119 (1974).
155. E. F. Paulus, D. Kobelt, and H. Jensen, *Angew. Chem. Int. Ed.*, **8**, 990 (1969).
156. L. M. Trefonas, R. L. Flurry, Jr., R. Majeste, E. A. Meyers, and R. F. Copeland, *J. Amer. Chem. Soc.*, **88**, 2145 (1966).
157. R. Majeste and L. M. Trefonas, *J. Heterocycl. Chem.* **5**, 663 (1968).
158. D. Bender, H. Rapoport, and J. Bordner, *J. Org. Chem.*, **40**, 3208 (1975).
159. L. A. Paquette, M. J. Broadhurst, C. Lee, and J. Clardy, *J. Amer. Chem. Soc.*, **94**, 630 (1972).
160. T. R. Allman and T. Debaerdemaeker, *Cryst. Struct. Commun.*, **3**, 365 (1974).
161. E. P. Abraham, *Topics Pharm. Sci.*, **1**, 1 (1968).
162. R. M. Sweet and L. F. Dahl, *J. Amer. Chem. Soc.*, **92**, 5489 (1970).
163. E. H. Flynn, "Cephalosporins and Penicillins," in *Chemistry and Biology*, Academic Press, New York, 1972.
164. B. Lee, *J. Mol. Biol.*, **61**, 463 (1971).
165. G. L. Simon, R. B. Morin, and L. F. Dahl, *J. Amer. Chem. Soc.*, **94**, 8557 (1972).
166. S. Abrahamsson, D. C. Hodgkin, and E. N. Maslen, *Biochem. J.*, **86**, 514 (1963).
167. M. O. Chaney and N. D. Jones, *Cryst. Struct. Commun.*, **2**, 367 (1973).
168. P. J. Cox, R. J. McClure, and G. A. Sim, *J. Chem. Soc. Perkin II*, **1974**, 360.
169. P. Blanpain and F. Durant, *Cryst. Struct. Commun.*, **5**, 83 (1976).
170. P. Blanpain, M. Melebeck, and F. Durant, *Acta Cryst.*, **B33**, 580 (1977).
171. M. O. Boles, and R. J. Girven, *Acta Cryst.*, **B32**, 2279 (1976).
172. I. Csöregh and T. B. Palm, *Acta Cryst.*, **B33**, 2169 (1977).
173. P. Blanpain, G. Laurent, and F. Durant, *Bull. Soc. Chim. Belg.*, **86**, 767 (1977).
174. P. Blanpain and F. Durant, *Cryst. Struct. Commun.*, **6**, 711 (1977).
175. T. B. Palm and I. Csöregh, *Acta Cryst.*, **B34**, 138 (1978).
176. M. O. Boles, R. J. Girven, and P. A. C. Gane, *Acta Cryst.*, **B34**, 461 (1978).
177. L. M. Lovell and N. A. Perkinson, *Cryst. Struct. Commun.*, **7**, 7 (1978).
178. D. D. Dexter and J. M. van der Veen, *J. Chem. Soc., Perkin II*, **1978**, 185.
179. P. Domiano, M. Nandelli, A. Balsamo, B. Macchia, and F. Mala, *Acta Cryst.*, **B35**, 1363 (1979).
180. D. C. Hodgkin and E. N. Maslen, *Biochem. J.*, **79**, 393 (1961).
181. D. Kobelt and E. F. Paulus, *Acta Cryst.*, **B30**, 1608 (1974).
182. E. F. Paulus, *Acta Cryst.*, **B30**, 2915 (1974).
183. E. F. Paulus, *Acta Cryst.*, **B30**, 2918 (1974).
184. J. P. Declercq, G. Germain, C. Moreaux, and M. Van Meerssche, *Acta Cryst.*, **B33**, 3868 (1977).
185. J. M. Dereppe, J. P. Declercq, G. Germain, and M. Van Meerssche, *Acta Cryst.*, **B33**, 290 (1977).
186. K. Neupert-Laves and M. Dobler, *Cryst. Struct. Commun.*, **6**, 153 (1977).
187. P. Domiano, M. Nardelli, A. Balsamo, B. Macchia, F. Macchia, and G. Mineardi, *J. Chem. Soc., Perkin II*, **1978**, 1082.

188. A. F. Cameron, J. McElhatton, M. M. Campbell, and G. Johnson, *Acta Cryst.,* **B35,** 1263 (1979).
189. M. Van Meerssche, G. Germain, J. P. Declercq, B. Coene, and L. Moreaux, *Cryst. Struct. Commun.,* **8,** 287 (1979).
190. C. Wentrup and H-W. Winter, *J. Amer. Chem. Soc.,* **102,** 6161 (1980).
191. T. Kametani, S-P. Huang, S. Yokohama, Y. Suzuki, and M. Ihara, *J. Amer. Chem. Soc.,* **102,** 2060 (1980).
192. M. Aratani and M. Hashimoto, *J. Amer. Chem. Soc.,* **102,** 6171 (1980).
193. E. M. Gordon, H. W. Chang, C. M. Cimarusti, B. Toeplitz, and J. Z. Gougoutas, *J. Amer. Chem. Soc.,* **102,** 1690 (1980).
194. C. Calvo, P. C. Ip, N. Krishnamachari, and J. Warkentin, *Canad. J. Chem.,* **52,** 2613 (1974).
195. C. J. Fritchie and J. L. Wells, *Chem. Commun.* **1968,** 917.
196. H. Ruben, H. Bates, A. Zalkin, and D. Templeton, *Acta Cryst.,* **B30,** 1631 (1974).
197. K. Prout, V. S. Stothard, and D. J. Watkin, *Acta Cryst.,* **B34,** 2602 (1978).
198. N. Kuhn, W. Schwarz, and A. Schmidt, *Chem. Ber.,* **110,** 1130 (1977).
199. B. Schuckmann, H. Fuess. O. Mosinger, and W. Ried, *Acta Cryst.,* **B34,** 988 (1978).

6 DIPOLE MOMENTS OF HETEROCYCLIC COMPOUNDS

WORTH E. VAUGHAN

*Department of Chemistry
University of Wisconsin
Madison, Wisconsin*

6.1.	Introduction	355
6.2.	Experimental Dipole Moments	356
	6.2.1. Determination of Dipole Moments of Heterocyclic Compounds, 357	
6.3.	Calculated Dipole Moments	358
6.4.	Applications	359
	6.4.1. Dipole Moments in Structure Determination, 360	
	6.4.2. Dipole Moments and Electronic Structure, 362	
	6.4.3. Conformational Analysis, 364	
	6.4.4. Theoretical Calculations of Dipole Moments, 365	
6.5.	Summary	366
References		367

6.1. INTRODUCTION

The determination of dipole moments of organic compounds has a long history. Heterocycles are invariably dipolar (except for a few symmetrical molecules), and the dipole moment is an indicator of structure and especially stereochemistry. Early applications of the dipole moment method were illuminating. For example, the dipole moment of 1,4-dioxane is about 0.4 debye (1 D = 3.33 x 10^{-30} coulombmeter) and hence the molecule exists primarily in the chair conformation. A number of other physicochemical methods have been introduced since for the determination of structure and conformation, and it is often the case that the dipole moment is a minor contributor to the information furnished by a comprehensive study using several physical methods. Other available methods include NMR, microwave spectroscopy, X-ray and electron diffraction, IR-Raman spectroscopy, and photoelectron spectroscopy. The problem is that the dipole moment is a property of the molecule as a whole, whereas we wish to infer from it details about the distribution of electronic charge in space. Despite this shortcoming the number of papers in which dipole moment determinations are a major part is increasing. Furthermore, the application in which dipole moments are most useful is conformational analysis, and heterocycles show particularly interesting and complex conformational effects. Fortunately, access to the voluminous literature is straightforward.

6.2. EXPERIMENTAL DIPOLE MOMENTS

Microwave spectroscopy via the Stark effect yields the dipole moment components along the principal axes with high accuracy. However, the determination of the dipole moment is really an afterthought, as the molecular structure (and conformation) is obtained (if a sufficient number of isotopically substituted species are studied) and the dipole moment is not used in the structure determination (1, 2). The dipole moment components do give an indication of the electron charge distribution. Molecules with large rings and/or flexible side groups yield spectra that are very complex, making line assignments and data reduction difficult and time consuming. Small rings (especially four-membered rings) are amenable to attack, and microwave structure-conformation investigations appear regularly in the literature.

There are beam methods that yield extremely precise dipole moment values, but these methods are restricted mainly to diatomic molecules.

The dipole moment may be found from the amplitude of the anomalous dispersion of the (liquid) dielectric permittivity (3). Measurements of the frequency dependence of the dielectric permittivity are motivated by a desire to characterize the dynamics of orientational and conformational change and are an inefficient path to the dipole moment, which is a static property.

Most dipole moments are obtained from solution measurements of the dielectric constant or from the dielectric constant of the neat liquid. The value of the dipole moment is only as good as the theory that relates it to the measured quantities (dielectric constant, density, index of refraction). The last 0.1 D is hard to pin down, and analysis of the problem of dipole moment determination in liquids continues (4). The factors involved are the molecular shape and the difference in the polarizability of the solute and the solvent. Good descriptions of common working equations are found in the literature (3, 5, 6). The variation in the calculated value of the dipole moment as a function of the working equation does not affect the use of the dipole moment in the structural-conformational analysis of heterocycles.

Dipole moments may be found from the temperature dependence of the orientation polarization in the vapor phase. This has the advantage of a secure formalism but is insensitive, since the amount of polar material in the system is small. Many heterocycles have sufficient vapor pressure to make the method at least possible.

Values of dipole moments are easy to find in the literature. McClellan has prepared two tables (7), and there is a tabulation due to Osipov (8). Compilations appeared yearly in the *Digest of Literature on Dielectrics* through the 1978 literature (9). This series has been discontinued. *Chemical Abstracts* may be searched under the name index entry "dipole moments."

The measurement of the dielectric constant of a (nonconducting) liquid is straightforward and employs a two- or three-terminal cell in conjunction with an audiofrequency bridge or as part of a heterodyne beat circuit (3, 10, 11). Two-terminal cells require calibration of the leads capacitance.

Special methods have been used to derive the dipole moments of electronically excited states. McRae (12) employed perturbation theory to calculate the effects of electric dipole interactions on the frequencies of electronic transitions (both emission and absorption). The frequency shift depends on the index of refraction and the static dielectric constant of the solvent and on the dipole moments of the states involved. If the molecule is considered to occupy an Onsager cavity (sphere, point dipole in the center), the frequency shifts may be related to the difference of the dipole moment in the ground and excited states. This procedure has been used to investigate the electronic structure of heterocycles.

6.2.1. Determination of Dipole Moments of Heterocyclic Compounds

In the vapor phase the Stark effect (electric field dependence of the microwave spectrum) provides a means of obtaining the dipole moment components along the principal axes of the molecule. Generating the relation between frequency shifts and the molecular interaction (especially electric dipole-electric field) with applied (static) electric fields is a considerable exercise in quantum mechanics, because of the complex structures encountered with heterocycles. Systematic developments of the theory are given by Townes and Schawlow (1) and by Wollrab (2). The basic spectrometer design and the underlying electromagnetic theory are described in references 1 and 2. Commercial units are available. The main source of error (apart from assignment of the spectrum) is determination of the electric field strength. Calibration with OCS is usual, as the dipole moment of OCS is known to high accuracy. Extracting dipole moments of heterocycles from the microwave spectrum is time consuming and requires expensive instrumentation.

By far the most direct route to dipole moments is the measurement of the dielectric constant of dilute solutions of the heterocycle in nonpolar solvents. Using a convenient frequency in the audio-radio region (1 kHz to 1 MHz), the dielectric constant is simply the ratio of the capacitance of a cell (leads capacitance subtracted if necessary) filled with the liquid (gas, solid) to that in vacuo. The capacitance is easily measured with bridges or obtained by difference if the cell is placed in parallel with a variable precision capacitor in the tank circuit of a heterodyne beat circuit. Methods for reducing errors in the capacitance determination have undergone extensive development. Excellent bridges are available commercially at relatively low cost. The cell design is an important consideration. Care must be taken to have an accurate description of the electric field distribution in the cell, as this determines how the cell capacitance varies as a function of the dielectric constant of the sample. Three-terminal cells employ a guard electrode, which produces a well-defined field distribution, but two-terminal cells can be used with proper calibration. Reviews of these methods for measuring dielectric constants are found in the literature (10, 3).

The relation between the solute dipole moment and the solution dielectric constant is somewhat subtle. The molar polarization of the solute at infinite dilution is estimated by using limiting forms of the Debye equation for the system polarization in conjunction with measurements of density and/or index of refraction as a function of solute concentration. The distortion polarization of the solute is estimated and the dipole moment calculated from the remaining orientation polarization. The solute is presumed spherical; there is a "solvent effect" even in the absence of specific interactions that are often present with commonly used solvents (carbon tetrachloride, benzene, dioxane). A critical comparative analysis of common procedures has appeared recently (6). Variations of 0.1 D or more in reported dipole moments from solution data are common (7). Fortunately, this accuracy is sufficient for most applications involving heterocycles. The determination of solution dielectric constants and reduction of the data to obtain the dipole moment of the solute is a common experiment in the undergraduate physical chemistry laboratory.

6.3. CALCULATED DIPOLE MOMENTS

If the molecular structure is known or postulated, the dipole moment may be calculated by vector addition of appropriate group or bond moments or by *ab initio* or semiempirical quantum-mechanical methods. Exner (13) and Minkin, Osipov, and Zhdanov (14) have devoted substantial portions of their books to the calculation of dipole moments. The calculation for heterocycles is more difficult than for many other molecules because the introduction of the heteroatom distorts the ring geometry, making it difficult to use model compounds whose structure is known (cyclohexane, for example). The effect of substitution on geometry has been investigated extensively. Riddell (15) gives a short review in his book. Of more concern is the fact that heterocycles often have numerous polar parts whose bond (or group) moments interact by induction and destroy simple vector addition of moments. Atoms with lone pairs require special treatment. Moments of conjugated systems are also nonadditive. Since these interaction effects depend on structure and conformation, they impede the implementation of a method whereby the conformation is deduced by comparison of experimental and calculated dipole moments. Despite the difficulties the vector addition method appears successful in its applications. Exner (13) provides a table of bond and group moments with corrections for mesomeric effects. The method is most successful when restricted to a closely related series of compounds in a single solvent. Some members of the series are used to establish the moments of bonds, groups, and (better) fragments, and this information is used to calculate the moments of the remainder as a function of conformation (structure).

Quantum-mechanical calculations of dipole moments operate on three levels, *ab initio*, full valence electron treatments, and σ-π separation. The situation as of a decade ago was described by Pople and Beveridge (16). With the great advances in computing power one might have expected quantum calculations to displace empirical methods. This appears not to have happened yet. *Ab initio* calculations have been very successful in obtaining the electron distribution and structure of small molecules but have yet to impact seriously on heterocycles. The σ-π separation has given way largely to the more rigorous treatment employing all valence electrons simultaneously. The method of Pople, Santry, and Segal (17) with a particular parameterization (18) (the method is called CNDO/2) has been used extensively on large systems. A number of related procedures have been developed and tested in comparative calculations of dipole moments among other properties. These semiempirical methods appear to predict dipole moments of about the quality of those of the vector addition scheme but with somewhat more effort. Although this is somewhat discouraging, the quality of the quantum calculations is certain to improve whereas the vector addition scheme appears to have stagnated.

6.4. APPLICATIONS

The applications described below are examples of current practice of the use of dipole moments to determine molecular structure, electron distribution, and conformation. They are selective and reflect a current survey of the literature (post-1972). Comprehensive tabulations can be found elsewhere (7–9, 13–15).

An illustration of the use of the dipole moment method in structure determination is given in a recent review of the tautomerism and acid-base properties of tetrazoles (19). Tetrazole (and *C*-deutero- and *N*-deuterotetrazoles) can exist in two tautomeric forms, 1-*H* and 2-*H*, which are in equilibrium with each other (Fig. 6.1). The dipole moments of the 1-*H* and 2-*H* forms computed from two vector addition schemes are 5.26 or 5.16 D and 2.04 or 2.38 D, respectively. *Ab initio* calculations gave 5.17 and 2.54 D. The dipole moments of tetrazole found experimentally in benzene and 1,4-dioxane are 5.10 and 4.96 D. Hence tetrazole in solution exists mainly in the 1-*H* form. NMR evidence in the literature is contradictory on this point. The dipole moments of tetrazole, *C*-deuterotetrazole, and *N*-deuterotetrazole from microwave spectroscopy (20) are 2.19, 5.30, and 2.14 D, respectively. The conclusion is that in the gas phase, tetrazole and *N*-deuterotetrazole exist primarily in the 2-*H* form, whereas *C*-deuterotetrazole is mainly in the 1-*H* form. These results seem both definitive and surprising. The two tautomers are close in energy, and the equilibrium constant appears very sensitive to solvation and (more surprisingly) to isotopic substitution.

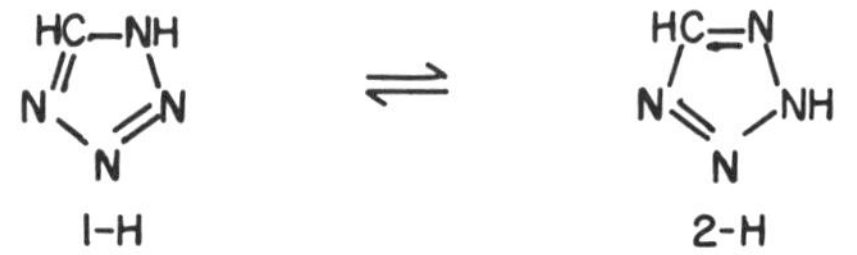

Fig. 6.1. Tautomer equilibrium in tetrazole.

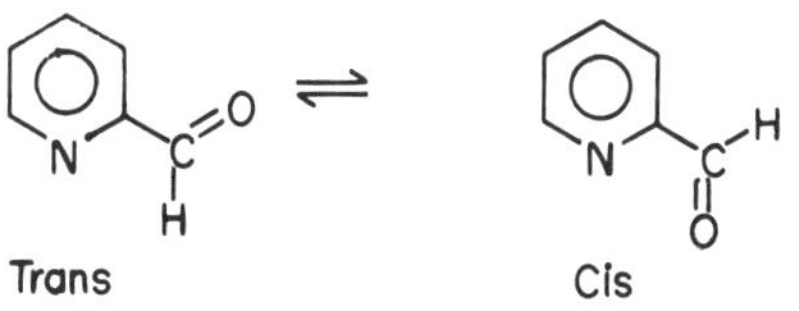

Fig. 6.2. cis-trans isomers of 2-pyridinecarbaldehyde.

Numerous conformational studies have been described (13). Riddell (15) gives a very systematic and complete discussion of heterocycle conformational analysis. However, other physical methods are generally used, and the dipole moment study highlighted is given as an example of the pitfalls inherent in the technique. The dipole moment method can be definitive. The microwave spectrum of 2-pyridinecarbaldehyde was determined by Kawashima and coworkers (21). Molecules of this type sometimes exist as a single conformer and sometimes as a mixture of (planar) *cis* and *trans* forms (Fig. 6.2). NMR results suggested an equilibrium slanted toward the O–N *s-trans* conformer, whereas only one torsional frequency was seen in the far-IR spectrum, suggesting a single conformer. The inertial defect obtained from the microwave spectrum indicated that the molecule is planar, and the rotational constants fit the *s-trans* structure well and the *s-cis* structure poorly. However, the analysis was clouded by the use of a model structure containing uncertain parameters. Dipole moment analysis removed the ambiguity. The dipole moment of 2-pyridinecarbaldehyde is 3.35 D in benzene solution. Vector addition provided estimates of 3.51 D for the *s-trans* conformer and 5.05 D for the *s-cis*. In the vapor Kozima found $\mu_a = 3.48$ D, $\mu_b = 0.76$ D, $\mu = 3.56$ D, $\tan^{-1}(\mu_a/\mu_b) = 12.3°$. Comparison values for the *cis* and *trans* conformers (in the vapor) could be obtained by adding pyridine and benzaldehyde fragment moments. The *s-trans* gave $\mu_a = 3.83$ D, $\mu_b = 0.78$ D, $\tan^{-1}(\mu_a/\mu_b) = 11.5°$, which is in substantial agreement with the observed values, while the *s-cis* gave $\mu_a = 4.47$ D, $\mu_b = 2.84$ D, and $\tan^{-1}(\mu_a/\mu_b) = 32.4°$, which lies well outside the uncertainties in the scheme.

6.4.1. Dipole Moments in Structure Determination

In an exposition of the method, Exner (13) points out that knowledge of the dipole moment is rarely of much use in structure determination. There are exceptions, however, even when microwave spectroscopy has been used. The case of 2-oxazoline (22) is an example. The dipole moment along a principal axis was found to be zero, which indicated that the heavy atom skeleton was planar.

The microwave spectrum was observed for 3,6-dihydro-2H-pyran (23) (Fig. 6.3). This ring system occurs in sugar residues of molecules of biological importance. Since only three rotational constants were derived from the data, it was not possible to do a complete structural determination. A model structure was adopted in which variable degrees of freedom were a ring-bending angle about the C_3–C_6 axis, a ring twist involving C_2–O about the perpendicular axis and the C_5–C_6–O angle. Some other parameters were taken from cyclohexene, and certain linear combinations of parameters were constrained to render the fitting problem determinate.

Fig. 6.3. 3,6-Dihydro-2H-pyran.

To increase the experimental information used in the structure determination, the (conformation-dependent) ratios μ_b/μ_a and μ_c/μ_a were estimated from group moments and included in the least squares refinement. The result was a reasonably definitive structure assignment.

From the fact that ground-state dipole moments were large and excited state dipole moments were small it was concluded that indolinespirobenzopyrans had a zwitterionic ground-state structure and a quinoid structure in the excited state (24).

Dipole moments and dielectric absorption at 100 GHz were found for diphenylene dioxide (I); Fig. 6.4 and analogs [phenoxathiin (II), thianthrene (III), xanthene, and phenothiazine] (25) to resolve a controversy in the literature as to whether diphenylene dioxide is planar or not and to investigate the structure of its analogs. Diphenylene dioxide exhibits a small dipole moment in benzene solution, but uncertainties in the magnitude of the distortion polarization cloud the conclusion that a nonzero moment implies a nonplanar structure. The other molecules have much larger moments, but their moments arise largely from a vector contribution from one heteroatom to the other. Uncertainties in bond moments (plus distortion polarization) preclude a quantitative estimate of the moment that would arise from bending of the plane of the molecule. An exception is thianthrene, whose moment (1.37 D) arises solely from distortion of the molecular geometry. The uncertainties are removed by the observation of dielectric loss at 100 GHz. This is reasonably assigned to a motion in which the bending angle changes and which requires a nonplanar molecular configuration. The magnitude of the dielectric loss permits an estimate of the maximum bend angle, and the relaxation time an estimate of the barrier for inversion.

Fragment dipole moments were used to calculate the dipole moments of four arylidene derivatives that were part of a structural study on α,β-unsaturated ketones (26). The *trans* structure of the conjugated chain was established from the dipole moments. The proton NMR spectra were also reported, but they did not yield definitive information as to structure.

Two possible values for the interfacial angle between the two pyridine nuclei in 2,7-diazathianthrene were found from a comparison of the measured dipole moment and its calculated value based on fragment moment differences (27). UV spectral evidence distinguished between the two geometries.

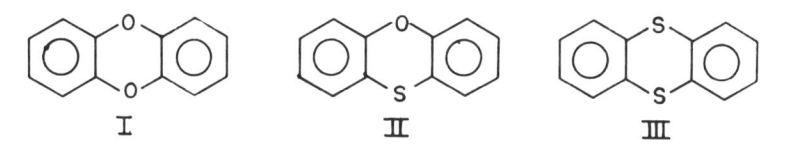

Fig. 6.4. Diphenylene dioxide analogs.

In an investigation aimed at characterizing the mechanism of cycloaddition of isothiocyanates and dicyclohexylcarbodiimide, Exner and coworkers (28) determined the moments of a series of 1:1 cycloadducts substituted in the *para* position of the aromatic moiety. Only one structure of the adduct was consistent with the data, and the stereochemistry of the addition process was inferred from the product structure.

6.4.2. Dipole Moments and Electronic Structure

A number of microwave spectra have been obtained with the aim of characterizing barriers to internal rotation. Dipole moments found as part of the studies yield nonspecific information on the charge distribution in the molecule. Compounds studied were 3-methylthiophene (29, 30), 3-methylfuran (31), and 2-methylpyrimidine (32).

In an effort to determine the distribution of electronic charge, experimental dipole moments have been compared to quantum calculations (mostly at the CNDO level). Agreement between experimental and calculated moments is taken to indicate that the quantum-mechanical calculation is adequate and thereby provide a rationalization of the moment value and a prediction of the electron distribution. For the procedure to be meaningful, the (generally rigid) structure must be known in advance. This method has been used for phenanthrolines (33), diazepines (34), diketopyrimidines (35), cytosine derivatives (36), heterocyclic-diketones (37), acridines (38), and triazines (39).

Many papers, often in the form of series, have systematically attacked the question of the role of heteroatoms and/or substituents in determining electronic structure. The motivation is the generation of a workable vector addition scheme or the identification of specific effects (mesomerism, induction) when the vector scheme fails to reproduce the experimental dipole moments. Two papers on azoles (40, 41) are numbers 24 and 27 in a series. An investigation of thienopyridines (42) is the twentieth of a series.

As part of an investigation employing several physicochemical techniques, the dipole moments of some 1,2-dithiolylium-4-olates and isothiazolium-4-olates were determined (43). Dipole moments of similar compounds have been rationalized in terms of resonance between zwitterionic structures with a negative charge on the oxygen (Fig. 6.5). All these structures have larger estimated moments than those observed for the compounds studied here. The author concludes that a third zwitterionic resonance form in which the charges are found on the carbons adjacent to the C=O group (estimated moment 0.35 D) is a contributor to the electronic structure. Other properties indicate anomalous electronic structure of the C=O group and support the interpretation. However, a totally coherent picture was not found.

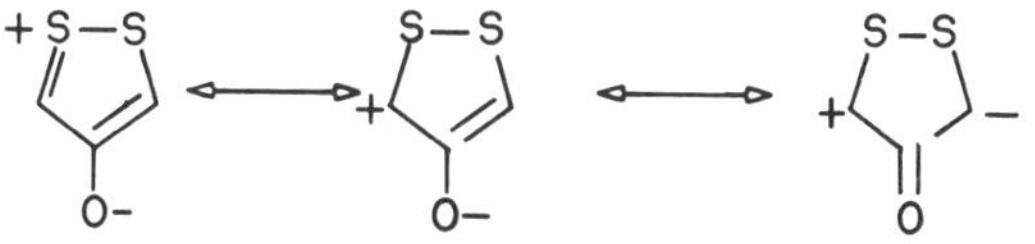

Fig. 6.5.   Resonance forms of 1,2-dithiolylium-4-olate.

Dipole moments of some heterocyclic cyclobutane derivatives were reproduced by vector addition (44).

Vector addition schemes based on bond and group moments require supplementary parameters to be workable. The role of the position of the nitrogen atom or *N*-oxide group in azaphenanthrenes (oxides) was investigated (45). Methyl-substituted pyrones and pyranpyrandiones were analyzed (46). Mesomeric effects were characterized for pyrazines and quinoxalines (47), benzoxadiazoles (thia, selena) (48), fused ring derivatives of thiophene and analogous compounds (49), and substituted benzazoles (50). Substituent effects in 1,2,4-triazoles were systematized (51).

Occasionally the value of the dipole moment itself is informative even without a supporting theoretical (semiempirical) formalism. The temperature dependence of the dipole moments of silatranes was related to the degree of transannular interaction (52). The dipole moments of a series of cyclic sulfides indicated no electronic effects in the main ring upon addition of side saturated rings (53). The dipole moment of 5-cycloheptatrienyliden-2(5H)-thiophene suggested olefinic character (54). Substituent effects on a series of pyrrolidinones, sulfonamides, sulfinamides, and sulfenamides were cataloged (55). A series of six-membered cyclic sulfones indicated the electronic interaction of the sulfonyl group (56).

Dipole moments have been used extensively to identify the predominant tautomeric form of a molecule or, in favorable cases, to characterize the tautomer equilibrium. Examples of the former are substituted benzoazolinethiones (57), benzotriazole (58), benzoxazolinones (59), derivatives of *s*-triazol(4,3-*b*)-*s*-triazole (60), imidazolinin-2(3H)-ones and related compounds (61), acetylaminopyrimidines (62), nitro derivatives of indazole (63), imidazo(4,5-*b*)pyridine (64), 5-*p*-tolyltetrazole (65), and tetrazole (20).

As part of a study to identify synthetic routes to five-membered heterocycles, the dipole moment of 2,4,6-tri-*tert*-butyl-7,8,9-dithiazabicyclo[4.3.0]nona-2,4,9-triene was derived (66). This molecule exists in equilibrium with the open ring structure, and the equilibrium constant is known from the NMR spectrum. The measured (squared) dipole moment of the equilibrium mixture was used to estimate the dipole moment of the heterocycle by the expedient of setting the dipole moment of the open-ring compound equal to that of its 6-methyl derivative in which ring closure is prevented. That the dipole moment of the ring compound (3.0 D) is greater than that of the open-ring compound (1.5 D) is confirmed by (polar) solvent effects on the equilibrium as observed in NMR spectra.

Equilibria were found in 4-pyridone derivatives (67), derivatives of 4-phenylthiazole (68), azoles and benzazoles (69), and substituted pyrimidines (70). An entire volume has been devoted to the tautomerism of heterocycles as studied by the entire range of physical and chemical methods (71).

Exner (13) has reviewed some techniques used to obtain dipole moments in electronically excited states. Although containing some approximations difficult to evaluate, the method of solvent shifts (12) has been a popular choice.

The method of solvent effects on absorption spectra was used to estimate dipole moments of excited states (singlet π-π^*) of some 1,3,5-triazaphosphorines

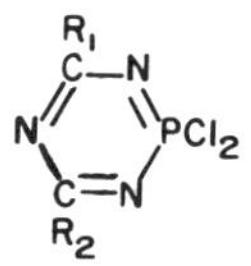

Fig. 6.6. Substituted 1,3,5,2λ^5-triazaphosphorins.

(72) (Fig. 6.6). The motivation was to characterize changes in conjugation in the nitrogen-phosphorus-nitrogen fragment. Since the phosphorus 3d orbitals are involved in the first three singlet π-π^* transitions, excited state dipole moments should be informative. The experimental study was supported by theoretical calculations in the Pariser-Parr-Pople approximation. The authors concluded that excitation to the first singlet state is accompanied by a displacement of π-electron density into the triazene fragment.

Excited state dipole moments have recently been reported for 1-methyl-2-pyridone (73) and for substituted 1,3-diphenyl-2-pyrazolines (74). Additional values can be found in the tabulations (7-9).

The dipole moments of singlet and triplet excited states of 1,3-diazaazulene were derived from the pseudo-Stark effect of molecular crystals on diazaazulene in a naphthalene host (75).

6.4.3. Conformational Analysis

The determination of conformation is the most effective current application of dipole moments. A classification of patterns of the conformational problem and a series of applications have been given by Exner (13). A very thorough discussion of the conformational analysis of heterocyclic compounds in which dipole moments play a supporting role is due to Riddell (15). We have attempted to avoid duplication with these monographs, which are excellent sources of information. If the conformational problem extends beyond the identification of a single structure or the estimation of the equilibrium constant between two conformers, the dipole moment method becomes uninformative. This has not prevented more ambitious studies, but the results are generally ambiguous. We confine our review to recent studies where the conformational problem is well defined.

Cis or *trans* isomers were identified from the dipole moments for heterocyclic aldonitrones (76), substituted *N*-alkylpyrroles (77), perhydropyridooxazines (78), 2-aryl-3-aroylaziridines (79), 2-benzoylthiophenes (80), phosphorinane derivatives (81), and 2-pyridinecarbaldehyde (21). *Syn* conformers were observed for *N*-allylphthalimide oxide and sulfide (82) and for 3,4-dimethylbicyclohep-3-eneoxides (83). Anhydrides derived from 3,6-epoxytricycloundecene had the *exo* configuration of the anhydride (or imide) ring (84). Axial or equatorial substituent orientations were deduced for tetrahydro-1,2-oxazines and related acyclic hydroxylamines (85), substituted oxazines and benzoxazines (86), and phenyl diazaphosphorinanes (87).

To establish the conformations of a series of *P*-substituted 5,5-dimethyl-2-oxo-1,3,2-dioxaphosphorinanes (88), dipole moments were obtained as part of a multi-

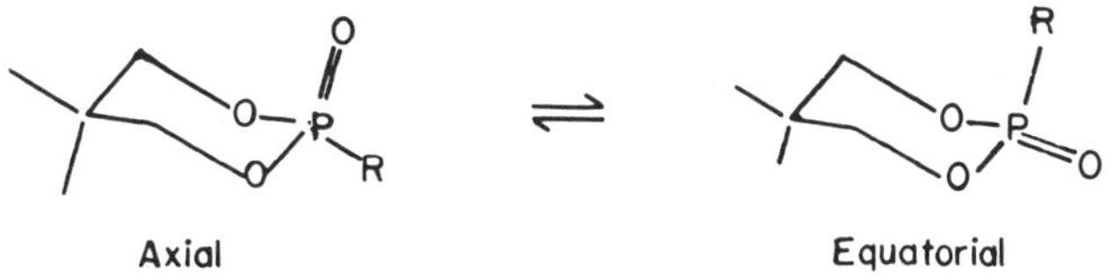

Fig. 6.7. Conformations of 2-oxo-1,3,2-dioxaphosphorinane.

technique study. CNDO/2 calculations indicated a strong preference for P=O equatorial except for the bulky *tert*-butyl and dimethylamino substituents, which were predicted to have P=O axial (gas-phase) (Fig. 6.7). These predictions corresponded to IR and NMR (solution) results. Dipole moments for the molecules with bulky substituents were in accord with values calculated for the axial form and not with values for the equatorial form. Dipole moments for the other compounds gave results that were in poor agreement with either the equatorial or axial form. However, the discrepancies were systematic compared to the equatorial series and were rationalized as an error in calculating the electron distribution on phosphorus.

Dihedral angles were determined in cyclic esters of sulfur, selenium, and arsenic (89), in *N*-arylsuccinimides and *N*-arylphthalimides (90), and in some phenylbenzimidazoles (91).

As part of a continuing study on *s-cis-trans* conformational equilibria in five-membered heterocycles (92), the dipole moments and Kerr constants of *N*-substituted pyrrole-2-carboxaldehydes were determined. Without a substituent, intramolecular hydrogen bonding causes a planar *O,N-cis* structure. The *N*-methyl compound is 80% *cis*, which is consistent with NMR data. When phenyl (heteryl) substituents are introduced, the *trans* form was found exclusively and the phenyl group was perpendicular to the pyrrole ring. This was interpreted as a stronger conjugation of the pyrrole ring with the carbonyl as opposed to the phenyl group. Clearly, steric effects have a strong influence on the geometry.

Equilibria between *cis* and *trans* conformers were characterized for thiophenecarboxaldehydes (93), 2-furaldehyde (94), diformylpyridines (95), and carbonyl derivatives of isozazole (96). Axial-equatorial equilibria were identified in 3-alkytetrahydro-1,3-thiazines (97), *N*-substituted *trans*-decahydroquinazolines (98), ketals of α-halocyclohexanones (99), and some substituted phenothiazines (100). Mixtures of axial and equatorial half-chair conformers were found for 1,2,3,6-tetrahydropyridine (101). Mixtures of chair and twist conformers were observed on some 1,3-dioxa-5-cycloheptenes (102) and 2-arylbenzo-1,3-dioxepanes (103).

As can be seen from the references, Osipov and Katritzky and their coworkers have contributed many papers in the area of conformational analysis by the dipole moment method. Many of these papers are in serial form, and each series is readily traced backwards from the latest citation.

6.4.4. Theoretical Calculations of Dipole Moments

A system that has attracted *ab initio* efforts recently is pyridine and its 2- and 4-monosubstituted derivatives (especially the hydroxy derivative). With the motivation of using the hydroxypyridines as model systems for the tautomerism of biological purines and pyrimidines, Kwiatowski (104) computed the dipole moments of pyridine and the lactim and lactam forms of the hydroxy derivatives. Reasonable agreement between the calculated and experimental moments was found (the tautomers were methylated to preserve the structure). Del Bene (105) calculated the dipole moments of a number of 2- and 4-substituted pyridines as part of a study to systematize trends in equilibrium structures, relative stabilities, and structural effects on electron distributions. Her dipole moments for the (lactim) hydroxy derivatives are in fair agreement with Kwiatkowski's.

CNDO/2 calculations of dipole moments continue to appear in the literature. The dipole moments of perimidine and naphthimidazoles were derived from the computed electron distribution and compared with experiment (106). This application was a strong test of the method as the perimidine system exhibits unusual electronic structure. The dipole moments of a series of uracil analogs were determined and correlated with the biological activity of the molecules. The most stable tautomer of 6-aminothymine was found by calculating the total energies of the 17 possible forms (107).

Deumié, Viallet, and Chalvet (108) employed the CNDO/S parameterization, which was developed for spectroscopic calculations (109), to obtain the electronic transition energies of carbazoles and azacarbazoles. Comparative dipoles in the S_0 state were obtained, and the results were consistent with experimental determinations of solvent shifts from which dipole moment changes upon excitation were derived (12).

Dewar and coworkers (110) carried out MNDO and MINDO/3 calculations on a series of fluoropyridines (which exhibit striking structural changes upon fluorine substitution) with the principal motivation of testing the ability of these procedures to reproduce the structures and dipole moments as found from microwave spectroscopy. MNDO was found to be the superior method for dipole moments, and this was attributed to a better treatment of the formal negative charge on the nitrogen. In an addendum to a comparative evaluation of current molecular orbital methods, Dewar and Ford (111) applied MNDO and MINDO/3 to a set of test data. These methods were superior to CNDO/2 for the calculation of dipole moments.

MINDO/3 was used to calculate the dipole moments of tautomeric forms of pyrimidine bases (112). Fair agreement with experimental dipole moments was found. A similar analysis on pyrrole derivatives was carried out (113).

Several new SCF MO methods have recently been used to calculate dipole moments (114–118).

6.5. SUMMARY

This chapter describes various uses of dipole moment measurements on heterocycles. References to recent work citing applications to structure determination, electron distribution, and conformational analysis are given. The methodology is not described in detail, as this information is available in depth in the general references. The papers cited are a selected subset of a larger class assembled from the tabular sources given in the references. The literature prior to 1971 has been selectively reviewed by Kraft and Walker (119, 120). There is also a review of the microwave spectroscopy of heterocycles (121). Minkin (122) has written a review with 233 references on applications of dipole moments to the stereochemistry of organic compounds. The dipole moments of heterocycles have been put to other uses, notably intermolecular hydrogen bonding and complex formation. These applications are generally oblique and are not discussed here. We hope this chapter will assist the reader in accessing the voluminous literature in the area of dipole moments of heterocyclic compounds.

REFERENCES

1. C. H. Townes and A. L. Schawlow, *Microwave Spectroscopy*, McGraw-Hill, New York, 1955.
2. J. E. Wollrab, *Rotational Spectra and Molecular Structure*, Academic Press, New York, 1967.
3. N. Hill, A. H. Price, W. E. Vaughan, and M. Davies, *Dielectric Properties and Molecular Behaviour*, Van Nostrand-Reinhold, London, 1969.
4. A. B. Myers and R. R. Birge, *J. Chem. Phys.*, 74, 3514 (1981).
5. C. J. F. Böttcher, O. C. Van Belle, P. Bordewijk, and A. Rip, *Theory of Electric Polarization*, 2nd ed., Vol. 1, *Dielectrics in Static Fields*, Elsevier, Amsterdam, 1973.
6. C. J. F. Böttcher and P. Bordewijk, *Theory of Electric Polarization*, 2nd ed., Vol. 2, *Dielectrics in Time Dependent Fields*, Elsevier, Amsterdam, 1978.
7. A. L. McClellan, *Tables of Experimental Dipole Moments*, Freeman, San Francisco, 1963; Vol. 2, Rahara Enterprises, El Cerrito, CA, 1974.
8. O. A. Osipov, V. I. Minkin, and A. D. Garnovoskit, *Spravochnik po dipolnym momentum*, 3rd ed., Vysshaya Shkola, Moscow, 1971.
9. For example, *Digest of Literature on Dielectrics*, Vol. 42, National Academy of Sciences, Washington, D.C., 1980. This is the final volume in the series (covering the 1978 literature).
10. C. P. Smyth, "Determination of Dipole Moments," in A. Weissberger and B. W. Rossiter, Eds., *Physical Methods of Chemistry* (*Techniques of Chemistry*, Vol. 1), Part 4, Wiley-Interscience, New York, 1972.
11. W. E. Vaughan, C. P. Smyth, and J. G. Powles, "Determination of Dielectric Constant and Loss," in A. Weissberger and B. W. Rossiter, Eds., *Physical Methods of Chemistry* (*Techniques of Chemistry*, Vol. 1), Part 4, Wiley-Interscience, New York, 1972.
12. E. G. McRae, *J. Phys. Chem.*, 61, 562 (1957).
13. O. Exner, *Dipole Moments in Organic Chemistry*, George Thieme, Stuttgart, 1975, Ch. 3.
14. V. I. Minkin, O. A. Osipov, and Yu. A. Zhdanov, *Dipole Moments in Organic Chemistry*, Plenum, New York, 1970, Ch. 3.
15. F. G. Riddell, *The Conformational Analysis of Heterocyclic Compounds*, Academic Press, London, 1980.
16. J. A. Pople and D. L. Beveridge, *Approximate Molecular Orbital Theory*, McGraw-Hill, New York, 1970.
17. J. A. Pople, D. P. Santry, and G. A. Segal, *J. Chem. Phys.*, 43, 5129 (1965).

18. J. A. Pople and G. A. Segal, *J. Chem. Phys.*, **44**, 3289 (1966).
19. G. I. Koldobskii, V. A. Ostravskii, and B. V. Gidasop, *Khim. Geterotsikl. Soedin.*, **7**, 867 (1980).
20. W. D. Krugh and L. P. Gold, *J. Mol. Spectrosc.*, **49**, 423 (1974).
21. Y. Kawashima, M. Suzuki, and K. Kozima, *Bull. Chem. Soc. Japan*, **48**, 2009 (1975).
22. J. R. Durig, S. Riethmiller, and Y. S. Li, *J. Chem. Phys.*, **61**, 253 (1974).
23. J. A. Wells and T. B. Thomas, Jr., *J. Chem. Phys.*, **60**, 3987 (1974).
24. D. Lapienis-Grochowska, M. Kryszewski, and B. Nadolski, *J. Chem. Soc., Faraday Trans. 2*, **75**, 312 (1979).
25. Y. Kogg, H. Takahashi, and K. Higasi, *Bull. Chem. Soc. Japan*, **46**, 3359 (1973).
26. V. D. Orlov, I. A. Borovoi, and V. F. Lavrushin, *Zh. Struk. Khim.*, **17**, 691 (1976).
27. K. Krowicki and P. Nantka-Namirski, *Rocz. Chem.*, **51**, 2435 (1977).
28. O. Exner, V. Jehlicka, and A. Dondoni, *Coll. Czech. Chem. Commun.*, **41**, 562 (1976).
29. T. Ogata and K. Kozima, *J. Mol. Spectrosc.*, **42**, 38 (1972).
30. L. N. Gunderova, A. A. Shapkin, and N. M. Pozdeev, *Opt. Spektrosk.*, **34**, 1211 (1973).
31. T. Ogata and K. Kozima, *Bull. Chem. Soc. Japan*, **44**, 2344 (1971).
32. W. Caminati, G. Cazzoli, and D. Troiano, *Chem. Phys. Lett.*, **43**, 65 (1976).
33. H. H. Perkampus, P. Mueller, and J. V. Knop, *Z. Naturforsch. B*, **26**, 83 (1971).
34. V. I. Minkin, I. D. Sadekov, L. L. Popov, and Yu. V. Kalodyazhnyi, *Zh. Obshch. Khim.*, **40**, 1865 (1970).
35. I. Kulakowska, M. Geller, B. Lesyng, and K. L. Weirzchowski, *Biochem. Biophys. Acta*, **361**, 119 (1974).
36. I. Kulakowska, M. Geller, B. Lesyng, and K. L. Weirzchowski, *Biochim. Biophys. Acta*, **407**, 420 (1975).
37. V. Galasso and G. C. Pappalardo, *J. Chem. Soc. Perkin Trans. 2*, **1976**, 574.
38. A. M. Galy, J. P. Galy, R. Faure, and J. Barbe, *Eur. J. Med. Chem.-Chim. Ther.*, **15**, 179 (1980).
39. G. Vernin, M. Meyer, L. Bouscasse, J. Metzger, and C. Parkanyi, *J. Mol. Struct.*, **68**, 209 (1980).
40. S. B. Bulgarevich, V. S. Bolotnikov, V. N. Sheinkes, O. A. Osipov, and A. D. Garnovskii, *Zh. Org. Khim.*, **12**, 197 (1976).
41. V. S. Bolotnikov, T. V. Lifintseva, S. B. Bulgarevich, V. N. Sheinkes, O. A. Osipov, and A. D. Garnovskii, *Zh. Org. Khim.*, **12**, 416 (1976).
42. L. H. Klemm and R. D. Jacquot, *J. Heterocycl. Chem.*, **12**, 615 (1975).
43. D. Barillier, *Phosphorus Sulfur*, **8**, 79 (1980).
44. M. Sanesi, *Z. Naturforsch. A*, **34**, 257 (1979).
45. A. Susharda-Sobczyk, L. Sobczyk, J. Mlochowski, and A. Koll, *Rocz. Chem.*, **48**, 1265 (1974).
46. H. H. Huang, S. F. Jan, and T. H. Tjia, *J. Heterocycl. Chem.*, **13**, 609 (1976).
47. H. Lumbroso, J. Curé, T. Konakahara, and Y. Takagi, *J. Mol. Struct.*, **68**, 293 (1980).
48. F. L. Tobiason, L. Huestis, C. Chandler, S. E. Pedersen, and P. Peters, *J. Heterocycl. Chem.*, **10**, 773 (1973).
49. M. G. Gruntfest, Yu. V. Kolodyazhni, V. Udre, M. G. Voronkov, and O. A. Osipov, *Khim. Geterosikl. Soedin.*, **6**, 448 (1970).
50. A. M. Galy, J. Llinares, J. P. Galy, and J. Barbe, *Compt. rend. C*, **287**, 459 (1978).
51. M. A. Pervosvanskaya, V. V. Mel'nikov, M. S. Pevzner, and B. V. Gidaspov, *Zh. Org. Khim.*, **11**, 1974 (1975).
52. P. Hencsei, G. Zsombok, L. Bihatsi, and J. Nagy, *Ric. Clin. Lab.*, **9**, 185 (1979).

53. E. N. Kharlamova, E. N. Gur'yanova, and V. G. Kharchenko, *Zh. Strukt. Khim.*, **12**, 637 (1971).
54. K. Takahashi, T. Sakae, and K. Takase, *Chem. Lett.*, **1980**, 179.
55. P. Ruostesuo, *Acta Univ. Oulunsis, A*, **1978**, 66.
56. V. V. Puchkova, E. N. Gur'yanova, V. G. Kharchenko, and A. A. Rassudova, *Zh. Org. Khim.*, **9**, 1531 (1973).
57. V. A. Granzhan, N. L. Poznanskaya, N. J. Shvetsov-Shilovski, and S. K. Laktionova, *Zh. Fiz. Khim.*, **48**, 2099 (1974).
58. P. Mauret, J. P. Fayet, M. Fabre, J. Elguero, and J. DeMendoza, *J. Chim. Phys. Physicochim. Biol.*, **71**, 115 (1974).
59. V. A. Granzhan, N. A. Poznanskaya, N. I. Shetsov-Shilivskii, S. K. Laktionova, and M. I. Koleskik, *Zh. Fiz. Khim.*, **45**, 1 (1971).
60. R. M. Claramunt, J. P. Fayet, M. C. Vertut, P. Mauret, and J. Elguero, *Tetrahedron*, **31**, 545 (1975).
61. C. W. N. Cumper and G. D. Pickering, *J. Chem. Soc., Perkin Trans. 2*, **1972**, 2045.
62. M. T. Mussetta, M. Selim, and N. Q. Trinh, *Compt. Rend. C*, **277**, 1279 (1973).
63. M. A. Pervozvanskaya, M. S. Pevzner, L. N. Gribanova, V. V. Mel'Nikov, and B. V. Gidaspov, *Khim. Geterosikl. Soedin.*, **1977**, 1669.
64. Yu. M. Yutilov, N. R. Kal'nitskii, and R. M. Bystova, *Khim. Geterotsikl. Soedin.*, **7**, 1436 (1971).
65. H. Lumbroso, J. Cure, and R. N. Butler, *J. Heterocycl. Chem.*, **17**, 1373 (1980).
66. Y. Inagaki, R. Okazaki, and N. Inamoto, *Heterocycles*, **9**, 1613 (1978).
67. B. D. Batts and A. J. Madeley, *Aust. J. Chem.*, **25**, 2605 (1972).
68. M. T. Mussetta, M. M. Selim, and N. Q. Trim, *C. R. Hebd. Seances Acad. Sci. C*, **281**, 3 (1975).
69. P. Mauret, J. P. Fayet, and M. Fabre, *Bull. Soc. Chim. France*, **1975**, 1675.
70. M. T. Musetta, M. Selim, and N. Q. Trinh, *Compt. Rend. C*, **276**, 1341 (1973).
71. J. Elguero, C. Marzin, A. R. Katritzky, and P. Linda, "The Tautomerism of Heterocycles," in A. R. Katritzky and A. J. Boulton, Eds., *Advances in Heterocyclic Chemistry*, Suppl. 1., Academic, New York, 1976.
72. E. A. Romanenki, S. V. Iksanova, and Yu. P. Egorov, *Teor. Eksp. Khim.*, **16**, 308 (1980).
73. A. Fujimoto and K. Inuzuka, *Bull. Chem. Soc. Japan*, **52**, 1816 (1979).
74. H. Gusten, G. Heinrich, and H. Fruhbeis, *Ber. Bunsenges. Phys. Chem.*, **81**, 810 (1977).
75. J. R. Braun, T. S. Lin, F. P. Burke, and G. J. Small, *J. Chem. Phys.*, **59**, 3595 (1973).
76. E. G. Merinova, V. N. Sheinker, O. A. Osipov, and V. I. Piven', *Zh. Obshch. Khim.*, **46**, 1191 (1976).
77. D. M. Bertin, G. Guiberry, C. Liegeois, and H. Lumbroso, *Bull. Soc. Chim. France*, **1976**, 1393.
78. I. D. Blackburne, A. R. Katritzky, D. M. Read, P. J. Chivers, and T. A. Crabb, *J. Chem. Soc. Perkin Trans. 2*, **1976**, 418.
79. V. D. Orlov, F. G. Yaremenko, Yu. N. Surov, and V. F. Lavrushin, *Dopov. Akad. Nauk Ukr. RSR, Ser. B: Geol., Khim Biol. Nauki*, **1979**, 931.
80. V. N. Sheinker, A. S. Kuzharov, V. F. Lavrushin, N. F. Pedchenko, and O. A. Osipov, *Khim. Geterotsikl. Soedin.*, **10**, 1327 (1979).
81. I. J. Patsanovskii, E. A. Ishmaeva, A. P. Logunov, Yu. G. Bosyakov, B. M. Butin, S. K. Shishkin, and A. N. Pudovik, *Zh. Obshch. Khim.*, **50**, 527 (1980).
82. B. A. Arbuzov, L. K. Aleksandrova, V. A. Mullin, A. N. Zolotov, S. G. Vul'fson, and A. N. Vereshchagin, *Izv. Acad. Nauk SSSR, Ser. Khim.*, **1979**, 664.

83. Z. G. Isaeva, G. Sh. Bikbulatova, and G. J. Kovylyaeva, *Izv. Akad. Nauk SSSR, Ser. Khim.*, **1980**, 1443.
84. N. A. Alekperov, R. D. Mishiev, and M. S. Salakhov, *Zh. Org. Khim.*, **16**, 770 (1980).
85. R. A. Y. Jones, A. R. Katritzky, S. Saba, and A. J. Sparrow, *J. Chem. Soc. Perkin Trans. 2*, **13**, 1554 (1974).
86. R. A. Y. Jones, A. R. Katritzky, and S. Saba, *J. Chem. Soc. Perkin Trans.*, 2, **14**, 1737 (1974).
87. B. A. Arbuzov, O. A. Erastov, G. N. Nikonov, T. A. Zyablikova, P. R. Arshinova, and R. A. Kadyrov, *Izv. Akad. Nauk SSSR, Ser. Khim.*, **1980**, 1571.
88. K. Faegri, Jr., T. Gramstad, and K. Tjessem, *J. Mol. Struct.*, **32**, 37 (1976).
89. B. A. Arbuzov, A. N. Vereshchagin, I. V. Anonimova, E. G. Yarkova, N. A. Chadaeva, and T. A. Lavrova, *Izv. Akad. Nauk SSSR, Ser. Khim.*, **1980**, 584.
90. B. A. Arbuzov, L. K. Aleksandrova, S. G. Vul'fson, and A. N. Vereshchagin, *Izv. Akad. Nauk SSSR, Ser. Khim.*, **1979**, 661.
91. A. S. Kuzharov, P. V. Tkachenko, I. I. Popov, and V. N. Sheinker, *Zh. Fiz. Khim.*, **52**, 2708 (1978).
92. A. S. Kuzharov, V. N. Shreinker, I. A. Kharizomenova, V. I. Shvedov, O. A. Osipov, and A. N. Grinev, *Zh. Obshch. Khim.*, **46**, 2570 (1976).
93. A. S. Kuzharov, V. N. Shreinker, E. G. Derecha, O. A. Osipov, and D. Ya. Movshovich, *Zh. Obshch. Khim.*, **44**, 2008 (1974).
94. V. N. Sheinker, A. S. Kuzharov, S. B. Bulgarevich, E. G. Derecha, O. A. Osipov, and V. I. Minkin, *Zh. Obshch. Khim.*, **44**, 175 (1974).
95. H. Lumbroso, D. M. Bertin, and G. C. Pappalardo, *J. Mol. Struct.*, **37**, 127 (1977).
96. V. N. Sheinker, T. V. Lifintseva, S. B. Bulgarevich, S. M. Vinogradova, S. D. Sokolov, A. D. Garnovskii, and O. A. Osipov, *Khim. Geterotsikl. Soedin.*, **9**, 1189 (1979).
97. L. Angiolini, A. R. Katritizky, and D. M. Read, *Gazzetta*, **106**, 111 (1976).
98. W. L. F. Armarego, R. A. Y. Jones, A. R. Katritzky, D. M. Read, and R. Scattergood, *Aust. J. Chem.*, **28**, 2323 (1975).
99. B. A. Arbuzov, E. N. Klimovitskii, and M. B. Timirbaev, *Izv. Akad. Nauk SSSR, Ser. Khim.*, **1980**, 1799.
100. C. Levayer, A. M. Galy, and J. Barbe, *Compt. Rend., Ser. C*, **287**, 569 (1978).
101. S. Chao, T. K. Avirah, R. L. Cook, and T. B. Malloy, Jr., *J. Phys. Chem.*, **80**, 1141 (1976).
102. B. A. Arbuzov, E. N. Klimovitskii, A. B. Remizov, V. V. Klochkov, A. V. Aganov, and M. B. Timirbaev, *Izv. Akad. Nauk SSSR, Ser. Khim.*, **1980**, 1794.
103. B. A. Arbuzov, E. N. Klimovitskii, A. B. Remizov, and G. N. Sergaeva, *Izv. Akad. Nauk SSSR, Ser. Khim.*, **1979**, 2031.
104. J. S. Kwiatkowski, *Acta Phys. Pol. A*, **55**, 923 (1979).
105. J. E. Del Bene, *J. Amer. Chem. Soc.*, **101**, 6184 (1979).
106. A. F. Pozharskii, E. N. Malysheva, A. N. Suslov, and L. L. Popova, *Khim. Geterotsikl. Soedin.*, **1979**, 692.
107. R. Hintsche, H. Sklenar, B. Preussel, D. Baerwolff, and J. Jaeger, *Abh. Akad. Wiss. DDR, Abt. Math., Naturwiss., Tech.*, **1978**, 221.
108. M. Deumié, P. Viallet, and O. Chalvet, *J. Photochem.*, **10**, 365 (1979).
109. J. Del Bene and H. H. Jaffé, *J. Chem. Phys.*, **50**, 1126 (1969).
110. M. J. S. Dewar, Y. Yamaguchi, S. Doraiswamy, S. D. Sharma, and S. H. Suck, *Chem. Phys.*, **41**, 21 (1979).
111. M. J. S. Dewar and G. P. Ford, *J. Amer. Chem. Soc.*, **101**, 5558 (1979).
112. R. Czermiński, B. Lesyng, and A. Pohorille, *Int. J. Quantum Mech.*, **16**, 605 (1979).

113. A. Karpten, P. Schuster, and H. Berner, *J. Org. Chem.*, **44**, 374 (1979).
114. R. M. Minyaev and V. I. Minkin, *Teor. Eksp. Khim.*, **16**, 659 (1980).
115. T. Kozar and I. Tvaroska, *Theor. Chim. Acta*, **53**, 9 (1979).
116. D. N. Nanda and K. Jug, *Theor. Chim. Acta*, **57**, 95 (1980).
117. K. Jug and D. N. Nanda, *Theor. Chim. Acta*, **57**, 107 (1980).
118. K. Jug and D. N. Nanda, *Theor. Chim. Acta*, **57**, 131 (1980).
119. J. Kraft and S. Walker, "Recent Applications of Electric Dipole Moments to Heterocyclic Systems," in A. R. Katritzky, Ed., *Physical Methods in Heterocyclic Chemistry*, Vol. 4, Academic Press, New York, 1971.
120. S. Walker, "Application of Dipole Moments to Heterocyclic Systems," in A. R. Katritzky, Ed., *Physical Methods in Heterocyclic Chemistry*, Vol. 1, Academic Press, New York, 1963.
121. J. Sheridan, "Microwave Spectroscopy of Heterocyclic Molecules," in A. R. Katritzky, Ed., *Physical Methods in Heterocyclic Chemistry*, Vol. 6, Academic Press, New York, 1974.
122. V. I. Minkin, *Stereochem., Fundam. Methods*, **2**, 1 (1977).

7 RAMAN SPECTROSCOPY OF HETEROCYCLIC COMPOUNDS

HELEN C. MACKIN, ELLEN A. KERR, AND NAI-TENG YU

School of Chemistry
Georgia Institute of Technology
Atlanta, Georgia

7.1.	Introduction	374
7.2.	Theory and Principles	375
	7.2.1. Classical Description, 375	
	7.2.2. Quantum Description, 377	
7.3.	Experimental Techniques	380
	7.3.1. Multichannel Detectors: Intensified Vidicon/Reticon, 380	
	7.3.2. Excitation Sources: Lasers, 383	
	7.3.3. A Multichannel Laser Raman System, 384	
	7.3.4. Measurement of the Depolarization Ratio, 387	
	7.3.5. Intensity Correction for Self-Adsorption: 90° versus 180° Geometry, 387	
	7.3.6. Sample-Handling Techniques, 389	
	7.3.7. Discrimination against Fluorescence: Nonlinear CARS and RIKES Techniques, 390	
	7.3.7.1. Coherent Anti-Stokes Raman Scattering (CARS), 390	
	7.3.7.2. Raman-Induced Kerr Effect Spectroscopy (RIKES), 392	
	7.3.8. Surface-Enhanced Raman Scattering (SERS), 393	
	7.3.9. High-Precision Raman Difference Spectroscopic (RDS) Technique for the Measurement of Very Small Frequency Shifts, 394	
	7.3.10. Raman Microprobe Analysis, 395	
7.4.	Applications	397
	7.4.1. Metalloporphyrins and Hemoproteins, 397	
	7.4.1.1. Axial Ligand-Associated Vibrations, 397	
	7.4.1.2. A Spectroscopic Ruler for Measuring the Size of the Heme Central Hole, 404	
	7.4.1.3 Time-Resolved Resonance Raman Studies, 406	
	7.4.2. Dioxygen Binding, Non-Heme Heterocyclic Compounds, 407	
	7.4.3. Chlorophylls, 409	
	7.4.4. Flavins and Flavoproteins, 410	
	7.4.5. Heterocyclic Enzyme-Substrate/Inhibitor Complexes, 412

7.4.6. Nucleic Acid Derivatives, 413
    7.4.6.1. Kinetics of Hydrogen-Deuterium Exchange in Purines, 413
    7.4.6.2. Ultraviolet Resonance Raman Effects, 415
    7.4.6.3. Metal Ion-Nucleotide Interactions, 416
7.4.7. Heterocyclic Free Radicals, 418

References   419

7.1. INTRODUCTION

The study of molecular vibrations by infrared and Raman techniques has played an important role in structural investigations of heterocyclic compounds. Infrared spectroscopy involves direct absorption of IR radiation at frequencies appropriate for molecular vibrations (see Balaban and coworkers, Chapter 1). Raman spectroscopy, on the other hand, employs a visible or UV laser (single-wavelength) beam as the excitation source and analyzes the light scattered from the samples. The great virtue of Raman scattering is that the excitation wavelength can be deliberately tuned to match an electronic transition to achieve selective enhancement (10^3–10^5) of certain vibrational modes; thus, information pertaining to vibronic couplings and excited state geometry may be obtained via resonance Raman excitation profiles (plots of Raman line intensities versus excitation wavenumber) (1–4). This is particularly important when one is interested in the structural behavior of heterocyclic molecules (e.g., heme, flavin, and enzyme substrate/inhibitor) embedded in a biological matrix such as a protein. The Raman scattering method is flexible and versatile with respect to the types of samples that can be studied (5, 6). Of particular significance is its ability to obtain structural information from dilute aqueous solutions (10^{-3}–$10^{-6}M$), which is difficult, if not impossible, by IR spectroscopy. Because of the small size of a focused laser beam (1–30 μm), Raman spectra of a sample as small as several picograms have been obtained with a microprobe Raman technique (7, 8).

The Raman effect was discovered by the Nobel laureate C. V. Raman in 1928. But only in the past few years has Raman spectroscopy become such a powerful and dynamic structural probe, thanks largely to the advances in laser technology and optical/electronic devices. The recent introduction of highly sensitive multichannel detectors (intensified vidicon/Reticon) (9–16) has helped to set the stage for a new cycle of Raman spectroscopy. It is now possible to record resonance Raman spectra of short-lived reaction intermediates even in the picosecond region (17). The enhanced detection capability made possible by multichannel detectors promises to generate exciting new structural information from biologically active heterocyclic molecules.

The purpose of this chapter is to review the important advances that have been made during the past decade in the applications of Raman spectroscopy to chemical and biochemical aspects of heterocyclic compounds, including metalloporphyrins/hemoproteins, non-heme heterocyclic oxygen carriers, chlorophylls, flavin, flavoproteins, heterocyclic enzyme substrates/inhibitors, nucleic acid derivatives, and heterocyclic free radicals.

7.2. THEORY AND PRINCIPLES

7.2.1. Classical Description

A classical description of the Raman effect serves to give a conceptually good picture of the process. It explains the appearance of Raman lines but fails to give the correct intensities. For this, and to explain resonance Raman scattering, a quantum-mechanical description must be employed.

When an oscillating electromagnetic field **E** interacts with matter, the induced dipole **μ** is related to **E** by

$$\mu = \alpha E = \alpha E_0 \cos(2\pi \nu t) \tag{1}$$

where α is the polarizability tensor. Equation (1) can be written in terms of the spatial components as

$$\mu_x = \alpha_{xx} E_x + \alpha_{xy} E_y + \alpha_{xz} E_z \tag{2}$$
$$\mu_y = \alpha_{yx} E_x + \alpha_{yy} E_y + \alpha_{yz} E_z$$
$$\mu_z = \alpha_{zx} E_x + \alpha_{zy} E_y + \alpha_{zz} E_z$$

The polarizability α is dependent upon the electron distribution that follows a nuclear oscillation. Thus $\alpha_{\rho\sigma}$ ($\rho, \sigma = x, y,$ or z) may be expanded in a power series in Q_k, the vibrational normal coordinate, about the equilibrium configuration Q_k° as

$$\alpha_{\rho\sigma} = (\alpha_{\rho\sigma})_0 + \left(\frac{\partial \alpha_{\rho\sigma}}{\partial Q_k}\right)_0 Q_k^\circ \cos(2\pi \nu_k t) + \cdots \tag{3}$$

and the dipole moment may be expressed as

$$\mu = \alpha_0 E_0 \cos(2\pi \nu_0 t) + \tfrac{1}{2} Q_k^\circ \left(\frac{\partial \alpha}{\partial Q_k}\right)_0$$
$$\times E_0 [\cos 2\pi(\nu_0 + \nu_k)t + \cos 2\pi(\nu_0 - \nu_k)t] \tag{4}$$

An oscillating dipole will radiate with intensity I (18), given by

$$I = \frac{16\pi^4 \nu^4}{3c^3} (\mu)^2 \tag{5}$$

Expansion of equation (5) yields

$$I = \frac{16\pi^4}{3c^3} \left\{ \left[\alpha_0^2 \nu_0^4 E_0^2 \cos^2[2\pi \nu_0 t] \right. \right.$$
$$+ \tfrac{1}{4}(\nu_0 + \nu_k)^4 Q_k^{\circ 2} \left(\frac{\partial \alpha}{\partial Q_k}\right)_0^2 E_0^2 \cos^2[2\pi(\nu_0 + \nu_k)t]$$
$$\left. \left. + \tfrac{1}{4}(\nu_0 - \nu_k)^4 Q_k^{\circ 2} \left(\frac{\partial \alpha}{\partial Q_k}\right)_0^2 E_0^2 \cos^2[2\pi(\nu_0 - \nu_k)t] \right] + \cdots \right\} \tag{6}$$

Equation (6) gives a qualitatively correct picture of the Raman effect. The first term is the Rayleigh scattering term. There is no change in frequency, and the intensity is related to the square of the polarizability. The second term is responsible for anti-Stokes scattering with frequency $v_0 + v_k$. The third term is responsible for Stokes scattering with frequency $v_0 - v_k$. The intensity of a Raman line depends on a change in polarizability, that is, $(\partial \alpha / \partial Q_k)_0 \neq 0$.

Classical theory predicts that the ratio of intensities of Stokes and anti-Stokes lines will be

$$\frac{I_{St}}{I_{anti\text{-}St}} = \frac{(v_0 - v_k)^4}{(v_0 + v_k)^4} \qquad (7a)$$

However, at room temperature most molecules are in the ground state and so the ratio of Stokes to anti-Stokes intensities should be greater than 1. The actual ratio must take into account the Boltzmann distribution factor. Thus,

$$\frac{I_{St}}{I_{anti\text{-}St}} = \frac{(v_0 - v_k)^4}{(v_0 + v_k)^4} (e^{hv_k/kT}) \qquad (7b)$$

where h is Planck's constant, k is the Boltzmann constant, and T is the absolute temperature.

Besides intensities, classical theory can give information about the polarization of the Raman lines. Consider the coordinate system shown in Figure 7.1, with the electromagnetic vector traveling in the y direction in the xy plane. The scattering is observed along the z axis. Since a dipole does not radiate in the direction of its axis, $\mu_z = 0$. The intensity can be expressed as

$$dI = \frac{4\pi^3 (v_0 \pm v_k)^4}{c^3} (|\mu_x|^2 + |\mu_y|^2) \, d\Omega \qquad (8)$$

Thus, $I_x \propto |\mu_x|^2$ and $I_y \propto |\mu_y|^2$. The ratio of intensities in the x and y directions is called the *depolarization ratio* ρ,

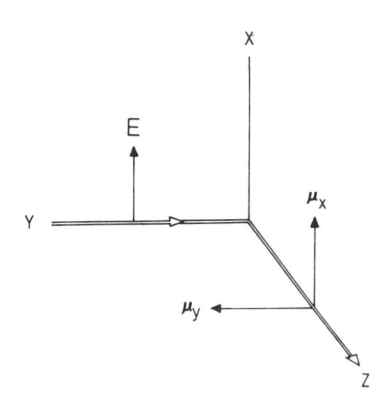

Fig. 7.1. Coordinate system for observation of scattered light.

$$\rho = \frac{I_y}{I_x} = \frac{I_\perp}{I_\parallel} \qquad (9)$$

and the total intensity I_{total} is equal to the sum of $I_\perp$ and $I_\parallel$. Since the intensity observed during the Raman experiment is from molecules that are randomly oriented, equation (8) must be multiplied by the number of molecules and averaged over all possible orientations relative to the experimental coordinate system. Thus results (18, 19) are expressed as invariants of the polarizability tensor,

$$G^0 = \tfrac{1}{3}(\alpha_{xx} + \alpha_{yy} + \alpha_{zz})^2 = 3\bar{\alpha}^2$$

$$G^s = \tfrac{1}{3}[(\alpha_{xx} - \alpha_{yy})^2 + (\alpha_{xx} - \alpha_{zz})^2 + (\alpha_{yy} - \alpha_{zz})^2] \qquad (10)$$
$$+ \tfrac{1}{2}[(\alpha_{xy} + \alpha_{yx})^2 + (\alpha_{xz} + \alpha_{zx})^2 + (\alpha_{yz} + \alpha_{zy})^2]$$

$$G^a = \tfrac{1}{2}[(\alpha_{xy} - \alpha_{yx})^2 + (\alpha_{xz} - \alpha_{zx})^2 + (\alpha_{yz} - \alpha_{zy})^2]$$

G^0 is the isotropic component and is related to α for Rayleigh scattering or $\partial \alpha / \partial Q_k$ for Raman scattering. G^s is the quadrupole component and reflects the anisotropy of α or $\partial \alpha / \partial Q_k$. G^a is the antisymmetric contribution; it contributes only in resonance Raman scattering (20).

The total intensity per molecule, after integration over the solid angle Ω, is (6)

$$I = \frac{2^7 \pi^5 (\nu_0 \pm \nu_k)^4}{9c^4}[G^0 + G^s + G^a]I_0 \qquad (11)$$

The depolarization ratio is

$$\rho = \frac{3G^s + 5G^a}{10G^0 + 4G^s} \qquad (12)$$

For a 90° scattering geometry, this leads to a total intensity of

$$I_{total} = I_\perp + I_\parallel = \frac{2^3 \pi^4 (\nu_0 \pm \nu_k)^4 I_0}{15c^4}[7G^s + 5G^a + 10G^0] \qquad (13)$$

The value of the depolarization ratio can give information about the symmetry type of a vibration. For normal Raman scattering, G^a is zero. Thus, if G^0 is equal to zero, the maximum value that ρ may have is $\tfrac{3}{4}$. These vibrations are nontotally symmetric and are denoted as depolarized (dp). When neither G^0 nor G^s is equal to zero, ρ can have any value between zero and $\tfrac{3}{4}$. These are totally symmetric modes and are denoted as polarized (p). For resonance Raman scattering, G^a is not necessarily equal to zero and G^0 and G^s may equal zero. This will lead to a depolarization ratio of ∞. These are inversely polarized modes (ip). They arise from nontotally symmetric vibrations (e.g., A_{2g} modes in D_{4h} symmetry). A fourth type are called anomalously polarized (ap) (1), with ρ having any value between $\tfrac{3}{4}$ and ∞. They may result from polarized or depolarized modes overlapping inversely polarized modes or from symmetry reduction.

7.2.2. Quantum Description

A clearer understanding of Raman scattering requires a detailed description of the polarizability $\alpha_{\rho,\sigma}$. This can be accomplished by a quantum-mechanical description that includes the vibrational and electronic factors affecting the Raman lines.

For a Raman transition from a ground vibronic state $|g\rangle$ to a final vibronic state $|f\rangle$ caused by a perturbation by plane-polarized light of frequency ν_0 and intensity I_0, the total scattered light intensity is

$$I = \frac{2^7 \pi^5 I_0}{9 c^4} (\nu_0 \pm \nu_{gf})^4 \sum_{\rho,\sigma} |(\alpha_{\rho,\sigma})_{gf}|^2 \qquad (14)$$

with the scattered light having frequency $\nu_0 \pm \nu_{gf}$, where $\nu_{gf} = \nu_f - \nu_g$.

Following Tang and Albrecht (21), an expression for $(\alpha_{\rho,\sigma})_{gf}$ can be derived from the second-order perturbation of Kramers-Heisenberg-Dirac dispersion theory:

$$(\alpha_{\rho,\sigma})_{gf} = \sum_{e}' \left[\frac{\langle g|R_\sigma|e\rangle\langle e|R_\rho|f\rangle}{E_e - E_g - h\nu - i\Gamma_e} + \frac{\langle g|R_\rho|e\rangle\langle e|R_\sigma|f\rangle}{E_e - E_f + h\nu - i\Gamma_e} \right] \qquad (15)$$

The expression is summed over all eigenstates $|e\rangle$, not including the ground state. The energy and homogeneous linewidth of $|e\rangle$ are E_e and Γ_e, respectively. The electric dipole operators are denoted as R_σ and R_ρ.

The expression for $(\alpha_{\rho,\sigma})_{gf}$ can be expanded and simplified by considering the problem in terms of vibronic interactions. Following Albrecht (22), who introduced the expansion of the wavefunctions using the Born-Oppenheimer approximation, the vibronic wavefunctions may be rewritten as

$$|g\rangle = |g(q,Q)\rangle|i(Q)\rangle$$
$$|f\rangle = |g(q,Q)\rangle|j(Q)\rangle \qquad (16)$$
$$|e\rangle = |e(q,Q)\rangle|u(Q)\rangle$$

where q and Q represent the electronic and nuclear coordinates, respectively. The expansion results in the product of electronic and vibrational wavefunctions. The electronic wavefunctions $|g\rangle$ and $|e\rangle$ depend on both the electronic and nuclear coordinates, while the vibrational wavefunctions $|i\rangle$, $|j\rangle$, and $|u\rangle$ depend only on the nuclear coordinates. Combination of equations (15) and (16) yields

$$(\alpha_{\rho,\sigma})_{gi,gj} = \sum_{eu}' \left[\frac{\langle gi|R_\sigma|eu\rangle\langle eu|R_\rho|gj\rangle}{E_{eu} - E_{gi} - h\nu - i\Gamma_{eu}} + \frac{\langle gi|R_\rho|eu\rangle\langle eu|R_\sigma|gj\rangle}{E_{eu} - E_{gj} + h\nu - i\Gamma_{eu}} \right] \qquad (17)$$

A second approximation can be made, expanding the electronic wavefunctions using a first-order Herzberg-Teller series for small coordinate displacements through first-order perturbation. Hence, $|g\rangle$ and $|e\rangle$ become

$$|g\rangle = |g^0\rangle + \sum_{k} \sum_{t \neq g} \frac{(h_k)_{gt}^0 \Delta Q_k}{E_g^0 - E_t^0} |t^0\rangle$$

$$|e\rangle = |e^0\rangle + \sum_{k} \sum_{s \neq e} \frac{(h_k)_{es}^0 \Delta Q_k}{E_e^0 - E_s^0} |s^0\rangle \qquad (18)$$

The term $(h_k)^0_{es} = \langle e^0|(\partial H/\partial Q_k)_0|s^0\rangle$ is the vibronic coupling term for normal mode k evaluated at $Q_k = 0$. The electronic wavefunction at $Q_k = 0$ is $|e^0\rangle$. The intermediate excited states are denoted as s and t. Hence, the electronic wavefunctions have been separated into two factors, each containing only electronic or vibronic factors.

Applying equation (18) to $(\alpha_{\rho,\sigma})_{gi,gj}$, the expanded version of $(\alpha_{\rho,\sigma})_{gi,gj}$ can be expressed as

$$(\alpha_{\rho,\sigma})_{gi,gj} = A_{\rho,\sigma} + B_{\rho,\sigma}$$

$$A_{\rho,\sigma} = \sum_e {}' \sum_u \frac{M_{ge}^\sigma M_{eg}^\rho}{E_{eu} - E_{gi} - h\nu_0 - i\Gamma_{eu}} \langle gi|eu\rangle\langle eu|gj\rangle$$

$$B_{\rho,\sigma} = \sum_e {}' \sum_u \sum_{s \neq e} \sum_k \frac{(h_k)^0_{es}}{(E_e^0 - E_s^0)(E_{eu} - E_{gi} - h\nu_0 - i\Gamma_{eu})} \quad (19)$$

$$\times [M_{ge}^\sigma M_{sg}^\rho \langle gi|eu\rangle\langle eu|Q_k|gj\rangle + M_{ge}^\rho M_{sg}^\sigma \langle gi|Q_k|eu\rangle\langle eu|gj\rangle]$$

where $M_{ge}^\sigma = \langle g^0|R_\sigma|e^0\rangle$ and coupling between ground and excited states is neglected.

Many features of the Raman effect can be understood through the examination of these terms. For nonresonance Raman scattering, the difference in energy between the excited intermediate state and the initial ground state is much greater than the energy of incidence $h\nu_0$. So, using Van Vleck's sum rule (23), and the harmonic oscillator approximation,

$$\sum_{e,u} \langle gi|eu\rangle\langle eu|gj\rangle = \langle gi|gj\rangle = \delta_{ij}$$

$$\sum_{e,u} \langle gi|eu\rangle\langle eu|Q_k|gj\rangle = \sum_{e,u} \langle gj|eu\rangle\langle eu|Q_k|gi\rangle = \langle gi|Q_k|gj\rangle$$

$$= \begin{cases} 0 & \text{if } V_k^j \neq V_k^i \pm 1 \\ \left(\frac{(V_k^i + 1)h}{8\pi^2\nu\mu}\right)^{1/2} & \text{if } V_k^j = V_k^i + 1 \\ \left(\frac{(V_k^i)h}{8\pi^2\nu\mu}\right)^{1/2} & \text{if } V_k^j = V_k^i - 1 \end{cases} \quad (20)$$

where V is the vibrational quantum number of normal coordinate k, with frequency ν and reduced mass μ.

For $i = j$, the B term vanishes and the A term contributes intensity. This is the Rayleigh scattering term. The B term contributes to the Raman intensity, since it is responsible for scattering of fundamentals ($V_k^j = V_k^i \pm 1$). Also, for the B term not to vanish, $(h_k)^0_{es}$ cannot be zero and M_{ge} and M_{gs} cannot be zero. This requirement leads to the selection rules for a Raman transition. The electronic states s and e must be upper states of an allowed electronic transition for $(h_k)^0_{es} \neq 0$.

If Γ_e and Γ_s are the symmetry species of states e and s, then both Γ_e and Γ_s must correspond to Γ_ρ ($\rho = x, y, z$) for a totally symmetric ground state. Also, the direct product $\Gamma_e \Gamma_{H'} \Gamma_s$ must contain the totally symmetric representation. $\Gamma_{H'}$ is the species symbol for the irreducible representation for which $(\partial H/\partial Q_k)_0$ is the basis (22); $\Gamma_{H'}$ is also the species symbol for the representation for which Q_k is the basis. Hence $\Gamma_k \Gamma_e \Gamma_s$ must also contain the totally symmetric representation, and thus Γ_k must also correspond to one of the species in $\Gamma_e \Gamma_s$ (x^2, xy, xz, y^2, yz, z^2). Therefore, a mode is Raman active if its irreducible representation contains x^2, y^2, z^2, xy, xz, or yz.

For resonance Raman scattering, with the damping factor $i\Gamma_{eu}$ included to prevent the denominator from becoming zero when $E_{eu} - E_{gi} = h\nu_0$, the sum rule no longer holds. Thus, if $i \neq j$, the A term need not vanish. The A term depends on Franck-Condon overlap factors between the ground and excited states. A-Term enhancement occurs when there is a change either in frequency or in internuclear separation upon transition from the ground to the excited state. This usually occurs in strongly allowed, totally symmetric vibrations. The A term has no dependence on vibronic mixing.

B-Term enhancement is due to vibronic borrowing of the resonant electronic transition and nearby electronic transition. The enhanced modes are usually non-totally symmetric.

Another term may be added to account for nonadiabatic coupling between excited states. This D term is given (24) by

$$D_{\rho\sigma} = \sum_{e} \sum_{u} \sum_{s \neq e}' \sum_{x} \sum_{k} \frac{(h_k)^0_{es}}{(E_e^0 - E_s^0)(E_{eu} - E_{gi} - h\nu_0 - i\Gamma_{eu})}$$

$$\times \left[M_{eg}^\sigma M_{gs}^\rho \langle gj|eu\rangle\langle eu|Q_k|sx\rangle\langle sx|gi\rangle (\delta_{u,x-1} - \delta_{u,x+1}) \frac{h\nu_s^k}{E_{eu} - E_{sx}} \right. \quad (21)$$

$$\left. - M_{eg}^\rho M_{gs}^\sigma \langle gj|sx\rangle\langle sx|Q_k|eu\rangle\langle eu|gi\rangle (\delta_{x,u-1} - \delta_{x,u+1}) \frac{h\nu_e^k}{E_{eu} - E_{sx}} \right]$$

7.3. EXPERIMENTAL TECHNIQUES

7.3.1. Multichannel Detectors: Intensified Vidicon/Reticon

A multichannel detector, like a photographic plate, can view a wide segment of a Raman spectrum in a single exposure; all channels acquire an optical signal simultaneously without a mechanical scanning of the monochromator. The detector responds to energy rather than power, so long-term data integration may be employed to minimize the preamplifier noises and enhance detection capability (15, 16). This enhanced detection capability allows one to extract extremely weak Raman signals and to obtain resonance Raman spectra of photolabile samples with low laser power. It also has the advantage over a photomultiplier tube in the measurement of relative intensities (resonance Raman

Table 7.1. Comparison of Intensified Vidicon (SIT) and Intensified Reticon (SPD) Detectors.

	SIT	SPD
1. Detector area	12.5 × 12.5 mm^2	18.0 × 2.5 mm^2
2. Quantum efficiency	< 10% except near 400 nm	10–20% (200–550 nm)
3. Resolution	40 channels/mm (500 total)	38 pixels/mm (700 total)
4. Dynamic range	16,000 : 1	16,000 : 1
5. Gating speed	50 ns	5 ns
6. Readout mechanism	Scanning electron beam; considerable lag	FET switches; practically lag-free
7. Cooling method	Dry ice ($-60°C$)	Peltier thermoelectric cooler ($-20°C$)

excitation profiles) because the multichannel detector is insensitive to fluctuations in laser intensity and sample scattering. Furthermore, the high speed of the detector permits the recording of resonance Raman spectra of ultra-short-lived transient species.

Although there are many types of multichannel detectors available, at least two have been demonstrated to be well suited for Raman spectroscopic work.

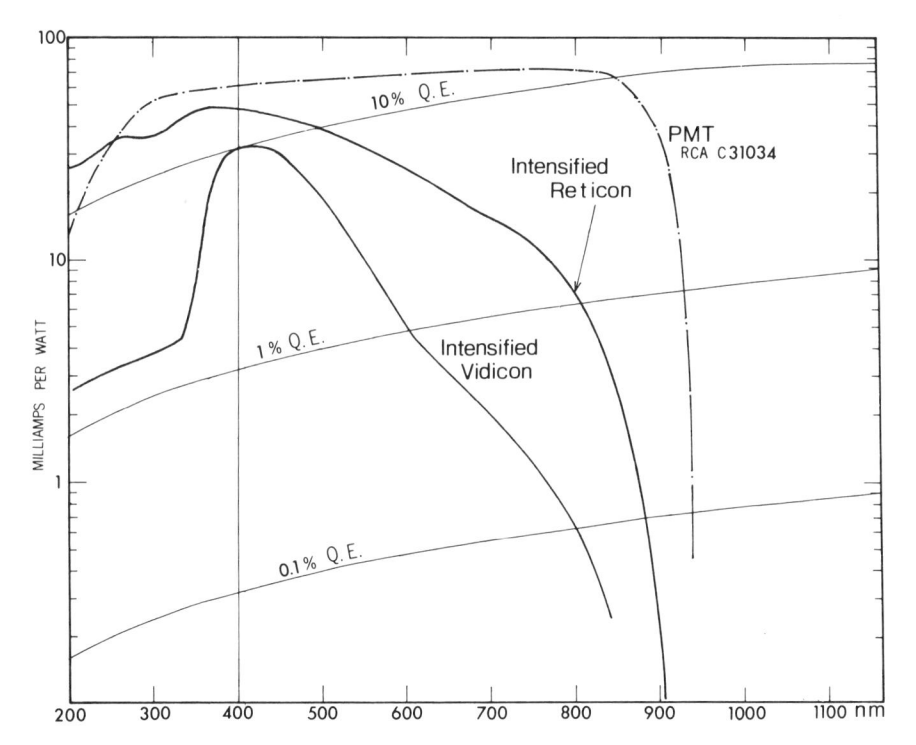

Fig. 7.2. Comparison of the spectral responses of three detectors: photomultiplier tube (RCA C31034), intensified reticon (PAR 1420), and intensified vidicon (PAR 1254).

They are the "intensified vidicon" (silicon intensified target, SIT) and "intensified Reticon" (silicon photodiode array, SPD). Both can be cooled and have a large dynamic range (an important parameter for Raman applications). Their important features are compared in Table 7.1. Their spectral responses, along with that of a popular Raman photomultiplier (RCA C31034), are shown in Figure 7.2. Since these are relatively new, their principles of operation are briefly described here.

SIT Vidicon (PAR 1254).   As shown in Figure 7.3, the intensifier stage consists of a fiber-optic faceplate that is optically coupled to a photocathode. The photoelectrons emitted from the photocathode surface are focused to produce an electron image on the silicon target that corresponds to the optical image on the faceplate. Because the photocathode is maintained at a potential (9 kV) below that of the silicon target (ground potential), and photoelectrons emitted strike the

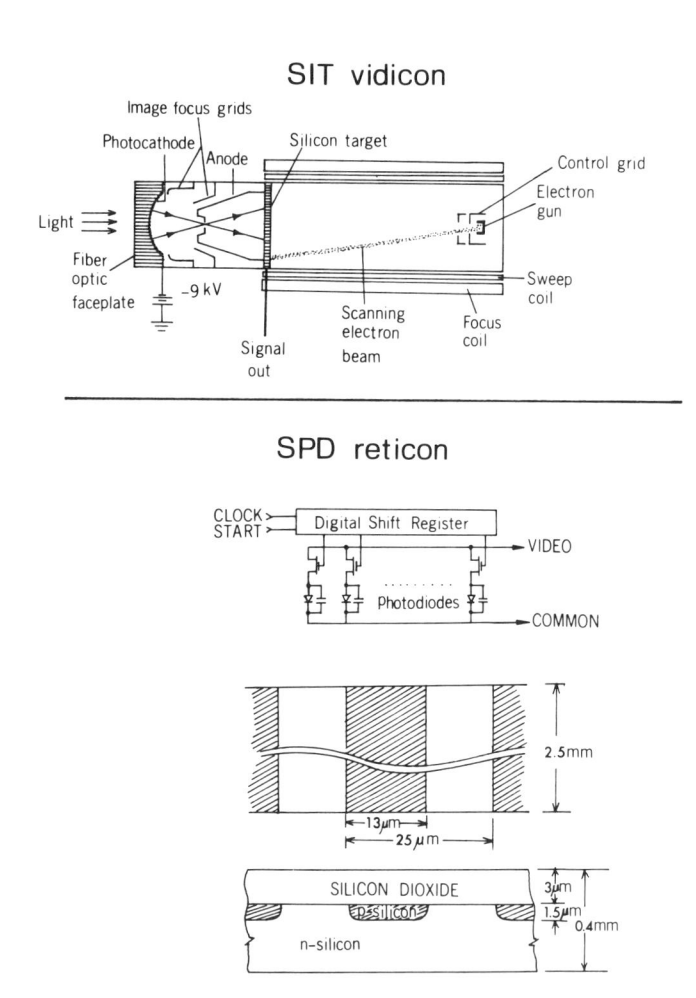

Fig. 7.3. Comparison of SIT vidicon and SPD reticon. From Y. Talmi and R. W. Simpson, *Appl. Opt.*, **19**, 1401 (1980) (with permission).

target with 9 keV of energy, producing electron-hole pairs on the target. The target consists of an array of *p*-type semiconductor islands grown on a wafer of *n*-type material. The electron beam, which is focused by focusing coils surrounding the tube, is emitted by an electron gun at one end of the tube and focused on the target. Sweep coils surrounding the tube permit the electron beam to scan the target. In operation, the electron beam is made to scan the target, charging the *p*-type islands to the potential of the electron gun. Each reverse-biased diode functions as a storage capacitor, because the negatively charged *p*-type material is separated from the *n*-type material at ground potential by an insulating depletion zone. These tiny capacitors may be discharged by electron-hole pair production in the *n*-type wafer. Thus there will be a reduced charge at locations where the target is exposed to light. At the end of an exposure, the electron beam scans the target again; electrons are redeposited on the *p*-type islands, producing a charging current that is amplified as the video signal. Normally, it takes many (10–100) scans to redeposit enough electrons to recharge the region back to the potential of the electron gun.

SPD Reticon (PAR 1420). This detector contains an ITT proximity focused channel intensifier coupled by fiber optics to a silicon photodiode array (SPD). The image intensifier consists of a photocathode, a microchannel plate, and an output phosphor screen. The photoelectrons emitted from the photocathode are accelerated by the microchannel plate (MCP) potential and cross the space between photocathode and MCP. The MCP consists of a bundle of fine glass tubes having conductive walls. The potential difference (~ 700 V) between the ends of the microchannels accelerates the electrons along the channels and causes them to collide with the walls. Thus the microchannels act as electron multipliers because each collision results in liberating additional electrons. The electrons exiting the MCP are accelerated toward the phosphor screen by the 5-kV potential. These electrons strike the phosphor, causing it to grow. In the SPD stage, each of the 700 photodiodes is effectively a charged capacitor, which is discharged by electron-hole pairs generated when light from the phosphor screen strikes a photodiode. As the photodiode is accessed by the array shift registers and multiplex switches, it is recharged by a video line (25, 26) (Fig. 7.3). This causes a change of voltage on that line, which is sensed by the preamplifier, digitized by the detector controller (PAR model 1218), and transmitted as data.

7.3.2. Excitation Sources: Lasers

Although nonresonance Raman spectra are independent of excitation wavelength, there is still a need to select a different wavelength in order to minimize the fluorescence interference due to impurities. For resonance Raman studies, one would like to have as many exciting wavelengths as possible so that wavelength dependence of Raman intensities (resonance Raman excitation profiles) can be measured.

The exciting radiation for both resonance and nonresonance Raman studies is normally provided by an argon-ion or krypton-ion laser. These two lasers generate

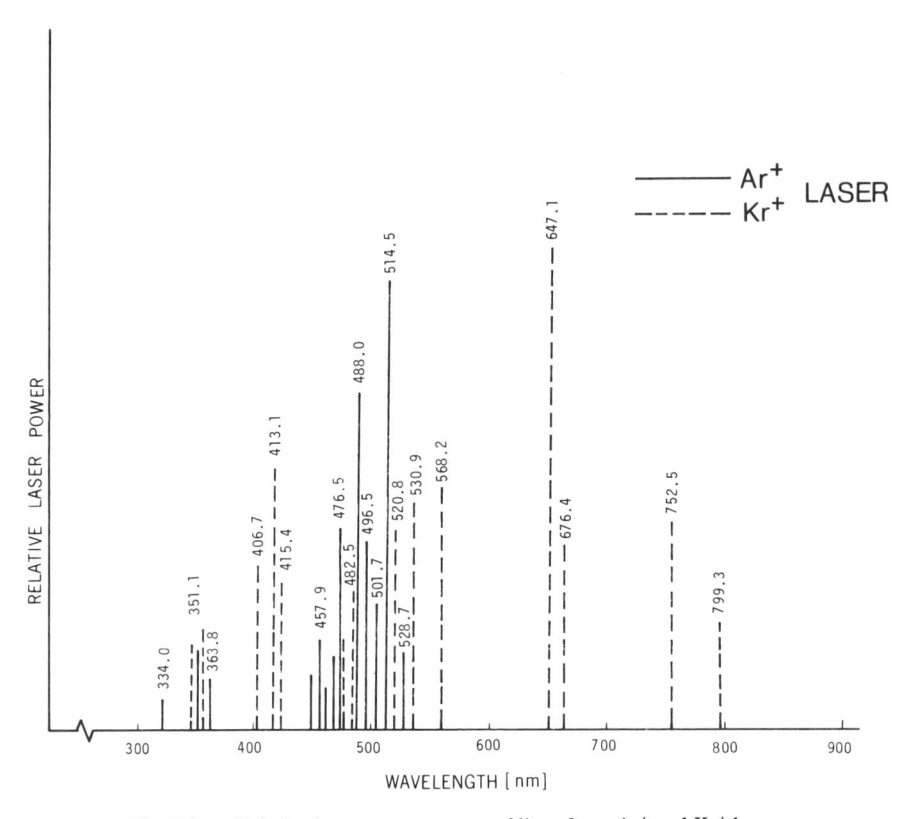

384   Raman Spectroscopy of Heterocyclic Compounds

Fig. 7.4. Relative laser output power of lines from Ar⁺ and Kr⁺ lasers.

a total of 27 discrete lines ranging from 334.0 to 799.3 nm (see Fig. 7.4 for relative output power). For continuous wavelength coverage in the visible region, a tunable dye laser (pumped by ion lasers) is needed. The typical laser output versus wavelength for various dyes from 400 to 875 nm is shown in Figure 7.5. In recent years there has been increasing interest in the 217–380-nm region, which is now provided by a second harmonic generation (frequency doubling) of a tunable coaxial flashlamp pumped dye laser with high average power (18–150 mW).

7.3.3. A Multichannel Laser Raman System

The selection of a monochromator is very important in setting up a multichannel-detected Raman system. Net dispersion, resolution, throughput, stray light level, and the quality of spectral image on the focal plane are among the factors to be considered. Different designs employing a single monochromator (9–14), a subtractive dispersion double monochromator (15), an additive dispersion double monochromator (16), or a triple monochromator (Spex 1807 Triplemate) have appeared. The complete system shown in Figure 7.6 has been used routinely by the present authors. It has excellent image on the detector, high throughput,

Dye	Pump Power	Pump Wavelength
CARB 165, C2, C1, C102	1.5 W	351 - 364 nm
C30	1.3 W	457.9 nm
C7	1.2 W	476.5 nm
Sodium Fluorescein, R110, R6G, RB, R101, Cresyl Violet	4 W	458 - 514 nm
Nile Blue	2 W	590 nm
Oxazine 1, DEOTC-P, HITC-P	4 W	647 - 676 nm

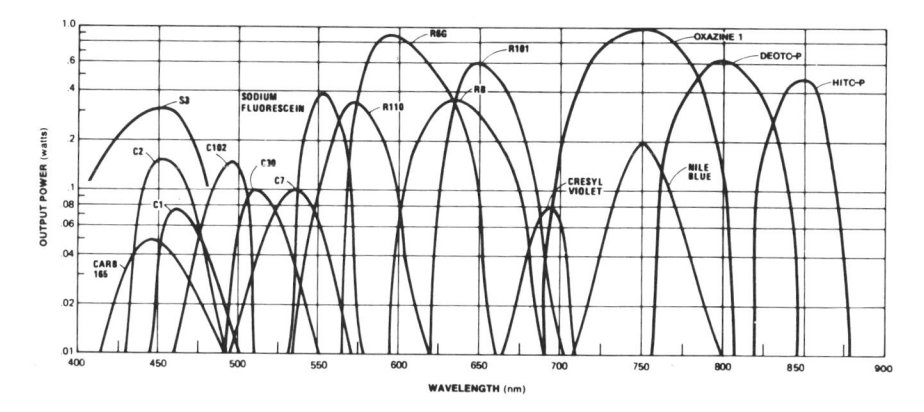

Fig. 7.5. Typical output power of the Spectra Physics model 375 dye laser with various dyes. The dyes, the pump laser power, and wavelength used to obtain the performance curves are listed: from "Spectra Physics CW Dye Lasers" Spectra Physics, Laser Instrumentation Division, Mountain View, California (with permission).

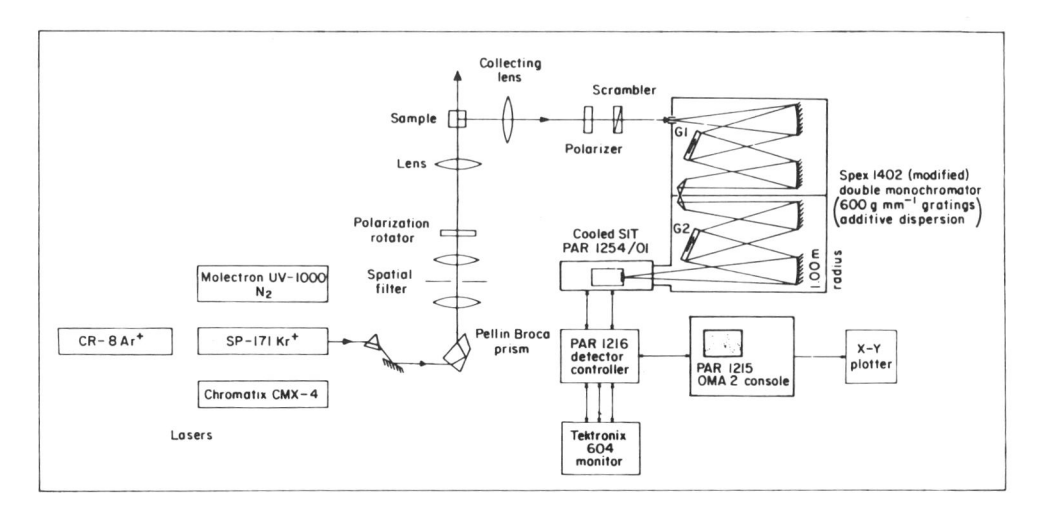

Fig. 7.6. Schematic diagram of a multichannel Raman system. From N. T. Yu and R. B. Srivastava, *J. Raman Spectroscopy*, 9, 166 (1980) (with permission).

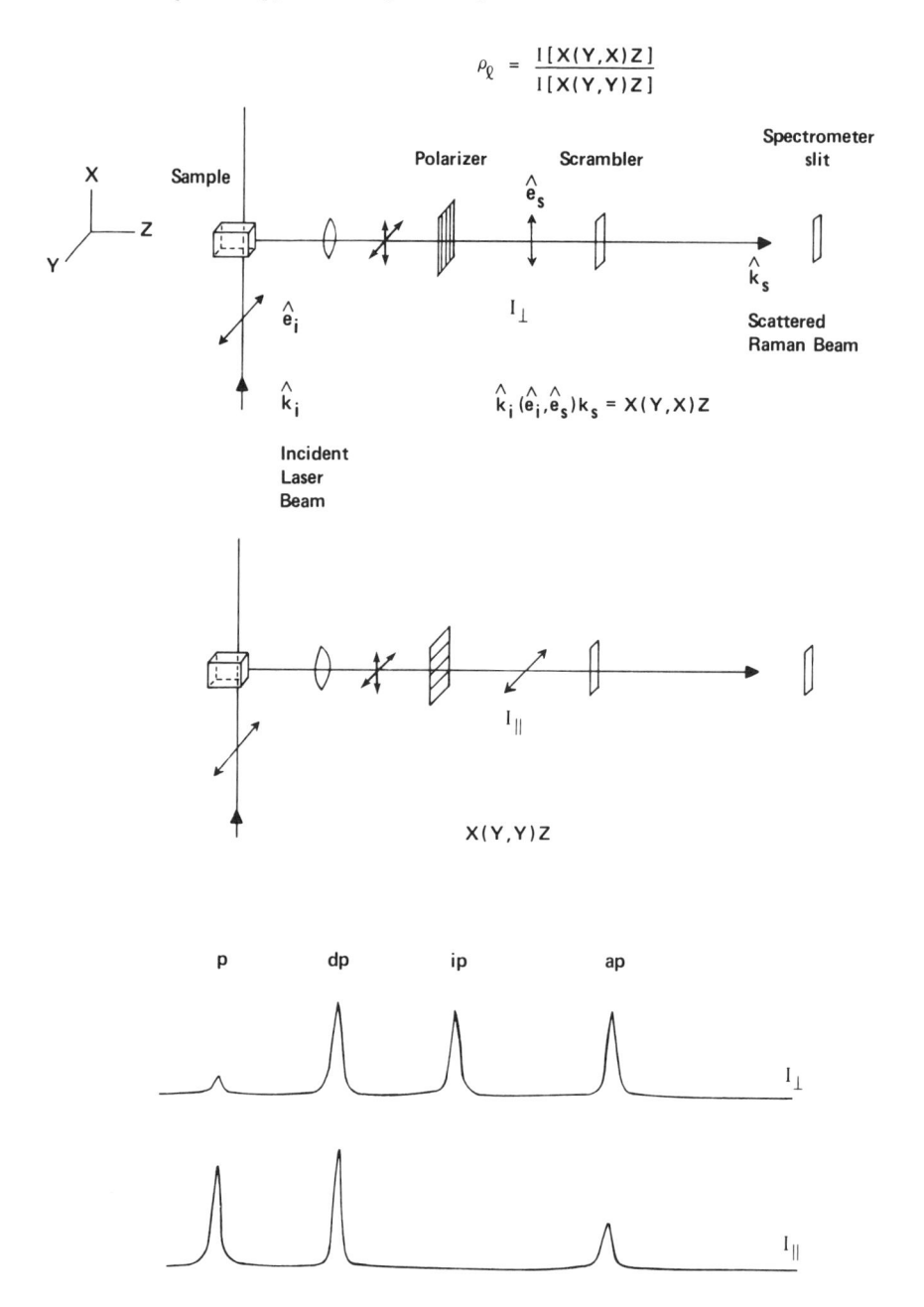

386 Raman Spectroscopy of Heterocyclic Compounds

$$\rho_\varrho = \frac{I[X(Y,X)Z]}{I[X(Y,Y)Z]}$$

$\hat{k}_i(\hat{e}_i,\hat{e}_s)k_s = X(Y,X)Z$

$X(Y,Y)Z$

Fig. 7.7. Measurement of depolarization ratio; schematic representation of *p*, *dp*, *ip*, and *ap* lines in polarized spectra ($I_\perp$ and $I_\parallel$).

adequate stray light rejection, and optimum bandpass and resolution (5–12 cm^{-1}) for most applications.

The system consists of excitation sources (CR-8 Ar$^+$, SP-171 Kr$^+$, Molectron UV-1000 N$_2$, and Chromatix CMX-4), a modified Spex 1402 double monochromator (two 600 grooves/mm classically ruled gratings in additive dispersion), a Spex 1419 sample illuminator (focusing lens, collecting lens, polarizer, and scrambler), a sample rotator with cell, a dry ice–cooled SIT detector (PAR 1254/01), a detector controller (PAR 1216), and an optical multichannel analyzer console (PAR OMA 2) with a Tektronix 604 monitor and an X-Y plotter. To eliminate plasma emission lines, a spatial filtering system (an equilateral prism, a Pellin Broca prism, a focusing lens, and a pinhole of 100–200 μm diameter) is employed. All the lenses and prisms are of Suprasil 1 grade quartz. For the 200–400-nm ultraviolet region, an intensified Reticon (PAR 1420) detector is far more sensitive in the entire spectral region (see Fig. 7.2). With this detector, the PAR 1216 detector controller is replaced by a PAR 1218 detector controller.

The spectral bandpass (in cm^{-1}) viewed by the two detectors is $\sim$ 750 cm^{-1} at 350 nm, $\sim$ 600 cm^{-1} at 400 nm, and $\sim$ 300 cm^{-1} at 600 nm for the SIT detector; and $\sim$ 1080 cm^{-1} at 350 nm, $\sim$ 850 cm^{-1} at 400 nm, $\sim$ 450 cm^{-1} at 600 nm for the SPD detector.

7.3.4. Measurement of the Depolarization Ratio

The scattering geometry is normally specified by the notation $\hat{k}_i(\hat{e}_i, \hat{e}_s)\hat{k}_s$, where the vector $\hat{k}_i$ ($\hat{k}_s$) refers to the direction of propagation of the incident (scattered) light, and the vector $\hat{e}_i$ ($\hat{e}_s$) refers to the direction of electric field polarization of incident (scattered) light. In Figure 7.7 the depolarization ratio is defined as

$$\rho_l = \frac{I[X(Y, X)Z]}{I[X(Y, Y)Z]}.$$

The subscript l indicates linearly polarized incident light. Two spectra are needed to determine this ratio: one with the polarizer oriented parallel to the X axis and the other with the polarizer oriented parallel to the Y axis. The scrambler between the polarizer and spectrometer slit is to eliminate the polarization dependence of the monochromator response. $I[X(Y, X)Z]$ and $I[X(Y, Y)Z]$ are often referred to as $I_\perp$ and $I_\parallel$, respectively. Four types of depolarization (p, dp, ip, and ap) are schematically shown in Figure 7.7.

7.3.5. Intensity Correction for Self-Absorption: 90° versus 180° Geometry

Resonance Raman spectroscopy of colored or opaque substances involves the problem of self-absorption. In quantitative Raman studies, self-absorption correction and concentration optimization are essential. Corrections may be made for the two basic illumination geometries, 90° transverse scattering (27) and 180° backscattering (28)(Fig. 7.8), based on appropriate intensity relationships.

Raman intensities are attenuated by absorption of the incident and scattered

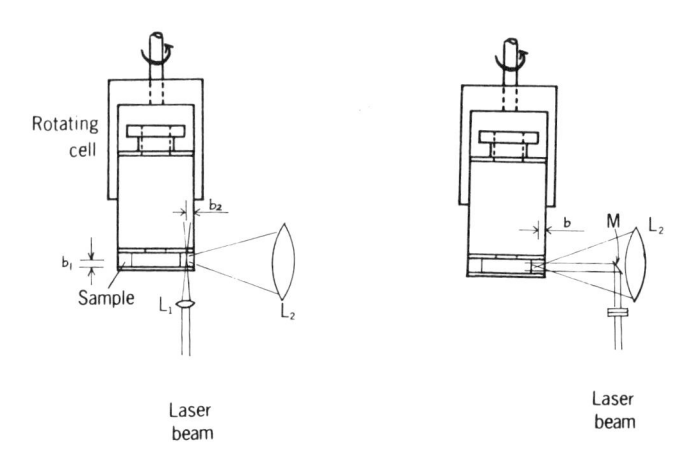

Fig. 7.8. Comparison of 90° and 180° scattering geometry.

radiation as an exponential function of the absorbing materials' concentration. There are fundamental differences in the effect of these absorption processes in the 90° and 180° geometries. For 90° transverse geometry, the observed intensity of a Raman line shows a maximum as a function of concentration. This optimum concentration is often quite low for intensely absorbing samples and found to correspond to an absorbance of about 13 per centimeter of path length. The observed Raman intensities are predicted by

$$I_{obs} = I_0 Jc \exp[-2.303c(b_1\epsilon_i + b_2\epsilon_s)] \qquad (22)$$

where I_0 is the laser intensity on entry into the sample, J is the molar scattering coefficient, c is the molar concentration of absorbing molecules, b_1 is the path length to the point of scattering, b_2 is the scattering path length, and ϵ_i and ϵ_s are the decadic molar absorptivity at the incident and scattering wavelengths, respectively. Studies by Strekas and coworkers (27) show that the position of maximum intensity is independent of absorptivity and of refractive index of the medium. Self-absorption corrections extend from expressing equation (22) as

$$\log R_{obs} = \log R_s - b_2 c \Delta\epsilon_s \qquad (23)$$

where $R_{obs} = I_{obs,1}/I_{obs,2}$, $R_s = J_1/J_2 \exp[-2.303 b_1 c(\epsilon_{i1} - \epsilon_{i2})]$, and $\Delta\epsilon_s = \epsilon_{s1} - \epsilon_{s2}$. With $\log R_{obs}$ plotted versus $c\Delta\epsilon_s$, extrapolation to zero concentration will give $\log R_s$, which eliminates the effect of self-absorption on measured scattering intensities. The slope of the line determines a value for b_2 that can be used with equation (23) to estimate the self-absorption correction at any given concentration for the same experimental arrangement.

For the 180° geometry, dependence of the Raman scattered intensity on the concentration of colored species is not as critical (28). Instead of reaching a maximum as seen before, an asymptotic limit is obtained with increasing concentrations. If the absorber is also the Raman scatter, the collected Raman intensity is given by

$$I_{obs} = \frac{0.434\alpha_s I_0}{\epsilon_s + \epsilon_i} (1 - 10^{-bc(\epsilon_s + \epsilon_i)}) \qquad (24)$$

where α_s is the effective differential cross section for a particular Raman line. The depth parameter b is experimentally determined. When the observed intensity is one-half the asymptotic maximum intensity, b is given by

$$b = \frac{0.301}{c(\epsilon_s + \epsilon_i)} \qquad (25)$$

For studies of a resonance Raman process, corrections for sample absorption are made with α_s, which is easily obtained for very strongly absorbing media from

$$\alpha_s \simeq \frac{2.303 I_{obs}}{I_0} (\epsilon_s + \epsilon_i)$$

Corrections can then be made from a knowledge of extinction coefficients. Other expressions are also known if the absorber and Raman scatterer are not the same species or if an internal reference is used.

For specific applications, certain advantages will select the design to be used. Self-absorption corrections require a simple plot of log intensity versus concentration for 90° but are even more straightforward for highly colored solutions by backscattering. Reproducible illumination is more easily accomplished at 180°, but extraneous signals from the container and the atmosphere as well as specular reflection from the sample cell into the collection lens are quite troublesome.

7.3.6. Sample-Handling Techniques

For resonance Raman studies of strongly absorbing samples, the sample cell should be spun at ~1000 rpm by a dc or ac motor. Figure 7.9 shows a rotating Raman cell of typical design that requires only ~0.2 ml of liquid sample. The cell can be sealed with a standard rubber septum. For colored solid samples, KBr pellets (~0.5 mg per 200 mg KBr) or crystalline powder evenly affixed to a rotating platform may be used to obtain resonance Raman spectra. The sample surface is oriented so that the incident beam makes a 10° angle with the surface.

For low-temperature experiments (25 to −80°C), which require sample spinning, the temperature may be controlled by a flow of adjustable cold N_2 gas through a dewar enclosing the rotating cell. The N_2 gas is cooled by flowing it through copper tubing immersed in a dry ice–isopropanol bath or a liquid N_2 dewar. For a rotating sample at liquid N_2 temperature, a novel design was recently developed by Braiman and Mathies (29).

For nonresonance Raman studies, the axial/transverse and transverse/transverse methods are commonly employed. In the axial/transverse method, the laser beam enters along the axis of a sample-containing capillary, and the scattered light is collected at 90° to the capillary wall. The capillary, which has a flat end, is held vertically so that its axis is parallel with the entrance slit of the spectrometer. In the transverse/transverse configuration, both incident beam and scattered light are

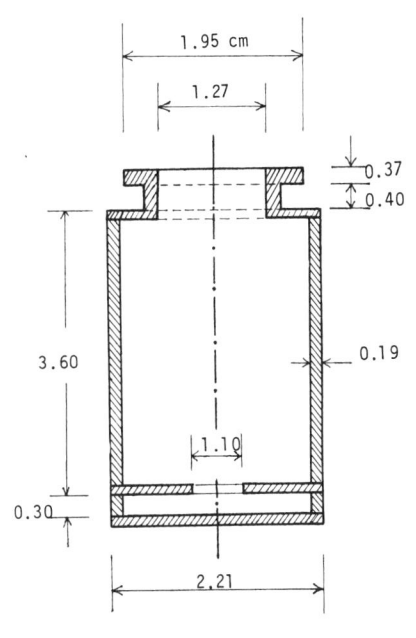

Fig. 7.9. Dimension of a typical quartz Raman rotating cell.

perpendicular to the capillary axis. Standard melting point capillary tubes of ~ 1.0 mm bore have been found to be adequeate for visible excitation. For ultraviolet wavelengths (below 350 nm), thin-wall quartz capillaries (Charles Supper Co., Natick, MA) are recommended. The sample volume that gives rise to Raman scattering for analysis is ~ $10^{-3} \mu l$. Variations of static sample temperature in the range -15 to $+100°C$ may be achieved by a thermostat (30) designed for the transverse/transverse capillary. In the range -196 to $+200°C$, the Harney-Miller variable-temperature device (available from Spex Industries, Inc.) is quite convenient.

The best method for handling solid powder (nonabsorbing) is to pack the sample into a conical depression at the end of an $\frac{1}{8}$-in. stainless steel rod, which may be held horizontally. The beam is then focused onto the sample at the grazing angle so that the scattering column is a strip on the powder surface, $\frac{1}{8}$ in. long and approximately 20 μm wide.

7.3.7 Discrimination against Fluorescence: Nonlinear CARS and RIKES Techniques

7.3.7.1. Coherent Anti-Stokes Raman Scattering (CARS)

The CARS technique was once considered exotic but impractical. However, its potential for biological applications has been greatly enhanced since the demonstration of its resonance enhancement by Hudson and coworkers (31), Nester and coworkers (32), and Carreira and coworkers (33). Because of the high discrimination against fluorescence, it allows one to obtain Raman spectra of fluorescent biological samples which are not obtainable by other means. Here we briefly discuss the principles and theoretical aspects of this nonlinear optical technique.

Consider the response of the medium to an electromagnetic field. In general, the induced electrical polarization can be written as a power series expansion:

$$P_i = \chi_{ij}^{(1)} E_j + \chi_{ijk}^{(2)} E_j E_k + \chi_{ijkl}^{(3)} E_j E_k E_l + \cdots$$

The coefficients $\chi^{(n)}$ are tensors of rank $n+1$, and the subscripts $ijkl$ specify different polarizations. The first coefficient, $\chi_{ij}^{(1)}$, is responsible for spontaneous Raman scattering; $\chi_{ijk}^{(2)}$ is related to optical second-harmonic generation (frequency doubling); and $\chi_{ijkl}^{(3)}$, the third-order nonlinear susceptibility, gives rise to the CARS signals as well as the RIKES signals (34).

In the CARS experiments, two laser beams at frequencies ω_1 and ω_2 ($\omega_1 > \omega_2$), are crossed in the sample at the phase-matching angle θ, and a new coherent beam at frequency $\omega_3 = 2\omega_1 - \omega_2$ is generated. The intensity of the CARS signal at ω_3 varies with the frequency difference $\omega_1 - \omega_2$ and reaches a maximum when $\omega_1 - \omega_2 = \omega_r$, the frequency of a Raman-active mode. The CARS spectra are usually obtained with ω_1 fixed while ω_2 is scanned.

The third-order nonlinear susceptibility may be expressed as $\chi^{(3)} = \chi_E^{(3)} + \chi_R^{(3)}$, where $\chi_E^{(3)}$ is the background or electronic susceptibility and $\chi_R^{(3)}$ is the Raman susceptibility. In the case of dilute solutions, the contributions to the background from both solute and solvent are included in $\chi_E^{(3)}$. When $\omega_1 - \omega_2 \to \omega_r$, the Raman susceptibility is given by (35)

$$\chi_{R, ijkl}(-\omega_3, \omega_1, \omega_1, -\omega_2) \propto \sum_r \frac{Q_r}{\omega_r - (\omega_1 - \omega_2) - i\Gamma_r}$$

where Q_r is proportional to

$$\sum_e \left(\frac{M_{ge}^i M_{er}^l}{\omega_{eg} - \omega_3 - i\Gamma_{eg}} + \frac{M_{ge}^l M_{er}^i}{\omega_{eg} + \omega_1 + i\Gamma_{eg}} \right) \sum_e \left(\frac{M_{re}^j M_{eg}^k}{\omega_{eg} - \omega_1 - i\Gamma_{eg}} + \frac{M_{re}^k M_{eg}^j}{\omega_{eg} + \omega_2 - i\Gamma_{eg}} \right)$$

where g stands for ground state $|g\rangle$; e, excited electronic state $|e\rangle$; r, the Raman vibrational level $|r\rangle$. M is the dipole moment operator, $\hbar\omega_{eg}$ is the energy difference between $|e\rangle$ and $|g\rangle$, $E_e - E_g$, and the factors Γ are phenomenological resonance widths. Since Q_r is a complex quantity it may be represented by $Q_r = R + iI$, where R and I are the real and imaginary parts, respectively. The intensity of a CARS signal at ω_3 is proportional to the square of $\chi^{(3)}$,

$$|\chi^{(3)}|^2 = \left| \chi_E^{(3)} + \frac{R + iI}{\delta - i\Gamma_r} \right|^2 = \left| B + \frac{R + iI}{\delta - i\Gamma_r} \right|^2$$

$$= B^2 + \frac{R^2 + I^2}{\delta^2 + \Gamma_r^2} + \frac{2BR\delta}{\delta^2 + \Gamma_r^2} - \frac{2BI\Gamma_r}{\delta^2 + \Gamma_r^2}$$

Here B is a slowly varying function of $\omega_1 - \omega_2$, and $\delta = \omega_r - (\omega_1 - \omega_2)$. When $\delta = 0$, $\omega_1 - \omega_2 = \omega_r$, representing the position of a Raman-active vibration.

The first term represents the background signal; the second term gives rise to a positive Lorentzian-shaped curve with its maximum at $\delta = 0$; the third term changes sign with δ, giving rise to a dispersion-shaped curve. The last term is small and negligible when ω_1 is away from the lowest electronic transition. Therefore, the CARS spectrum is the sum of the background, a Lorentzian curve, and a dispersion term. Near the one photon absorption region (i.e., resonance CARS), $\omega_1, \omega_3 \approx \omega_{eg}$ and simultaneously $\omega_1 - \omega_2 \approx \omega_r$. The resonance contribution to $\chi_R^{(3)}$ is proportional to (36)

$$\frac{\Gamma_{eg}^2}{(\omega_{eg} - \omega_3 - i\Gamma_{eg})(\omega_{eg} - \omega_1 - i\Gamma_{eg})} \frac{\Gamma_{rg}}{(\omega_{rg} - (\omega_1 - \omega_2) - i\Gamma_{rg})}$$

The resonance CARS lines were shown to change from positive to dispersion-shaped and finally to completely negative as ω_1 changes from the high-frequency to the low-frequency side of an electronic transition (ω_{eg}).

7.3.7.2. Raman-Induced Kerr Effect Spectroscopy (RIKES)

While the resonance CARS technique has been proven to be suitable for biological applications, the major drawback is the phase-matching requirement and the changes in line shapes, which make both experiments and interpretations somewhat difficult. The so-called RIKES technique, first demonstrated by Heiman and coworkers (37) appears to eliminate this drawback and is capable of generating Raman spectra of fluorescent biological samples.

In the RIKES experiments, two beams at ω_1 and ω_2 are crossed in the sample, as in the CARS technique. The circularly polarized pump beam at ω_1 induces an intensity-dependent birefringence, which changes the state of polarization of a second probe beam at ω_2. The change in polarization of the probe beam reaches a maximum when $|\omega_1 - \omega_2| = \omega_r$, the frequency of a Raman-active vibration in the sample. The birefringence induced by ω_1 causes the exiting polarized probe beam to be transmitted through a crossed polarizer that blocks it with ω_1 absent. The spectrum of the transmitted probe beam (ω_2 fixed, when ω_1 is varying) reflects the Raman spectrum.

Unlike the CARS technique, the RIKES method always conserves momentum and energy for any two frequencies ω_1 and ω_2 and for any two relative beam angles. The direction of the probe beam and the crossing angle can be fixed as one scans a spectrum.

Levenson and coworkers (38) introduced an optical heterodyne detection method to the RIKES technique and were able to extract signals linearly proportional to the spontaneous Raman cross section. When the detected signal corresponds to the imaginary part of the nonlinear susceptibility, the OHD RIKES (occasionally called the "rippled RIKES") trace reproduces the spontaneous Raman scattering spectrum.

The phenomenon of the Raman-induced Kerr effect can be described in terms of a third-order nonlinear susceptibility. If the probe beam (ω_2) is initially linearly polarized in the x direction, the effect of the birefringence is to produce a component polarized in the y direction.

$$E_y(\omega_2) = \frac{24\pi i \omega_2 l_{\text{eff}} \chi^{(3)}(-\omega_2, \omega_1, -\omega_1, \omega_2) E_x(\omega_1) E_y^*(\omega_1) E_x(\omega_2)}{cn_2}$$

where $E_x(\omega_1)$ and $E_y^*(\omega_1)$ are amplitudes of the pump wave components, $E_x(\omega_2)$ is the probe beam amplitude, n_2 is the refractive index at ω_2, and l_{eff} is the effective interaction length of the two beams. For a circularly polarized pump beam, the effective nonlinear susceptibility is

$$\chi_{\text{eff}}^{(3)} = i[\chi_{2211}^{(3)}(-\omega_2, \omega_1, -\omega_1, \omega_2) - \chi_{2121}^{(3)}(-\omega_2, \omega_1, -\omega_1, \omega_2)]$$

$$= \frac{i[B(\omega_1 - \omega_2) - 2A(\omega_1 - \omega_2)]}{24}$$

where $A(\Delta\omega)$ and $B(\Delta\omega)$ are the complex functions related to the cross sections for polarized and depolarized Raman scattering (37).

The intensity of the y polarized component of the RIKES output I_R is equal to $(n_2 c/8\pi)|E_y(\omega_2)|^2$. This intensity is proportional to the square of the spontaneous Raman scattering cross section, just as in the case of CARS. This makes weak vibrational modes disproportionately difficult to detect. In addition, static birefringence in the sample and optics produces a spurious background intensity.

In Levenson's OHD RIKES technique, a local oscillator field $E_{\text{LO}}(\omega_2)$ polarized in the y direction is purposely inserted at the detector. The local oscillator field adds coherently to the RIKES field, producing a total intensity at the detector of

$$I_T = \frac{n_2 c}{8\pi}|E_{\text{LO}} + E_R|^2 + I_B = I_{\text{LO}} + I_B + I_R + \frac{n_2 c}{8\pi}(E_R^* E_{\text{LO}} + E_{\text{LO}}^* E_R)$$

If $I_{\text{LO}} \gg I_R$, the cross term $H = n_2 c/8\pi(E_R^* E_{\text{LO}} + {}^*E_R)$ is larger than the former RIKES intensity. It is *linear* in the scattering cross section. It has been demonstrated (38) that the heterodyne term H can be extracted electronically from the other contributions to the total intensity at the detector.

The relative phase of the RIKES and local oscillator field is an important parameter determining the appearance of the spectrum. If a retardation plate is inserted in the probe beam to generate a local oscillator 90° out of phase with $E_x(\omega_2)$, the detected signal reproduces the spontaneous Raman spectrum.

The OHD RIKES technique is still in an infant stage of development. In principle, it should exhibit resonance enhancement with electronic transitions. While it appears attractive for biological applications, its real value remains to be demonstrated.

7.3.8. Surface-Enhanced Raman Scattering (SERS)

Raman scattering from a monolayer of nonresonant molecules is weak to the point of being unobservable. However, certain molecules when adsorbed to a metal surface show Raman scattering with enhancement of up to a factor of 10^6. The system that has attracted the most interest is pyridine adsorbed on a silver electrode (39). When the silver electrode undergoes anodization and pyridine adsorption, the pyridine Raman lines at 1006 and 1033 cm^{-1} are inordinately enhanced. Other systems show this SERS. Silver, gold, copper, nickel, and platinum have been used as the substrate in the form of wire, foil, single crystals, evaporated film, and colloidal particles. Adsorbed species such as CN⁻, benzoic acid, isonicotinic acid, crystal violet, CO_3, Cl⁻, and CO have shown SERS. Methods of adsorption include electrochemical and chemical deposition and vapor deposition from vacuum.

Although there still remains a considerable question as to the exact mechanism responsible for SERS, most agree that surface roughness of the substrate plays an important role (39, 40). Other proposals include (1) excitation in an electronic adsorption band that is present only in the adsorbed system (41–44); (2) electron-hole pair excitations (45–47); (3) mixing of surface plasmons with the adsorbate's molecular electronic states (43); and (4) image field enhancement (48). Studies have also been performed to determine the roles of laser excitation (49) and chemisorption (50). Recent studies show that silver colloids (51) may mimic the roughened surface of a silver electrode in the ability to show SERS.

Although much of the work done in SERS has been on aqueous electrolytic systems, the technique shows potential for further applications. The primary importance of SERS is that it can serve as a probe of the vibrational characteristics of the gas-solid or liquid-solid interface without the restriction of ultrahigh vacuum conditions. SERS may be applicable to the study of catalytic surfaces, corrosive surfaces, photovoltaic solar cells, and electrocatalytic reaction mechanisms and kinetics (39).

7.3.9. High Precision Raman Difference Spectroscopic (RDS) Technique for the Measurement of Very Small Frequency Shifts

The detection of frequency differences of less than 1 cm⁻¹ between two samples is normally considered difficult if not impossible. However, the Raman difference technique using rotating cells, originally introduced by Kiefer (52, 53), has now been demonstrated (54–58) to be capable of measuring frequency differences of ~ 0.1 cm⁻¹, or as small as 0.02 cm⁻¹ under favorable conditions (56). The schematic diagram shown in Figure 7.10 represents the instrumentation developed at Bell Laboratories (59). It employs a cylindrical rotating cell with a central partition so that samples 1 and 2 are alternately irradiated by the laser beam. For each monochromator setting (advanced by a stepping motor) two samples are irradiated for a preset time period (usually 0.5 s to several seconds). The amplified signal after the photomultiplier is sent to the inputs of two linear gates. With proper adjustment of the gate delay times and window widths, only the signal from sample 1 passes through gate 1 and only the signal from sample 2 passes through gate 2.

The shutters of the gates are controlled by a synchronous signal from the rotator. After data integration in their respective counters for a preset period, the stepping motor is advanced by a preset increment and data integration restarts. Up to 4096 data points can be collected for each of the two spectra.

Here, intensities in the difference spectrum are used to determine frequency differences. Analysis of line-shape functions reveals the following relationships in the limit of small frequency differences:

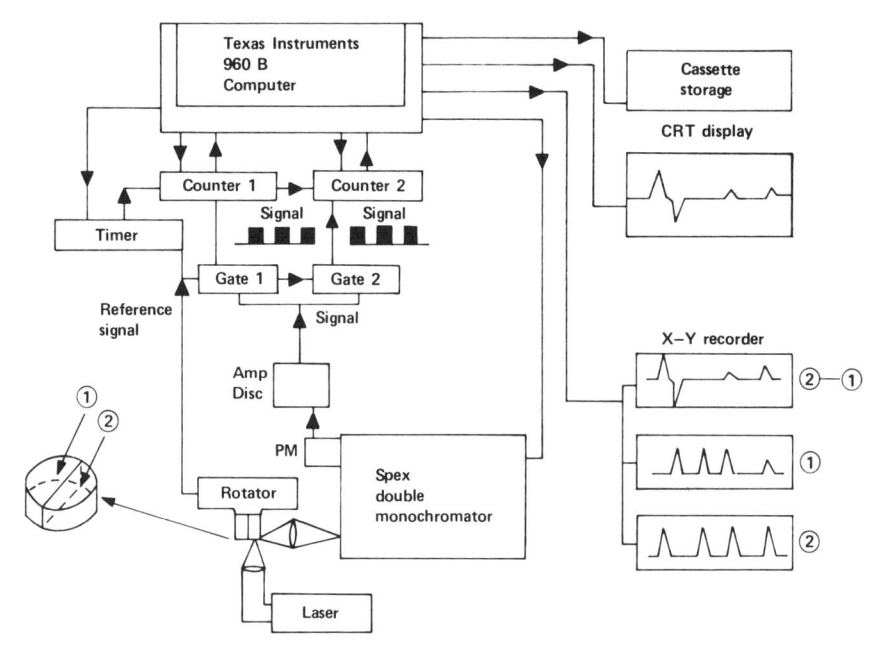

Fig. 7.10. Schematic diagram of a precision Raman difference spectroscopy system at Bell Laboratories. From D. L. Rousseau, *J. Raman Spectroscopy*, **10**, 94 (1981) (with permission).

$$\frac{\Delta \nu}{\Gamma} = 0.38 \frac{I_D{}^0}{I^0} \quad \text{(Lorentzian)} \tag{26}$$

$$\frac{\Delta \nu}{\Gamma} = 0.35 \frac{I_D{}^0}{I^0} \quad \text{(Gaussian)} \tag{27}$$

where $\Delta \nu$ is the frequency difference, Γ is full width at half-height of a Lorentzian or Gaussian line, $I_D{}^0$ is the measured intensity difference at the maximum and minimum in the difference spectrum, and I^0 is the center intensity of a Lorentzian or Gaussian line. When $\Delta \nu$ is very much smaller than Γ, frequency differences may be determined directly from the intensity measurements by using equations (26) and (27).

As the difference between two lines becomes smaller, the frequency spacing of the maximum and minimum reach a limiting value but the intensity continues to drop. The lower limit of a frequency difference is limited by the signal-to-noise ratio of the spectrum. It is also necessary to isolate the system from vibrations, stabilize the laser intensity, and minimize the wobbling of the scattering column.

7.3.10. Raman Microprobe Analysis

Direct detection and identification of molecular components in microsamples was not possible until the introduction of micro-Raman spectroscopy into the field of instrumental microanalysis. Raman microprobe analysis permits nondestructive compound-specific molecular microanalysis from single microparticles or sample regions of micrometer dimensions.

The extension of laser-excited Raman spectroscopy into microprobe spectroscopy, suggested first by Hirschfeld (60), was done by a simple modification of a conventional monochannel Raman spectrometer optimized for microparticle measurement. As illustrated by NBS instrumentation (61), Raman microprobes today employ a mixed gas Ar^+/Kr^- laser to irradiate the sample, which is mounted on a substrate. With the aid of a microscope objective the exciting beam is focused on a small area of the sample through a circular opening at the center of an ellipsoid mirror. With the sample positioned at the near focus of this mirror, the scattered radiation is collected at the far focus and transferred into a double monochromator employing concave holographic gratings. A single-channel, cooled photomultiplier tube (PMT) and photon-counting processing electronics detect the signal. In more recent years, Raman microprobes have come to employ multichannel optical detection, which can offer the advantage of increased data-handling capability (62).

Another major advance relevant to Raman microanalysis is the Raman microprobe/microscope developed at the Centre National de la Recherche Scientifique in Lille, France, and now commercially offered as MOLE (63, 64). This system has two independent modes of operation. One is a spectral mode that functions as a molecular microprobe. Using a monochannel detector, classical Raman spectra are recorded. In the multichannel spectral mode, spectra in spectrographic form are observed on an oscilloscope in real time. The other is an imaging mode that functions as a molecular microscope. Using a characteristic Raman frequency of a specific component in the sample, one can obtain a micrographic image showing the distribution of this species in the sample or sample area. The basic design consists of (1) a laser, (2) a conventional optical microscope for sample stage and part of the imaging system, (3) an optical filter, and (4) monochannel and multichannel detection systems. Observation of the micrographic image is possible from a TV monitor linked to the multichannel system.

Analytically useful Stokes-Raman spectra with good molecular specificity and detection limits in the picogram range have been obtained for various scattering species. Sample types include single microparticles, collection of particles (airborne particles and atmospheric aerosols), fluid inclusions in minerals, and biological and pathological materials (65, 66). Specific details on sample preparation and measurement procedures have also been discussed (65). The observed micro-Raman spectra are equivalent to or in some cases better than corresponding spectra from macroscopic samples of the same materials.

Raman microanalysis still presents inherent limitations and problems. Absorption of the focused laser radiation can cause sample heating and possibly sample decomposition. Suitable substrate materials such as high-quality sapphire partially eliminate this problem by serving as efficient heat sinks. Even more recently, spectra of intensely colored, highly opaque microsamples have been obtained using a method to further enhance thermal sinking into the substrate (67). Fluorescence from sample impurities or contaminants can also cause severe limitations since "cleanup" of microsamples is virtually impossible. Further aspects still under study include quantitation and effective sampling volume and spatial resolution (65, 68).

7.4. APPLICATIONS

7.4.1. Metalloporphyrins and Hemoproteins

7.4.1.1. Axial Ligand-Associated Vibrations

Resonance Raman spectroscopy has clearly emerged as the technique of choice for directly probing the nature of ligand bonding in hemoproteins and synthetic model hemes (69–93). While infrared spectroscopy has been useful in providing information such as $\nu(C-O)$ (94–97), $\nu(O-O)$ (98–101), $\nu(N-O)$ (102), $\nu(N=N=N)$ (103, 104), and $\nu(C\equiv N)$ (105) frequencies, its ultimate usefulness has been limited by strong water absorption and lack of sensitivity. Resonance Raman spectroscopy, on the other hand, can provide more detailed and specific information regarding the bonding geometry of ligands and the direct interactions between metal and ligands. Many ligand-associated vibrations not observable by IR absorption have now been detected by resonance Raman scattering (69–93). For example, in carbon monoxide complexes of hemoproteins, IR spectroscopy can measure only the $\nu(C-O)$ frequencies, whereas resonance Raman spectroscopy detects not only the $\nu(C-O)$ but also the $\nu(Fe-CO)$ bond stretching and the $\delta(Fe-C-O)$ angle bending. Other iron–axial ligand vibrations of hemoproteins detected by resonance Raman include $\nu(Fe^{III}-N_3)$ (75), $\nu(Fe^{III}-CN)$ (75a), $\nu(Fe^{III}-NO)$ (75b), $\nu(Fe^{III}-OH)$ (79), $\nu[Fe^{II}-N_\epsilon(His)]$ (87), and $\nu(Fe^{III}-F)$ (76). Axial ligand-associated vibrations in cobalt- and manganese-substituted hemoproteins are also resonance Raman visible (69, 74).

The ability to detect these vibrations in both hemoproteins and model hemes places resonance Raman spectroscopy in a very good position to critically test the applicability of various hypotheses for the mechanism of protein control of heme reactivity: the proximal base tension effect (106–109), the proximal imidazole deprotonation/H-bonding effect (110–114), the distal side steric effects (96, 115, 116), and the π-donor/acceptor interactions between the heme macrocycle and aromatic amino acids (54, 117, 118).

Identification of the $\nu(Fe-CO)$ stretch in hemoproteins by resonance Raman has been difficult because carbon monoxide dissociates from heme readily upon illumination by laser light, generating deoxy species which interfere with the observation of signals from unphotolyzed complexes. The quantum yield for photodissociation of carbonmonoxy Mb is 1.0 at 21°C (118). However, Tsubaki and coworkers (70), employing a multichannel detector, were able to obtain resonance Raman spectra of carbonmonoxy Hb and Mb (Fig. 7.11), which contain negligible contributions from photolyzed deoxy species. With excitation at 406.7 nm they observed lines at 507 (512), 578 (577), and 1951 (1944) cm^{-1} in HbCO (MbCO) sensitive to CO isotope substitution. Based on normal coordinate calculations and the pattern of observed isotope ($^{13}C^{16}O$, $^{12}C^{18}O$, $^{13}C^{18}O$) shifts, it was established (70) that the intense line at 507 (512) cm^{-1} is from the ν(Fe–CO) stretching mode and the weaker one at 578 (577) cm^{-1} is due to the δ(Fe–C–O) bending mode. The frequency at 1951 (1944) cm^{-1} are the ν(C–O) vibration in HbCO (MbCO), in agreement with those observed by IR spectroscopy (94, 97). It had been assumed (96) that in hemoproteins the $sp^2 \to \pi^*$ donation from the distal histidine would weaken the C–O bond [hence reducing the ν(C–O) stretching frequency] and oppose $d_\pi \to \pi^*$ donation from the iron to the CO, weakening the Fe–C bond. Thus, one might expect a decrease in the ν(Fe–CO) frequency as the ν(C–O) frequency decreases. However, in comparing MbCO and HbCO, it was surprising to note that the ν(Fe–CO) frequency in MbCO (512 cm^{-1}) is higher than that in HbCO (507 cm^{-1}), whereas the ν(C–O) frequency is lower in MbCO ($\sim$ 1944 cm^{-1}) than in HbCO (1951 cm^{-1}). In CO complexes of simple ferrous porphyrins, where there is no $sp^2 \to \pi^*$ donation, the ν(C–O) frequency should

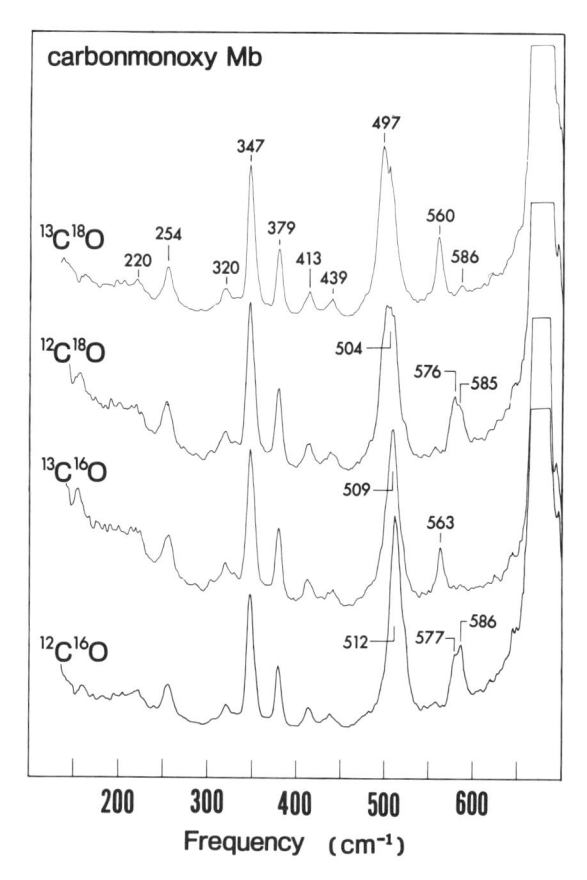

Fig. 7.11. Carbon monoxide isotope effects on low frequency region (100–700 cm^{-1}) spectra of carbonmonoxy Mb. The spectra are arranged in the order of decreasing masses of carbon monoxide isotopes. Excitation wavelength 406.7 nm. From M. Tsubaki, R. B. Srivastava, and N. T. Yu, *Biochemistry*, 21, 1132 (1982) (with permission).

appear higher (1970–1980 cm^{-1} versus ~1950 cm^{-1}) as actually observed. One would also expect the ν(Fe–CO) frequency to be much higher in simple ferrous hemes than in hemoproteins. Quite interestingly, Kerr and coworkers (71) found that the ν(Fe–CO) frequency in Fe(II) (TPP) (N-MeIm) (CO), as well as in Fe(II) (TpivPP) (N-MeIm) (CO), is ~25 cm^{-1} lower than that in MbCO despite the much higher affinity in the former. This appears to indicate a definite change in CO binding geometry in going from the simple heme–CO complex to MbCO. Normal mode analysis (71) showed that the ν(Fe–CO) frequency would increase in going from perpendicular linear to tilted linear or bent geometry, due to different degrees of mixing with other coordinates even without a change in force constants. Correlations between CO-associated vibrations and detailed bonding geometry are currently unknown. Since carbon monoxide binds to almost any ferrous hemes and hemoproteins and is incapable of oxidizing the heme, the detection of ν(Fe–CO) and δ(Fe–C–O) by resonance Raman opens up new possibilities for gaining important insights into the exact nature of ligand bonding in hemoproteins. Resonance Raman spectroscopy may serve to bridge the gap left by the extensive work of Caughey and associates on infrared ν(C–O) frequency of hemoproteins (94–97).

The frequency of bound O–O stretching vibration in hemoproteins has been a puzzle for some time. With infrared difference spectroscopy, Caughey and coworkers (98–100) reported the ν(O–O) frequencies at 1107 (oxy HbA), 1103

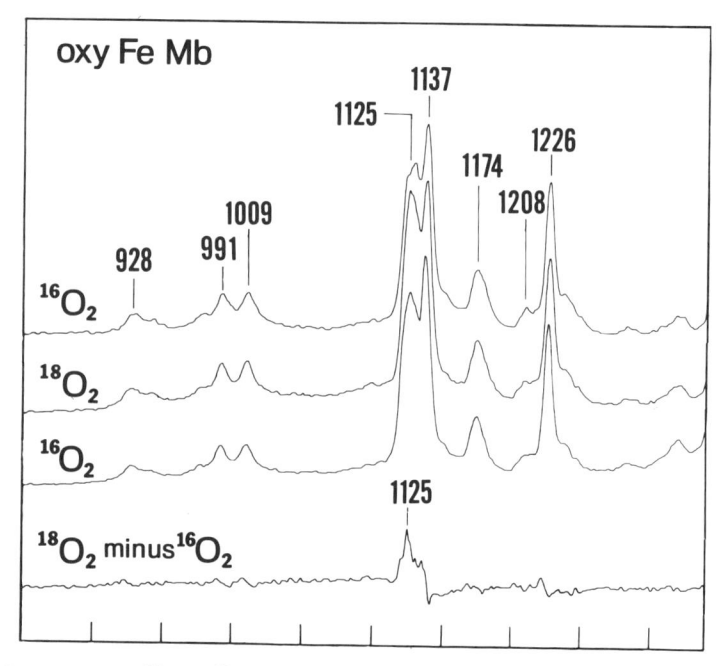

Fig. 7.12. Effect of $^{16}O_2 \rightarrow {}^{18}O_2$ isotope substitution in oxymyoglobin on Raman intensity at 1125 cm^{-1}. Excitation wavelength 406.7 nm. From M. Tsubaki and N. T. Yu, *Proc. Nat. Acad. Sci. USA*, **78**, 3581 (1981) (with permission).

(oxy Mb), and 1105 cm^{-1} (oxy Co-deuteroporphyrin IX-substituted HbA). However, these frequencies are quite different from the $\sim$ 1150–1160 cm^{-1} values in the oxygen adducts of Fe and Co "picket fence" porphyrins (119). Alben and coworkers (120) reported an additional infrared ν(O–O) stretch at 1156 cm^{-1} in oxy HbA. The splitting of the ν(O–O) mode into two IR bands at 1107 and 1156 cm^{-1} in oxy HbA was interpreted (120) as due to Fermi resonance with the first overtone of the ν(Fe–O) stretch at $\sim$ 570 cm^{-1} observed by Brunner (81), using resonance Raman spectroscopy. However, Tsubaki and Yu (69) observed an anomalous Raman intensity increase at 1125 cm^{-1} in oxy FeMb (Fig. 7.12) upon $^{16}O_2 \rightarrow {}^{18}O_2$ isotope substitution. This was not readily interpretable until the detection of three isotope-sensitive lines at 1103 (1107), 1137 (1137), and 1153 (1152) cm^{-1} in the resonance Raman spectrum of oxy CoMb (or oxy CoHbA) (69) (Fig. 7.13). It now turns out that the kind of Fermi resonance suggested by Alben and coworkers (120) may not be operative in both the Fe and Co systems.

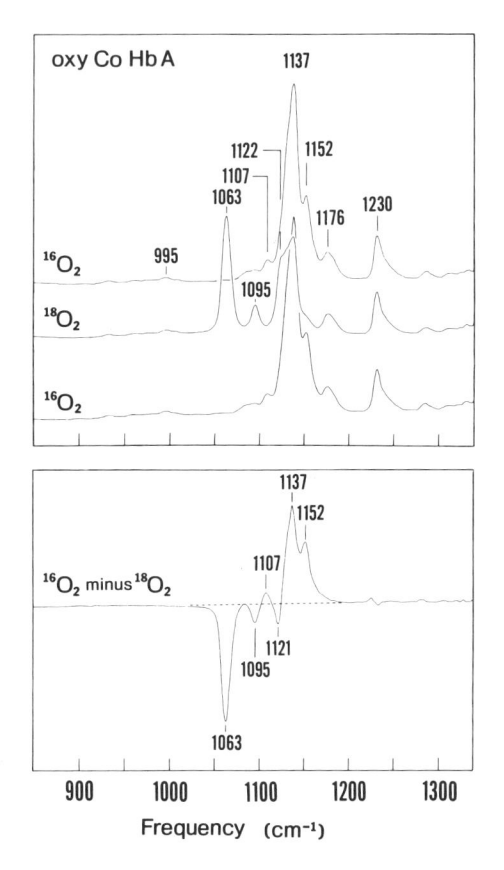

Fig. 7.13. Resonance Raman spectra of oxy-CoHbA (*upper*) and difference spectrum (CoHbA $^{16}O_2$ minus CoHbA $^{18}O_2$) (*lower*) in the 900–1300 cm^{-1} region. Excitation wavelength 406.7 nm. From M. Tsubaki and N. T. Yu, *Proc. Nat. Acad. Sci. USA*, 78, 3581 (1981) (with permission).

It was proposed (69) that the first two frequencies arise from resonance interaction between a ν(O–O) mode at $\sim 1122\,\text{cm}^{-1}$ and an accidentally degenerate porphyrin ring mode at $1123\,(1121)\,\text{cm}^{-1}$, whereas the third represents an "unperturbed" ν(O–O) vibration from a different conformer. The ν(Co–O) stretch was detected at $\sim 538\,\text{cm}^{-1}$, which is lower than the ν(Fe–O$_2$) frequency at $\sim 570\,\text{cm}^{-1}$ in oxy FeMb and oxy FeHbA. The Co–O bond appears to be longer and weaker than the Fe–O bond. Resonance Raman enhancement of both ν(O–O) and ν(Co–O) indicates the existence of a charge-transfer transition underlying the Soret band, which has been assigned as $\pi^*(\pi_g^* O_2/d_{xz}) \to \sigma^*(d_{z^2}\text{Co}/\pi_g^*)$. The bound O$_2$ is in a bent end-on configuration which may have two allowed orientations differing in the extent of $sp^2(N_\epsilon) \to \pi^*(O_2)$ donation from distal histidine. The observed ν(O–O) mode at $1152\,\text{cm}^{-1}$ in oxy CoMb and oxy CoHbA is close to the modes of dioxygen adducts of Fe or Co "picket fence" porphyrin complexes (~ 1150–$1160\,\text{cm}^{-1}$) in which there is no donation to bound dioxygen from the distal site.

The effect of a weakening in the Fe–N$_\epsilon$ (imidazole) bond upon the ν(Fe–O$_2$) stretching frequency in model hemes is of considerable interest because of possible implications for hemoglobin cooperativity. Walters and coworkers (121) reported a $4\,\text{cm}^{-1}$ difference between Fe(II)(TpivPP)(1-MeIm)(O$_2$)/CH$_2$Cl$_2$ ($568\,\text{cm}^{-1}$) and Fe(II)(TpivPP)(1,2-Me$_2$Im)(O$_2$)/CH$_2$Cl$_2$ ($564\,\text{cm}^{-1}$) at 25°C. On the other hand, Hori and Kitagawa (92) found the ν(Fe–O$_2$) stretch to be insensitive to steric hindrance of the 2-methyl group in the bound imidazole. They located the ν(Fe–O$_2$) frequency at $568\,\text{cm}^{-1}$ in both complexes. Recently, Kerr and coworkers (71) confirmed the observation of Walters and coworkers (121) and found a $6\,\text{cm}^{-1}$ difference in ν(Fe–O$_2$) between the two complexes in benzene solution. In the case of CO complexes it is rather unexpected (71) that the ν(Fe–CO) stretching frequency increases upon a decrease in affinity of Fe(II)(TpivPP)(1-MeIm)(CO) to Fe(II)(TpivPP)(1,2-Me$_2$Im)(CO). The difference in CO binding free energy between these two derivatives is definitely not localized in the Fe–C bond. The free energy associated with the weakening of other bonds, particularly the C–O bond, must also be considered.

Novel resonance Raman enhancement of nontotally symmetric internal azide vibrations was reported by Yu and Tsubaki (74) for the Mn(III)Mb·N$_3$ complex. Herzberg-Teller vibronic couplings between a charge-transfer and $\pi \to \pi^*$ Soret states are indicated. In contrast, the one-state Franck-Condon scattering mechanism, via resonance with the Fe(d_π) $\to$ pyridine (π^*) transition, may be responsible for the resonance enhancement of only totally symmetric internal pyridine vibrational modes in bis-(pyridine)iron(II)mesoporphyrin(IX)dimethyl ester (122). In Mn(III)Mb·N$_3$, the enhancement of bound azide vibrations at $650\,\text{cm}^{-1}$ (dp, bending) and $2039\,\text{cm}^{-1}$ (dp, antisym stretch) upon excitation at 400–460 nm suggests the existence of a new charge-transfer transition (Figs. 7.14 and 7.15) which has been assigned as azide (π) $\to$ porphyrin (π^*).

The azide complexes of metmyoglobin and methemoglobin exhibit a temperature-dependent spin equilibrium. An absorption band at $\sim 640\,\text{nm}$ has been assigned to an x, y-polarized high-spin charge-transfer transition (123). When azide metmyoglobin is converted completely to the low-spin form (at 77 K) the absorp-

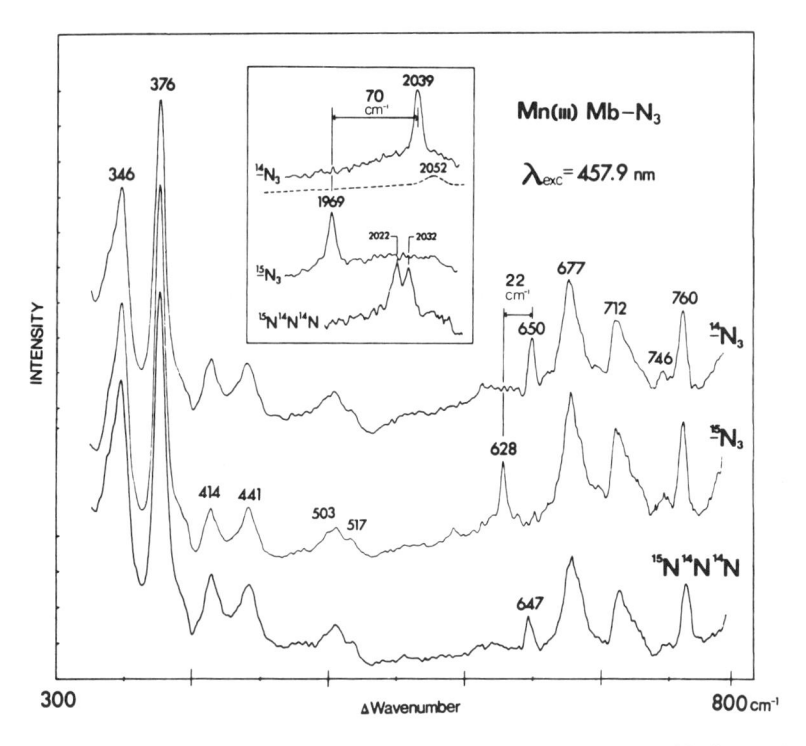

Fig. 7.14. Resonance Raman spectra of Mn(III)Mb–N$_3$ with λ_{exc} = 457.9 nm, showing isotope-sensitive lines at 650 and 2039 cm^{-1}. From N. T. Yu and M. Tsubaki, *Biochemistry*, **19**, 4647 (1980) (with permission).

tion spectrum still shows a weak broad band centered at ~650 nm, which was suggested by Eaton and Hochstrasser (123) as a charge-transfer band (z-polarized) corresponding to a transition of the low-spin form of metmyoglobin azide. By excitation into the 640-nm absorption region, Asher and coworkers (79) observed the selective enhancement of a Raman mode at 413 cm^{-1} in methemoglobin azide and assigned it to the ν(Fe–N$_3$) stretch (without isotope evidence). Later, Desbois and coworkers (86) assigned a 570-cm^{-1} mode (excited near Soret band) in metmyoglobin azide to the ν(Fe–N$_3$) stretch. This 570-cm^{-1} line shifts to 554 cm^{-1} upon ^{15}N$_3$ isotope substitution. To explain the discrepancy between these two assignments, Asher and Schuster (77) suggested the the 413-cm^{-1} mode was from the high-spin form and the 570-cm^{-1} mode from the low-spin form. However, Tsubaki and coworkers (75a) obtained evidence that suggested different assignments. They demonstrated that with decreasing temperature the intensity of the 413-cm^{-1} mode increases in spite of the decrease in high-spin component (hence the absorption at ~640 nm). It appears that the 413-cm^{-1} mode derives its intensity from the weak low-spin charge-transfer band underlying the stronger high-spin charge-transfer band at ~640 nm. Based on temperature studies, isotope substitution, depolarization measurements, and normal coordinate analysis, they assigned

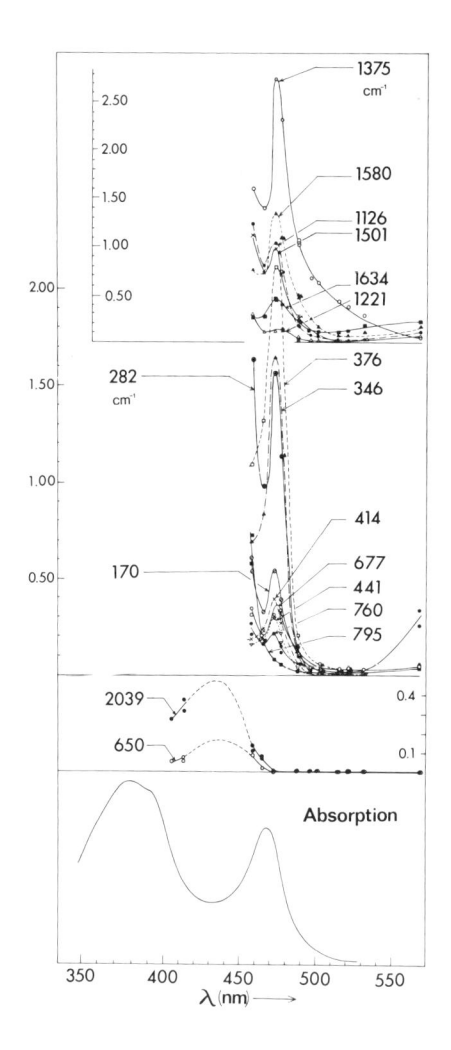

Fig. 7.15. Excitation profiles of major Raman lines of Mn(III)Mb·N$_3^-$ near band V and band VI absorption maxima. From N. T. Yu and M. Tsubaki, *Biochemistry*, **19**, 4647 (1980) (with permission).

the 413 and 570 cm^{-1} modes as Fe-azide stretching and azide internal bending, respectively, both in the low-spin state. In addition, they detected resonance enhancement of bound azide anti-symmetric stretching from both high- and low-spin forms with excitation at 406.7 nm.

Cytochromes P-450 are a unique class of hemoproteins that catalyze the hydroxylation of a wide variety of organic compounds through the activation of molecular oxygen. The presence of a sulfur ligand (mercaptide) in P-450 has been repeatedly suggested (124–129), but no direct evidence was available. The search for the ν(Fe–S) stretching mode in the resonance Raman spectrum of P-450 began in 1977 when Champion and Gunsalus (130) reported a low-frequency vibration at 351 cm^{-1} in the spectrum of cytochrome P-450$_{cam}$ excited at 363.8 nm. This 351-cm^{-1} line, also observed (6) at ~550 nm, was suggested (6) as the ν(Fe–S) stretch on the basis of excitation profiles, although it was assigned (130) to a ring mode involving Fe–N stretching. Recently, Champion and coworkers (131), employing isotopically labeled (^{54}Fe and ^{34}S) samples of the oxidized P-450$_{cam}$ substrate complex, conclusively demonstrated the existence of an Fe–S bond. Indeed, the 351-cm^{-1} Raman line suggested by Felton and Yu (6) is the ν(Fe–S) stretching vibration. It exhibits a -4.6 ± 0.3 cm^{-1} downshift upon the ^{34}S isotope substitution and a $+2.4 \pm 0.2$ cm^{-1} upshift upon ^{54}Fe substitution.

Identification of the ν[Fe–N$_\epsilon$(His)] stretch in hemoproteins is important, as it provides crucial information for understanding the mechanism of protein control of heme reactivity. Initial assignments of this mode to a 411-cm^{-1} line (86) or a 375-cm^{-1} line (132) have not been substantiated by recent data (133–137). Kitagawa and coworkers (134–137), based on isotope and model compound studies, have assigned the Raman mode at 220 cm^{-1} in deoxy Hb (excited near Soret) as the iron-histidine stretching frequency. It was found (135, 136) that large differences in both frequency and line shape of this mode exist between T-state HbA and R-state NES des-Arg Hb at room temperature. In T-state valency hybrid hemoglobins, the α chains display a large frequency shift (15 cm^{-1}) from their R-state values than do the β chains (~4 cm^{-1}) (136). Recently, Ondrias et al. (138) detected a Raman line at ~230 cm^{-1} in deoxy Hb at 77°K, which was thought to be due to the iron-histidine stretch, but later identified as an ice lattice mode (93).

7.4.1.2. *A Spectroscopic Ruler for Measuring the Size of the Heme Central Hole*

The distance between the center of the porphyrin core and the pyrrole nitrogen (Fig. 7.16), denoted as $d(C_t$–N), may be used to indicate the size of the heme central hole. This $d(C_t$–N) distance of metalloporphyrins and hemoproteins in solution could not be directly measured until the discovery of a "core-expansion"

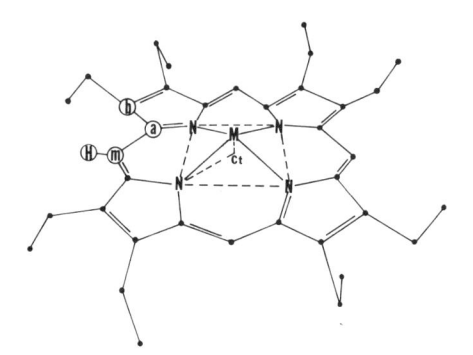

Fig. 7.16. Structure of a porphinato core, showing the Ct–N distance. From L. D. Spaulding, R. C. C. Chang, N. T. Yu, and R. H. Felton, *J. Amer. Chem. Soc.*, **97**, 2517 (1975) (with permission).

Table 7.2. Correlation of the Anomalously Polarized Frequency with C_t–N Distance[a]

Compound	$\bar{\nu}$ (cm^{-1})	C_t–N (Å)	Structure
SnOEPCl$_2$	1545	2.082 (2)	SnOEPCl$_2$
AgOEP	1550	~2.08	AgTPP
MgOEP	1558	2.055 (6)	MgTPP(H$_2$O)
ZnOEP	1565	2.047 (2)	PyZnTPyP
RhEtio (DMA)$_2$Cl	1578	2.038 (6)	RhEtio (DMA)$_2$Cl
(FeOEP)$_2$O	1570	2.027	(FeTPP)$_2$O
FeOEPCl	1568	2.019 (3)	Average hemin
FeProtoDMECl	1575	2.007 (5)	FeProtoCl
VOEtio	1574	2.01 (4)	VODPEP
PdOEP	1585	2.009 (9)	PdTPP
Fe^{II}Etio (Py)$_2$	1587	2.004 (4)	Fe^{II}TPP(Im)$_2$
CuOEP	1585	2.000 (5)	CuTPrP
Fe^{III}OEP(Im)$_2$Cl	1590	1.989 (5)	Fe^{III}TPP(Im)$_2$Cl
FeOEP(NO)	1593	1.990	FeTPP(NO)
MnOEPCl	1591	1.99	MnTPPCl
CuOEP	1605	1.981 (7)	CuTPP
CoOEP(NO)	1605	1.976 (3)	CoTPP(NO)
NiOEP (D_{4h})	1609	1.958 (2)	NiOEP
NiOEP (D_{2d})	1590	1.929 (3)	NiOEP

[a] Py = pyridine, Im = imidazole, TPyP = tetra(4-pyridyl)porphinato, DMA = dimethylamine, Proto = protoporphyrin IX, DPEP = deoxyphylloerythroetioporphinato, TPrP = tetra(n-propyl)porphinato, DME = dimethoxy ester.

correlation (139). Spaulding and coworkers (139) found that an ap line between 1540 and 1610 cm^{-1} (observable only with Q-band excitation) correlates well with $d(C_t$–N) in a series of metalloporphyrin complexes with known X-ray structural data. The original data used in establishing the correlation are shown in Table 7.2 and plotted in Figure 7.17, which includes recent data of Spiro and coworkers (140).

In addition to this ap (band IV) frequency, Huong and Pommier (141) proposed a similar inverse linear relationship for band V (dp), which is also plotted in Figure 7.17. Useful correlations have also been found for band II (140), band VI (142), and other skeletal modes (above 1450 cm^{-1}) (143).

The "core-expansion" correlations may be used for assessment of the out-of-plane displacement of the iron from the plane of the four pyrrole nitrogens, *if the iron-to-pyrrole nitrogen distance, d(Fe–N), can be estimated.* For example, in HbO$_2$, a d(Fe–N) = 2.00 Å is anticipated for the low-spin Fe(II) complex. Since the ap line at 1585 cm^{-1} for HbO$_2$ implies a $d(C_t$–N) of 2.00 Å, it is concluded that oxyhemoglobin contains in-plane iron. The application of the core-expansion correlation to deoxy Hb yields a value of 0.40 Å for the out-of-plane displacement of high-spin Fe(II), which is a good agreement with the X-ray data (144, 145) and Mössbauer experiments (146).

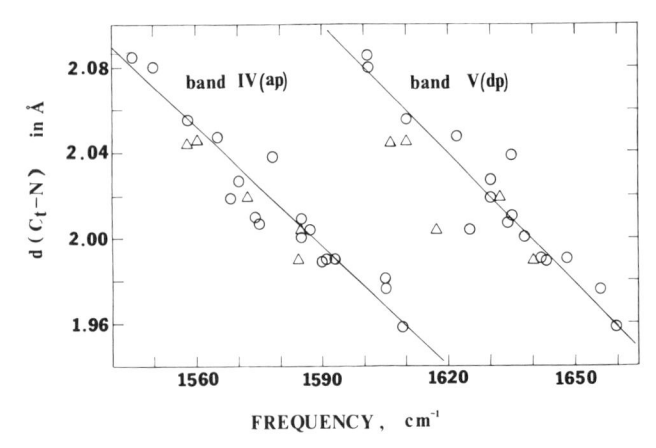

Fig. 7.17. A plot of the frequency of Raman bands IV (*ap*) and V (*dp*) vs. d(Ct–N). From A. Lanir, N. T. Yu, and R. H. Felton, *Biochemistry*, 18, 1656 (1979) (with permission).

7.4.1.3. Time-Resolved Resonance Raman Studies

The structural events immediately following the photolysis of carbonmonoxy Hb/Mb may be monitored by pulse laser resonance Raman techniques (17, 147, 148). Resonance Raman spectra contain structural information related to iron-ligand bond length (134–137), porphyrin core expansion (139–143), and electron donation into the porphyrin π^* orbitals (149–151), and thus are more informative than absorption spectra. In two-pulse experiments in which the initial high-power pulse photolyzes the CO ligand, followed by a certain delay by a probe pulse to excite the Raman spectra, Friedman and Lyons (147a) were able to obtain data in the 20 ns to 1 ms region. Time-resolved resonance Raman spectra may also be obtained in single-pulse experiments in which each pulse serves both to photolyze the sample and to excite the Raman spectrum. The resolution is limited only by the pulse duration. Recently, Terner and coworkers (17) recorded resonance Raman spectra of COHb photoproduct on the picosecond time scale by employing the 30-ps pulses of a synchronously pumped mode-locked cavity-dumped dye laser. The spectrum of the photoproduct was found to be similar to that of deoxy Hb, but the frequencies at 1603, 1552, and 1542 cm^{-1} were 2–4 cm^{-1} lower than those of deoxy Hb. The frequency shifts relative to deoxy Hb persisted even when the laser pulses were increased to 20 ns. This slow relaxation may be associated with the globin quaternary structure change.

Continuous applications of pulse laser time-resolved resonance Raman techniques to probe the structural events associated with photolysis and recombination may ultimately establish the relationship between the heme structural changes and the tertiary and quaternary protein structural changes.

Several excellent review articles (6, 76, 152) dealing with resonance Raman spectroscopy of metalloporphyrins and hemoproteins have appeared and may serve as important references for those who desire further information.

7.4.2. Dioxygen Binding, Non-Heme Heterocyclic Compounds

Certain transition metal heterocyclic complexes are capable of binding oxygen. Studies of these compounds are of interest since they can serve as models for biologically important substances such as vitamin $B_{12}r$, hemerythrin, and hemocyanin. When dioxygen binds to these compounds, the dioxygen can take on superoxo (O_2^-) or peroxo (O_2^{2-}) character. The most commonly formed complex involves a cobalt complex with dioxygen. The complexes form as 1:1 cobalt-dioxygen, 2:1 cobalt-dioxygen with dioxygen being the only bridging species, or as 2:1 cobalt-dioxygen with a second bridging species such as hydroxyl. These complexes usually have a charge-transfer transition involving O_2 and cobalt(III) ($O_2^- \rightarrow Co^{3+}$ or $O_2^{2-} \rightarrow Co^{3+}$). This transition usually lies in the UV or visible region, making resonance Raman spectroscopy an excellent tool for observing both the $\nu(O-O)$ and $\nu(Co-O)$.

Szymanski and coworkers (153) were the first to observe the O—O stretching frequency of a 1:1 cobalt-dioxygen complex by resonance Raman spectroscopy. The complex, 7,8,15,16,17,18-hexahydro-3,12-dinitrobenzo[e,m][1,4,8,11]tetra-azocyclotetradecinatocobalt(II), Co(NO$_2$Cyen), in pyridine-dimethylformamide (10% v/v) solution, gave an absorption maximum at 4900 Å. Thus, a 4880 Å laser line was used as the exciting wavelength. At $-80 \pm 0.1°C$, oxygenation of the complex produced a new resonance Raman line at 1137 cm^{-1}, which shifted to 1078 cm^{-1} upon substitution with $^{18}O_2$. The observation of $\nu(O-O)$ at 1137 cm^{-1} is in agreement with the Co^{3+}-O_2^- formalism of a superoxo complex. One anomalous observation is that the integrated intensity of the $\nu(^{18}O-^{18}O)$ is only 21% of that of $\nu(^{16}O-^{16}O)$, although oxygenation was checked by ESR and the possibility of contamination by $^{16}O_2$ was ruled out.

The equilibrium between a 1:1 and 1:2 dioxygen-cobalt complex was studied by Kozuka and coworkers (154). The cobalt(II) chelate, Co(II) (J-en) in methylene chloride solution containing 1% pyridine, was examined by resonance Raman spectroscopy using 514.5 nm excitation. At $-45°C$. the oxygenated complex revealed oxygen isotope–sensitive lines at 1144 and 836 cm^{-1} ($^{16}O_2$). The 1144-cm^{-1} line is in the superoxo region, indicating a 1:1 complex. The 836-cm^{-1} line is in the peroxo region, indicating a 1:2 complex, At $-75°C$, the 836-cm^{-1} line nearly disappears, while the 1144 cm^{-1} line gains intensity, showing that at this low temperature the 1:1 complex predominates. Raising the temperature back to $-45°C$ gives back the original spectrum, showing the reversibility of the equilibrium in this temperature range.

Another system of interest is the Co(salen) complex. This compound binds dioxygen in the solid state, forming the dinuclear, monobridged complex [Co(salen)]$_2$O$_2$. Both Hester and Nour (155a) and Suzuki and coworkers (156) have recently examined this complex via resonance Raman spectroscopy. Hester and Nour assigned the $\nu(O-O)$ to 1012 cm^{-1} and the $\nu(Co-O)$ to 376 cm^{-1} using 568.2-nm excitation. Using an excitation wavelength of 579 nm, Nakamoto and coworkers found the $\nu(O-O)$ at 1011 cm^{-1} and $\nu(Co-O)$ at 533 cm^{-1}. Also, another $\nu(Co-O)$ was found at ≈ 370 cm^{-1} using a 514.5-nm laser line. These workers assigned the 533-cm^{-1} line to the symmetric Co–O stretch and the 370-cm^{-1} line to the antisymmetric stretch and also concluded that the Co–O–O–Co atoms were very nearly transplanar. Both groups determined that the $O_2^{2-} \rightarrow Co^{3+}$ charge-transfer transition was located at ∼ 570–580 nm.

Hester and Nour (155a) also studied [(L)(salen)CoO$_2$Co(salen)(L)] with L = DMSO, py, DMF, pyO. In all cases, the ν(O–O) (875–911 cm^{-1}) were lower than the corresponding ν(O–O) in the complexes without L. The ν(Co–O) were observed in the 525–545 cm^{-1} region. The effect of the transaxial ligand on the ν(O–O) and ν(Co–O) values is explained in terms of σ and π donor effects through the metal. An excitation profile of the ν(O–O) of the DMF complex resulted in assignment of the ∼ 370-nm band of the electronic absorption spectrum to the $O_2^{2-} \rightarrow Co^{3+}$ charge transfer transition.

The complex bis(dimethylglyoximato)cobalt(II) undergoes reactions similar to cobalamine, vitamin B$_{12}$r, and so the complex is named cobaloxime(II). Hester and Nour (155b) have studied the μ-peroxo dinuclear complex of cobaloxime(II), [(L)(D$_2$H$_2$)CoO$_2$Co(D$_2$H$_2$)(L)], where D$_2$H$_2$ is dimethylgloxime and L = py, (C$_2$H$_5$)$_3$N, (CH$_3$)$_2$S, (C$_6$H$_5$)$_3$P, (C$_6$H$_5$)$_3$As, or (C$_6$H$_5$)$_3$Sb. At $-60°$C, the ν(O–O) were observed in the 814–831 cm^{-1} region and the ν(Co–O) ranged from 555 to 574 cm^{-1} using the 568.2-nm laser line. The ν(O–O) and ν(Co–O) varied with L, but to a lesser extent than the salen complexes. Also, the salen complexes revealed higher ν(O–O) and lower ν(Co–O). This was explained in terms of the higher conjugation of the salen ligand, resulting in delocalization of cobalt d electrons via the π-molecular orbitals of the salen. This results in a weakening of the Co–O bond.

Resonance Raman studies of μ-peroxo and μ-peroxo-μ-hydro dicobalt(II) ethylenediamine and L-histidine complexes were performed by Hester and Nour (155c). For the ethylenediamine complexes, [(en)$_2$CoO$_2$Co(en)$_2$]$^{4+}$ and [(en)$_2$CoO$_2$(OH)Co(en)$_2$]$^{3+}$, the ν(O–O) were found at 799 and 795 cm^{-1}, respectively. The ν(Co–O) were found at ∼ 590 cm^{-1} using λ_{exc} = 406.7 nm. The 360 nm band of the absorption spectrum is assigned to the $O_2^{2-} \rightarrow Co^{3+}$ charge-transfer transition. Also, the ν(Co–N), N being from the ethylenediamine, was found at 494 and 534 cm^{-1} for the monobridged complex and at 530 cm^{-1} for the dibridged complex. The L-histidine complexes, [(L-his)$_2$CoO$_2$Co(L-his)$_2$] and [(L-his)$_2$CoO$_2$(OH)Co(L-his)$_2$] showed ν(O–O) at 811 and 805 cm^{-1}, respectively, and ν(Co–O) at 600 and 590 cm^{-1}. The ν(Co–N) were found at 530 and 493 cm^{-1} using λ_{exc} = 457.9 nm. The $O_2^{2-} \rightarrow Co^{3+}$ charge-transfer transition for the dibridged complex was assigned to the 366 nm band. For the monobridged complex, either the 385 or 329 nm band may be responsible, but short-wavelength laser excitation was unavailable. In the L-histidine complexes, the difference between the mono- and dibridged complexes lies in the values for ν(Co–N), indicating involvement of L-histidine in the resonance effect. Thus, changes in geometry of the ligands that occur on changing from mono- to dibridged may be reflected in the ν(Co–N).

7.4.3. Chlorophylls

Chlorophylls are the photosynthetic pigments found in green plants, algae, and photosynthetic bacteria that function in absorbing and transporting photons of light (antenna chlorophyll) and converting the light energy into electrical charge separation (reaction center chlorophyll). The chlorophyll molecule contains a magnesium porphyrin with intense electronic absorption bands in the visible region. Hence, resonance Raman spectroscopy can be used to probe the chromophore to gain information about binding interactions.

In the past few years, Lutz and coworkers have done resonance Raman studies of chlorophylls *in vitro*, in chlorophyll-protein complexes, algae and bacterial antenna chlorophyll, and bacterial reaction centers (157).

Resonance Raman spectra of chlorophyll-protein complexes from green and blue-green algae and higher plants (158) showed that these complexes are very similar to bulk antenna chlorophyll *in vivo*, that is, the same multiplicity of binding stites for ketone carbonyl groups of Chl a. It was concluded that the binding sites are not magnesium atoms from other chlorophyll molecules and that the magnesium atoms bind one single external ligand. The resonance Raman spectra of chlorophyll b–protein complexes were similar to the intact membrane complexes except for partial rearrangement of one of two environmental subspecies of Chl b.

Bacteriochlorophyll is thought to be the primary electron donor in photosynthetic bacteria reaction centers. The light-induced, one-electron oxidation of BChl *in vivo* produces a radical cation similar to the *in vitro* BChl$^+$ π-cation. Ionization weakens the carbon-carbon pyrrole and methine bridge bonds, while the magnesium-nitrogen and carbon-nitrogen bonds are basically unaffected as shown by resonance Raman studies (158). Previous ESR and ENDOR studies (159, 160) showed the unpaired electron to be delocalized over two bacteriochlorophyll molecules; however, Raman studies showed the electron to be localized on a single bacteriochlorophyll molecule by virtue of the shorter time scale of the scattering process ($\approx 10^{-13}$ s).

Selective excitation of bacterial reaction centers can give resonance Raman spectra of a particular chromophore (161). A reaction center consists of a three-subunit protein containing one cartenoid, two bacteriopheophytin, and four bacteriochlorophyll molecules, two quinone, and one nonheme iron. The reaction-center (RC) bound carotenoids were found to be in a (di) *cis* conformation, whereas bulk carotenoids assume an all-*trans* conformation. Carotenoidless mutants of the R26 strain bind bulk carotenoids, causing a flip from all-*trans* to (di) *cis* upon binding (161, 162). Thus RC proteins contain a specific binding site for the carotenoid.

Excitation in the 528–535- and 545–550-nm regions can selectively enhance the Raman spectra of either one of two bacteriopheophytin molecules (BPheo) (161). The spectra show that the two BPheo share different environments in the reaction center, particularly through differences in the stretching frequency of the nine C=O ketone groups and the two C=O acetyl groups (163).

Excitation of the Q_x band (580–610 nm) can selectively enhance the four bacteriochlorophyll molecules (161). It was concluded that the nine C=O and two C=O of all six BPheo and BChl molecules are bound to either protein or water and no intermolecular bonding involving magnesium atoms exists.

Magnesium isotope studies (161) indicated that the band at 290 cm^{-1} is sensitive to isotope substitution. Based on other studies of hexa- and pentacoordinated magnesium complexes, the BChl complexes have been assigned to be 5-coordinate (161, 163).

7.4.4. Flavins and Flavoproteins

Flavins and flavoproteins are essential molecules of the biological process of oxidoreduction. The flavins, flavomononucleotide and flavin adenine dinucleotide, are always found covalently bound to a protein, and thus the holoenzyme is called a flavoprotein. The flavin ring system, isoallozazine (Fig. 7.18), can undergo reversible oxidation and reduction in two consecutive one-electron transfers. Flavins are natural chromophores and are therefore good candidates for resonance Raman studies. However, they fluoresce very strongly, serving to obscure the weaker Raman signals. Recently, three techniques to overcome the fluorescence problem have proven to be successful in obtaining the resonance Raman spectra of flavins and flavoproteins.

Making use of the fact that fluorescence is observed on the Stokes side of the

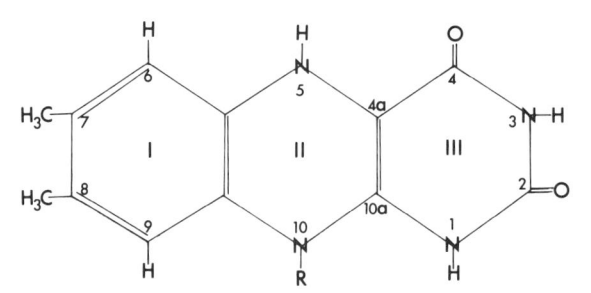

Fig. 7.18. Chemical structures of oxidized and reduced flavin ring system.

exciting frequency, Spiro and coworkers have studied flavins using coherent anti-Stokes Raman scattering (CARS). This nonlinear Raman process has produced high-quality spectra of flavin adenine dinucleotide (FAD), riboflavin binding protein, and glucose oxidase (164). Several vibrational modes were observed and identified by analogy with uracil. Of interest was the ~ 1295-cm^{-1} line that was sensitive to the degree of hydrogen bonding of the N_3 atom, shown by shifts observed in riboflavin binding protein and the complete disappearance in glucose oxidase. Resonance CARS studies of the semiquinone and oxidized forms of flavodoxin (165) excited in a longer wavelength absorption band showed small lowering of vibrations compared to flavomononucleotide (FMN). Three bands were assigned to environmental effects of protein binding site. At this excitation, the semiquinone and oxidized forms differed markedly. At shorter wavelength excitation, both forms showed only one enhanced band at 1584 and 1616 cm^{-1} for the oxidized and semiquinone, respectively. Thus, the shorter wavelength absorption band is the same $\pi \to \pi^*$ transition in both forms. Other work by this group includes line-shape studies of the CARS spectrum of FAD (166) and MNDO-MOCIC calculations for the interpretation of flavin vibrational spectra (167).

Another technique used to minimize the fluorescence problem is addition of KI to act as a collision quencher. McFarland and coworkers have studied 8-mercaptoriboflavin in the neutral and anionic forms (168). Comparison of the two spectra revealed that when protein binds, the flavin is in the 8-SH form rather than the S—S form. Shifts in the 1257-cm^{-1} band are indicative of weaker hydrogen bonding between flavin and riboflavin binding protein than between flavin and water. Studies of protonated, deuterated, and partially reduced flavin complexes (169) showed marked differences in the Raman spectra. However, charge-transfer complexation, as observed in fatty acyl CoA dehydrogenase showed no significant covalent binding effects. The only major spectral change occurred in the region sensitive to hydrogen bonding. It was recently suggested (170) that in the charge-transfer complex of acetoacetyl-CoA dehydrogenase with substrate, the electron transfer takes place at the N_5 and C_{4a}. This was based on the strong coupling between the charge transfer band and the 1586 cm^{-1} Raman line which is associated with the $\nu(C=N)$ at N_5 and C_{4a}.

Raman studies of a Ag$^+$ complex of flavin mononucleotide (171) showed that certain spectral changes occur in the flavin upon complexation in the $\nu(C_{4a}=N_5)$ region and the δN—N—H bending at N_3. Similar changes were observed in an Ru^{2+}-FMN complex, even though the electronic spectra are very different. Thus, the spectral changes are a result of metal complexation at N_5 and the oxygen at C_4 of flavin rather than the vibronic interactions that give rise to the resonance enhancement.

Kitagawa and coworkers have studied flavins by resonance Raman spectroscopy by virtue of the fact that when flavins are bound to proteins the fluorescence is nearly completely quenched. Isotope studies of riboflavin bound to egg white flavoprotein (172) revealed, at 488.0-nm excitation, a line at 1252 cm^{-1} indicative of N_3—H • • • protein interaction. At 370-nm excitation, a line at 1584 cm^{-1} indicates N_5 displacement, and at 450-nm excitation, the 1355-cm^{-1} line involved displacement of all ring III carbon atoms. These last two lines can be used to probe the electronically excited states of isoalloxazine. Charge-transfer interactions (173, 174) between old yellow enzyme, which contains one FMN per monomer, and phenols, which are enzyme inhibitors, have been investigated. The resonance Raman spectra support the model that the charge transfer involves electron donation from phenolate oxygen to FMN of old yellow enzyme, adopting an oxidized-like state. The spectra contained primarily in-plane modes of bound phenol, including several FMN modes. Most noticeable were the 1585 and 1550 cm^{-1} lines, which are vibrational displacements of N_5 and C_{4a} of isoalloxazine. Ring III modes were moderately enhanced. The conclusion drawn was that the π electrons of the benzene ring interact as a whole with ring III of isoalloxazine in a parallel arrangement of molecular planes with the phenolate oxygen above the N_5–C_{4a} bond.

Two separate resonance Raman studies of flavocytochrome C_{552}, a multicenter redox molecule containing heme and flavin centers, have recently been performed. Ondrias and coworkers (175) found lines corresponding to porphyrin ring modes, and the presence of the flavin was noted by fluorescence background. Variations in pH and conditions of inhibitor binding showed no direct interaction between flavin and heme. It was proposed that the two groups interact via protein-mediated interactions.

Kitagawa and coworkers did find direct flavin-heme interaction (176). In the intact flavocytochrome, both flavin and heme were photoreduced in the presence of EDTA. However, in the split unit, the flavin was photoreduced but the heme was not, even with EDTA. Therefore, it was proposed that the flavin is reduced first with EDTA, followed by electron transport to the heme.

7.4.5. Heterocyclic Enzyme-Substrate/Inhibitor Complexes

Resonance Raman spectroscopy can be used as a unique probe of the active site of an enzyme. Because most enzymes are not chromophores, a technique called resonance Raman labeling (177) may be used. This involves use of a chromophoric substance that mimics the biological substrate. Hence, enhancement of the resonance Raman spectra of the chromophore can reveal information about the nature of the active site of bound enzyme. Many of the chromophoric substrates contain heterocycles such as furans. Carey and coworkers (178) have used this technique to probe enzymes, enzyme-inhibitor complexes, and enzyme-substrate complexes. Two main approaches are taken (179). The first involves the use of substrates that are chromophores due to a delocalized π-electron system. Upon binding, a dipole moment is induced by the enzyme into the substrate's π system, causing spectral changes in quantities such as $\nu(C=C)$ and λ_{max}. The second approach involves study of a dithio ester ($-\overset{\overset{S}{\|}}{C}-S-$) formed at catalysis. Enhancement of $\nu(C=S)$ and $\nu(C-S)$ allows observation of these bonds as they undergo enzymolysis.

Resonance Raman studies (180) of a series of furylacryloyl and thienylacryloyl chymotrypsins at pH 3.0 revealed that the acyl group adopts two conformations having different environments about the carbonyl group. Thus, spectral values such as λ_{max} and $\nu(C=C)$ are treated as weighted values, $\langle\lambda_{max}\rangle$ and $\langle\nu(C=C)\rangle$. The correlation between $\langle\lambda_{max}\rangle$ and $\langle\nu(C=C)\rangle$ is explained in terms of a π-electron polarization model.

Higher pH values studies of 4-NH_2-3-NO_2-cinnamoyl and 5-CH_3-thienylacryloyl chymotrypsin with 350.7-nm laser excitation (181) allows observation of the enzyme at optimal activity rather than just the stable bound enzyme complex. Use of UV excitation greatly improves the resonance Raman spectra of the unstable species. Three acyl enzymes, 3-CH_3-thienylacryloyl, α-CH_3-5-CH_3-thienylacryloyl, and α-CH_3-5-CH_3-furylacryloyl chromotrypsins, were studied and showed low rates of deacylation and anomalous $\langle\nu(C=C)\rangle$ versus $\langle\lambda_{max}\rangle$ correlations. This was interpreted as due to steric hindrance at the active site caused by the 3-CH_3 or α-CH_3 groups.

Comparative electronic absorption and resonance Raman studies of rabbit furylacryloyl glyceraldehyde-3-phosphate dehydrogenase (FA-rabbit enzyme) and FA-sturgeon enzyme (182) show that the FA-rabbit enzyme exists in two forms while the FA-sturgeon enzyme exists in only one form. One form of FA-rabbit enzyme and the FA-sturgeon enzyme shows a red shifted λ_{max} and a decrease in $\nu(C=C)$ upon binding NAD^+, indicating a polarization of π-electrons of the FA-chromophore. It was concluded that this polarization does not activate the carbonyl group solely by polarization of the C–O moiety. For a series of FA-derivatives of the type

where X = H, N, O, there is a clear correlation between λ_{max} and $\nu(C=C)$. However, when X = S, the correlation breaks down, possibly due to through-space $d\pi$-$p\pi$ overlap between the ethylene bond and the d orbitals of sulfur.

Resonance Raman spectra of the dithioacyl-enzyme intermediate produced during the papain-catalyzed hydrolysis of methyl thionohippurate (183) showed enhancement of $\nu(C=S)$ and $\nu(C-S)$. Since the natural reaction involves the formation of the $-\overset{\overset{O}{\|}}{C}-S-$ group, the dithioester is a simple one-atom replacement of sulfur for oxygen. This is a more direct probe of the ester bonds making up the enzyme-substrate complex.

7.4.6. Nucleic Acid Derivatives

7.4.6.1. Kinetics of Hydrogen-Deuterium Exchange in Purines

Raman spectroscopy can be used to monitor the kinetics of the hydrogen-deuterium exchange of the 8-CH groups of purine nucleotides. The main advantages that Raman spectroscopy has over other methods of measuring these kinetics are (184):

1. Structural changes and concentration can be measured simultaneously.
2. In principle, it is possible to discriminate between simultaneous exchange on different purines of a polynucleotide or oligonucleotide.
3. Compared with some techniques, such as tritium labeling, Raman spectroscopy is a relatively simple technique.

Hydrogen-deuterium exchange kinetics have been measured for adenine nucleotides (185, 186) and guanine nucleotides (184). The results of these studies show that exchange at C_8 in guanine nucleotides occurs about 2.5 times faster than in adenine nucleotides. At low temperature ($\sim 30°C$), each cyclic purine nucleotide (guanine or adenine 3',5' monophosphate) exchanges about 1.9 times faster than the corresponding 5' nucleotide. At higher temperatures, the rates become nearly equal, as shown in Figure 7.19 for the guanine nucleotides. The enhanced lability of the 8-CH bond in guanine nucleotides compared to adenine nucleotides is explained in terms of different electronic structures of the purines rather than due to different intermolecular interactions. This is because these interactions would decrease with temperature, which is inconsistent with the results. But the differences in exchange rates between 5' and cyclic nucleotides of a given purine do diminish with temperature, suggesting that inter- or intramolecular interactions are important.

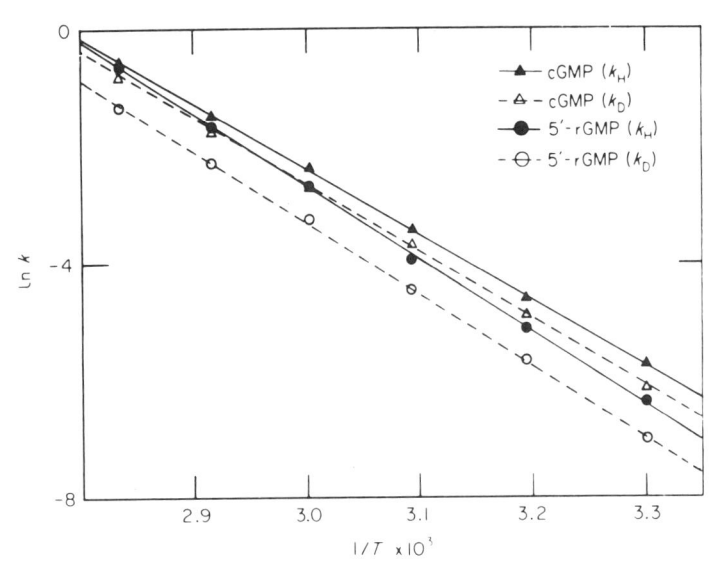

Fig. 7.19. Plots of ln k vs. $1/T$ for 5'-rGMP and cGMP. Each linear plot is a least squares fit to the data points shown. From G. J. Thomas and M. J. Lane, *J. Raman Spectroscopy*, 9, 134 (1980) (with permission).

7.4.6.2. Ultraviolet Resonance Raman Effects

Recent resonance Raman studies of nucleic acid derivatives using UV excitation have provided valuable information concerning the nature of these substances. DNA–polar peptide complexes form in a selective way, depending on the "specific recognition" of the peptide for a base of the DNA. This specific interaction was monitored for DNA-lysine and DNA-arginine complexes (187). The UV resonance Raman spectra (Fig. 7.20) showed different interactions for the two complexes. Observation of structural differences in the nucleic acid led to the conclusion that electrostatic effects with the phosphate groups are not the only interactions. For the arginine-DNA complexes, a modification of the Raman line due to adenine at 1582 cm^{-1} shows recognition of arginine for adenine. The unperturbed guanine Raman line at 1490 cm^{-1} is contrary to a hypothesis that arginine "recognizes" guanine. The lysine-DNA complexes provided no clear evidence for recognition of a specific base by lysine.

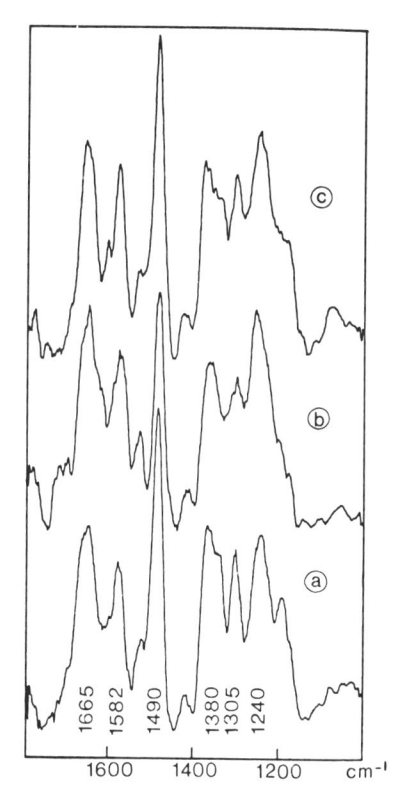

Fig. 7.20. Raman spectra of DNA and DNA-polar peptide complexes, 1000–1800 cm^{-1} range, excitation 300 nm. (*a*) DNA NaCl, 1 M, pH 7, aqueous solution. (*b*) DNA-lysine methylester complex. (*c*) DNA-arginine methylester complex. From A. Laigle, L. Chinsky, and P. Y. Turpin in W. F. Murphy, Ed., *Proc. VIIth International Conference on Raman Spectroscopy*, North-Holland Publishing Co., Amsterdam, 1980, p. 586 (with permission).

The order-disorder transition, that is, the conformational changes of a nucleic acid from a random coil at high temperatures to an ordered structure with stacked nucleic bases at low temperatures, can be monitored by Raman line intensities. Upon base stacking, the molar absorptivity of an electronic transition deccreases. This affects the intensities of resonance Raman lines, an effect called *Raman hypochromism*. When Raman excitation is in the preresonance region (285–320 nm) of the UV electronic transition in nucleic acids, this effect is much enhanced. Thus, the order-disorder transition can be understood if the Raman hypochromism excitation profile is known. The first theoretical approach to understanding this effect was taken by Chinsky and Turpin (188) and verified by studies of polyribouridylic acid. Raman excitation profiles of adenine derivatives have been examined by Lytle and coworkers (189) in the 285–320-nm region. They pointed out that changes in preresonance Raman intensities can also be due to shifts in the excited state, and thus the behavior of excited state energies must be known before definitive conclusions about hypochromism can be made.

A semiempirical method for calculating resonant Raman excitation profiles was recently developed and applied to Raman lines of two pyrimidines (cytidine 5′-monophosphate and uridine 5′-monophosphate) from the UV excited spectra by Blazej and Peticolas (190). The excitation profile was calculated for a given vibration, knowing only the absorption spectrum, yielding a unique profile for each vibration. Hence, excited state geometries can be better understood.

The enhancement of the intensity of a carbonyl stretching vibration can be used to probe the π-electron state of the carbonyl. When the Raman spectrum is excited in a $\pi \to \pi^*$ transition of a conjugated double-bond system, $\nu(C-O)$ will be enhanced if the carbonyl bond is involved in the conjugated system. This has been used in a study of nicotinamide adenine dinucleotide (NAD) and its reduced form, NADH (191). NAD has an absorption band at 265 nm, With 300-nm excitation, the ring-"breathing" line at 1033 cm^{-1} is enhanced, but the C=O stretching is not. However, for NADH, a new absorption band appears at 340 nm. The carbonyl stretching frequency at 1693 cm^{-1} is enhanced with 351.1-nm excitation. Hence, in NADH the carbonyl bond is involved in the conjugated double-bond system. This is not true in NAD.

7.4.6.3. Metal Ion–Nucleotide Interactions

Most adenine nucleotides require divalent cations as cofactors in enzymatic reactions. In order to probe the nucleotide binding sites, Raman studies of metal ion–ATP complexes can complement present data from NMR, ESR, etc. and also provide new information. Two recent Raman studies involve divalent cation-ATP complexes (192) and Co(III)-ATP complexes (193).

The divalent cations Ca^{2+}, Mg^{2+}, Co^{2+}, Cu^{2+}, and Hg^{2+} were complexed with ATP and studied over pH 3–12. There are two types of interactions possible: metal-adenine and metal-triphosphate. The metals coordinate the triphosphate throughout the entire pH range. For Ca^{2+}- and Mg^{2+}-ATP complexes, the triphosphate frequency (~ 1118 cm^{-1}) is pH dependent, with a transition near neutrality. At low pH, evidence was found for interaction with the adenine ring. The interaction was proposed to be through a bridging water molecule. For Cu^{2+}-ATP complexes, six different species were identified in the range of pH 3–12. At neutral pH, the Raman data indicate Cu^{2+} binding to both N_7 and the amino group of the adenine ring, but at pH 10–11 the copper binds at N_3 as indicated by enhancement of the 760 and 1360 cm^{-1} lines. In the Hg^{2+}-ATP complex at neutral pH, the Hg^{2+} ion coordinates directly to N_1. At lower pH, N_1 is protonated, so Hg^{2+} does not bind to the adenine moiety.

Co(III)-ATP complexes were examined both with and without 1,10-phenanthroline and superoxide. A preresonance effect was observed in the excitation profiles of some of the ATP Raman lines of Co(III)-(phen)-ATP-O_2^-, indicating a charge-transfer transition between 300 and 400 nm. The Raman spectra of Co(III)-(phen)-ATP-O_2^- (prepared by oxidation with molecular oxygen), Co(III)-(phen)-ATP-CN$^-$, and Co(III)-(phen)$_2$ are presented in Figure 7.21. The most dramatic differences are the enhancement of the 734 and 1430 cm^{-1} lines in Co(III)-(phen)-

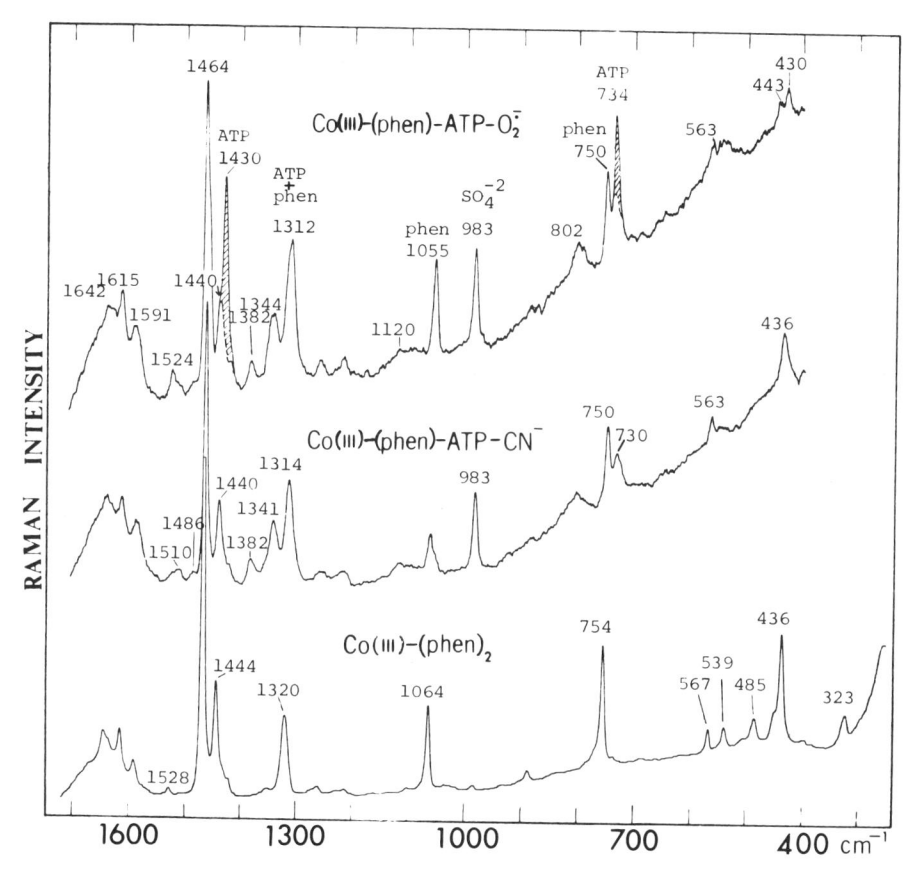

Fig 7.21. Raman spectra of Co(III) complexes. (a) Co(III)-phen-ATP–O_2^-; (b) Co(III)-phen-ATP–CN$^-$; (c) Co(III)-(phen)$_2$. From N. -T. Yu, A. Lanir, and M. M. Werber, *J. Raman Spectroscopy*, 11, 150 (1981) (with permission).

418 Raman Spectroscopy of Heterocyclic Compounds

ATP-O$_2^{\cdot -}$, which are assigned to adenine ring modes. This is interpreted as interaction between Co(III) with the six-membered ring of adenine, probably through binding to the 6-amino group. Broadening of the 1118 cm^{-1} phosphate stretching line relative to the sharp line in free ATP reflects the nature of Co(III)-O-P binding. The binding of two nitrogens in the phenanthroline to a superoxide ion and the amino group of adenine results in a distortion of the triphosphate group to allow additional binding of Co(III) with two phosphate groups. The enhancement of the 734 and 1430 cm^{-1} lines is lost when cyanide is added. Thus, cyanide replaces both the superoxide and the adenine amino group on the cobalt.

7.4.7. Heterocyclic Free Radicals

Resonance Raman studies of free radicals can yield structural information about excited state geometries of the parent compound. Two heterocyclic free-radical systems recently studied are the diazobicyclooctane radical cation (194) and the phenothiazine, 10-methylphenothiazine, and phenoxazine radical cations (195).

The radical cation of diazabicyclooctane, as shown in Figure 7.22, can be generated by oxidation of the parent compound with ClO$_2$ or OCl$^-$. The free radical shows strong absorption in the visible region, making it suitable for study by resonance Raman spectroscopy. By comparison of the force constants of the parent and radical, the electronic structure of the radical can be evaluated. Photoelectron (196) and ESR (197) studies showed that oxidation results in removal of an electron from the A_1' orbital, which has N lone-pair and C–C π-bond character. The symmetry of both the parent and the radical cation is D_{3h}. Only totally symmetric modes were observed. However, in the resonance Raman spectra more polarized bands than expected were observed in the 0–1800 cm^{-1} region. This was explained in terms of Fermi resonance playing an important role. Calculations involving establishment of a reasonable force field for the parent compound and modification for the radical cation, along with more extensive INDO and MINDO/2 calculations of the force constants, adequately described the resonance Raman spectra. On going from parent to radical, calculations showed weakening of C–C bonds and strengthening of C–N bond strengths, but to a lesser extent.

Phenothiazine, 10-methylphenothiazine, and phenoxazine, shown in Figure 7.23, and the corresponding radical cations are of interest in the elucidation of the tranquillizing properties of phenothiazine-based drugs. Phenothiazine and its derivatives are easily oxidized, and this property is thought to play a key role in the tranquillizing ability. Resonance Raman spectroscopy can help evaluate

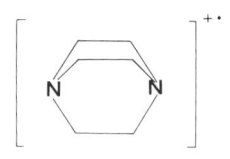

Fig. 7.22. Chemical structure of the diazabicyclooctane radical cation.

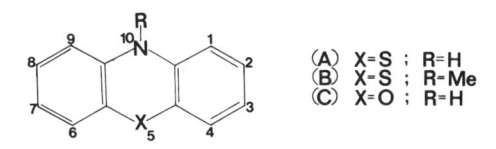

Fig. 7.23. Chemical structures of (a) phenothiazine; (b) 10-methylphenothiazine; (c) phenoxazine.

structural changes that occur upon oxidation. In addition, excitation profiles of certain Raman lines can help to unravel the complicated visible absorption spectra. A partial assignment of the spectra was made on the basis of approximations using data from smaller, related molecules and previous Raman data on phenothiazines (198). Changes in skeletal deformation modes, $\delta_d(CNC)$ and $\delta_d(CSC)$, provide the most information about structural changes that occur upon radical formation. Excitation profiles showed that most resonance-enhanced bands are maximal at the wavelength of maximal absorption. However, three lines show maxima at lower wavelength. One of the three coincides with a separate absorption band. It was proposed that Franck-Condon coupling between ground and excited states can be achieved through several routes, as evidenced by the other two excitation profiles (195).

ACKNOWLEDGEMENTS

We thank Drs. S. A. Asher, G. T. Babcock, P. R. Carey, J. M. Friedman, M. Lutz, J. T. McFarland, D. L. Rousseau, T. G. Spiro, R. P. Van Duyne and W. H. Woodruff for kindly communicating their results prior to publication. Our resonance Raman work at Georgia Tech was supported by the National Institutes of Health (GM 18894).

REFERENCES

1. T. G. Spiro and T. C. Strekas, *Proc. Nat. Acad. Sci. US*, 69, 2622 (1972).
2. J. A. Shelnutt, D. C. O'Shea, N. T. Yu, L. D. Cheung, and R. H. Felton, *J. Chem. Phys.*, 64, 1156 (1976).
3. J.A. Shelnutt, L. D. Cheung, R. C. C. Chang, N. T. Yu, and R. H. Felton, *J. Chem. Phys.*, 66, 3387 (1977).
4. L. D. Cheung, N. T. Yu, and R. H. Felton, *Chem. Phys. Lett.*, 55, 527 (1978).
5. N. T. Yu, *CRC Crit. Rev. Biochem.*, 4, 229 (1977).
6. R. H. Felton and N. T. Yu, in D. Dolphin, Ed., *The Porphyrins*, Vol. 3, Academic Press, New York, 1978, p. 347.
7. G. J. Rosasco, in R. J. H. Clark and R. E. Hester, Eds., in *Advances in Infrared and Raman Spectroscopy*, Vol. 7, Heyden, London, 1980, p. 223.
8. E. S. Etz and J. J. Blaha, in R. H. Geiss, Ed., *Microbeam Analysis*, San Francisco Press, San Francisco, 1979, p. 173.
9. M. Bridoux and M. Delhaye, in R. J. H. Clark and R. E. Hester, Eds., *Advances in Infrared and Raman Spectroscopy*, Vol. 2, Heyden, London, 1976, p. 140.
10. Y. Talmi, *Anal. Chem.*, 47, 685A (1975).

11. W. H. Woodruff and S. Farquharson, *Anal. Chem.*, **50**, 1389 (1978).
12. R. B. Srivastava, M. W. Schuyler, L. R. Dosser, F. J. Purcell, and G. H. Atkinson, *Chem. Phys. Lett.*, **56**, 595 (1978).
13. J. Sturm, R. Savoice, M. Edelson, and W. Peticolas, *Indian J. Pure Appl. Phys.*, **16**, 327 (1978).
14. J. Terner, C. L. Hsieh, A. R. Burns, and M. A. El-Sayed, *Proc. Nat. Acad. Sci. US*, **76**, 3046 (1979).
15. R. Mathies and N. T. Yu, *J. Raman Spectrosc.*, **7**, 349 (1978).
16. N. T. Yu and R. B. Srivastava, *J. Raman Spectrosc.*, **9**, 166 (1980).
17. (a) J. Terner, J. D. Stong, T. G. Spiro, M. Nagumo, M. Nicol, and M. A. El-Sayed, *Proc. Nat. Acad. Sci. US*, **78**, 1313 (1981). (b) J. Terner, T. G. Spiro, M. Nagumo, M. F. Nicol, and M. A. El-Sayed, *J. Amer. Chem. Soc.*, **102**, 3238 (1980).
18. J. A. Koningstein, *Introduction to the Theory of the Raman Effect*, D. Reidel Publ. Co., Dordrecht, Holland, 1972.
19. E. B. Wilson, J. C. Decius, and P. C. Cross, *Molecular Vibrations*, McGraw-Hill, New York, 1955, p. 43.
20. J. Behringer, *Mol. Spectrosc.*, **2**, 100 (1974).
21. J. Tang and A. C. Albrecht, *J. Chem. Phys.*, **49**, 1144 (1968).
22. A. C. Albrecht, *J. Chem. Phys.*, **34**, 1476 (1961).
23. H. J. Van Vleck, *Proc. Nat. Acad. Sci. US*, **15**, 754 (1929).
24. M. Mingardi and W. Siebrand, *J. Chem. Phys.*, **62**, 1074 (1975).
25. Y. Talmi and R. W. Simpson, *Appl. Opt.*, **19**, 1401 (1980).
26. R. W. Simpson, *Rev. Sci. Instrum.*, **50**, 730 (1979).
27. T. C. Strekas, D. H. Adams, A. Packer, and T. G. Spiro, *Appl. Spectrosc.*, **28**, 324 (1974).
28. D. F. Shriver and J. B. R. Dunn, *Appl. Spectrosc.*, **28**, 319 (1974).
29. (a) M. Braiman and R. Mathies, *Proc. Nat. Acad. Sci. US*, **79**, 403 (1982); (b) M. Braiman and R. Mathies, *Methods Enzymol.*, **88**, 77 (1982).
30. G. J. Thomas, Jr., and J. R. Barylski, *Appl. Spectrosc.*, **24**, 463 (1970).
31. B. Hudson, W. Heatherington III, S. Kramer, I. Chabay, and G. K. Klauminzer, *Proc. Nat. Acad. Sci. US*, **73**, 3798 (1976).
32. J. R. Nestor, T. G. Spiro, and G. K. Klauminzer, *Proc. Nat. Acad. Sci. US*, **73**, 3329 (1976).
33. L. A. Carreira, T. C. Maguire, and T. B. Malloy, Jr., *J. Chem. Phys.*, **66**, 2621 (1977).
34. M. D. Levenson, *Phys. Today*, May 1977, p. 44.
35. H. Lotem, R. T. Lynch, Jr., and N. Bloembergen, *Phys. Rev.*, **A14**, 1748 (1976).
36. R. T. Lynch, Jr., H. Lotem, and N. Bloembergen, *J. Chem. Phys.*, **66**, 4250 (1977).
37. D. Heiman, R. W. Hellwarth, M. D. Levenson, and G. Martin, *Phys. Rev. Lett.*, **36**, 189 (1976).
38. G. L. Eesley, M. D. Levenson, and W. M. Tolles, *IEEE J. Quantum Electron.*, **QE-14**, 45 (1978).
39. R. P. Van Duyne, in C. B. Moore, Ed., *Chemical and Biological Applications of Lasers*, Vol. 4, Academic Press, New York, 1979, p. 101.
40. E. Burstein and C. Y. Chen, in W. F. Murphy, Ed., *Proc. VIIth International Conference on Raman Spectroscopy*, North-Holland Publ. Co., Amsterdam, 1980, p. 345.
41. F. W. King and G. C. Schatz, *Chem. Phys.*, **38**, 254 (1979).
42. M. R. Philpott, *J. Chem. Phys.*, **62**, 1812 (1975).
43. R. M. Hexter and M. G. Albrecht, *Spectrochim. Acta*, **35A**, 233 (1978).

References

44. B. Pettinger, V. Wenning, and D. Kolb, *Ber. Bunsenges. Phys. Chem.*, **82**, 1326 (1978).
45. E. Burstein, Y. J. Chen, and S. Lundquist, *Bull. Amer. Phys. Soc.*, **23**, 30 (1978).
46. F. R. Aussenegg and M. Lippitsch, *Chem. Phys. Lett.*, **59**, 214 (1978).
47. E. Burstein, Y. J. Chen, C. Y. Chen, S. Lundquist, and E. Tosatti, *Solid State Commun.*, **29**, 567 (1979).
48. F. W. King, R. P. Van Duyne, and G. C. Schatz, *J. Chem. Phys.*, **69**, 4472 (1978).
49. (a) J. A. Creighton, C. G. Blatchford, and M. G. Albrecht, *J. Chem. Soc. Faraday II*, **75**, 790 (1979): (b) M. Moskovits, *J. Chem. Phys.*, **69**, 4159 (1978).
50. C. A. Murray, D. L. Allara, and M. Rhinewine, in W. F. Murphy, Ed., *Proc. VIIth International Conference on Raman Spectroscopy*, North-Holland Publ. Co., Amsterdam, 1980, p. 406.
51. (a) D. S. Wang, H. Chew, and M. Kerker, *Appl. Opt.*, **19**, 2256 (1980); (b) M. Kerker, O. Süman, L. A. Bumm, and D.-S. Wang, *Appl. Opt.*, **19**, 3253 (1980); (c) M. Kerker, D.-S. Wang, and H. Chew, *Appl. Opt.*, **19**, 4159 (1980).
52. W. Kiefer, *Appl. Spectrosc.*, **27**, 253 (1973).
53. W. Kiefer, in R. J. H. Clark and R. E. Hester, Eds., *Advances in Infrared and Raman Spectroscopy*, Vol. 3, Heyden, London, 1977, p. 1.
54. J. A. Shelnutt, D. L. Rousseau, J. K. Dethmers, and E. Margoliash, *Proc. Nat. Acad. Sci. US*, **76**, 3865 (1979).
55. J. A. Shelnutt, D. L. Rousseau, J. M. Freidman, and S. R. Simon, *Proc. Nat. Acad. Sci. US*, **76**, 4409 (1979).
56. D. L. Rousseau, J. A. Shelnutt, and S. R. Simon, *FEBS Lett.*, **111**, 235 (1980).
57. D. L. Rousseau, J. A. Shelnutt, E. R. Henry, and S. R. Simon, *Nature*, **285**, 49 (1980).
58. J. A. Shelnutt, D. L. Rousseau, J. K. Dethmers, and E. Margoliash, *Biochemistry*, **20**, 6405 (1981).
59. D. L. Rousseau, *J. Raman Spectrosc.*, **10**, 94 (1981).
60. T. Hirschfeld, *J. Opt. Soc. Amer.*, **63**, 476 (1973).
61. G. J. Rosasco and E. S. Etz, *Res. Develop.*, **28**, 20 (1977).
62. E. S. Etz, "New Raman Microprobe with Multichannel Optical Detector", *NBS Dimensions* (Nov. 1980), p. 16.
63. M. Delhaye and P. Dhamelincourt, *J. Raman Spectrosc.*, **3**, 33 (1975).
64. P. Dhamelincourt, F. Wallart, M. Leclercq, A. T. N'Guyen, and D. O. London, *Anal. Chem.*, **51**, 414A (1979).
65. E. S. Etz and J. J. Blaha, in K. F. J. Heinrich, Ed., *Characterization of Particles*, NBS Special Publication 533, Washington, 1980, p. 153.
66. J. L. Abraham and E. S. Etz, *Science*, **206**, 716 (1979).
67. E. S. Etz and J. J. Blaha, in *Technical Activities NBS Center for Analytical Chemistry* (Oct. 1980), p. 197.
68. E. S. Etz in R. H. Geiss, Ed., *Microbeam Analysis*, San Francisco Press, San Francisco, 1981, p. 73.
69. M. Tsubaki and N.-T. Yu, *Proc. Nat. Acad. Sci. US*, **78**, 3581 (1981).
70. M. Tsubaki, R. B. Srivastava, and N.-T. Yu, *Biochemistry*, **21**, 1132 (1982).
71. E. A. Kerr, H. C. Machin, and N.-T. Yu, *Biochemistry* (1983) in press.
72. M. Tsubaki and N.-T. Yu, *Biochemistry*, **21**, 1140 (1982).
73. H. C. Mackin, M. Tsubaki, and N.-T. Yu, *Biophys. J.*, **41**, 349 (1983).
74. (a) N.-T. Yu and M. Tsubaki, *Biochemistry*, **19**, 4647 (1980); (b) N.-T. Yu, E. A. Kerr, B. Ward and C. K. Chang, *Biochemistry* (1983) in press.
75. (a) M. Tsubaki, R. B. Srivastava, and N.-T. Yu, *Biochemistry*, **20**, 946 (1981); (b) B. Benko and N.-T. Yu, *Proc. Nat. Acad. Sci. US*, (in press).

76. S. A. Asher in E. Antonini, L. Rossi-Bernardi, and E. Chicanone, Eds., *Methods in Enzymology*, Vol. 76, Academic Press, New York, 1981, p. 371.
77. S. A. Asher and T. M. Schuster, *Biochemistry*, **18**, 5377 (1979).
78. S. A. Asher and T. M. Schuster, *Biochemistry*, **20**, 1866 (1981).
79. S. A. Asher, L. E. Vickery, T. M. Schuster, and K. Sauer, *Biochemistry*, **16**, 5849 (1977).
80. S. A. Asher and T. M. Schuster, in C. Ho, W. A. Eaton, J. P. Collman, Q. H. Gibson, J. S. Leigh, Jr., E. Margoliash, J. K. Moffat, and W. R. Scheidt, Eds., *Interactions between Iron and Proteins in Oxygen and Electron Transport*, Elsevier, Amsterdam, 1981.
81. H. Brunner, *Naturwiss.*, **61**, 129 (1974).
82. J. D. Stong, J. M. Burke, P. Daly, P. Wright, and T. G. Spiro, *J. Amer. Chem. Soc.*, **102**, 5815 (1980).
83. G. Chottard and D. Mansuy, *Biochem. Biophys. Res. Commun.*, **77**, 1333 (1977).
84. T. G. Spiro and J. M. Burke, *J. Amer. Chem. Soc.*, **98**, 5482 (1976).
85. P. G. Wright, P. Stein, J. M. Burke, and T. G. Spiro, *J. Amer. Chem. Soc.*, **101**, 3531 (1979).
86. A. Desbois, M. Lutz, and R. Banerjee, *Biochemistry*, **18**, 1510 (1979).
87. K. Nagai, T. Kitagawa, and H. Morimoto, *J. Mol. Biol.*, **136**, 271 (1980).
88. J. Kinciad, P. Stein, and T. G. Spiro, *Proc. Nat. Acad. Sci. US*, **76**, 4156 (1979).
89. J. Kincaid, P. Stein, and T. G. Spiro, *Proc. Nat. Acad. Sci. US*, **76**, 549 (1979).
90. T. Kitagawa, K. Nagai, and M. Tsubaki, *FEBS Lett.*, **104**, 376 (1979).
91. K. Nagai and T. Kitagawa, *Proc. Nat. Acad. Sci. US*, **77**, 2033 (1980).
92. H. Hori and T. Kitagawa, *J. Amer. Chem. Soc.*, **102**, 3608 (1980).
93. M. R. Ondrias, D. L. Rousseau, and S. R. Simon, *J. Biol. Chem.*, **258**, 5638 (1983).
94. J. O. Alben and W. S. Caughey, *Biochemistry*, **7**, 175 (1968).
95. W. S. Caughey, J. O. Alben, S. McCoy, S. H. Boyers, S. Charache, and P. Hathaway, *Biochemistry*, **8**, 59 (1969).
96. P. W. Tucker, S. E. V. Phillips, M. F. Perutz, R. A. Houtchens, and W. S. Caughey, *Proc. Nat. Acad. Sci. US*, **76**, 1076 (1978).
97. M. W. Makinen, R. A. Houtchens, and W. S. Caughey, *Proc. Nat. Acad. Sci. US*, **76**, 6042 (1979).
98. C. H. Barlow, J. C. Maxwell, W. J. Wallace, and W. S. Caughey, *Biochem. Biophys. Res. Commun.*, **55**, 91 (1973).
99. J. C. Maxwell, J. A. Volpe, C. H. Barlow, and W. S. Caughey, *Biochem. Biophys. Res. Commun.*, **58**, 166 (1974).
100. J. C. Maxwell and W. S. Caughey, *Biochem. Biophys. Res. Commun.*, **60**, 1309 (1974).
101. W. S. Caughey, C. H. Barlow, J. C. Maxwell, J. A. Volpe, and W. J. Wallace, *Ann. N.Y. Acad. Sci.*, **244**, 1 (1975).
102. J. C. Maxwell and W. S. Caughey, *Biochemistry*, **15**, 388 (1976).
103. S. McCoy and W. S. Caughey, *Biochemistry*, **9**, 2387 (1970).
104. J. O. Alben and L. Y. Fager, *Biochemistry*, **11**, 842 (1972).
105. D. H. O'Keefe, R. E. Ebel, J. A. Peterson, J. C. Maxwell, and W. S. Caughey, *Biochemistry*, **17**, 5845 (1978).
106. M. F. Perutz, E. J. Heidner, J. E. Lander, J. G. Bettlestone, C. Ho, and F. Slade, *Biochemistry*, **13**, 2187 (1974).
107. J. L. Hoard, in B. Chance, R. W. Estabrook, and T. Yonetani, Eds., *Hemes and Hemoproteins*, Academic Press, New York, 1966, p. 9.

108. M. Rougee and D. Brault, *Biochemistry,* **14**, 4100 (1975).
109. J. Geibel, J. Cannon, D. Campbell, and T. G. Traylor, *J. Amer. Chem. Soc.,* **100**, 3575 (1978).
110. H. B. Dunford, *Physiol. Veg.,* **12**, 13 (1974).
111. J. Peisach, *Ann. N.Y. Acad. Sci.,* **244**, 187 (1975).
112. M. Morrison and R. Schonbaum, *Ann. Rev. Biochem.,* **45**, 861 (1976).
113. T. Mincey and T. G. Traylor, *J. Amer. Chem. Soc.,* **101**, 765 (1979).
114. J. C. Schwartz, M. A. Stanford, J. N. Moy, B. M. Hoffman, and J. S. Valentine, *J. Amer. Chem. Soc.,* **101**, 3396 (1979).
115. J. P. Collman, J. I. Brauman, K. M. Doxsee, T. R. Halbert, and K. S. Suslick, *Proc. Nat. Acad. Sci. US,* **75**, 564 (1978).
116. D. K. White, J. B. Cannon, and T. G. Traylor, *J. Amer. Chem. Soc.,* **101**, 2443 (1979).
117. E. H. Abbott and P. A. Rafson, *J. Amer. Chem. Soc.,* **96**, 7378 (1974).
118. M. A. Stanford, J. C. Swartz, T. E. Phillips, and B. M. Hoffman, *J. Amer. Chem. Soc.,* **102**, 4492 (1980).
119. J. P. Collman, J. I. Brauman, T. R. Halbert, and K. S. Suslick, *Proc. Nat. Acad. Sci. US,* **73**, 3333 (1976).
120. J. O. Alben, G. H. Bare, and P. P. Moh, in W. S. Caughey, Ed., *Biochemical and Clinical Aspects of Hemoglobin Abnormalities,* Academic Press, New York, 1978, p. 607.
121. M. A. Walters, T. G. Spiro, K. S. Suslick, and J. P. Collman, *J. Amer. Chem. Soc.,* **102**, 6857 (1980).
122. P. G. Wright, P. Stein, J. M. Burke, and T. G. Spiro, *J. Amer. Chem. Soc.,* **101**, 3531 (1979).
123. W. A. Eaton and R. M. Hochstrasser, *J. Chem. Phys.,* **49**, 985 (1968).
124. D. Y. Copper, O. Rosenthal, R. Snyder, and C. Witmer, *Cytochromes P-450 and b5,* Plenum Press, New York, 1975, p. 1.
125. C. K. Chang and D. Dolphin, *Proc. Nat. Acad. Sci. US,* **73**, 3338 (1976).
126. J. P. Collman and T. N. Sarrell, *J. Amer. Chem. Soc.,* **97**, 4133 (1975).
127. J. O. Stern and J. Piesach, *J. Biol. Chem.,* **249**, 7495 (1974).
128. R. Tsai, C. A. Yu, I. C. Gunsalus, J. Peisach, W. Blumberg, W. H. Orme-Johnson, and H. Beinert, *Proc. Nat. Acad. Sci. US,* **66**, 1157 (1970).
129. S. C. Tang, S. Koch, G. C. Papaefthymiou, S. Foner, R. B. Frankel, J. A. Ibers, and R. H. Holm, *J. Amer. Chem. Soc.,* **98**, 2414 (1976).
130. P. M. Champion and I. C. Gunsalus, *J. Amer. Chem. Soc.,* **99**, 2000 (1977).
131. P. M. Champion, B. R. Stallard, G. C. Wagner, and I. C. Gunsalus, *J. Amer. Chem. Soc.,* **104**, 5469 (1982).
132. J. Kincaid, P. Stein, and T. G. Spiro, *Proc. Nat. Acad. Sci. US,* **76**, 549 (1979).
133. J. Kincaid, P. Stein, and T. G. Spiro, *Proc. Nat. Acad. Sci. US,* **76**, 4156 (1979).
134. T. Kitagawa, K. Nagai, and M. Tsubaki, *FEBS Lett.,* **104**, 376 (1979).
135. K. Nagai, T. Kitagawa, and H. Morimoto, *J. Mol. Biol.,* **136**, 271 (1981).
136. K. Nagai and T. Kitagawa, *Proc. Nat. Acad. Sci. US,* **77**, 2033 (1980).
137. H. Hori and T. Kitagawa, *J. Amer. Chem. Soc.,* **102**, 3608 (1980).
138. M. R. Ondrias, D. L. Rousseau, and S. R. Simon, *Proc. Nat. Acad. Sci. US,* **79**, 1511 (1982); *Science,* **213**, 657 (1981).
139. L. D. Spaulding, R. C. C. Chang, N.-T. Yu, and R. H. Felton, *J. Amer. Chem. Soc.,* **97**, 2517 (1975).
140. T. G. Spiro, J. D. Stong, and P. Stein, *J. Amer. Chem. Soc.,* **101**, 2648 (1979).
141. P. V. Huong and J. C. Pommier, *Compt. Rend. C,* **285**, 519 (1977).

142. P. M. Callahan and G. T. Babcock, *Biochemistry,* **20**, 952 (1981).
143. S. Choi, T. G. Spiro, K. C. Langry, K. M. Smith, D. L. Budd, and G. N. LaMar, *J. Amer. Chem. Soc.,* **104**, 4354 (1982).
144. G. Germi, *J. Mol. Biol.,* **97**, 237 (1975).
145. T. Takano, *J. Mol. Biol.,* **110**, 569 (1977).
146. H. Eicher, D. Dade, and F. Parak, *J. Chem. Phys.,* **64**, 1446 (1976).
147. (a) J. M. Friedman and K. B. Lyons, *Nature,* **284**, 570 (1980); (b) J. M. Friedman, D. L. Rousseau, M. R. Ondrias and R. A. Stepnoski, *Science,* **218**, 1244 (1982).
148. K. B. Lyons and J. M. Friedman, in C. Ho, W. A. Eaton, J. P. Collman, Q. H. Gibson, J. S. Leigh, Jr., E. Margoliash, J. K. Moffat, and W. R. Scheidt, Eds., *Interaction between Iron and Proteins in Oxygen and Electron Transport,* Elsevier, Amsterdam 1981.
149. T. G. Spiro and T. C. Strekas, *J. Amer. Chem. Soc.,* **96**, 338 (1974).
150. T. Kitagawa, T. Iizuka, M. Saito, and Y. Koyogku, *Chem. Lett.,* **8**, 849 (1975).
151. T. G. Spiro and J. M. Burke, *J. Amer. Chem. Soc.,* **98**, 5482 (1976).
152. T. G. Spiro, in A. B. P. Lever and H. B. Gray, Eds., *Iron Porphyrins,* Part II, Addison Wesley, Reading, Massachusetts, 1983, p. 89.
153. T. Szymanski, T. Cape, R. VanDuyne, and F. Basolo, *J. Chem. Soc. Chem. Commun.,* **1979**, 5.
154. M. Kozuka, M. Suzuki, Y. Nishioa, S. Kida, and K. Nakamoto, *Inorg. Chim. Acta,* **45**, L11 (1980).
155. (a) R. E. Hester and E. M. Nour, *J. Raman Spectroscopy,* **11**, 49 (1981). (b) *ibid.,* p. 59; (c) *ibid.,* p. 64.
156. M. Suzuki, T. Ishiguro, M. Kozuka, and K. Nakamoto, *Inorg. Chem.,* **20**, 1933 (1981).
157. For a recent review, see P. R. Carey and V. R. Salares in R. J. H. Clark and R. E. Hester, Eds., *Advances in Infrared and Raman Spectroscopy,* Vol. 7, Heyden, London, 1980, p. 19.
158. M. Lutz, J. S. Brown, and R. Reimy, *Ciba Foundation Symp.,* **61** (*Chlorophyll Organ. and Energy Transfer in Photosynthesis*), 1979, p. 105.
159. M. Lutz and J. Kleo, *Biochim. Biophys, Acta,* **546**, 365 (1979).
160. W. M. Parson and R. J. Cogdell, *Biochem. Biophys. Acta,* **416**, 105 (1973).
161. M. Lutz, in W. F. Murphy, Ed., *Proc. VIIth Int. Conf. on Raman Spectrosc.,* North-Holland, Amsterdam, 1980, p. 520.
162. I. Agalidis, M. Lutz, and F. Reiss-Husson, *Biochim. Biophys. Acta,* **589**, 264 (1980).
163. M. Lutz, *Proc. 5th Int. Congr. Photosynthesis, Kallithea, Greece,* 1980.
164. P. K. Dutta, J. Nestor, and T. G. Spiro, *Biochem. Biophys. Res. Commun.,* **83**, 209 (1978).
165. P. K. Dutta and T. G. Spiro, *Biochemistry,* **19**, 1590 (1980).
166. P. K. Dutta and T. G. Spiro, *J. Chem. Phys.,* **69**, 3119 (1978).
167. W. D. Bowman and T. G. Spiro, *J. Chem. Phys.,* **73**, 5482 (1980).
168. J. Schmidt, M. Y. Lee, and J. T. McFarland, *Arch. Biochem. Biophys.,* **215**, 22 (1982).
169. M. Benecky, T. Y. Li, J. Schmidt, F. Frerman, K. L. Watters, and J. T. McFarland, *Biochemistry,* **18**, 3471 (1979).
170. J. Schmidt, J. Reinsch and J. T. McFarland, *J. Biol. Chem.,* **256**, 11667 (1981).
171. M. Benecky, T. J. Yu, W. L. Watters, and J. T. McFarland, *Biochem. Biophys. Acta,* **626**, 197 (1980).
172. T. Kitagawa, Y. Nishina, Y. Kyogoku, T. Yamano, N. Ohishi, A. Takai-Suzuki, and K. Yagi, *Biochemstry,* **18**, 1804 (1979).

173. T. Kitagawa, Y. Nishina, K. Shiga, H. Watari, Y. Matsumura, and T. Yamano, *J. Amer. Chem. Soc.,* **101**, 3376 (1979).
174. Y. Nishina, T. Kitagawa, K. Shiga, H. Watari, and T. Yamano, *J. Biochem.,* **87**, 831 (1980).
175. M. R. Ondrias, E. W. Findsen, G. E. Leroi, and G. T. Babcock, *Biochemistry,* **19**, 1723 (1980).
176. T. Kitagawa, Y. Fukumori, and T. Yamanaka, *Biochemistry,* **19**, 5721 (1980).
177. P. R. Carey and H. Schneider, *Acc. Chem. Res.,* **11**, 122 (1978).
178. P. R, Carey and V. R. Salares, in R. J. H. Clark and R. E. Hester, Eds., *Advances in Infrared and Raman Spectroscopy,* Vol. 7, Heyden, London, 1980, p. 27.
179. P. R. Carey, *Canad. J. Spectrosc.,* **26**, 134 (1981).
180. B. A. E. MacClement, R. G. Carriere, D. J. Phelps, and P. R. Carey, *Biochemistry,* **20**, 3438 (1981).
181. D. J. Phelps, H. Schneider, and P. R. Carey, *Biochemistry,* **20**, 3447 (1981).
182. A. C. Storer, D. J. Phelps, and P. R. Carey, *Biochemistry,* **20**, 3454 (1981).
183. A. C. Storer, W. F. Murphy, and P. R. Carey, *J. Biol. Chem.,* **254**, 3163 (1979).
184. M. J. Lane and G. J. Thomas, Jr., *Biochemistry,* **18**, 3839 (1979); G. J. Thomas and M. J. Lane, *J. Raman Spectrosc.,* **9**, 134 (1980).
185. G. J. Thomas and J. Livramento, *Biochemistry,* **14**, 5210 (1975).
186. G. J. Thomas and J. Livramento, *J. Amer. Chem. Soc.,* **96**, 6529 (1975).
187. A. Laigle, L. Chinsky, and P. Y. Turpin, in W. F. Murphy, Ed., *Proc. VIIth International Conference on Raman Spectroscopy,* North-Holland Publ. Co., Amsterdam, 1980, p. 586.
188. L. Chinsky and P. Y. Turpin, *Biopolymers,* **19**, 1507 (1980).
189. T. H. Bushaw, F. E. Lytle, and R. S. Tobias, *Appl. Spectrosc.,* **34**, 521 (1980).
190. D. C. Blazej and W. L. Peticolas, *J. Chem. Phys.,* **72**, 3134, (1980).
191. Y. Nishimura and M. Tsuboi, *Science,* **210**, 1358 (1980).
192. A. Lanir and N. -T. Yu, *J. Biol. Chem.,* **254**, 5882 (1979).
193. N. -T. Yu, A. Lanir, and M. M. Werber, *J. Raman Spectrosc.,* **11**, 150 (1981).
194. E. E. Ernstbrunner, R. B. Girling, W. E. L. Grossman, and R. E. Hester, *J. Chem. Soc. Faraday Trans. 2,* **74**, 501 (1978).
195. R. E. Hester and K. P. J. Williams, *J. Chem. Soc. Perkin II,* **5**, 852 (1981).
196. E. Heilbronner and K. A. Musckat, *J. Amer. Chem. Soc.,* **92**, 3818 (1970).
197. G. W. Eastland and M. C. R. Symons, *Chem. Phys. Lett.,* **45**, 422 (1977).
198. B. Kure and M. D. Morris, *Talanta,* **23**, 398 (1976).

8 ELECTROCHEMICAL BEHAVIOR OF HETEROCYCLIC COMPOUNDS

J. ARMAND

Laboratoire de Physicochimie des Solutions
Université Pierre et Marie Curie
4, Place Jussieu
Paris, France

J. PINSON

Laboratoire d'Electrochimie
Université Paris 7
2, Place Jussieu
Paris, France

8.1. Experimental Procedures and Electroanalytical Techniques   430
    8.1.1. Experimental Procedures for Carrying Out an Organic Electrochemical Experiment, 430
       8.1.1.1. Cells, 431
       8.1.1.2. Diaphragms, 434
       8.1.1.3. Reference Electrodes, 434
       8.1.1.4. Working and Counter Electrodes, 436
       8.1.1.5. Solvents, 436
       8.1.1.6. Supporting Electrolytes, 440
       8.1.1.7. Electronic Devices, 440
    8.1.2. The Use of Electroanalytical Techniques in the Investigation of Organic Reactions, 442
       8.1.2.1. Polarography, 447
       8.1.2.2. Rotating Disk Electrode (RDE), 450
       8.1.2.3. Cyclic Voltammetry, 453
       8.1.2.4. Coulometry and Preparative Electrolysis, 463
       8.1.2.5. Spectroscopic Methods, 465

8.2. Reduction of Heterocyclic Compounds   468
    8.2.1. Reduction of Heterocyclic Systems, 469
       8.2.1.1. Compounds with One Nitrogen Atom, 469
       8.2.1.2. Compounds with One Oxygen Atom, 475
       8.2.1.3. Compounds with One Sulfur Atom, 477
       8.2.1.4. Compounds with Two Nitrogen Atoms, 478

428 Electrochemical Behavior of Heterocyclic Compounds

        8.2.1.5. Compounds with One Nitrogen and One Oxygen Atom, 487
        8.2.1.6. Compounds with One Nitrogen and One Sulfur Atom, 488
        8.2.1.7. Compounds with Two Phosphorus Atoms, 489
        8.2.1.8. Compounds with Two Sulfur Atoms, 490
        8.2.1.9. Compounds with Three Nitrogen Atoms, 491
        8.2.1.10. Compounds with Two Nitrogen Atoms and One Oxygen Atom, 494
        8.2.1.11. Compounds with Two Nitrogen Atoms and One Sulfur Atom, 495
        8.2.1.12. Compounds with Four Nitrogen Atoms, 496
        8.2.1.13. Compounds with Five Nitrogen Atoms, 500
    8.2.2. Reduction of Substituted Heterocycles, 500
        8.2.2.1. Compounds with a Substituent that has an Oxygen Atom Attached to the Ring, 500
        8.2.2.2. Compounds with a Substituent that has a Nitrogen Atom Attached to the Ring, 516
        8.2.2.3. Compounds with a Substituent that has a Sulfur Atom Attached to the Ring, 522
        8.2.2.4. Halogen Derivatives, 524
        8.2.2.5. Compounds with a Substituent that has a Carbon Atom Attached to the Ring, 526

8.3. Oxidation of Heterocylic Compounds   530
    8.3.1. Oxidation of Heterocyclic Systems, 530
        8.3.1.1. Compounds with One Nitrogen Atom, 530
        8.3.1.2. Compounds with One Oxygen Atom, 549
        8.3.1.3. Compounds with One Sulfur Atom, 553
        8.3.1.4. Compounds with Two Nitrogen Atoms, 556
        8.3.1.5. Compounds with One Nitrogen and One Oxygen Atom, 568
        8.3.1.6. Compounds with One Nitrogen and One Sulfur Atom, 568
        8.3.1.7. Compounds with Two Oxygen Atoms, 571
        8.3.1.8. Compounds with One Oxygen and One Sulfur (Selenium, Tellurium) Atom, 574
        8.3.1.9. Compounds with Two Sulfur Atoms, 574
        8.3.1.10. Compounds with Two Nitrogen Atoms and One Phosphorus Atom, 580
        8.3.1.11. Compounds with Three Sulfur Atoms, 580
        8.3.1.12. Compounds with Four Nitrogen Atoms, 581
        8.3.1.13. Compounds with Four Sulfur Atoms, 590
        8.3.1.14. Compounds with Four Nitrogen and Two Sulfur Atoms, 590
    8.3.2. Oxidation of Substituted Heterocycles, 591
        8.3.2.1. Substituents with Oxygen Attached to the Ring, 591
        8.3.2.2. Substituents with Nitrogen Attached to the Ring, 595
        8.3.2.3. Substituents with Sulfur Attached to the Ring, 597
        8.3.2.4. Substituents with Carbon Attached to the Ring, 598

References   600

SYMBOLS

A	Area of the electrode
C_O^*	Bulk concentration of Ox
C_R^*	Bulk concentration of Red
D_O	Diffusion coefficient of Ox
D_R	Diffusion coefficient of Red
$E_{1/2}$	Measured half-wave potential in polarography
E_p	Peak potential in cyclic voltammetry
$E^{\circ\prime}$	Formal electrode potential
E°	Standard electrode potential
F	The faraday
f	Rotation rate
i	Current
i_d	Diffusion-limited current
i_p	Peak current in cyclic voltammetry
k°	Standard heterogeneous rate constant
k_f, k_b	Homogeneous rate constants
m	Mercury flow rate at the dropping mercury electrode
n	Number of electrons involved in a reduction or an oxidation
Ox	Oxidized species
Red	Reduced species
R	Gas constant
T	Absolute temperature
$t_{\max}$	Drop time at the dropping mercury electrode
v	Potential sweep rate
α	Transfer coefficient
ν	Kinematic viscosity
ω	Angular frequency of rotation

Electrochemistry has been used in organic chemistry and particularly in heterocyclic chemistry to prepare new compounds, to unravel the mechanism of reactions, or in an analytical way to titrate electroactive products in a complex mixture. This chapter deals with the electrochemical reduction and oxidation of heterocyclic compounds and their mechanistic analysis.

Several reviews and books have already been published that deal specifically with heterocyclic electrochemistry (1–4) or encompass organic electrochemistry in a general manner (5–12). "Specialist Periodical Reports" published by the Chemical Society (13), the recent compilation of organic electrochemical data by Meites and Zuman (14), and *Ascatopic Electrochemistry* (15) are also useful publications.

In preparing this chapter we have consulted the books and reviews quoted above and have scanned *Chemical Abstracts* up to Volume 96 (1982).

430 Electrochemical Behavior of Heterocyclic Compounds

8.1. EXPERIMENTAL PROCEDURES AND ELECTROANALYTICAL TECHNIQUES

This section describes how organic electrochemical experiments are carried out experimentally and then how electroanalytical techniques are used to investigate reaction mechanisms.

8.1.1. Experimental Procedures for Carrying Out an Organic Electrochemical Experiment

A typical but simplified electrochemical experiment is shown in Figure 8.1. It is composed of three electrodes: the working electrode on which the desired reduction or oxidation takes place, the reference electrode of which the potential does not vary with the content of the solution, and the counter electrode. With such an assembly the potential of the working electrode will be monitored by the potentiostat through a high-impedance feedback loop, and the current will flow between the working and counter electrodes. Thus each electrode plays a well-defined role: the reaction to be carried out takes place at the working electrode; the reference electrode serves as a standard; the counter electrode closes the circuit, and another electrochemical reaction which is very often disregarded takes place at this electrode. The potential between the working and reference electrodes can be measured by a voltmeter. The current flowing through the cell is indicated by an ammeter, which must be included between the counter electrode and the potentiostat; if it were

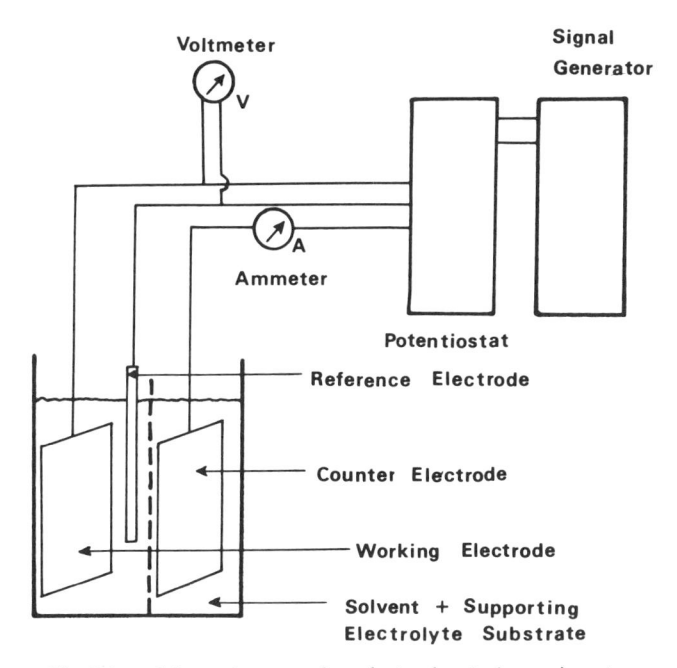

Fig. 8.1. Schematic setup of an electrochemical experiment.

placed between the working electrode and the potentiostat it would introduce an ohmic drop, and the potential of the working electrode would not be the desired one. The potentiostat is an electronic device that maintains constant potential between the working and reference electrodes irrespective of the current flowing in the cell and the resistance of the solution. In a preparative electrolysis or a coulometric experiment the potential will remain constant, but in many electroanalytical techniques (polarography, voltammetry, potential step technique) the potential will be varied with time; this is achieved by a signal generator, which pilots the potentiostat.

The solution obviously contains the solvent and the compound to be reduced or oxidized, but such a solution would generally be too resistive. Hence one has to add a supporting electrolyte, a salt that will dissociate in the solvent and allow the current to flow through the cell. Another utility of this supporting electrolyte is to suppress the migration of charged species toward the electrodes under the influence of the electric field and thus to simplify the analysis of the electroanalytical experiments. Finally, in the case of preparative electrolysis, where large amounts of products are electrolyzed, a diaphragm should be placed between the working and counter electrodes to prevent species formed at the electrodes from mixing and interacting.

In most cases, particularly when reductions are performed, the solution must be deoxygenated by bubbling nitrogen or argon through it, as oxygen is easily reduced and interferes with the process under investigation.

This schematic setup is the basis for most of the electrochemical experiments that deal either with large-scale electrolysis with large electrodes or electroanalytical measurements with microelectrodes and small amounts of solution. However, it should be noted that in some cases (chronopotentiometry) it is not the potential of the working electrode that is monitored, but the current flowing through the cell, and the variable that is observed is the potential between the working and reference electrodes. In some industrial processes only two electrodes are used, the current is maintained at a fixed value, and the reactant is fed continuously to the solution.

The various parts of the experimental setup are discussed in detail in the following pages.

8.1.1.1. *Cells*

Different kinds of cells are used for different electroanalytical experiments, and the type selected for a particular experiment depends on the nature of the process.

The cell designed for electrochemical experiments that is shown in Figure 8.2 can be used with as little as 5 ml of solution. An important feature of these cells is that the reference electrode must be placed as close as possible to the working electrode to minimize *iR* drop phenomena. Cells for even smaller volumes have been designed as electrochemical detectors (16–18) for high performance liquid chromatography (HPLC).

In the typical cell for preparative laboratory reductions shown in Figure 8.3, a few grams of compound in 150 ml of solvent can be reduced. In preparative

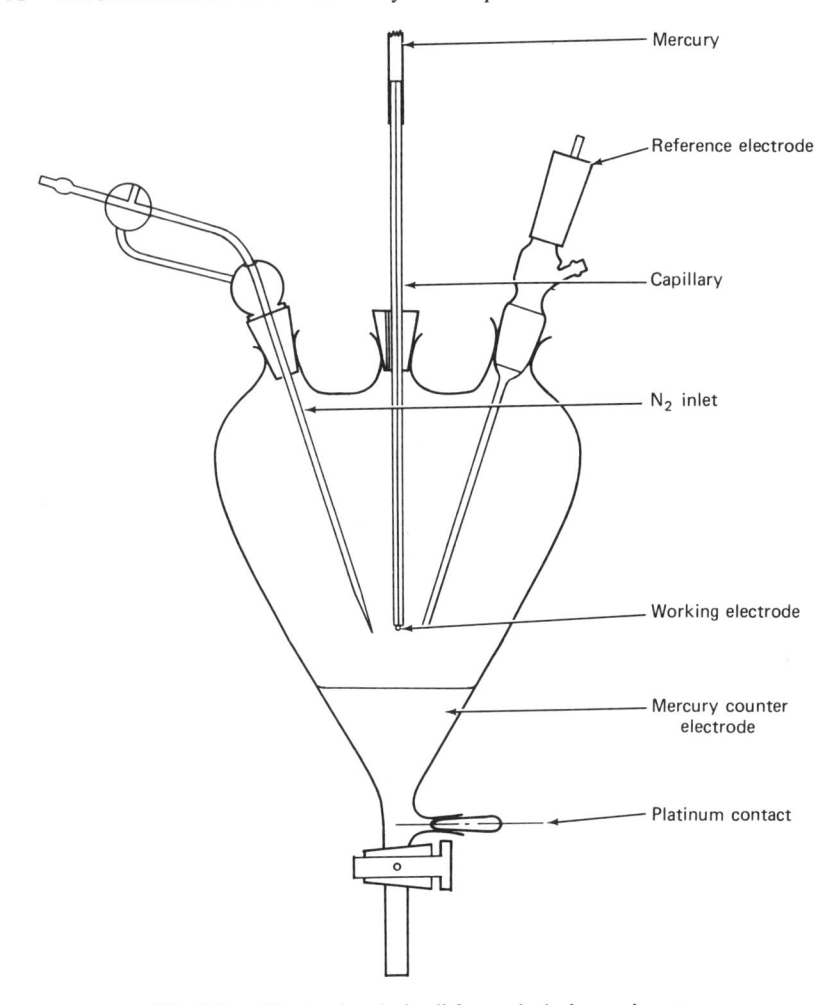

432 Electrochemical Behavior of Heterocyclic Compounds

Fig. 8.2. Electrochemical cell for analytical experiments.

cells it is mandatory to separate the catholyte from the anolyte to prevent mixing of the products formed at each electrode. In such a cell it is desirable that the current be uniformly distributed on the working electrode to prevent iR drop phenomena on some points of this electrode; this is generally achieved by placing two electrodes of similar size face to face.

A cell that is very convenient for reductions or oxidations on carbon is shown in Figure 8.4. Its main component is a vitreous carbon crucible that serves as both the container and the working electrode.

Several examples of such laboratory cells have been described along with cells designed for special purposes, as for ESR and UV spectrometry and for high-pressure work (19).

Industrial cells (20, 21) are generally flow cells in which anolyte and catholyte of fixed composition are fed continuously and equivalent quantities are withdrawn

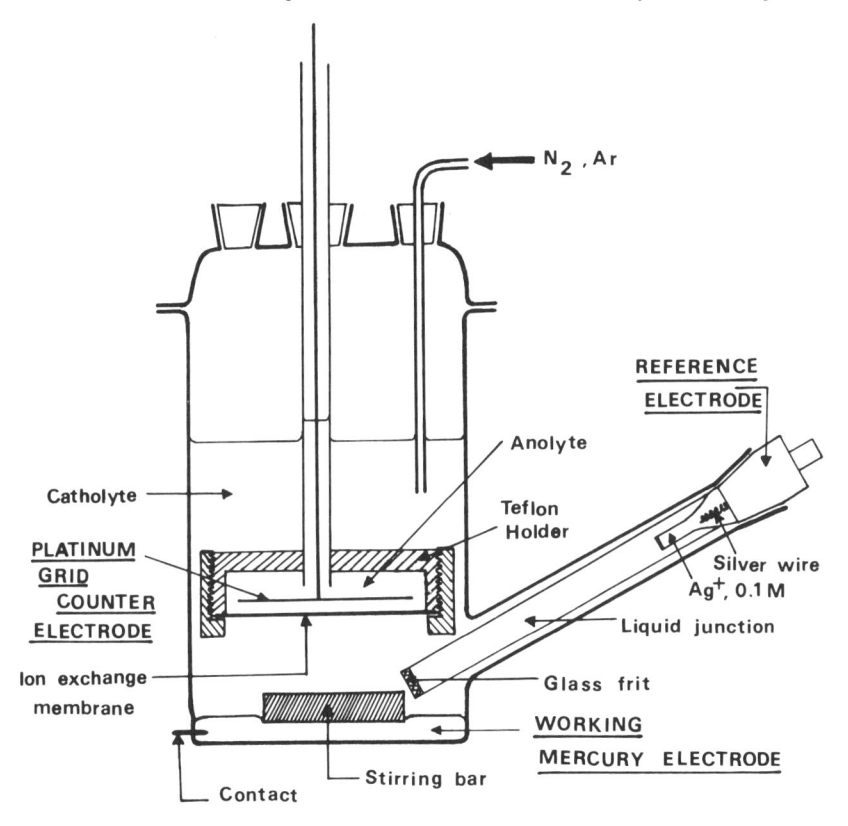

Experimental Procedures and Electroanalytical Techniques 433

Fig. 8.3. Laboratory cell for preparative electrolysis.

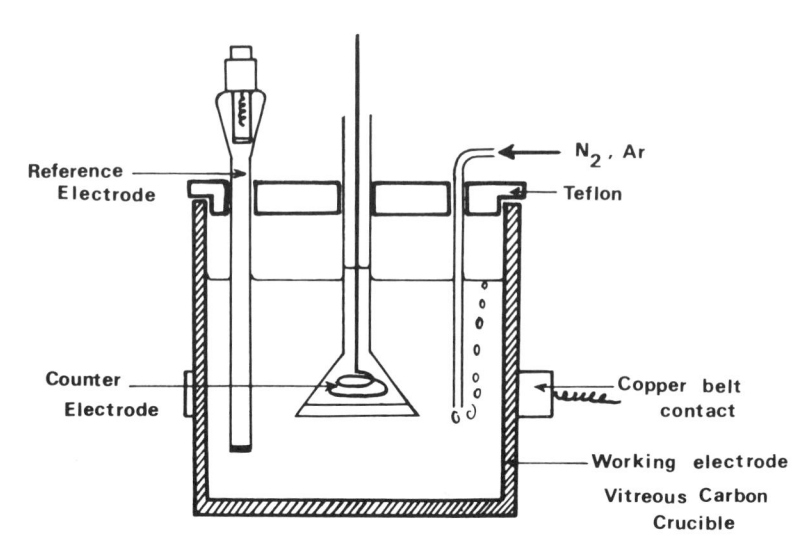

Fig. 8.4. Vitreous carbon laboratory cell for electrolysis.

for further workup. Industrial cells (22) include filter press cells consisting of stacks of anodes, diaphragms, and cathodes (23), capillary gap cells with an extremely narrow spacing between anode and cathode (100–125 μm) (24), the Swiss-Roll cell (25) where anode and cathode are wrapped cylindrically as in a paper condenser, and the Pump Cell (26) where the stator and the rotor of a pump constitute the two electrodes.

8.1.1.2. Diaphragms

Diaphragms should prevent the products of the catholyte and anolyte from mixing and should not be too resistive. Two kinds of diaphragms are used in laboratory work, porous materials such as glass frits and unglazed porcelain and ion-exchange membranes. These permselective membranes efficiently prevent the diffusion of neutral compounds from one compartment to the other.

8.1.1.3. Reference Electrodes

The potential of the working electrode is measured with reference to a half-cell, the potential of which is stable over time and is not changed by flow of a small current (unpolarizable electrode). The half-cell to which thermodynamic standard potentials are referred is the normal hydrogen electrode (NHE), but it is very inconvenient for practical purposes and other reference electrodes are used. The most convenient and widely used is the saturated calomel electrode (SCE), which is $Hg/Hg_2Cl_2/KCl$ saturated in water. It is commercially available as a rod electrode with a glass joint that fits in the cell and a glass frit that separates the saturated KCl solution from the solution under investigation. There is nearly no restriction to its use in aqueous solutions only in the case when chloride ions should be avoided it should be replaced by the mercurous sulfate electrode Hg/Hg_2SO_4/saturated K_2SO_4. However, in nonaqueous solvents there is some diffusion of water in the solvent as well as potassium and chloride ions, which may bring some changes in the electroanalytical curves. Besides this, some side reactions between the components of SCE and some nonaqueous solvents may result in potential shifts. A very simple way to prevent these phenomena is to use a liquid junction together with a fine glass frit.

Other electrodes are used in nonaqueous solvents such as the silver–silver chloride electrode Ag/AgCl(s), KCl(s), and the silver–silver ion electrode Ag/Ag^+ (0.01 M). The former is prepared by electrodepositing a layer of AgCl on a silver wire and then placing it in a saturated KCl solution, and the latter by placing a silver wire in a 0.01 M Ag^+ solution. Both electrodes are separated from the solution by a liquid junction. Neither is very stable, especially the silver–silver chloride electrode, and it is desirable to check the electrode potential against a SCE before each series of experiments.

The potentials of the different electrodes are reported in Figure 8.5. A review of electrode potentials in nonaqueous solvents has been published (27).

Let us consider now the exact meaning of imposing the potential between the working and reference electrodes. The equivalent circuit of the cell is given

Experimental Procedures and Electroanalytical Techniques   435

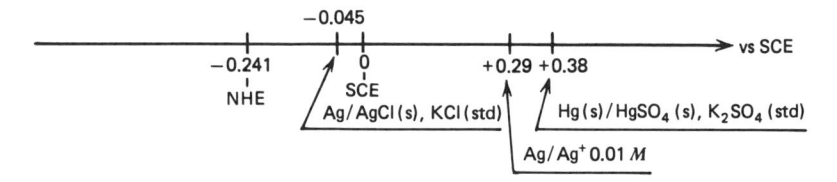

Fig. 8.5.   Relative potentials of different reference electrodes.

in Figure 8.6, and it is clear that a potential (E_{appl}) imposed between the reference and working electrodes by the potentiostat forces a current i through the cell and the potential of the working electrode is $E = E_{appl} - R_u i$. By lowering R_u, E can be brought near to E_{appl}. This can be achieved in preparative electrolysis by positioning the reference electrode in the close vicinity of the working electrode. In electroanalytical experiments, specially designed potentiostats compensate most of R_u. This also explains why two-electrode systems where the reference electrode also serves as the counter electrode can be used only in supporting electrolyte-solvent systems of very low resistivity (generally buffered aqueous solutions).

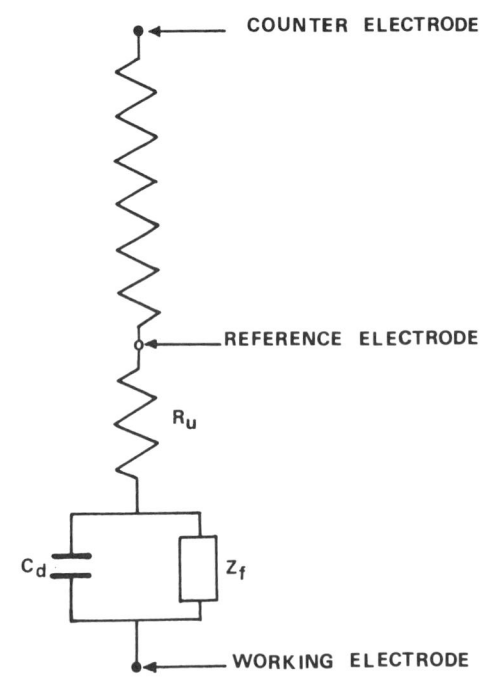

R_u: Uncompensated Resistance
C_d: Double layer capacitance
Z_f: Faradic Impedance

Fig. 8.6.   Equivalent electrical circuit of an electrochemical cell.

8.1.1.4. Working and Counter Electrodes

Ideally an electrode should play a single role, to transfer electrons to the substrates, and should not reduce or oxidize the solvent and supporting electrolytes, so that a large potential range will be available for both the reduction and oxidation sides. With a wider usable potential range it is possible to investigate a large number of compounds. However, this usable potential range depends not only on the electrodes but also on the solvent and the supporting electrolyte (28). No material fulfills all requirements.

Mercury is widely used as a cathode, mostly in laboratory practice. In industry, its use has been restricted by environmental regulations. In both aqueous and nonaqueous solvents it offers a wide cathodic range, but it cannot be used as an anode (except for very oxidizable compounds) as its oxidation takes place at about 0 V/SCE in all solvents. Another advantage of mercury is its ease of purification. Besides this, polarography, one of the most widely used electroanalytical techniques, entirely relies on the liquid state of mercury. In some cases, however, mercury reacts with the compounds under investigation or with some intermediates (29, 30).

Lead can also be used as a cathode material. It sometimes shows specific effects leading to products different from those observed on mercury (31, 32).

Platinum can be used as either a cathode or anode, although its cathodic range in aqueous solvents is somewhat less than that of mercury. Indeed, in laboratory experiments, it is one of the most widely used anodes (both as a working macro- or microelectrode or as a counter electrode). *Gold* behaves like platinum. *Carbon* is used under different forms, either as vitreous carbon or pyrolitic graphite or as carbon paste. One of the most important advantages is that it has a quite wide potential range both on the anodic and cathodic side, and hence it allows the study of both the oxidation and reduction of a compound (33, 34). For industrial purposes, lead, carbon, steel, copper, and nickel are used as cathodes. Lead and some of its alloys, carbon, and platinum, as well as dimensionally stable anodes (DSA) (where a non-noble metal is covered by a noble metal, e.g., platinum on titanium) are used as anodes.

8.1.1.5. Solvents

An electrochemical solvent should fulfill the following requirements:

1. It should have a wide accessible potential range; that is, a solvent for oxidation should be oxidized at potentials as high as possible, and conversely a solvent for reduction should be reduced at potentials as negative as possible.
2. It should not be too resistive. This depends on its ability to dissociate the salts used as supporting electrolytes, which in turn depends on its dielectric constant.
3. It should dissolve correctly the compounds to be investigated (although it

is often possible to carry out preparative electrolysis with reactants in suspension) and the supporting electrolyte.
4. In the case of preparative work, the ease of recovery of the products should be considered. For example, it is quite difficult to recover compounds from dimethylsulfoxide (DMSO), which has a high boiling point (bp_{760} 189°C) and is highly soluble in water, while acetonitrile (MeCN) is easily evaporated (bp_{760} 82°C).

Besides, many reaction mixtures are analyzed by chromatography, and the behavior of the solvent should be taken into consideration. For example, DMSO gives very broad peaks on polar VPC columns, as does dimethylformamide (DMF). Some solvents may absorb during the UV detection in HPLC (for example, DMSO absorbs below 250 nm, while MeCN is clear down to 200 nm).

Electrochemical solvents are classified as protic or aprotic. Reduction leads to negatively charged species such as radical anions and anions, which will have a tendency to protonate. In an aprotic solvent, these species will have a smaller tendency to react with the solvent and will be stabilized. On the other hand, in a protic solvent many protonation steps intervene in the reaction. Scheme 8.1 includes the example of 3-phenyl pyrido[3,2-e]as-triazine (35), where in MeCN the radical anion is stable, while in aqueous solution the electronic transfer takes place on a protonated species and the resulting radical undergoes further reduction and protonation. Oxidations are less sensitive to the protic or aprotic character of the solvent; in this case positively charged species (radical cations) are produced that are prone to nucleophilic attacks, and therefore the important parameter is the nucleophilicity of the medium. For example, diphenylanthracene (Scheme 8.2) gives a stable radical cation in MeCN, and a nucleophilic substitution occurs (36) in the presence of pyridine. Another characteristic of the solvent is its hydrogen atom donating ability. Hydrogen atom abstractions from neutral radicals have often been postulated to explain the formation of products and unambiguous evidence has been reported in some cases (37) (Scheme 8.3).

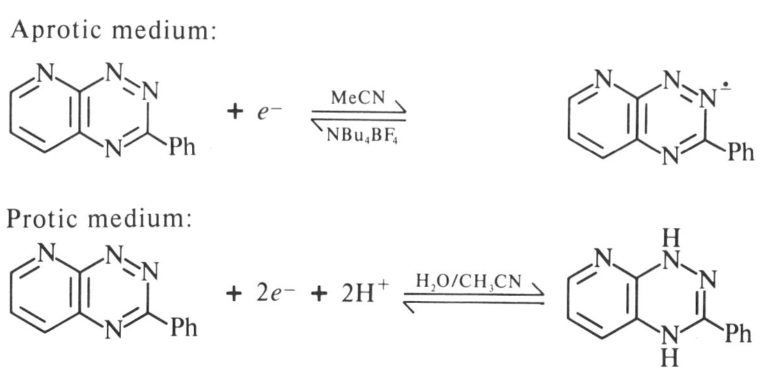

Aprotic medium:

[structure] + e^- $\underset{NBu_4BF_4}{\overset{MeCN}{\rightleftarrows}}$ [structure]

Protic medium:

[structure] + $2e^-$ + $2H^+$ $\underset{}{\overset{H_2O/CH_3CN}{\rightleftarrows}}$ [structure]

Scheme 8.1

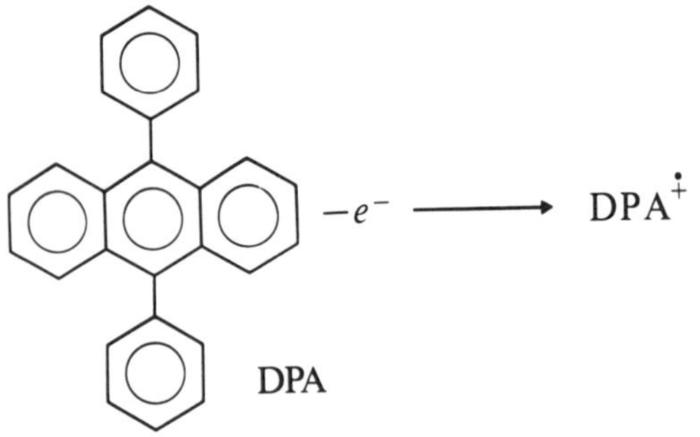

438 Electrochemical Behavior of Heterocyclic Compounds

$$DPA \xrightarrow{-e^-} DPA^{+\cdot}$$

$$DPA - e^- \xrightarrow[\substack{\text{Pyridine} \\ \text{NEt}_4\text{ClO}_4}]{\text{CH}_3\text{CN}} DPA\text{-}Py^{+\cdot} \xrightarrow{-e^-} DPA^+\text{-}Py^+$$

Scheme 8.2

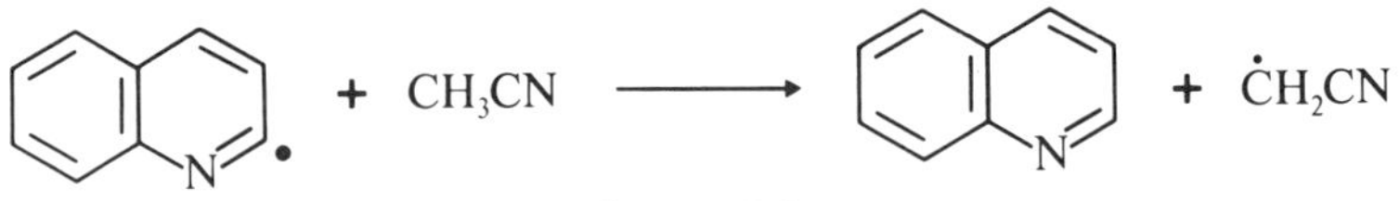

Quinoline$^{\cdot}$ + CH$_3$CN $\longrightarrow$ Quinoline + $\dot{\text{C}}$H$_2$CN

Scheme 8.3

Several reviews (8, 19, 38, 39) have appeared on the physicochemical properties and useful potential range of different solvent electrolyte systems, as well as on purification procedures. IUPAC-recommended methods of solvent purification have also been published for acetonitrile (40), pyridine (41), propylene carbonate (42), N-methylacetamide (43), dimethylsulfoxide (44), N-methylpropionamide (45), dimethylformamide (46), and sulfolane (47). The compilation of electrochemical reactions classified according to the use of solvents has been published by Meites and Zuman (14).

8.1.1.5.1. Protic Solvents. The most widely used protic solvent is water, the proton availability of which can be modified through the use of appropriate buffers. Its use is really trouble-free, but its main drawback is that organic compounds are scarcely soluble in this medium. This can be overcome through the use of hydroorganic mixtures (methanol-water, acetonitrile-water, dimethylformamide-water). The recovery of products from such mixtures is fairly easy: evaporation of the solvent, and precipitation of the products by addition of water or extraction with an immiscible organic solvent. Electrolyzed solutions in such hydroorganic solvents cannot be analyzed by VPC (except on Porapak columns), but they are readily quantized by reverse-phase HPLC, where the same hydroorganic solvents (methanol-water, acetonitrile-water) can be used as eluant. The usable potential range of water strongly depends on the pH. In acidic medium at pH 1 it goes from $\sim +1.5$ to ~ -1.1 V/SCE and at pH 13 from ~ -0.2 to ~ -2.0 V. A problem that arises in the use of hydroorganic solvents is the determination of the pH: one can either use pH scales that have been published for such media (48) or simply use the direct reading obtained with a glass electrode, specifying the standardization conditions. In the latter case one does not measure an actual pH, but the measurement can be reproduced.

Other protic solvents include alcohols, particularly methanol; acids, such as acetic acid, which has been used to carry out acetoxylation reactions through chemical oxidations (49); and Kolbe reactions (50, 51). Hydrogen fluoride has been used to study the oxidation of bromoalkanes and hydrocarbons (52, 53). Fluorosulfonic acid (FSO_3H) has been used to carry out electrochemical isomerization, oligomerization, and cracking of n-alkanes (54, 55).

8.1.1.5.2. Aprotic Solvents. Ordinarily, aprotic solvents contain some residual water (for example, ~ 0.3% in distilled MeCN, 0.15% in distilled DMSO). These low water contents are far from negligible, as they correspond to contrations (0.15% water in DMSO ~ 2×10^{-2} M) that are higher than the concentrations of substrates used in electroanalytical experiments (up to 5×10^{-3} M). The water content can be measured either by Karl Fischer titration or by VPC on Porapak columns (56). This water content is decreased through the use of activated neutral alumina added directly in the solution or used in a column on which the solvent-supporting electrolyte is recycled several times (57, 58). Conversely, an aprotic medium can be gradually transformed into a protic one by the addition of proton donors such as phenol, acetic acid, or alcohol. This slow transition from an aprotic to a protic solvent is useful for the analysis of reactions.

Acetonitrile (MeCN) is a widely used solvent (although somewhat acidic in nature, its conjugated base $^-CH_2CN$ is not a very strong base) for oxidations and reductions. It is easily removed by distillation in preparative work, its retention time on most of the VPC columns is quite small, and its peak is generally not too broad (the peaks of the products can be clearly observed). In HPLC the injection of acetonitrile solutions on reverse-phase columns is a particularly recommended technique. It is well suited for spectroelectrochemical experiments, being transparent down to 200 nm. Dimethylformamide (DMF) is also of general use in electrochemistry although it is more easily oxidized than acetonitrile. It is less convenient for preparative work. Products may be recovered by precipitating with water, by adding water and extracting with ether, or by distilling off the solvent under reduced pressure. Dimethylsulfoxide (DMSO) is also a good solvent for reduction, but the extraction of products is difficult and its UV cutoff is at 250 nm. Care must be taken in handling DMSO (59), which although it is not toxic itself easily allows other toxic substances to be transported through the skin. Methylene chloride is a solvent of choice for oxidation because of its ability to stabilize radical cations and its low nucleophilicity (60, 61).

Among basic solvents, liquid ammonia (62, 63) is particularly useful, because it is at the same time a poor proton donor and a poor hydrogen atom donor. Even dianions can be stable in this medium (64–66). Some special equipment is obviously needed due to its low boiling point (bp_{760} $-33°C$) (62, 63). Other aprotic solvents include pyridine (67), hexamethylphosphorotriamide (HMPA) (68), butyronitrile (69), 1,2-dimethoxyethane (70), sulfolane (71), propylene carbonate (66), tetrahydrofuran (72), and liquid sulfur dioxide (73).

8.1.1.6. Supporting Electrolytes

In water and hydroorganic solvents, the usual buffers (Britton-Robinson, Clark-Lubs) can be used. Their concentration must be high enough, particularly when performing large-scale electrolysis, to avoid buffering problems. For the electrolysis of 2×10^{-2} M solutions it is usual practice to use 0.2 M buffers. There are some solubility problems with phosphate buffers, particularly when large amounts of organic cosolvents are used. In such cases phosphate buffers are replaced by buffers made from carboxylic acids and organic amines.

In aprotic solvents, tetraalkylammonium perchlorates, fluoborates, and nitrates are generally used both for oxidation and reduction. They are now commercially available in reagent grade purity and are very soluble in the usual electrochemical solvents (up to about 1 M). In reduction the tetrabutylammonium cation gives the widest potential range (74). Since explosions have been observed with perchlorates, the use of tetrafluoborates seems safer (75). The problems are that in the presence of a strong base they can undergo Hoffmann elimination and that they can give hydrogen atoms to neutral radicals (37). When these problems become too serious, lithium or sodium perchlorate is used as a supporting electrolyte, and in such cases the accessible cathodic range is reduced (for example, in acetonitrile the supporting electrolyte discharge occurs at about -2.7 V/SCE with NBu_4BF_4 but at only -2.0 V/SCE with lithium perchlorate).

8.1.1.7. Electronic Devices

The most important part of the electronic equipment is the potentiostat, which maintains constant potential between the working and reference electrodes. The electronic principles on which the potentiostat is based are found in references 4 and 76. The characteristics of the potentiostat are not the same for preparative electrolysis as for electroanalytical techniques. In the former case, large currents (up to 1 A) are necessary, while the time taken to reestablish the potential after a perturbation (the response time) is not too important. In analytical techniques where fast voltage changes are applied (potential steps or voltage ramps), fast-rise wide-bandpass potentiostats are needed. An additional feature necessary in a fast-rise potentiostat is a positive feedback loop, which allows one to compensate for the residual resistance R_u between the working and reference electrodes (77-79). Figure 8.7 includes a voltammogram recorded with and without *iR* drop compensation. All kinds of potentiostats for large-scale electrolysis or analytical work are available commercially.

When performing electrolysis at constant potential, one does not need a signal generator, but a voltmeter is needed to check the potential between the reference and working electrodes, and an ammeter between the counter electrode and potentiostat. A coulometer is also important; in series with the ammeter it integrates the current and gives the total amount of charge passed through the cell (a datum of prime importance for the establishment of the mechanism).

When performing polarographic, voltammetric, chronopotentiometric, etc.

Experimental Procedures and Electroanalytical Techniques 441

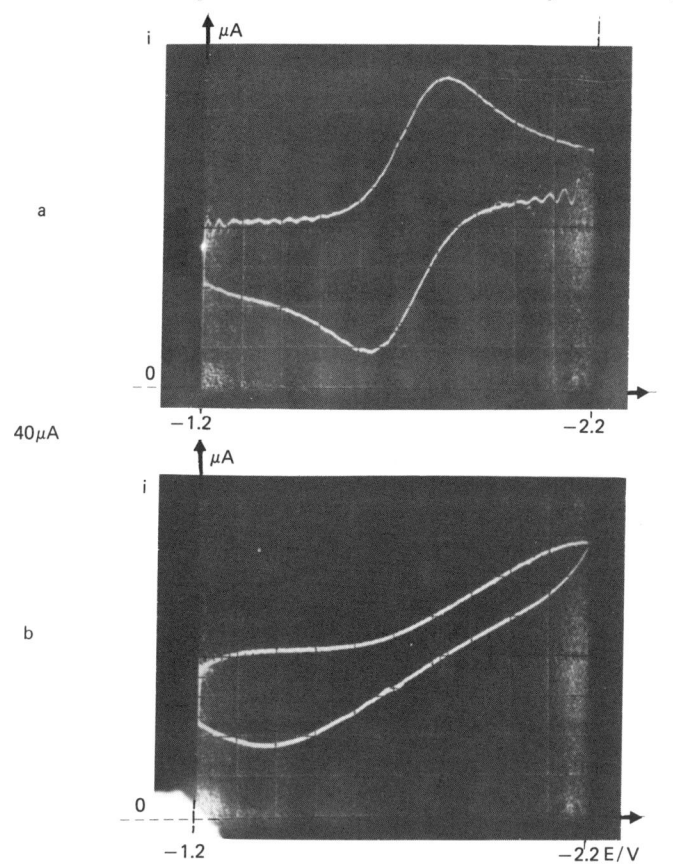

Fig. 8.7. Cyclic voltammogram of benzo[c]cinnoline in DMF ($c = 10^{-3}\,M$; supporting electrolyte, NBu_4BF_4, $10^{-1}\,M$) at $v = 1000\,V\,s^{-1}$ (a) with and (b) without ohmic drop compensation, reference SCE.

experiments, it is necessary to monitor the potential as a function of time. This is done by a signal generator which pilots the potentiostat. Voltage ramps (Fig. 8.8a) and potential single steps (Fig. 8.8b) or double steps (Fig. 8.8c) can be applied to the electrode. Finally, the current is recorded as a function of time on a chart recorder in the case of slow signals and on a memory oscilloscope in the case of transient signals. A digital acquisition system must be used for fast and precise quantitative measurements.

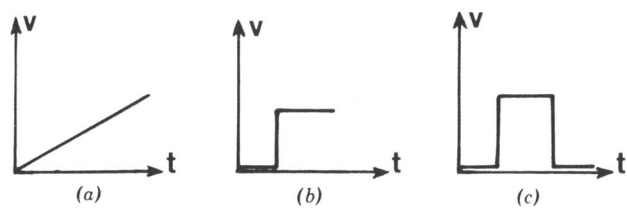

Fig. 8.8 (a) Voltage ramps (b) potential single step, (c) double steps, used in electroanalytical experiments.

8.1.2. The Use of Electroanalytical Techniques in the Investigation of Organic Reactions

Different electrochemical techniques and the basic concepts of electrochemistry have been dealt with in a number of books (80–92) and particularly in a recent and comprehensive treatise by Bard and Faulkner (76, 93). Reviews have also been published on the use of electrochemical methods for the elucidation of reaction mechanisms (94–96). The main features of electroanalytical techniques and their applications in the study of reaction mechanisms are discussed in a qualitative manner. Scheme 8.4 includes an example of reduction.

The compound Y to be investigated is dissolved in an electrochemical solvent supporting electrolyte system. The following steps may be involved:

1. Y may react with the solvent (may be protonated). This step symbolizes all the reactions that can occur before any electrochemical event to the starting compound.
2. The compound resulting from step **1** or the starting compound will have to be transported to the electrode (mass transfer) to be electronated. This can occur in three ways:
   (*a*) By migration of charged species in the electrical field produced by the electrodes, which is prevented by the use of high concentrations of supporting electrolyte.
   (*b*) By convection, that is, by stirring the solution or by rotating the electrode.

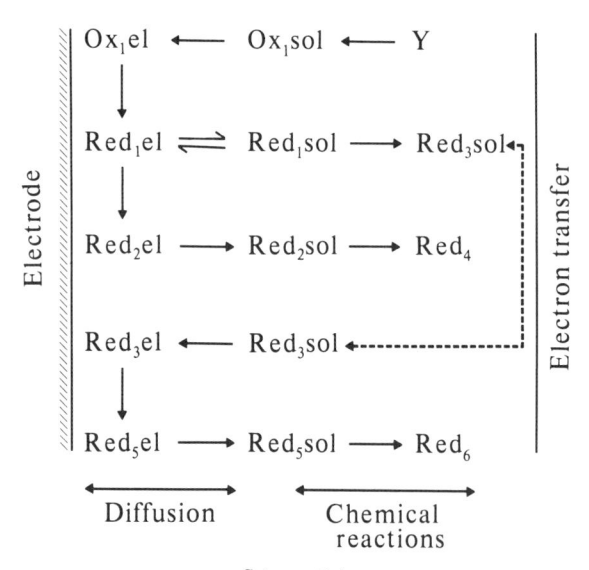

Scheme 8.4

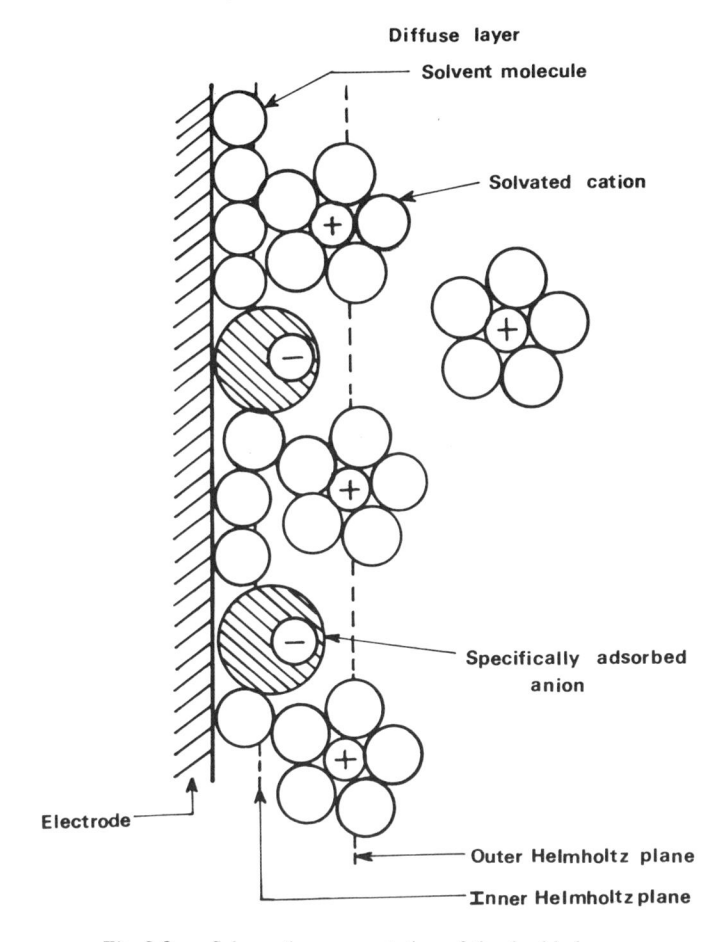

Fig. 8.9. Schematic representation of the double layer.

(c) By diffusion in the concentration gradient. (Obviously, the concentration of Ox_1 must be smaller, even equal to zero near the electrode where it is reduced, than in the bulk of the solution).

What makes the problem complicated is that this concentration gradient depends on the electrode potential and on the time. Electroanalytical techniques have been devised to render the problem as simple as possible and then diffusion is the only mode of *mass transfer*. In preparative electrolysis, convection is an important way of bringing compounds to the electrode.

3. Compound Ox_1 is now near the electrode in a region termed *double layer*, where the structure of the solution is different from that of the bulk solution (Fig. 8.9). This particular structure is electrically equivalent to a capacitor (it is represented as C_d in Fig. 8.6). An electron is added to Ox_1 while it is in the outer Helmholtz plane, or it is specifically adsorbed on the

electrode; in the latter case there is a supplementary adsorption-desorption step in the mechanism. There is general agreement that electrons are added one by one, but in the case where two separate identical functions exist on the molecule two electrons can be transferred at the same potential (97). By addition of one electron, Ox_1 is reduced to Red_1, which in turn can be specifically adsorbed on the electrode. As in a chemical reaction, the transfer of an electron is characterized by a rate constant k_s (which, being heterogeneous, is expressed in cm s^{-1}), and the rate of electron transfer depends on the potential.

4. Red_1 will now diffuse back to the bulk of the solution.
5. Red_1 can undergo further chemical reactions: protonation, electrophilic attack, dimerization, etc., to give Red_3, which is very often more easily reducible than the starting Ox_1.
6. It diffuses toward the electrode, where it undergoes a second electronic transfer to give Red_5.
7. Red_5 diffuses toward the bulk of the solution and undergoes chemical reaction to give Red_6.
8. At a more negative potential, Red_1 may also undergo a second electron transfer and further chemical reactions.

Besides this, homogeneous redox reactions can occur between species in different oxidation steps as shown:

$$Ox_1 + e^- \longrightarrow Red_1 \text{ sol} \qquad ArX + e^- \longrightarrow ArX^{\cdot -}$$

$$Red_1 \text{ sol} \longrightarrow Red_3 \qquad ArX^{\cdot -} \longrightarrow Ar^{\cdot} + X^-$$

$$Red_3 + Red_1 \rightleftharpoons Ox_1 + Red_5 \text{ sol} \qquad Ar^{\cdot} + ArX^{\cdot -} \rightleftharpoons Ar^- + ArX$$

$$Red_5 \text{ sol} \longrightarrow Red_6 \qquad Ar^- + H_2O \longrightarrow ArH + OH^-$$

where ArX = 2-chloroquinoline. The homogeneous electron transfer leads to the same product Red_5 as a second electron transfer on Red_3.

$$Red_3 + e^- \longrightarrow Red_5 \qquad Ar^{\cdot} + e^- \longrightarrow Ar^-$$

With such reaction schemes, efforts are directed to finding out the possible intermediates, the reactions involved, the rate-determining step, and the rate constants. Efforts are made to study the electrochemical curves and their modification with the different parameters together with nonelectrochemical methods such as spectroscopy. In some cases thorough analysis of the reaction is possible, but sometimes it is possible only to suggest the most likely reaction path.

Each electroanalytical technique has a characteristic time, which must be of the same order as the lifetime of the intermediates to be observed. There is, for example, no hope of observing an intermediate with a lifetime of 1 s by performing a coulometric procedure that lasts $\frac{1}{2}$ h. Thus the time characteristic is an important factor in choosing a method in the search for transient intermediates.

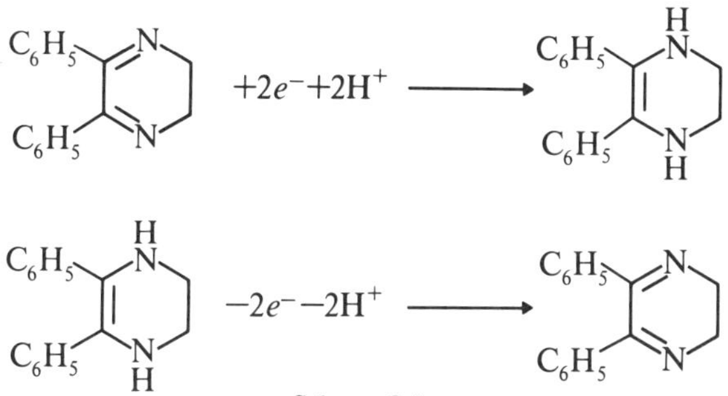

Scheme 8.5

An important concept that is often referred to in the analysis of chemical reactions is that of reversibility and irreversibility. Unfortunately, these terms are used with several different meanings, which may be quite confusing for the reader. The first meaning of chemical reversibility is that there is one pathway from the reactant to the product and another, reverse pathway from the product to the reactant (98) (Scheme 8.5). The reduction is carried out at $E = -0.95$ V/SCE at pH 13, while the electrochemical oxidation takes place at $E = -0.45$ V/SCE. There is no reason to think that the reoxidation takes place by way of the reverse path of the reduction.

A completely irreversible reaction is shown in Scheme 8.6 (2). There is no way to change the potential to reverse the course of reaction, and the irreversibility is due to the irreversible character of bond cleavage.

The second meaning is related to thermodynamic reversibility. For a simple $Ox + e^- \rightleftarrows Red$ system at equilibrium, a small negative change of the potential will transform some Ox to Red, and conversely for a small positive change of the potential. But the observation of this thermodynamic reversibility relies on kinetic conditions: if the rate of change of the potential is fast and if the time during which the observation takes place is short, is it possible to observe the formation of Red or Ox? Yes, if the system is kinetically fast; no, if the system is kinetically slow. Therefore, the terms *fast* and *slow electron transfer* should be preferred in this case to *reversibility* and *irreversibility*. For example, the system included in Scheme 8.7 appears fast (reversible) for slow variations of the potential (0.1 V s^{-1}) but slow (irreversible) for fast variations of the potential (50 V s^{-1}). It is shown in Figure 8.10, where the reduction and oxidation peaks shift apart upon increasing the sweep rate.

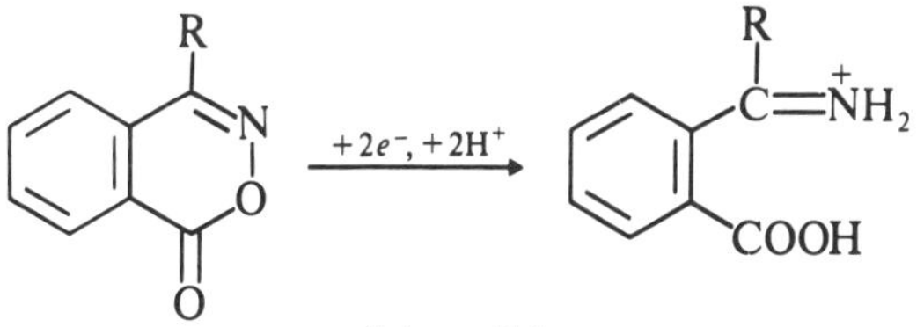

Scheme 8.6

446   Electrochemical Behavior of Heterocyclic Compounds

[Scheme 8.7: nitrodurene + e⁻ → nitrodurene radical anion]

Scheme 8.7

Some of the electroanalytical techniques and their applications for the analysis of reaction mechanisms will be discussed now. It is common practice to classify the different possible reaction schemes in which chemical reactions are associated with electron transfers by using E for an electron transfer and C for a chemical reaction. A CE mechanism is

$$Y \longrightarrow Ox_1$$
$$Ox_1 + e \longrightarrow Red_1$$

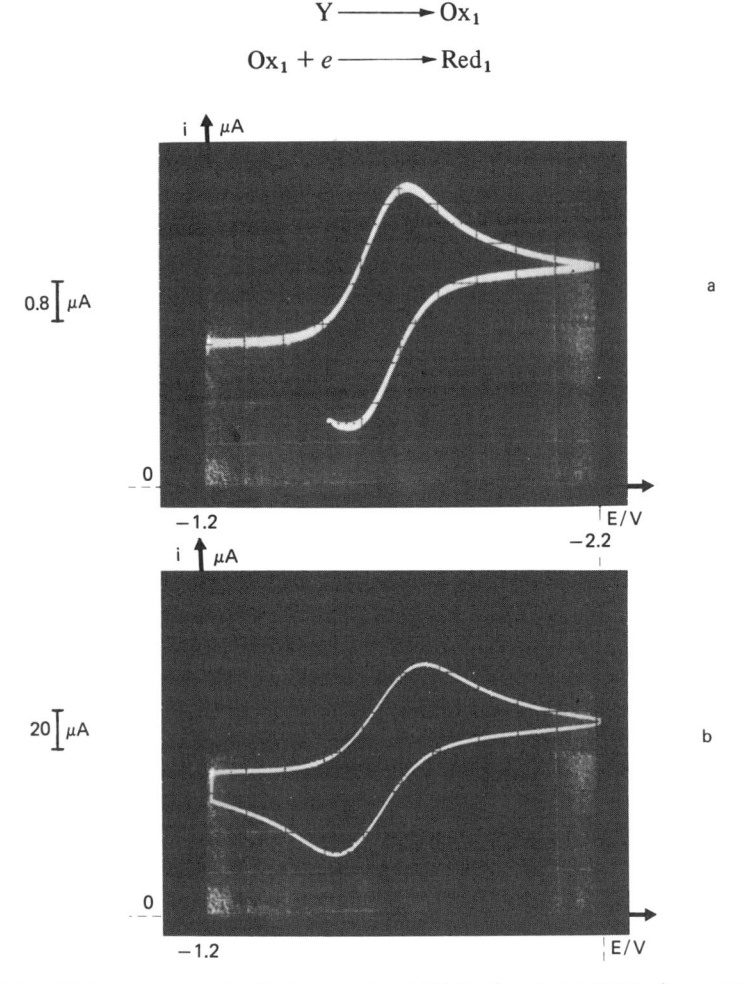

Fig. 8.10. Voltammogram of nitrodurene at (a) $0.1\,\mathrm{V\,s^{-1}}$ and (b) $50\,\mathrm{V\,s^{-1}}$ ($c = 10^{-3}\,M$ in DMF, NBu_4BF_4 $10^{-1}\,M$, reference SCE).

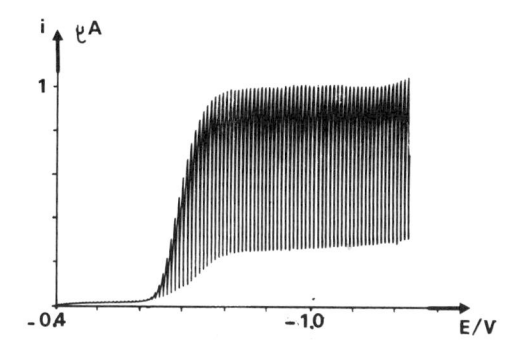

Fig. 8.11. Polarogram of 2,3-diphenyl quinoxaline ($c = 10^{-3}$ M, 50/50 water methanol, pH 7, reference SCE).

while an EC mechanism will be

$$Ox_1 + e \longrightarrow Red_1$$
$$Red_1 \longrightarrow Red_3$$

Mechanisms are also termed ECE, DISP (involving a disproportionation reaction), or DIM (involving a dimerization reaction).

8.1.2.1. Polarography

Polarography, a technique discovered by Heyrovsky in 1922, has been for a long time the overwhelming dominant electroanalytical technique for both analytical purposes and the study of reaction mechanisms. The working electrode of a polarograph is a dropping mercury electrode (DME); the drops of mercury are obtained from a glass capillary connected to a mercury reservoir maintained at a certain height above the end of the capillary (30–100 cm). The advantage of this technique is that each drop is a new clean electrode, and as the fall of the preceding one has stirred the solution this new electrode is borne in a fresh solution. A slow voltage ramp is applied to this electrode. The main limitation of this technique is that it can be used only at negative potentials (i.e., mainly in reduction), mercury being oxidized at about 0 V/SCE in all solvents. The time parameter of this method is the drop time t_{max}, which can vary from 1 to 10 s, that is, in a rather small range compared with other methods. Figure 8.11 shows the polarogram of diphenylquinoxaline (99); the wave corresponds to the reaction shown in Scheme 8.8.

Scheme 8.8

448   Electrochemical Behavior of Heterocyclic Compounds

The simplest data that can be obtained from a polarogram are:

1. The half-wave potential $E_{1/2}$, the potential at which the current reaches half its limiting value. For a given solvent supporting electrolyte system it is a characteristic of the product. For the identification of a product, an $E_{1/2}$ is just as good as its melting point; that is, the identity of half-wave potentials does not necessarily imply the identity of two compounds.

2. The limiting current of the plateau i_d. In this region the current is limited by diffusion; that is, the potential (by reference to the standard thermodynamic potential) is negative enough that the rate of electron transfer becomes large and the diffusion becomes rate-determining. This limiting current for the $Ox + ne^- \rightarrow Red$ reaction, measured just before the drop falls is given by the Ilkovic equation:

$$i_d = 708 n D_O^{1/2} C_O^* m^{2/3} t_{max}^{1/6}$$

where n is the number of electrons involved in the reduction, D_O is the diffusion coefficient of the reagent in $cm^2\ s^{-1}$, C_O^* is the concentration of Ox in $mol\ cm^{-3}$, m is the flow rate of mercury in $mg\ s^{-1}$, t_{max} is the drop time in seconds, and i_d is given in amperes. This equation forms the basis for the analytical use of polarography; for a given compound, D_0 and n are known and m and t_{max} are characteristic of the capillary, so by measuring i_d, C_O^* can be determined or, alternatively, calibration curves can be used. For the analysis of reaction mechanisms, the Ilkovic equation is used to determine n. The diffusion coefficients D_O do not vary very much from one organic compound to another (mostly for compounds with similar structures). Thus by using the same capillary to compare the height of the wave of the compound investigated at a C_O^* concentration with that of a compound of similar structure at the same concentration for which n is known, n for the unknown process is determined. n is very often an integer (1, 2, 3, ...).

Several other data can be obtained from a polarogram. In the case of a simple $Ox + ne^- \rightarrow Red$ reaction, if the electrode reaction is reversible (the electron transfer is always very rapid on the time scale of polarography), $E_{1/2}$ is given by the relation

$$E_{1/2} = E^{\circ\prime} + \frac{RT}{nf} \ln \frac{D_R^{1/2}}{D_0^{1/2}}$$

and as $D_R \simeq D_O\ E_{1/2} = E^{\circ\prime}$ (the formal potential of the Ox/Red system). In this case, the equation of the polarographic wave is given by

$$E = E_{1/2} + \frac{RT}{nF} \ln \frac{i_d - i}{i}$$

where R is the gas constant, T is absolute temperature, F is the Faraday constant ($= 96500\ C$), and i is current on the wave at potential E. For such a fast electron transfer, a straight line with a slope of $59.1/n$ mV at 25° is obtained by plotting E versus $\ln(i_d - i)/i$. In the case of a fast electron transfer, the diffusion is rate-determining and nothing can be known about the kinetics of electron transfer except the value of $E^{o'}$. In the case of an irreversible electron transfer (i.e., electron transfer that is slow and rate-determining on the time scale of polarography) E is given by the relation

$$E = E^{o'} + \frac{0.059}{\alpha n_a} \log\left(1.349 \frac{k^\circ t^{1/2}}{D_O^{1/2}}\right) + \frac{0.0542}{\alpha n_a} \log\left(\frac{i_d - i}{i}\right)$$

where α is the electrochemical transfer coefficient (100), whose value is generally near 0.5, k° is the standard heterogeneous rate constant of reduction, and n_a is the number of electrons involved in the rate-determining step. In this case, E vs. $\log[(i_d - i)/i]$ should be linear with a slope of $54.2/\alpha n_a$ mV, and $E_{1/2}$ is given by the relation

$$E_{1/2} = E^{o'} + \frac{0.059}{\alpha n_a} \log\left(1.349 \frac{k^\circ t^{1/2}}{D_O^{1/2}}\right)$$

If chemical reactions are associated with electron transfers, it is possible to characterize these reactions in some cases:

1. If the chemical reaction takes place after the first electronic transfer, the current on the plateau of the wave should be (as in a simple reaction $Ox + e^- \rightarrow Red$) controlled by the diffusion. In such cases i_d should vary linearly with $\sqrt{h}$ (the square root of the effective height of the mercury reservoir above the tip of the capillary). The existence of this followup reaction can be characterized only by the shift of the half-wave potential toward positive potentials.
2. In the case of kinetic currents a preceding reaction (101) is rate-determining. In Scheme 8.9, the free aldehyde is the only reducible form, and the height of the wave in acidic medium (between $H_0 = -2$ and $+2$) is limited by the rate of the dehydration steps; it depends on the concentration of the reagent C_O^* but is independent of the mercury height. If the rate of the preceding reaction becomes very fast or very slow, such kinetic currents will no longer be observed and the height of the wave becomes diffusion controlled.
3. With catalytic currents such as in the reduction of protons catalyzed by the 1,4-dihydroquinoxalinium ion (Scheme 8.10) (102), i_d increases with C_O^* until a certain concentration and then remains constant.

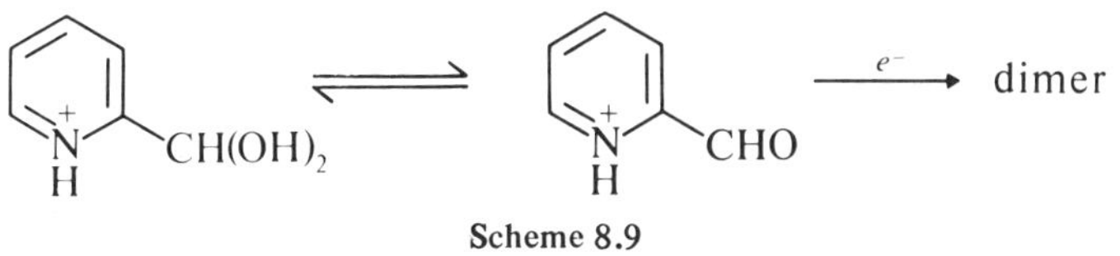

Scheme 8.9

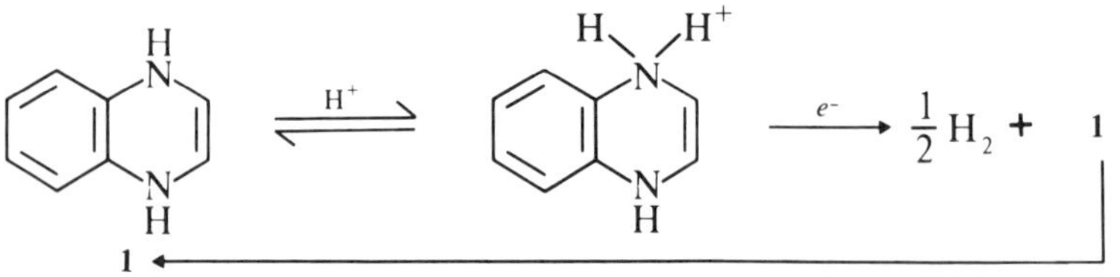

450 Electrochemical Behavior of Heterocyclic Compounds

Scheme 8.10

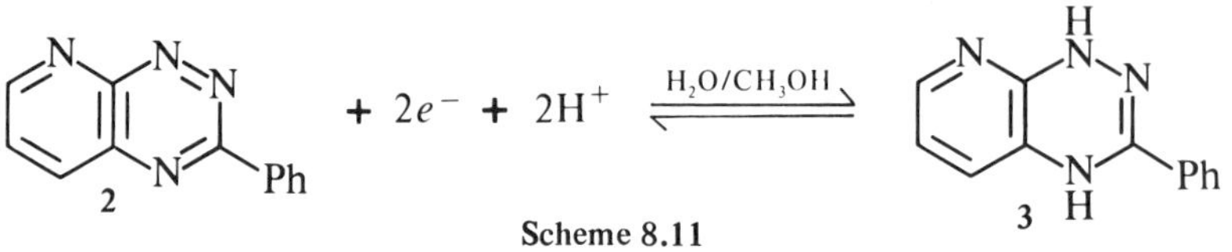

Scheme 8.11

4. Adsorbtion of the reactants or of the products has been the subject of many polarographic investigations (103).
5. In aqueous media, protonations are important reactions. For a reaction such as

$$Ox + ne^- + mH^+ \rightleftharpoons Red$$

where both the electron and proton transfers are fast, the half-wave potential depends on the pH:

$$E_{1/2} = E_{1/2}(\text{pH } 0) - 0.059 \frac{m}{n} \text{pH}$$

Many other protonation schemes have been investigated (104). As an example (35), protons are involved in the reduction of 3-phenylpyrido[3,2-e]triazines (Scheme 8.11), in which one observes the following pH dependence:

pH < 4.4 $E_{1/2} = 0.28 - 0.090 \text{ pH}$

pH > 4.4 $E_{1/2} = 0.15 - 0.060 \text{ pH}$

which corresponds to the reactions

$2 + 2e + 3H^+ \rightleftharpoons [3H^+]$ pH < 4.4

$2 + 2e + 2H^+ \rightleftharpoons 3$ pH > 4.4

The pK of the reduced compound **3** is 4.4.

8.1.2.2. Rotating Disk Electrode (RDE)

A schematic representation of an RDE is shown in Figure 8.12. A platinum or gold wire or a small carbon rod is embedded in Teflon or epoxy resin and ground smooth to the insulating material and perpendicular to the axis of the electrode. The electrode is rotated perpendicularly to its axis with a rotation rate between 100 and 10,000 rpm. An electrical connection is provided through a brush contact.

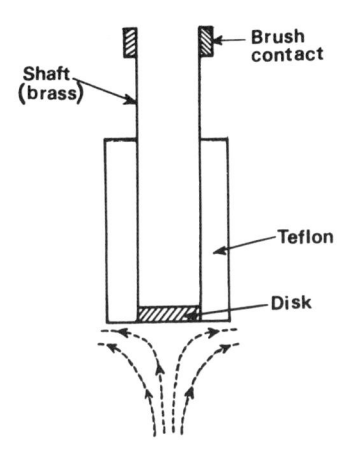

Fig. 8.12. Schematic representation of a rotating disk electrode.

Such devices can be constructed in the laboratory but are also commercially available. In addition to references 76 and 8, which give a detailed treatment of the RDE, the reviews by Riddiford (105) and Opekar and Beran (106) should be consulted.

The time parameter of this technique is $1/\omega = (2\pi f)^{-1}$, where ω is the angular velocity in rad s^{-1} and f is the frequency in s^{-1}. The variation range of this time parameter is much larger (10^{-2}–10^{-4} s) than in polarography (1–10 s), and the characterization of transient intermediates with much smaller lifetimes will be possible.

Rotation of the electrode causes the liquid under it to rise, and in the vicinity of the electrode it is thrust horizontally, creating a laminar layer under the electrode. Everything takes place as if there were two regions: a diffusion layer of thickness δ near the electrode and a layer in which mass transfer occurs only by convection. While the electrode is rotated, a slow voltage ramp is applied. The resulting curve is shown in Figure 8.13. The shape of the curve is similar to that

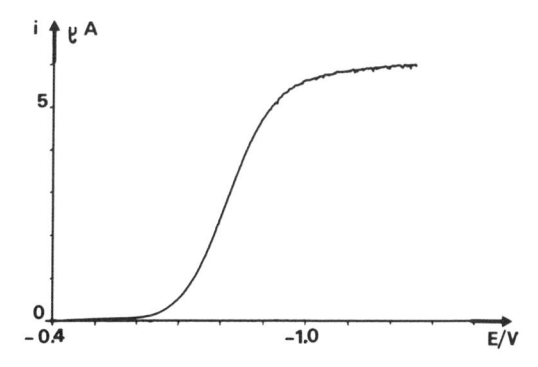

Fig. 8.13. Rotating disk voltammogram of 2,3-diphenyl quinoxaline ($c = 10^{-3} M$, 50/50 water-methanol, pH 7, reference SCE).

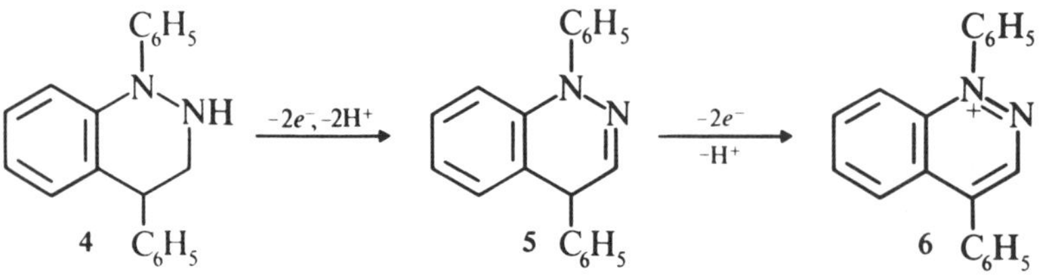

452 Electrochemical Behavior of Heterocyclic Compounds

Scheme 8.12

of a polarogram and it can be qualitatively interpreted in the same manner. In contrast to polarograms, RDE voltammograms can be recorded at positive potential for oxidation studies. The limiting current on the plateau of the wave is given by the Levich equation:

$$i_l = 0.620 nFAD_O^{2/3} \omega^{1/2} \nu^{-1/6} C_O^*$$

where n is the number of electrons involved, A is the area of the electrode in cm², ω is the rotation speed in rad s⁻¹, ν is the kinematic viscosity in cm² s⁻¹, C_O^* is in millimoles, and i_l is in amperes.

For a fast electron transfer in a simple Ox + $ne \rightleftarrows$ Red reaction, the equation of the wave is

$$E = E_{1/2} + \frac{RT}{nF} \ln\left(\frac{i_l - i}{i}\right)$$

with

$$E_{1/2} = E^{o\prime} + \frac{RT}{nF} \ln\left(\frac{D_R}{D_O}\right)^{2/3}$$

and the shape of the wave does not depend on ω. As the limiting current depends on $\omega^{1/2}$, the same dependence should be observed for i at any potential on the wave. Therefore, for a fast electron transfer, the plot of i versus $\omega^{1/2}$ should be a straight line passing through the origin (Levich's criterion). Any deviation from linearity indicates the existence of a slow step in the investigated process. The oxidation of tetrahydrocinnolines (107) in MeCN is investigated with the help of RDE (Scheme 8.12).

A first wave is observed at the Pt RDE ($E_{1/2} = +0.13$ V vs. Ag/Ag⁺ 10⁻² M). Its height corresponds to only 0.66 F mol⁻¹ because the protons released in the oxidation protonate the starting tetrahydrocinnoline 4, and the protonated form is oxidizable only at more positive potentials. The addition of Na₂CO₃ results in a 2e⁻ oxidation, which gives 5. Compound 5 presents a 2-F wave at $E_{1/2} = +0.77$ V corresponding to its oxidation into the cinnolinium cation 6. The latter shows at −0.55 V a one-electron reversible reduction wave leading to the cinnolinyl radical (Scheme 8.13), which presents an anodic wave at the same potential.

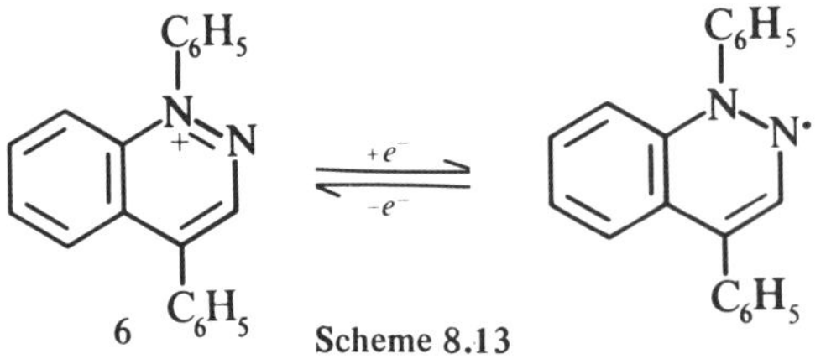

Scheme 8.13

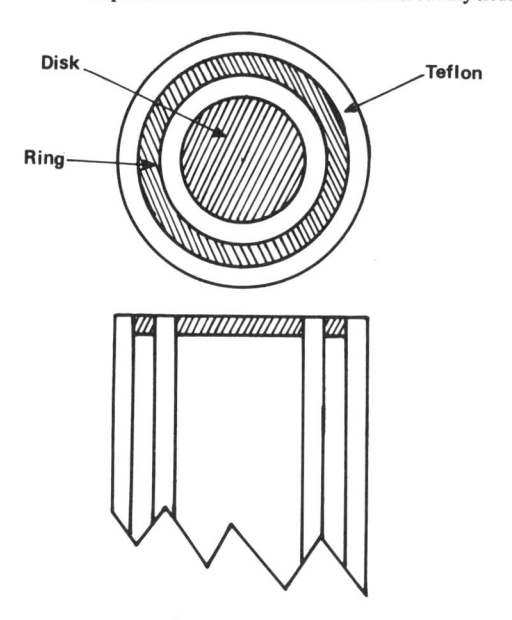

Fig. 8.14. Schematic representation of a rotating ring disk electrode.

Deriving from the RDE is the rotating ring disk electrode (RRDE) in which (in addition to the disk electrode) a concentric ring electrode is inserted in the insulator (Fig. 8.14). The ring and the disk, which can be of the same or of different materials, are independently potentiostatted and the currents are also recorded independently. The rotation of the electrode thrusts the species formed at the disk toward the ring, where they can be evidenced by reduction or oxidation.

8.1.2.3. Cyclic Voltammetry

This technique makes use of a stationary microelectrode to which a triangular voltage sweep is applied as shown in Figure 8.15, with the current recorded as a function of potential. The time parameter of the method is RT/Fv, where v is the sweep rate in $V\,s^{-1}$. As v can be varied from $20 \times 10^{-3}\,V\,s^{-1}$ to $10^3\,V\,s^{-1}$, the parameter RT/Fv varies from 15 to 2.5×10^{-5} s. Very short-lived intermediates are observed, and diagnostic criteria of reaction schemes can be applied with a wide variation of the parameter, that is, much more safely than with slower techniques such as polarography. A simple reversible voltammogram is shown in Figure 8.7. The potential scan is started at the potential where no reaction occurs, and the current increases up to i_p, which corresponds to the potential E_p. After the peak, the voltage sweep is reversed and an anodic peak is observed on the reverse scan. At the start of the voltage sweep, no reaction occurs, and as the potential becomes more and more negative the current increases until the concentration of the oxidant at the electrode $C_O(0, t) = 0$ at E_p. At potential beyond E_p, the gradient of concentration extends farther and farther in solution; that is, less and less Ox available

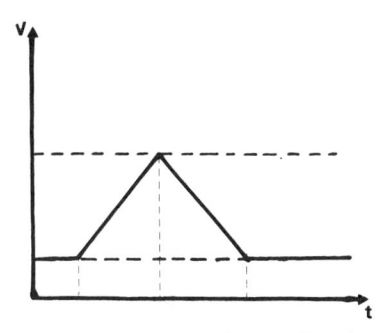

Fig. 8.15. Triangular voltage sweep used for a cyclic voltammetry experiment.

for reduction in the vicinity of the electrode and the current decreases. If R (the radical anion of benzo[c]cinnoline) is stable, when the potential scan is reversed it will still be available for reoxidation in the vicinity of the electrode (diffusion layer) and will give rise to the anodic peak. At higher sweep rates a sudden increase of the current at the very beginning of the voltage sweep is due to a nonfaradaic process (the charging of the double layer, which does not decrease to zero because the potential is constantly increased). As stated before, it is important to compensate the ohmic drop when using a high sweep rate because as i increases with the sweep rate $R_u i$ becomes large (see Fig. 8.7).

For a reversible (fast electron transfer) system such as that of Figure 8.7 the peak current is given by

$$i_p = (2.69 \times 10^5) n^{3/2} A D_O^{1/2} v^{1/2} C_O^*$$

with A in cm^2, D_O in cm^2 s^{-1}, C_O^* in mole cm^{-3}, and v in V s^{-1}, and i_p in amperes. For such a system, the cathodic and anodic peaks should be separated by $59/n$ mV (at 25°C) and the peak potential should not vary with the scan rate. As long as the two peaks do not shift apart on increasing the sweep rate, mass transfer is rate-determining and the electron transfer can be considered as fast.

For an irreversible (slow electron transfer) system (Fig. 8.10), the peak current also depends on $v^{1/2}$ and C_O^*, but the potential of the cathodic peak is shifted to negative potentials by $30/\alpha n_a$ mV (at 25°C) (n_a is the number of electrons involved in the rate-determining step) for each tenfold increase in v.

From Figure 8.10 it is obvious that it is much easier with this than with other techniques to distinguish between a reversible (fast electron transfer) and irreversible (slow electron transfer) process. Figure 8.10 also shows that by increasing the sweep rate (i.e., the diffusion), the process can be transformed to give a pattern in which the electron transfer appears to be slow.

A drawback of cyclic voltammetry as compared to polarography of RDE voltammetry is that it is quite difficult to measure the height of the second wave. Its height must be measured from the decreasing part of the current of the first wave.

A particularly interesting feature of cyclic voltammetry is that by increasing the sweep rate it is possible to outrun chemical reactions that occur after the electron transfer. Such an example is given by the reduction of 7-bromo-2,3-

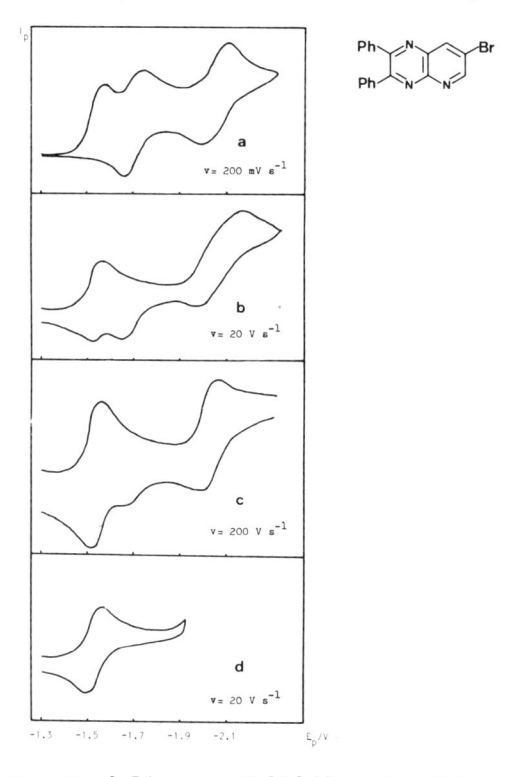

Experimental Procedures and Electroanalytical Techniques 455

Fig. 8.16. Voltammogram of 7-bromopyrido[2,3-*b*]pyrazine. (Solvent MeCN, supporting electrolyte NEt$_4$ClO$_4$; reference Ag/Ag$^+$ 0.01 *M*.)

diphenylpyrido[2,3-*b*]pyrazine in acetonitrile (108) (Fig. 8.16, Scheme 8.14). At low sweep rate $v = 200$ mV s^{-1} (Fig. 8.16*a*), three peaks are observed of which the first irreversible one corresponds to $7 \rightarrow 7^{\div}$ followed by the cleavage of the bromine atom with concomitant formation of **8**. The last two reversible peaks correspond to the formation of $8^{\div}$, and 8^{2-}. At a high scan rate (200 V s^{-1}) (Fig. 8.16*c*) it is possible to outrun the cleavage of $7^{\div}$, and reversible peaks occur, the first corresponding to $7 \rightarrow 7^{\div}$ and the second to $7^{\div} \rightarrow 7^{2-}$. Some decomposition occurs at the level of 7^{2-} as a small anodic peak corresponding to the reoxidation of $8^{\div}$ can still be observed. In Figure 8.16*b* [recorded at an intermediate sweep rate

Scheme 8.14

456 Electrochemical Behavior of Heterocyclic Compounds

Scheme 8.15

($v = 20$ V s^{-1})], the decomposition of 7^{2-} can take place during a longer time and the anodic peak corresponding to $8^{\cdot-} \to 8$ is higher. Another example is given by the oxidation of uric acid in water at pH 4.7 on a pyrolytic graphite electrode (109, 110). The oxidation mechanism is described in Scheme 8.15.

The voltammogram recorded at 4.6 V s^{-1} shows an anodic peak I_a on the first anodic scan, and two peaks I_c and II_c on the cathodic scan (Fig. 8.17). At 0.1 V s^{-1}, peak I_c does not appear, and therefore peak I_c corresponds to the formation of an

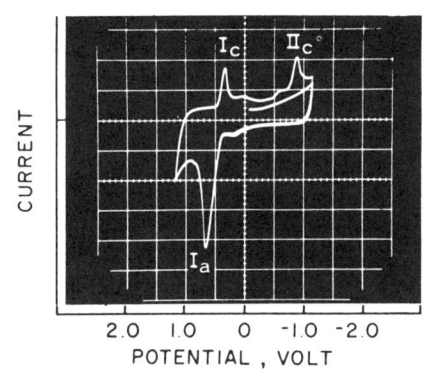

Fig. 8.17. Cyclic voltammogram at a clean pyrolytic graphite electrode in acetate buffer pH 4.7, 0.075 mM uric acid. Scan pattern 0.00 → −1.20 V → 1.30 V → −1.20 V → 0.00 V; scan rate 5 V s^{-1}.

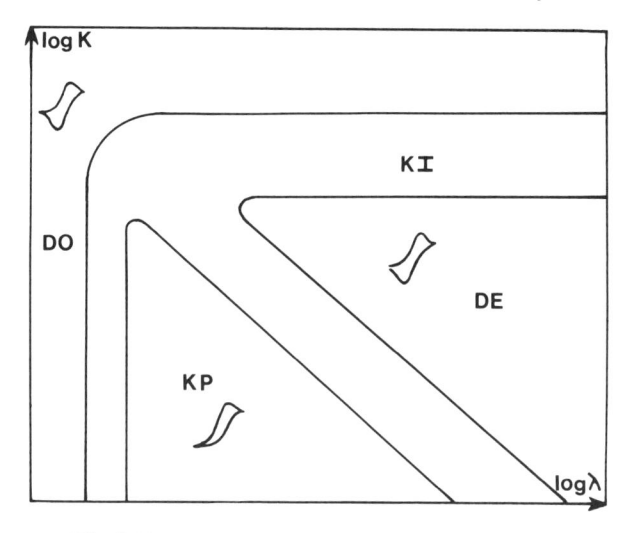

Fig. 8.18. Zone diagram for a CE reaction scheme.

unstable species to which the structure of the diimine **10** has been assigned. The hydration of **10** is proposed to take place in two steps, and peak II$_c$ would correspond to the reduction of **11**.

Cyclic voltammetry is also particularly useful for the determination of the kinetics of reactions associated with electron transfer. This is related to the large variations of sweep rates available with this technique. The variations of peak potentials or peak heights are recorded as a function of the different operational parameters (scan rate, concentrations, etc.) and are compared with those predicted for the different reaction schemes (111). The precision of these criteria can be improved by using not only the peak but all the points of a voltammogram, which is possible through the use of a digital acquisition system connected to a computer that processes the data. Many reaction schemes have been investigated.

CE (112–114).   A reversible reaction followed by a reversible electron transfer is considered.

$$Y \underset{k_b}{\overset{k_f}{\rightleftarrows}} Ox \qquad K = \frac{k_f}{k_b}$$

$$Ox + ne^- \rightleftarrows Red$$

The behavior of this system depends on two parameters $K = k_f/k_b$ and

$$\lambda = \frac{k_f + k_b}{v} \frac{RT}{nF}$$

and it is represented by the zone diagram as shown in Figure 8.18. In zone DO, the voltammogram looks like a diffusion-controlled reversible voltammogram for a simple electron transfer. This happens under two conditions: (1) K is large so that the starting material is essentially present as Ox, which diffuses toward the electrode and (2) λ is small (k_f and k_b are small), everything takes place as if the preceding reaction were frozen and as if only the Ox originally present in the solution were diffusing toward the electrode. In this case the peak height is smaller than in the first case.

In zone DE, λ is large (i.e., k_f and k_b are large, the equilibrium is fast, and Ox and Y are always at equilibrium). The voltammogram is also reversible, but it is shifted on the potential axis (by reference to the Ox $+ e^- \rightleftarrows$ Red system), by an amount that depends on $\ln K$. It can be expressed in a simple manner. For a fast electron transfer,

$$E = E^{o\prime} + \frac{RT}{nF} \ln \frac{C_O(x=0)}{C_R(x=0)}$$

where $C_O(x=0)$ and $C_R(x=0)$ are the concentrations of Ox and Red at the electrode; because of the preceding reaction, $C_O(x=0)$ is smaller than it would be without it, the logarithm decreases, shifting the potentials (and the peak potential) to negative values.

In zone KP, K is small and λ is large so that in a reaction layer adjacent to the electrode there is a mutual balance between the formation of Ox by the chemical equilibrium and its consumption at the electrode. In this case the concentration of Ox near the electrode becomes time-independent. These conditions are referred to as "pure kinetic conditions." In this case the voltammogram has a peculiar shape indicating a kind of plateau. In this zone the peak potential is shifted by 30 mV in the positive direction for a tenfold increase of v. The shapes of the voltammogram for different values of K and λ have been dealt with in reference 112, and KI is a zone of intermediate behavior. The case where the electron transfer is slow has also been treated (113).

EC (113, 114). A fast electron transfer is followed by an irreversible first-order reaction:

$$\text{Ox} + e^- \rightleftarrows \text{Red}$$
$$\text{Red} \xrightarrow{k} \text{X} \quad \text{(first order in Red)}$$

The behavior of the system depends on a single parameter $[\lambda = k/v(RT/nF)]$. If λ is small (k small, zone DO), the reaction is slow and does not occur to any appreciable extent during the electrochemical experiment, and the reversible voltammogram is of a simple fast electron transfer. If λ is large (zone KP), pure kinetic conditions are operative, and any R produced at the electrode is immediately consumed, so that on the reverse scan no anodic peak occurs, and the voltammogram becomes irreversible. In this region the peak potential shifts toward negative values by 30 mV (at 25°C) for a tenfold increase of v and toward positive potentials (by the same amount) for a tenfold increase of k. The plot of E_p as a function of $\log v$, a straight line with 30 mV slope is in the KP region, and a horizontal straight line (E independent of v) is in the DO region. Between these two straight lines, a curve is obtained that corresponds to the intermediate KI zone. The intercept of the two straight lines determines v_i, from which the rate constant of the chemical reaction k is determined: $k = 0.52(nF/RT)v_i$. An example

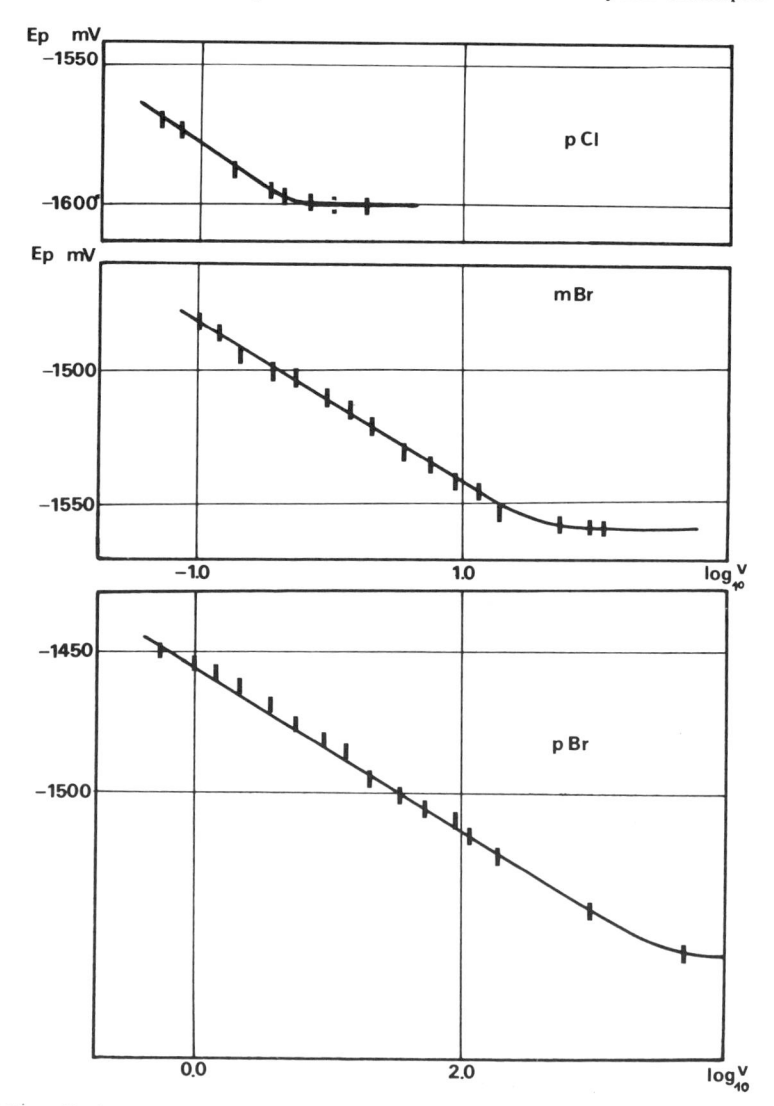

Fig. 8.19. Peak potential $-\log v$ diagram for p-chloro, m-bromo, and p-bromo benzophenone. (Solvent, dimethylformamide; supporting electrolyte NEt_4ClO_4, 10^{-1} M, reference SCE.)

is given in Figure 8.19 for the reduction of m-bromobenzophenone in DMF (Scheme 8.16), a value of $k_1 = 740 \pm 200$ s^{-1} (115) is obtained.

DIM (116, 117a, b). This corresponds to the following dimerization reactions:

$$A^+ + e^- \rightleftharpoons A \qquad E_1^\circ$$

$$2A \rightleftharpoons A-A \qquad \text{equilibrium constant } K_1$$

$$A + A^+ \rightleftharpoons A-A^+ \qquad \text{equilibrium constant } K_2$$

$$A-A^+ + e^- \rightleftharpoons A-A \qquad E_2^\circ$$

460 Electrochemical Behavior of Heterocyclic Compounds

Scheme 8.16

where two processes are operative, a radical-radical coupling (DIM 1) or a radical-ion coupling (DIM 2). The zone diagram is shown in Figure 8.20 for DIM 1, assuming fast electron transfers. Two parameters are involved: an equilibrium parameter ($\chi = K/C_O^*$) and a parameter describing the competition between diffusion and chemical reactions $\lambda = (RT/F)(kC_O^*/v)$, where k is the rate constant of the dimerization reaction. In the DO region, the chemical reaction is slow in comparison to the mass-transfer process. The voltammogram is diffusion controlled and reversible. In the DE zone, the dimerization reaction is much faster than the mass transfer; when the concentrations in solution are modified by diffusion of the electroactive species, the dimerization equilibrium shifts in favor of the dimer. Zone DE is also representative of a purely diffusion-controlled process and the voltammogram is reversible (shifted toward positive potentials). In zone KP "pure kinetic conditions" are achieved and the voltammogram is completely irreversible. In the intermediate KI zone, the voltammogram follows a partial reversibility.

According to the relative values of E_1°, E_2°, K_1, K_2, and C_O^* one or two peaks are observed on the voltammogram. Under "pure kinetic conditions" simple diagnostic criteria are obtained: (1) in the single-peak (DIM 1) case, a linear variation of the peak potential with the logarithm of initial concentration and with sweep rate (19.7 mV per decade at 25°C), anodically for an increase in the first parameter and cathodically for an increase in the second. (2) In the single-peak (DIM 2)

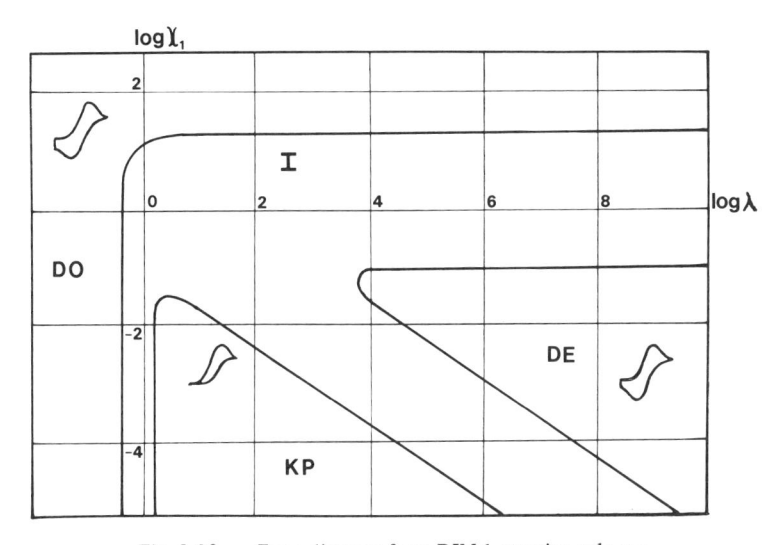

Fig. 8.20. Zone diagram for a DIM 1 reaction scheme.

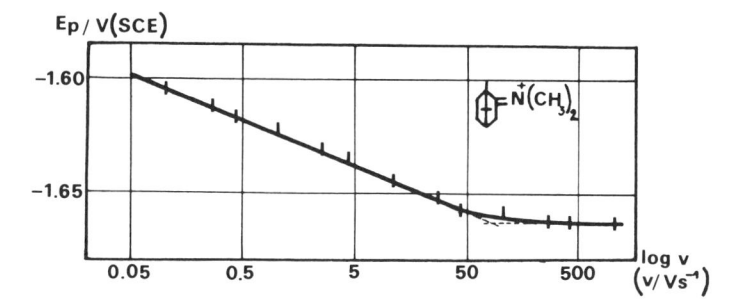

Experimental Procedures and Electroanalytical Techniques 461

Fig. 8.21. Peak potential — log v diagram of **14** in MeCN, 0.4 M NEt$_4$ClO$_4$, reference Ag/Ag$^+$ 10^{-2} M.

case, the peak shifts in the same direction as in the first case, but the rate of variation is different (29.6 mV per decade at 25°C).

The rate constant of the dimerization reaction can be obtained by increasing the sweep rate until the peak potential remains constant. It allows determination of v_i as shown in Figure 8.21 on the particular example of the iminium salt **14** (117a, b). Rate constants are given by

$$k = \frac{0.8 F v_i}{RTC_O^*} \quad \text{and} \quad k = \frac{0.27 F v_i}{RTC_O^*}$$

for DIM 1 and DIM 2, respectively.

More complex reactions involving dimerizations and protonations (118, 119) as well as cyclizations (120) have been investigated.

ECE-DISP. The corresponding reactions follow:

ECE: $\quad$ Ox$_1$ + e$^-$ ⇌ Red$_1$ $\quad$ $E_1^{o'}, E_{1/2}(1)$

$\quad\quad\quad$ Red$_1$ ⇌ Ox$_2$

$\quad\quad\quad$ Ox$_2$ + e$^-$ ⇌ Red$_2$ $\quad$ $E_2^{o'}, E_{1/2}(2)$

DISP: $\quad$ Ox$_1$ + e$^-$ ⇌ Red$_1$ $\quad$ $E_1^{o'}, E_{1/2}(1)$

$\quad\quad\quad$ Red$_1$ ⇌ Ox$_2$

$\quad\quad\quad$ Ox$_2$ + Red$_1$ ⇌ Red$_2$ + Ox$_1$

Both the mechanisms yield the same final product and can only be distinguished because in the ECE mechanism the second electron transfer takes place at the electrode, while in the DISP mechanism it takes place by electron exchange in solution. They are important mechanisms because they explain why two electron

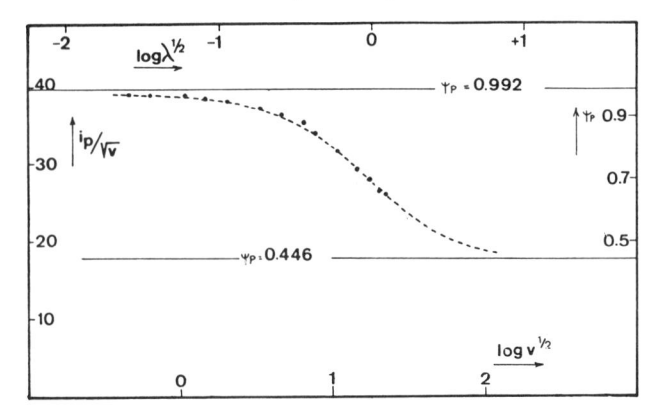

Fig. 8.22. Determination of the rate constant for the cleavage of 2-chloroquinoline radical anion. Variation of the normalized peak current in cyclic voltammetry vs. scan rate.

waves are observed in organic electrochemistry, although electrons are known to be transferred one by one. It occurs when $E_{1/2}(2) \gg E_{1/2}(1)$, that is, when the species produced by the chemical reaction is more easily reduced than the starting compound. An example of such mechanisms is given by the reduction of 2-chloro- and 2-iodoquinoline in liquid ammonia (121):

$$ArX + e^- \rightleftharpoons ArX^{\cdot -}$$

$$ArX^{\cdot -} \xrightleftharpoons{k_1} Ar^{\cdot} + X^-$$

$$Ar^{\cdot} + e^- \longrightarrow Ar^-$$

and/or

$$Ar^{\cdot} + ArX^{\cdot -} \xrightleftharpoons{k_D} Ar^- + ArX$$

$$Ar^- + H_2O \longrightarrow ArH + OH^-$$

Different limiting cases depending on the homogeneous (DISP) or heterogeneous (ECE) nature of the second electron transfer and on the reversibility of the chemical reaction have been investigated: ECE irreversible (122, 123) where the chemical reaction is irreversible; ECE reversible (124); DISP 1 when the homogeneous electron transfer follows an irreversible chemical reaction that is the rate-determining step (124); and DISP 2 when the homogeneous electron transfer is the rate-determining step and occurs after a fast reversible chemical reaction (124). These four mechanisms can take place simultaneously for numerous substrates and reactive media. The general case when all the reactions can take place has been thoroughly investigated (125). When the chemical step is irreversible (rate constant k_1), the ECE-DISP 1 problem depends on a single parameter p given by

$$p = \frac{k_D C_O^*}{k_1^{3/2}} \left(\frac{Fv}{RT}\right)^{1/2}$$

For substrates giving rise to fast electron transfers, k_D can be considered equal to the diffusion limit rate constant k_{dif} (126, 127). Then p depends on k_1 only.

The reduction of 2-haloquinolines is cited as an example to provide a physical explanation: If k_1 is large (p small) the radical $Ar^{\cdot}$ is formed close to the electrode

Scheme 8.17

and is able to diffuse back to the electrode, where it undergoes a second electron transfer (ECE). If k_1 is small (p large), $ArX^{\cdot-}$ has time to diffuse in the solution before decomposing, and $Ar^{\cdot}$ is produced in the solution and undergoes a homogeneous electron transfer (DISP).

The rate constants for ECE or DISP mechanisms can be obtained by fitting the experimental data (peak height or peak potential) on the working curves, which have been published (122–124). An example of a rate constant determination is shown in Figure 8.22 for the reduction of 2-chloroquinoline (Scheme 8.17). The peak height (as i_p/i_{pd}, where i_{pd} is the height of the peak in the case of reversible one-electron voltammogram at the same sweep rate) is plotted as a function of $\log v^{1/2}$, and the curve obtained is fitted on the theoretical working curve for a DISP mechanism. This provides a value of $k_1 = 1.7 \pm 0.2 \times 10^4 \text{ s}^{-1}$.

8.1.2.4. Coulometry and Preparative Electrolysis

In organic electrochemistry, coulometry is a very important technique that is used to measure the number of electrons involved in a reduction or oxidation. For this purpose a solution (generally 10^{-3}–10^{-2} M) is electrolyzed at constant potential, and the number of electrons is measured by an electronic integrator. There is generally no problem in hydroorganic media, where the final current decreases to very low values (compared to the initial current), but in organic solvents basic species can be generated that attack the solvent and the current always remains at higher values. In such a situation, a modification of the reaction medium should be considered. The number of electrons measured by coulometry may or may not be in agreement with those obtained from the height of the polarographic wave. Interesting conclusions may be drawn from such discrepancies. In the case of 4-aminoquinazoline (128) (Scheme 8.18), a two-electron polarographic wave is

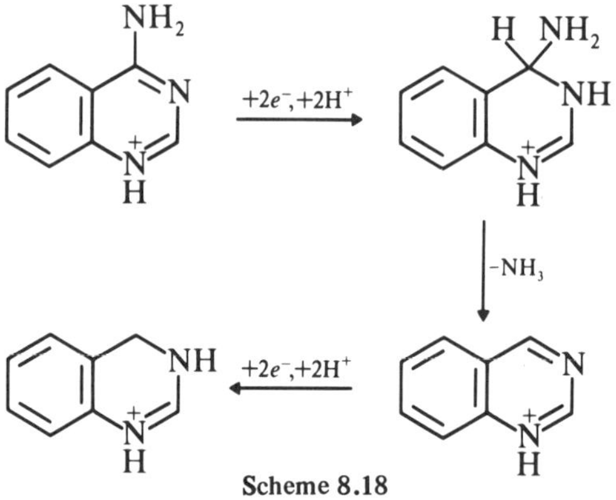

Scheme 8.18

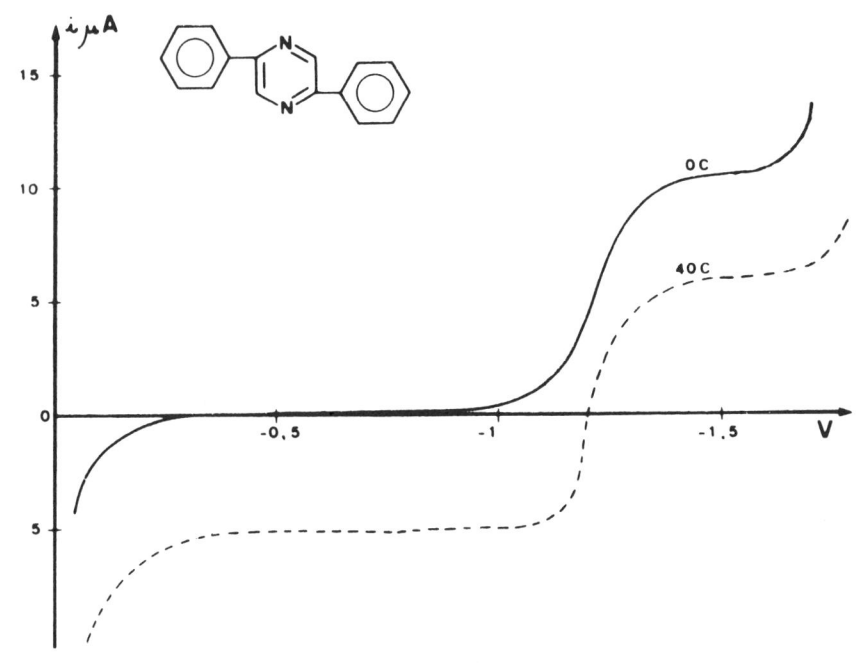

Fig. 8.23. Electrolysis of **15** in dilute solution; 0.5 N NaOH; $c = 2 \times 10^{-3}$ M; $V = 200$ ml; 50% DMF; $E = -1.45$ V. Polarograms before (OC) and during electrolysis.

observed while 4e per molecule are measured by coulometry. This occurs because the deamination step is slow on the time scale of polarography (i.e., it does not occur to any appreciable extent during the lifetime of a drop), but it has time to go to completion during the time required for coulometry.

It is also possible to obtain evidence on slow-reacting intermediates by recording polarograms or voltammograms during coulometry. For example, in the reduction of 2,5-diphenylpyrazine **15** (129), an anodic wave appears during the electrolysis (Fig. 8.23) that corresponds to the formation of the 1,4-dihydro derivative **16** (Scheme 8.19), and this wave slowly disappears at the expense of new anodic and cathodic waves corresponding to the 1,6-dihydro compound **17**.

A preparative electrolysis should be carried out in every electroorganic investigation to ascertain the nature of the final products. In ordinary laboratory practice, amounts of products ranging from 50 mg to several grams are exhaustively electrolyzed, the solutions are analyzed, and the products are isolated by the usual organic

Scheme 8.19

techniques. The analysis of electrolyzed solutions is carried out by gas chromatography (if the products are amenable to such a technique), which when coupled with mass spectroscopy allows a fast and easy determination and identification of the products. When electrolyzed solutions are injected directly on VPC one should be aware of the possible appearance of peaks resulting from the decomposition of supporting electrolytes. High performance liquid chromatography (HPLC) is used mostly on reverse-phase columns with water-methanol or water-acetonitrile elution mixtures in which the electrolyzed solutions are injected directly in an organic or hydroorganic solvent.

Preparative electrolysis is also used for the determination of mechanisms and rate constants both for slow reactions taking place in the bulk of the solution (130) after generation of an intermediate at the electrode and for fast reactions taking place in a thin layer close to the electrode: ECE-DISP mechanism, ECC-ECE-DISP competition (as in the example of aromatic S_{RN1} substitutions) (131), hydrogen atom transfer from neutral aromatic radicals (132), and catalytic aromatic S_{RN1} reactions (133).

8.1.2.5. *Spectroscopic Methods*

Several spectroscopic techniques are used with electrochemical methods either to characterize transient intermediates or to study reaction kinetics.

8.1.2.5.1. Electron Spin Resonance (ESR). Radical anions (119) and radical cations (135, 136) can be characterized by ESR (134). Cells have been reported (137–139) that allow the generation of such species directly in the cavity of a spectrometer. The radical anions or cations must have a certain stability to be detected. The observation of an ESR spectrum does not necessarily imply that this species is an intermediate in the investigated reaction and it can only be considered an intermediate if its kinetics of formation and/or consumption fit the overall reaction kinetics.

Radicals can be identified by comparison of the hyperfine coupling constants of their ESR spectra with that obtained from molecular orbital calculations.

In the case of very unstable radicals, the spin trapping technique can be used (140) to obtain an ESR spectrum. In Scheme 8.20 (136) the unstable radical reacts with a typical spin trap (phenyl-*tert*-butylnitrone) to give a stable radical, the ESR spectrum of which can be recorded.

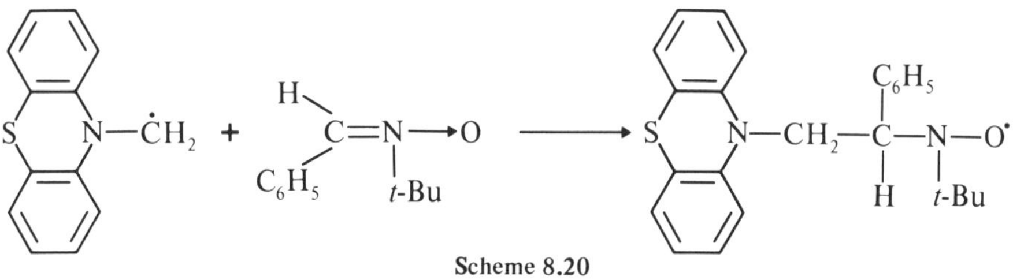

Scheme 8.20

8.1.2.5.2. UV and Visible Spectroscopy. In order to record UV or visible spectra during electrolysis, platinum, gold, or even a vitreous carbon minigrid (141) is placed in the thin cell of a spectrophotometer. It is thus possible to record a spectrum while performing coulometric analysis. This kind of spectroelectrochemical cell is well suited to long-term electrochemical experiments, so that the intermediate to be observed has enough time to diffuse in the spaces of the grid.

Another possibility involves the use of optically transparent electrodes OTE (142, 143) obtained by the deposition of a semiconductor (SnO_2 or In_2O_3) or a metal (Au, Pt) on glass or quartz. This transparent electrode is used as one of the optical windows of the cell, which is placed in the beam of the spectrophotometer. It is possible to apply a potential step to the electrode and at the same time to scan the wavelength to obtain the spectra. The use of rapid-scan spectrophotometers that can record a spectrum (in as little as 10^{-3} s) allows observation of the spectra of unstable intermediates. For example, in the case of **18** it was possible to record the spectrum of the radical cation which dimerizes (with a rate constant of $5 \times 10^4 \ M^{-1} \ s^{-1}$) by averaging a large number of spectra obtained after a potential step from a region where it is not oxidized to a potential behind its oxidation peak (144).

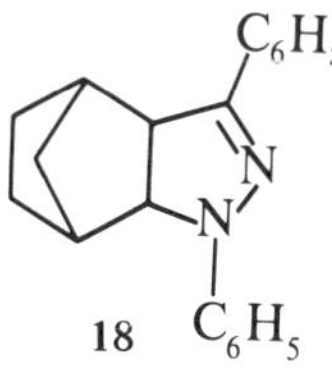

It is also possible to follow the absorbance of the solution at a given wavelength as a function of time after a potential step. The variations of absorbance, which is proportional to the total amount of product formed, as a function of time have been established for different reaction schemes (145). In this way it was possible to establish the reaction mechanism of the radical cation of phenoxathine with anisole (135) (Scheme 8.21). The concentration of $P^{\ddot{+}}$ was followed through the variation of optical density at its maximum absorbance, while the OTE undergoes a potential step followed by relaxation (i.e., the electrode is electronically disconnected).

$$P - e^- \longrightarrow P^{\ddot{+}}$$

$$P^{\ddot{+}} + AnH \underset{k_2}{\overset{k_1}{\rightleftharpoons}} PAnH^{\ddot{+}} \quad k_1 = 160 \ M^{-1} s^{-1}; k_2 = 10 \ M^{-1} s^{-1}$$

$$PAnH^{\ddot{+}} + P^{\ddot{+}} \overset{k_3}{\longrightarrow} PAn^+ + H^+ + P \quad k_3 = 700 \ M^{-1} \ s^{-1}$$

Scheme 8.21

Th = Thianthrene

$$Th \longrightarrow Th^{\cdot +} + e^-$$

$$Th^{\cdot +} + Th^{\cdot +} \rightleftharpoons Th + Th^{2+}$$

$$Th^{2+} + H_2O \longrightarrow ThO + 2H^+$$

Scheme 8.22

Optically transparent electrodes can also be used in internal reflection spectroelectrochemical experiments (142, 146). The light beam enters the OTE through a prism and is reflected several times at the electrode solution interphase, and a second prism allows it to exit from the OTE. At each reflection at the solution-electrode interface, the light beam penetrates for a short distance into the solution, where it can interact with compounds existing close to the electrode surface or adsorbed on it. Thus any modification in the vicinity of the electrode will affect the intensity of the light beam (147, 148). This technique was applied to the investigation of the reduction of methylviologen MV^{2+} in which the monocation radical was monitored at 650 nm. The potential was stepped to a region where MV^0 is produced at the electrode by a diffusion-controlled two-electron reduction of MV^{2+}. A coproportionation reaction occurs:

$$MV^{2+} + MV^0 \underset{k_b}{\overset{k_f}{\rightleftharpoons}} 2MV^{\cdot +}$$

With this technique, it was possible to measure k_f by a curve-fitting method, and a very high value ($k_f > 3 \times 10^9 \, M^{-1} \, s^{-1}$) was obtained (149).

In specular reflectance spectroscopy (150), a beam of monochromatic polarized light is directed (with an incidence of perhaps 45°) onto a mercury electrode or a carefully polished solid electrode. The potential of the electrode is generally modulated, and a lock-in detection and amplification of the reflected light are achieved. The results are expressed as $\Delta R/R$ (R is the reflectance) as a function of time or wavelength. In this manner, it was possible to investigate the oxidation mechanism of thianthrene by monitoring the concentration of its cation radical at 540 nm (151). The results do not support a disproportionation mechanism (152) (Scheme 8.22, Th = thianthrene), but do not allow discrimination between the other two possible mechanisms (Scheme 8.23).

$$Th^{\cdot +} + H_2O \longrightarrow Th + H^+ + \tfrac{1}{2}H_2O_2$$

$$Th + \tfrac{1}{2}H_2O_2 \longrightarrow ThO + H_2O$$

and $\quad Th^{\cdot +} + H_2O \longrightarrow ThOH^{\cdot} + H^+$

$$ThOH^{\cdot} + Th^{\cdot +} \longrightarrow ThOH^+ + Th$$

$$ThOH^+ \longrightarrow ThO + H^+$$

Scheme 8.23

8.1.2.5.3. Raman Spectroscopy. The use of lasers, which provide a high-intensity source of monochromatic light, together with the finding that the intensity of a Raman spectrum can be increased by a factor of 10^4 to 10^6 by using an excitation wavelength corresponding to an electronic absorption band of the product, rendered possible the use of Raman spectroscopy (153) for the detection of electrochemical intermediates. This appears to be a powerful and promising technique because of the detailed information contained in a Raman spectrum, which allows safe identification of an intermediate. It has been used for the characterization of the tetracyanoquinodimethane (TCNQ) radical anion and dianion (154, 155) as well as for the identification *in situ* of the reaction product of $TCNQ^{2-}$ with oxygen (156) as

$$TCNQ^{2-} + O_2 \longrightarrow O^- - C \equiv N + (N \equiv C)_2 - \bar{C} - \langle \bigcirc \rangle - \overset{O}{\underset{\|}{C}} - C \equiv N$$

The reduction mechanism of 5-nitroimidazole (157) as well as the oxidation of bacteriochlorophyll to its radical cation (158) have been investigated by this technique.

The increased efficiency of Raman scattering (by a factor up to 10^6) observed for the molecules adsorbed on the electrodes (159) is also of interest. The spectra of pyridine, pyrazine, and *N*-methylimidazole adsorbed on Ag electrodes have been obtained (160, 161). This technique was also used to study the conformation of nucleic acids adsorbed on the Ag electrode (162).

8.2. REDUCTION OF HETEROCYCLIC COMPOUNDS

Let us consider a heterocyclic compound that will undergo an overall $2e^-$ transfer. Depending on the protic or aprotic solvent which is used, and on the addition of electrophilic species different reaction paths can be followed leading to various final products.

Aprotic Solvents (DMF, MeCN, DMSO, and Others)

If the starting compound is electrically neutral, the first electron transfer will lead to a radical anion that may have a certain stability or can decompose or dimerize or react with residual water to give a neutral radical. This species can also be obtained in the presence of proton donors such as phenol or react with purposely added electrophiles [e.g., $(CH_3CO)_2O$, alkyl halides, or CO_2] to give a substituted neutral radical.

The second electron transfer on the radical anion will take place at a more negative potential, while on a neutral radical it will generally take place at a more positive potential. In the first case, a very basic dianion will be formed, which will undergo one or two protonations (the protons being furnished by the residual water, the solvent, or the supporting electrolyte). In the second case, the anion formed will react with the electrophilic species present in the solution. The reactions can be summarized as:

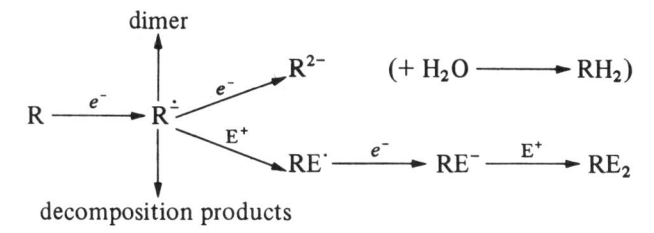

$$R \xrightarrow{e^-} R^{\pm} \begin{cases} \text{dimer} \\ R^{2-} \quad (+ H_2O \longrightarrow RH_2) \\ RE^{\cdot} \xrightarrow{e^-} RE^- \xrightarrow{E^+} RE_2 \\ \text{decomposition products} \end{cases}$$

with E^+ branch from $R^{\pm}$.

The reaction path $R \to RE_2$ involves two electron transfers (which can take place at the electrode or by electron exchange in the solution) and two chemical reactions. Such a mechanism will be called ECEC.

When the starting compound is a cation (e.g., pyrilium or pyridinium, the first electron transfer gives a neutral radical, which can be stable or dimerize. The second electron transfer leads to an anion.

Aqueous Solutions (H_2O or Hydroorganic Medium)

In acidic media (and often in neutral media), the first electron transfer will occur on a protonated species:

$$R \xrightarrow{H^+} RH^+ \xrightarrow{e^-} RH^{\cdot} \xrightarrow{e^-, H^+} RH_2$$
$$\downarrow \text{dimerization}$$

In basic media, most often the first electron transfer takes place on the unprotonated molecule:

$$R \xrightarrow{e^-} R^{\pm} \xrightarrow{H_2O} RH^{\cdot} \xrightarrow{e^-, H_2O} RH_2$$
$$\downarrow \text{dimerization}$$

Remark. The formation of dimers may take place by reaction of two identical species ($R^{\pm}$ or $RH^{\cdot}$) as well as of two different species (e.g., $R^{\pm}$, $RH^{\cdot}$).

8.2.1. Reduction of Heterocyclic Systems

In this section, the reduction of heterocyclic compounds in which the nucleus undergoes a reduction is discussed. Substituted compounds are included in which the substituent remains unreduced.

One advantage of electrochemical reduction in heterocyclic systems lies in the possibility of preparing a partly reduced system in most cases from π-electron-deficient compounds. It is probably the most general method for effecting such partial reductions.

8.2.1.1. Compounds with One Nitrogen Atom

8.2.1.1.1. Pyrroles, Indoles, and Carbazoles. In acidic media pyrrole is reduced to pyrroline and further to pyrrolidine (163, 164). Under similar

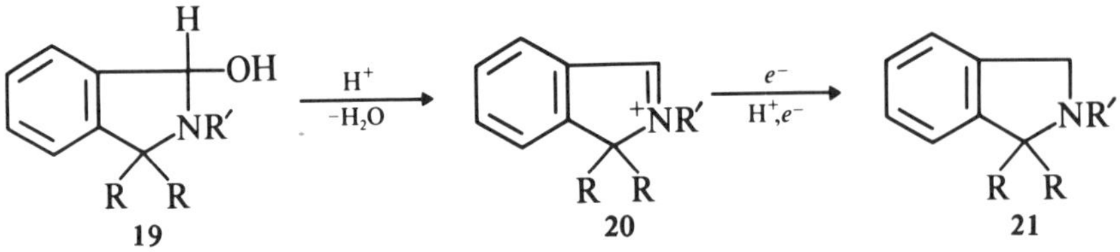

Scheme 8.24

conditions indoles are reduced to indolines (165–168) or to dimerized products (169). In acidic media, hydroxyisoindolines **19** are dehydrated into the cations **20**, which are reduced to isoindolines **21** (170) (Scheme 8.24).

In MeCN the α,ω-diazopolyenes **22** show two reversible one-electron steps (171) (Scheme 8.25). The coproportionation constant K of the intermediate cation radical defined at 298 K by $E_2^\circ - E_1^\circ = 0.059 \log K$ falls from $K = 7 \times 10^5$ when $n = 1$ to $K \simeq 3$ for $n = 3$, showing a marked dependence on the length of the polyene chain.

8.2.1.1.2. Pyridine Derivatives.

At pH values close to 7, pyridine gives rise to a catalytic hydrogen reduction wave at $E_{1/2} \sim -1.7$ V (172). In acidic media, it can be reduced to piperidine (173, 174); side products are 2,2'- and 4,4'-dipiperidyl. Electron-attracting substituents facilitate the reduction.

In MeCN, reduction of 2,6-dimethyl-3,5-dicarbethoxy-1,4-dihydropyridine yields a radical that dimerizes into a mixture of 2,4'- and 4,4'-dimers (175).

In aqueous solutions, nicotinamide **23** is reduced to a radical that dimerizes to the 6,6'-dimer and/or is further reduced to 1,6-dihydronicotinamide (176). A similar mechanism has been proposed in MeCN and DMSO (177). In alkaline solution isonicotinamide derivative **24** is reduced in 1,4-dihydro derivatives (178).

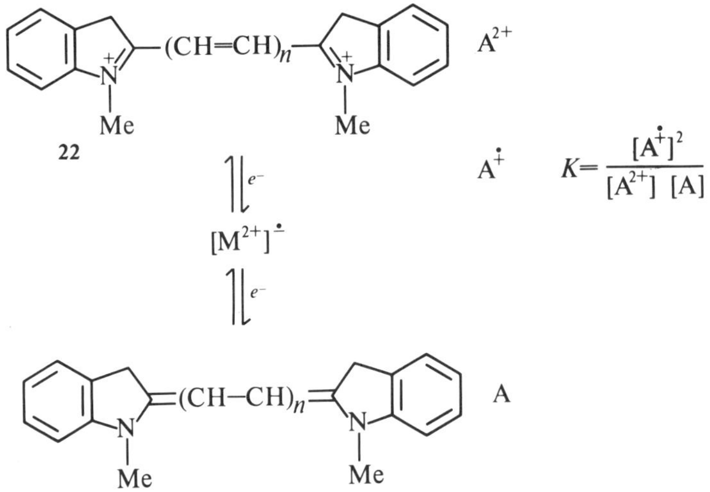

Scheme 8.25

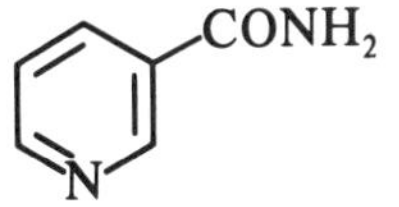

Reduction of Heterocyclic Compounds 471

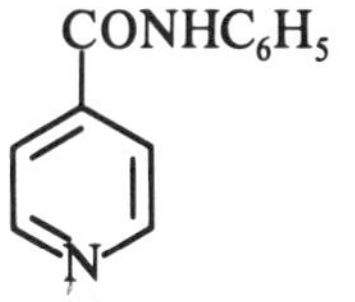

[Structures 23 (nicotinamide) and 24 (N-phenyl isonicotinamide pyridinium)]

In aqueous solutions *N*-alkylpyridine cations are polarographically reduced in a one-electron process (179, 180, 180a). In the case of *N*-methylpyridinium cation **25** (180, 180a), the reduction furnishes a radical that dimerizes, followed by elimination of hydrogen, leaving methylviologen **26** (which undergoes further reduction) (Scheme 8.26).

In aqueous solutions the 3-cyano-1-methylpyridinium cation is reduced into a radical that dimerizes into four dimeric products identified as two diastereoisomeric pairs (181). 4-Benzoylpyridinium derivatives are reduced in a one-electron step to a fairly stable radical (182). In the case of 1-ethyl-4-carbomethoxypyridinium cation, the equilibrium constant of the disproportionation equilibria $2R^\cdot \rightleftarrows R^+ + R^-$ has been calculated (183, 184). Reduction of 1-benzyl-3-carbamoylpyridinium cation gives tetrahydropyridine dimers and dihydropyridine (184a).

In MeCN, 1-alkylpyridinium cations without substituents on the ring yield the 4,4'-dimer on reduction (185). In MeCN with phenol as a proton donor, *N*-alkyl-2,3,4,6-tetraarylpyridinium cations are reduced to 5,6-dihydro derivatives (186).

In DMF the reduction products of pyridinium cations depend upon the substituents (187). Simple *N*-alkyl compounds yield dimers; the 2,4,6-trimethyl-*N*-alkyl compounds reversibly produce free radicals, whereas 2-methyl-4,6-diphenyl-*N*-alkyl pyridinium cations are reduced to cyclic dienes in two steps. In DMF the reduction of 1-ethyl-4-methoxy carbonyl pyridinium cation involves a two-electron transfer, yielding an anion that reacts with alkylated halides to give alkylated dihydropyridine (188); a mechanism involving electron transfer in solution has been proposed. The reduction of quaternary derivatives of nicotinamide has been the subject of various studies as they are NAD$^+$ model compounds (189). In aqueous media, 1-methylnicotinamide cation reversibly forms a free radical, which can form the 6,6'-dimer or be reduced further to give 1,6-dihydro derivative (190). Nicotinamide mononucleotide behaves analogously (190, 191). 1-Benzyl-3-carbamoylpyridinium cation is reversibly reduced into a radical that dimerizes to a 4,4'-dimer instead of a 6,6'-dimer. At more negative potential the radical is reduced further into a mixture of 1,4- and 1,6-dihydro derivatives (192, 193). The reduction of NAD$^+$ at -1.1 V produces a radical that reacts further into dimers that were identified, and it was found that three stereoisomers of the 4,4'-dimer

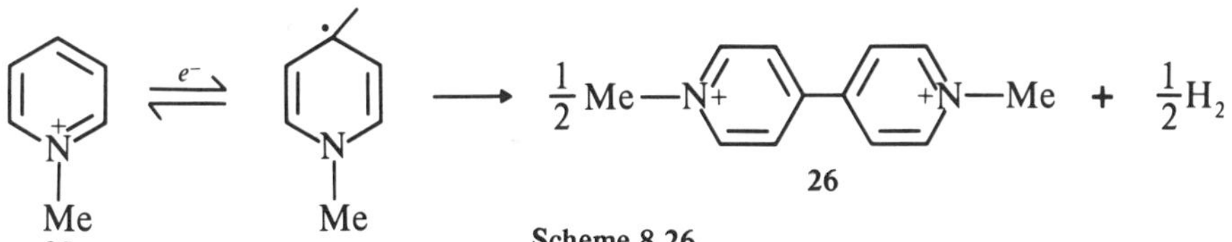

[Scheme 8.26: N-methylpyridinium cation 25 ⇌ (e⁻) radical → ½ Me–N⁺⟨=⟩⟨=⟩N⁺–Me (26) + ½ H₂]

Scheme 8.26

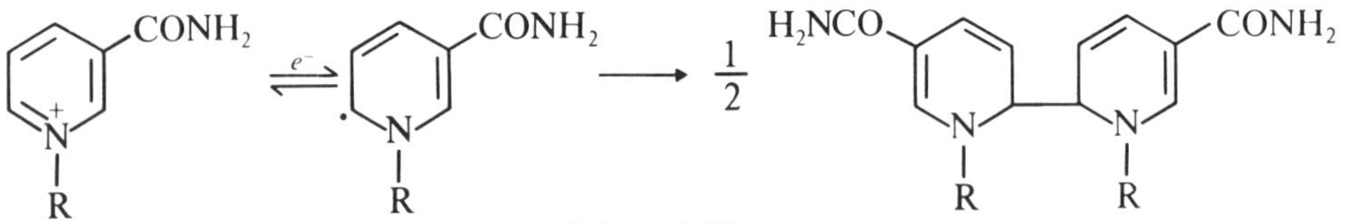

Scheme 8.27

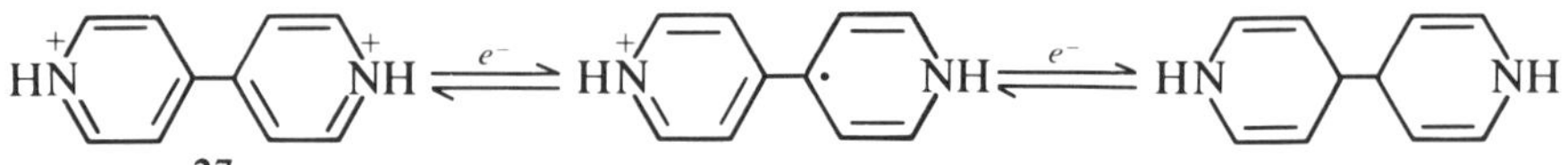

Scheme 8.28

account for 90% of the dimer mixture and three 4,6'-dimers are responsible for the remaining 10%. The reduction at −1.8 V produced 1,4-NADH (50%), 1,6-NADH (30%), and dimers (20%) (193a). Electrolysis of 1-phenylnicotinamide cation in D_2O yields a mixture of 1,4- and 1,6-dihydro derivatives fully deuterated at C-4, showing that a 1,6-dihydro product is formed via the 1,4-dihydro derivative (194).

In MeCN and DMSO, various biologically important 1-substituted nicotinamides are reduced to free radicals that dimerize rapidly at the 6-position (195) (Scheme 8.27). Dimerization rate constants have been determined by cyclic voltammetry.

In aqueous solutions, 2,2'-bipyridyl is reduced reversibly, according to a two-electron, two-proton process, probably to an N,N'-dihydro derivative, which undergoes a fast irreversible reaction to 1,4- or 1,6-dihydro compound (196). 4,4'-Bipyridyl **27** is reduced in acid solution in two stages (197) (Scheme 8.28).

The reduction of N-ethyl-4,4'-pyridylpyridilium cation proceeds similarly (198). Bipyridilium compounds are reduced in the same manner (199–201) the cation radicals have a considerable stability. In MeCN, voltammetry of N-substituted 4,4'-bipyridyls yields two reversible one-electron steps in most cases (202); the difference in the redox potentials $E_1^°$ and $E_2^°$ can be correlated with the Taft substituent constants and absorption maximum of the radical cations. The reduction of paraquat dimer molecules has been found to be a potential direct bielectronic mediating system (202a).

Diquaternary salts of phenanthrolines, 2,7-diazopyrene, and diazoniapentaphenes are reduced to cation radicals that are fairly stable in aqueous solutions (203). The same is true for the reduction of compounds of the types **28**, **29**, and **30**.

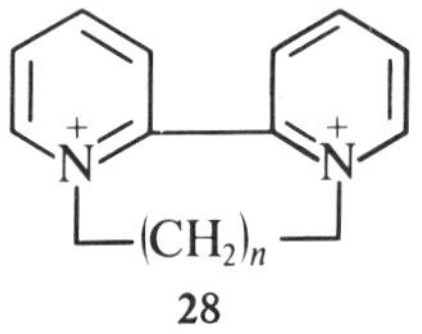

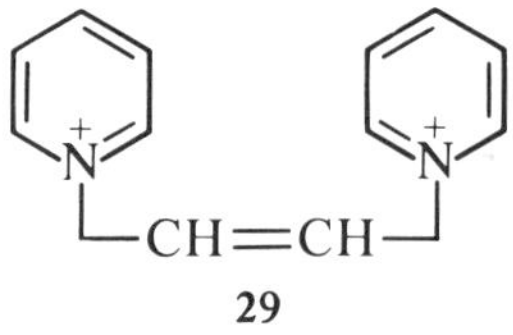

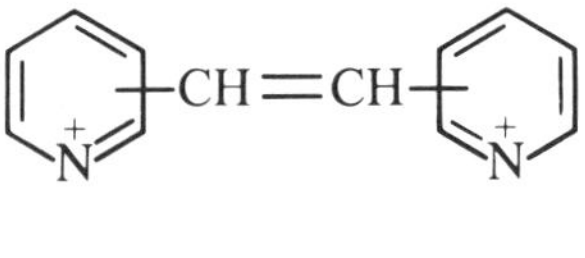

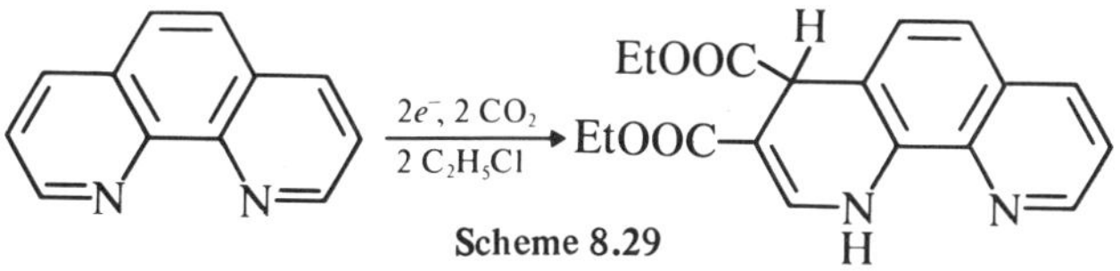

Reduction of Heterocyclic Compounds   473

Scheme 8.29

8.2.1.1.3. Quinoline, Isoquinoline, and Acridine.

In DMF the reduction of 4-substituted-2-phenylquinolines **31** gives two one-electron steps leading reversibly to the radical anion and irreversibly to the dianion (204). In the presence of proton donors, a two-electron process gives 1,4-dihydroquinolines. In anhydrous ammonia, quinoline is reduced in two steps to yield the radical anion and dianion, and both dimerize (205). In the presence of alkyl bromides, reductive alkylation proceeds via an ECEC mechanism to yield a mixture of 1,2-dihydro-1,2-dialkyl and 1,4-dihydro-1,4-dialkyl derivatives. In DMF, quinolines and isoquinolines in the presence of *t*-butyl chloride give various *t*-butylated dihydro and tetrahydro derivatives (206). In DMF, quinolines, phenanthridine, and *o*-phenanthroline in the presence of 1-bromo-adamantane furnish adamantylated product (207); for example, 2-methoxyquinoline leads to 7-(1-adamantyl)-7,8-dihydro-2-methoxyquinoline. The proposed mechanism involves a reduction of heterocycles to dianions that react with 1-bromoadamantane. In DMF, isoquinolines, *o*-phenanthroline, and benzo[*h*]-quinoline (208) in the presence of CO_2 undergo reductive carboxylation via an ECEC mechanism (Scheme 8.29). In aqueous solutions, 3-carbethoxyquinoline is reduced in two one-electron steps. The first step corresponds to the reduction to a radical, which dimerizes. The second step corresponds to the reduction of the radical into 1,4-dihydro-3-carbethoxyquinoline (209), and 3-cyanoquinoline behaves similarly (210).

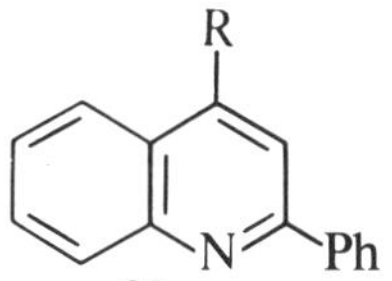

31

In acidic media quinoline yields dihydro and tetrahydro derivatives (211). In aqueous solution, 2-methylquinoline gives a 4,4'-dimer that transforms into a self-cyclized product (212). Reduction of 2-methylisoquinolinium cation in alkaline medium gives a dimer (213). Depending on the conditions, 1,2,3,4-tetrakis(methoxycarbonyl)quinolizinium cation **32** produces several hydroquinolizines (214).

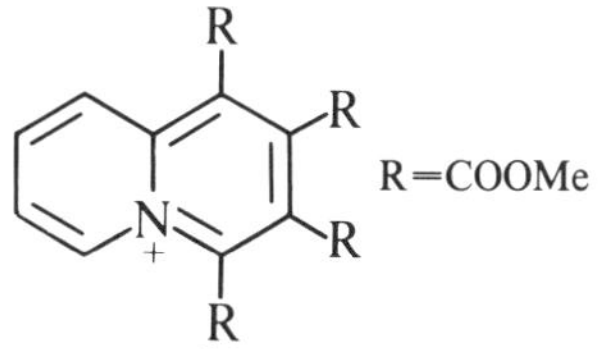

R=COOMe

32

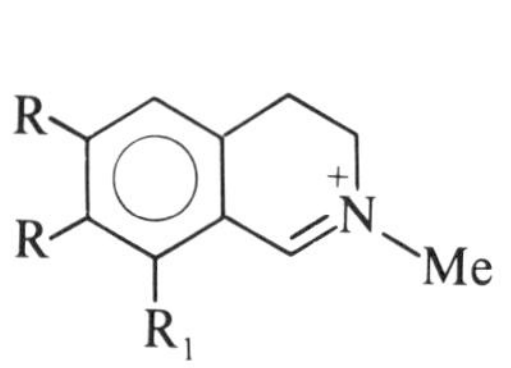

33

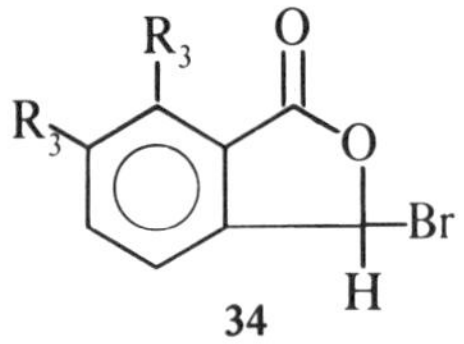

34

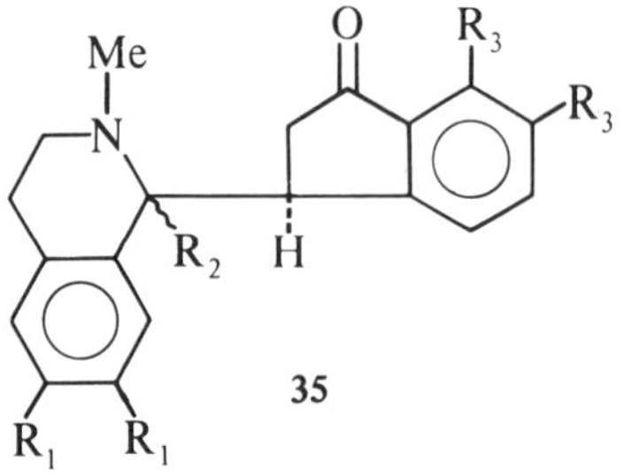

35

In DMF, reduction of the cation **33** in the presence of bromo derivative **34** gives the phthalid alkaloid **35** (215). The mechanism involves probably a two-electron reduction of **33** to give an anion that reacts with **34**. Two similar reactions yielding interesting polycyclic compounds have been described. Thus, in MeCN the iodo benzoquinolinium cations **36** are reduced to compounds **37** via an ECEC mechanism (216) (Scheme 8.30). Hydrogenation of **37** leads to interesting aporphines. In a similar manner the cation **38** gives the compound **39** (217) (Scheme 8.31).

Acridine **40** is polarographically reducible in both acid and alkaline solutions (218). In strongly acidic solution a single one-electron wave is found, and compounds dimerized at C-9 have been isolated (219) in preparative electrolysis. In less acidic and in alkaline solution, two one-electron waves are found (Scheme 8.32). In

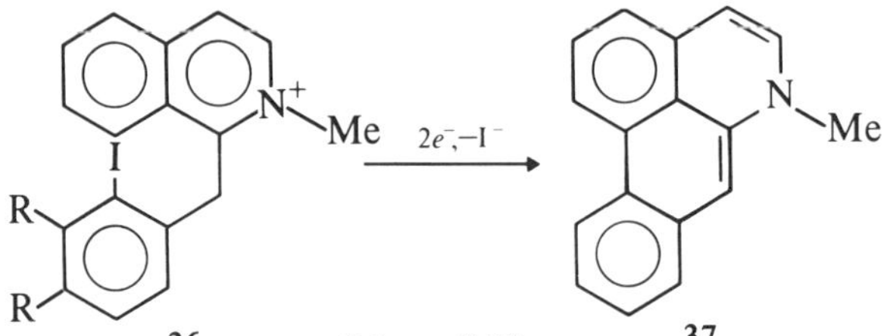

36 **Scheme 8.30** **37**

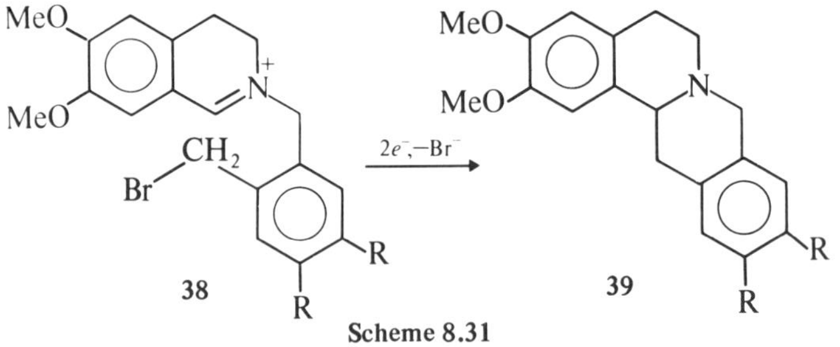

38 **39**

Scheme 8.31

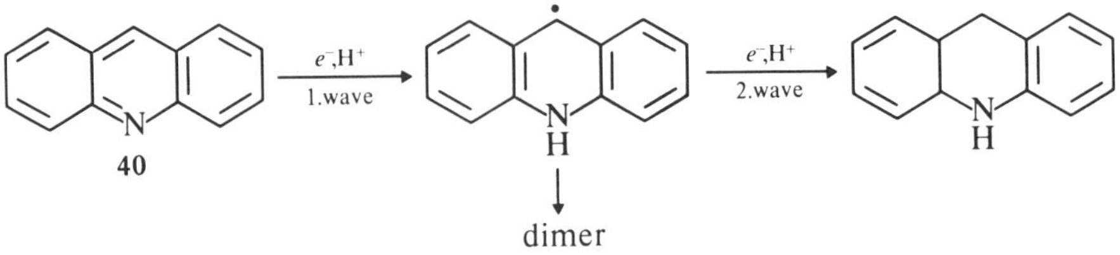

Scheme 8.32

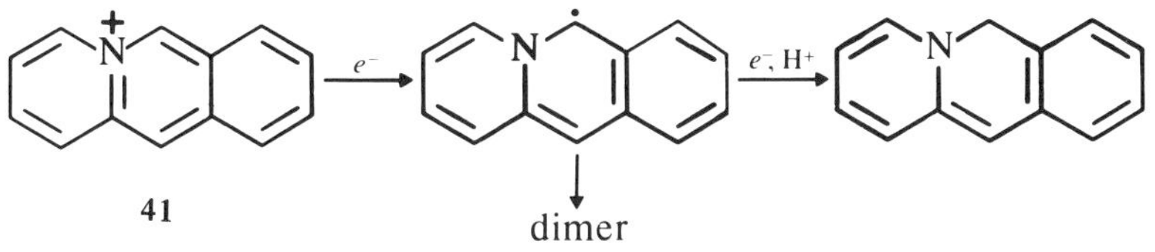

Reduction of Heterocyclic Compounds 475

Scheme 8.33

DMF, acridine gives two one-electron polarographic waves corresponding to the formation of an anion radical and dianion. In the presence of phenol, the acridanyl radical and then acridan were formed (220). The acridizinium cation **41** is reduced in acidic solution (Scheme 8.33) (221).

8.2.1.1.4. Azocines. In DMF, 3,8-dimethyl-2-methoxyazocine **42** involves a slow reduction to the radical anion **43** followed by a fast reduction to the dianion **44** and reaction of **42** and **44** in solution to give **43** (222).

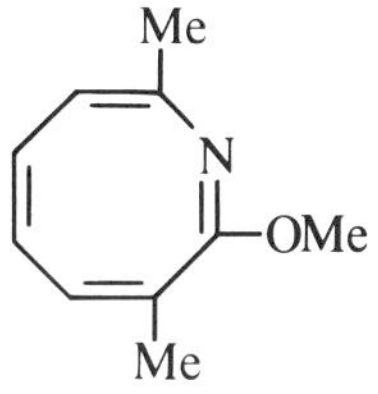

8.2.1.2. Compounds with One Oxygen Atom

8.2.1.2.1. Furan Derivatives. In DMF, cyclic voltammetry shows that aryl-substituted isobenzofurans give two one-electron peaks (223).

8.2.1.2.2. Pyrilium Derivatives. In MeCN, the reduction of substituted pyrilium compounds **45** gives the radical, which dimerizes (224, 225). In MeCN in the presence of alkyl halide R_1X, 2,4,6-triaryl pyrilium cation **46** is reduced via a two-electron process to the anion, which reacts with the alkyl halide to give 2,4,6-triaryl-4-alkyl-4*H*-pyranes **47** (226). 1-Veratryl-3-methyl-4-methyl-6,7-dimethoxyisobenzopyrilium cations **48** are reduced to the corresponding isobenzopyran by a mechanism that involves a slow disproportionation of the intermediate radical (227) (Scheme 8.34). In acidic solution 5,6,7,8-tetrahydro-2,4-diphenyl-1-benzopyrilium cation **49** undergoes a reversible electron transfer to form a radical that dimerizes (228) (Scheme 8.35).

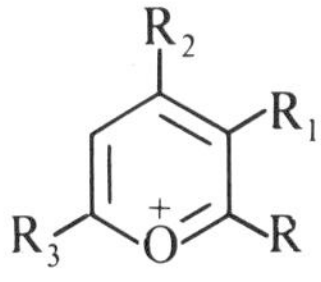

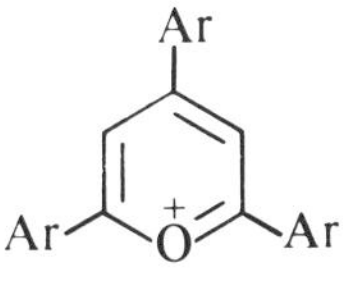

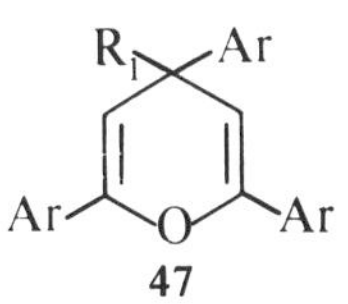

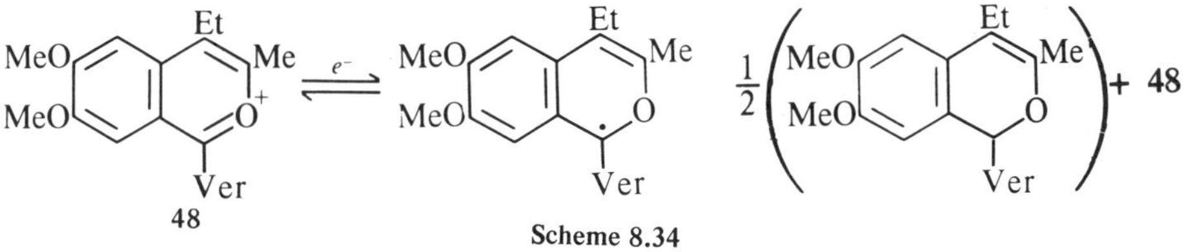

476 Electrochemical Behavior of Heterocyclic Compounds

Scheme 8.34

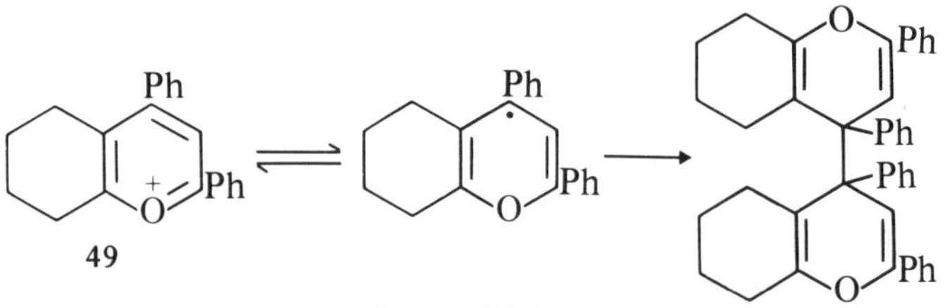

Scheme 8.35

8.2.1.2.3. Pyrones (229) and Coumarines. Substituted pyrones have been reduced in aqueous solutions; in some cases "double dimers" have been obtained (229a). 6-Phenyl-4-hydroxy-2-pyrone gives a 5,6-dihydro derivative at $E = -2.05$ V and 5-phenyl-3-hydroxypentanoic acid at $E = -2.2$ V. The methoxy derivative **50** is reduced as in Scheme 8.36. The esters of 4-hydroxy derivatives are first reduced to the parent -4-hydroxy derivative (230). The reduction of 2-acyl-5,6-dihydro-4H-pyrans **51** in acid solution leads to 2-acyltetrahydropyrans, whereas in alkaline medium 5,6-dihydro-4H-pyran-2-yl carbinol are obtained (231).

In aqueous solution, the reduction of coumarins **52** (232–234) affords the *meso* and *d,l* forms of a product that are dimerized at C-4. In the presence of tertiary amines, which promote catalytic evolution of hydrogen, 3,4-dihydrocoumarins are formed. Asymmetric amines, for example, alkaloids, cause asymmetric induction. 3,4-Dihydro-3-methylcoumarin obtained by reduction of 4-methylcoumarin had an optical purity up to 17% depending on the alkaloid employed (235). In DMF, coumarins give two polarographic waves, the first corresponding to the formation of a radical and a dimer and the second corresponding to the reduction to the dianion (236).

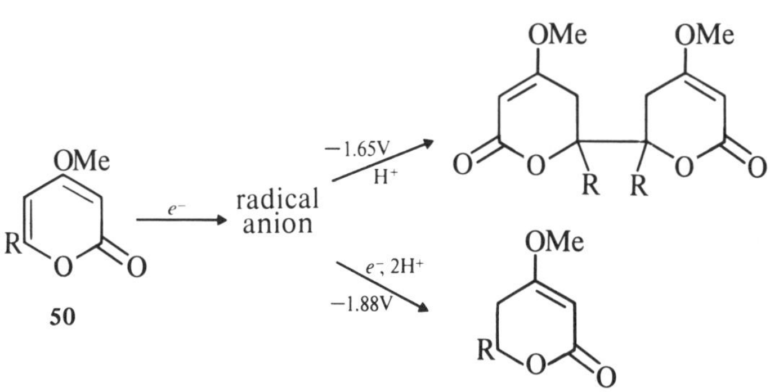

Scheme 8.36

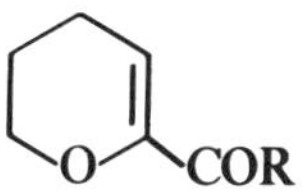
51

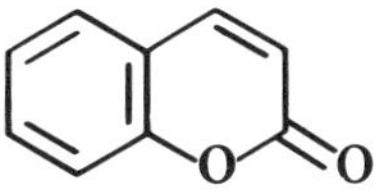
52

8.2.1.3. *Compounds with One Sulfur Atom*

8.2.1.3.1. Thiophene Derivatives. In aqueous solutions the first polarographic wave of benzo[*b*]thiophene **53** corresponds to a two-electron reduction to the 2,3-dihydro derivative and the second to hydrogen evolution following decomposition of a dimeric complex (237). In DMF-H_2O, thiophene has not been reduced directly; it can be reduced by indirect electrolysis. In the presence of biphenyl as electron-transfer agent, thiophene leads to 2,5-dihydro and tetrahydro derivatives (238). In acetate buffer, benzothiophene-1,1-dioxide **54** is probably reduced as shown in Scheme 8.37 (239). *S*-Methylthiophenium cations such as **55** and **56** exhibit two one-electron polarographic waves (240).

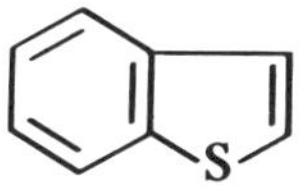
53

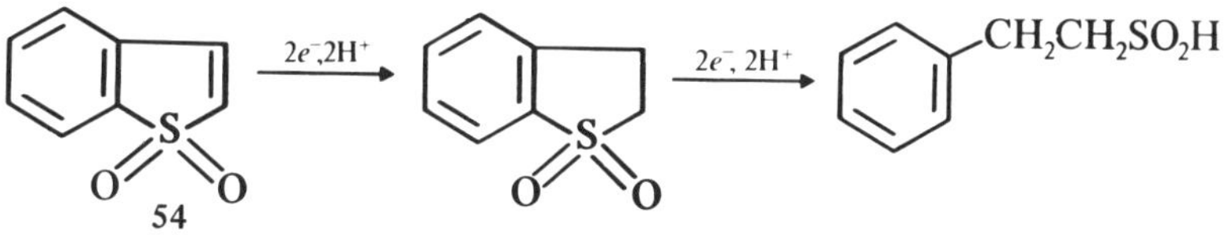

54

Scheme 8.37

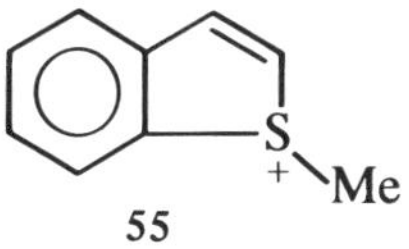
55

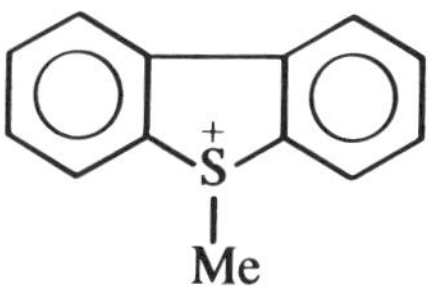
56

8.2.1.3.2. Sulfones. In DMF or MeOH, cyclic sulfones are reduced (241) according to the reaction

$$-CH_2SO_2- \xrightarrow{2e^-, H^+} -CH_3 + -SO_2^-$$

The position of cleavage is shown in the case of some sulfones in Scheme 8.38.

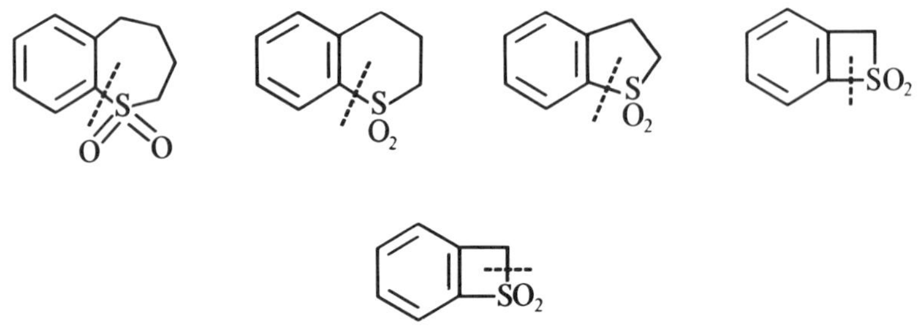

Scheme 8.38

8.2.1.4. *Compounds with Two Nitrogen Atoms*

8.2.1.4.1. ***Diazirines.*** Diazirines **57** are reducible in acid solution in a four-electron reaction and in alkaline medium in a two-electron reaction (242), whereas diaziridines **58** are reducible only in acid solution (243) (Scheme 8.39). In DMF or DMSO, di-*t*-butyldiaziridone **59** is reduced to di-*t*-butylurea (244) (Scheme 8.40).

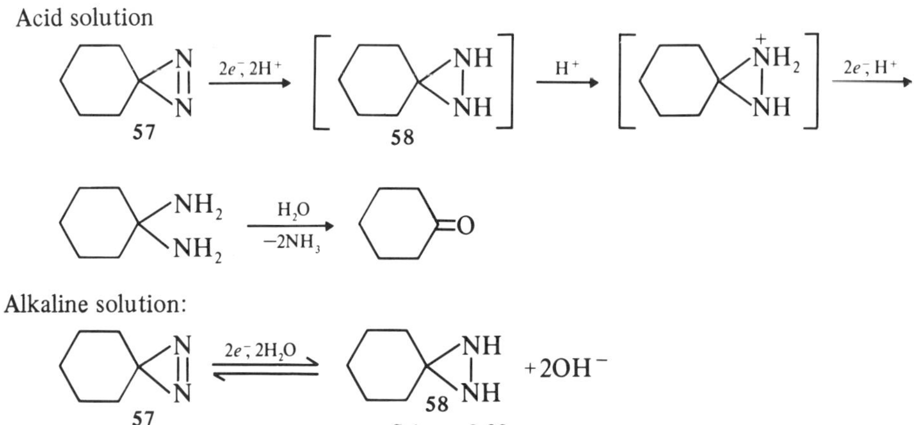

Scheme 8.39

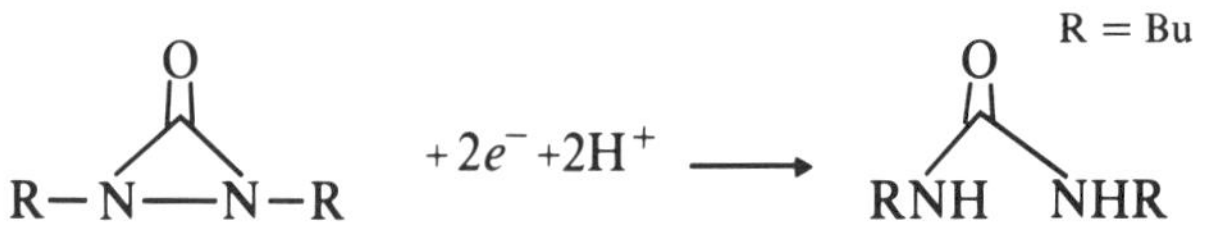

Scheme 8.40

8.2.1.4.2. ***Pyrazole and Imidazole Derivatives.*** Substituted *N*-phenyl pyrazoles **60** can be reduced in DMF-H_2O to 4,5-dihydro derivatives (245) (Scheme 8.41). 1,2′-Diimidazoles and 1,4′-diimidazoles undergo reductive bond cleavage via ECE mechanisms from which imidazole anions result (246).

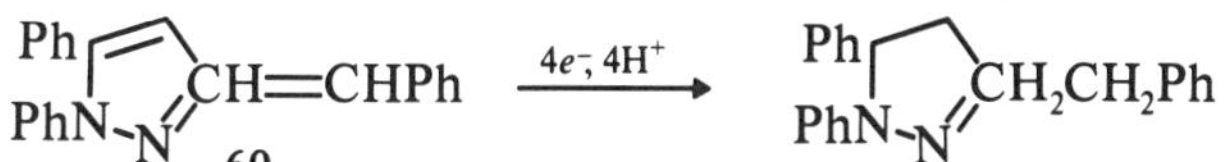

Reduction of Heterocyclic Compounds 479

[Scheme 8.41: Ph/PhN-N pyrazole with CH=CHPh (60) → 4e⁻, 4H⁺ → Ph/PhN-N with CH₂CH₂Ph]

Scheme 8.41

8.2.1.4.3. Pyridazines.

In aqueous solutions pyridazines are reduced polarographically stepwise (247, 248). In acid solution, pyridazine is reduced chiefly to the 1,2-dihydro derivative (249). For substituted pyridazines the first reduction step generally yields different tautomeric forms of dihydropyridazine, the most stable forms usually being 1,4- and 4,5-dihydropyrazines. For example, 3-phenyl-6-dimethylaminopyridazine **61** yields the 4,5-dihydro derivative (247) (Scheme 8.42).

3-Phenyl-6-methoxypyridazine and 3-methyl-6-chloropyridazine are reduced in a similar way, but the dihydro derivatives are unstable and lose methanol or hydrogen chloride, forming the corresponding 4,5-dihydropyridazinone (247). In alkaline solutions 3,6-diphenylpyridazine **62** gives two polarographic waves; preparative electrolysis at the potential of the second wave yields 3,6-diphenyl-2,3,4,5-tetrahydropyridazine (248). A mechanism has also been proposed (Scheme 8.43).

Reductions at the potential of the second wave are generally complicated because the different tautomeric forms may be reduced differently at different potentials and the rate of transformation of a given tautomer to the most easily reducible one may or may not be fast compared with further reduction.

As pointed out by Lund (250), the reduction in acid solution of di- and tetrahydropyridazines is often best understood when the resemblance with hydrazones is considered: in hydrazones the primary two-electron, two-proton reduction

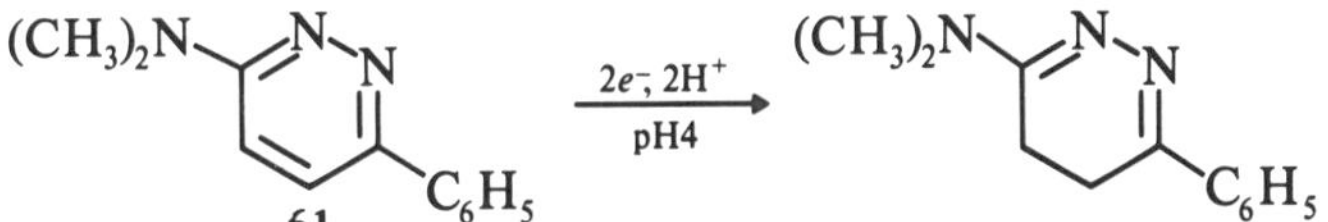

Scheme 8.42

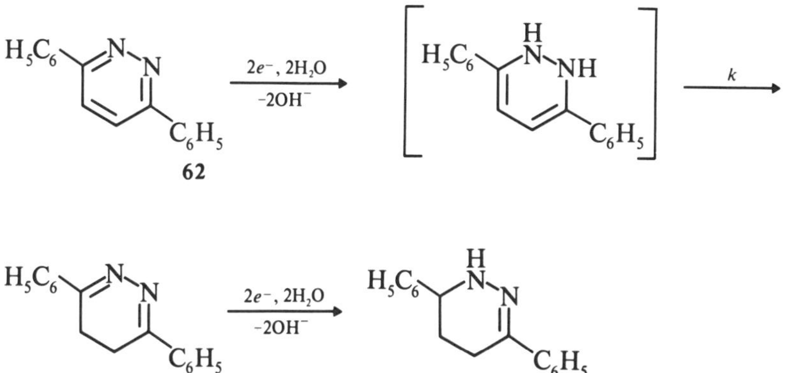

Scheme 8.43

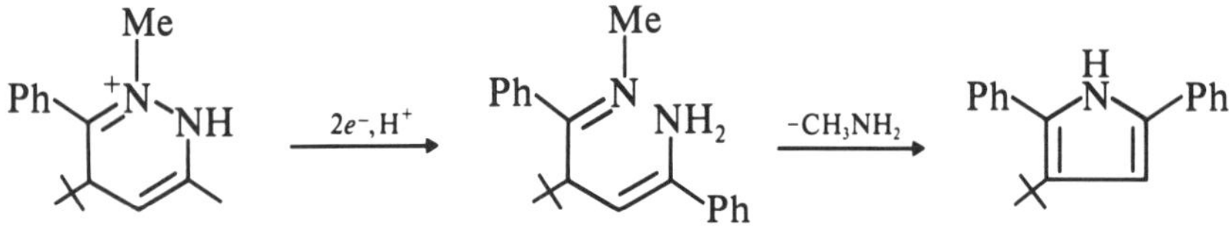

480 Electrochemical Behavior of Heterocyclic Compounds

Scheme 8.44

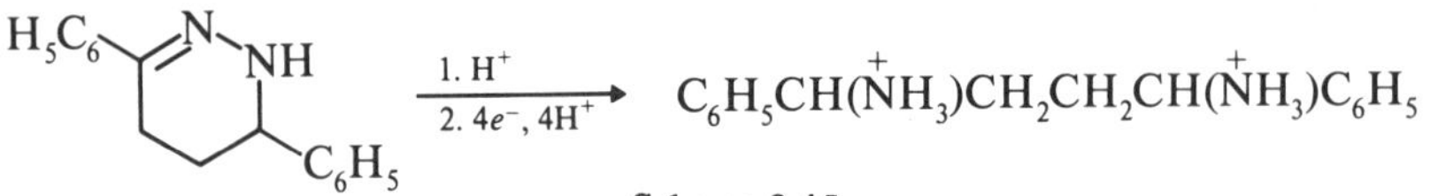

Scheme 8.45

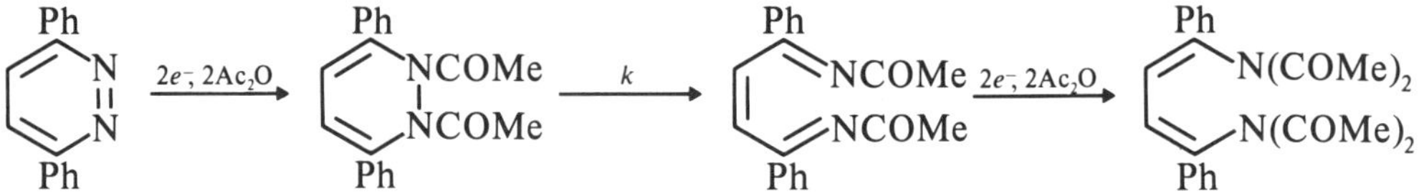

Scheme 8.46

step corresponds to the hydrogenolysis of the N—N bond. Two examples (247, 251) are given in Schemes 8.44 and 8.45.

In DMF, 3,6-diphenylpyridazine **63** in the presence of acetic anhydride is reduced according to Scheme 8.46 (252).

8.2.1.4.4. Cinnolines. Cinnolines are generally reducible (253) in several steps. The first step in the reduction of 3-aryl- or 3-alkyl-substituted cinnolines is the formation of 1,4-dihydrocinnolines (Scheme 8.47). 1,4-Dihydrocinnolines may be regarded as cyclic phenylhydrazones (254). The first reduction step is hydrogenolysis of an N—N bond, and further reduction or ring contraction to indoles depends on the nature of the substituent R (Scheme 8.47). 1-Methyl-3-

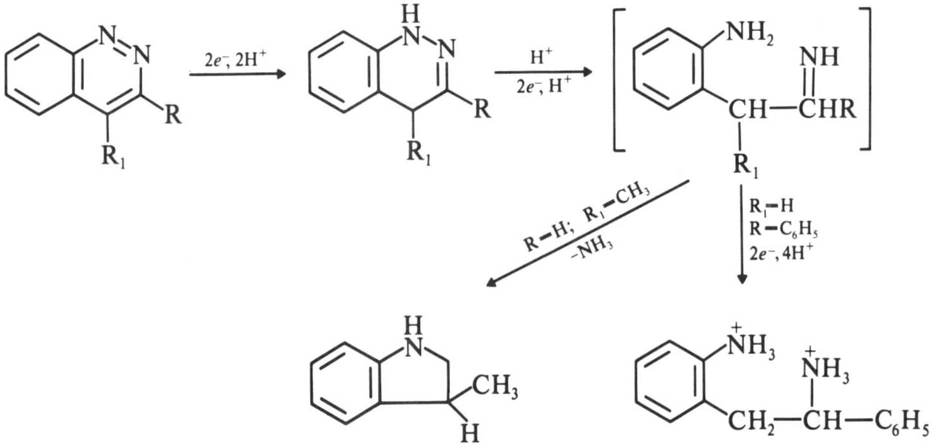

Scheme 8.47

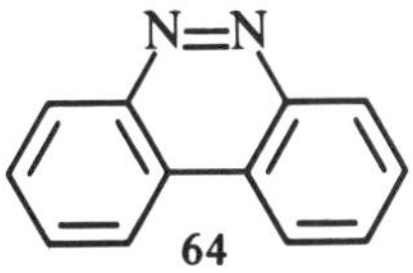

Reduction of Heterocyclic Compounds 481

phenylcinnolinium cation is similarly reduced (253, 255). 1,4-Diphenylcinnolinium cation is reversibly reduced into 1,4-dihydro-1,4-diphenylcinnoline (256, 257). Benzo[c]cinnoline **64** is reduced to the 5,6-dihydro derivative (258).

In DMF, 4-methylcinnoline in the presence of acetic anhydride yields the 1,2-diacetyl-1,2-dihydro derivative, which can be reduced into a tetraacetylated diamine (251). In the same manner, benzo[c]cinnoline gives the 1,2-diacetyl-1,2-dihydro derivative (259).

8.2.1.4.5. Phthalazines. In alkaline solution, phthalazine **65** is reduced (Scheme 8.48) (260). In acidic solution the reduction yields isoindoline and o-xylene-α,α'-diamine; two routes were postulated depending on acidity (260). The reduction course of substituted phthalazines depends on the nature of the substituents. Thus 1-methylphthalazine in alkaline solution forms the 3,4-dihydro derivative (260), whereas 4-dimethylamino- and 4-methoxy-1-phenylphthalazine are reduced to the 1,2-dihydro derivative (261). The reduction of dihydro compounds in acidic medium proceeds like that of hydrazones, the first step being the hydrogenolysis of the nitrogen-nitrogen bond (260, 261).

Dimerized products have been obtained from the reduction of 2-methylphthalazinium cation (260).

8.2.1.4.6. Pyrimidines. The electrochemistry of pyrimidine derivatives has been reviewed (262). In acid solution pyrimidine **66** is reduced polarographically in two one-electron waves; two two-electron waves are observed in neutral media, whereas a four-electron reduction occurs in alkaline solution. The mechanism of reduction (263) is summarized in Scheme 8.49. Pyrimidine substituted with an amino or methyl group gives a one-electron wave like pyrimidine (264). In

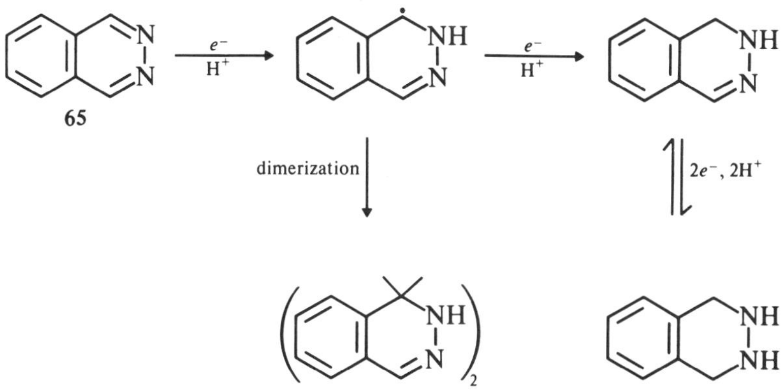

Scheme 8.48

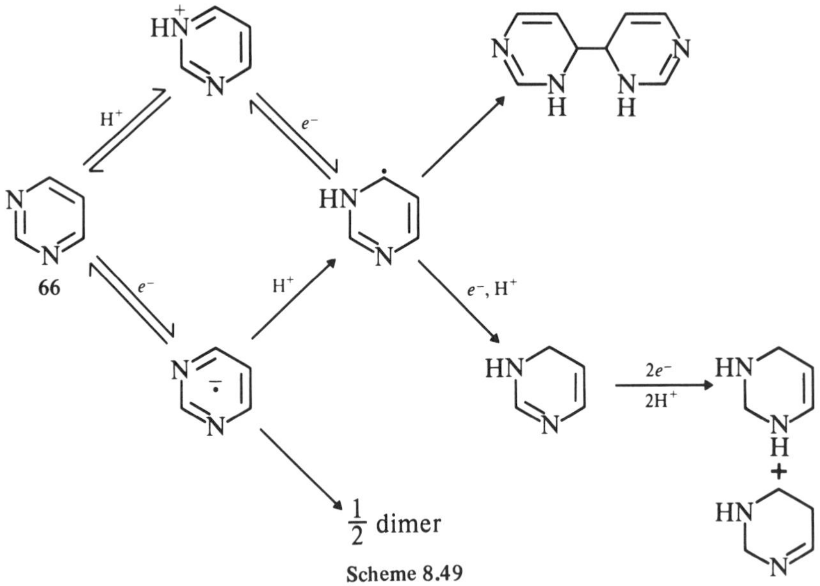

Scheme 8.49

aqueous media the reduction of 4-aminopyrimidine is accompanied by isomerization, ring cleavage, and/or deamination of the primary reduction products and the resulting secondary chemical products (264a).

In acid solution, 2-phenylpyrimidines are reduced to 2-phenylpyrroles with ring contraction in a four-electron reaction. The presence of a phenyl group at the 2-position changes the reduction of the pyrimidine ring and gives the 1,6- instead of the 1,2-dihydro derivative (265). 4-Phenylpyrimidine is reduced in a one-electron step into a dimeric, 1,6-dihydro derivative (265).

8.2.1.4.7. Quinazoline. The polarographic behavior of quinazoline is complicated by the hydration of the quinazolinium cation. The hydrated ion is not reducible. The reduction of quinazoline resembles that of pyrimidine (266, 267).

8.2.1.4.8. Pyrazines. In aqueous solution, pyrazine is reversibly reduced to 1,4-dihydropyrazine (268–270). In acid solution the reduction proceeds in two one-electron steps; the same behavior is observed with N-methylpyrazinium cation (271). In neutral and alkaline media, substituted pyrazines are reduced to 1,4-dihydro derivatives, which isomerize into 1,2- or 1,6-dihydro compounds, which can be further reduced (272, 273). For example, the reduction of 2,3-diphenylpyrazine 67 is summarized in Scheme 8.50. In DMF or MeCN, 2,3-diphenylpyrazine in the presence of acetic anhydride yields, 1,4-diacetyl-1,4-dihydro-2,3-diphenylpyrazine (259), probably via an ECEC mechanism (Scheme 8.51). As 1,4-dihydropyrazines are too oxidizable to be isolated, this easy preparation of diacetyl derivatives must be pointed out, especially as the same procedure can be used with other derivatives containing the pyrazine ring such as quinoxalines, phenazines, pyrido[2,3b]pyrazines. In aqueous

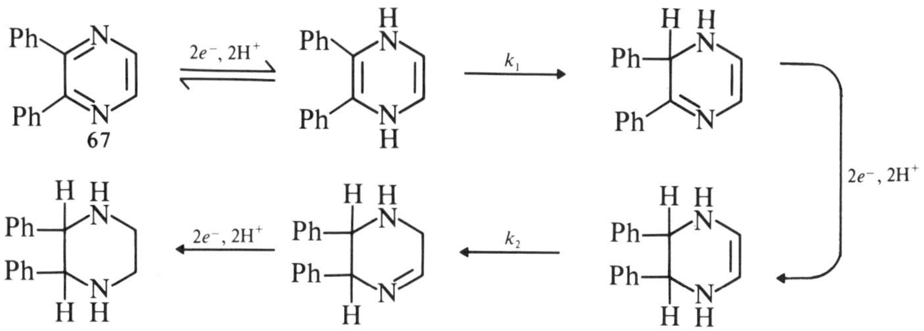

Scheme 8.50

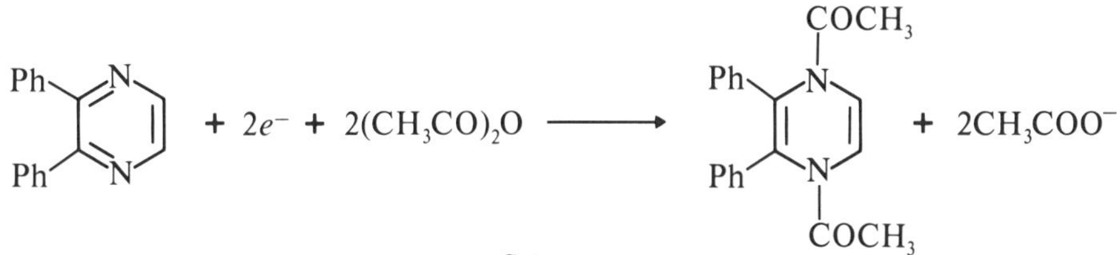

Scheme 8.51

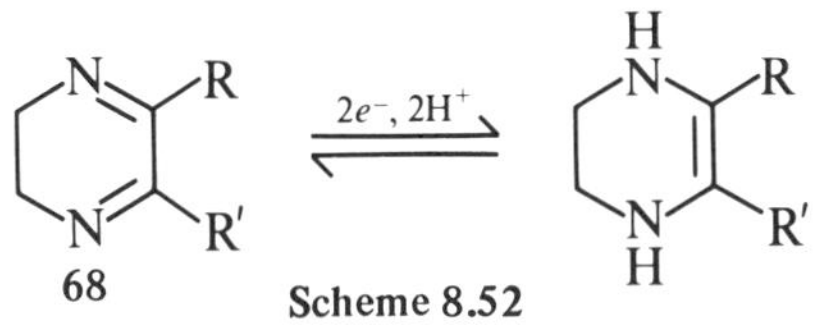

Scheme 8.52

media, the electrochemical reduction of 2,3-disubstituted-5,6-dihydropyrazines **68** yields 1,4,5,6-tetrahydropyrazines (274) (Scheme 8.52); the diimine-enediamine system is thus analogous to the dione-enediol system.

8.2.1.4.9. Quinoxalines. In aqueous media, quinoxaline (275–277) is reduced reversibly to 1,4-dihydroquinoxaline. In neutral or alkaline media, substituted quinoxalines **69** are reduced to 1,4-dihydroquinoxalines, which rearrange into 1,2- or 3,4-dihydro compounds. The latter compounds can be reduced to 1,2,3,4-tetrahydroquinoxalines (277, 278) (Scheme 8.53).

In acidic media 2,3-diphenyl-1-methyl quinoxalinium cation **70** is reduced in two one-electron steps (279). The first step leads to a rather stable protonated radical and the second to the 1,4-dihydro-1-methyl derivative, which rearranges into 1,2-dihydro-1-methyl-2,3-diphenyl quinoxaline (Scheme 8.54).

In DMF or MeCN, quinoxaline in the presence of acetic anhydride is reduced to 1,4-diacetyl-1,4-dihydroquinoxaline (252). In DMF, quinoxaline in the presence of CO_2 is reduced via an ECEC mechanism to a 1,4-dicarboxylate dianion 1,4-dihydroquinoxaline, which reacts with CH_3Cl to give the corresponding dimethyl ester (208).

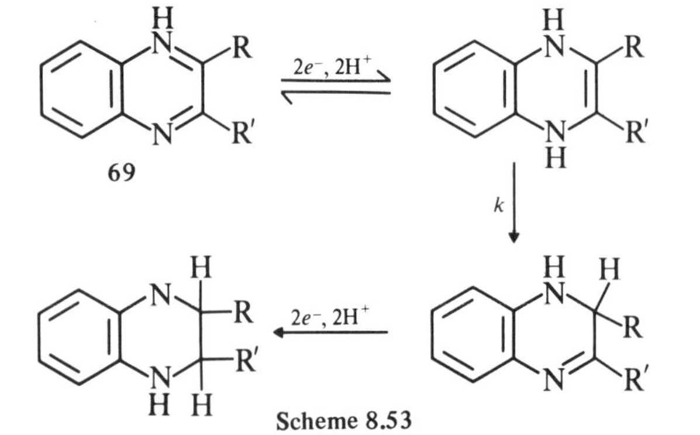

484 Electrochemical Behavior of Heterocyclic Compounds

Scheme 8.53

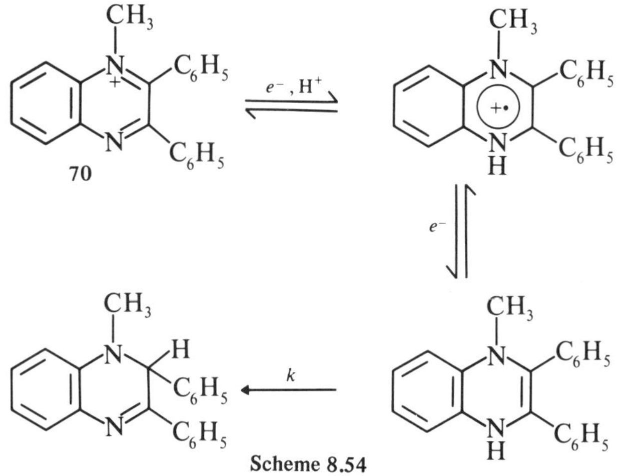

Scheme 8.54

8.2.1.4.10. Phenazines. In aqueous solutions, phenazine **71** is reversibly reduced into 5,10-dihydrophenazine via the stable semiquinone radical (280–286). In DMF, phenazine in the presence of acetic anhydride is reduced to 5,10-diacetyl-5,10-dihydrophenazine (252). In MeCN, in the presence of alkyl halides or dialkyl sulfates, phenazine is reduced to 5,10-dihydro-5,10-dialkylphenazine (287). In MeCN, 5-methylphenazinium cation **72** undergoes two reversible one-electron reductions (288). Electrolysis at the level of the first wave yields 5-methylphenazyl free radical. After addition of acid, two one-electron steps are also evidenced; electrolysis at the first step gives 5-methyl-10-hydrophenazine cation radical. Pyocyanine **73** has a similar behavior.

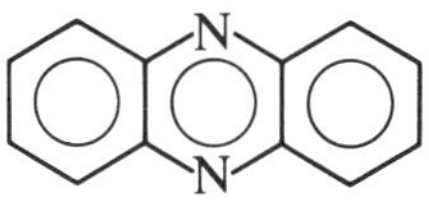

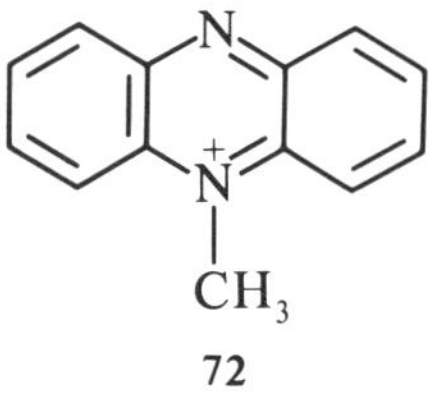

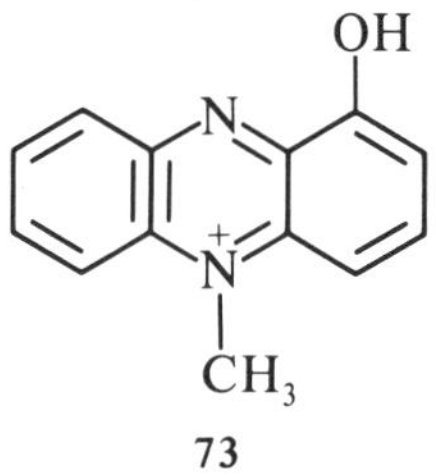

At pH 5–9 in aqueous solutions, 5-methylphenazinium cation 72 is reversibly reduced to 5,10-dihydro-5-methylphenazine (289).

8.2.1.4.11. Diazepines and Benzodiazepines.

In DMF, reduction of 5,7-diphenyl-2,3-dihydro-1,4-diazepinium cation 74 occurs in two steps. The first step leads to a radical that disproportionates to the dihydro derivative 75 and to 74 and the second leads to the anion of 74 (290). Reduction of 76 unexpectedly yields pyrrolodiazepine 77 (291). The reduction of 1,4-dibenzyl-6-phenyl-2,3-dihydro-1,4-diazepinium cation gives the *meso* and racemic isomers of 7,7'-bis(1,4-diphenyl-6-phenyl-1,2,3,4-tetrahydro-1,4-diazepine) and (1Z,3E)-1,4-bis(1,3-dibenzyl-2-imidazolidymyl)-1,4-diphenylbutadiene (292).

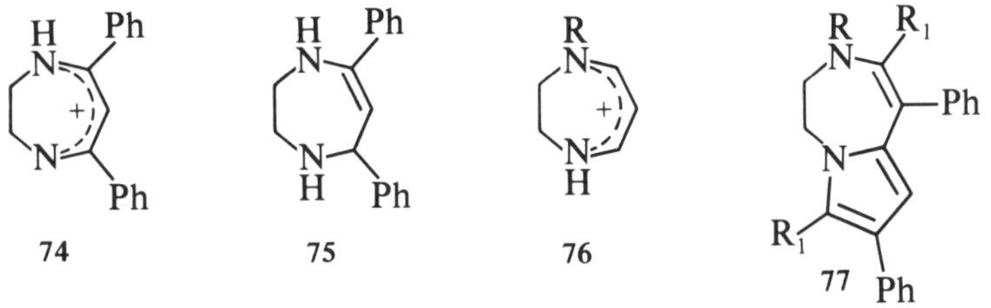

In aqueous solutions, derivatives of dibenzo[c,f]-1,2-diazepine are reduced to 1,2-dihydro derivatives (293, 294). The polarography of 1,4-benzodiazepines has been investigated in view of analytical applications (295–299). The elucidation of the mechanism of reduction of 7-chloro-2-methylamino-5-phenyl-3H-1,4-benzodiazepine 79 has been made from the study of its 4-oxide 78 (300, 301).

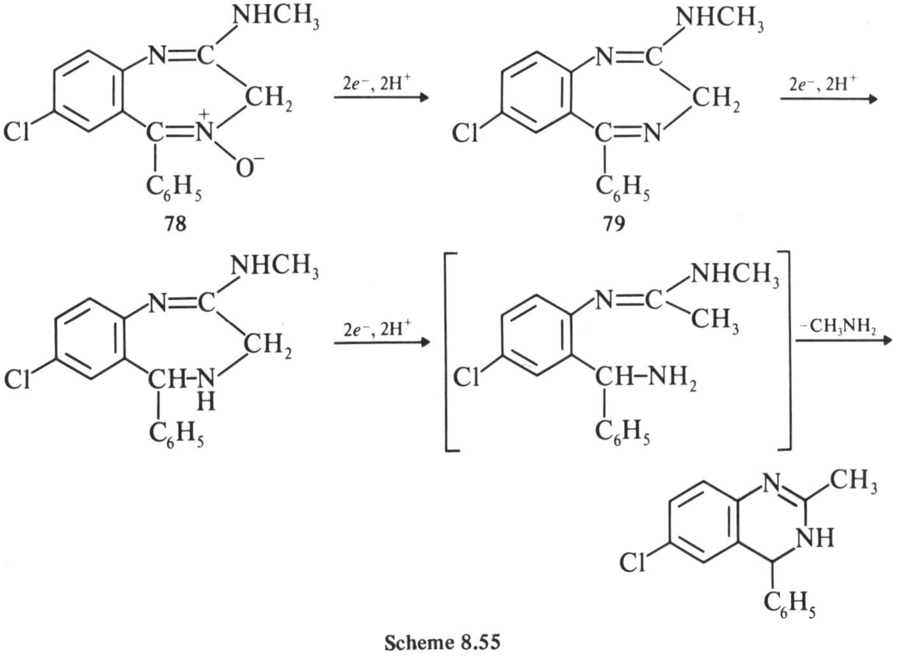

Scheme 8.55

This compound is reduced in acid solution in three steps: the first step corresponds to reduction of the $\overset{+}{N}-O^-$ bond, the second to saturation of the C_5-N_4 double bond, and the third to the formation of dihydroquinazoline (Scheme 8.55).

In aqueous solution, 1,5-benzodiazepines **80** are polarographically reducible (302, 303). The two two-electron waves correspond to the successive reduction of the $C_4=N_5$ and $N_1=C_2$ double bonds. In MeCN the reduction of benzodiazepines **81** gives the corresponding barbaralanes **82** (304); in MeCN, bridged 1,5-benzodiazepines, in the presence of acetic anhydride, are reduced to substituted barbaralanes (304a).

8.2.1.4.12. Naphthyridines. In acid media, 1,5-, 1,7-, 2,6-, and 2,7-naphthyridines **83–86** are reduced to an anion radical (stable in the case of N,N'-dimethyl derivative of 1,5-naphthyridine), which generally isomerizes into a second radical. The first and the second radicals react and give a dimer (305, 306).

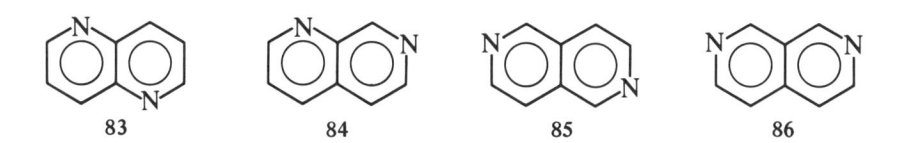

8.2.1.4.13. Imidazopyridinium and Pyrrolopyridinium Cations. In aqueous solution, 1-methyl-3-methylimidazo[1,2-a]pyridinium cation **87** is reduced (307) in two two-electron steps according to Scheme 8.56. In the case of 1-methyl-2-phenylimidazo[1,2-a]pyridinium cation, dimerization occurs after addition of one electron per molecule (307).

Scheme 8.56

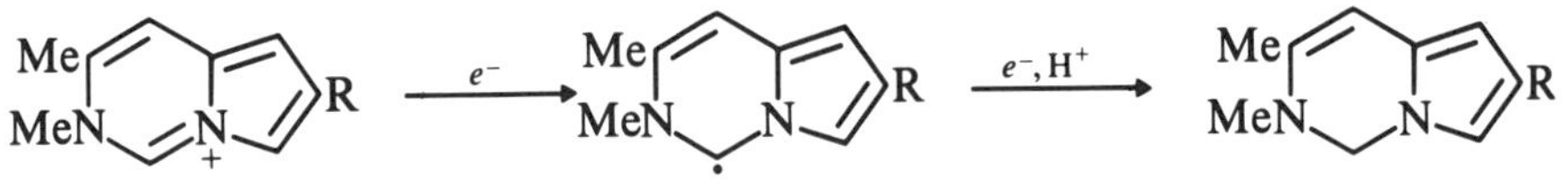

Reduction of Heterocyclic Compounds 487

Scheme 8.57

Pyrrolo-pyridinium cations **88** are reduced in two one-electron stages (307) (Scheme 8.57).

8.2.1.5. Compounds with One Nitrogen and One Oxygen Atom

8.2.1.5.1. Oxaziridines. In aqueous media, oxaziridines are very easily reduced (308). The cleavage of the N–O bond occurs as shown in Scheme 8.58 in the case of compounds **89**.

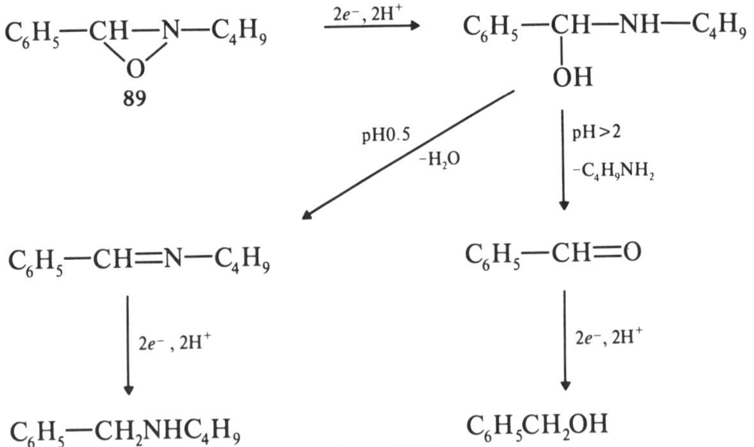

Scheme 8.58

8.2.1.5.2. Anthranils. 3-Methylanthranil **90** is reduced (309) in acidic solution to o-aminoacetophenone (Scheme 8.59).

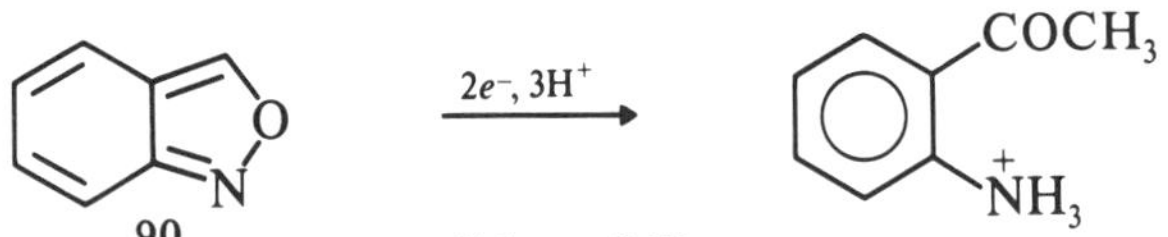

Scheme 8.59

8.2.1.5.3. Benzoxazines. Lund (310) suggested that benzoxazines are cyclic oximes and are reduced correspondingly in aqueous solution; the nitrogen-oxygen bond is hydrolyzed before the saturation of the carbon-nitrogen double bond (310). The reduction of 4-(4′-methoxyphenyl)-2,3-benzoxazin-1-one **91** is described in Scheme 8.60. It is possible to isolate unstable ketimine **92**.

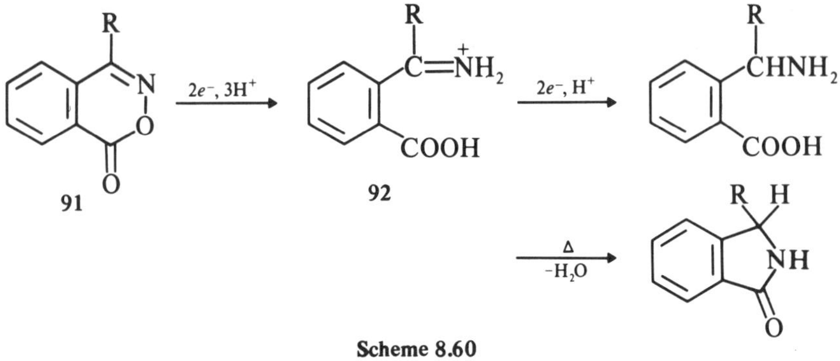

488   Electrochemical Behavior of Heterocyclic Compounds

Scheme 8.60

8.2.1.5.4. Oxazoles. In DMF, 2,5-diphenyloxazole **93** gives two one-electron waves (311, 312), the first corresponding to formation of a stable anion radical and second presumably to that of the dihydro derivative.

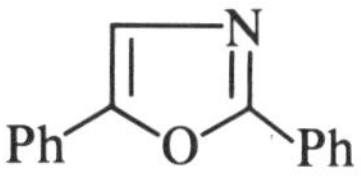

$$\underset{93}{\text{Ph}-\text{[oxazole]}-\text{Ph}}$$

In aqueous solutions, 2,5-diaryl-substituted oxazoles are reduced in most cases with consumption of six electrons (313). The reduction is suggested to proceed as shown in Scheme 8.61.

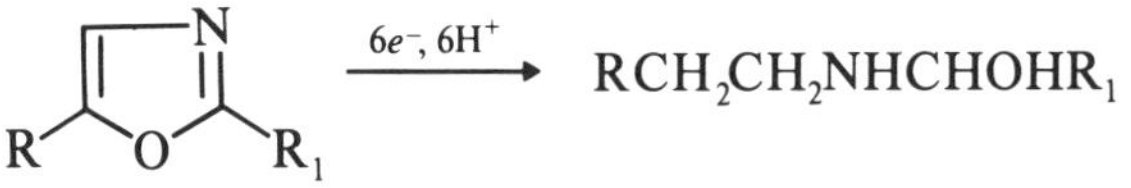

$$\text{oxazole} \xrightarrow{6e^-,\, 6H^+} RCH_2CH_2NHCHOHR_1$$

Scheme 8.61

8.2.1.6. Compounds with One Nitrogen and One Sulfur Atom

8.2.1.6.1. Thiazole Derivatives. Thiazole and its simple alkyl and aryl derivatives are not polarographically reducible in aqueous solutions. Derivatives of thiazole-2-carboxylic acid are reduced in alkaline solution, and the reduction is assumed to take place in the nucleus (314). The reduction of 3-methylbenzothiazolium cation **94** (315) gives a mixture of dihydro derivative and dimer (Scheme 8.62). An aryl group in the 2-position hinders dimerization (316).

A voltammetric study of 2-methyl-3-ethylbenzothiazolium cation in aqueous solution (317) led to the proposal of a mechanism of reduction. In the first one-electron process a dimerization mechanism takes place with a radical-substrate reaction as a rate-determining step. In the second one-electron step the neutral radical produced through the first electron transfer furnishes a dihydro derivative after addition of a proton and electron.

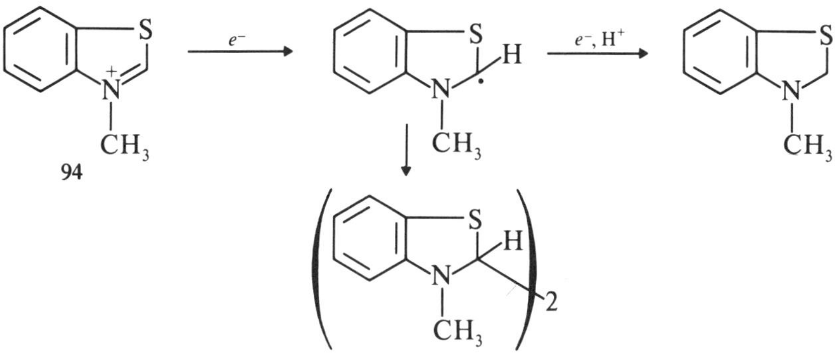

Scheme 8.62

8.2.1.6.2. Isothiazole Derivatives. In alkaline solution, saccharin **95** is reduced as in Scheme 8.63 (318). In acid solution, the primary reduction step concerns the carbonyl group (319, 320).

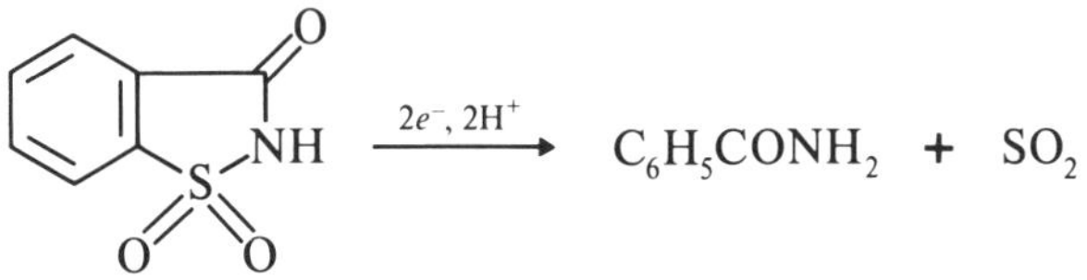

$$\text{95} \xrightarrow{2e^-, 2H^+} C_6H_5CONH_2 + SO_2$$

Scheme 8.63

8.2.1.6.3. Phenothiazine Derivatives. The polarographic reduction of methylene blue **96** to leuco-methylene blue **97** in aqueous solution (Scheme 8.64) has been carefully investigated (321), particularly concerning the adsorption phenomena.

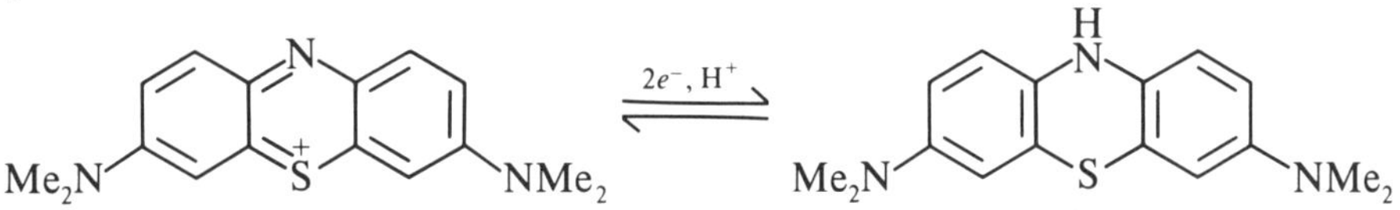

Scheme 8.64

8.2.1.7. Compounds with Two Phosphorus Atoms

The reduction of 1,4-diphosphoniacyclohexa-2,5-dienes **98** in aqueous medium furnishes the saturated products **99** (322) (Scheme 8.65).

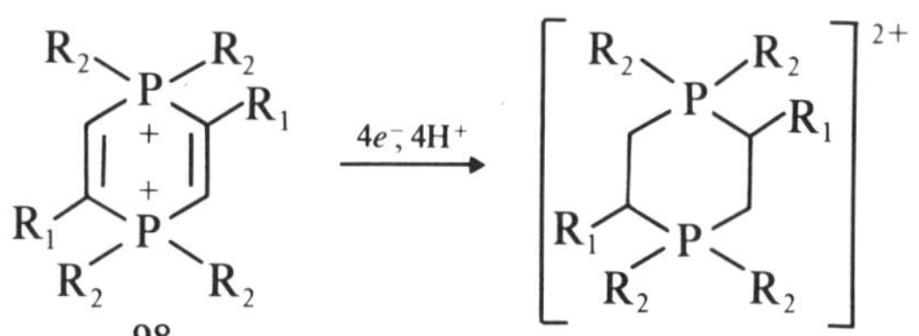

Scheme 8.65

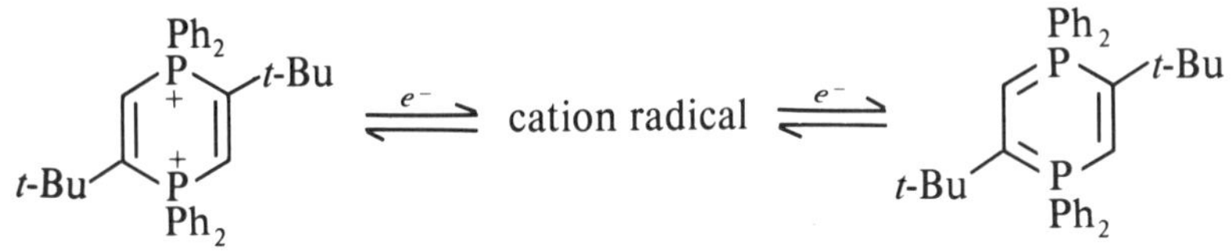

490 Electrochemical Behavior of Heterocyclic Compounds

$$100 \underset{e^-}{\overset{e^-}{\rightleftarrows}} \text{cation radical} \underset{e^-}{\overset{e^-}{\rightleftarrows}} 101$$

Scheme 8.66

In DMF, diphosphobenzenes **101** can be obtained by reduction of the corresponding diphosphonium salts **100** (323) (Scheme 8.66).

8.2.1.8. Compounds with Two Sulfur Atoms

In MeCN or CH$_2$Cl$_2$, reduction of 3,5-disubstituted-1,2-dithiolylium ions **102** produce 1,2-dithiolylium radicals **103**, which equilibrate with the dimers **104**.

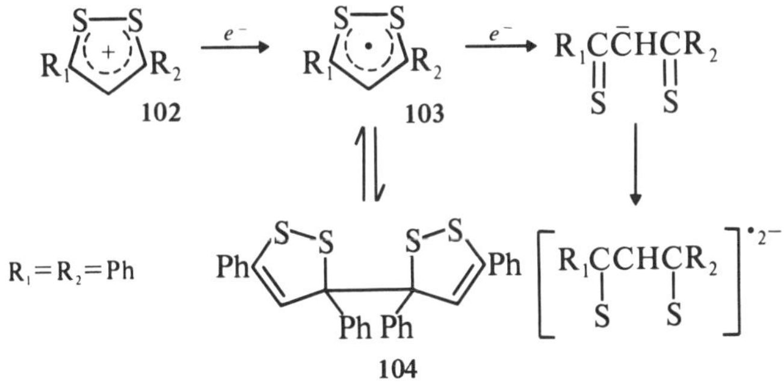

Scheme 8.67

Further reduction of the radicals causes ring opening between the sulfur atoms, with the formation of 1,3-dithioketonate anion (324) (Scheme 8.67). In MeCN, 5-methyl-1,2-dithiole-3-thione **105** is reduced stepwise (325) (Scheme 8.68). Reduction of 2-thioethoxy-1,3-dithiolium ions **106** in MeCN leads to the formation of orthothiooxalate **107** (326) (Scheme 8.69).

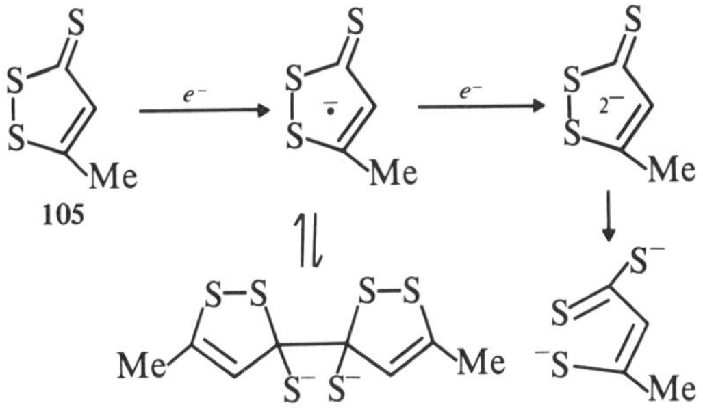

Scheme 8.68

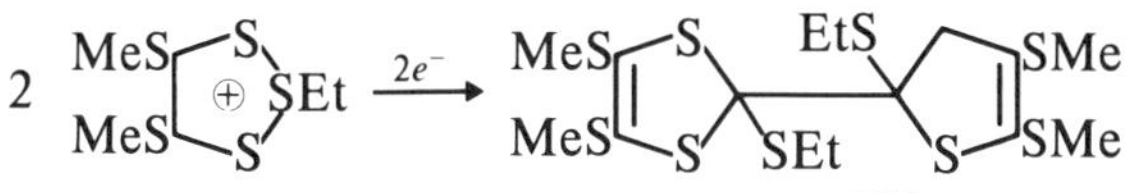

Reduction of Heterocyclic Compounds 491

$$2 \; \underset{106}{\text{MeS-S}\overset{\oplus}{\underset{\text{MeS-S}}{\bigg]}}\text{SEt}} \xrightarrow{2e^-} \underset{107}{\text{MeS}\underset{\text{MeS}}{\bigg\lfloor}\overset{\text{S}}{\underset{\text{S}}{\bigg\rfloor}}\overset{\text{EtS}}{\underset{\text{SEt}}{\bigg\rceil}}\overset{\text{SMe}}{\underset{\text{SMe}}{\bigg\rceil}}}$$

Scheme 8.69

In MeCN, **108** undergoes three one-electron reduction steps without decomposition, the first two of which are reversible (327). Radical **109** was prepared and is monomeric at temperature $< 18°C$.

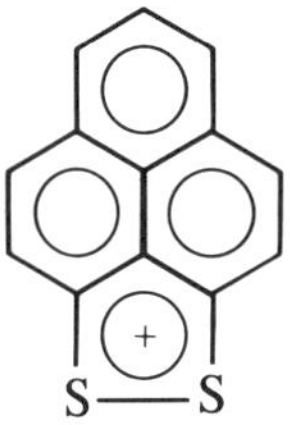

[Structures 108 and 109: pyrene-based disulfide cation and radical]

8.2.1.9. Compounds with Three Nitrogen Atoms

8.2.1.9.1. Benzotriazoles. Benzotriazoles are reducible in acid solution, and only 2-substituted derivatives have been found to be reducible in alkaline solution (328). In acid media, the reduction of benzotriazole **110** proceeds as in Scheme 8.70. In alkaline solution, 2-methylbenzotriazole is reduced reversibly to 1,3-dihydro derivative (328).

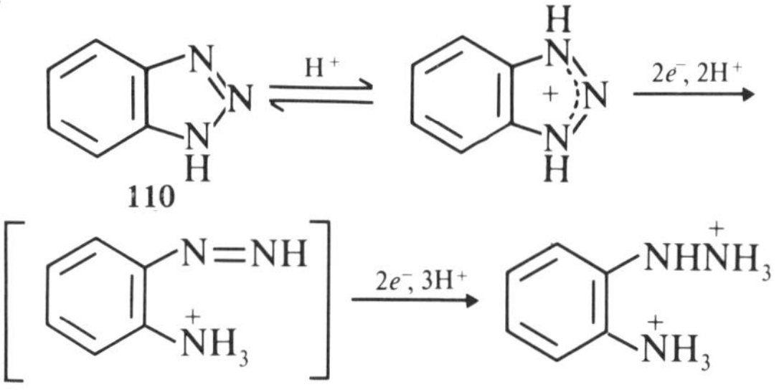

[Scheme 8.70: benzotriazole 110 reduction mechanism]

Scheme 8.70

8.2.1.9.2. Triazines and Benzotriazines. In aqueous solution, 3,4-dihydro-3-phenyl-benzo-1,2,3-triazine **111** is reduced to indazoline **112** through a four-electron, four-proton process (329) (Scheme 8.71). In aqueous solution, 3-phenylbenzo-1,2,4-triazine **113** is reversibly reduced into the dihydro-1,4 derivative (330), which can be reduced to benzimidazole **114** (Scheme 8.72).

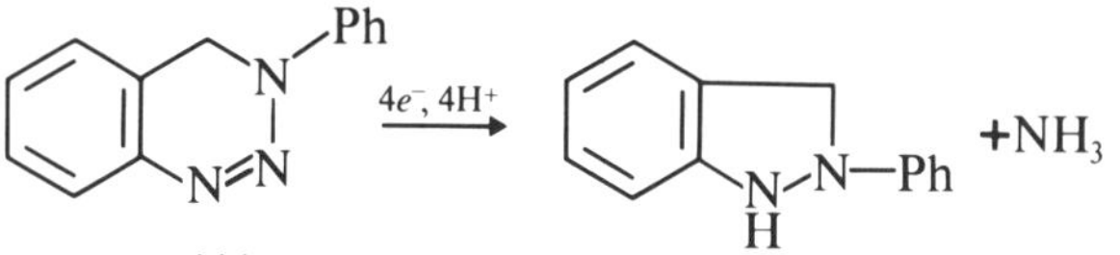

[Scheme 8.71: 111 → 112 + NH₃ via 4e⁻, 4H⁺]

Scheme 8.71

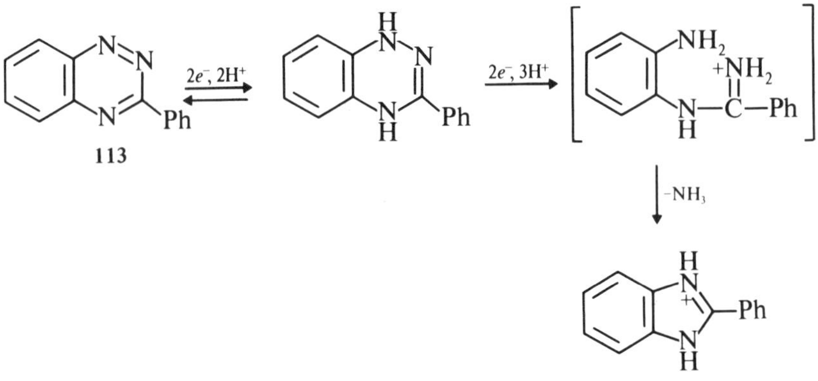

Scheme 8.72

In aqueous solution, 3,5,6-triphenyl-1,2,4-triazine 115 is polarographically reduced in two two-electron waves (331). Reduction at a potential corresponding to the first plateau furnishes the 1,4-dihydro derivative, which isomerizes into 1,2- and 4,5-dihydro compounds, and the latter can be reduced to the 1,4,5,6-tetrahydro derivative. Reduction at the level of the second plateau gives a mixture of 2,4,5-triphenylimidazole 116 and 1,4,5,6-tetrahydro-3,5,6-triphenyl-1,2,4-triazine. The reactions are summarized in Scheme 8.73.

In acid solution 1,3,5-triazines are polarographically reduced through a two-electron process (332).

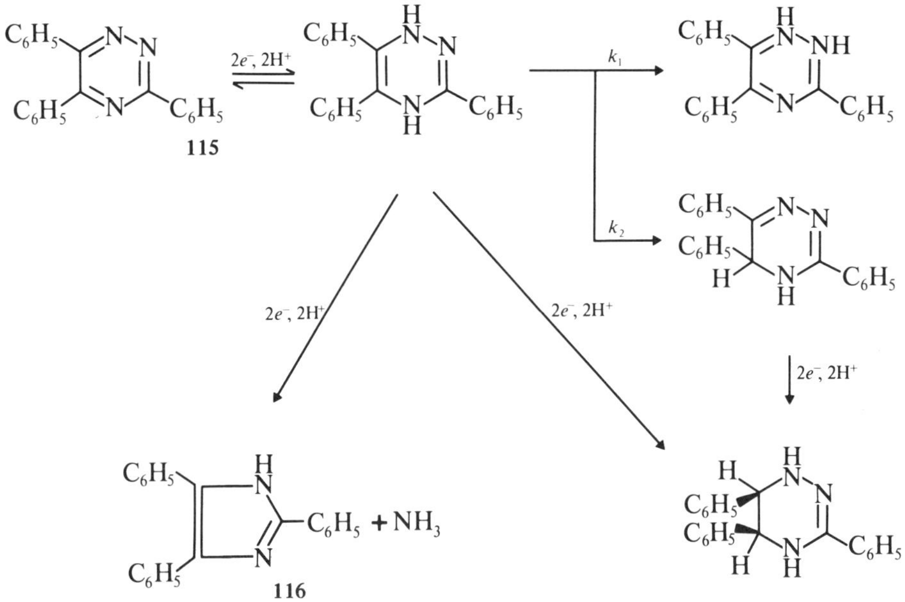

Scheme 8.73

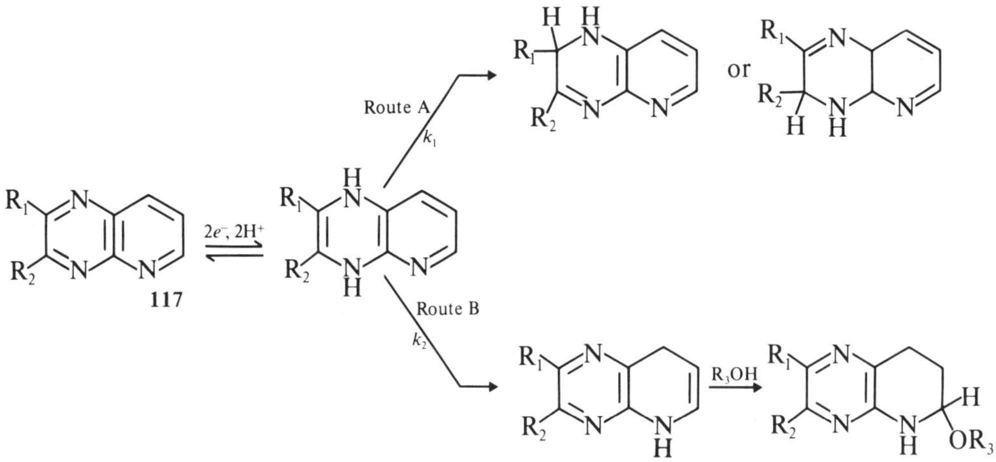

Scheme 8.74

8.2.1.9.3. Pyridopyrazines. In water-methanol or water-ethanol solution, pyrido[2,3b]pyrazines **117** (333) are reversibly reduced into 1,4-dihydro derivatives, which isomerize into 1,2-, or 3,4- and 5,8-dihydro derivatives, and the latter covalently adds a molecule of alcohol (Scheme 8.74).

In MeCN, in the presence of phenol, 5,8-dihydro derivatives are obtained. In MeCN, in the presence of acetic anhydride, phenyl-3-pyrido[2,3b]pyrazine furnishes the 1,4-diacetyl-1,4-dihydro derivative, whereas diphenyl-2,3-pyrido[2,3b]pyrazine gives the 5-acetyl-5,8-dihydro derivative.

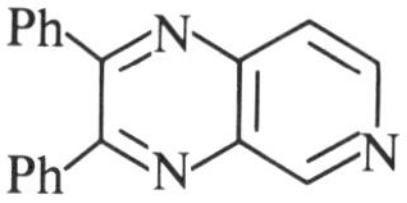

In aqueous solution, 2,3-diphenylpyrido[3,4b]pyrazine **118** is reversibly reduced into the 1,4-dihydro derivative, which isomerizes into a 1,2- or 3,4-dihydro compound that can be further reduced to the 1,2,3,4-tetrahydro derivative (333).

8.2.1.9.4. Pyrazolo[1,5a]Pyrimidines. In acid solution, 3,6-diphenylpyrazolo[1,5a]pyrimidine **119** is reduced into a 4,5-dihydro derivative (334) (Scheme 8.75).

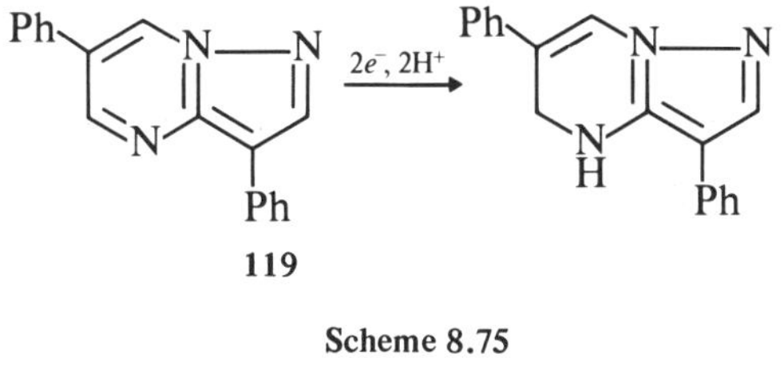

Scheme 8.75

8.2.1.10. Compounds with Two Nitrogen Atoms and One Oxygen Atom

8.2.1.10.1. Oxadiazoles. In aqueous solution, aryloxadiazoles are reducible at rather negative potentials; both four- and six-electron reactions occur (312, 313, 335). In the case of 2-(1-naphthyl)-5-(*m*-tolyl)-1,3,5-oxadiazole **120**, the six-electron reduction is believed to proceed as in Scheme 8.76.

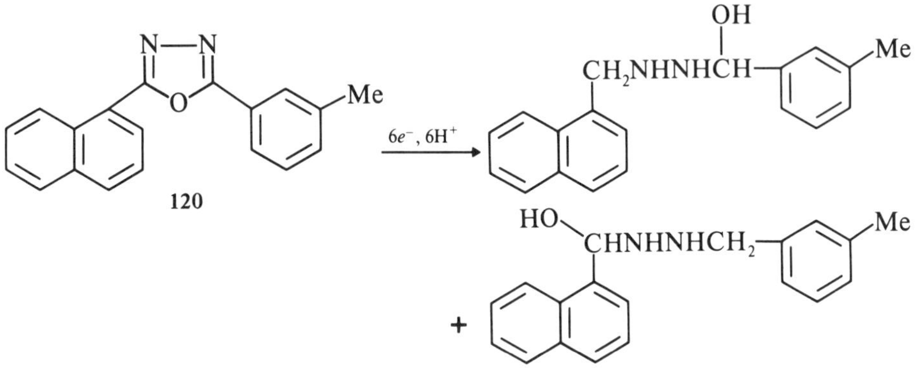

Scheme 8.76

8.2.1.10.2. Furazans and Furoxans. Benzofurazan **121** is polarographically reducible in aqueous solutions to *o*-phenylenediamine in a six-electron reduction

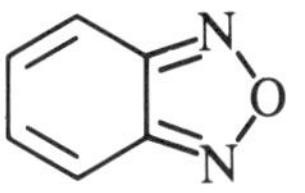

(336). The reduction of benzofuroxan **122** in neutral or basic solutions consumes 2 F mol^{-1} and gives *o*-benzoquinonedioxime, which can be further reduced (337) (Scheme 8.77).

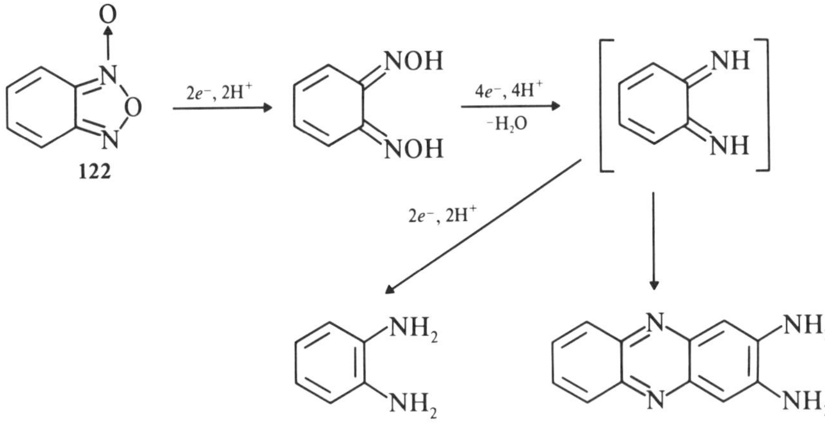

Scheme 8.77

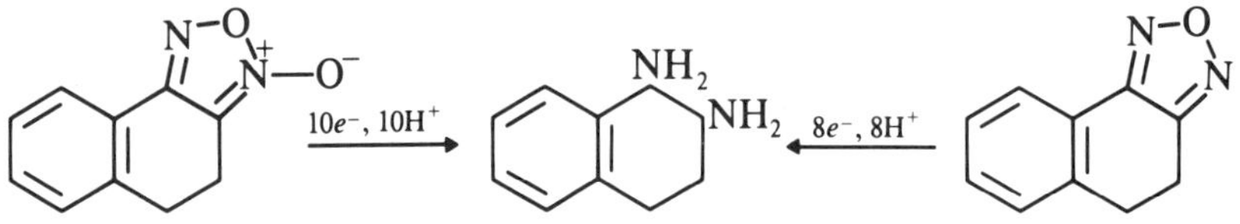

124 123

Scheme 8.78

Naphthofurazan and naphthofuroxan derivatives **123**, **124** have been reduced to diaminotetralin (338) in aqueous solution and DMF (Scheme 8.78).

8.2.1.10.3. ***Sydnones.*** 3-Phenylsydnones **125** are reducible in aqueous solution (339). In acid media, six electrons are consumed per molecule, whereas in alkaline solutions only four electrons are consumed (Scheme 8.79).

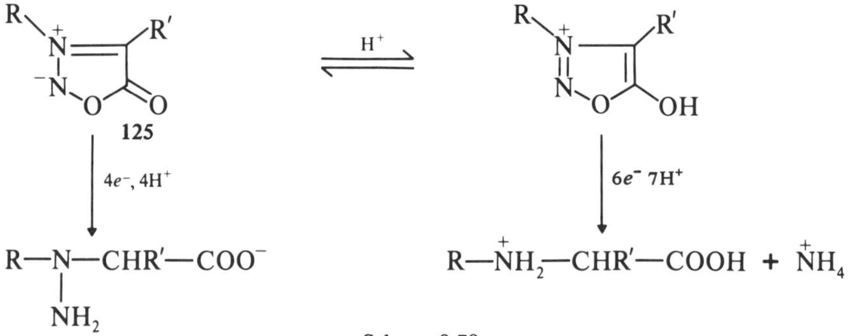

Scheme 8.79

8.2.1.11. Compounds with Two Nitrogen Atoms and One Sulfur Atom

8.2.1.11.1. ***2,1,3-Thiadiazoles and Benzo Derivatives.*** 2,1,3-Thiadiazole **126** is reduced in a six-electron reaction as in Scheme 8.80 (340). The behavior of benzo-2,1,3-thiadiazole **127** and benzo-2,1,3-selenodiazole **128** is similar to that of benzofurazan. The six-electron reduction furnishes *o*-phenylenediamine and hydrogen sulfide or selenide (341, 342) (Scheme 8.81).

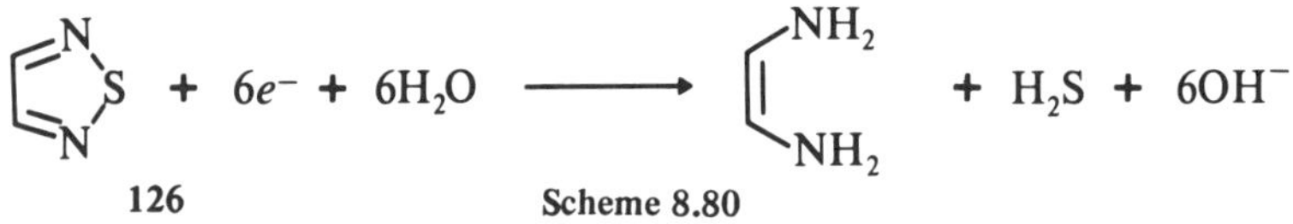

126

Scheme 8.80

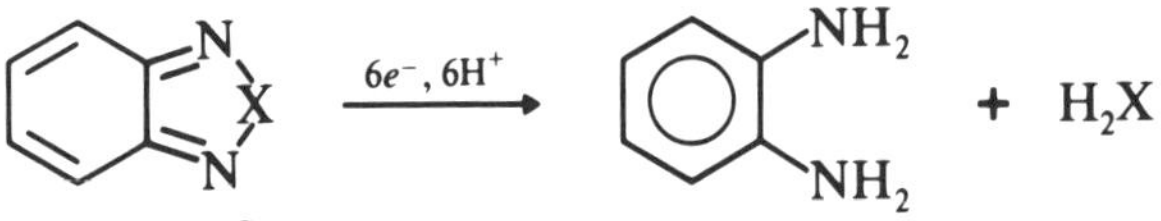

127 X = S
128 X = Se

Scheme 8.81

496 Electrochemical Behavior of Heterocyclic Compounds

8.2.1.11.2. 1,3,4-Thiadiazoles. These compounds are often polarographically reducible (343, 344). In acid solution, 2,5-diphenyl-1,3,4-thiadiazole **129** is reduced in a two-electron reaction to the dihydro derivative that is probably hydrolyzed to benzaldehyde-thiobenzoylhydrazone.

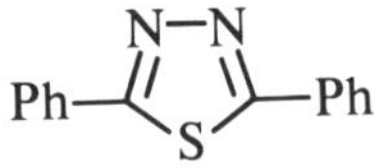

$$Ph-\overset{N-N}{\underset{S}{\diagdown\diagup}}-Ph$$
129

8.2.1.12. Compounds with Four Nitrogen Atoms

8.2.1.12.1. Tetrazolium Cations. In "super dry" MeCN (with activated Al_2O_3 in suspension), the tetrazolium cation **130** gives a reversible one-electron peak in voltammetry corresponding to the formation of a free radical (345). In MeCN containing "residual" H_2O ($10^{-3} M < C < 5 \times 10^{-3} M$), a two-electron irreversible peak is evidenced. Electrolysis furnishes compound **131** (Scheme 8.82) via an ECCE mechanism. The first electron transfer gives a free radical, which is then protonated, and the cycle is cleaved before addition of a second electron. Solution electron transfer between **130** and **131** may also occur.

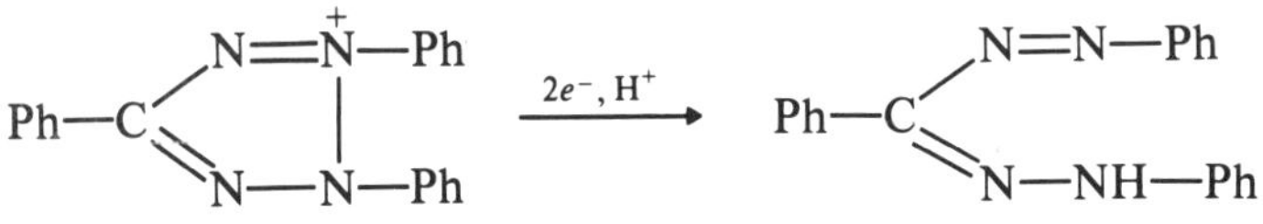

130 $\xrightarrow{2e^-, H^+}$ **131**

Scheme 8.82

In aqueous solution, **130** is irreversibly reduced to **131**, which decomposes into phenylbenzamidrazone, aniline, and triphenylformazan (346).

8.2.1.12.2. Purines. The electrochemistry of purines has been reviewed (347, 348). The reduction of purine **132** takes place only in acid solution and thus differs from that of pyrimidine and quinazoline. The reduction proceeds in two two-electron steps (263) (Scheme 8.83). The tetrahydro derivative obtained **133** is further hydrolyzed. In DMF, MeCN, or DMSO, 6-substituted purines are reduced to

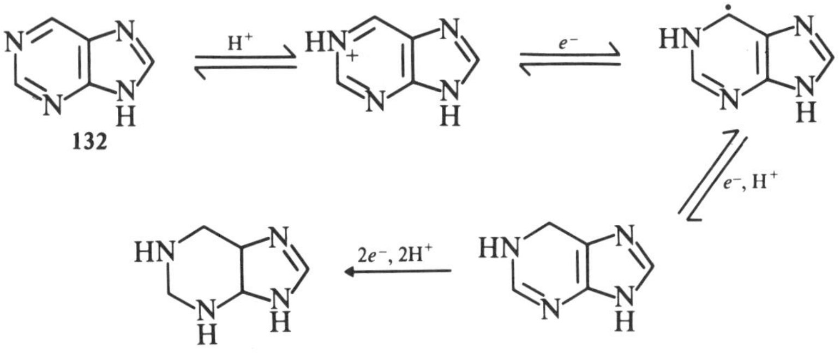

Scheme 8.83

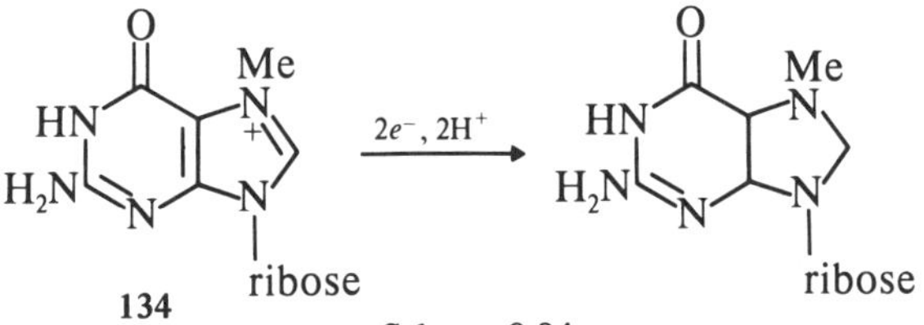

Reduction of Heterocyclic Compounds 497

Scheme 8.84

a dimer (349). In the presence of a proton donor, purine and 6-methylpurine each gives a dihydro derivative at less negative potentials and the tetrahydro derivative at more negative potentials (349). 6-Dimethylaminopurine, 6-methoxypurine, and adenine yield the tetrahydro derivative in a single step in the presence of a proton donor (350). 7,9-Dimethylpurinium cations are reduced progressively to 1,6-dihydro and 1,2,3,6-tetrahydro derivatives (351). In the reduction of 7-methylguanosine **134** (352) in acidic media, the imidazole ring (Scheme 8.84) is reduced.

8.2.1.12.3. Pteridines. The polarography of pteridine **135** is complicated because of its instability in aqueous solution (353). Pteridine is reversibly reduced into the 5,8-dihydro derivative (354) (Scheme 8.85). The monohydrated form of pteridine (3,4-dihydro-4-hydroxypteridine) is also reduced to 5,8-dihydropteridine, but in an irreversible process. The 5,8-dihydropteridine reacts with pteridine, producing a dimer that can be reduced to 7,8-dihydropteridine. The reduction of pterin and 7,8-dihydropterin has been investigated (354a).

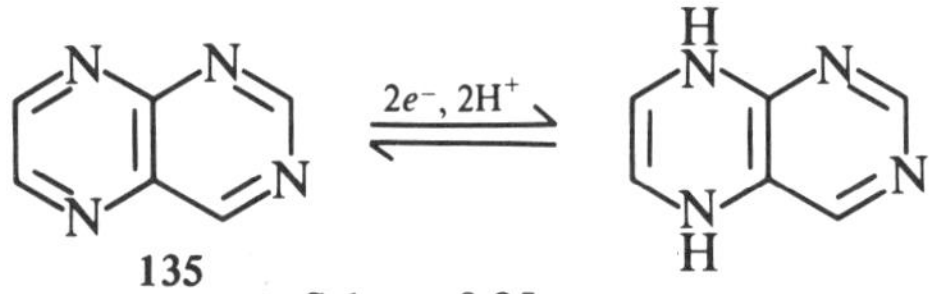

Scheme 8.85

8.2.1.12.4. Pyridotriazines. In aqueous solutions, pyrido[3,2e]-*as*-triazines **136** and pyrido[3,4e]-*as*-triazines **137** are reversibly reduced to the 1,4-dihydro derivatives (355) (Scheme 8.86). The reduction of 3-phenylpyrido[3,4e]triazine in the presence of acetic anhydride furnishes a mixture of 1,4-diacetyl-1,4-dihydro and 1,2-diacetyl-1,2-dihydro derivatives (355), probably via an ECEC mechanism.

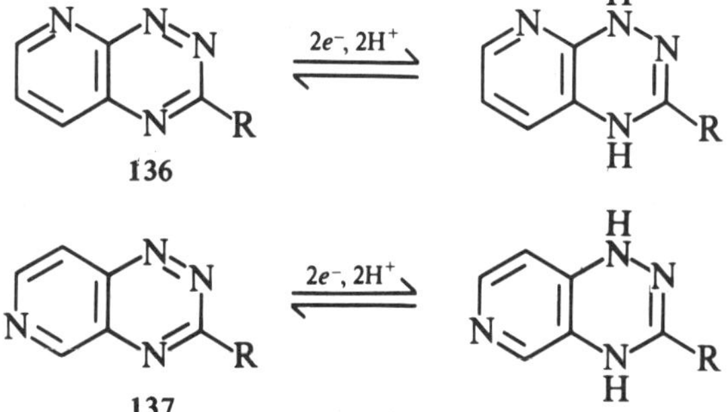

Scheme 8.86

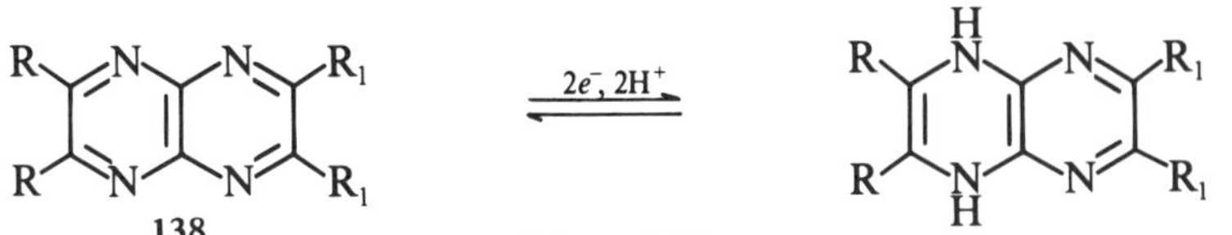

498 Electrochemical Behavior of Heterocyclic Compounds

Scheme 8.87

8.2.1.12.5. Pyrazino[2,3-b]pyrazines. In aqueous solution these compounds **138** are reversibly reduced to 1,4- or 5,8-dihydro derivatives (356) (Scheme 8.87).

8.2.1.12.6. Pyrazino[2,3-b]quinoxalines. In aqueous solution phenyl-2- and 2,3-dimethylpyrazino[2,3-b]quinoxalines **139** are reversibly reduced to the 5,10-dihydro derivative (357) (Scheme 8.88).

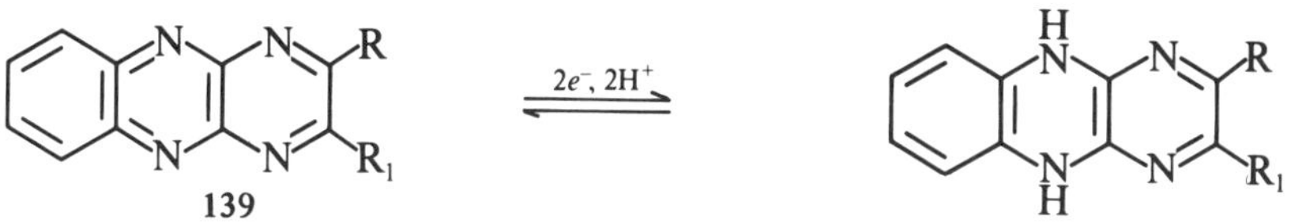

Scheme 8.88

8.2.1.12.7. Quinoxalino[2,3-b]quinoxaline 140. A polarographic study with an erroneous interpretation of polarograms has been reported (358). In aqueous solution, **140** is reversibly reduced to the 5,12-dihydro derivative (359), which can be further reduced in acidic media to 2-(o-aminoanilinomethyl)benzimidazole (Scheme 8.89).

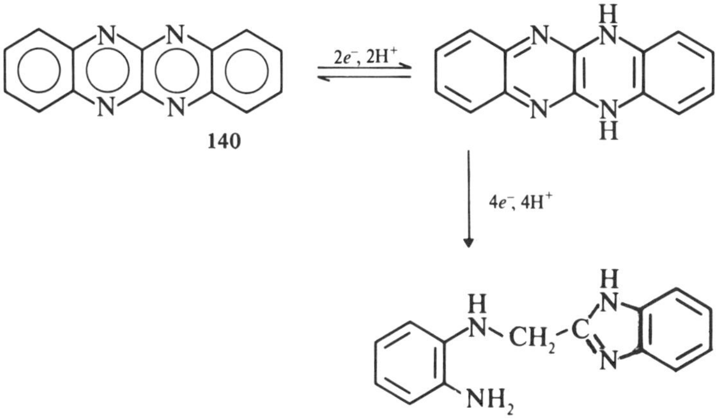

Scheme 8.89

In MeCN, **140** is reversibly reduced to the anion radical and then to the dianion (359). In MeCN, in the presence of acetic anhydride, 5,12-diacetyl-5,12-dihydroquinoxalino[2,3-b]quinoxaline **141** is reduced to the 5,6,11,12-tetraacetyl-5,6,11,12-tetrahydro derivative (Scheme 8.90).

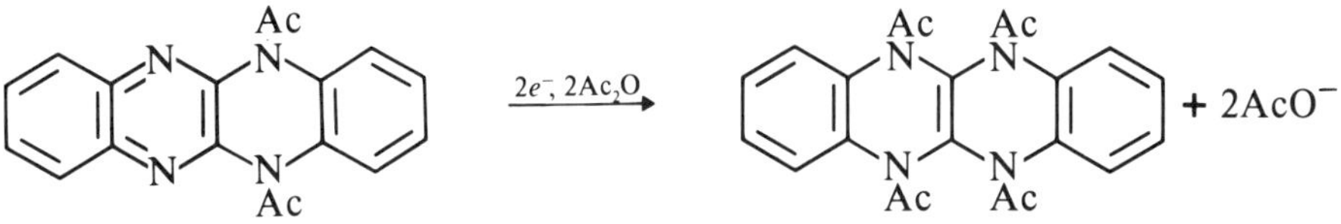

Reduction of Heterocyclic Compounds 499

Scheme 8.90

8.2.1.12.8. Tetraazabinaphthylene. In DMF, cyclobuta[1,2-b:3,4-b'] diquinoxaline **142** is reversibly reduced (360) through two one-electron processes to an anion radical and then to a dianion. In MeNO$_2$ the dication **143** is reversibly reduced in two one-electron steps to give a cation radical and then **144**.

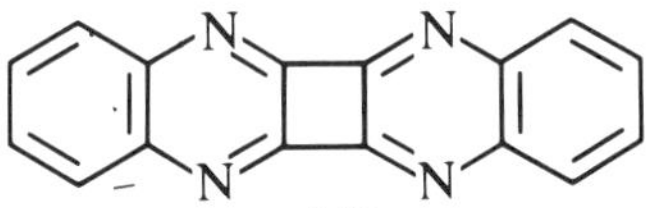

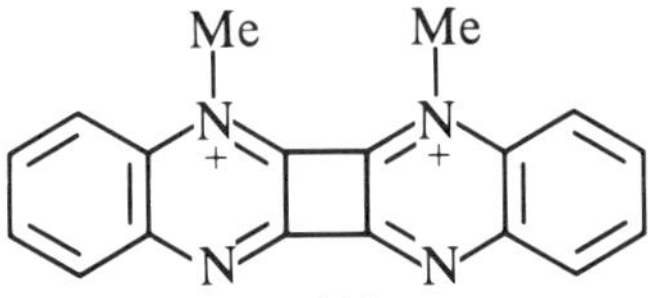

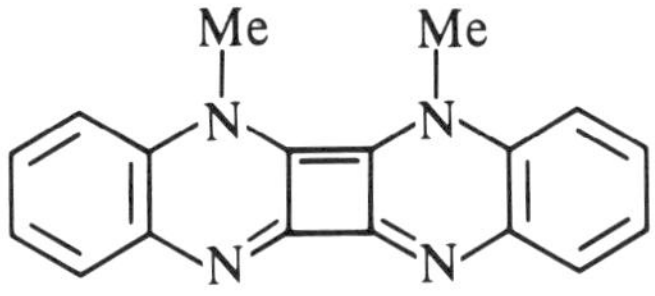

8.2.1.12.9. Porphyrins. The electrochemistry of porphyrins has been reviewed (361, 362). Woodward (363) has pointed out that the reduction of porphyrins **145** by addition of two electrons and two protons can result in the formation of two types of isomeric dihydroporphyrins, chlorins **146** and phlorins **147**. The chlorins are stable to air oxidation, but phlorins are unstable. The two-electron, two-proton reduction of these dihydroporphyrins yields three types of tetrahydroporphyrins: chlorin-phlorins **148**, bacteriochlorins **149**, and porphomethenes **150**.

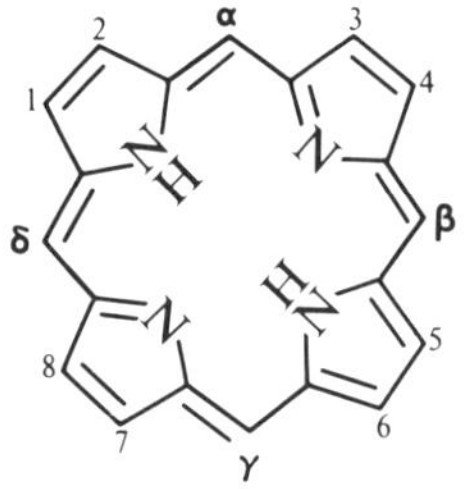

Porphyrins
145

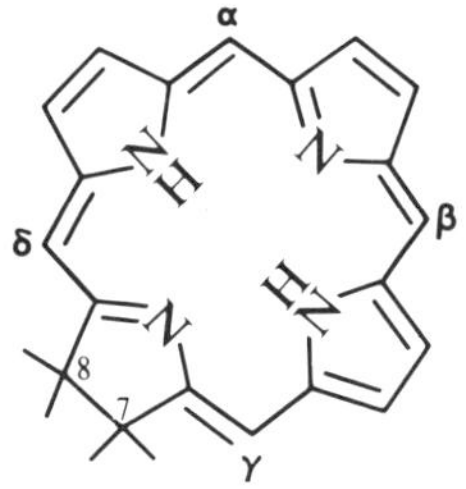

Chlorins
146

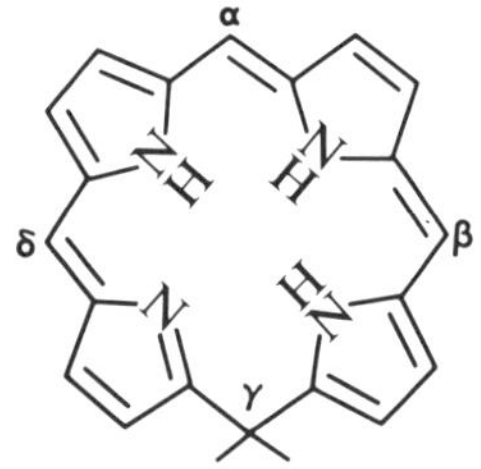

Phlorins
147

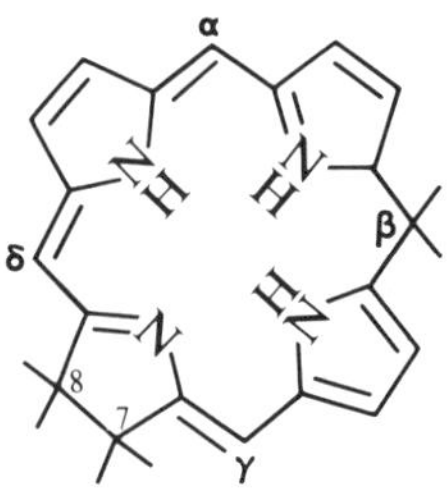

500   Electrochemical Behavior of Heterocyclic Compounds

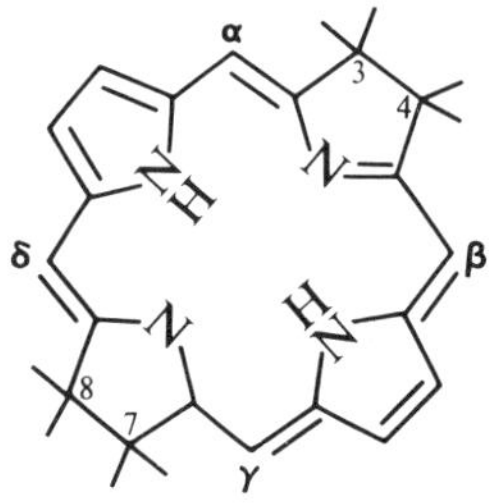

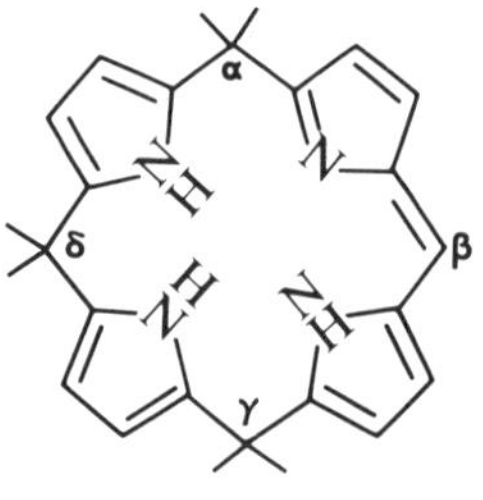

|Chlorin-phlorins|Bacteriochlorins|Porphomethenes|
|148|149|150|

Reduction of mesoporphyrin dimethyl ester yields a phlorin rather than a chlorin (364). The reduction of chlorin-e_6 trimethyl ester gives rise to the corresponding chlorin-phlorin (364). The electrochemistry in DMF of tetraphenylporphyrin, tetraphenylchlorin, and tetraphenylbacteriochlorin has been studied using cyclic voltammetry (365).

8.2.1.13. Compounds with Five Nitrogen Atoms

Pyrimidotriazole **151** is reduced in aqueous solution to the 5,6-dihydro derivative (366) (Scheme 8.91).

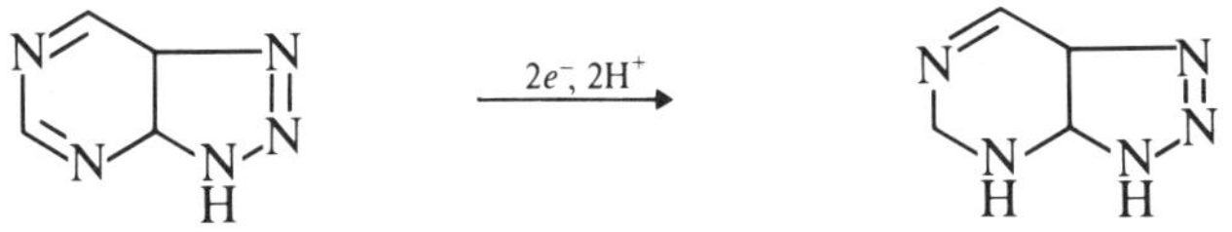

151 $\xrightarrow{2e^-, 2H^+}$

Scheme 8.91

8.2.2. Reduction of Substituted Heterocycles

In this section, the electrochemical reduction of substituted heterocyclic compounds is discussed for cases when (1) the substituent plays an essential role in determining the course of the reaction, (2) the electrode reaction involves the substituent directly, and/or (3) the electrode reaction may be different in some way from the analogous reaction in the carbocyclic series.

8.2.2.1. Compounds with a Substituent that has an Oxygen Atom Attached to the Ring

8.2.2.1.1. C-Hydroxy and Oxo Derivatives. Heterocyclic compounds having a hydroxyl substituent attached to the carbon atom of the ring exist in two tautomeric forms (367), the hydroxy and oxo. In many cases the study of the tautomeric equilibrium in solution has allowed to determine the form which is reduced at the electrode which is most often the oxo form.

Reduction of Heterocyclic Compounds   501

Derivatives of Pyrrole. In acidic solution, succinimide **152** is reduced to pyrrolidone at the Pb cathode (368) (Scheme 8.92), whereas some pyrrolidine is also formed at the amalgam electrode (369). *N*-Substituted succinimides (370) and maleimides (371) have been reduced to the corresponding pyrrolidones. In MeCN or DMF the reduction of succinimide yields the anion of succinimide and hydrogen (371).

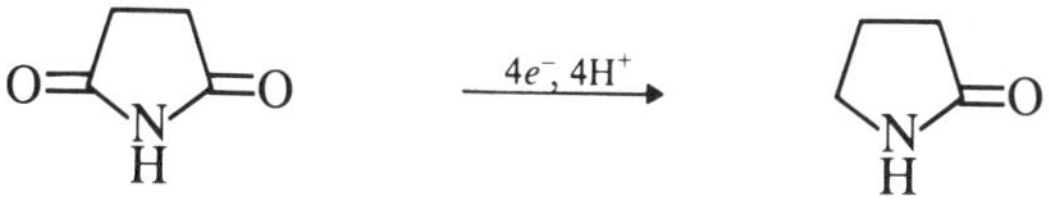

152 → (4e⁻, 4H⁺) → pyrrolidone

Scheme 8.92

Derivatives of Isoindole. In acidic solution phthalimide **153** gives a two-electron polarographic wave and in slightly alkaline media two one-electron waves (372–376). In acid media (Scheme 8.93), the reduction gives hydroxyphthalimidine **154**, and at low pH **154** is dehydrated to 3-oxoisoindole **155**, which is reduced to phthalimidine **156**. The latter can be further reduced to isoindoline **157**. Substituted phthalimidines behave similarly (377).

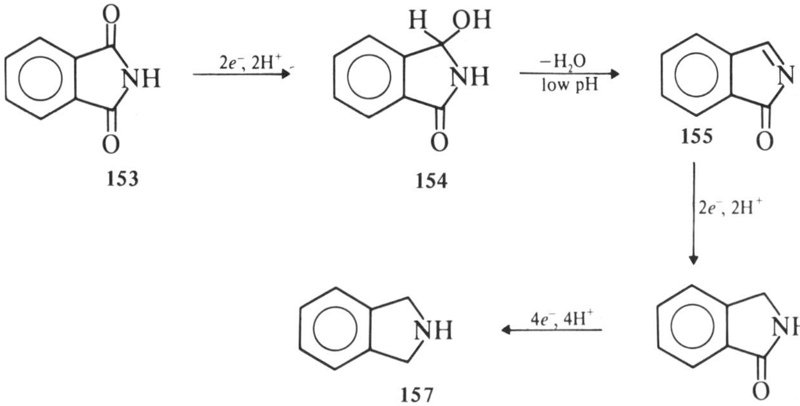

Scheme 8.93

In alkaline solution, electrolysis at the level of the plateau of the first polarographic wave gives a radical anion that rapidly dimerizes into an epoxide dimer (376).

In DMF, 2-phenyl-3-oxo-3(*H*)indole **158** is reduced in two one-electron steps to a radical anion and a dianion (378). In acidic media **158** is reversibly reduced to 3-hydroxy-2-phenylindole **159** (Scheme 8.94).

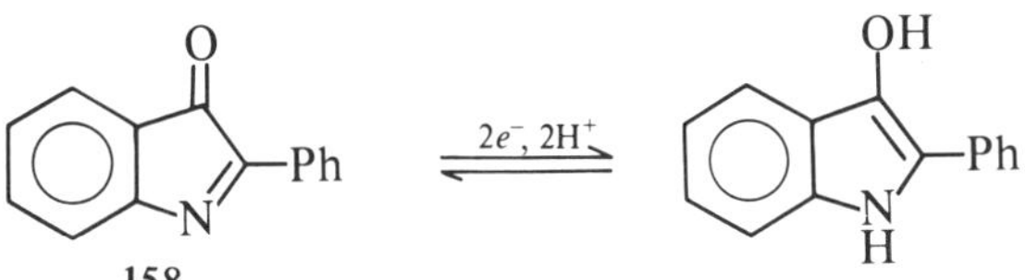

Scheme 8.94

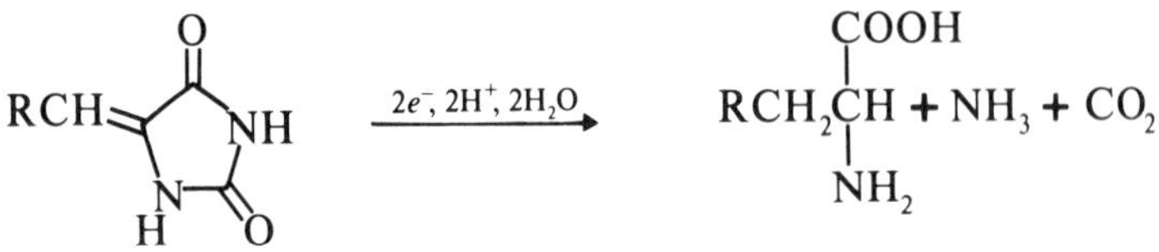

160

Scheme 8.95

Derivative of Imidazolidine. In aqueous solutions, reduction of hydantoins **160** yields amino acids (379) (Scheme 8.95). In aqueous solutions, parabanic acid **161** and derivatives are reduced in a two-electron, two-proton process to 5-hydroxyhydantoins **162**, which are slowly further hydrolyzed to a complex mixture (380) (Scheme 8.96).

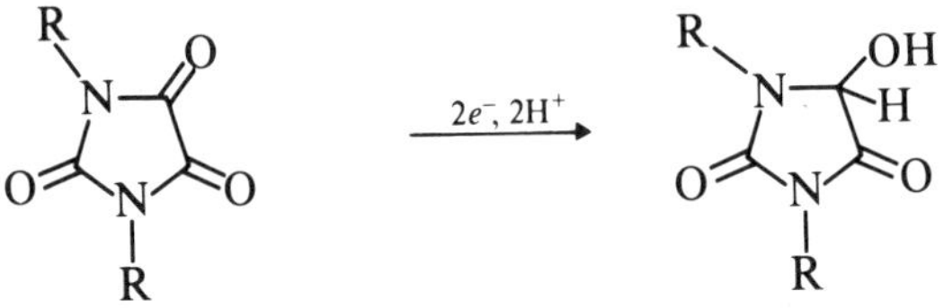

161 Scheme 8.96 162

Derivatives of Acridine. Acridone is reduced in DMF-EtOH to acridane (381) (Scheme 8.97).

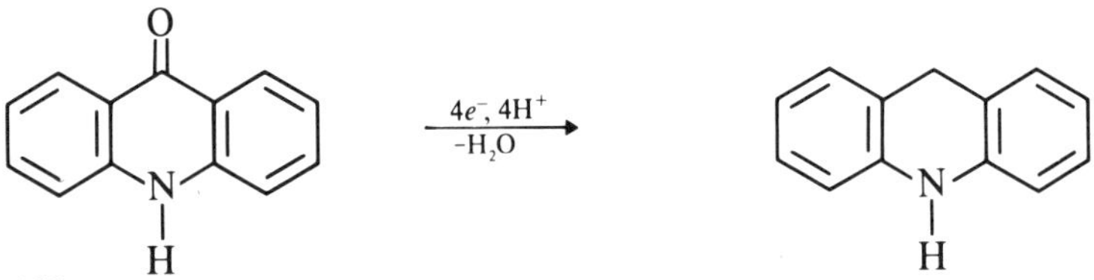

163

Scheme 8.97

Derivatives of Phenoxazine and Phenothiazine. In DMF, 3*H*-phenoxazin-3-one **164a** is reduced (382) in two one-electron stages, the first of which is reversible. Upon addition of phenol, a two-electron, two-proton process is evidenced, leading to the dihydro derivative **165**. 3*H*-Phenothiazine-3-one **164b** has similar behavior (Scheme 8.98).

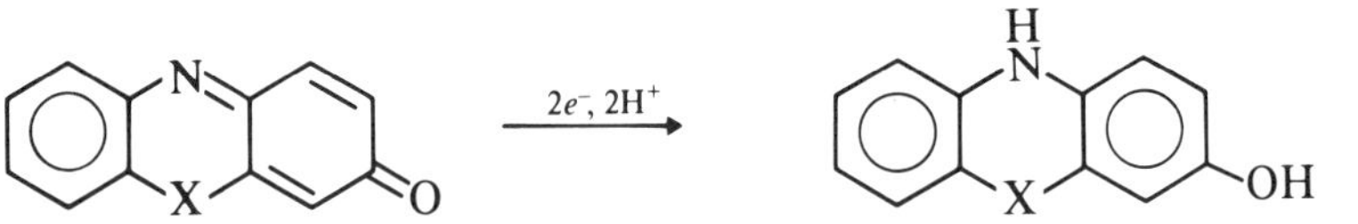

164a X = O
164b X = S

165

Scheme 8.98

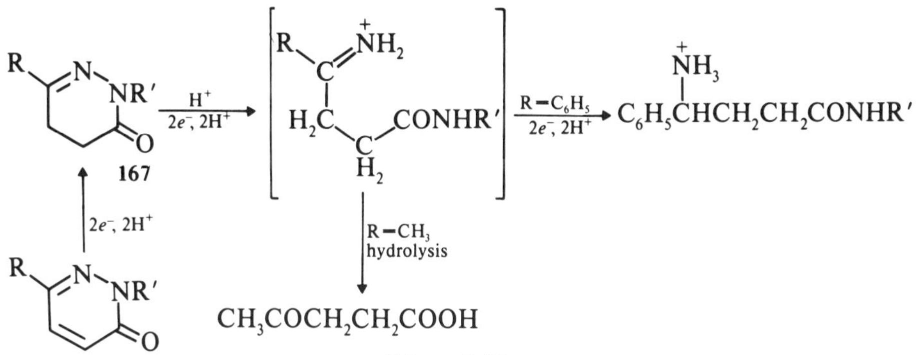

Reduction of Heterocyclic Compounds 503

Scheme 8.99

Pyridazinones. In aqueous solutions, in the reduction of both 3-pyridazinones **166** (247, 383) and maleic hydrazide (3-hydroxy-pyridazine-6-one) (384), the first step is the saturation of 4,5-double bond, and further reduction of **167** depends on the substituents at the 6-position (Scheme 8.99).

Hydroxycinnolines and Cinnolinone. 3-Hydroxycinnoline **168** (385) is reduced in strongly acid solution to 1-aminooxindole **169**, and in neutral media to 3-keto-1,2,3,4-tetrahydrocinnoline **170** (Scheme 8.100), which can be easily oxidized to 3-hydroxycinnoline. 4-Keto-1,4-dihydrocinnoline **171** is reduced (385) in acidic solution in two two-electron steps to dihydro **172** and further to tetrahydro derivatives **173**, which can be easily oxidized anodically in alkaline media (Scheme 8.101). As **171** is easily accessible, the electrochemical preparation of cinnoline **174** from **171** is the method of choice (385). In alkaline media, the reduction of **172** leads to dimeric products (385).

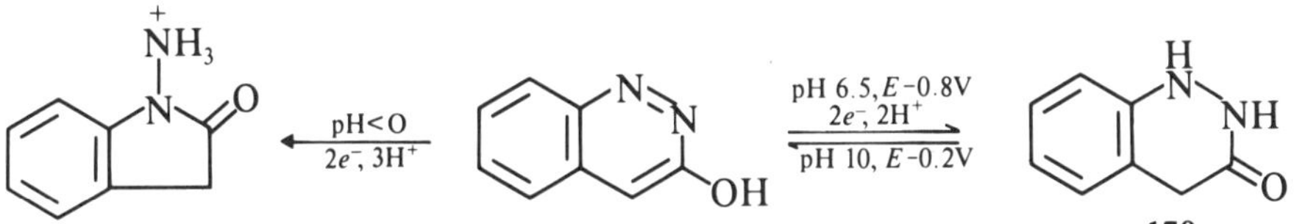

Scheme 8.100

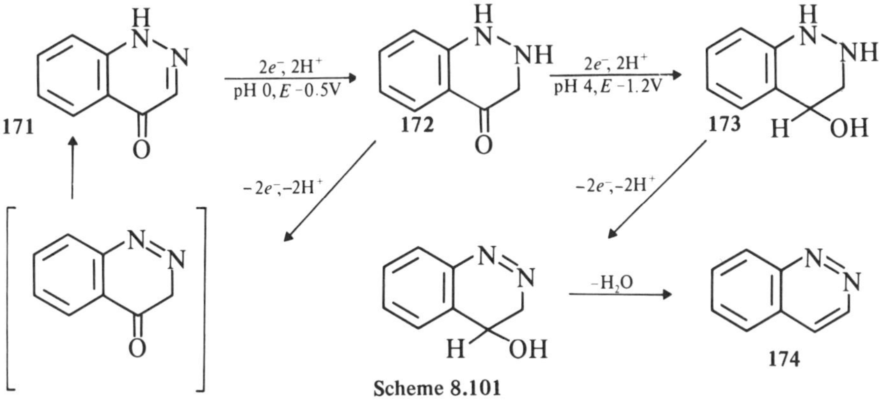

Scheme 8.101

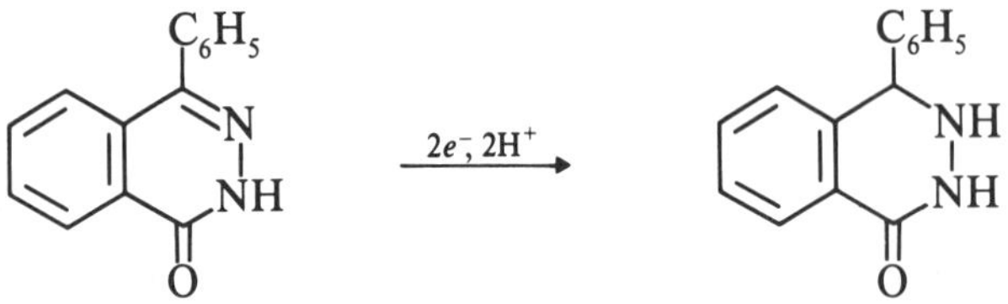

504 Electrochemical Behavior of Heterocyclic Compounds

Scheme 8.102

Phthalazinones and Phthalizinediones. 1(2H)-Phthalazinones **175** are reduced in neutral and alkaline media to the corresponding 3,4-dihydro derivative **176** (386, 387) (Scheme 8.102). 4-Oxo-furo[2,3d]pyridazines **177** are similarly reduced (388).

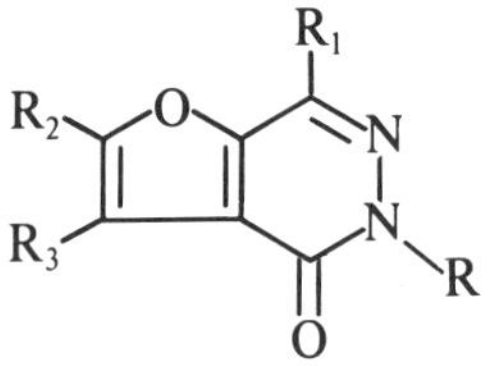

177

In acid solution, 3,4-dihydro-4-methyl-1(2H)phthalazinone **178** can be reduced in a two-electron reduction, with the cleavage of the nitrogen-nitrogen bond and ring closure to phthalimides (386) (Scheme 8.103).

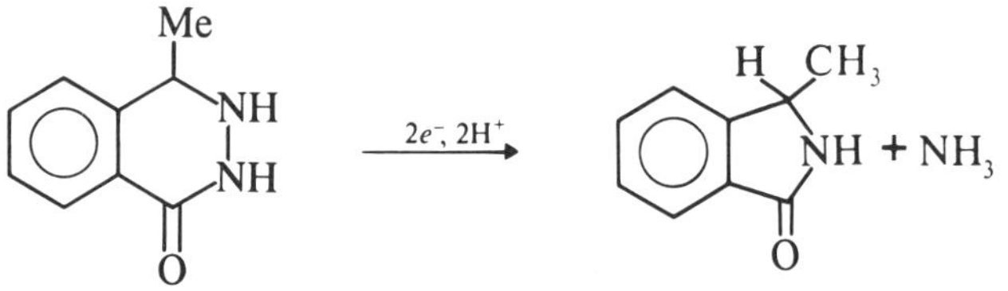

Scheme 8.103

Phthalazinediones (386) such as 2,3-dihydro-2,3-dimethyl-1,4-phthalazinedione **179** are reduced at low pH to phthalimidine **180** in a six-electron reduction. In alkaline solution, 4-hydroxy-3,4-dihydro-2,3-dimethyl-phthalazinone **181** is obtained in a two-electron reduction. The latter cannot be further reduced in alkaline media, but in acidic media the pseudo base loses water and gives a phthalazinonium cation **182** which can be further reduced (Scheme 8.104).

Derivatives of Pyrimidine. In aqueous solutions, 2-pyrimidones such as **183** are reduced with a one-electron process to a radical that rapidly dimerizes into a 6,6' dimer (389, 390) (Scheme 8.105).

In DMSO, 2-pyrimidone **183** is reversibly reduced to the radical anion, which dimerizes more slowly than its attack on unreduced **183** to abstract a proton (father-son reaction), producing the neutral free radical, which dimerizes more rapidly than the radical anion (391). Barbituric acid, 5,5-dialkylbarbiturates,

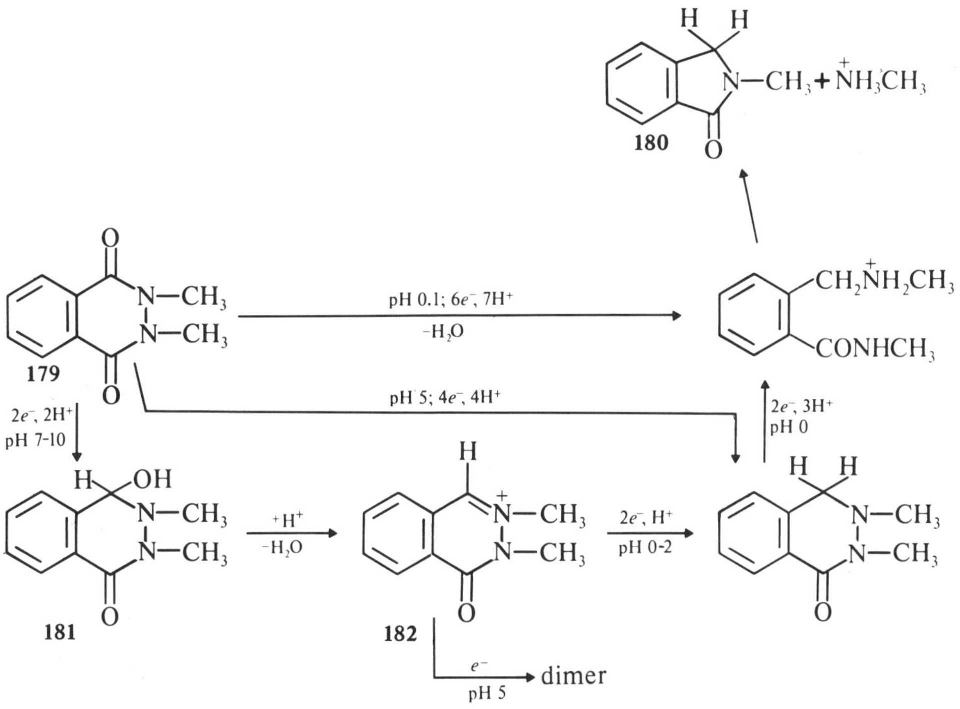

Reduction of Heterocyclic Compounds   505

Scheme 8.104

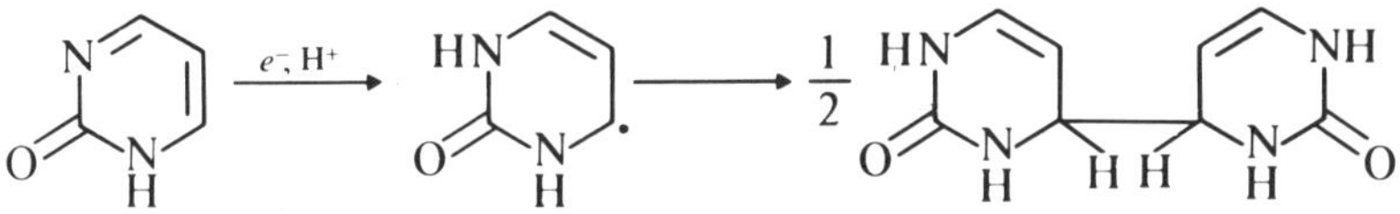

Scheme 8.105

uracil, and thymine are not polarographically reducible in aqueous solution. In MeCN, 5,5-dialkylbarbituric acid **184** gives (392) a one-electron cathodic peak in voltammetry due to the reduction (Scheme 8.106).

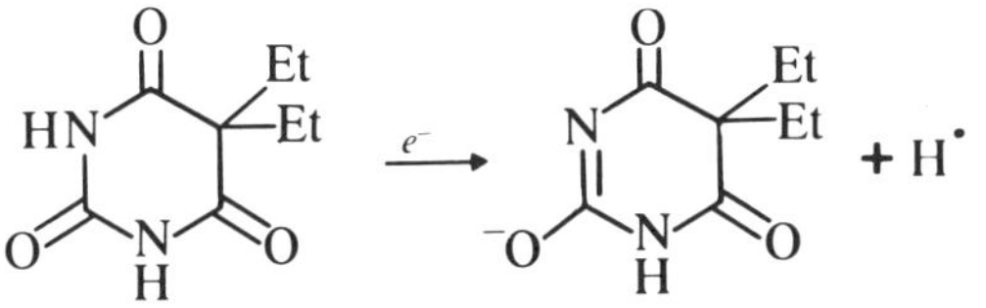

Scheme 8.106

In DMSO, uracil **185** gives a one-electron step corresponding to the formation of a radical anion protonated by unreacted uracil (393). Thymine **186** has similar behavior (394).

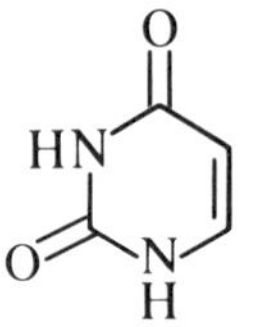

506   Electrochemical Behavior of Heterocyclic Compounds

185, 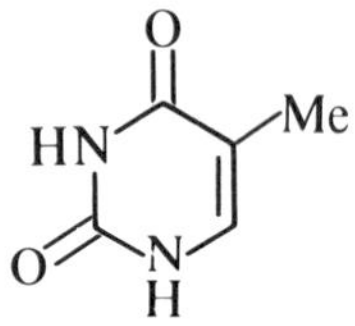 **186**

In aqueous solutions, alloxans (395) **187** exhibit three cathodic waves I_a, I_b, and II, corresponding to the reduction of alloxans or their hydrated forms (Scheme 8.107).

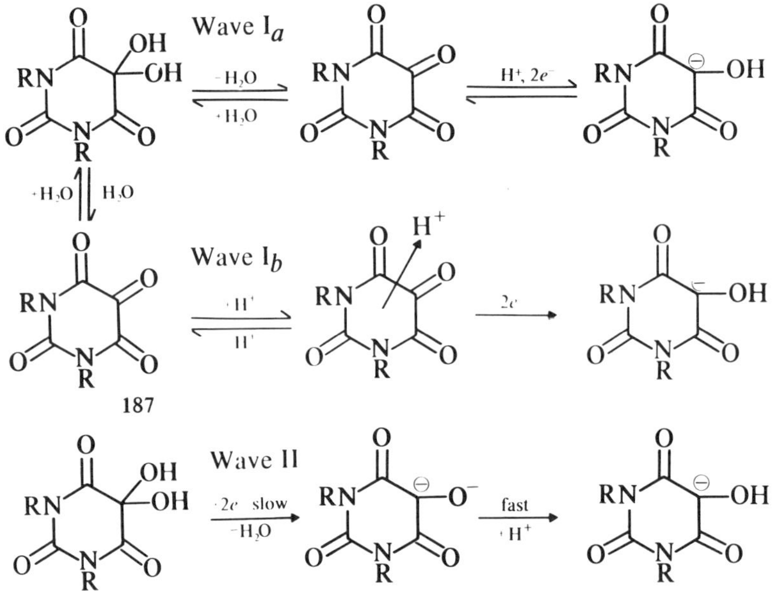

Scheme 8.107

Derivatives of Quinazoline. In aqueous solution, 4(3H)-quinazolinone **188** is reduced as in Scheme 8.108 (396).

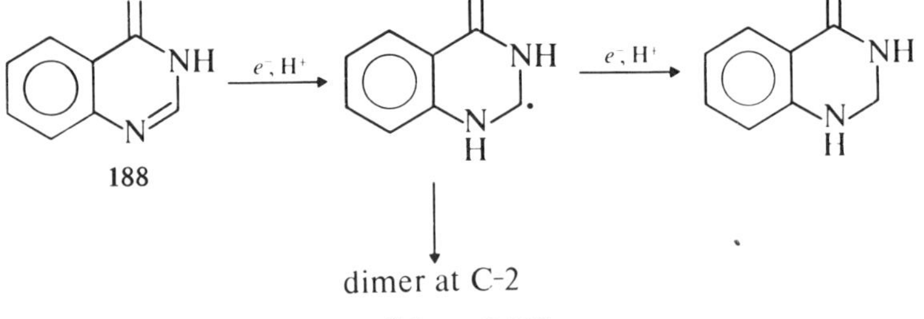

dimer at C-2

Scheme 8.108

Reduction of Heterocyclic Compounds 507

Derivatives of Pyrazine and Piperazine. In aqueous solutions, 5,6-diphenyl-2(1*H*)pyrazinone and its *N*-methyl derivative are reduced (397) to the 3,4-dihydro derivative, which isomerizes to 3,6-dihydro derivatives. The reduction of the latter compound yields 3,4,5,6-tetrahydro derivatives (Scheme 8.109).

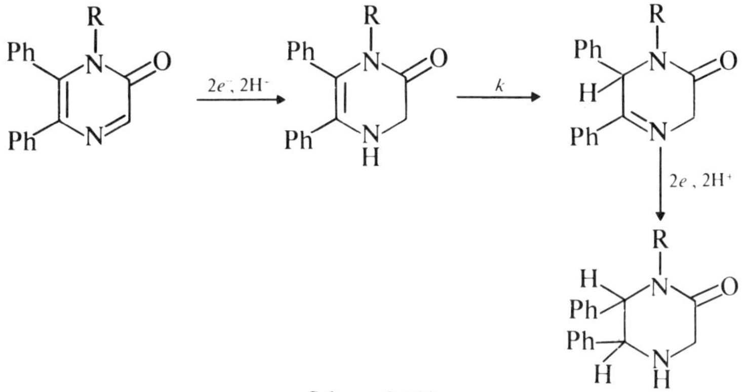

Scheme 8.109

2,3-Diphenyl-5-methoxypyrazine is reduced to the 4,5-dihydro derivative, which isomerizes to the 2,5-dihydro derivative (397). The 1,4-dihydro compound does not seem to be an intermediate in the reduction process, in contrast with what occurs with alkyl and arylpyrazines (273). In aqueous solution, tetraketopyrazine **189** is reduced (398) through a two-electron, two-proton process to 2-hydroxy-triketopiperazine, which undergoes loss of water to give triketopyrazine **190**. The latter is reduced through a six-electron process to 2,5-diketopiperazine **191**.

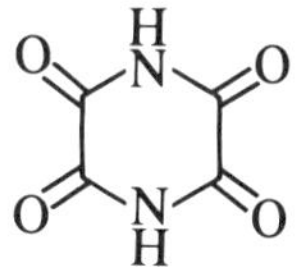

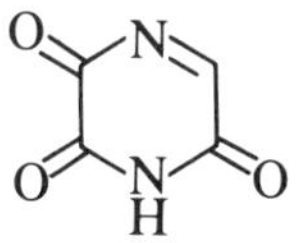

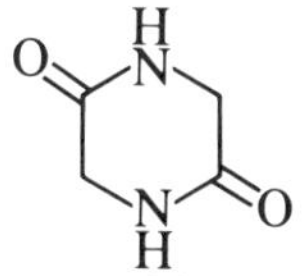

Derivatives of Quinoxaline. In aqueous solutions, 3-aryl quinoxaline(1*H*)-2-ones are reduced to 3,4-dihydro derivatives, whereas 3-alkylated compounds give 3,4-dihydro derivatives and a product dimerized at N-4 (399) (Scheme 8.110).

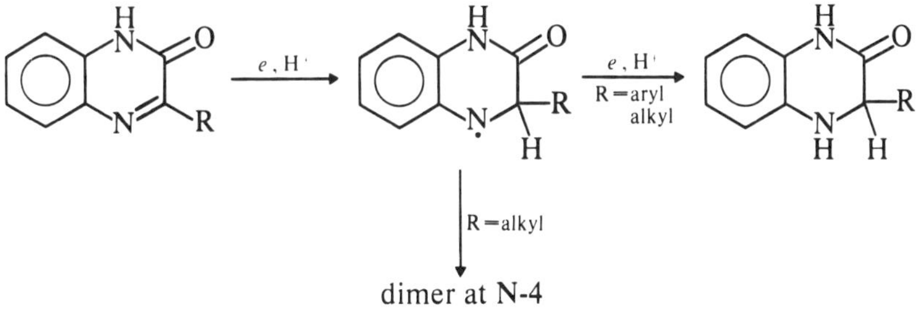

Scheme 8.110

2,3-dimethyl-5,8-quinoxalinedione **192** gives two two-electron polarographic waves at pH 7. The first corresponds to the reduction of the *p*-quinone system and the second to the reduction of the pyrazine nucleus (400) (Scheme 8.111).

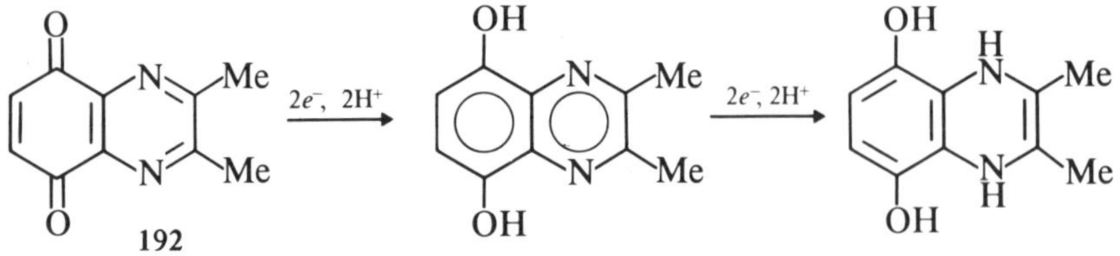

Scheme 8.111

In aqueous solutions, 2,3-dioxo-1,2,3,4-tetrahydroquinoxalines (401–404) are reducible. As an example, 1,4-dimethyl-2,3-dioxo-1,2,3,4-tetrahydroquinoxaline **193** is reduced as in Scheme 8.112. In aqueous solutions, 2,3-dimethoxyquinoxaline is reduced to the 1,4-dihydro derivative, which rearranges into the 1,2-dihydro derivative (402). The latter compound can be further reduced to the 1,2,3,4-tetrahydro compound.

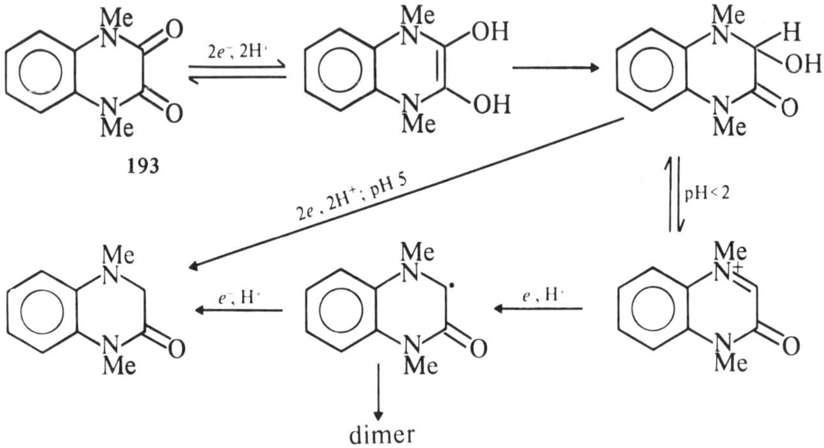

Scheme 8.112

In DMF, cyclobuta[*b*]quinoxaline-1,2-dione **194** is reduced in two one-electron steps to the radical anion and then the dianion (405). In the presence of a proton donor, two-electron, two-proton process leads to the formation of dihydro derivative **195** (Scheme 8.113).

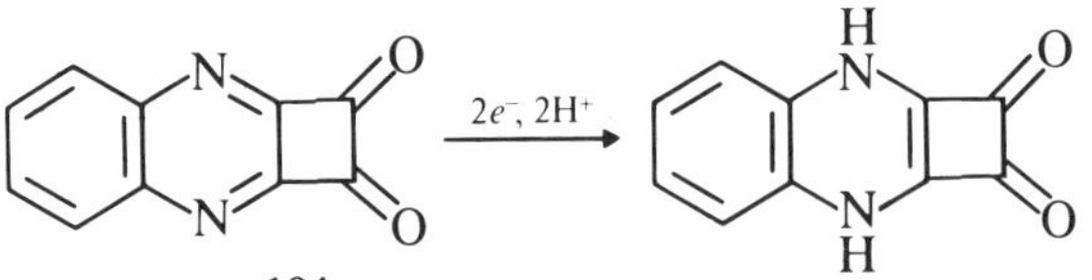

Scheme 8.113

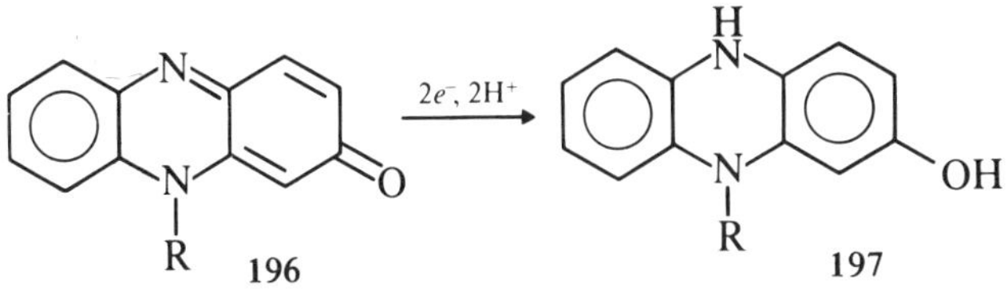

Reduction of Heterocyclic Compounds   509

Scheme 8.114

Derivatives of Phenazine. In DMF, 5-methyl and 5-phenyl-3(5*H*)-phenazinones **196** are reduced (406) in two one-electron stages, the first being reversible. Upon addition of phenol, a two-electron, two-proton process is evidenced, leading to the formation of dihydro derivative **197** (Scheme 8.114).

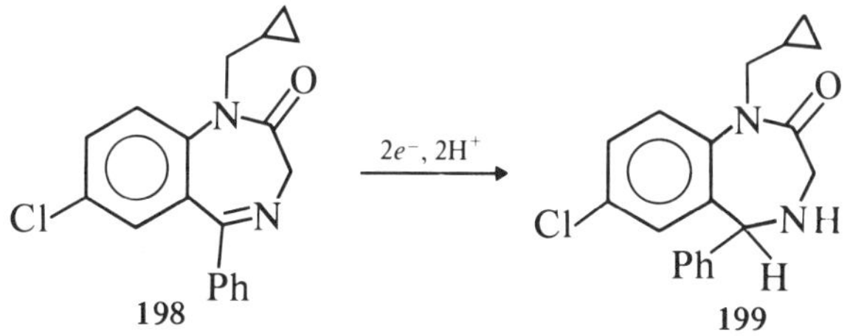

Scheme 8.115

Derivatives of Benzodiazepine. In aqueous solutions, prazepam **198** can be selectively reduced to dihydro derivative **199** (407) (Scheme 8.115). In acid solution, the reduction of benzo-1,4-diazepine derivatives **200** yields compounds **201** after uptake of four electrons per molecule (408) (Scheme 8.116).

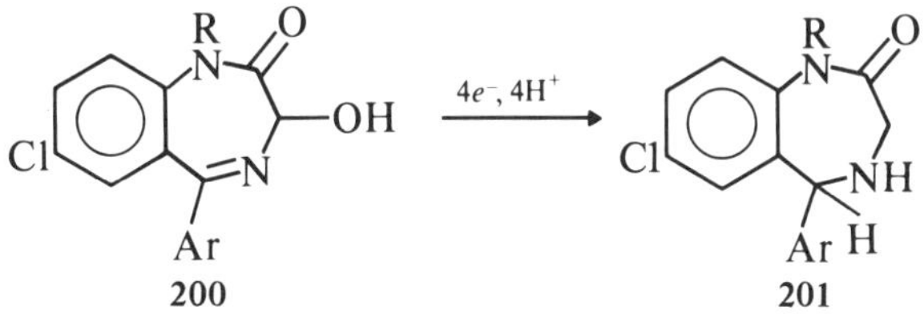

Scheme 8.116

1,3-Dithio Derivatives. In DMF, the reduction of 4,5-bis(alkylthio)-1,3-dithio-2-ones **202** followed by alkylation gives tetrathioethylenes **203** (409) (Scheme 8.117).

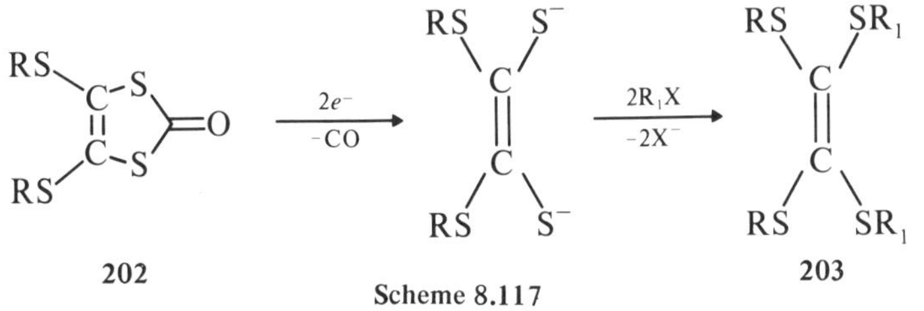

Scheme 8.117

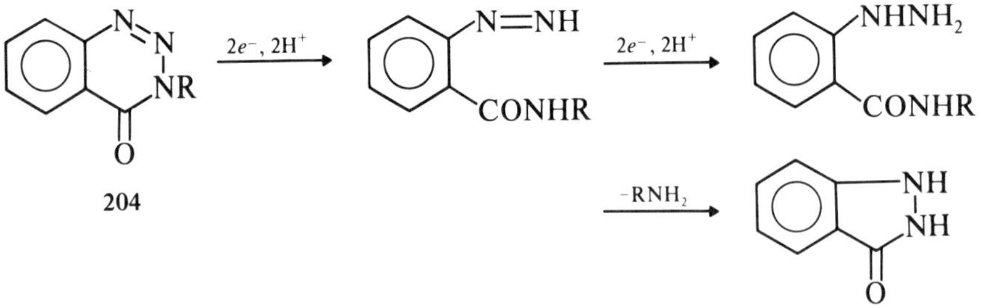

204

205

Scheme 8.118

Derivatives of Triazines. The reduction of benzo-1,2,3-triazin-4(3*H*)-one **204** in acidic solution yields indazolinone **205** (Scheme 8.118) (410). In aqueous solutions, 5,6-diphenyl-1,2,4-triazine-3-one **206** is reduced as in Scheme 8.119 (331). In aqueous media, 1,2,4-triazine-3,5-diones **207** are reduced to dihydro derivative **208** (366) (Scheme 8.120).

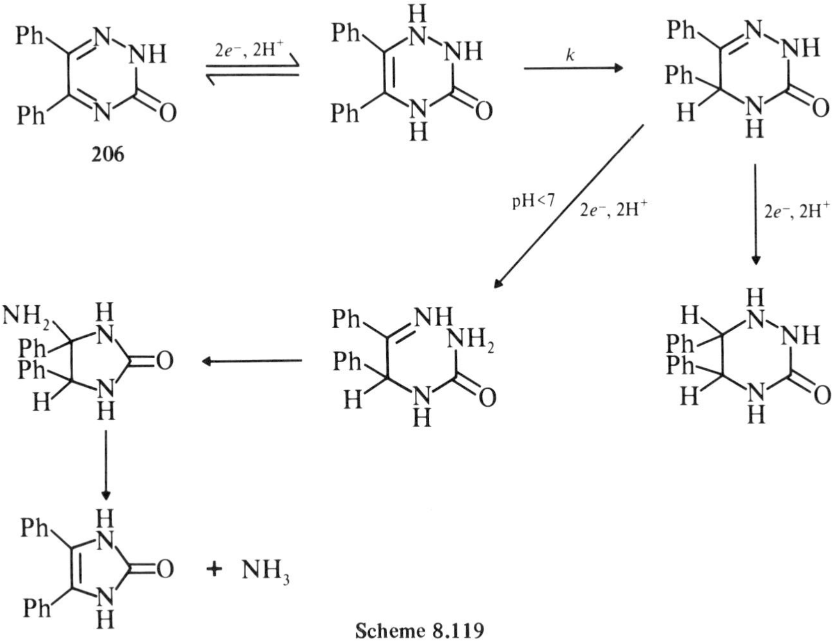

206

Scheme 8.119

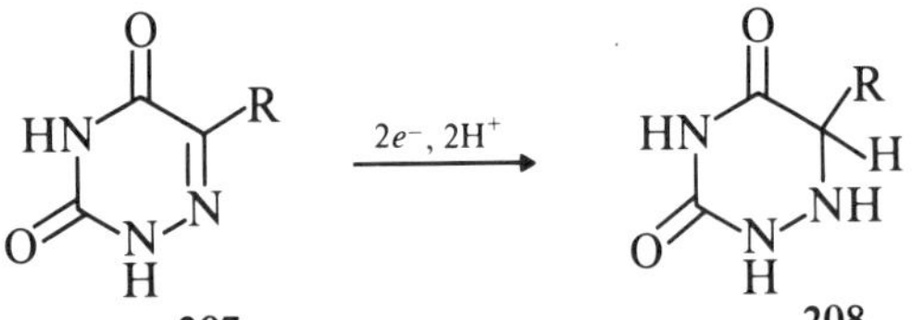

207

208

Scheme 8.120

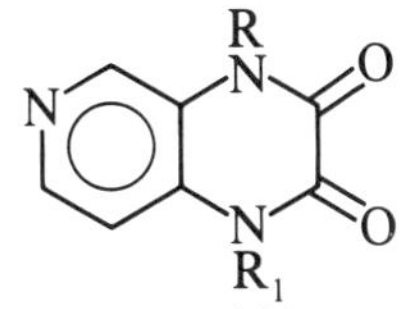

Reduction of Heterocyclic Compounds 511

209

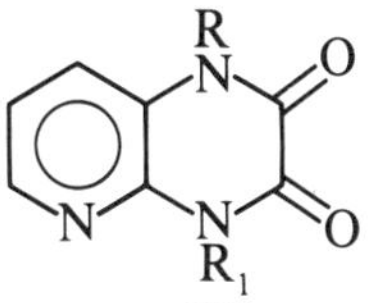

210

Derivatives of Pyridopyrazines. 2,3-Dioxotetrahydropyridopyrazines **209** and **210** are reduced similarly to 2,3-dioxo-1,2,3,4-tetrahydroquinoxaline (see under "Derivatives of Quinoxaline"), and dihydro and tetrahydro derivatives are obtained along with small amounts of dimeric products (411).

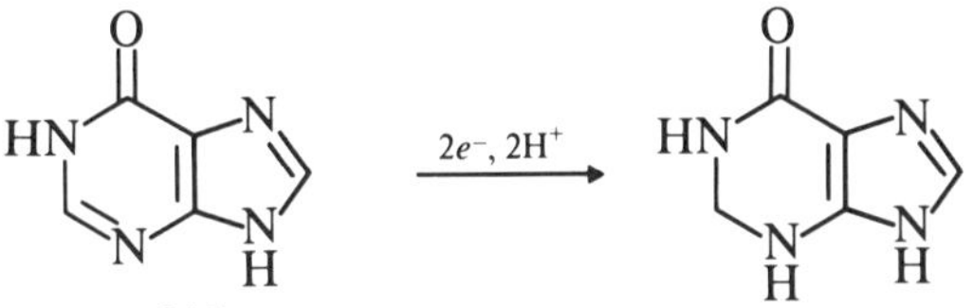

211 Scheme 8.121

Derivatives of Purine. The electrochemistry of purine derivatives has been reviewed in references 347 and 348. In aqueous solutions, hypoxanthine **211** gives a two-electron polarographic wave as in Scheme 8.121 (412). The cation **212** is reduced to dihydro derivative **213** (Scheme 8.122) (413).

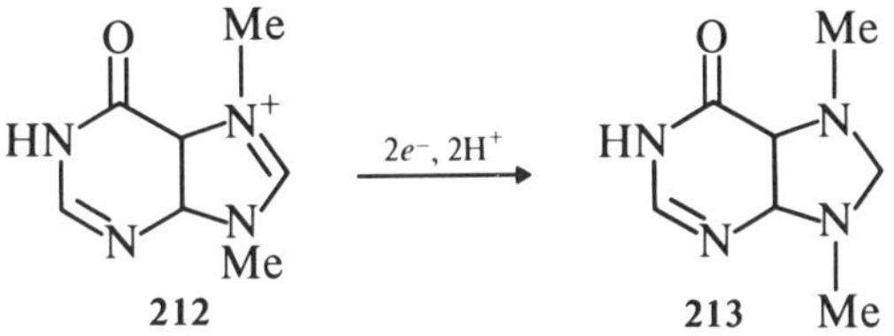

212 **213**

Scheme 8.122

In aqueous solutions, allopurinol **214** is reduced in a two-electron, two-proton process at pH > 2 and in a four-electron, four-proton path at pH < 2 (414) (Scheme 8.123). In aqueous solutions, 2-oxopurine **215** is reduced via two

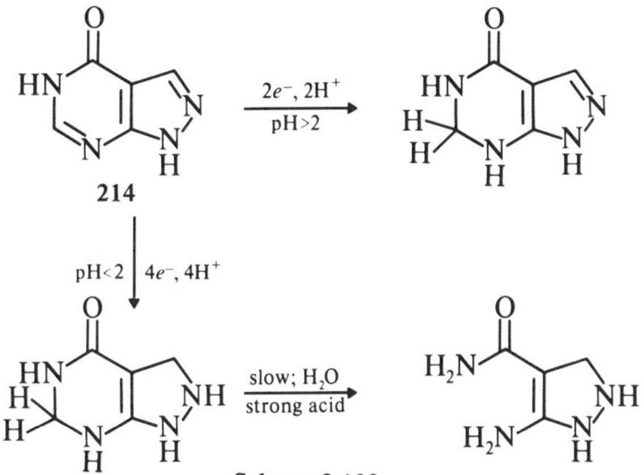

214

Scheme 8.123

one-electron steps, the first one leading to a radical that dimerizes into 6,6'-bis(1,6-dihydro-2-oxopurine) and the second to 1,6-dihydro-2-oxopurine (415) (Scheme 8.124).

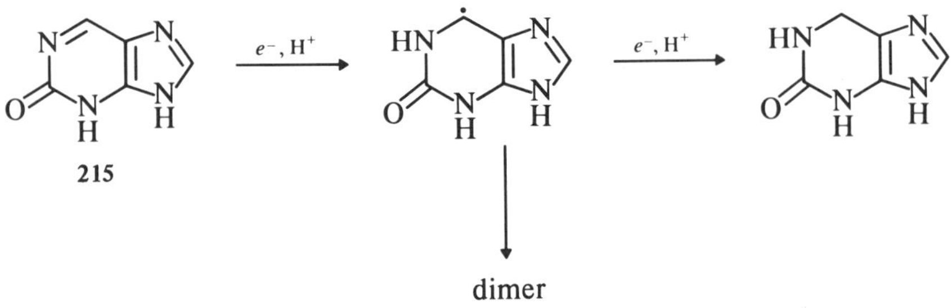

Scheme 8.124

Derivatives of Pteridines. The electrochemistry of pteridine derivatives has been reviewed (416). Pteridine-6(5H)-one 216 and pteridine-7-(8H)-one 217 are reduced in aqueous solution to the dihydro derivative (417) (Scheme 8.125).

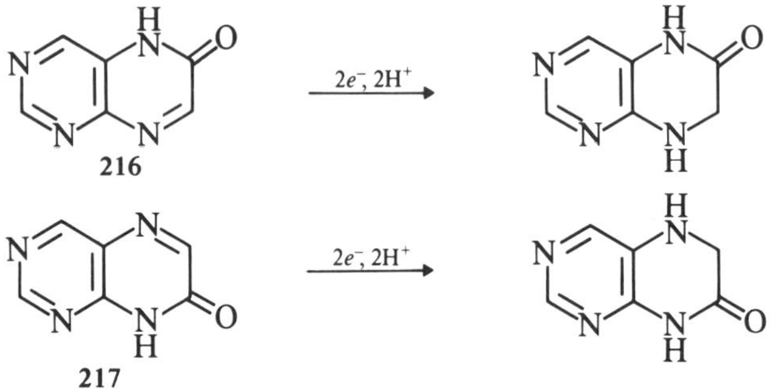

Scheme 8.125

Many amino-2-pteridine 4-(3H)-ones 218 have been studied (418, 419) in aqueous solutions. The reduction is similar to that of quinoxalines (see Sec. 8.2.1.4.9) in neutral and alkaline solutions. A two-electron, two-proton reversible process gives 5,8-dihydro derivatives that isomerize into 7,8-dihydro derivatives and are further reduced to 5,6,7,8-tetrahydro derivatives. In the case of folic acid (419), the reduction of the 7,8-dihydro derivative in acid solution cleaves the C_9-N_{10} bond.

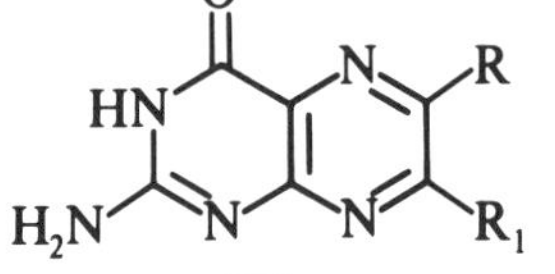

218

In aqueous solution, the reduction of 6,7-dioxo-5,6,7,8-tetrahydropteridine 219 (411) is similar to that of 2,3-dioxo-1,2,3,4-tetrahydroquinoxaline (see discussion on derivatives of quinoxaline) (Scheme 8.126).

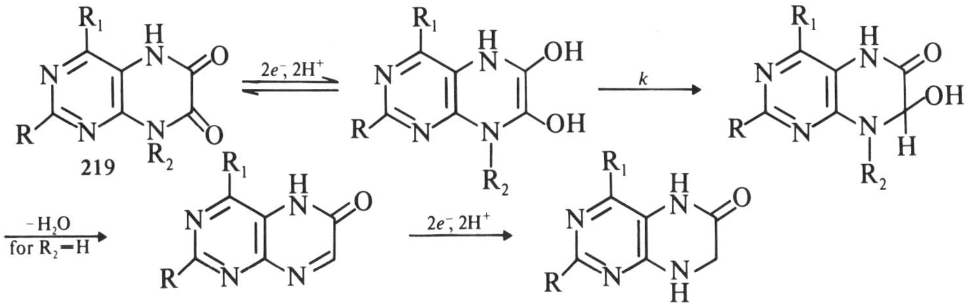

Scheme 8.126

Flavins (Isoalloxazines). The electrochemistry of flavins has been reviewed (420). In aqueous solutions, flavin **220** and its dihydro-1,5 derivatives **221** form a redox system (Scheme 8.127).

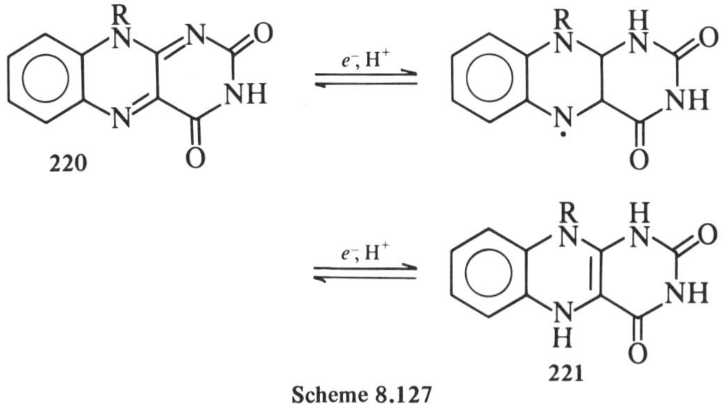

Scheme 8.127

The oxidation-reduction chemistry of the N^5-ethyl-3-methyllumiflavinium ion (FlEt⁺), its reduction products, its hydroperoxide (4a-FlEtOOH), and the adduct formed by the combination of the N^5-ethyl-3-methyllumiflavo radical (FlEt·) and superoxide ion ($O_2^{\dot{-}}$) in dimethylformamide has been determined by cyclic voltammetry, controlled potential coulometry, UV-visible spectroscopy, and ESR spectroscopy (420a). The FlEt⁺ cation exhibits four reduction steps with electron stoichiometries for the first three steps of 0.5, 1.0, and 1.5 electrons per molecule, respectively, and a single reversible one-electron oxidation at +1.04 V. Electrochemical measurements indicate that several binuclear adducts are formed during electrolytic reduction of FlEt⁺; these include (FlEt⁺)(FlEt·), (FlEt·)₂, and (FlEt·)(FlEt⁻). Combination of FlEt and $O_2^{\dot{-}}$ results in the transient formation of the N^5-ethyl-3-methyllumiflavoperoxide anion (4a-FlEtCOO⁻). The species 4a-FlEtOOH and 4a-FlEtOO⁻ are effective reaction mimics for flavo mono- and dioxygenases.

8.2.2.1.2. *N-Oxides, Nitroxides, and N-Hydroxy Derivatives.* The electrochemistry of *N*-oxides and *N*-hydroxy derivatives has been reviewed (421).

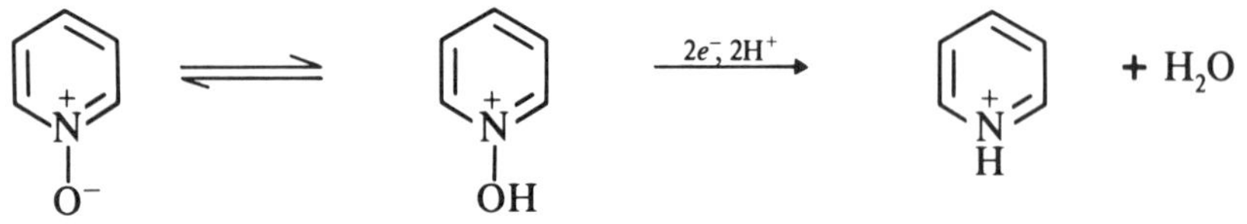

514 Electrochemical Behavior of Heterocyclic Compounds

Scheme 8.128

Heterocycles with One Nitrogen Atom. In aqueous solution, pyridine *N*-oxide 222 is reduced to pyridine (422–424); the species reduced at the electrode is the protonated form (Scheme 8.128). Other pyridine *N*-oxides behaves similarly (424). In DMF, pyridine *N*-oxide is also reduced to pyridine via an ECEC mechanism (425). In DMF, the reduction of substituted pyridine *N*-oxides gives the anion radical (426). In DMF, an extensive study of the mechanism of reduction of 4,4′-azobispyridine-1,1′-dioxide 223 has been reported (427). In DMF, 4-substituted-2-phenylquinoline-1-oxides 224 (428) are reduced in two one-electron steps in which the first step leads to a stable anion radical and the second involves an electron transfer followed by a chemical reaction to give the corresponding quinoline. In protic media, a two-electron step is observed that yields the corresponding quinoline. The polarography of *N*-oxides of quinolines (429) and acridines (430, 431) has been investigated in aqueous solution.

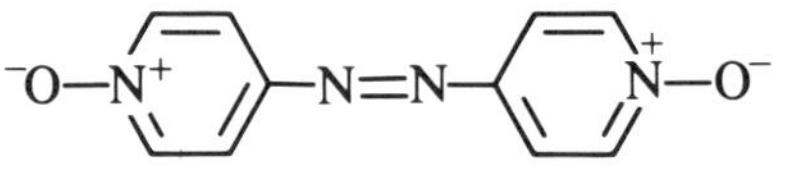

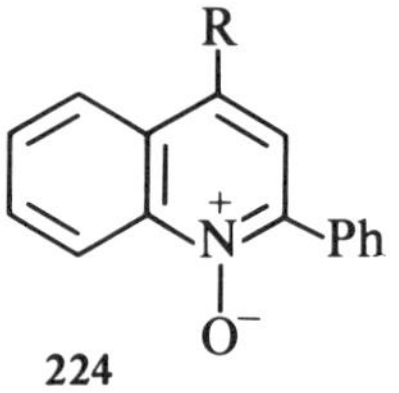

In DMF, in the presence of strong proton donors, 2-phenyl-3-arylimino indolenine-1-oxides 225 are reduced as in Scheme 8.129 (432).

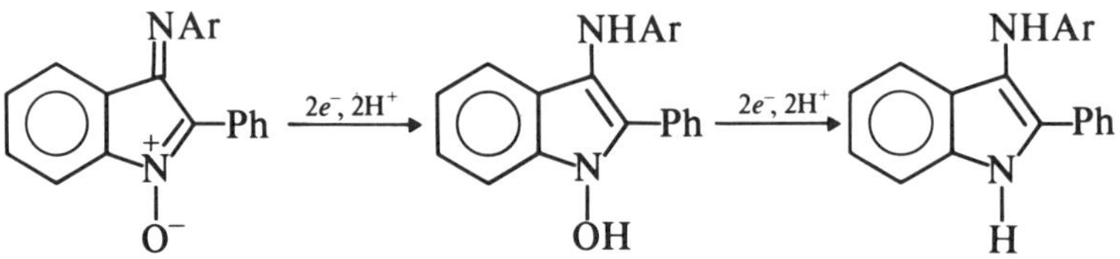

Scheme 8.129

In DMF, 1,1′-dioxy-2,2′-diphenyl-$\Delta^{3,3'}$-bi[3*H*]indole 226 is reduced to a stable radical anion (433). In the presence of proton donors the reduction gives dihydro derivative 227 (Scheme 8.130).

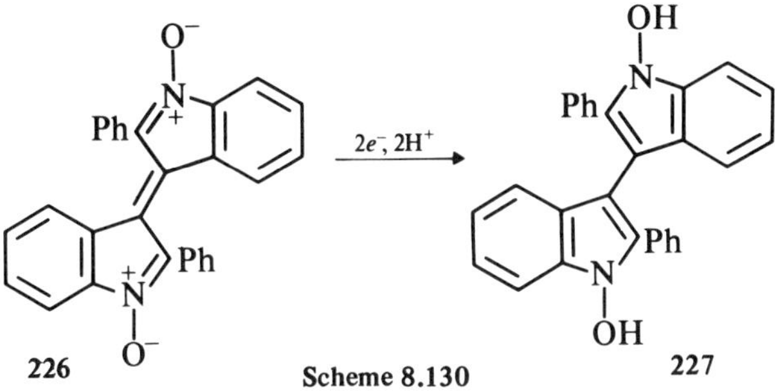

Reduction of Heterocyclic Compounds 515

Scheme 8.130

In aqueous media, ether derivatives of 1-hydroxypyridinium ion (343, 422–424) **228** are reduced to give pyridine. In aqueous solution, 2,2,6,6-tetramethylpiperidine nitroxide **229** is reduced through a one-electron, one-proton process to the *N*-hydroxy derivative (434).

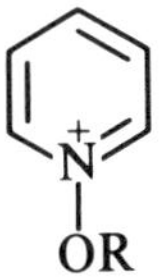

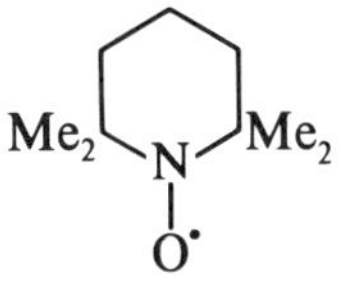

Heterocycles with Two or More Nitrogen Atoms. In aqueous media when the heterocyclic ring of heterocyclic *N*-oxides contains more than one nitrogen atom, the initial reduction involves either the nucleus or simultaneously the nucleus and *N*-oxide. The latter behavior has been observed in the case of quinazoline-3-*N*-oxide **230** (266) and cinnoline-2-*N*-oxide **231** (435). At pH < 1.5, *N*-oxides like cinnoline-1-*N*-oxide (385) and 3,6-diphenylpyridazine-*N*-oxide (343) give a two-electron wave followed by the reduction waves of the parent compound.

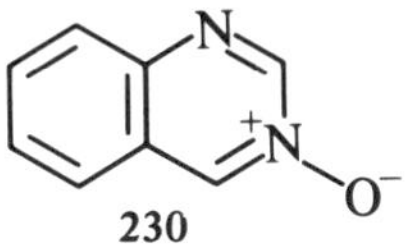

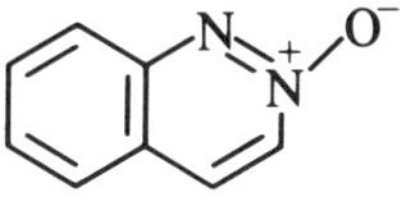

Benzo[*c*]cinnoline dioxide **232** has been investigated by polarography (436) in aqueous media. Two waves are observed, the first due to the reduction to monoxide via a two-electron, two-proton process, and the second to the reduction into 5,6-dihydrobenzo[*c*]cinnoline via a four-electron, four-proton process (Scheme 8.131).

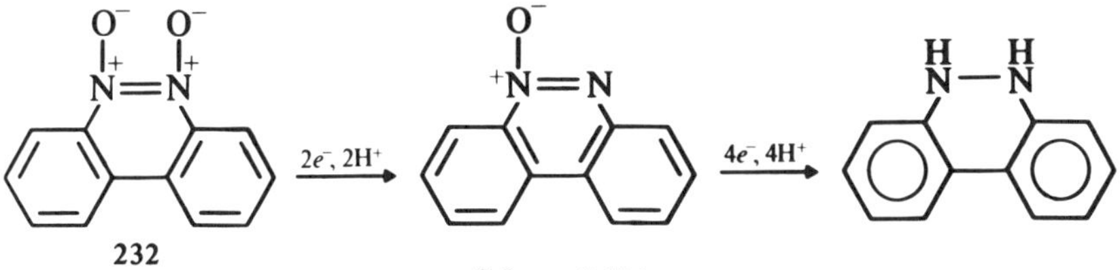

Scheme 8.131

In DMF, pyrazine-*N*-oxide **233** is reduced to pyrazine via an EECC mechanism (437). In aqueous solution, **233** is also reduced to pyrazine (437, 438). In DMF, quinoxaline-1,4-di-*N*-oxide **234** is reduced to a relatively stable anion radical (439), whereas in acid aqueous solution it is reduced via four-electron, four-proton process to quinoxaline (422). The same behavior has been reported for phenazine-5,10-di-*N*-oxide **235** in acidic aqueous solution (440). In neutral solution, 6-methoxy-1-phenazinol-5,10-di-*N*-oxide is reduced stepwise to monoxide and then to 6-methoxy-1-phenazinol (441); at pH < 3 and pH > 9, the two two-electron polarographic waves merge into one. In aqueous solutions, adenine-*N*-oxide is reduced to adenine (442). Furoxans are *N*-oxides of furazans, and their behavior has already been described in Section 8.2.1.10.2. Ethers derived from 1-hydroxybenzo[*d*]-1,2,3-triazole **236** are reduced in a two-electron, two-proton process to give the parent compounds (443).

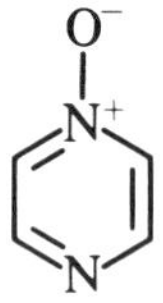

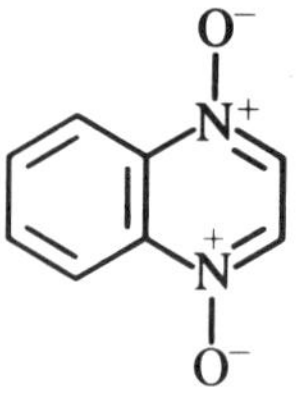

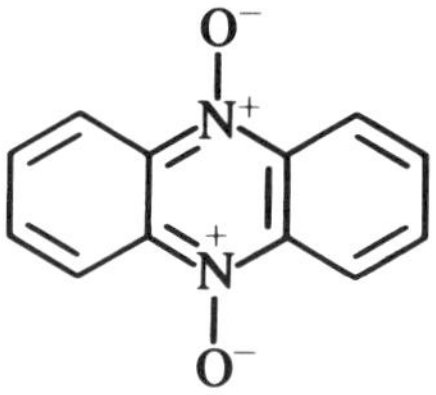

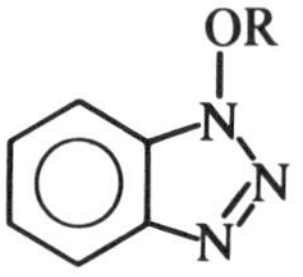

8.2.2.2. Compounds with a Substituent that has a Nitrogen Atom Attached to the Ring

8.2.2.2.1. Amino Derivatives.
Frequently, the presence of an amino group does not modify the reduction mechanism of the heterocycles, but in some cases the amino group is removed by the reduction, either by the cleavage of the C–N bond (route *A*) or by the elimination of ammonia following the reduction step (route *B*).

2-Pyridiltrimethylammonium cation **237** is reduced (343) in acid solution (route *A*) (Scheme 8.132).

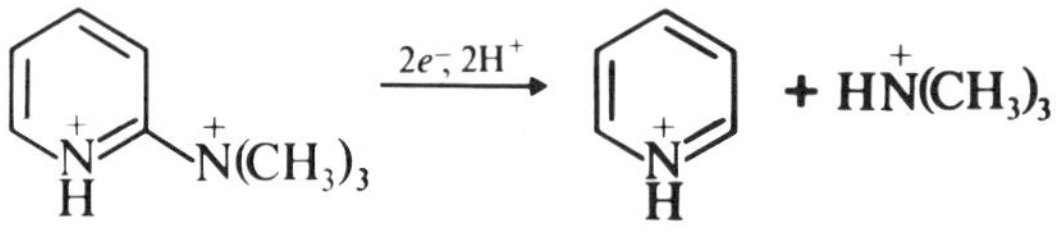

Scheme 8.132

The reduction of adenine **238** in aqueous media (444) involves the elimination of ammonia (route *B*, Scheme 8.133).

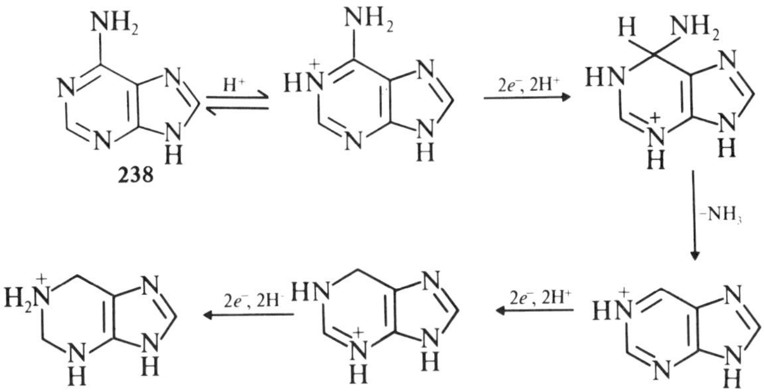

Reduction of Heterocyclic Compounds 517

Scheme 8.133

In another mechanism it has been proposed (445) that the elimination of ammonia occurs after the uptake of four electrons and four protons, but the mechanism in Scheme 8.133 appears to be more probable. Like adenine, 4-aminoquinazoline **239** is reduced in acid solution to dihydro-3,4-quinazoline (446) (Scheme 8.134).

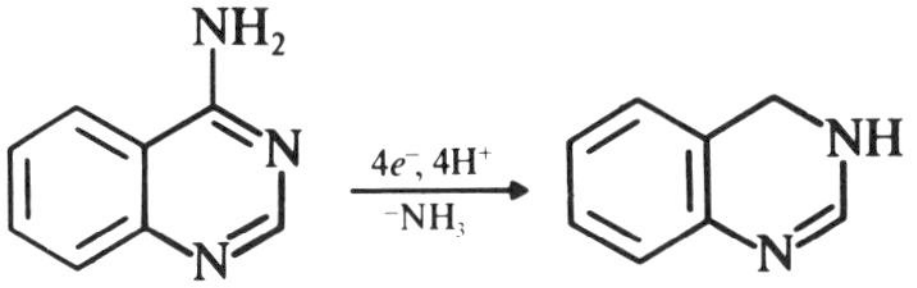

Scheme 8.134

Cytosine **240** and cytidine are also reduced in aqueous solution to a dihydro derivative that loses ammonia to form the parent compound, which then undergoes further reduction (447). In aqueous solution, the aminotriazinones **241** are reduced to dihydro derivative **242** (366) (Scheme 8.135) and aminopyrimidinotriazole **243** gives dihydro derivative **244** (366) (Scheme 8.136).

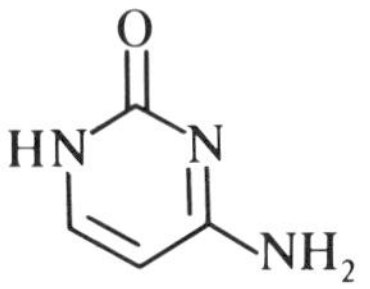

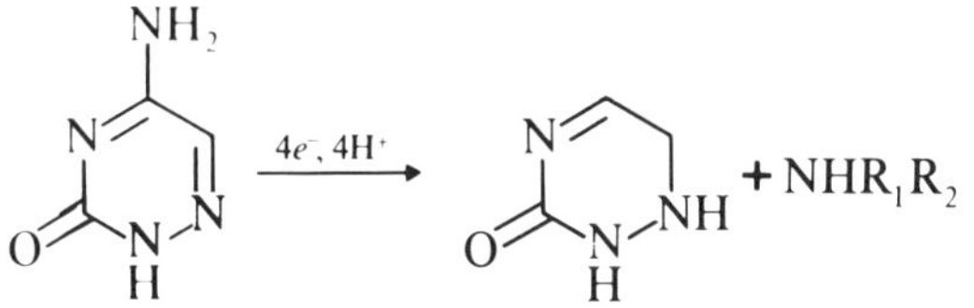

Scheme 8.135

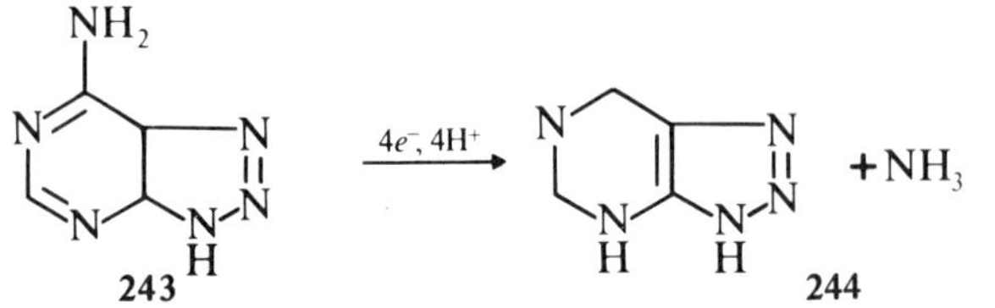

518 Electrochemical Behavior of Heterocyclic Compounds

Scheme 8.136

In acid aqueous solutions, 7-amino-6-phenylpyrazolo[1,5-*a*]pyrimidines **245** are reduced (334) to dihydro derivative **246** (Scheme 8.137). Experimental results have not allowed determination of whether route *A* or route *B* is involved in the reduction mechanism. The electrochemical reduction of **245** is a key step in obtaining new pyrazolo[1,5-*a*]pyrimidines.

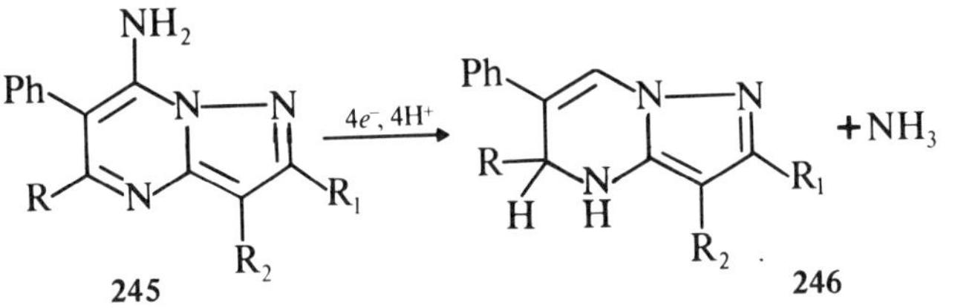

Scheme 8.137

In aqueous solution, *N*-aminopyridine is reductively cleaved to give pyridine and ammonia (see Sec. 8.2.2.2.3).

8.2.2.2.2. Nitroso Derivatives. The study of the reduction of 2-, 3-, and 4-nitropyridines showed that the corresponding nitrosopyridines are reduced reversibly to hydroxylamino derivatives (Scheme 8.138) (448, 449). This behavior is similar to that of aromatic nitroso derivatives (450). In aqueous solution, nitroso-2-quinoxaline **247** is reduced to quinoxalin-2-one oxime **248** (451) (Scheme 8.139).

In aqueous media, 4-nitrosoantipyrine **249** (452, 453) gives a four-electron polarographic wave corresponding to the reduction into 4-aminoantipyrine **251** as in Scheme 8.140. The mechanism of reduction involves (453) the dehydration

$$\text{Py—NO} + 2e^- + 2\text{H}^+ \rightleftharpoons \text{Py—NHOH}$$

Scheme 8.138

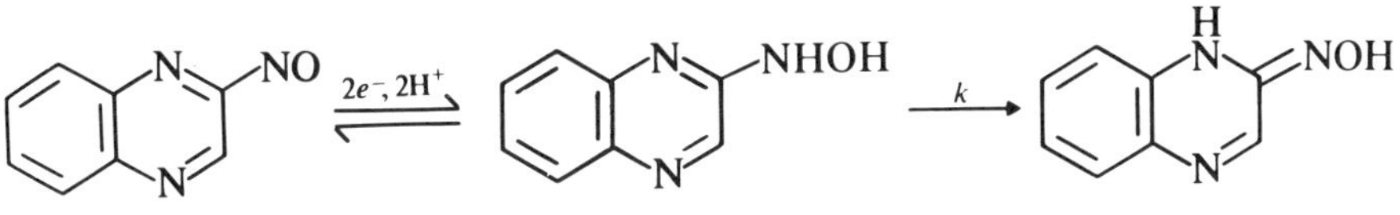

Scheme 8.139

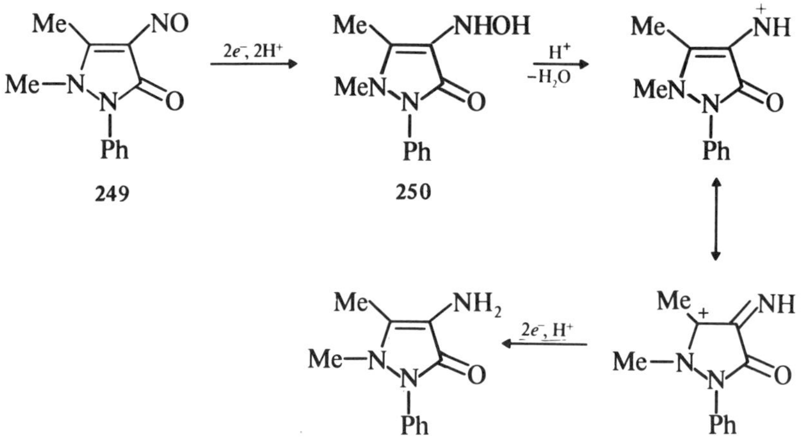

Scheme 8.140

of hydroxylamino derivative **250** to a readily reducible immonium ion. Substituted 5-nitrosouracils **252** are reduced in a similar way to 5-amino derivatives **253** (454–455a).

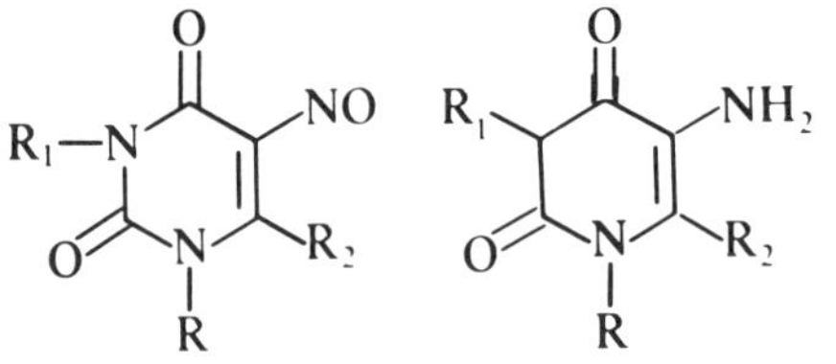

N-Nitroso derivatives of saturated heterocycles such as N-nitrosopiperidine, N-nitrosomorpholine, and N,N'-dinitrosopiperazine (456–458) behave like their open-chain analog (450). The reduction in acid solution leads, after the uptake of four electrons and four protons, to the formation of hydrazine (Scheme 8.141).

$$RR_1N-N=O \xrightarrow{4e^-, 4H^+} RR_1N-NH_2 + H_2O$$

Scheme 8.141

8.2.2.2.3. Nitro Derivatives. The reduction of aromatic and heteroaromatic nitro compounds are similar (459). In aprotic solvents, a one-electron polarographic wave corresponding to the reversible reduction to radical anion is observed first, generally followed by a second irreversible wave, but in *superdry* solvent (with activated alumina in suspension) there is a reversible reduction to a dianion (460). In aqueous solution (459), a four-electron wave corresponding to the reduction to hydroxylamino derivative is generally observed. The suggested reduction mechanism involves the intermediate formation of the nitroso derivative (Scheme 8.142).

520 Electrochemical Behavior of Heterocyclic Compounds

$$\text{RNO}_2 \xrightarrow[(1)]{2e^-, 2H^+} \text{R—N(OH)}_2 \xrightarrow{-H_2O} \text{R—NO} \underset{(2)}{\overset{2e^-, 2H^+}{\rightleftharpoons}} \text{R—NHOH}$$

Scheme 8.142

In the case of 2-, 3-, and 4-nitropyridines (448, 449) a study has shown that step 2 is slow enough at low temperature and permits reversible two-electron, two-proton reduction as in step 1.

4-Nitroantipyrine 254 is reduced to 4-aminoantipyrine via a six-electron, six-proton process (453). The suggested mechanism involves two-electron, two-proton reduction to 4-nitrosoantipyrine, which is further reduced as in Scheme 8.140. The corresponding saturated nitro compound 255 is reduced (453) in acid solution in a four-electron, four-proton process to the corresponding hydroxylamino derivative, which shows that the C_3–C_4 double bond is essential to induce dehydration of hydroxylamino derivative 250, probably because of the stabilization of the resulting immonium ion. In alkaline solution, aliphatic nitro compounds of the type RR_1CHNO_2 give a nitronate anion ($RR_1C=NO_2^-$), which is generally not reducible (459). However, the nitronate anion of 255 is reduced through a six-electron, six-proton process to the corresponding amino derivative (453).

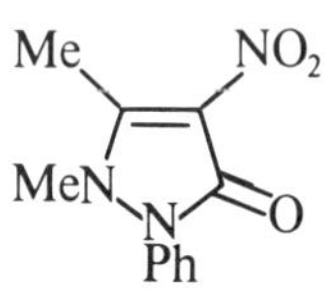

254

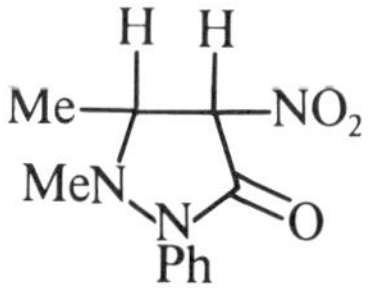

255

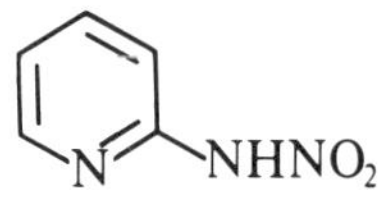

256

N-Nitro derivatives generally behave (459, 461–463) like their open-chain analogs. In acid solution, the hydrazines are produced in a six-electron reduction, and in alkaline media the two-electron reduction results in the formation of N-nitroso derivatives, which can be further reduced. N-Nitropyrazole behaves slightly differently, as the reduction in acid solution involves the cleavage of the N–N bond (464). 2-Nitraminopyridine 256 is reduced in acid solution to the hydrazine, but secondary chemical reactions occur, yielding 2-aminopyridine and 2-chloropyridine (in 2 M HCl solution) (465). In contrast with the nitramines derived from primary amines, the anion of 256 is reducible in alkaline medium. Pyridine-1-nitroimide cation 257 is reduced to pyridine (Scheme 8.143) (465).

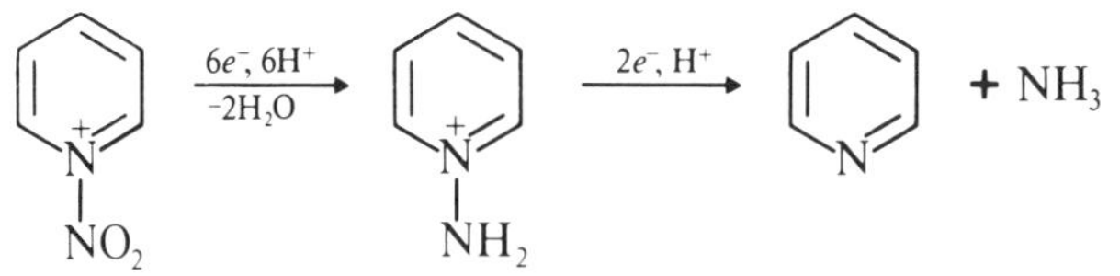

257

Scheme 8.143

Reduction of Heterocyclic Compounds 521

$$R-N=N-R \underset{}{\overset{e^-}{\rightleftharpoons}} [R-N=N-R]^{\cdot -} \rightleftharpoons [R-N=N-R]^{2-}$$
$$[R-N=N-R]^{2-} + HS \longrightarrow R-NH-\bar{N}-R + S^-$$

Scheme 8.144

8.2.2.2.4. Azo Derivatives.

Aromatic and heteroaromatic azo derivatives behave similarly (465). In aprotic solvent, two one-electron waves are observed. The first one is due to the reversible one-electron transfer, which yields the radical anion, and the second is often irreversible because the dianion is a strong base that reacts with any proton donors present in the medium (SH = solvent, residual water, etc.), presumably to give a monoanion (Scheme 8.144). Sometimes, as in

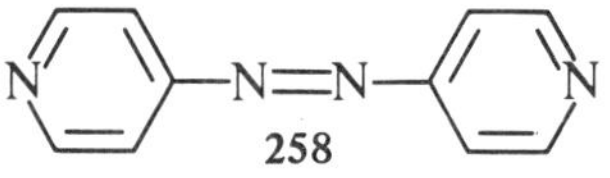

[Structure 258: 4,4'-azopyridine]

the case of 4,4'-azopyridine **258** (466) the second wave is reversible. Due to delocalization of the negative charges, the dianion is a weak base that cannot abstract a proton from a medium. In aqueous solutions, azo derivatives are reversibly reduced to the corresponding hydrazo derivatives (465) (Scheme 8.145).

$$R-N=N-R \underset{}{\overset{2e^-,\, 2H^+}{\rightleftharpoons}} R-NH-NH-R$$

Scheme 8.145

Benzo[c]cinnoline **259**, cinnoline **260**, pyridazine **261**, and phthalazine **262** may be considered as cyclic azo compounds (253); it was reported that in 1 M HCl the corresponding half-wave potentials are -0.17, -0.29, -0.51, and -0.71 V. The contribution of the N=N double bond (B) to the Kekule structure is 0.80, 0.67, 0.55, and 0.33, respectively, which correlates linearly with half-wave potential values ($E_{1/2} = -1.09 + 1.17$ B), and this is corroborated by the fact that azobenzene has $E_{1/2} \approx +0.05$ V.

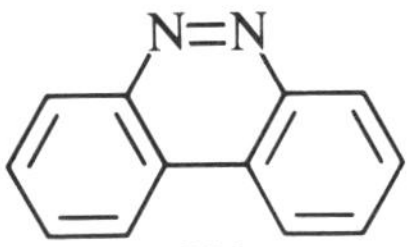

[Structures 259, 260, 261, 262]

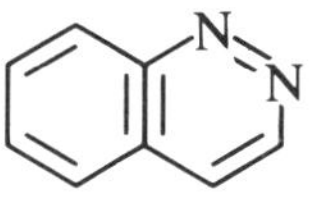

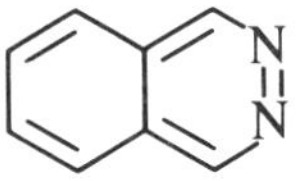

8.2.2.2.5. Other Substituents.

2-Phenyl-3-imino-3H-indole **263** is reversibly reduced in aqueous solution to 2-phenyl-3-amino-1H-indole **264** (466) (Scheme 8.146). This behavior is similar to that of α-diimines, which are reduced to

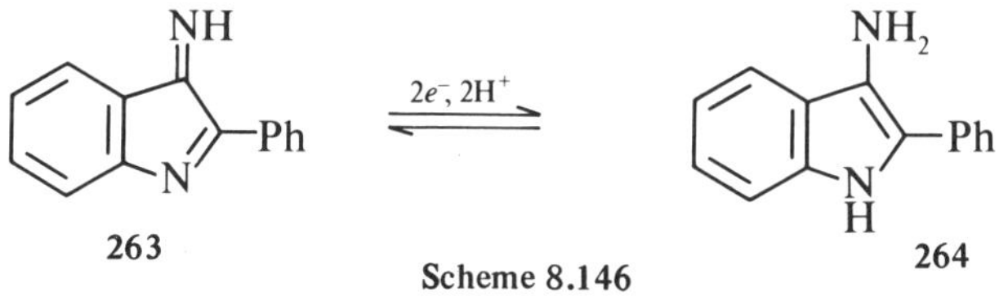

[Structures 263 and 264 with 2e⁻, 2H⁺ equilibrium arrow]

Scheme 8.146

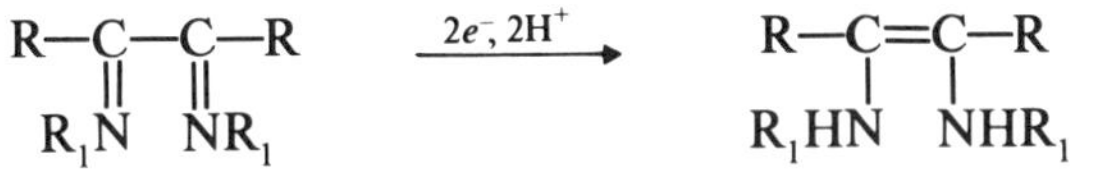

522 Electrochemical Behavior of Heterocyclic Compounds

$$\underset{R_1N\ NR_1}{R-\underset{\|}{C}-\underset{\|}{C}-R} \xrightarrow{2e^-,\ 2H^+} \underset{R_1HN\ NHR_1}{R-\underset{|}{C}=\underset{|}{C}-R}$$

Scheme 8.147

enediamines (**274**) (Scheme 8.147). The monoxime **265** is reduced to imine **263** and then to amine **264** (466).

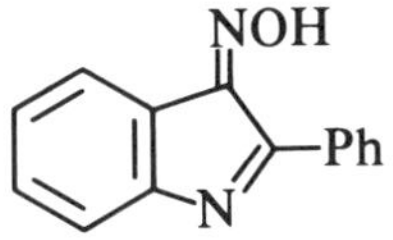

265

In DMF, bis-(2-phenyl-3-indolinone) azine **266** (467) is reduced in two one-electron steps, the first leading to the radical anion and the second to the dianion. With proton donors the two waves merge into one two-electron wave, and electrolysis gives the dihydro derivative **267** (Scheme 8.148).

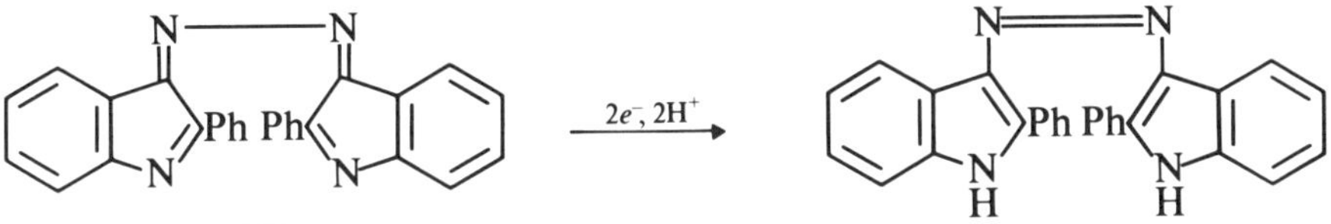

266 → **267**

Scheme 8.148

8.2.2.3. Compounds with a Substituent that has a Sulfur Atom Attached to the Ring

Aromatic and heteroaromatic disulfides behave similarly (for a review, see ref. 468). In aqueous media, 4,6-dimethyl-2-thiopyrimidine **268** and its 1-methyl derivative undergo a one-electron, one-proton process to give a radical that rapidly dimerizes

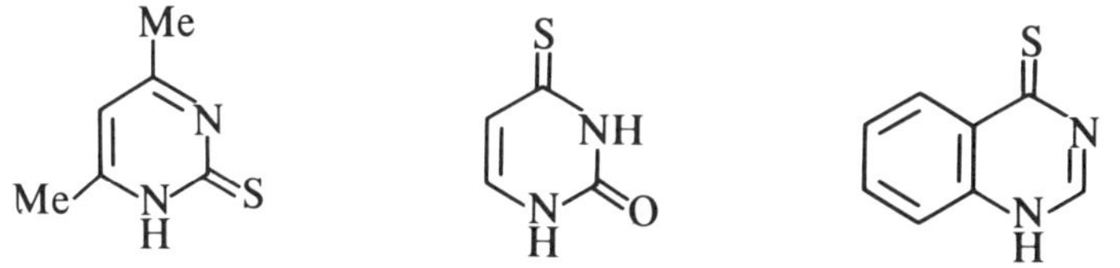

268   **271**   **272**

to give 4,4'-dimers (469). In aqueous solution, cinnoline-4-thione **269** gives a two-electron wave, but an electrolysis consumes four electrons and produces 1,4-dihydrocinnoline **270** (385) (Scheme 8.149). Similarly, reductions involving a two-electron, two-proton reduction step followed by a chemical elimination reaction take place in 2-keto-4-thiouracil **271** (470), quinazoline-4-thione **272** (446, 471), purine-2,6-dithione **273** (472), purine-6-sulfonic acid **274** (473), and

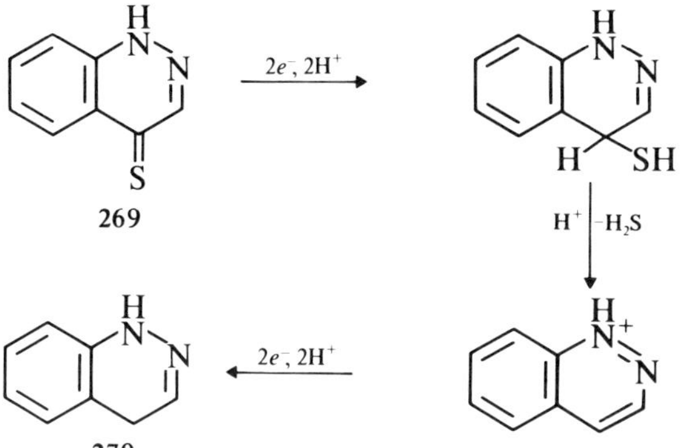

Reduction of Heterocyclic Compounds   523

Scheme 8.149

purine-6-sulphonic acid **275** (474). Purine-6-thione **276** is also reduced in a similar way, whereas purine-2-thione **277** is reduced without cleavage of the C–S bond (475) (Scheme 8.150).

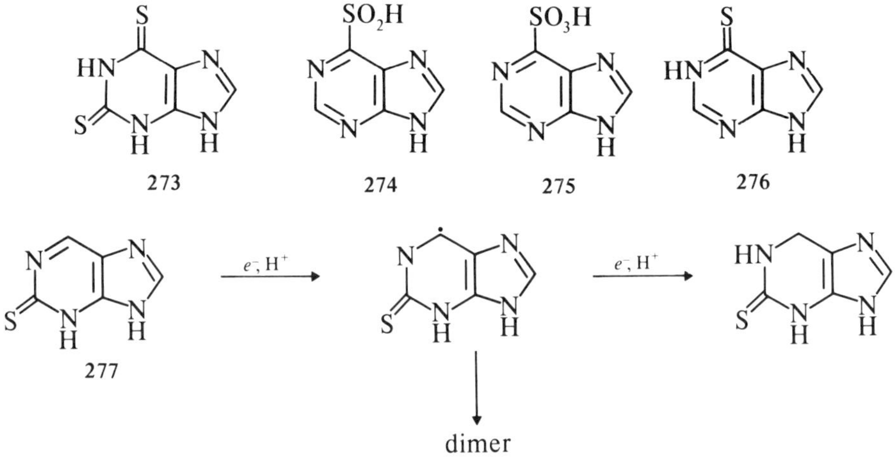

Scheme 8.150

In DMF, the thiazine thione **278** undergoes stepwise reduction to the corresponding stable radical anion and then to the dianion (475). The dianion further reacts to yield bithiazymilidene **279**. In aqueous media, 4-thiomethyluracyl and derivatives **280** are reduced (476) to the corresponding 2-pyrimidones **281**, which are further reducible (Scheme 8.151).

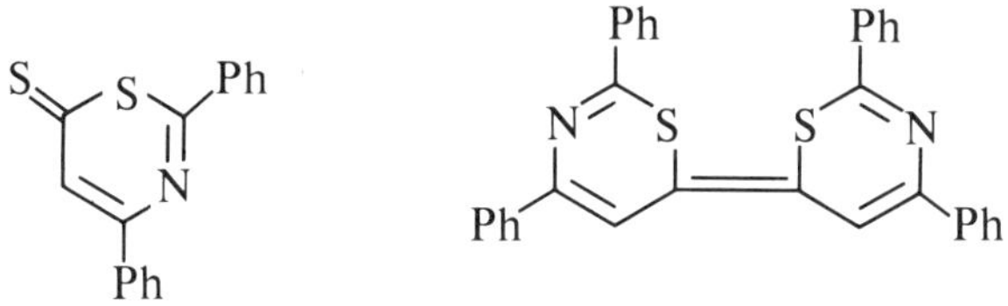

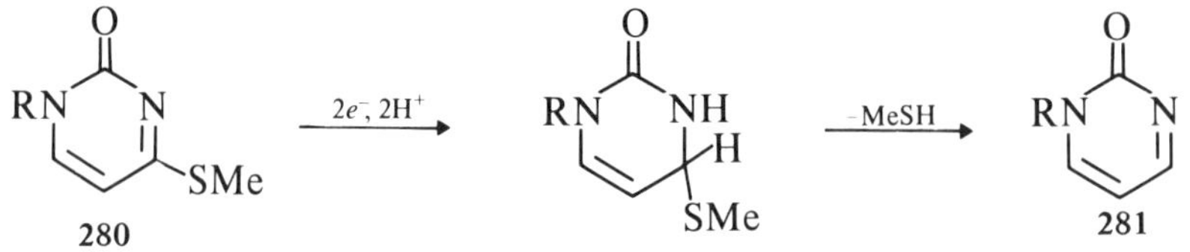

524   Electrochemical Behavior of Heterocyclic Compounds

Scheme 8.151

8.2.2.4. Halogen Derivatives

The reducibility of a halogenated heteroaromatic system depends on the heteroaromatic ring and the nature of the halogen and its position in the ring. Iodides are more easily reducible than bromides, which in turn are more easily reduced than chlorides and fluorides. Halogen substituents in positions activated toward nucleophilic attack are preferentially reduced.

When the electrode reaction is only a reductive cleavage of the carbon-halogen bond, the mechanisms are similar to that of other organic halides, which have been reviewed (477). The electrode reaction may be a reduction of the nucleus or a cleavage of the carbon-halogen bond and reduction of the nucleus.

The selectivity of electrochemical reduction frequently allows selective dehalogenation of organic halides, as in the case of 4-chloroquinazoline **282** (266), 2-amino-4-chloro-pyrimidine **283** (478), 4-iodophthalazines **284** (247), and 7-bromopyrido[2,3b]pyrazine **285** (333) and is of preparative value.

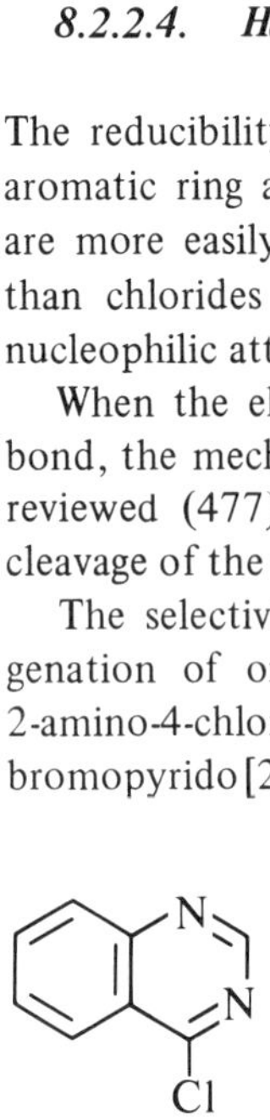

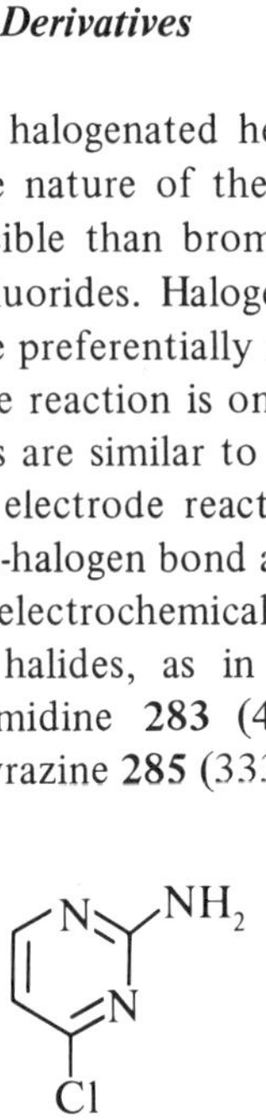

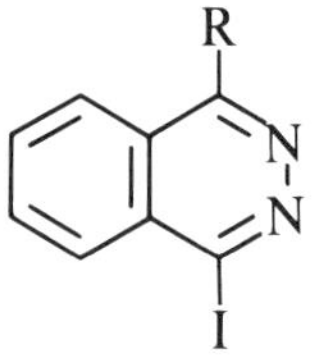

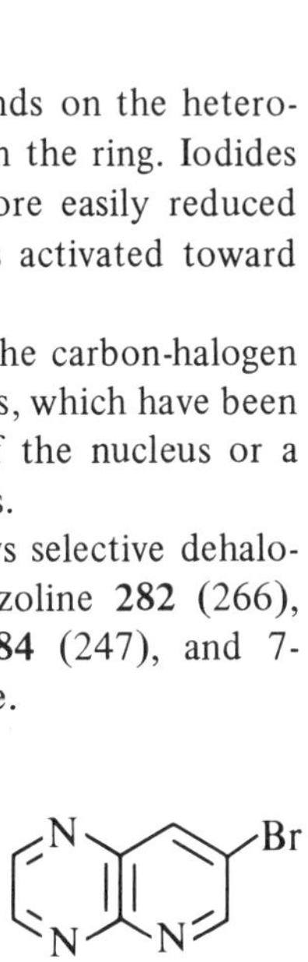

In aprotic solvents and in liquid ammonia, 2- and 6-haloquinolines **286**, 6-haloquinoxaline **287**, and 2-halophenazine **288** are reduced to anion radicals, which decompose to give halide ion and the neutral heteroaromatic radical R˙

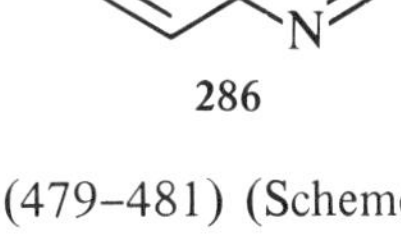

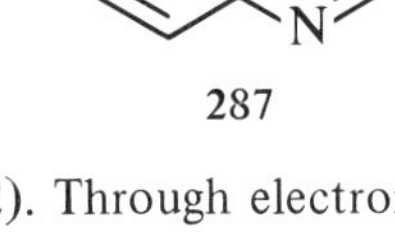

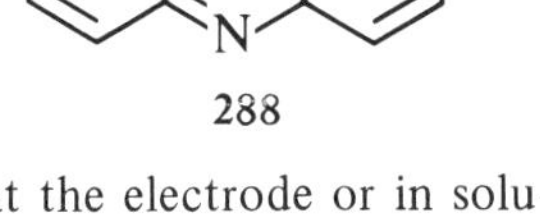

(479–481) (Scheme 8.152). Through electron transfer at the electrode or in solution and protonation, or through hydrogen atom transfer from the solvent, this neutral radical yields the parent heterocycle. In the presence of nucleophiles (480, 481), the electrochemically catalyzed aromatic nucleophilic ($S_{RN}1$) substitution takes place. The rate constant for the cleavage of the radical anion of 2-chloroquinoline has been measured by cyclic voltammetry (481).

$$R-X \xrightarrow{e^-} R-X^{\cdot -} \xrightarrow{k} X^- + R^{\cdot} \quad (+SH \longrightarrow RH + S^{\cdot})$$
Scheme 8.152   SH = solvent

In aqueous media, halopyridines are reduced to pyridine (482), and 5-chloro-, 5-bromo-, and 5-iodouracil are reduced to uracil, whereas in the case of 5-fluorouracil **289** a four-electron process is observed, showing that the reduction involves the substituent and the ring (483).

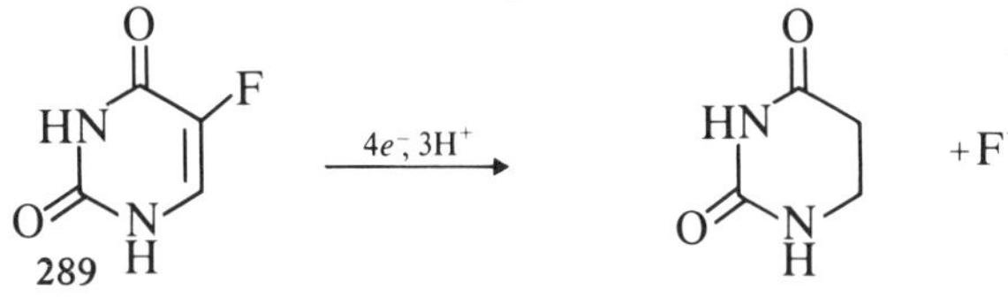

In the reduction of 4,6-dichloro-2-diethanolamino-s-triazine **290**, one chlorine is first removed and then the reduction in the ring takes place (484). The first step in the reduction of 6-chloro-7,9-dimethylpurinium cation **291** in acid solution is the removal of chlorine (351). 3-Chloro-6-methylpyridazine **292** is reduced in a two-electron reaction; the isolated product is 4,5-dihydro-6-methyl-pyridazinone. This indicates that the reduction takes place in the ring and the chlorine is lost by hydrolysis (247).

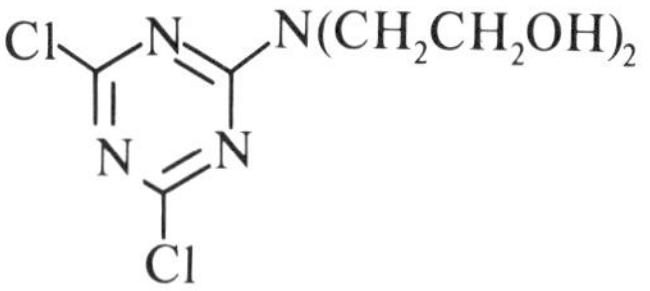

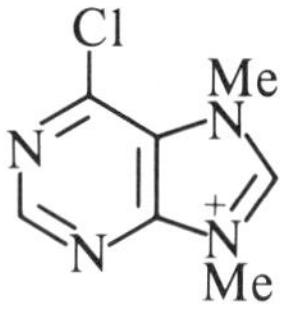

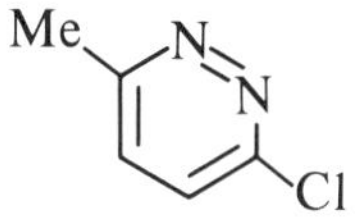

In aqueous media, 7-bromo-2,3-dimethylpyrido[2,3-b]pyrazine **293** is reversibly reduced to the 1,4-dihydro derivative, which in neutral media isomerizes to the 1,2-dihydro (or 3,4-dihydro) derivative, whereas in alkaline media it decomposes to give 2,3-dimethylpyrido[2,3-b]pyrazine, which undergoes further reduction (333) (Scheme 8.153).

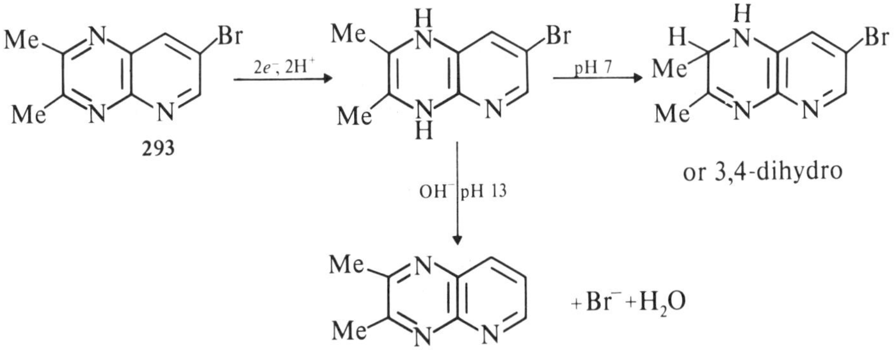

8.2.2.5. Compounds with a Substituent that has a Carbon Atom Attached to the Ring

8.2.2.5.1. Alcohols. The hydroxyl groups are reducible only when they are activated by an electron-attracting group. A pyridine ring activates its substituents particularly in the 2- and 4-positions. This is illustrated by the reduction of pyridoxol **294** in aqueous solution where only one of the hydroxyl groups is removed (485) (Scheme 8.154). The mechanism of reduction of 1-(2-pyridil) and 1-(4-pyridil) alkanols to the corresponding alkyl pyridines has been examined by electrolyzing several optically active derivatives in aqueous acid solution (486). As

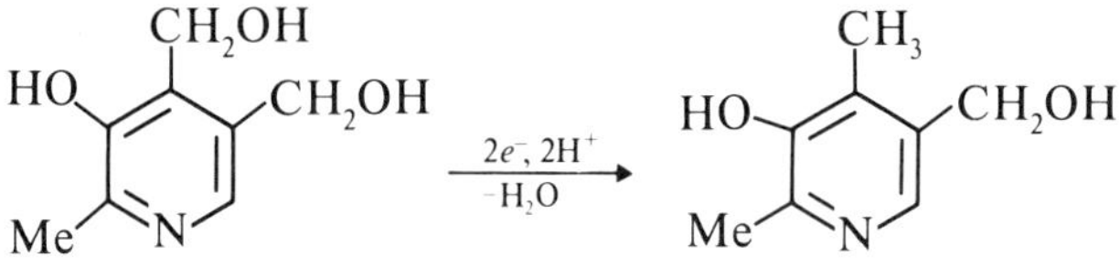

Scheme 8.154

hydroxyl groups are difficult to reduce, it is not surprising that in 2-[α-hydroxybenzyl]-3-phenylquinoxaline **295** reduction occurs in the ring to give the 1,4-dihydro derivative (487).

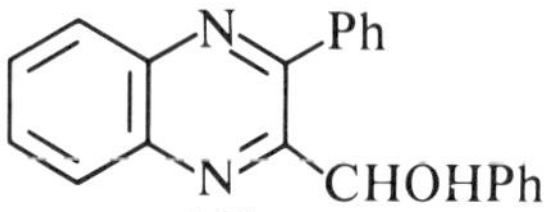

8.2.2.5.2. Aldehydes, Ketones, and Derivatives. Aldehydes and ketones are reduced to alcohols and/or dimeric 1,2-diols (Scheme 8.155). The reduction of aromatic and heteroaromatic derivatives is similar and has been reviewed (488, 489).

$$R-CO-R' \xrightarrow{2e^-, 2H^+} R-CHOH-R'$$
$$2R-CO-R' \xrightarrow{2e^-, 2H^+} R-\underset{OH}{C(R')}-\underset{OH}{C(R')}R$$

Scheme 8.155

When aldehydes bear an electron-withdrawing group, they are partially hydrated in aqueous solutions. For example, pyridine carboxaldehydes (490, 491) are hydrated to a high degree in solution, which influences their polarographic behavior because the hydrated form, a *gem*-diol, is not polarographically reducible. Hydration equilibrium constants, dissociation constants, and dehydration rate constants have been evaluated for the equilibrium (491) (Scheme 8.156). Since both heterocyclic ring and carbonyl groups are reducible, the reduction concerns one or the other

$$R-CH(OH)_2 \rightleftharpoons R-CHO + H_2O$$

Scheme 8.156

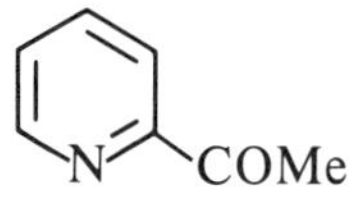

Reduction of Heterocyclic Compounds 527

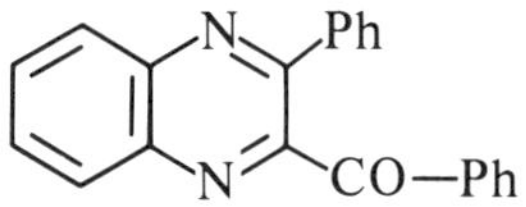

296: 2-acetylpyridine (N-COMe)

297: 2-benzoyl-3-phenylquinoxaline

depending upon their relative reducibility. In 2-acetylpyridine **296** (490, 491) the carbonyl group is first reduced, whereas in 2-benzoyl-3-phenylquinoxaline **297** the ring is first reduced to the 1,4-dihydro derivative (487). The reduction of azomethine derivatives of aromatic and heteroaromatic aldehydes and ketones, such as imines, oximes, and hydrazones, is similar and has been reviewed (488, 489). In aqueous solution, imines are reduced to amines. Oximes and hydrazones are reduced to amines via imines (Scheme 8.157). Monoimines of 1,2-diketones (2-pyridil, benzil) are reduced to eneaminols, which rearrange to α-aminoketones (492)

$$RR_1C=NOH \xrightarrow[-H_2O]{2e^-, 2H^+} RR_1C=NH \xrightarrow{2e^-, 2H^+} RR_1CHNH_2$$

$$RR_1C=N-NHR_2 \xrightarrow[-R_2NH_2]{2e^-, 2H^+} RR_1C=NH \xrightarrow{2e^-, 2H^+} RR_1CHNH_2$$

Scheme 8.157

(Scheme 8.158). The same mechanism is involved in the reduction of monoximes of α-diketones (493). In the case of dioximes of 1,2-diketones such as α-furil dioxime **298** (493, 494), the reduction gives 1,2-bis(2-furyl)ethylenediamine **299**.

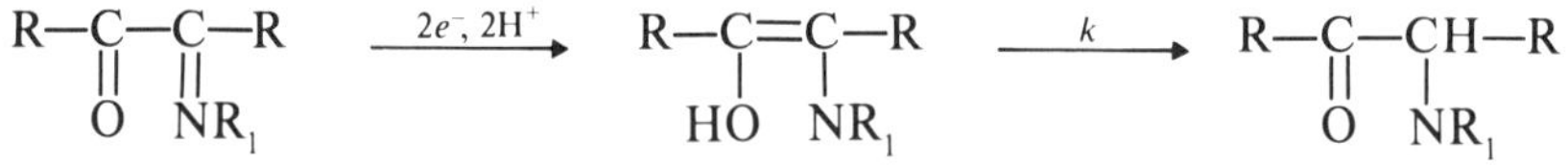

$$\underset{O\;\;NR_1}{R-C-C-R} \xrightarrow{2e^-, 2H^+} \underset{HO\;\;NR_1}{R-C=C-R} \xrightarrow{k} \underset{O\;\;NR_1}{R-C-CH-R}$$

Scheme 8.158

Two mechanisms of the reduction have been suggested. The first assumes the formation of hydroxylamino derivative **300** after the uptake of six-electrons

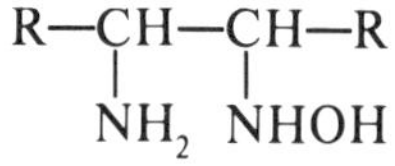

$$\underset{NH_2\;\;NHOH}{R-CH-CH-R}$$

300

(494), and the second (493) involves the formation of α-diimine **301**, which undergoes reduction as in Scheme 8.159. This assumption seems supported by the reduction of the dioxime of 1,2-dihydropyrroziline-1,2-dione **302** (495) to enediamine **303** through a six-electron, six-proton process (Scheme (8.160).

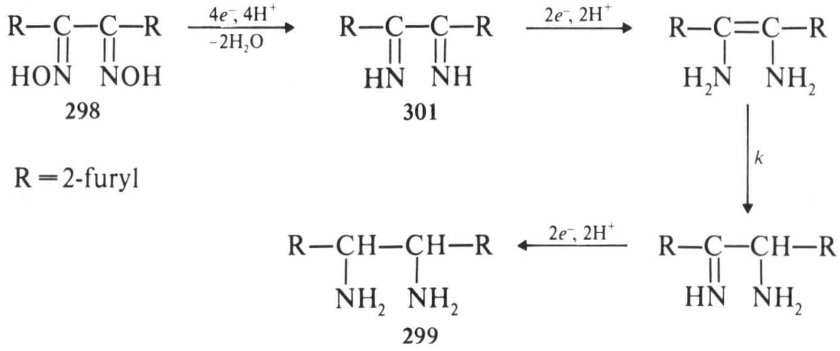

528 Electrochemical Behavior of Heterocyclic Compounds

$$R-\underset{\underset{HON}{\parallel}}{C}-\underset{\underset{NOH}{\parallel}}{C}-R \xrightarrow[-2H_2O]{4e^-,\ 4H^+} R-\underset{\underset{HN}{\parallel}}{C}-\underset{\underset{NH}{\parallel}}{C}-R \xrightarrow{2e^-,\ 2H^+} R-\underset{\underset{H_2N}{|}}{C}=\underset{\underset{NH_2}{|}}{C}-R$$

298   301

R = 2-furyl

$$R-\underset{\underset{NH_2}{|}}{CH}-\underset{\underset{NH_2}{|}}{CH}-R \xleftarrow{2e^-,\ 2H^+} R-\underset{\underset{HN}{\parallel}}{C}-\underset{\underset{NH_2}{|}}{CH}-R$$

299

Scheme 8.159

302 $\xrightarrow{6e^-,\ 6H^+}$ 303

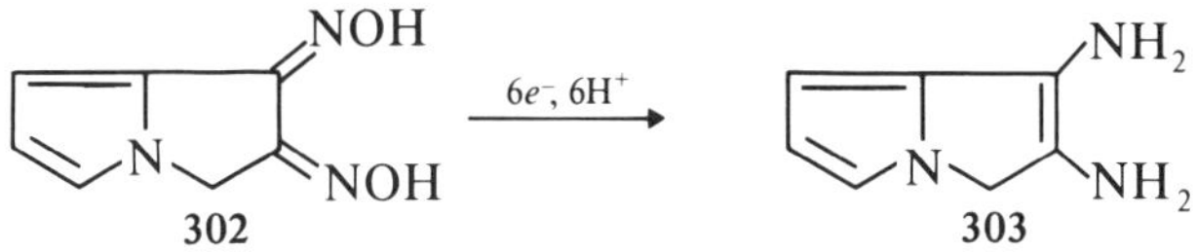

Scheme 8.160

8.2.2.5.3. Carboxylic Acids and Derivatives. Carboxylic acids and most of their derivatives are difficult to reduce and require activation by a strong electron-attracting group. The mechanisms are similar for aromatic and heteroaromatic derivatives and have been reviewed (496, 497). In the case of heteroaromatic derivatives the reduction will involve the substituent or the ring, depending on their relative reducibility, which may sometimes depend on pH. In acidic media, N-phenyl-4-pyridine carboxamide **304** gives a mixture of 4-pyridinemethanol **305** and 4-anilinomethylpyridine **306** (Scheme 8.161), whereas reduction in alkaline media gives the 1,4-dihydro derivative **307** (178) (Scheme 8.162).

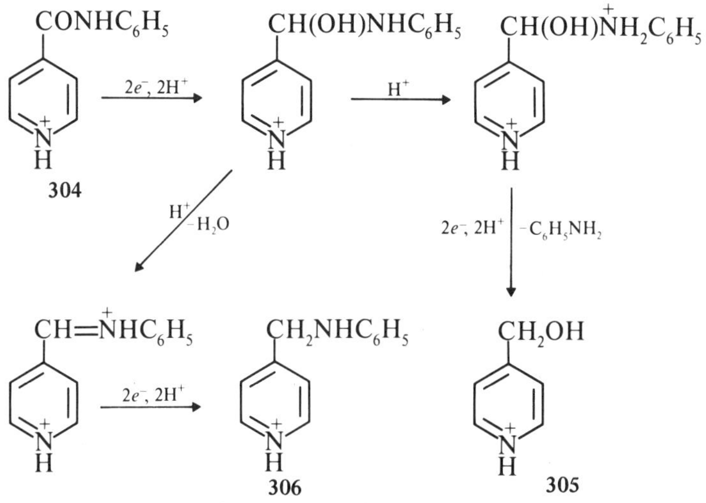

Scheme 8.161 (Acidic medium)

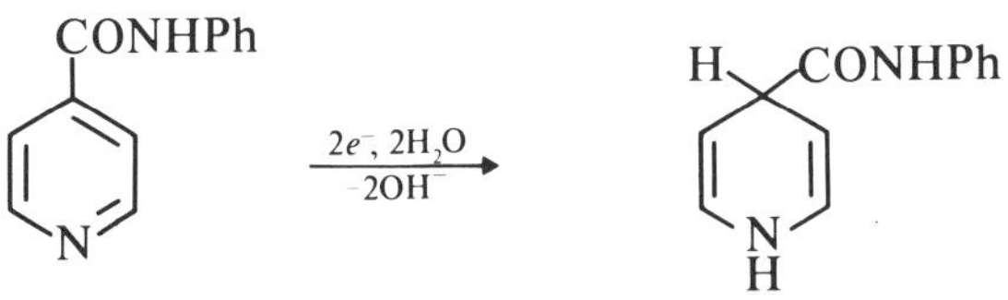

Reduction of Heterocyclic Compounds 529

Scheme 8.162 (Alkaline medium)

304 → 307 (2e⁻, 2H₂O, −2OH⁻)

The reduction of carboxylic acids yields the aldehydes, which are reduced at less negative potential, and the isolated reduction product is generally the corresponding alcohol (497). If the aldehyde is hydrated to a high degree and if the rate constant k is low enough, the aldehyde is trapped (Scheme 8.163).

$$R-COOH \xrightarrow{2e^-, 2H^+} R-CH(OH)_2 \underset{}{\overset{k}{\rightleftharpoons}} RCHO + H_2O$$

$$R-CHO \xrightarrow{2e^-, 2H^+} R-CH_2OH$$

Scheme 8.163

The reduction of esters involves a similar scheme; aldehyde or carbinol is isolated (496), and in strongly acidic media the final product is an ether (Scheme 8.164).

$$R-COOR' \xrightarrow{2e^-, 2H^+} R-CH(OH)OR' \underset{}{\overset{k}{\rightleftharpoons}} RCHO + HOR'$$

$$R-COOR' \xrightarrow{4e^-, 4H^+} RCH_2OR'$$

Scheme 8.164

Amides may also be reduced to aldehydes and then to alcohols or amines (496) (Scheme 8.165).

The corresponding aldehydes have been obtained by reduction of several carboxylic acids and their derivatives. In aqueous acidic media the reduction of several carboxylic acids and their derivatives, for example, pyridine carboxylic acids (498), imidazole-2-carboxylic acid (499), thiazole-2-carboxylic acid (500), and the ethyl esters (178, 499, 500) and amides (178, 499, 500) of these acids have been reported to yield the corresponding aldehydes, whereas the reduction of N-phenyl-4-pyridine carboxamide **304** yields only a mixture of 4-pyridine methanol and 4-anilinomethylpyridine.

$$R-CONH_2 \xrightarrow[NH_3]{2e^-, 2H^+} R-CHO \underset{}{\overset{H_2O}{\rightleftharpoons}} R-CH(OH)_2$$

$$R-CONH_2 \xrightarrow[H_2O]{4e^-, 4H^+} R-CH_2NH_2$$

Scheme 8.165

530 Electrochemical Behavior of Heterocyclic Compounds

Scheme 8.166

The reduction of nitriles depends on the pH. In the case of 4-cyanopyridine (501) a four-electron reduction is observed in acid media, leading to 4-pyridilmethylamine, whereas in alkaline media the carbon-carbon bond is cleaved to give pyridine and cyanide ion (Scheme 8.165). 2-Cyanopyridine has a similar behavior (502). In strongly acid media, the reduction of 2-cyano-1-methylpyridinium gives the corresponding aminomethyl derivative (502a).

8.3. OXIDATION OF HETEROCYCLIC COMPOUNDS

The reactions observed during oxidation of organic compounds are the counterparts of those observed during their reduction. As far as the reaction medium is concerned, the difference is between nucleophilic and nonnucleophilic media rather than protic and aprotic. The oxidation reactions can be summarized as:

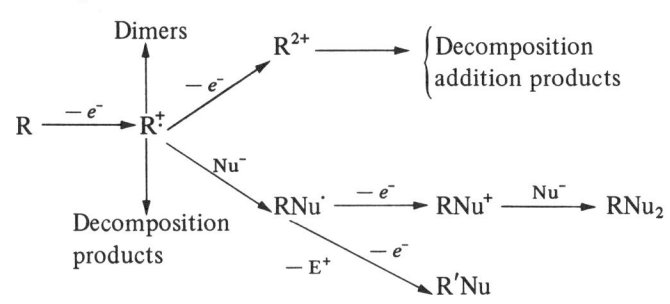

As in the case of reductions, the second electronic transfer can take place at the electrode or in solution.

8.3.1. Oxidation of Heterocyclic Systems

8.3.1.1. Compounds with One Nitrogen Atom

8.3.1.1.1. Aziridines. These compounds such as **308** are oxidized to the azaalkyl cation **309** in methanol, which reacts with the solvent to give **310** (503), which in turn undergoes further solvolytic and electrochemical reactions.

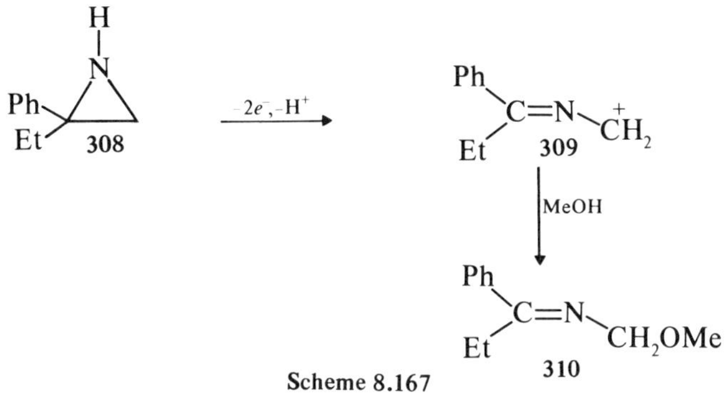

Oxidation of Heterocyclic Compounds   531

Scheme 8.167

On electrochemical oxidation, N-benzylaziridines give a macrocyclic tetramer (Scheme 8.168) (504), the cyclization mechanism of which is of particular interest. The consumption of electricity is very low (down to 0.05 F mol^{-1}) and suggests an electrochemically catalyzed reaction (Scheme 8.169).

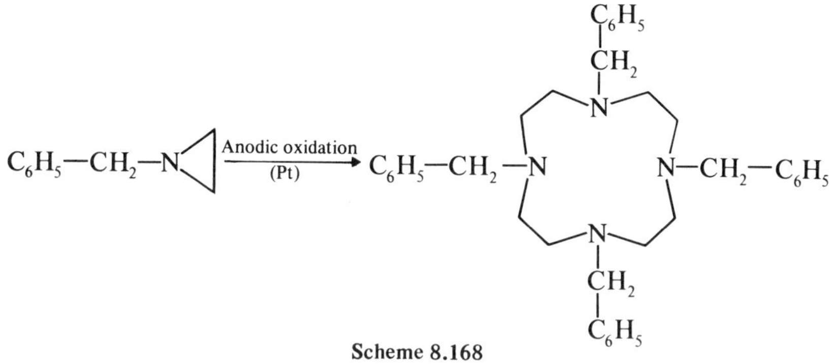

Scheme 8.168

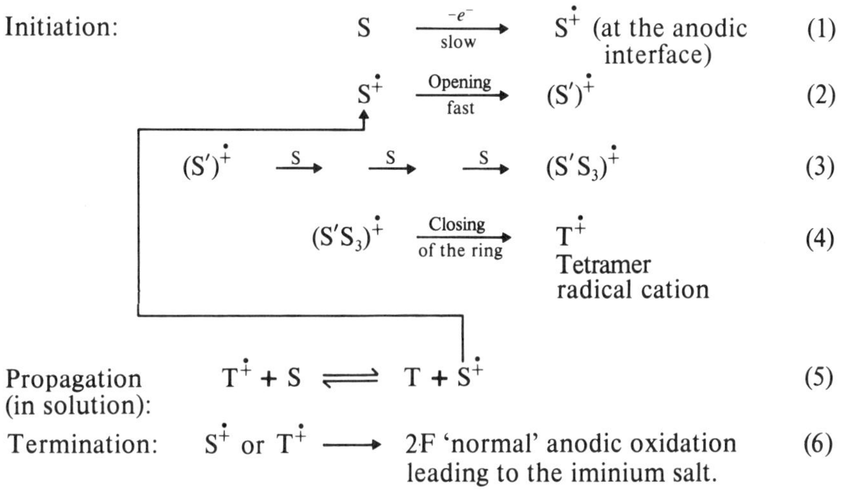

Initiation:

$$S \xrightarrow[\text{slow}]{-e^-} S^{\dot{+}} \text{ (at the anodic interface)} \quad (1)$$

$$S^{\dot{+}} \xrightarrow[\text{fast}]{\text{Opening}} (S')^{\dot{+}} \quad (2)$$

$$(S')^{\dot{+}} \xrightarrow{S} \xrightarrow{S} \xrightarrow{S} (S'S_3)^{\dot{+}} \quad (3)$$

$$(S'S_3)^{\dot{+}} \xrightarrow{\text{Closing of the ring}} T^{\dot{+}} \quad (4)$$

Tetramer radical cation

Propagation (in solution): $T^{\dot{+}} + S \rightleftharpoons T + S^{\dot{+}}$ (5)

Termination: $S^{\dot{+}}$ or $T^{\dot{+}} \longrightarrow$ 2F 'normal' anodic oxidation leading to the iminium salt.   (6)

Scheme 8.169

532   Electrochemical Behavior of Heterocyclic Compounds

8.3.1.1.2. Pyrroles. The electrochemical oxidation of pyrrole in 0.1 N H$_2$SO$_4$ or 0.1 N KOH at the Pt electrode gives 4-pyrrolin-2-one, 5-(2-pyrrolyl)-2-pyrrolidinone, 2,5-bis-(5-oxopyrrolidin-2-yl)pyrrole, and pyrrole black (505). In acetonitrile, at the Pt electrode, pyrrole has been reported to form a tar covering the electrode (506), but recently (507a, b, c; 508a–g) it has been shown that in aprotic solvents at the Pt electrode the one-step oxidation of pyrrole results in the formation of a flexible metallic polymer with *p*-type conductivity. The black films stripped from the electrode are thermally stable up to 300°C. Similarly, films of poly(*N*-methyl or *N*-phenyl)pyrrole and *N*-methylpyrrole–pyrrole copolymers have been obtained and characterized (508). However, the oxidation of α,α'-disubstituted pyrroles gives only soluble products (508c).

Pentaphenylpyrroles give fairly stable cation radicals in MeCN when the phenyl group is *p*-substituted with –H, –CH$_3$, or –OCH$_3$, and their UV-visible and ESR spectra have been studied (509). 2,3,4,5-Tetrasubstituted pyrroles have been oxidized in acidic, nonbuffered, or basic media, and the conditions have been established for the generation of cation radicals, dications, nitrenium ions, and pyrryl radicals (510). In the case of 2,5-bis(*p*-methylaminophenyl)-3,4-dimethyl pyrrole, the four species have been observed.

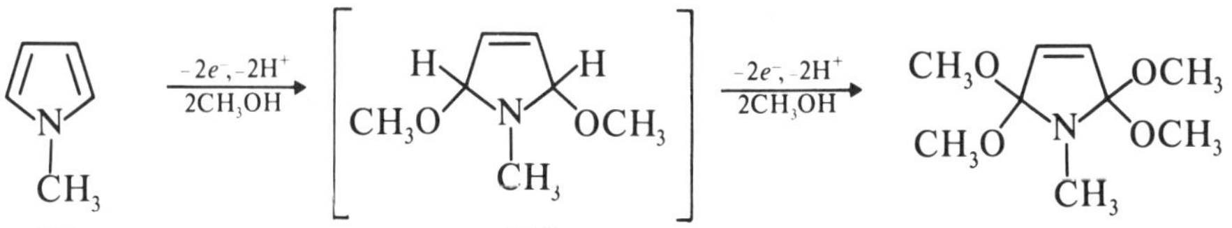

Scheme 8.170

In methanolic solutions, 1-methylpyrrole **311** forms 1-methyl-2,2,5,5-tetramethoxypyrroline **312** (511) (Scheme 8.170). 2,3,4,5-Tetraphenylpyrrole **313** shows two oxidation waves in MeCN. The radical cation, corresponding to the first wave disproportionates to give a dication, which reacts with the nucleophiles present in the solution to give 2-hydroxy-2,3,4,5-tetraphenylpyrrole **314–315** or its methyl ethyl ether (Scheme 8.171) (512). In nitromethane at low temperature the same kind of behavior is observed (513), but at higher temperatures (> 40°C) a

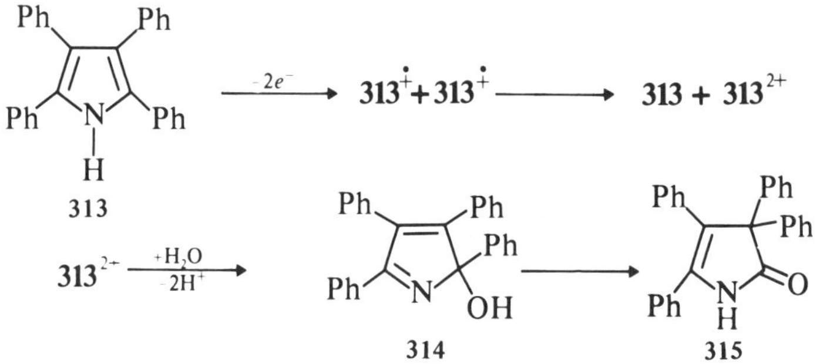

Scheme 8.171

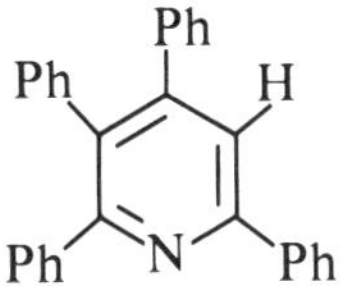
316

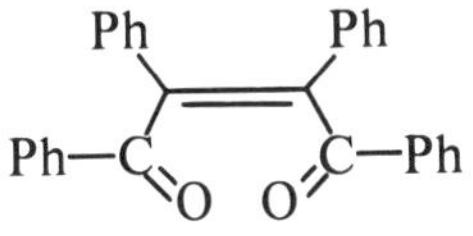
317

nitromethane molecule is incorporated in the ring to give **316**, and the nitrite ion released reacts with the starting pyrrole to give **317**. Similar results have been obtained with 2,3,4,5-tetraanisylpyrrole (514), 2,3,4,5-tetratolylpyrrole (515), 2,3,4,5-tetra(biphenylyl)pyrrole (516), and tetraarylpyrroles substituted by Me, CD_3, or Et groups (517) at the nitrogen.

1,2,3,5-Tetraphenylpyrrole **318** dimerizes in acetonitrile (518) (Scheme 8.172), but on addition of cyanide (as Et_4CN) to the solution, 1,2,3,5-tetraphenyl-2,5-dicyano-Δ^3-pyrrolline **319** is formed.

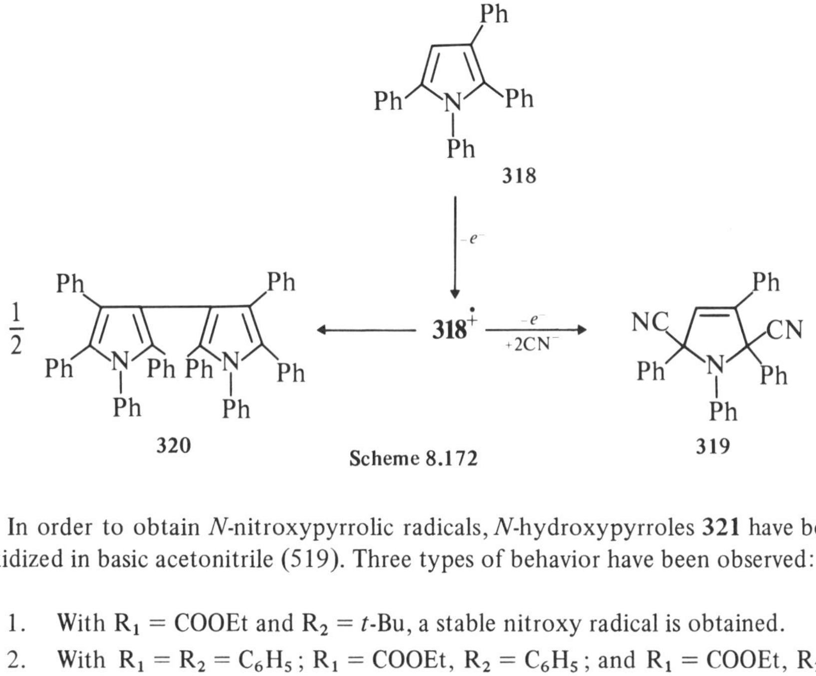

Scheme 8.172

In order to obtain *N*-nitroxypyrrolic radicals, *N*-hydroxypyrroles **321** have been oxidized in basic acetonitrile (519). Three types of behavior have been observed:

1. With R_1 = COOEt and R_2 = *t*-Bu, a stable nitroxy radical is obtained.
2. With $R_1 = R_2 = C_6H_5$; R_1 = COOEt, $R_2 = C_6H_5$; and R_1 = COOEt, R_2 = *p*-$MeOC_6H_4$; mixtures of the radicals and their dimers are obtained.
3. With R_1 = COOEt and R_2 = CH_3 or COOEt, only dimers are observed.

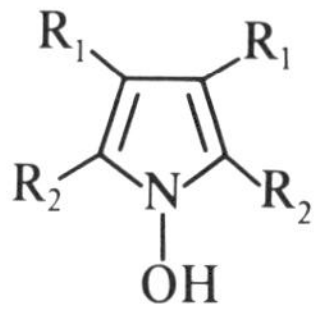
321

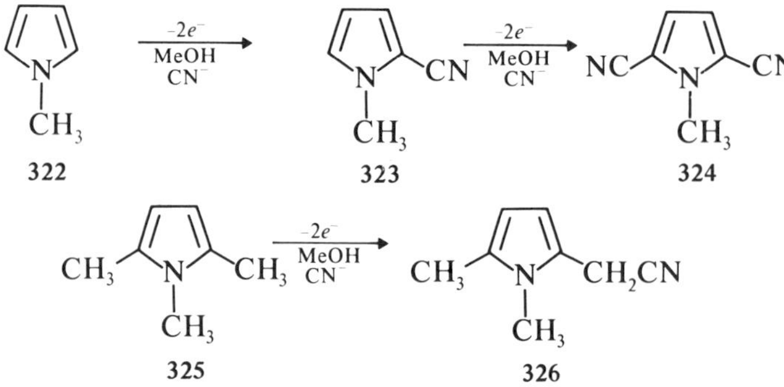

Scheme 8.173

Cyanation of 1-alkylpyrrole 322 can be achieved in methanolic NaCN (520a, b). Substitution takes place exclusively at the vacant 2- or 5-position, while 1,2,5-trialkyl derivatives 325 undergo cyanation at the side chain to form dialkylpyrrole 2-acetonitrile 326; no isocyanide or methoxylated products are obtained. Cyanation of indoles takes place exclusively at the pyrrole ring. The mechanism of ring substitution by CN⁻ has been proposed to take place by direct attack of CN⁻ on the radical cation, while the substitution at the chain involves deprotonation of the cation radical (520). Both mechanisms have been disputed by Eberson (521). In the case of ring cyanation, electron transfer would take place between the radical cation and cyanide ion to give a cyanide radical:

$$HArCH_3^{+\cdot} + CN^- \longrightarrow HArCH_3 + CN^{\cdot}$$

$$HArCH_3^{+\cdot} + CN^{\cdot} \longrightarrow H_3CAr^+ \begin{matrix} H \\ \diagdown \\ CN \end{matrix}$$

This explains why the cyanide ion, a relatively strong base, does not deprotonate the methyl group of the radical cation, while acetate ion, a considerably weaker base, does. The mechanism of the side-chain substitution has been proposed to be as in Scheme 8.174. This is supported by the isolation and characterization of the 2,5-adduct 327 obtained from the electrolysis of 325 in a divided cell where no MeO⁻ is present that could act as a base catalyst for the decomposition of 327.

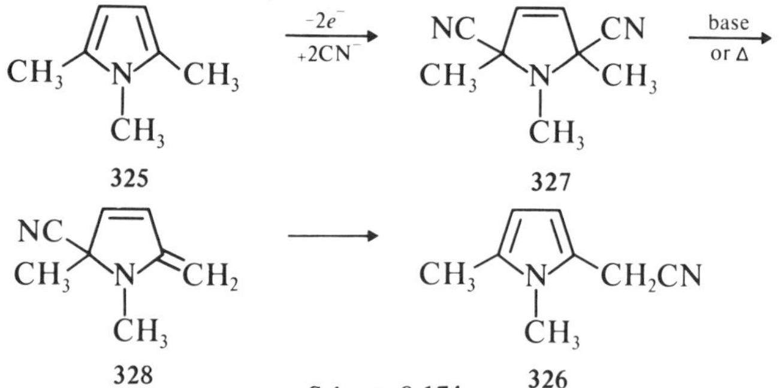

Scheme 8.174

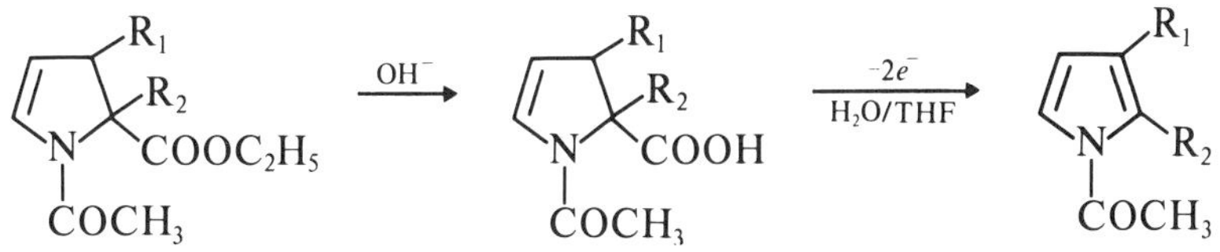

N-Acetyl-2,3-disubstituted pyrroles **331** (522) of physiological interest are synthesized from the readily available N-acetyl-2,3-substituted-Δ^4-pyrroline-2-carboxylic acid **330** (Scheme 8.175) by anodic oxidation under Kolbe electrolysis conditions. N-Formylproline **332** and derivatives can be methoxylated under Kolbe type conditions as shown in Scheme 8.176 (523).

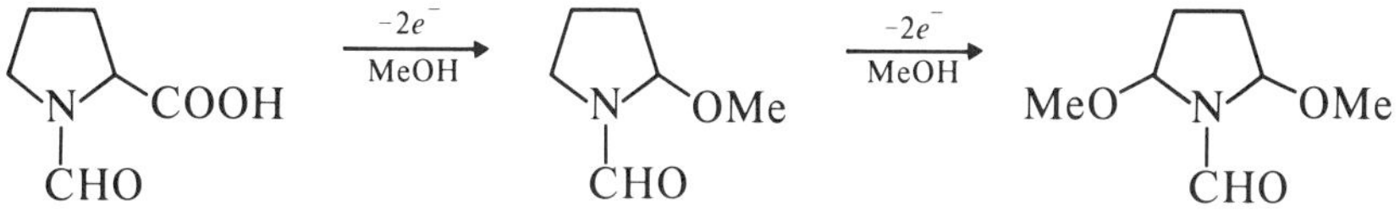

In the presence of benzaldehyde, pyrrole is oxidized to tetraphenyl porphine (506). The oxidation of tetraphenylporphirin iron(III) chloride in CH_2Cl_2/MeOH yields the methoxylated derivative **333** (524). In DMSO, bilirubin **334** is oxidized (2 F mol^{-1}) at the gold anode to biliverdin **335** (525a, b) (Scheme 8.177).

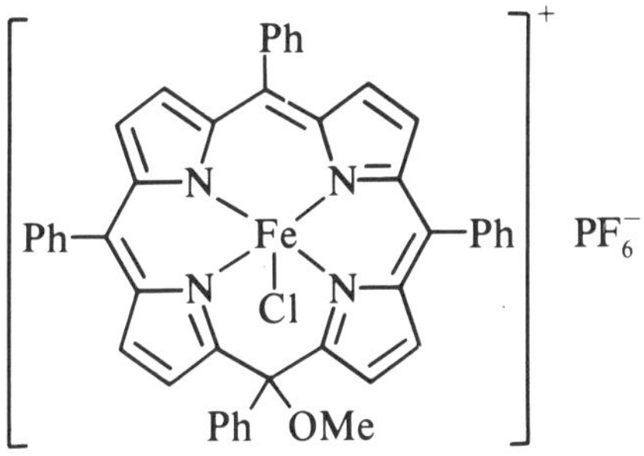

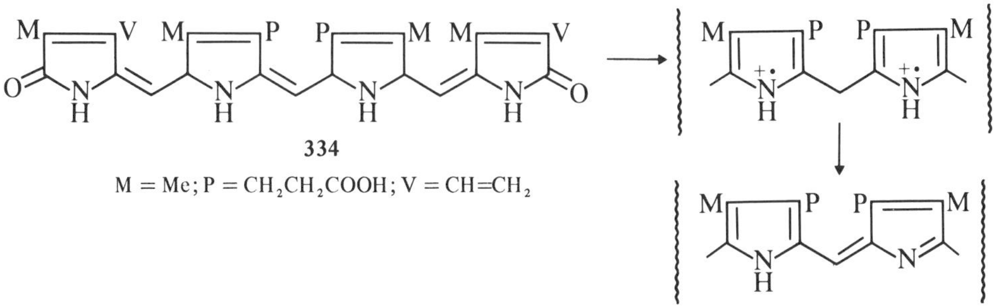

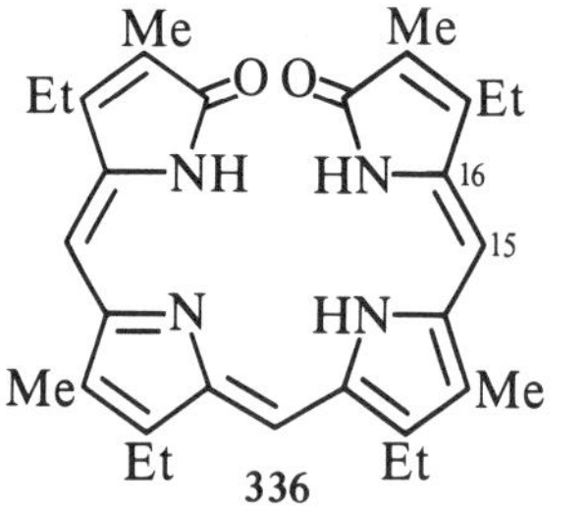

536 Electrochemical Behavior of Heterocyclic Compounds

Preparative electrolysis of bilindione **336** in a MeOH/CH$_2$Cl$_2$ solution gives the 15,16-dimethoxy derivative (526, 527). Compounds such as **337–339**, which are bile pigment models, are oxidized in MeCN at the Pt rotating disk electrode (528).

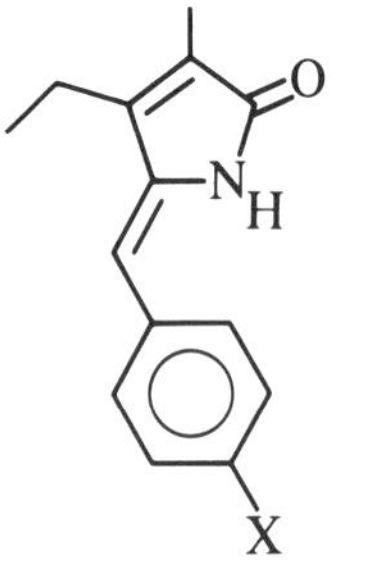

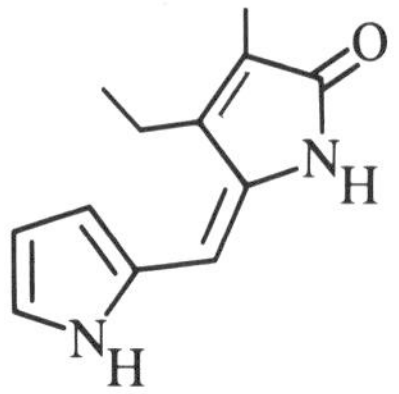

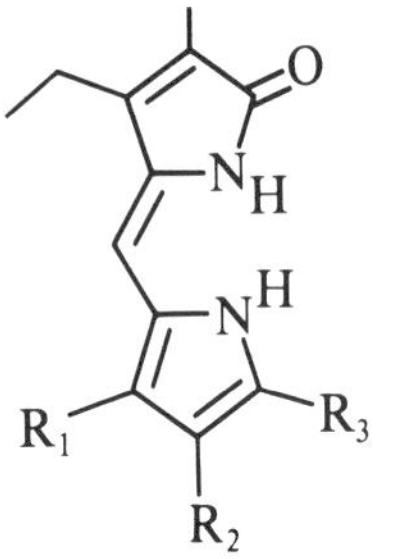

8.3.1.1.3. Pyridines. The oxidation of pyridine with perchlorate as the supporting electrolyte yields onium ion, which reacts with excess pyridine to give a brown material. The latter has been assigned the structure of *N*-(2-pyridyl)-pyridinium (529). Methoxy groups facilitate oxidation (511). In alkaline methanol, 2,6-dimethoxy pyridine **340** gives mainly 2,3,5,6-tetramethoxypyridine **341** and azaquinone **342** (Scheme 8.178). On electrochemical fluorination in HF (530), 3-chloropyridine gives very low yields of perfluorinated compounds, while 2,6-dimethylpyridine **343** (531) gives a 26% yield of perfluoro-(*N*-fluoro-2,6-dimethyl)-pyridine **344**.

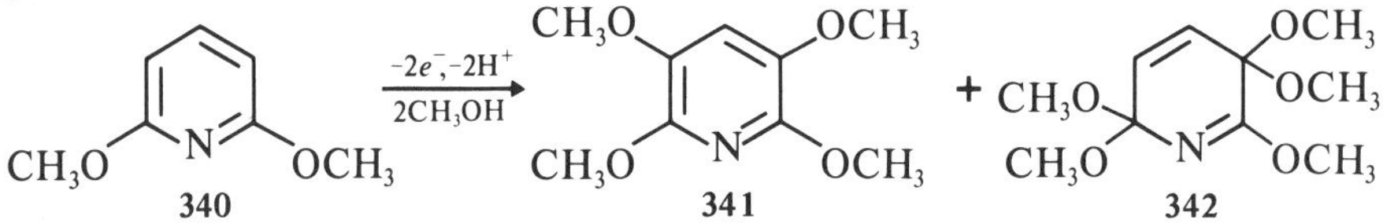

Scheme 8.178

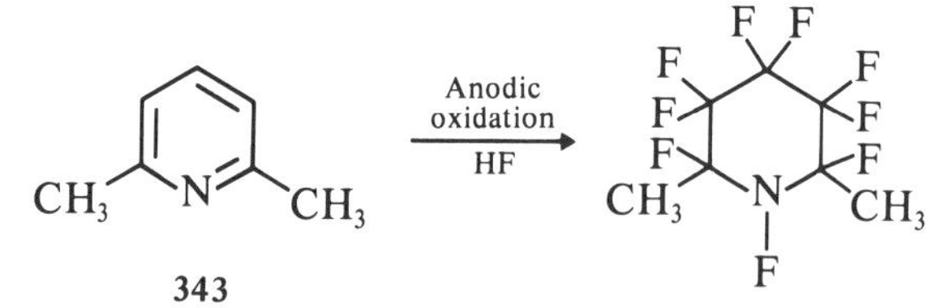

8.3.1.1.4. Dihydropyridines.

1,4-Dihydropyridines are important compounds because they possess the same basic skeleton as 1,4-dihydronicotinamide, the basic structural block of the coenzyme NADH, which takes part in the electron transfer chain as the NAD⁺/NADH couple **345, 346**.

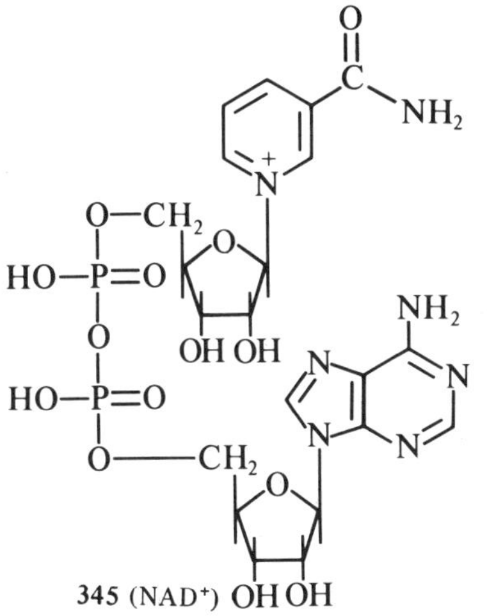

345 (NAD⁺)

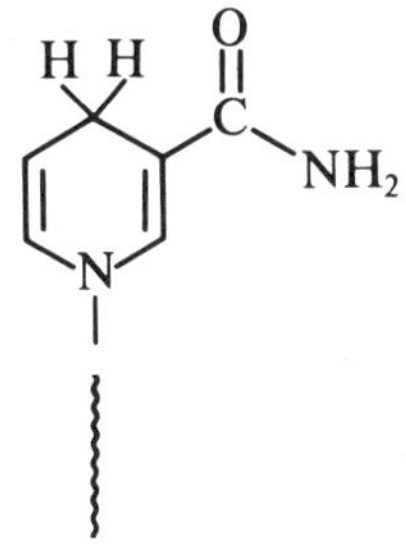

346 (NADH)

Hydride ion transfer mechanisms have been proposed for the chemical oxidation of NADH (532), but several reports indicate that NADH can act as an electron-transfer reagent (533–537). This is why there has been sustained interest in the electrochemical oxidation of NADH and model compounds.

The electrochemical investigation of model compounds such as **347a** (R = Me, Pr, Bz) and **347b** (R = 2,6-dichlorobenzyl) has been carried out in MeCN (538). It has been proposed that in buffered acetonitrile **347** undergoes one-electron transfer to give a radical cation **347$^{\ddot{+}}$**.

The radical cation can disproportionate to give **348** (the oxidized NAD⁺ form of the model compound) and the protonated form of **347**. The latter species undergoes further chemical reactions, which accounts for the nonintegral number of electrons observed in the coulometric experiments. In the presence of a base, **347$^{\ddot{+}}$** is deprotonated to give a neutral radical that undergoes a further electron transfer (ECE mechanism) leading to the overall two-electron reaction

$$347 + B \longrightarrow 348 + 2e^- + BH^+$$

In MeCN, it is possible to obtain the ESR spectrum of **349$^{\ddot{+}}$** (no deprotonation can occur at the 4-position) produced by the electrochemical oxidation at the Pt electrode (539). The investigation of the oxidative reaction mechanism of several 1,4-dihydropyridines by electrogenerated chemiluminescence suggests an ECE mechanism, the deprotonation of the radical cation (analog of NADH⁺) by the dihydropyridine itself and the oxidation of the neutral radical (analog of NAD·) (Scheme 8.179) (554). The electrochemical oxidation of other dihydropyridine derivatives in MeCN has been investigated (Scheme 8.180) (540, 541), and the electrochemical preparation of pyridines **350** from dihydropyridines has been patented in 45–68% yield (542a, b, c).

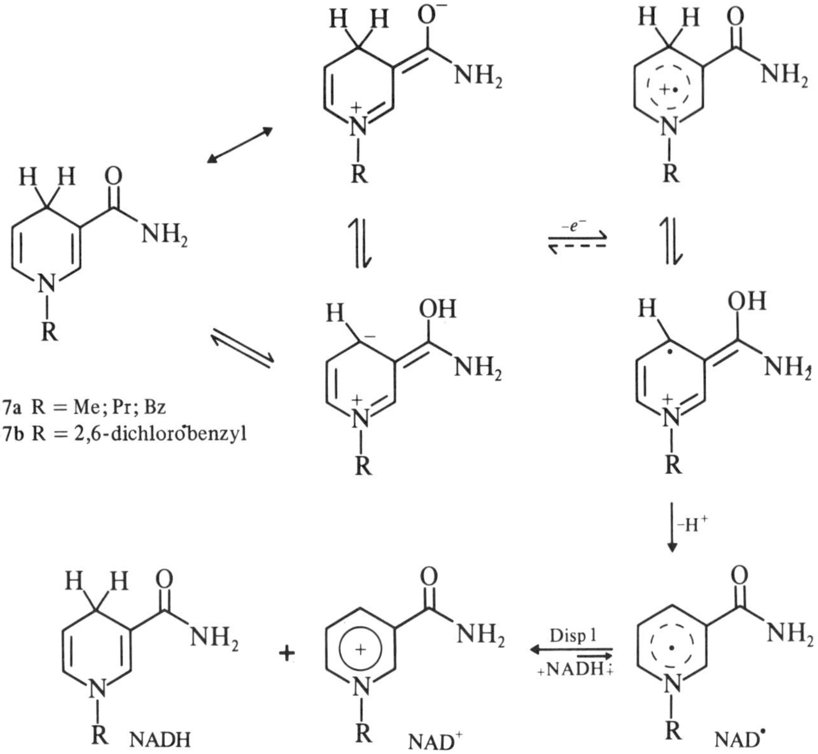

538   Electrochemical Behavior of Heterocyclic Compounds

347a R = Me; Pr; Bz
347b R = 2,6-dichlorobenzyl

Scheme 8.179

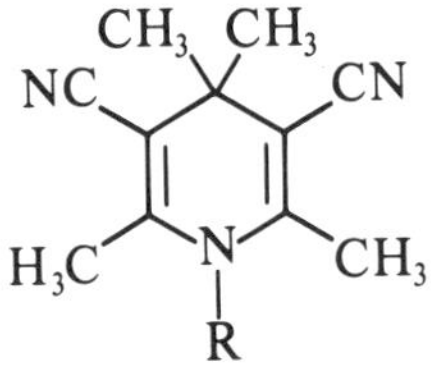

349

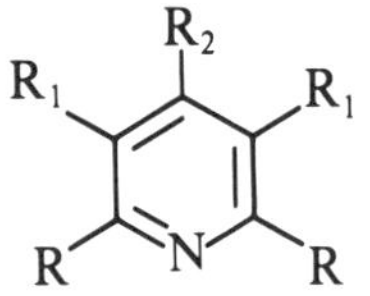

350

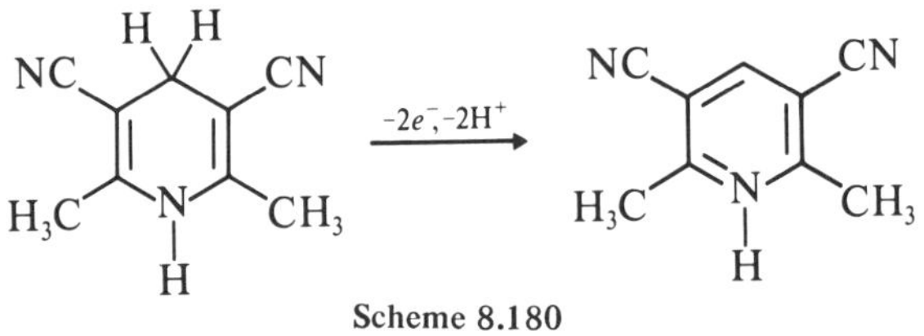

Scheme 8.180

Peak potentials in cyclic voltammetry were correlated linearly with Taft σ^* constants for a large number of 4-substituted 1,4-dihydropyridines (543), and the electrochemical oxidation of differently substituted 1,4-dihydropyridines has been examined (544a–546).

NADH model compounds have also been investigated in aqueous media: 1-benzyl- and 1-propyl-1,4-dihydronicotinamide have been oxidized to a cation in buffered aqueous solutions at the Pt, Au, or vitreous carbon electrodes following a two-electron irreversible process, independent of pH (547).

The electrochemical oxidation of NADH itself is difficult, since the current potential curves are poorly defined (548a, b) and electrode fouling occurs at NADH concentrations of 0.1–1 mM. Attempts have been made to circumvent this phenomenon by the use of mediators (549) and steady voltammetry (550). It was also found that much of the difficulty associated with the voltammetric determination of NADH at the carbon electrodes (glassy carbon or pyrolytic graphite) can be eliminated by covering the electrode surface with adsorbed electrochemically generated NAD$^+$. Such an electrode can be used for NADH measurements without pretreatment (551, 552). This technique was used to determine the electrochemical oxidation mechanism of NADH in aqueous media at the glassy carbon electrode. The oxidation takes place in a single irreversible two-electron stage that is marginally pH dependent, involving removal of two electrons and one proton to form NAD$^+$ (553). The first heterogenous electron transfer is slow and yields a cation radical NADH^{+}, which loses a proton (first-order reaction rate constant k of the order of $60\,\text{s}^{-1}$ at pH 6.10) to form the neutral radical NAD$^{\cdot}$, which participates in a second homogeneous electron transfer with NADH^{+} (DISP1 mechanism) to give NAD$^+$. A heterogeneous second electron transfer at the electrode has been proposed as an alternative pathway but does not seem likely in view of the value of the deprotonation rate constant (Scheme 8.179). A similar reaction path is operative with different electrodes and in DMSO. At the Pt electrode NADH 346 is oxidized to NAD$^+$ in 100% yield at pH 7.4 (555).

The NADH 346 oxidation in aqueous media at both Pt and C electrodes proceeds with considerable overpotential, which can be reduced by some conditioning pretreatment of the electrode (550). Electrochemical catalysis by mediators also permits reduction of this overpotential: at a vitreous carbon electrode (pH 7), NADH 346 can be oxidized to NAD$^+$ 345 at about 0.6 V/SCE. Electroactive aromatics, such as 1,4-diaminobenzene, catalyze this reaction, allowing it to be performed at potentials as low as 0.1 V. The catalytic mechanism involves a two-step process. The diamine is oxidized to the diimine at the electrode. The diimine or another related oxidant then attacks NADH in the solution. Controlled potential electrolysis showed that NAD$^+$ 345 is the product and $n = 2$. It is proposed that the mechanism involves the transfer of the 4-hydrogen of NADH to a protonated diimine (561a). Another method for reducing the overpotential is the construction of a chemically modified electrode where o-quinones derived from dopamine and 3,4-dihydroxy benzylamines are covalently bound to C electrodes, in order to catalyze the oxidation of NADH as in Scheme 8.181 (561b).

540 Electrochemical Behavior of Heterocyclic Compounds

$$\text{\textit{4}-QH}_2 \longrightarrow \text{\textit{4}-Q} + 2e^- + 2H^+$$

$$\text{\textit{4}-Q} + \text{NADH} + H^+ \longrightarrow \text{\textit{4}-QH}_2 + \text{NAD}^+$$

Scheme 8.181

where $\text{\textit{4}-QH}_2$ = surface-bound dihyroquinone
$\text{\textit{4}-Q}$ = surface-bound oxidized quinone

In the same way, a polymer from poly(methacryloyl chloride) and dopamine adsorbed on the surface of a vitreous carbon electrode decreases the overpotential by about 0.25 V and the surface coverage is improved (561c). Aromatics containing catechol functional units can be adsorbed on graphite to produce a modified electrode, which reduces the overpotential by 0.235 V, 30% of the original covering remaining after 30 min of electrochemical cycling at pH 7 (561d). Phenazine methosulfate and ethosulfate can also be immobilized on graphite electrodes and the overpotential reduced by 550 mV (561e).

Studies have been conducted to determine the best conditions for the regeneration of NAD⁺ **345** from NADH **346** *in situ* for use in alcohol dehydrogenase catalysis of EtOH to MeCHO. Up to 92–95% yields of NAD⁺ have been obtained, but the turnover number is far from reaching those of enzymic reactions (557). Similarly, a liver alcohol dehydrogenase to which NAD⁺ was covalently bound was immobilized on the surface of a glassy carbon electrode. The reduced coenzyme could be reoxidized, but only one cycle could be carried out due to the catalytic decomposition of the coenzyme at the electrode surface (558). A similar attempt involved the coulometric oxidation of NADH at the pretreated rotating Pt gauze electrode. NAD⁺ formed was reduced in a reactor containing immobilized alcohol dehydrogenase. Several recyclings were achieved with the same lot of NADH, and the overall conversion to enzymically active NAD⁺ showed a current efficiency of 99.3%. A continuous method of oxidation of alcohol with electrolytic regeneration of cofactor and removal of aldehyde by dialysis was tested (559).

As a model system of the NADH/FAD (FAD = flavine adenine dinucleotide) redox couple, the 1-benzyl-1,4-dihydronicotinamide **351** riboflavin **351A** system was investigated by polarography in DMSO. Riboflavin was reduced by **351**, and the mechanism was presumed to involve an hydride ion transfer (560). NADH model compounds (1-benzyl-1,4-dihydronicotinamide) have been used to reduce ketones in the presence of Zn^{2+}, which acts as a catalyst (556).

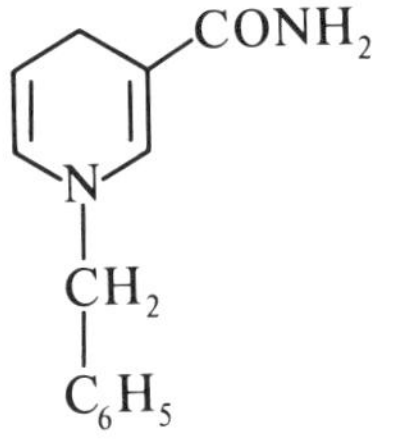

351

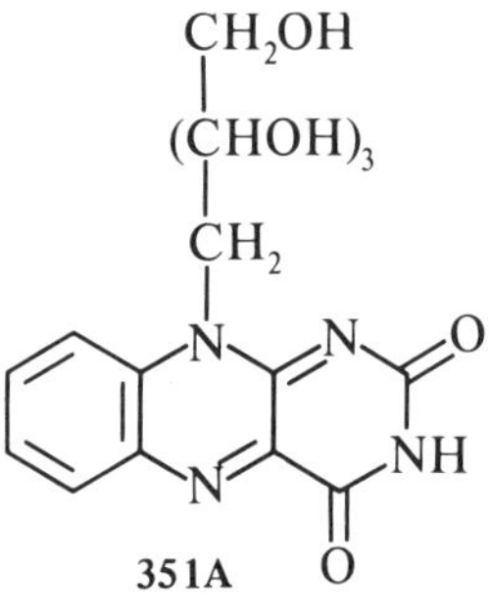

351A

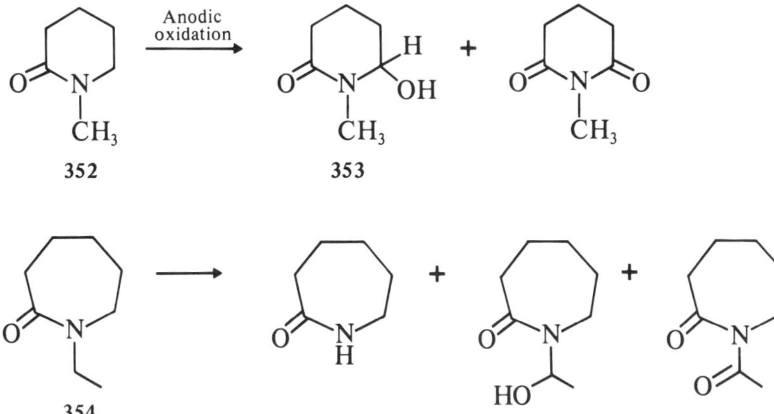

Oxidation of Heterocyclic Compounds 541

Scheme 8.182

8.3.1.1.5. Piperidine. Piperidine has been reported to undergo electrochemical oxidation in aqueous media, but the authors (562, 563) do not agree on the number of electrons involved and on the formation of final product, pyridine (562) or piperidine *N*-oxide (563). 1-Methylpiperidine-2-one **352** is oxidized in acetonitrile to give an α-hydroxylated derivative **353** as is 1-methylpyrrolidine-2-one, and the analogous seven-membered ring compound **354** yields hydroxylactams, imides, and dealkylation product (564) (Scheme 8.182).

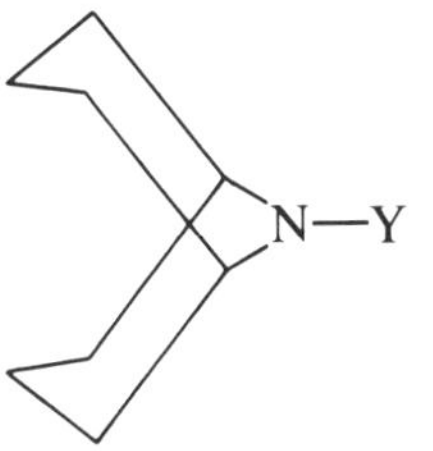

Y = CR$_3'$; NR$_2'$; OR; Cl; NMe$_3$

355a

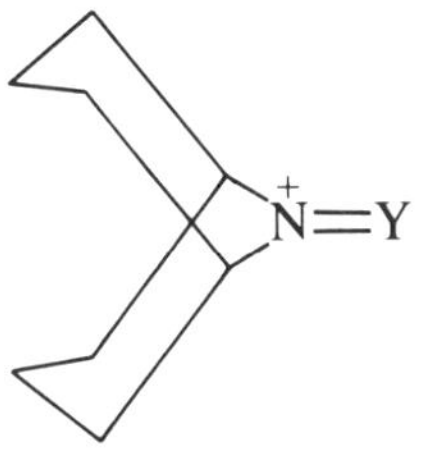

Y = CR$_1$R$_2$; NR; OR

355b

The stability of Bredt's rule, kinetically stabilized, nitrogen-centered radical cations and radicals has been investigated in 9-azabicyclo[3.3.1]nonyl systems **355** and **356** by measurements of E_0' (565a). The destabilization effect of a γ-keto group (565b) has been investigated. The results are discussed in terms of resonance stabilization and structural factors. Compound **356**, in contrast with compound **355**, shows an irreversible oxidation peak by voltammetry, indicating that the lifetime of **356**$^+$ is short. This instability is discussed in terms of the energy level of the orbitals obtained in photoelectronic spectra (566).

Various *N*-substituted pyrrolidines and piperidines have been perfluorinated electrochemically (567) (Scheme 8.183).

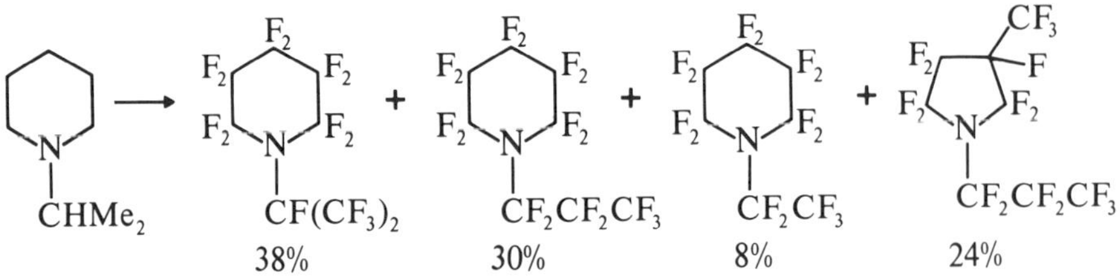

542 Electrochemical Behavior of Heterocyclic Compounds

Scheme 8.183

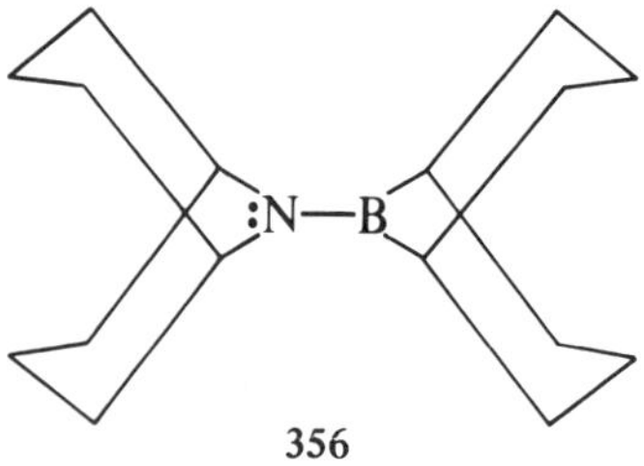

356

8.3.1.1.6. Indole. In the MeCN–NEtClO₄ system, indole is oxidized to **357** in 11% yield (568a); a conducting polymeric film can be obtained at a Pt electrode (568b). Indoline **358** is oxidized at constant current in AcOH, in the presence of triethylamine at the Pt electrodes, and gives 77% **359**, 2% **360**, and 3% 1-acetylindole

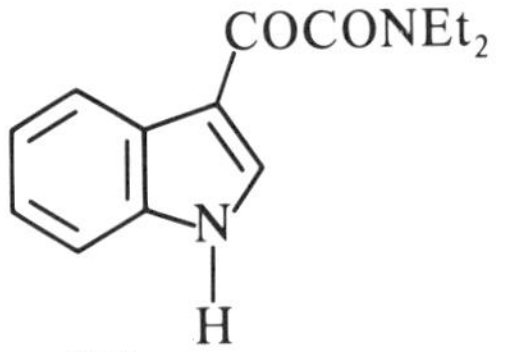

357 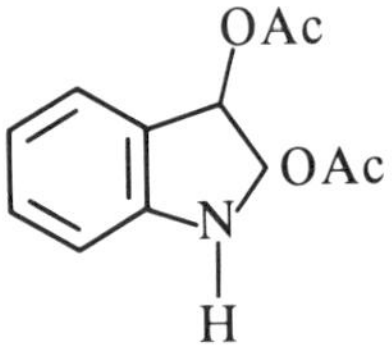
358 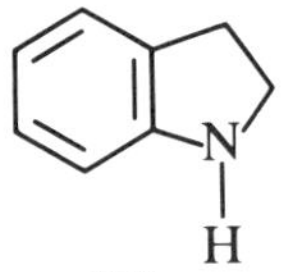
359 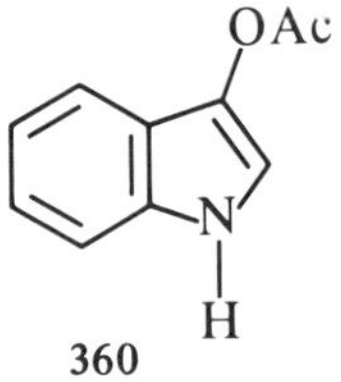
360

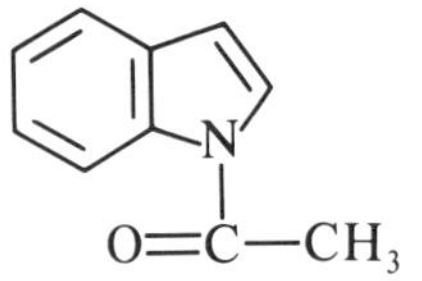

361

361 (569a). In MeOH with ammonium bromide as supporting electrolyte, 1-methylindole, 1,3-dimethylindole, and 1,2-dimethylindole are oxidized at the Pt electrode to a variety of products (569b). In aqueous AcOH, **359** is obtained in 48% yield. In MeCN, 2,3-diphenylindole **362** oxidizes to give a radical cation at the Pt electrode which dimerizes to **363** in 95% yield (570) (Scheme 8.184).

2-Phenyl-3-arylamino indoles **364** are oxidized in MeCN, DMF, and propylene carbonate at the rotating or pulsating Pt electrode. In unbuffered media, two one-electron steps are observed, the first leading to a radical cation (identified

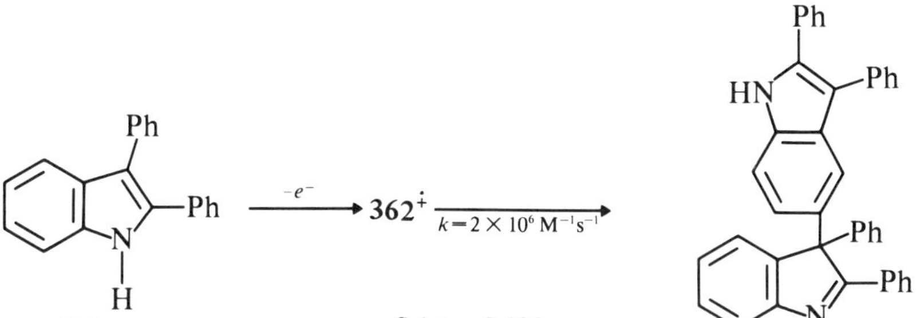

Oxidation of Heterocyclic Compounds   543

Scheme 8.184

by its ESR spectrum) and second followed by a very rapid deprotonation step (EEC mechanism). In the presence of a base (diphenylguanidine), **364** undergoes a two-electron oxidation to an imine (571, 572).

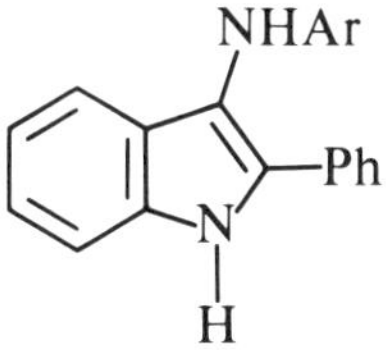

5,6-Dihydroxy-2-methylindole **365** in aqueous media undergoes a two-electron irreversible oxidation at the C paste electrode via an EC mechanism involving a short-lived intermediate, probably 2-methylindole-5,6-quinone, which is converted into electroinactive products (573) (Scheme 8.185). When $n = 1$, **366** is

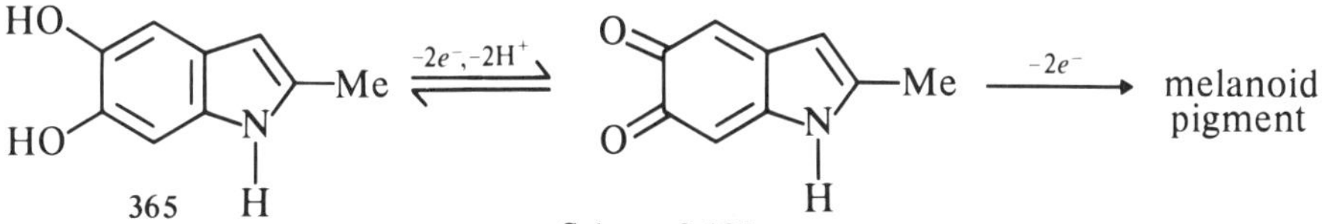

Scheme 8.185

oxidized to its stable radical cation, and when $n = 2$ the radical cation of **367** undergoes an initial attack at position 3 followed by bond cleavage and rearrangement to give an almost quantitative yield of **368** (574). The difference in behavior is attributed to the existence of a six-membered transition state in the case of **367**.

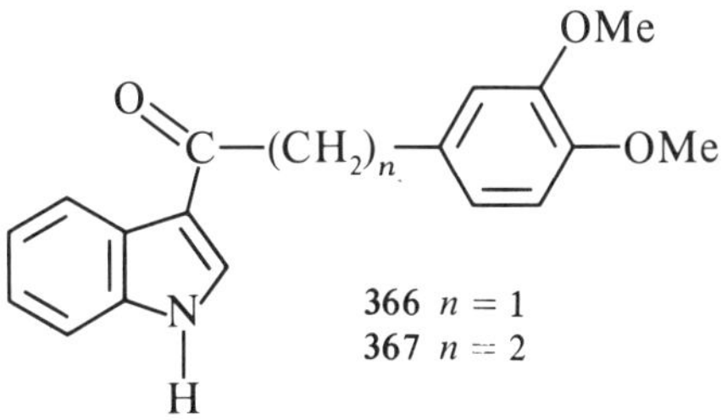

366 $n = 1$
367 $n = 2$

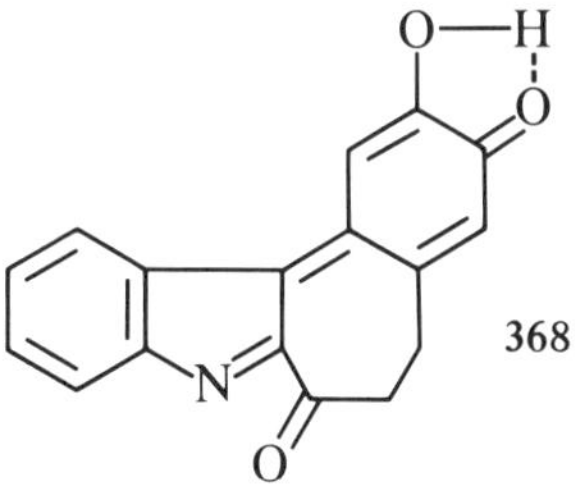

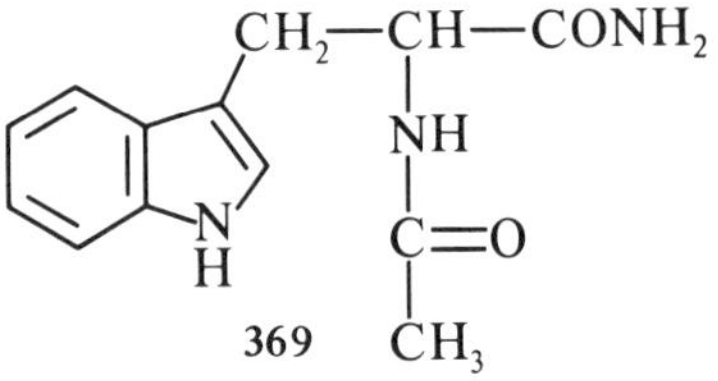

544 Electrochemical Behavior of Heterocyclic Compounds

N-Acetyl-L-tryptophan amide **369** is oxidized in aqueous media at a graphite paste electrode. The first oxidation step gives a monocarbocation that dimerizes, and in the second step the dimer is oxidized to the carbodication, which undergoes polymerization (575). In acetonitrile, indolinonic nitroxide radicals (2,2-disubstituted 3-oxoindolines-1-oxyls) are oxidized to the corresponding oxoammonium cations (576).

8.3.1.1.7. Quinolines. Anodic oxidation of quinoline at the Pt anode in 75% sulfuric acid yields quinolinic acid in 77% yield (577a). Other ring-substituted quinolines behave similarly, with the exception of 2- and 4-substituents, which yield tarry material (577b). 2-Methyl-8-hydroxyquinoline gives a polymeric film that can be made electroactive by complexing Cu^{2+} ions (577c, d). The solvation energies of 8-hydroxyquinoline in MeCN and dichloromethane have been obtained from half-wave potentials and energies of the highest molecular orbitals (578).

The electrochemical oxidation of **370** proceeds via an electron-electron-proton mechanism to **371** (579); isoquinoline, acridine, and phenanthridine behave similarly.

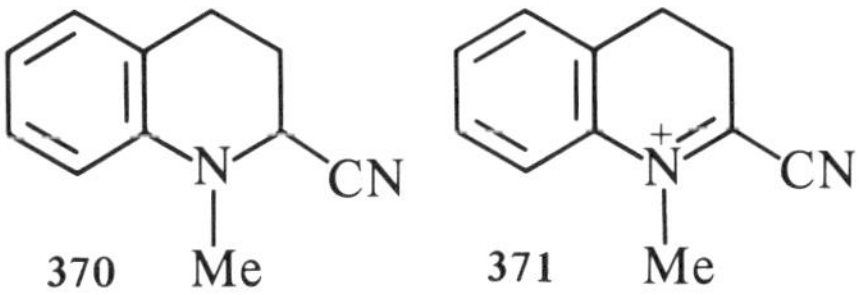

N-Amino-1,2,3,4-tetrahydroquinoline **372** is oxidized in MeCN to a diazenium cation **373**, which is fairly stable in acidic media but leads to a tetrazene **374** in basic media (580). The diazenium cation can also be condensed with olefins, for example, with acenaphthylene, to give **375** (Scheme 8.186).

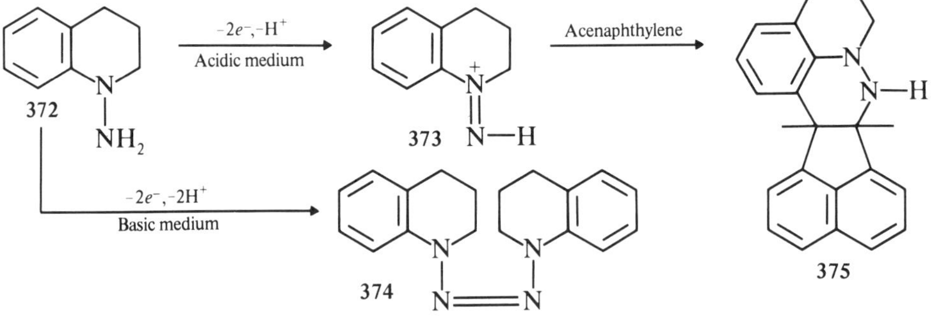

Scheme 8.186

Oxidation of Heterocyclic Compounds 545

Narcotine sulfate **376** is oxidized to opianic acid **377** and cotarnine **378** in 80% and 55% yields in an undivided cell with a nickel cathode using chlorides as protective additives. The undesired reduction of narcotine and its oxidation products is prevented by the formation of a layer of hydrated oxide at the cathode surface (581). The electrochemical oxidation of papaverine **379** has also been reported (582). The phenethylisoquinoline **380** is oxidized to dibenzoquinolizinium perchlorate **381**, while the *N,N*-dimethyl derivative yields a dimer **382** (583).

546   Electrochemical Behavior of Heterocyclic Compounds

In a series of papers, Miller and coworkers have reported an interesting synthesis of morphinanedienone alkaloids (584–587a, b) by the oxidative cyclization of 1-benzyltetrahydroisoquinoline. The reactions were carried out in acetonitrile with addition of Na_2CO_3 at 0° in a divided cell, with platinum anodes and NBu_4BF_4 or $LiClO_4$ as supporting electrolytes. ± Laudanosine (383; $R_1 = R_2 = R_3 = R_4 = Me$) is converted into o-methylflavinantine 384 (52% yield); o-benzyllaudanine (383; $R_3 = CH_2Ph$, $R_1 = R_2 = R_4 = Me$) to o-benzylisoflavinantine (43% yield); o-benzylcodanine (383; $R_1 = R_3 = R_4 = Me$, $R_2 = CH_2Ph$) to o-methylflavinantine (53% yield); and o-benzyl pseudolaudanine (383; $R_1 = CH_2Ph$, $R_2 = R_3 = R_4 = Me$) to 2,3-dimethoxy-6-benzyloxymorphinandienone (44% yield) (585). The yields obtained by electrochemical cyclization (586) are higher than those obtained by chemical methods. Even higher yields of alkaloids have been obtained by modifying the experimental procedure (589), that is, by carrying out the electrolysis in CF_3COOH in an undivided cell using HBF_4 as supporting electrolyte. ± Flavinantine, 386a, ± pallidine 386b, and ± annurine 386c have been obtained similarly from 385a, b, c.

385a $R_1 = OCH_2Ph$; $R_2 = OCH_3$
385b $R_1 = OCH_3$; $R_2 = OCH_2Ph$
385c $R_1 = OCH_3$; $R_2 = OAc$

386a $R_1 = OH$; $R_2 = OCH_3$
386b $R_1 = OCH_3$; $R_2 = OH$
386c $R_1 = R_2 = -OCH_2O-$

The mechanism of the cyclization involves initial oxidation of the amine moiety followed by coupling (587). It has been proposed that the amine group anchimerically assists coupling. This can be conceptualized either as an electrophilic attack on the isoquinoline aromatic ring by the aminium ion or homoconjugation between

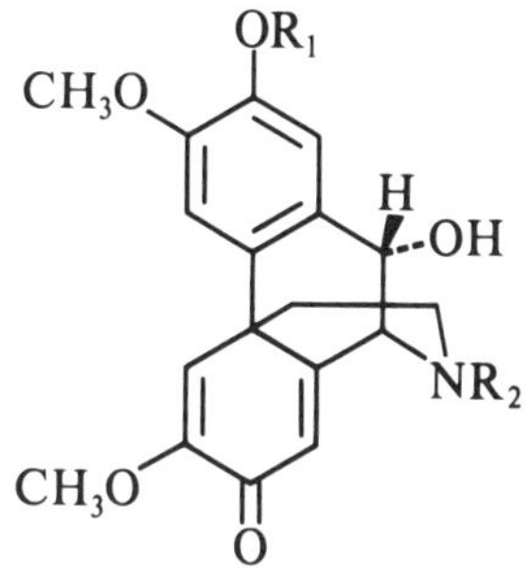

Oxidation of Heterocyclic Compounds 547

387a

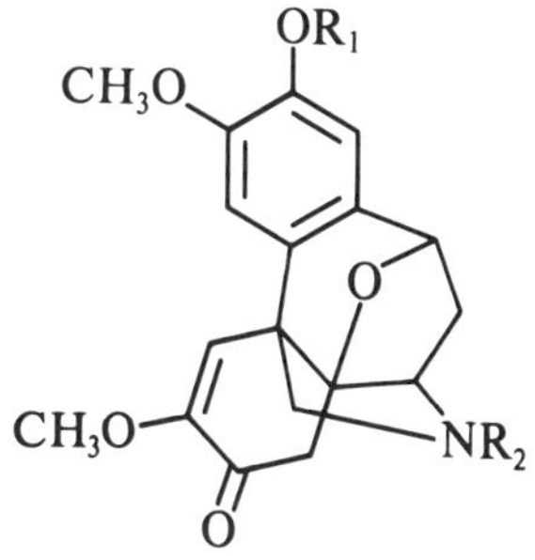

387b

the amine and aromatic ring. The morphinandienones can be oxidized in their turn in acetonitrile-fluoboric acid. (o-Methylflavinantine **384** gives *trans*-10-hydroxy-o-methylflavinantine **387a** in 38% yield together with an oxohomomorphinan containing an acetal function **387b** in 41% isolated yield.) O-Benzylpallidine and o-methylflavinine are similarly oxidized (588).

8.3.1.1.8. Carbazoles. In acetonitrile, at the Pt electrode the one-electron oxidation of carbazole **388** yields a radical cation that dimerizes predominantly at the 3-position to give dicarbazyl **389**. The latter is more easily oxidized than carbazole to the dication **390**. In basic media (containing pyridine), deprotonation of the NH group yields 9,9'-dicarbazyl **391** (590a) (Scheme 8.187). *N*-Vinylcarbazole

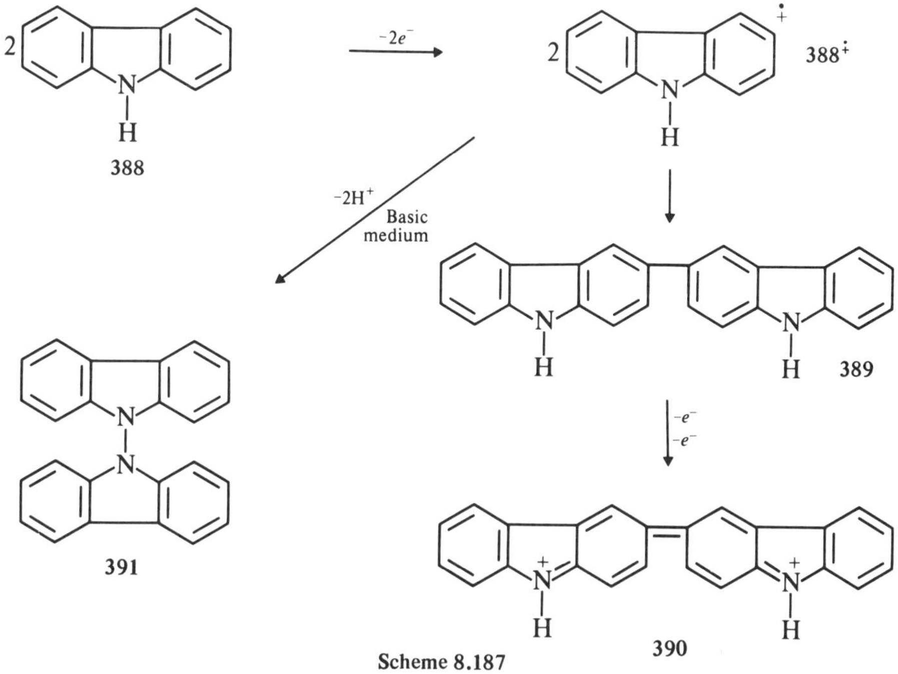

Scheme 8.187

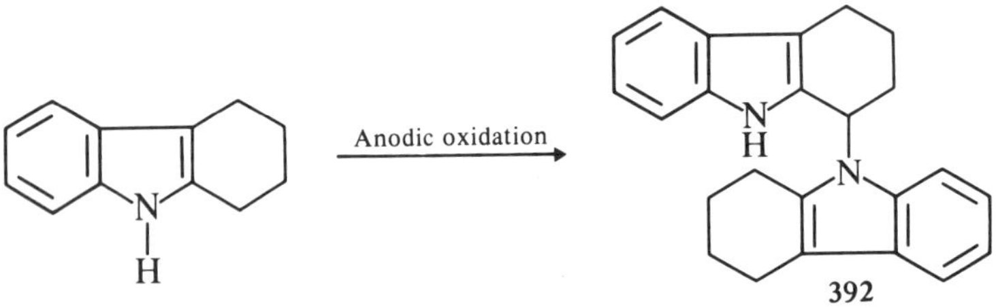

is oxidized to a film which adheres to Pt or Au electrodes (590b, c). When the 9-position is blocked by substituents, 3,3'-dimers are obtained, and if the 3-, 6-, and 9-positions are substituted dimerization does not occur and stable cation radicals are formed (590a, 591).

Tetrahydrocarbazole gives a dimer **392** on oxidation in CH_3CN/H_2O (592).

8.3.1.1.9. Acridines. The electrochemical oxidation of **393** involves the sequential loss of two electrons to give **394** via an intermediate radical-cation (593a, b). Acridine itself is oxidized in MeCN, showing two oxidation waves due to the neutral and protonated species. Preparative electrolysis yields, via an ECEC mechanism, a dication of a tetramer in equilibrium with acridyl acridinium cation radical (594a, b) (Scheme 8.188).

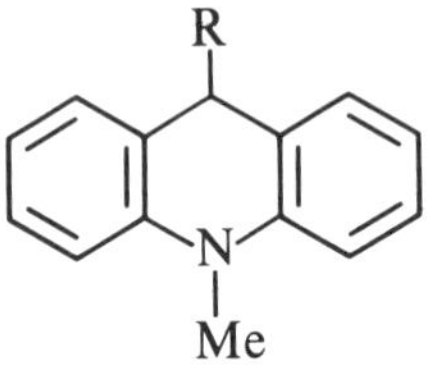

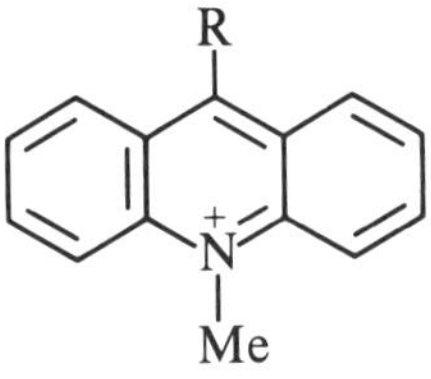

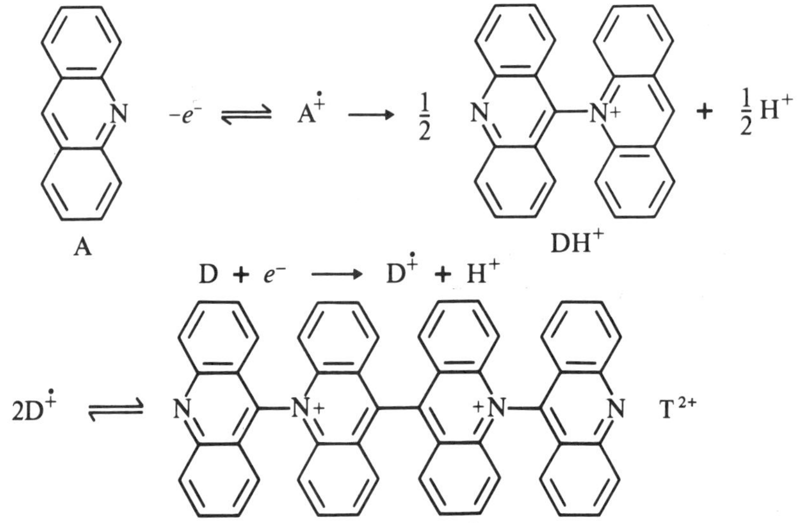

Scheme 8.188

8.3.1.2. Compounds with One Oxygen Atom

8.3.1.2.1. Furans. Furan is easily oxidized. In acetonitrile a conducting polymeric film can be obtained at a Pt electrode (568b). In aqueous solution (1 N H_2SO_4) at a PbO_2 anode, it is oxidized to 2,5-dihydroxy-2,5-dihydrofuran and β-formylacrylic acid (595a, b), whereas 2-methylfuran yields acetylacrylic acid (595a, b). The anodic substitution of furan and its derivatives has been extensively investigated. In methanol, 2,5-dimethoxy-2,5-dihydrofuran is obtained via an ECEC mechanism (596a, b) (Scheme 8.189). Similarly, furylacetate 395 is oxidized to 396 (597) (Scheme 8.190), and 2,5-dimethylfuran to 2,5-dimethoxy-2,5-dimethyldihydrofuran (598, 599). Anodic acetoxylation of furan is carried out in acetic acid containing acetate ions (598, 600), while benzoyloxylation takes place in DMF containing benzoic acid and lithium chloride (601).

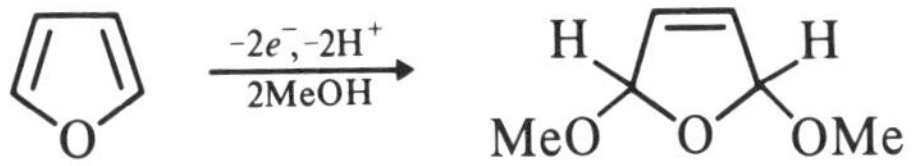

Scheme 8.189

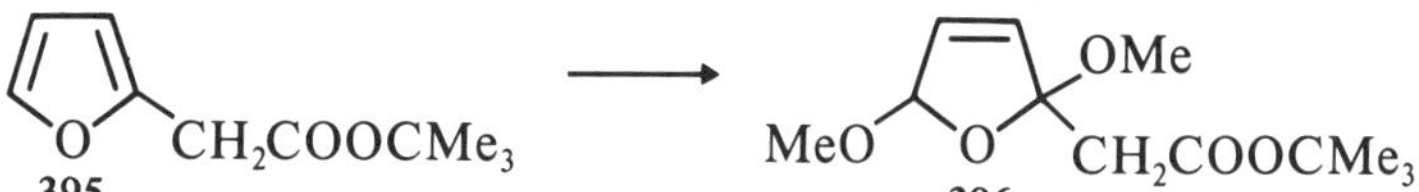

Scheme 8.190

Cyanation carried out in MeOH/NaCN leads to *cis*- and *trans*-2-cyano-5-methoxy-2,5-dimethyldihydrofuran along with a small amount of 2,5-dimethoxy-2,5-dimethyldihydrofuran. The proposed mechanism is similar to the classical ECEC mechanism of anodic aromatic substitution, where the chemical step is a nucleophilic attack (602). Furan can also be oxidized indirectly by electrogenerated bromine in methanol to 2,5-dihydro-2,5-dimethoxyfuran (603a–d). The electrolysis is considered to be indirect, since a film of bromine appears at the anode, the material of which is unimportant. In ethanol, 2,5-diethoxydihydrofurans (604) are formed. When the furan ring is substituted by an aromatic ring, 5-methoxyfurans (not the dihydro derivatives) are formed (605a, b) via an ECEC mechanism, in which the last chemical step is the loss of the proton at C-5 of the furan ring. The driving force of the whole process is conjugation of the substituent in position 2 with the arising furan heterocyclic system. Tetraphenylfuran 397 is oxidized to dibenzoylstilbene in buffered (Na_2CO_3) neutral nitromethane, while in unbuffered neutral nitromethane a dication is also formed, which, according to the authors, could be isolated as a perchlorate. The dication reacts with methanol to give 398 (606) (Scheme 8.191). In acidic nitromethane, a dimer is obtained for which structure 399 has been proposed.

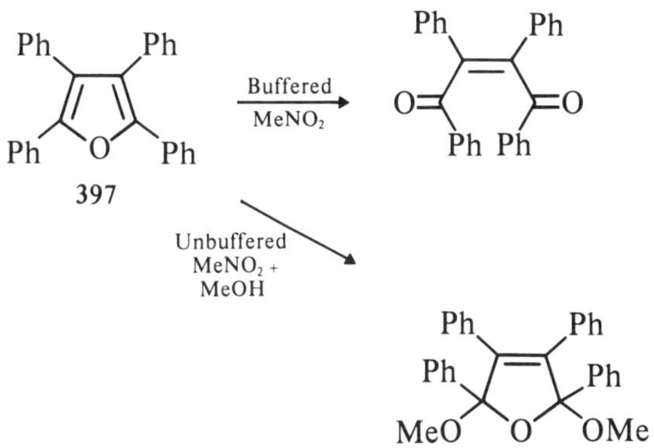

550   Electrochemical Behavior of Heterocyclic Compounds

Scheme 8.191

Furfural is oxidized at the PbO$_2$ anode (in 1 N H$_2$SO$_4$) to β-formylacrylic acid and maleic acid at +1.30 and +1.50 V (vs, SCE), respectively. In the presence of K$_2$Cr$_2$O$_7$ or V$_2$O$_5$ the oxidation is accelerated, the selectivity of the process is enhanced, and the yield of β-formylacrylic acid reaches 90% (607a). Similarly, MeCOCH=CH–COOH is obtained from 5-methylfurfural (607b).

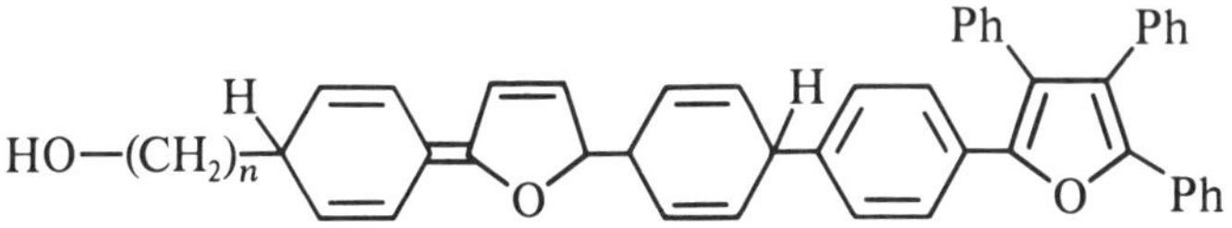

Alkyl-2-furoic acids **400** are oxidized in protic media (H$_2$O, MeOH, AcOH) at the Pt electrodes. The proposed reaction path is shown in Scheme 8.192. The intermediate **401** was proposed by analogy with the previously described electrochemical behavior of furan derivatives (608).

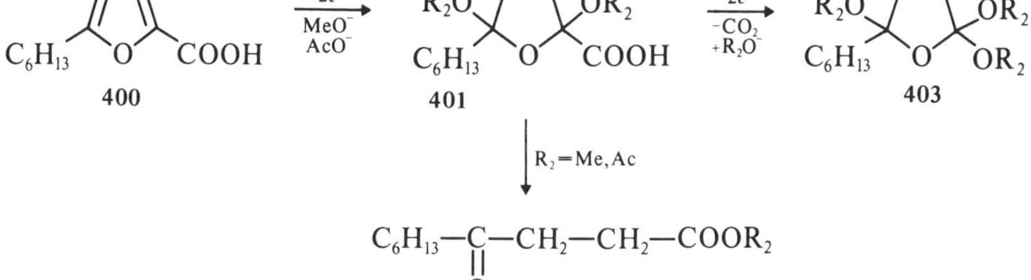

Scheme 8.192

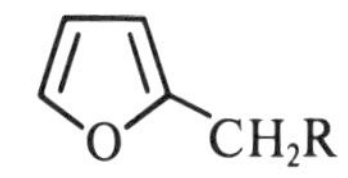

Oxidation of Heterocyclic Compounds 551

404 R = CN
405 R = COOCH₃

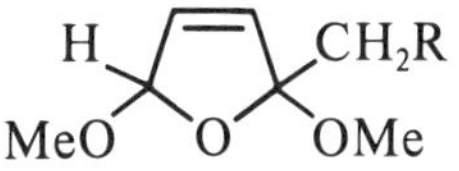

406 R = CN
407 R = COOCH₃

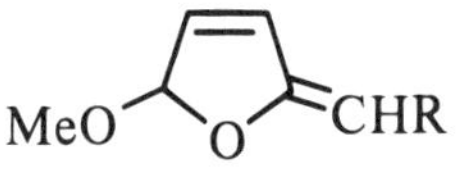

408 R = CN
409 R = COOCH₃

2-Furylacetic acid derivatives are oxidized in methanol at the Pt electrodes. With NaClO₄ as supporting electrolyte at −50°C, **404** gives 98% **406**, while with NH₄Br at 0°C it gives 98% **408**. Similar results are observed in the case of **405** (609).

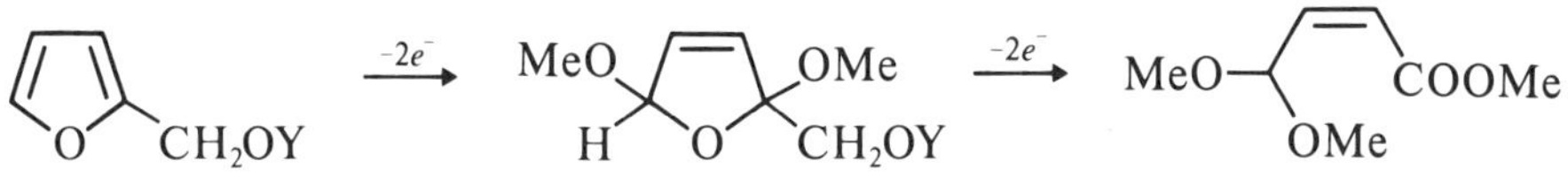

410a Y = H
410b Y = Ac

Scheme 8.193

In MeOH with Et₄NClO₄ or Et₄NClO₄/NEt₃, at the Pt electrodes using a constant current of 0.02 A cm⁻² 2-substituted furans **410** are oxidized to methyl Z-4,4-dimethoxybutenoate **411** after consumption of 4F mol⁻¹ (610) (Scheme 8.193).

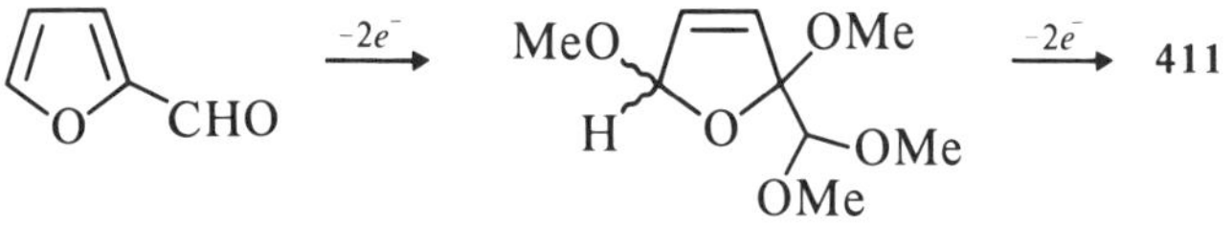

Scheme 8.194

Similarly, furfural yields **411** (Scheme 8.194), but 2-fuoric acid leads to **412** or **411** as in Scheme 8.195 (611). Anodic methoxylation of furans containing a ketonic group in the side chain such as **413** leads to normal 2,5-dimethoxy compounds **414**, but aldehydes undergo intramolecular cyclization to give **415** (yields over 70%) (Scheme 8.196) (611).

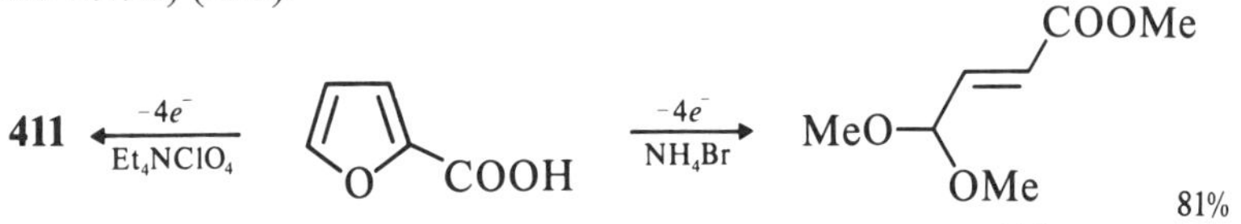

84%

Scheme 8.195

81%

413

414

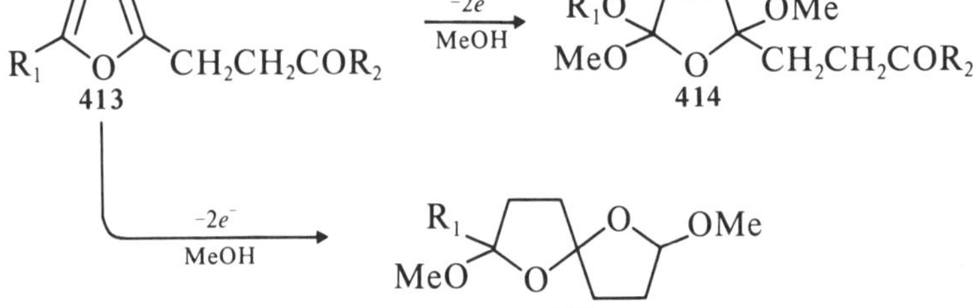

415

Scheme 8.196

8.3.1.2.2. Tetrahydrofuran.

Tetrahydrofuran (THF) is oxidized to γ-butyrolactone in KBr solution with Na_2SO_4 as supporting electrolyte at the Pt anode at pH 1. Anodic efficiency reaches 98% when a diaphragm is used. THF is oxidized indirectly by BrO_3^- formed on oxidation of Br^-. At pH 6-7 the oxidizing agent BrO^- is formed at the anode and the current efficiency decreases due to the simultaneous formation of oxybutyric acid (612). The oxidation of THF to succinic acid is carried out in aqueous media using Pt, C, or PbO_2 anodes. Intermediates include peroxides which dehydrate at the PbO_2 anodes to γ-butyrolactone, which is further oxidized to succinic anhydride. The dimer **416** can be formed in high yields at Pb anodes in aqueous H_2SO_4 by the oxidation of butyrolactone or THF (613). Similarly, tetrahydrofurfuryl alcohol gives mainly γ-butyrolactone and succinic acid (614). Electrochemical fluorination of THF leads to perfluorofuramidine in yield up to 51% (615).

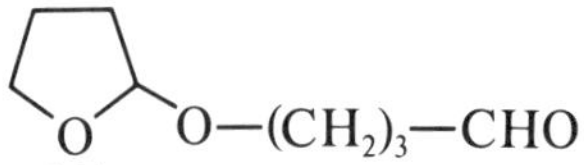

$$\underset{\mathbf{416}}{\text{[THF-ring]}-O-(CH_2)_3-CHO}$$

8.3.1.2.3. Pyrans.

2-Alkoxy-Δ^5-dihydropyrans **417** can be indirectly bromoalkylated by electrogenerated bromine, which attacks the C=C double bond, and the two alkoxy groups are added *trans* to each other (616) (Scheme 8.197). Electrochemical oxidation of 2,4,6-triphenyl-4H-pyran to 2,4,6-triphenylpyrilium perchlorate involves an ECE process (617).

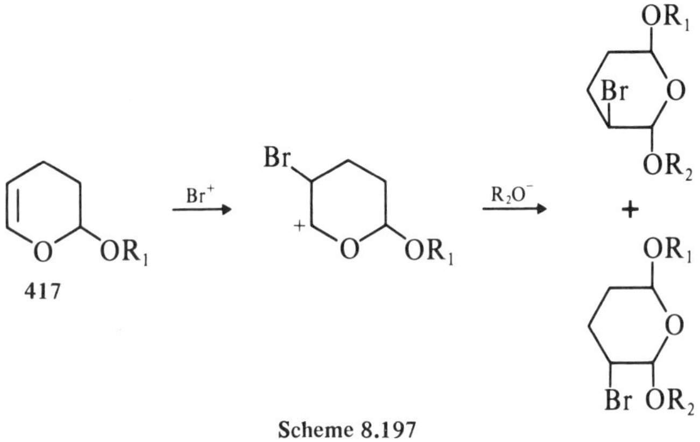

Scheme 8.197

The anodic oxidation of 4,4′-bipyran **418** has been reported (618). During the anodic oxidation of a mixture of a fluorescent hydrocarbon (perylene or rubrene) with 4,4′-bipyran in $MeCN-C_6H_6$, luminescent emission from the hydrocarbon has been observed. RDE and RRDE studies indicate that luminescence is caused by the electron transfer between the pyranyl radicals obtained from the bipyran and aromatic radical cation. Magnetic field effect and thermodynamic considerations indicate a triplet electrochemical luminescence mechanism (618).

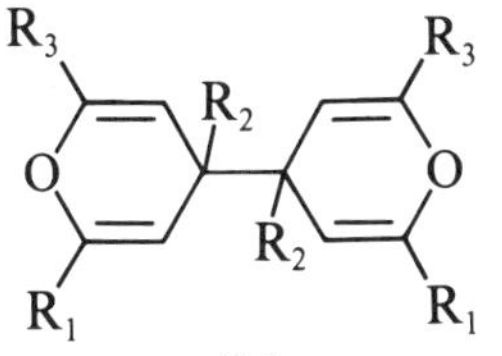

418

8.3.1.2.4. Benzofurans. The anodic oxidation of benzofurans **419** at the C anode in acidic methanol leads to 2,3-dialkoxydihydrobenzo[*b*]furans **420** in 40–90% yields (Scheme 8.198) (619). In DMF, electrogenerated electroluminescence of arylisobenzofurans **421** has been studied in mixed systems with aromatic compounds (620).

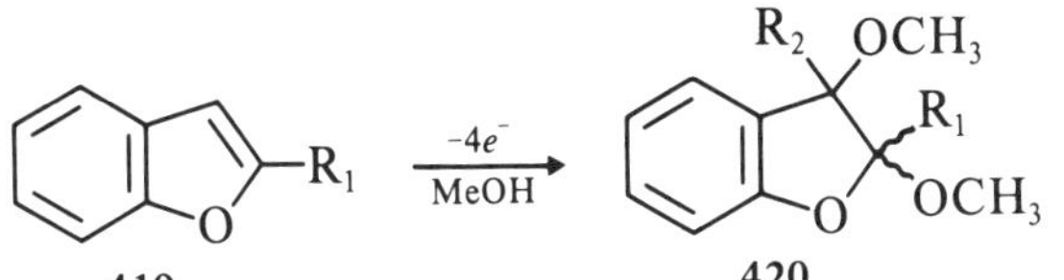

Scheme 8.198

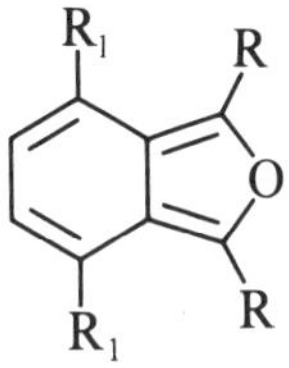

8.3.1.3. Compounds with One Sulfur Atom

8.3.1.3.1. Thiophenes. In MeCN, oxidation of thiophene at a Pt electrode leads to a metallic polymer film (568b). Anodic oxidation of thiophene and substituted derivatives in acidic methanol at low temperature (-20, $-30°C$) results in ring opening with the loss of SO_2. Thiophene yields butenedialdehyde tetramethylacetal, methoxy succinic dialdehyde tetramethylacetal, and a small amount of β-formyl propionate (621–624). In general, oxidation of thiophenes results in the formation of derivatives of α,β-unsaturated γ-dicarbonyl compounds or γ-ketoesters. Thiophenes can be brominated by anodic oxidation with NH$_4$Br as supporting electrolyte (the electrogenerated bromine being the brominating reagent) to give 2-bromo and 2,5-dibromo derivatives (625). 2-Methylthiophene gives 5-bromo-2-methylthiophene as the main product, 3,5-dibromo-2-methylthiophene, and 1,1,4,4-tetramethoxy-2-pentene (626). 3-Methylthiophene yields mainly 2-bromo-3-methylthiophene (626). 2,5-Dimethylthiophene results in the formation of 3-bromo-2,5-dimethylthiophene (627). If nonhalide electrolytes such as

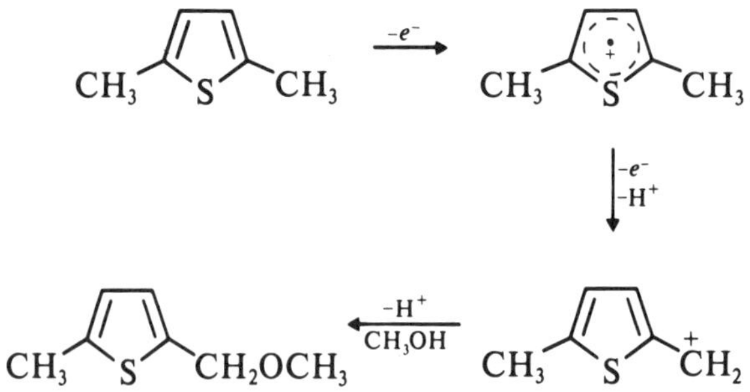

554 Electrochemical Behavior of Heterocyclic Compounds

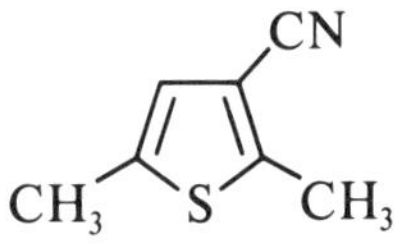

Scheme 8.199

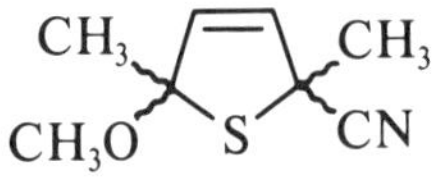

ammonium nitrate, sodium methoxide, acetate, or perchlorate are used, 2,5-dimethylthiophene yields 2-methoxymethyl-5-methylthiophene **422** (627) (Scheme 8.199). With methanolic sodium cyanide, 2,5-dimethylthiophene gives *cis*- and *trans*-2-cyano-5-methoxy-2,5-dimethyldihydrothiophene **423**, 3-cyano-2,5-dimethylthiophene **424**, and 2-methoxymethylthiophene **422** (627). Thiophene-2- and 3-carboxylic acids are oxidized at graphite anodes in acidic methanol to give products resulting from the ring cleavage (628) (Scheme 8.200). 2,3,4,5-Tetraphenylthiophene is oxidized in nitromethane to a radical cation, which reacts with the solvent or residual water to give 2-(*p*-nitrophenyl)-3,4,5-triphenylthiophene **425** and 1,2,3,4-tetraphenylbut-2-ene-1,4-dione **426** (629).

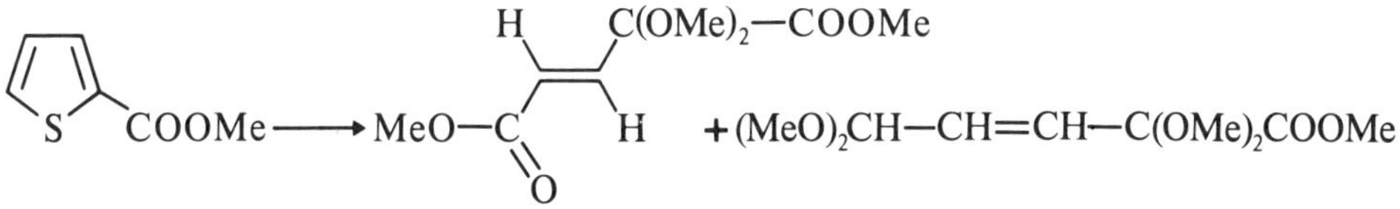

Scheme 8.200

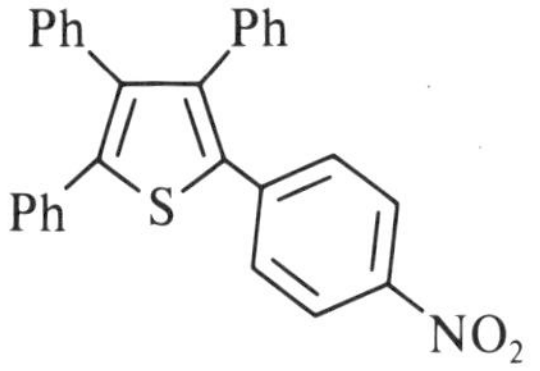

40% in neutral solution
425

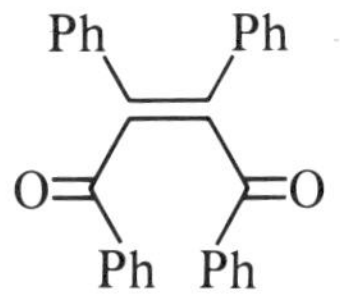

30% with Na$_2$CO$_3$
426

Oxidation of Heterocyclic Compounds 555

Electrochemical fluorination of tetrahydrothiophene and its 2- and 3-methyl derivatives as well as tetrahydropyran leads to perfluorinated compounds (630).

8.3.1.3.2. Benzothiophenes and Dibenzothiophenes. Electrochemical oxidation of benzothiophene in methanol using MeONa as supporting electrolyte gives a mixture of isomeric 2,3-dimethoxy-2,3-dihydrobenzothiophenes **427**, 2,3-dimethoxybenzothiophene **428**, and 4,7-dimethoxybenzothiophene **429**. 2-Methyl and 2,3-dimethylbenzothiophene behave similarly (631). It appears that these reactions are not very reproducible and are sensitive to the operating conditions.

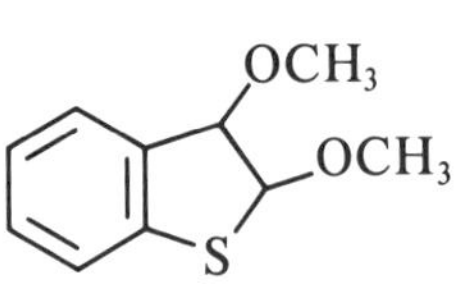

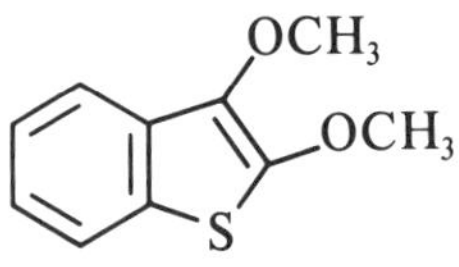

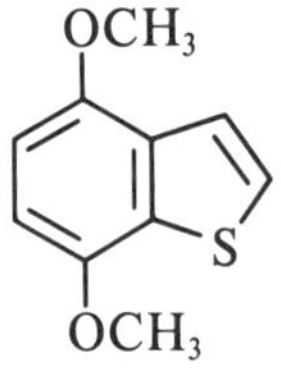

In 1% methanolic potassium at the Pt electrode, methoxylation takes place at the benzene ring (632) of 4-methoxybenzothiophene **430** (Scheme 8.201). At $-30°C$ **429** predominates, but in refluxing methanol **431** is obtained. The mechanism

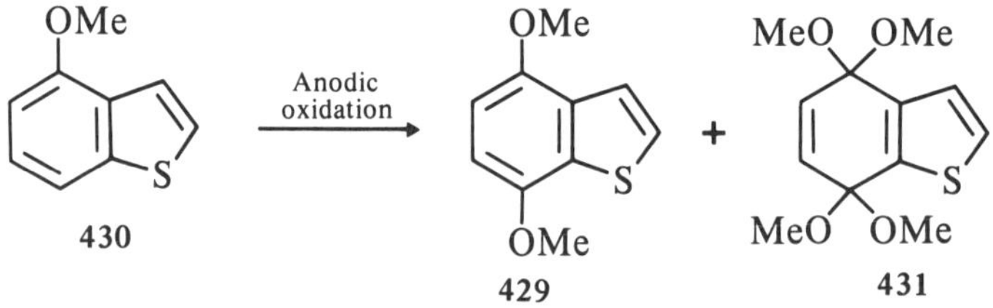

Scheme 8.201

involves the initial formation of **432** ($2e^-$) (which could be isolated with R = Me) followed by the loss of MeOH to **429** and further $2e^-$ oxidation to **431** (632). In acetonitrile at the Pt electrode, dibenzothiophene is oxidized to a radical cation that dimerizes into a sulfonium ion. The formed dimer can be further oxidized (Scheme 8.202) (633). In sulfuric or perchloric acid, dibenzothiophene is oxidized quantitatively to its *S*-oxide and in hydrochloric acid to its *S*-dioxide (634).

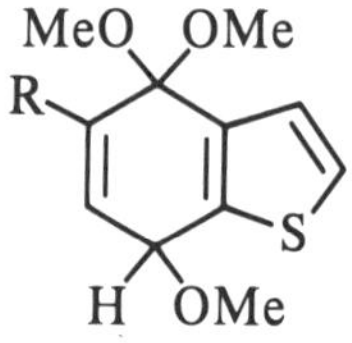

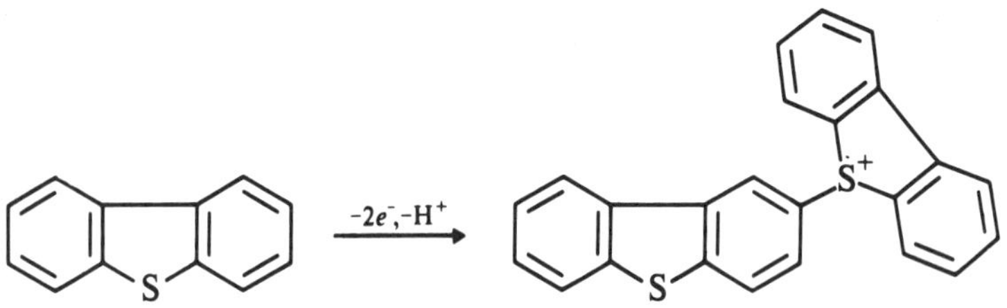

556   Electrochemical Behavior of Heterocyclic Compounds

Scheme 8.202

8.3.1.4. Compounds with Two Nitrogen Atoms

8.3.1.4.1. Diaziridines. 3,3-Pentamethylene diaziridine **433** is oxidized in alkaline solution to diazirine **434**, which in the same medium can be reduced to the former. Unlike their aromatic counterparts they do not form a reversible couple; about 1.2 V exists between the anodic half-wave potential of **433** and the cathodic half-wave potential of **434** (635) (Scheme 8.203).

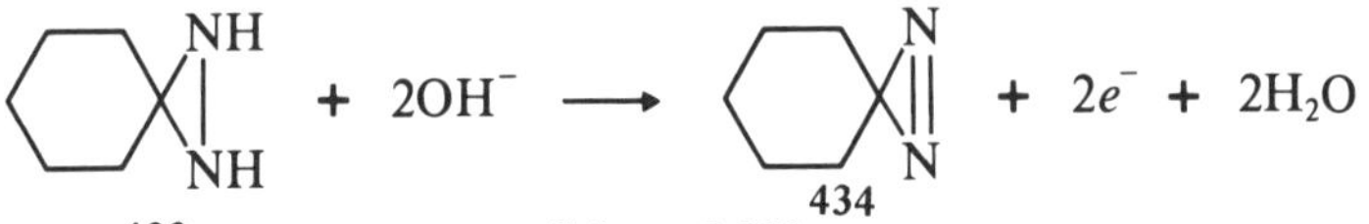

Scheme 8.203

8.3.1.4.2. Pyrazolines. 1-Phenyl-Δ^2-pyrazoline **435** is oxidized in acetonitrile to a radical cation that dimerizes into **436** (Scheme 8.204) (636). Compound **436** is oxidized to a dication. Highly substituted pyrazolines such as 1,3,5-triaryl-Δ^2-pyrazolines give biphenyl-type dimers like those obtained from **437** and its phenyl-substituted derivatives (637, 638) (Scheme 8.205) and 1-phenyl-3-amino-Δ^2-pyrazoline (639). Similar behavior is observed for 1,3,4,5-tetraphenyl-Δ^2-pyrazoline (640). However, when the phenyl group at the 1-position of pyrazoline is substituted by a nitro group, 1-p-nitrophenylpyrazole (639) and bipyrazoline are obtained (641). Kinetic measurements have been reported for the electrochemical oxidation of these triarylpyrazolines (642). In the presence of pyridine in MeCN, the electrochemical oxidation of 1,3,5-triaryl- (643) and 1,3,4,5-tetraphenyl-Δ^2-pyrazoline (640) yields the corresponding pyrazole **440** and pyrazolyl dimer **441**. The mechanism of these oxidations in the presence of pyridine has been investigated (644, 645). First electronic transfer furnishes a radical cation, which is rapidly deprotonated by pyridine and is then oxidized to a carbonium ion located

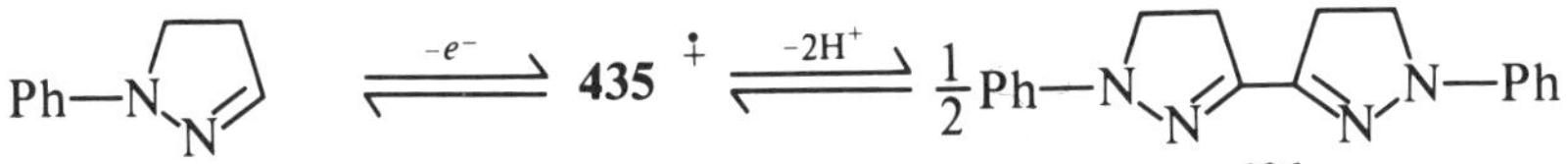

Scheme 8.204

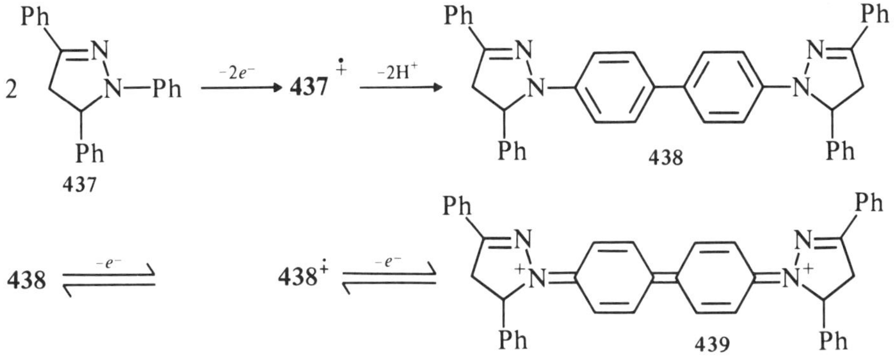

Scheme 8.205

at the 4-position of the pyrazoline ring, eliminating hydrogen from the 5-position to give a pyrazole. When the 5-position bears two substituents, one of them migrates to the 4-position and elimination of a proton yields pyrazole. In some cases, pyridine traps the intermediate carbonium ion by forming a quaternary ammonium salt, and the reaction does not occur with the more hindered bases (collidine). In the case of 1-p-tolyl-3,5-diphenyl-5-methyl-Δ^2-pyrazoline (646) a reversible oxidation peak is observed by voltammetry, which disappears on addition of propylamine. Electrolysis in MeCN leads to **442**, and the mechanism involves an anodic substitution and dehydration of the 5H-pyrazolium cation. Cyanide ion does not act as a nucleophile.

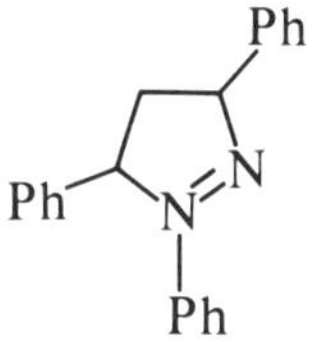

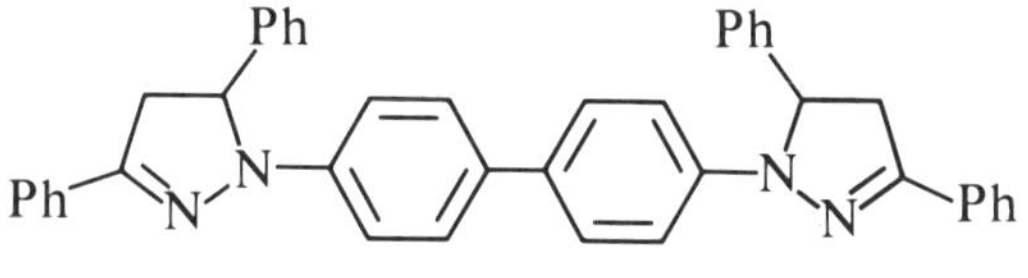

The mechanism of the electrochemical oxidation of 1-p-anisyl-3,5-diphenyl-Δ^2-pyrazoline (PH$_2$) has been investigated by classical electrochemical techniques (647) and spectroelectrochemistry (648, 649). In MeCN a stable radical cation is obtained, which is deprotonated by pyridine (644, 645). The resulting radical is oxidized in solution through a disproportionation mechanism (Scheme 8.206) (648, 649). The rate constants (649) and pK_a of the radical cation PH$_2^{\cdot+}$ (648)

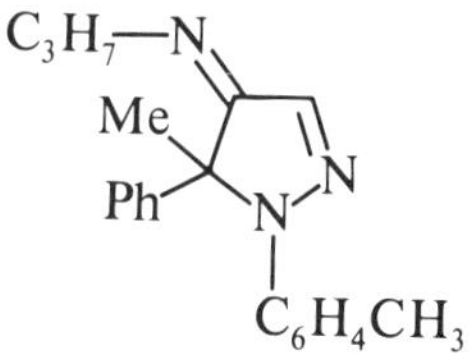

have been reported. The oxidation on the second wave has been shown (649) to follow a coproportionation mechanism:

$$PH_2 - 2e^- \longrightarrow PH_2^{2+}$$
$$PH_2^{2+} \rightleftharpoons PH^+ + H^+$$
$$PH^+ + PH_2 \rightleftharpoons PH^{\cdot} + PH^{\cdot+}$$
$$PH^{\cdot} + H^+ \rightleftharpoons PH_2^{\cdot+}$$

$$PH_2 - e^- \longrightarrow PH_2^{\cdot+}$$
$$PH_2^{\cdot+} + Py \rightleftharpoons PH^{\cdot} + PyH^+$$
$$PH^{\cdot} + PH_2^{\cdot+} \rightleftharpoons PH^+ + PH_2$$
$$PH^+ + Py \rightleftharpoons P + PyH^+$$

PH$_2$ = 1,3-diphenyl-2-(p-methoxyphenyl)-Δ²-pyrazoline (Ph, Ph, C$_6$H$_4$OCH$_3$)

P = 1,3-diphenyl-2-(p-methoxyphenyl)pyrazole (Ph, Ph, C$_6$H$_4$OCH$_3$)

Py = pyridinic base

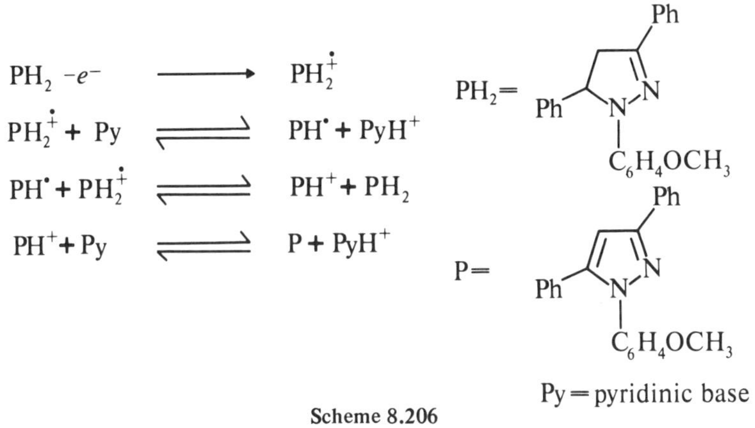

Scheme 8.206

1,3-Diphenyl-Δ²-pyrazoline **443** is oxidized in MeCN to a biphenyl dimer. The electrochemical and spectrochemical data indicate the dimerization of the radical cation but do not exclude the possibility of the reaction of the radical cation with the starting compound (650). Similar dimerization was also investigated for **443** chemically bonded to the electrode surface; the dimerization process is less important than for the dissolved species and is dependent on the sweep rate (651). Pyrazolines **444** and **445** have been investigated by cyclic voltammetry in MeCN solution and when bonded to an SnO$_2$ surface by silyl groups. For **444** reversible oxidation peaks are observed in both situations and **445** shows a reversible peak in solution, but the reversibility is greatly reduced in the bonded form. It is assigned to the low stability of bonded radical cations as compared to that of radical cations in solution (652). The methyl ester **446** shows a two-electron irreversible peak in MeCN and leads to **447**. The corresponding acid **448** shows a new, less cathodic

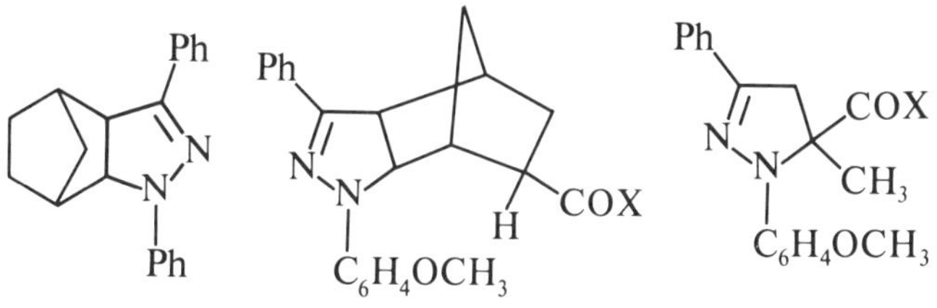

443   444   445

Oxidation of Heterocyclic Compounds 559

peak due to the carboxylate ion, as ionization is enhanced by the proximity of the pyrazolin π system (653). Electrolysis leads to **449**, and the oxidation of the carboxylate ion involves a radical stabilized by carboxylate group and π-system interactions.

1-Phenylpyrazolidin-3-one **450** (an antioxidant and photographic developer) has been oxidized in strong aqueous acid to dimeric and polymeric species (654, 655). In alkaline solutions, a reversible electron transfer on the pyrazolidinone anion gives a free radical, which disproportionates to 1-phenylpyrazolin-3-one **451** and **452** (656). In MeCN, two waves are observed by cyclic voltammetry, which are assigned to the protonated and neutral molecules. Controlled potential electrolysis resulted in the isolation of **451** (75%) and **452** (20%) (657). In the presence of an excess of chloride ions, a single two-electron irreversible wave is observed, and a quantitative yield of **452** is obtained via a rate-determining formation of **450$\overset{+}{\cdot}$** followed by rapid proton capture by Cl$^-$ (658).

The oxidation of various antipyrines has been studied at vitreous carbon anodes (659, 660). 4-Dimethylaminoantipyrine **453** gives a radical cation that disproportionates to **454** and/or **455** in addition to **453**.

8.3.1.4.3. Imidazoles.

The electrochemical oxidation of 2,4,5-triaryl imidazole has been investigated in benzonitrile (661) and nitromethane (662). 2,4,5-Triphenylimidazole **456** (662) is oxidized to a radical cation that deprotonates and dimerizes to **457**. In the course of preparative electrolysis, **457** isomerizes to **458**, but there is disagreement (661, 662) on the structure of the primary dimer (Scheme 8.207). At the higher temperature or in the presence of water, a ketone **459** is also obtained. The reaction was investigated (663, 664) with 2,4,5-tris-p-methoxyphenylimidazole (ImH) **460**, and the formation of the dimer was shown to follow an ECE mechanism.

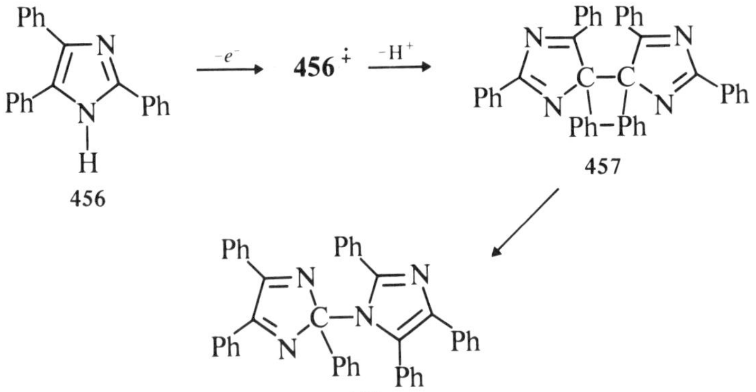

Scheme 8.207

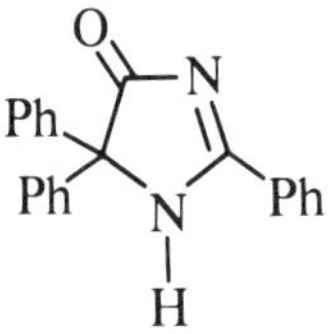

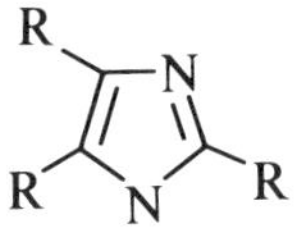

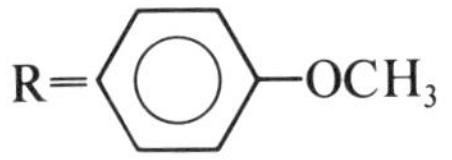

$$\text{ImH} \xrightarrow{-e} \text{ImH}^{\cdot+}$$

$$2\text{ImH}^{\cdot+} \rightleftarrows \text{Im}^{\cdot} + \text{ImH}_2^+$$

$$\text{Im}^{\cdot} \xrightarrow{-e} \text{Im}^+$$

$$\text{Im}^+ + \text{ImH} \longrightarrow \text{protonated dimer}$$

The electrochemical oxidation of **461** was also investigated (665).

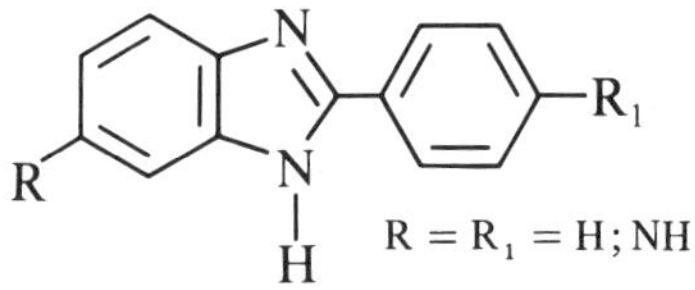

$R = R_1 = H; NH_2$

461

8.3.1.4.4. Pyridazine and Cinnoline. 1,4-Diphenyl-1,2,3,4-tetrahydrocinnoline is prepared by the oxidation of diphenylhydrazine in the presence of styrene (666) in acidic acetonitrile and is further oxidized to the corresponding cinnoline and cinnolinium cation (667) (Scheme 8.208).

Scheme 8.208

When a tetraalkylhydrazine cation radical is forced to twist at the N—N bond or is strongly pyramidal at nitrogen destabilization results. The degree of instability is estimated by measuring changes in $E^{o\prime}$ (the standard potential of the hydrazine-hydrazine cation radical in MeCN), which allows the determination of the energy gap between the hydrazine and its radical cation in strained **462** and unstrained **463** forms (668). It appears that bending a nitrogen a modest amount is not destabilizing at all (as in **462**), a great amount of bending seriously destabilizes a hydrazine radical cation (as in **464**). Great twisting of the radical cation is not, at least, as costly in energy but it does give the cation radical an extremely short lifetime.

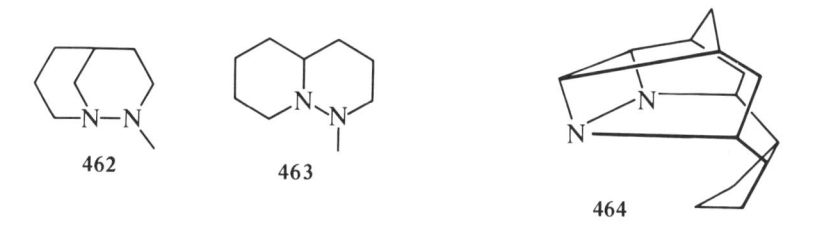

8.3.1.4.5. Pyrimidines. The electrochemical oxidation of pyrimidines (diaminopyrimidines, barbituric acid, and uracil) has been reviewed (669), and several papers dealing with barbituric acids and derivatives have been published. The oxidation of barbituric acid **465** and its 1-methyl and 1,3-dimethyl derivatives at pH 1 in the presence of chloride ions at a pyrolytic graphite electrode involves a single two-electron voltammetric peak (670). Three major products identified after electrolysis are 5,5′-dichlorohydurilic acid **466**, 5,5-dichlorobarbituric acid **467**, and

562 Electrochemical Behavior of Heterocyclic Compounds

alloxan **468**. The proposed mechanism is supported by the results of the electrochemical oxidation of 5-chlorobarbituric acid (an intermediate in the process) and explains the formation of **466**, **467**, and **468**.

465

466

467

468

The electrochemical oxidation of 1,3-dimethyl barbituric acid **469** has also been investigated in the absence of chloride ions (671). Two voltammetric peaks are observed at the pyrolytic graphite electrode. The first anodic voltammetric peak involves 1.33 electrons per molecule of **469** (Scheme 8.209), and an initial one-electron oxidation of **469** gives a neutral radical that dimerizes at the 5-position to a hydurilic acid. The latter compound is then further electrooxidized in a two-electron, one-proton process to give a carbonium ion that reacts further with barbituric acid to give an open-chain trimer **470**. The second voltammetric peak involves a somewhat more complex process. At this potential **470** is oxidized in a two-electron, two-proton reaction to give a cyclic trimer **471**. Direct oxidation of **469** at the level of the second peak involves two reaction routes (Scheme 8.209). The first proceeds via an initial one-electron process to a radical that dimerizes and ultimately gives **470**, which is further oxidized to **471**. The second route involves a further one-electron oxidation of the radical to give a carbonium ion, which is attacked by water to give dialuric acid. It is further oxidized to 1,3-dimethyl alloxan (Scheme 8.209).

The oxidation of **469** in 1 M acetic acid provides an easy synthetic route to **471** (47% yield) (672). 5-Alkylbarbituric acids are electrolyzed at the pyrolytic graphite electrode at pH 1–2.3 in the absence of Cl⁻, via a single one-electron voltammetric peak to form barbiturate radicals, which dimerize to 5,5′-dialkyl hydurilic acids **472** (673) (44–69% yield) (Scheme 8.210).

Hydurilic acid and its 1,1-dimethyl and 1,1′,3,3′-tetramethyl derivatives **472** have been oxidized at the pyrolytic graphite electrode (674). The first voltammetric peak involves a quasi-reversible two-electron, two-proton oxidation to dehydrohydurilic acid **473**, which reacts with MeOH to give **474**, and with water to give

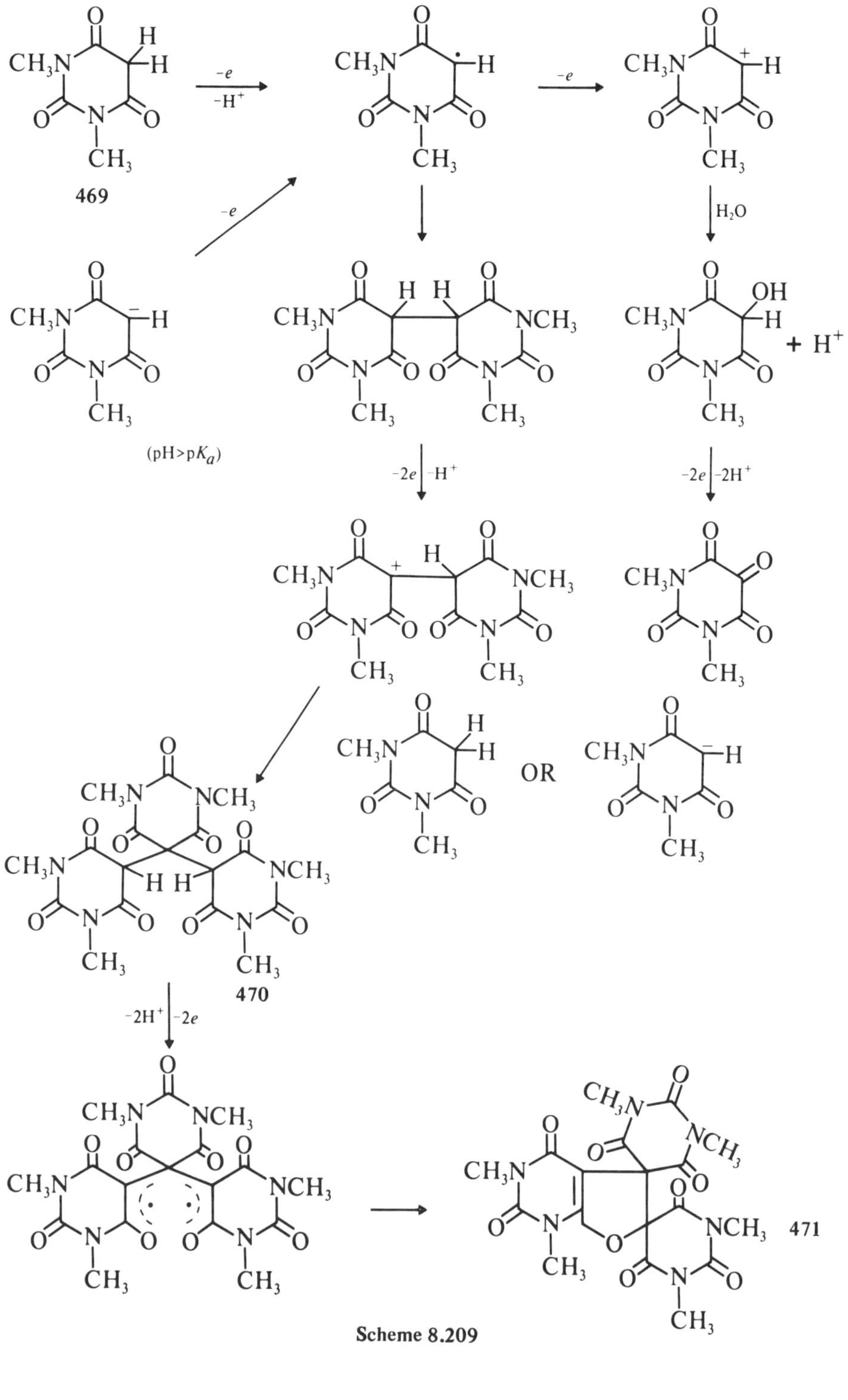

Oxidation of Heterocyclic Compounds 563

Scheme 8.209

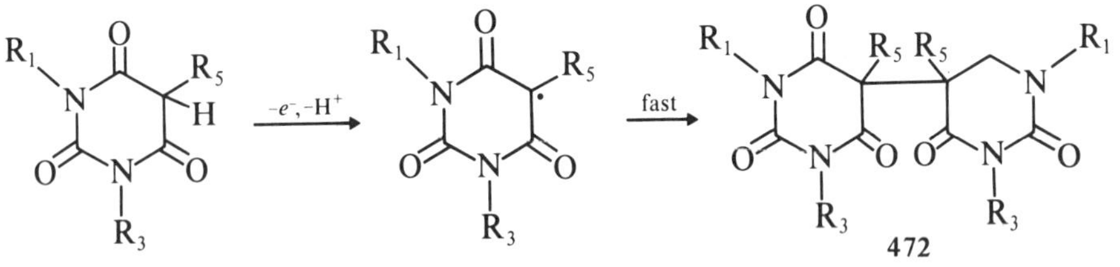

564 Electrochemical Behavior of Heterocyclic Compounds

Scheme 8.210

5-hydroxyhydurilic acid **475**, which decomposes to barbituric acid and alloxan. The kinetic constants have been measured (Scheme 8.211). Oxidation at the potential of the more positive peak appears to proceed via four-electron, four-proton process to give a dication that undergoes nucleophilic attack by water and/or chloride ion to give 5,5'-dichlorohydurilic acid, alloxantin, and 5-chloro-5'-hydroxyhydurilic acid. Further decomposition and electrochemical oxidation results in the formation of alloxan, 5,5-dichlorobarbituric acid, and 5,5-dichlorohydurilic acid (674).

5,6-Diaminouracil **476** has been investigated as a structural model of naturally occurring purines such as uric acid (675). At pH 0–4 the monocation of 5,6-diaminouracil is oxidized at the pyrolytic graphite electrode to the corresponding

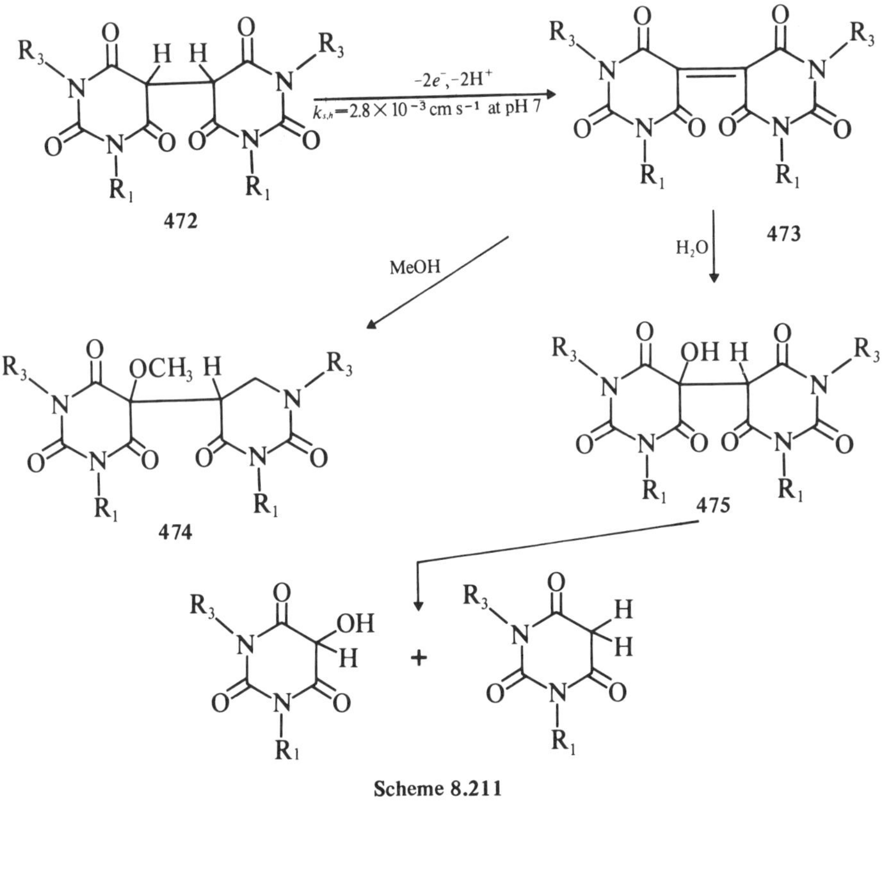

Scheme 8.211

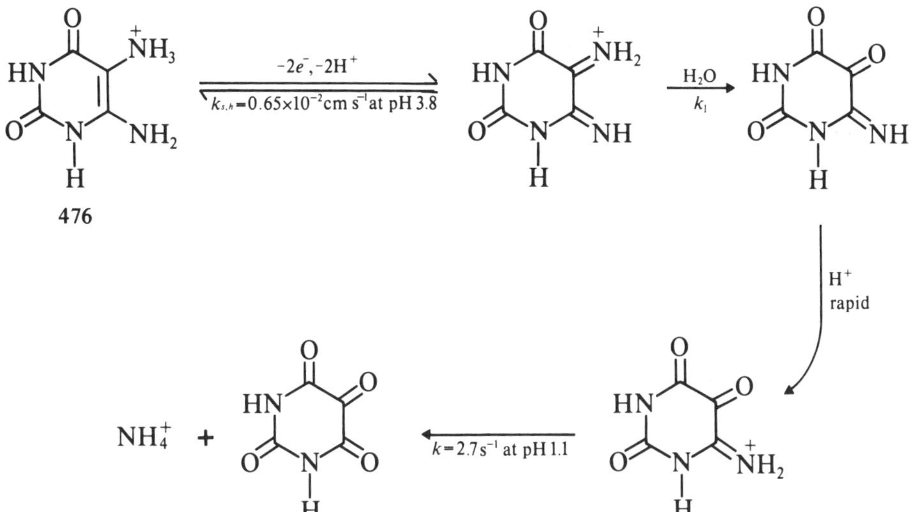

Scheme 8.212

diimine, which undergoes stepwise hydrolysis to quinone imine and then to alloxan. The characteristic rate constants of the EC mechanism have been measured (heterogeneous apparent electron transfer by cyclic voltammetry and hydrolysis by double potential step chronoamperometry). The major products of the reaction are alloxan and ammonia (675) (Scheme 8.212). The intermediacy of this important diimine (a similar diimine is a key intermediate during the oxidation of uric acid) and the value of the rate constant of hydrolysis were confirmed by thin layer spectroelectrochemistry (676). 5-Cyanouracil **477** has been prepared (95% yield) by the oxidation of uracil and tetrabutylammonium cyanide (677) (Scheme 8.213). 5-Fluorouracil has been prepared (60–90% yield) by anodic oxidation of uracil in HF (678) with Ni anodes and a current density of 15–30 mA cm^{-2} (679). The halogenation (Cl, Br, I) of uracil by anodic substitution reactions has also been investigated.

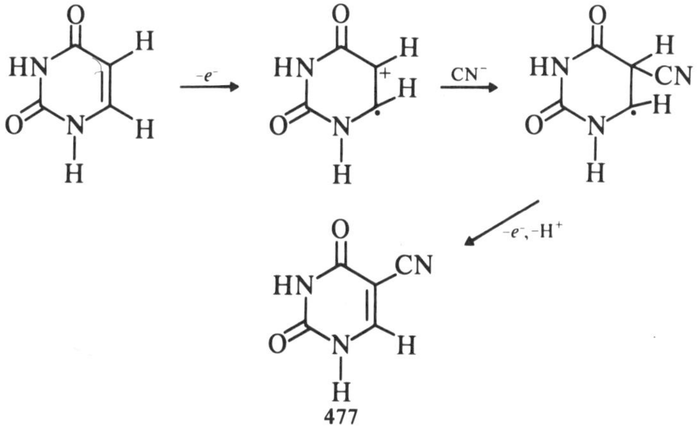

Scheme 8.213

8.3.1.4.6. Pyrazines, Quinoxalines, and Phenazines.

5,10-Dihydro-5,10-dimethylphenazine exhibits two successive one-electron steps in MeCN. In the presence of a nucleophile (or a nucleophilic solvent), the dication acts as a methylating agent toward the nucleophile, with the formation of 5-methylphenazinium cation (680).

The electrochemical oxidation of 5,10-diaryl-5,10-dihydrophenazines (PH$_2$) **478** in MeCN (681) gives rise to a stable radical cation and dication in the absence of nucleophile. In the presence of nucleophiles (Cl$^-$, AcO$^-$, pyridine), a substitution reaction involving the mechanism given in Scheme 8.214 occurs.

$$PH_2 - e^- \rightleftharpoons PH_2^{\cdot+}$$
$$PH_2^{\cdot+} - e^- \rightleftharpoons PH_2^{2+}$$
$$PH_2^{2+} + 2Nu^- \longrightarrow PHNu + HNu$$
$$PHNu - 2e^- \rightleftharpoons PHNu^{2+}$$
$$PHNu^{2+} + 2Nu^- \rightleftharpoons PNu_2 + HNu$$
$$PNu_2 - e^- \longrightarrow PNu_2^{\cdot+}$$

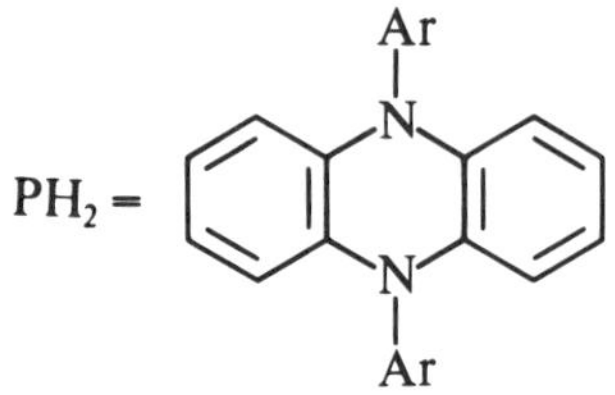

Scheme 8.214

Phenazine N,N'-dioxide **479** can be oxidized in benzonitrile, and the radical cation leads to a dimer (682, 683) (Scheme 8.215). The oxidation of 3,6-diisobutylpiperazine-2,5-dione **480** in acetonitrile leads to the formation of a new heterocyclic compound diimidazopyrazine dione **481** (Scheme 8.216) (684). The oxidation of N,N'-diformylpiperazine in MeOH/NBu$_4$BF$_4$ leads to the

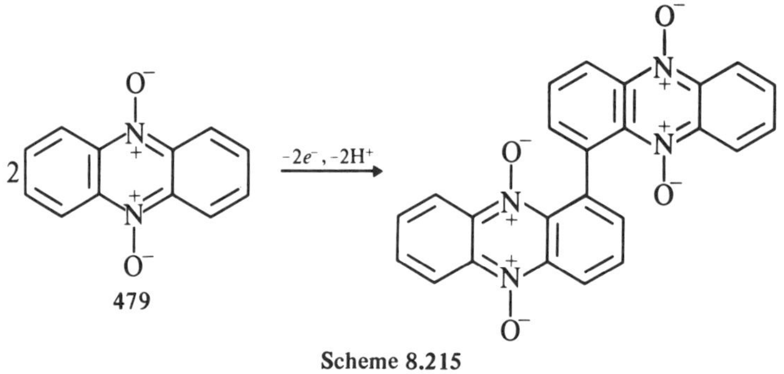

Scheme 8.215

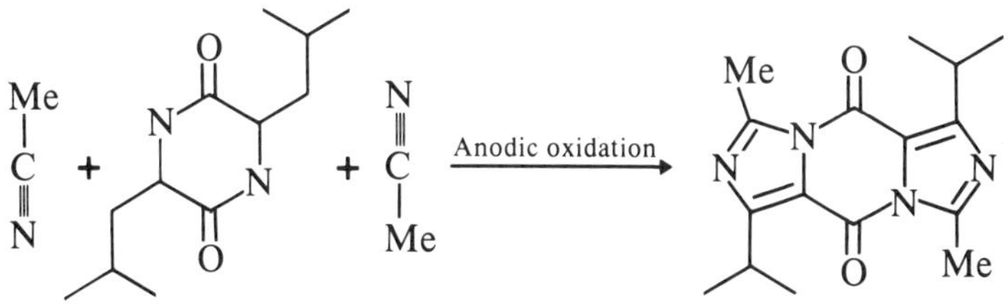

Oxidation of Heterocyclic Compounds 567

Scheme 8.216

formation of *N,N'*-diformyl-2-methoxypiperazine, which undergoes an acid-catalyzed elimination of methanol to give enamides (685, 686) (Scheme 8.217). Many other heterocyclic systems — pyrrolidine, piperidine, azacyclopentane, morpholine, and tetrahydroisoquinoline, undergo similar reactions.

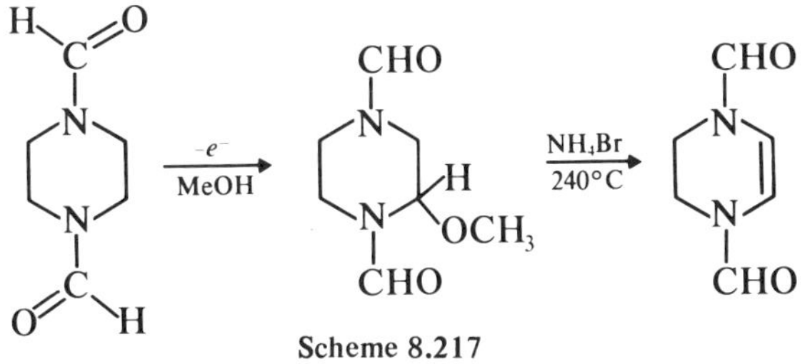

Scheme 8.217

Anodic oxidation of diacyldiazines such as **482** in DMF gives the acylium ion followed by the formation of the parent diazine (687). The reaction proceeds via an ECEC process, and the first chemical step (the deacylation of the radical cation) is greatly enhanced by the basic solvent.

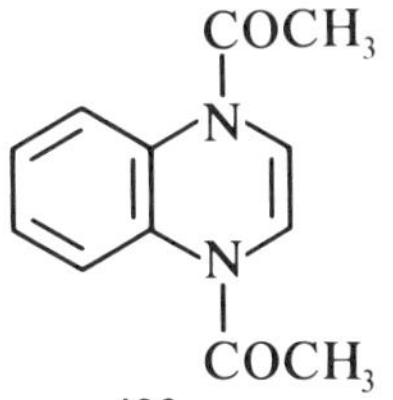

482

A large number of tertiary amines containing nitrogen atoms at 1,4 positions, mostly in cage structures, have been investigated by cyclic voltammetry (688). Most of them show an irreversible peak up to 5 V s^{-1} and a few, such as **483**, give reversible patterns (689). Due to favorable lone-pair–σ-CC interactions, through-bond (not through-space) interactions.

483

8.3.1.4.7. Other Heterocycles with Two Nitrogen Atoms.

The electrochemical behavior of 9-hydroxyellipticine **484**, an anticancer agent, has been investigated in aqueous media. The ellipticinium cation undergoes a one-electron transfer accompanied by a fast deprotonation, and the resulting neutral radical dimerizes but the dimer could not be characterized, due to its insolubility (690a, b). The dimerization rate constant has been measured.

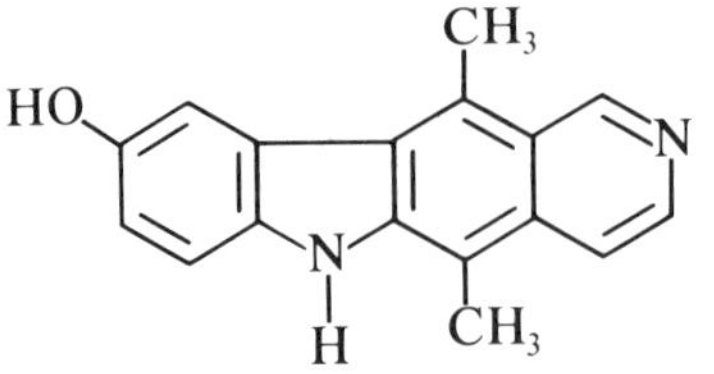

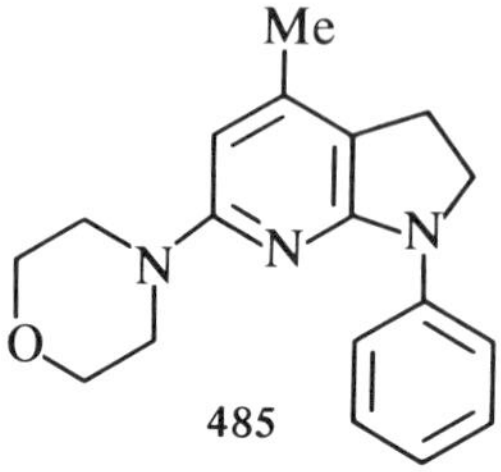

1-Phenyl-4-methyl-6-morpholino-7-azaindoline **485** is oxidized to a 7-azaindole and a biphenyl type of dimer (691). The macrobicyclic ligand [2.2.2]cryptate has been oxidized in propylene carbonate (692).

8.3.1.5. Compounds with One Nitrogen and One Oxygen Atom

Anodic oxidation of cycloserine **486** and its acetyl derivatives has been studied at the glassy carbon electrode in aqueous buffered solution. On electrolysis, **486** consumes 1 F mol^{-1} and yields about 50% serine, 10% NH$_3$, and a small amount of resinous products (693).

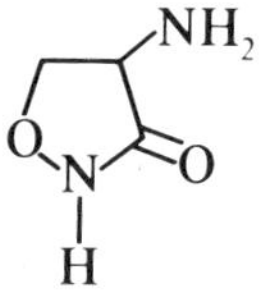

8.3.1.6. Compounds with One Nitrogen and One Sulfur Atom

8.3.1.6.1. Phenothiazines.

Tranquilizing drugs deriving from phenothiazine have been found to be oxidizable at the Au anode in dilute sulfuric acid (Scheme 8.218) (694a), and a DISP mechanism has been proposed in the case of chlorpromazine (694b) based on spectroelectrochemical measurements.

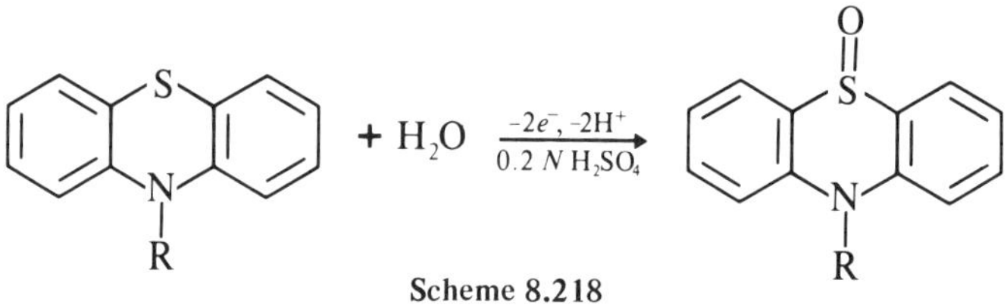

Scheme 8.218

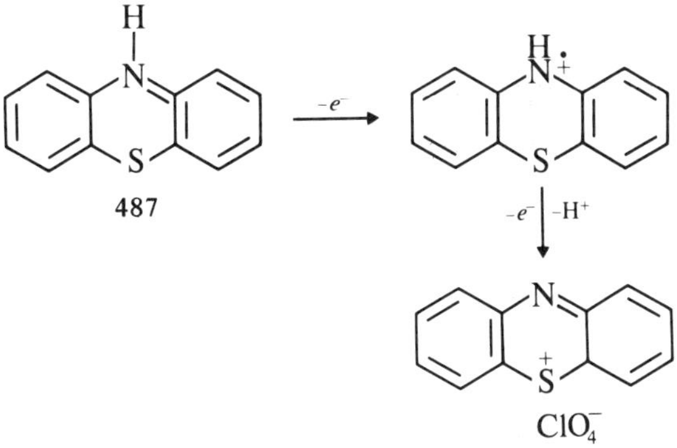

Scheme 8.219

In acetonitrile, phenothiazine **487** is oxidized in two one-electron waves. The first leads to the radical cation, and a second electron transfer and the loss of a proton (Scheme 8.219) yields phenothiazonium ion. This phenothiazonium ion is isolated as perchlorate (695–698). Cyclic voltammetry and coulometric studies of phenothiazine and thianthrene have been carried out in liquid SO_2. Both show two reversible one-electron waves corresponding to the formation of stable (for about 2 h) radical cations and dications (699).

The disproportionation equilibrium of the radical cation of 10-methylphenothiazine has been investigated in several solvent-supporting electrolyte systems by cyclic voltammetry (700a). Disproportionation is significantly influenced by association interactions between the cation and anion radicals of the supporting electrolyte and is favored by anions of greater ionic potentials. This anion effect is leveled by solvents of increased polarity. In the absence of ion association, disproportionation is favored by more polar solvents. The kinetics and mechanism of the reaction of 10-phenylphenothiazine dication with water in acetonitrile have been investigated (700b).

The electrochemical oxidation of 10-methyl phenothiazine (PCH_3) in MeCN in the presence of nonnucleophilic base collidine leads to the iminium cation (Scheme 8.220). In the presence of diethylphosphonate (RH), R-substituted compounds both on the ring **488** and on the methyl group **489** are obtained (701). Addition of R^- on the iminium cation PCH_2^+ and coupling of $R^{\cdot}$ and $PCH_3^{\dot{+}}$ ($R^{\cdot}$ being generated by electron transfer between R^- and $PCH_3^{\dot{+}}$) result in the formation of **488** and **489**, respectively. The oxidation of 10-methyl phenothiazine has also

$$PCH_3 \rightleftharpoons PCH_3^{\dot{+}} + e^-$$
$$PCH_3^{\dot{+}} + coll \rightleftharpoons PCH_2^{\cdot} + coll\,H^+$$
$$PCH_2^{\cdot} + PCH_3^{\dot{+}} \rightleftharpoons PCH_2^+ + PCH_3$$
$$PCH_2^{\cdot} \rightleftharpoons PCH_2^+ + e^-$$

Scheme 8.220

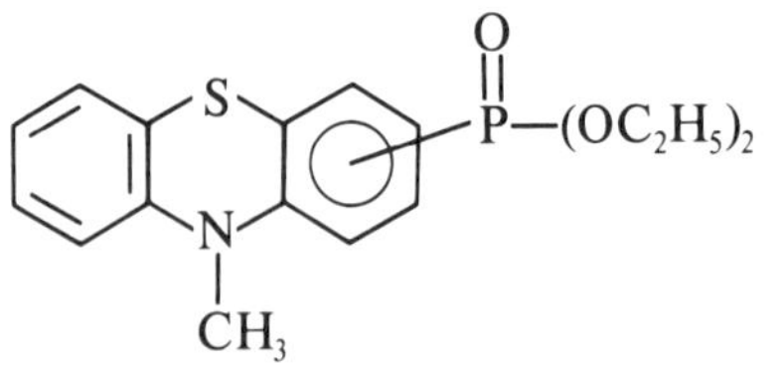

488

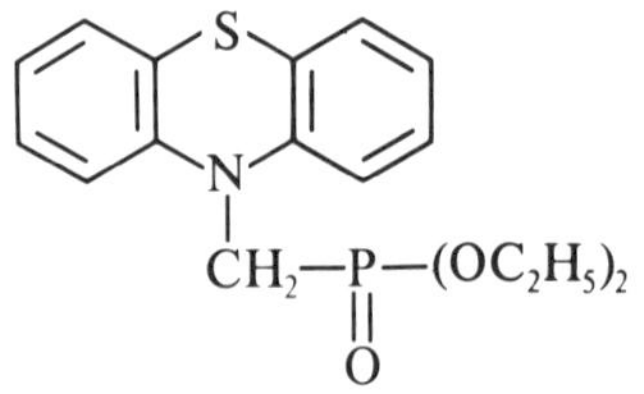

489

been investigated in micellar systems, and very small changes were observed in the stability of the radical cation compared with those in similar aqueous solutions (702, 703a, b). On oxidation, 3,7-dimethoxy phenothiazine **490** gives rise to four species that are stable in MeCN and have distinctive UV spectra. Therefore, this system has been used for the measurement of hydrogen ion concentration in MeCN

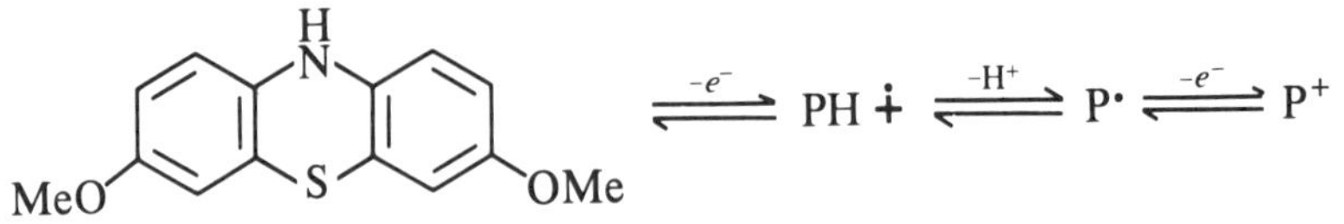

$$\text{490} \underset{}{\overset{-e^-}{\rightleftharpoons}} PH\overset{\cdot}{+} \underset{}{\overset{-H^+}{\rightleftharpoons}} P\cdot \underset{}{\overset{-e^-}{\rightleftharpoons}} P^+$$

(704). Other 3,7-disubstituted phenothiazines and *N*-methyl phenothiazines have been investigated in acetonitrile (705, 706). In acidic acetonitrile, the one- and two-electron oxidized species are characterized by electrochemistry and spectroscopy. In basic acetonitrile, 1,10 dimers **491** and 3,10 dimers **492** are obtained.

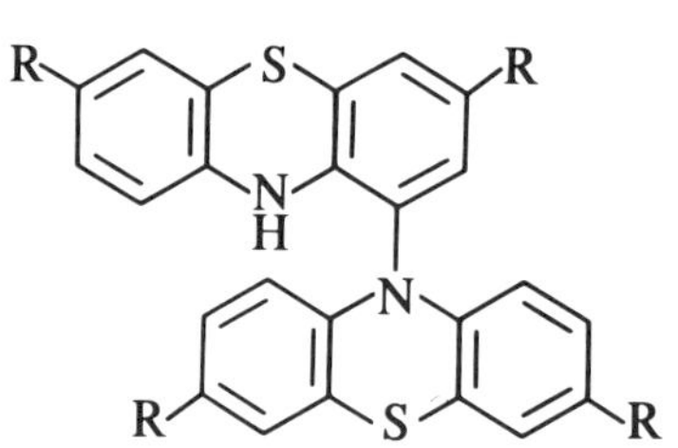

491

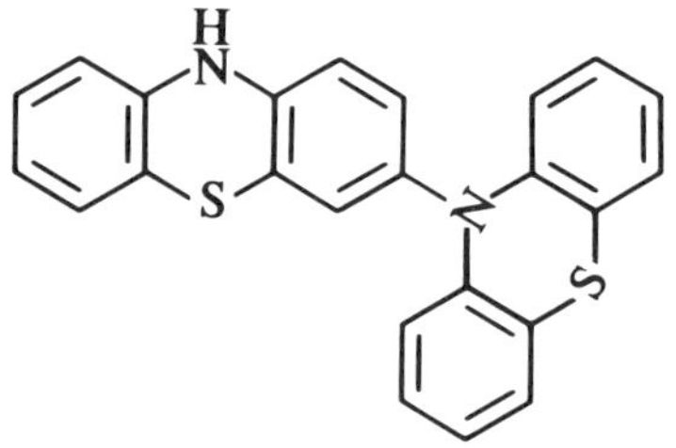

492

Hydroxy phenothiazines **493**, **494**, and **495**, which are metabolites of the drug chlorpromazine, are oxidized in aqueous solutions at low pH to stable protonated cation radicals (707). At slightly higher pH, these radical cations disproportionate to quinone imines, which undergo hydroxylation (Scheme 8.221).

A correlation has been established between the oxidation potential, drug activity, and side effects of several such tricyclic psychoactive drugs (708).

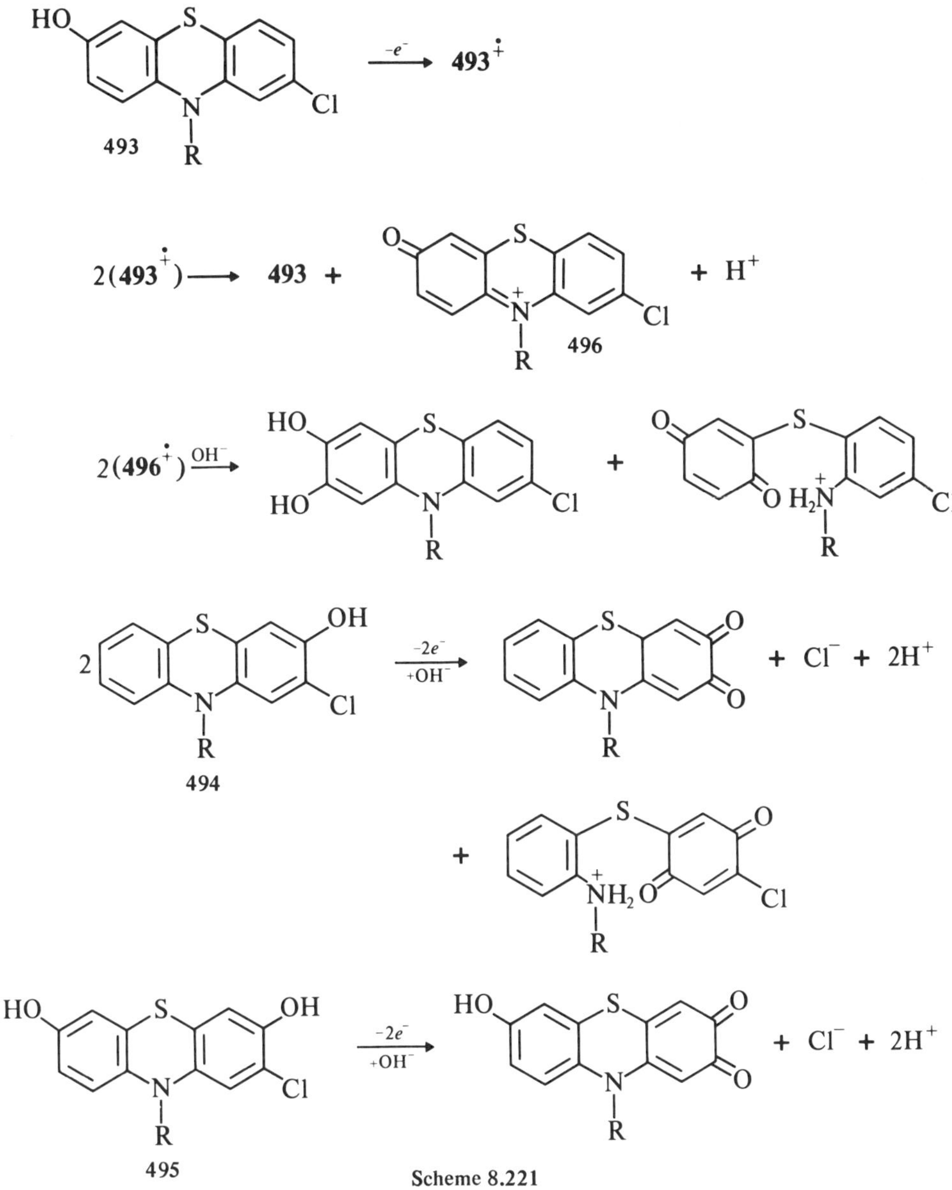

Scheme 8.221

8.3.1.7. Compounds with Two Oxygen Atoms

1,3-Dioxolane **497** is electrochemically oxidized in H_2SO_4 at noble metals and fuel cell electrodes. At low anodic potentials, partial oxidation occurs, giving CO_2 and ethylene glycol. At higher anodic potentials, $HOCH_2COOH$ and oxalic acid are obtained. The electrode-bound species C_2O_2 was identified as an intermediate in

497

the formation of oxalic acid (709). Oxidation of dioxene in methanol leads to C-methylated dioxanes (710a). 1,4-Dioxanes in MeCN gives a radical cation and a dication in two oxidation steps (710b). In a methanol/sodium methoxide system the electrochemical oxidation of 1,4-dioxane leads to a monomethoxylated compound (Scheme 8.222) (711). Use of ammonium nitrate as supporting electrolyte also gave the methoxylated product; lithium perchlorate does not yield any product.

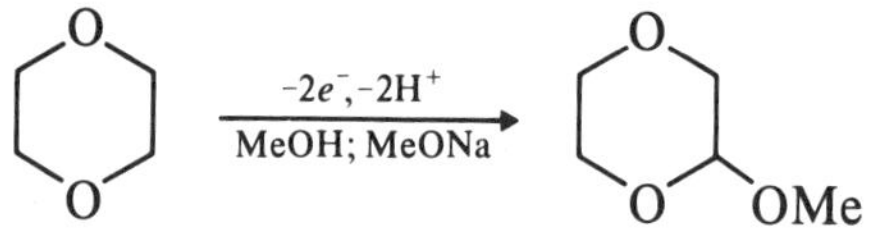

Scheme 8.222

The proposed mechanism is

$$\text{Supporting electrolyte} \xrightarrow{-e^-} S^{\cdot}$$

$$ROCH_2R' \xrightarrow{S^{\cdot}} RO\dot{C}HR' \xrightarrow{-e^-} RO\overset{+}{C}HR'$$

$$RO\overset{+}{C}HR' + MeOH \longrightarrow RO\underset{\underset{OMe}{|}}{C}HR' + H^+$$

Oxidation of benzodioxane in trifluoroacetic acid–tetrafluoborate–tetrabutylammonium yields the radical cation of its trimer (712) **498**.

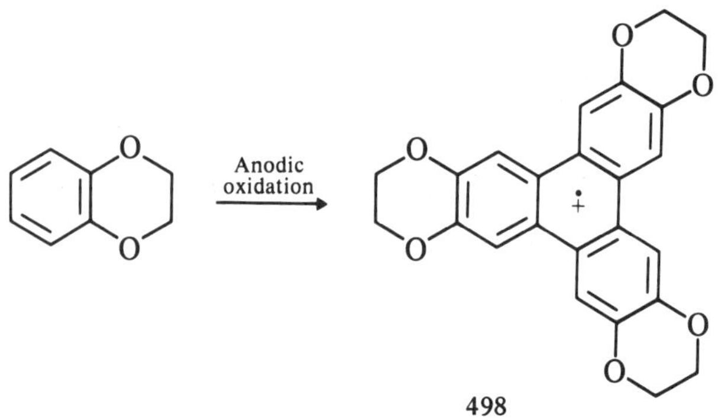

498

Benzodioxane behaves similarly to methoxy benzenes and is methoxylated at the Pt anode in methanolic KOH at constant current (713) (Scheme 8.223). In MeCN, dibenzodioxane is oxidized in two one-electron steps, the first yielding a fairly stable cation radical. In the presence of water, a more complex reaction is observed due to further oxidation of quinone obtained after the second electron transfer (Scheme 8.224) (714a, b).

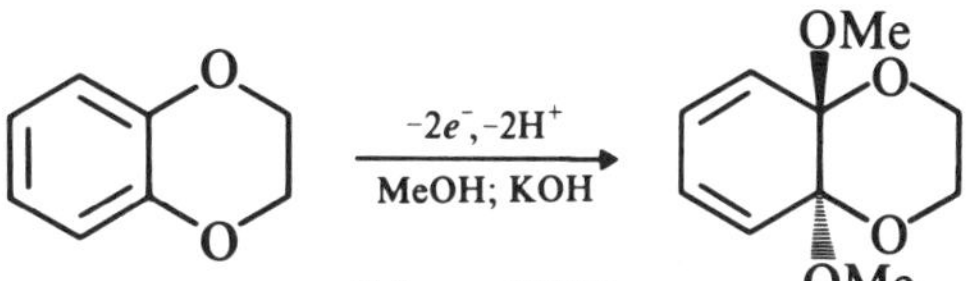

Oxidation of Heterocyclic Compounds 573

Scheme 8.223

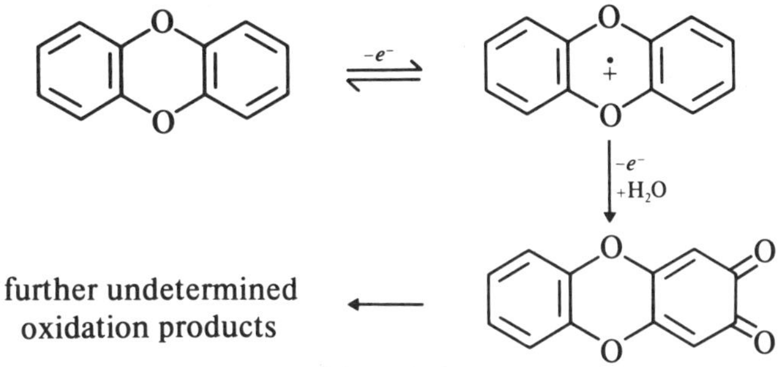

Scheme 8.224

The electrocatalytic oxidation of 1,1-diphenylethylene (DPE) by the radical cation of dibenzodioxane (DPO⁺) leads to the formation of 1,2,4,4-tetraphenyl-3-butene-1-one (TBPO) (715). The mechanism investigated by spectroelectrochemistry is shown in Scheme 8.225. Kinetic data concerning this process have been reported.

$$DPO \rightleftharpoons DPO^{\cdot+} + e^-$$
$$DPO^{\cdot+} + DPE \rightleftharpoons DPO + DPE^{\cdot+}$$
$$2DPE^{\cdot+} \longrightarrow TPB + 2H^+$$
$$\text{and/or } DPE^{\cdot+} + DPE \rightleftharpoons (DPE)_2^{\cdot+}$$
$$(DPE)_2^{\cdot+} + DPO^{\cdot+} \longrightarrow DPO + TPB + 2H^+$$
$$TPB + DPO^{\cdot+} \rightleftharpoons TPB^{\cdot+} + DPO$$
$$TPB^{\cdot+} + DPO^{\cdot+} \rightleftharpoons TPB^{2+} + DPO$$
$$TPB^{2+} + H_2O \longrightarrow TPBO + 2H^+$$

DPO = [dibenzodioxane structure]

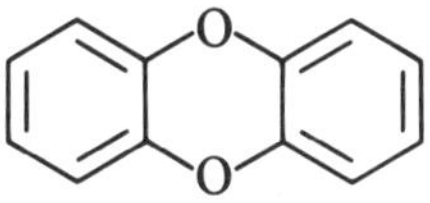

TPB = tetraphenylbutadiene

TPBO = Ph−C(=O)−CH(Ph)−CH=C(Ph)(Ph)

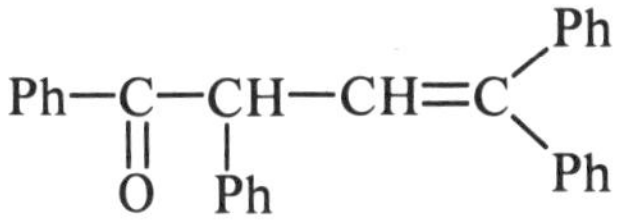

Scheme 8.225

8.3.1.8. Compounds with One Oxygen and One Sulfur (Selenium, Tellurium) Atom

Oxidative anisylation of phenoxathine **499** has been investigated by transmission spectroelectrochemistry involving a simple potential step followed by relaxation of the system (716). The reaction takes place by a half-regeneration mechanism. The first step involves a nucleophilic addition of anisole on the radical cation of phenoxathine (Scheme 8.226). The rate constants have been measured.

$$P - e^- \longrightarrow P^{\dot+}$$
$$P^{\dot+} + AnH \rightleftharpoons PAnH^{\dot+}$$
$$PAnH^{\dot+} + P^{\dot+} \longrightarrow PAn + H^+ + P$$

499   Scheme 8.226

Phenoxatellurin is oxidized at the Pt electrode in anhydrous MeCN/LiClO$_4$ to a complex consisting of two radical cations and one neutral molecule which is isolable as a perchlorate salt (717). Phenoxaselenin is also oxidized under these conditions to a radical cation that dimerizes or reacts with water (718).

8.3.1.9. Compounds with Two Sulfur Atoms

Tetrathiafulvalene **500** and its derivatives give rise to stable radical cations and dications on electrochemical oxidation in aprotic solvents. These species can be characterized by their UV spectra (719). The oxidation potentials of differently substituted tetrathiafulvalenes and dibenzotetrathiafulvalenes have been reviewed (720).

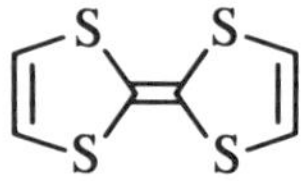

500

The electrochemistry of **500** and related donor molecules has been studied in the presence of Br$^-$ and MeCN. Cyclic voltammograms of tetraselenafulvalene **501** and benzotetrathiafulvalene **502** (721) provide evidence for the formation of a radical cation bromide salt. A linear phenoxytetrathiafulvalene polymer has been prepared and adsorbed on a metal substrate to prepare a modified electrode. The film has been characterized by optical and electrochemical methods (722).

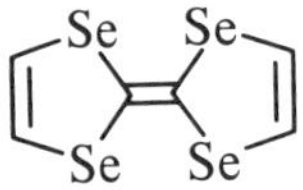

501 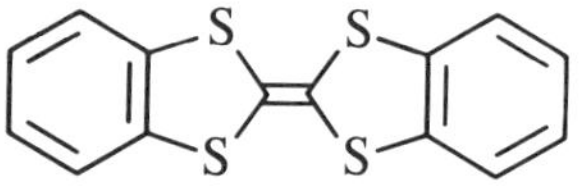   **502**

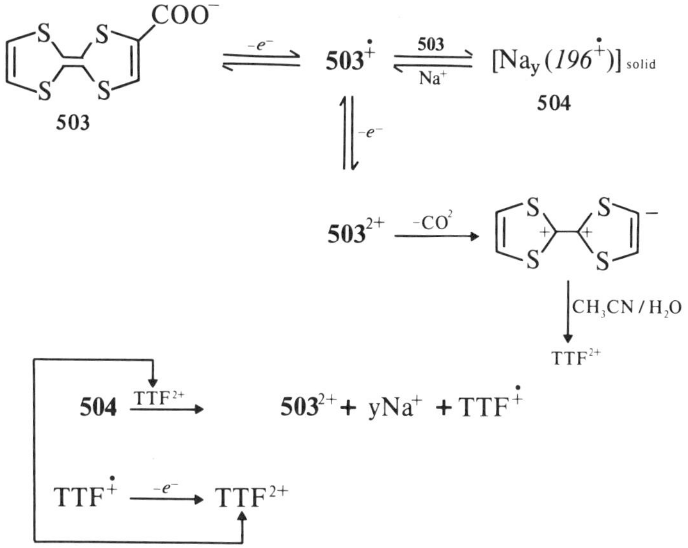

Scheme 8.227

The electrochemical oxidation of tetrathiafulvalene monocarboxylic acid **503** (723) and its conjugate base has been studied in MeCN by electrochemical and spectroscopic techniques. On the coulometric time scale, decarboxylation occurs for the dication of **503**. Nonstoichiometric salts **504** of the radical cation **503**$^{+\cdot}$ are formed during electrolysis and are further oxidized by the dication of tetrathiafulvalene (TTF^{2+}). The autocatalytic oxidation is shown in Scheme 8.227. In propylene carbonate, **505** (A) oxidizes to a dicationic dimeric species (A$_2^{2+}$), which can undergo a two-electron reduction to **506** (724).

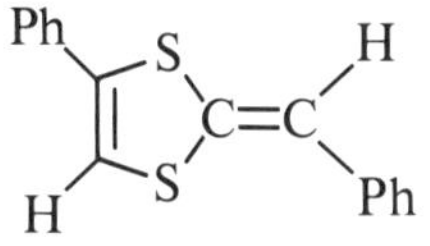

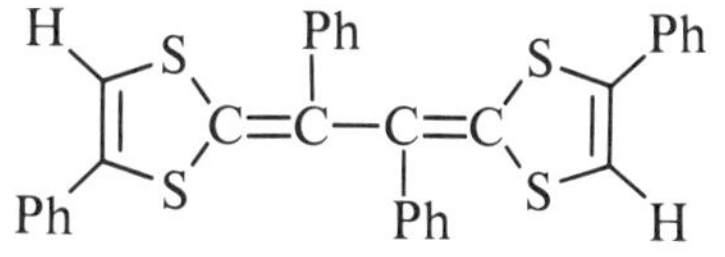

Tetrathioethylene **507** and thianthrene are oxidized at the tungsten anodes in AlCl$_3$–NaCl melts, and the dications formed are much more stable than in MeCN at room temperature (725, 726).

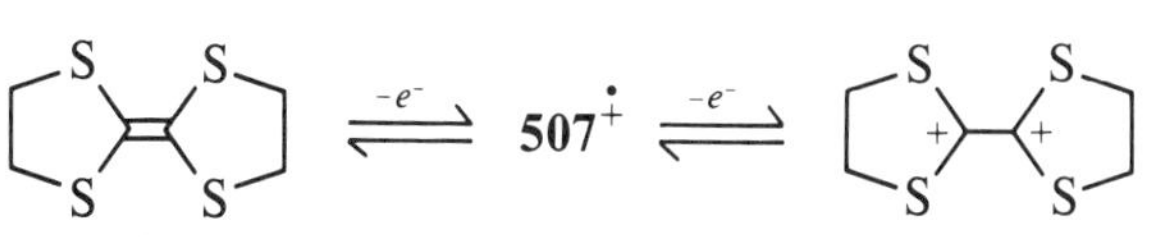

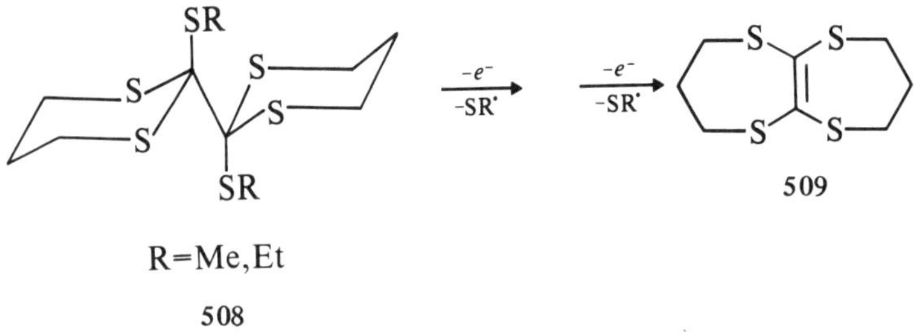

576 Electrochemical Behavior of Heterocyclic Compounds

R=Me,Et
508

Scheme 8.228

Cyclic trithiocarbonates undergo an irreversible two-electron oxidation in MeCN followed by reaction of water in the solvent to give cyclic dithiocarbonates and elemental sulfur (727).

The electrochemical oxidation of orthothiooxalates **508** containing the 1,3-dithian ring in MeCN proceed via rearrangement reactions of cationic intermediates to form the dication of an endocyclic tetrathioethylene **509** (Scheme 8.228) (728). Compounds **509** and **510** are oxidized to the radical cations and dications. At room temperature these dications undergo an interesting endocyclic-to-exocyclic rearrangement (Scheme 8.229) (728, 729). The driving force for the rearrangement is attributed to the minimization of coulombic interactions between positive sulfur atoms, which is possible by rotation about the central carbon-carbon bond in the exocyclic but not the endocyclic dication.

Noncyclic mono- and dithioethers undergo irreversible oxidation. The cyclic dithio ethers such as 1,5-dithiacyclooctane **511** undergo a facile and reversible oxidation with two closely spaced reversible peaks (20 mV). The oxidation yields a dimer (Scheme 8.230). The unusual ease of oxidation is due to a transannular interaction between two sulfur atoms that results in a three-electron bond. The second electron transfer is facilitated because the loss of the second electron is energetically favorable, the coulombic repulsion being more than offset by the removal of the unpaired electron from the antibonding orbital (730, 731).

Anodic oxidation of thianthrene **512** in acetic acid containing perchloric acid leads to the quantitative formation of the monoxide (732). At higher potential, either thianthrene or its monoxide forms the *cis*-dioxide (44%), *trans*-dioxide (28%), sulfone (13%), and trioxide and tetroxide (5%). The reactions of thianthrene

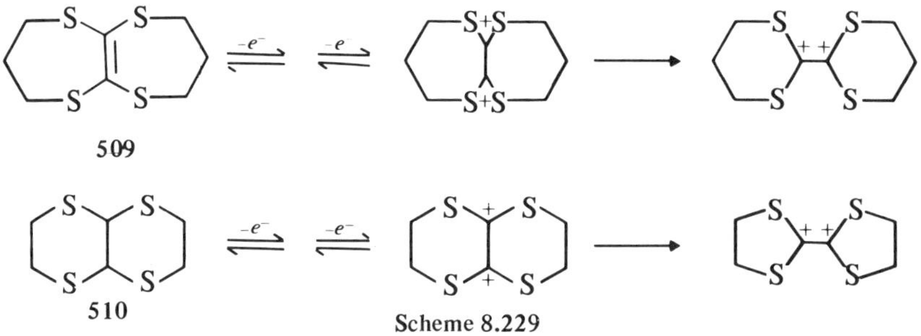

Scheme 8.229

$$DTCO \rightleftharpoons DTCO^{\cdot+} + e^-$$

$$DTCO^{\cdot+} \rightleftharpoons DTCO^{2+} + e^-$$

$$2DTCO^{\cdot+} \rightleftharpoons (DTCO)_2^{2+} \qquad K_d = 5000$$

$$(DTCO)_2^{2+} \rightleftharpoons 2DTCO^{\cdot+}$$

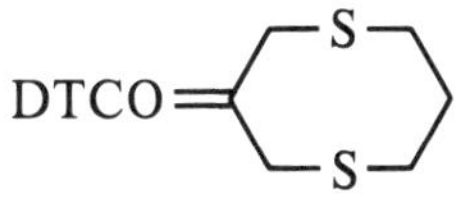

DTCO = [structure: ring with two S atoms] 511

Scheme 8.230

radical cation **513** and dication **514** have been the subject of controversial discussions on the reactivity of aromatic cation radicals. In a series of papers, Shine and coworkers have investigated the reactivity of chemically prepared cation radicals with different nucleophiles: water (733, 734), aromatics such as anisole (735), phenols (736), ketones (737), and amines (738). In the case of water (733, 734), equal amounts of thianthrene and thianthrene-5-oxide are formed. On the basis of experimental second-order kinetics, a disproportionation mechanism has been proposed (Scheme 8.231). However, an alternative half-regeneration mechanism (Scheme 8.232) has also been proposed by Parker and Eberson (739) on the basis of potentiostatic experiments.

$$2Th^{\cdot+} \rightleftharpoons Th + Th^{2+}$$
$$\quad 513 \qquad\quad 512 \quad 514$$

$$Th^{2+} + H_2O \longrightarrow ThO + 2H^+$$

512 = Th — thianthrene
ThO — thianthrene-5 oxide

Scheme 8.231

$$Th^{\cdot+} + H_2O \rightleftharpoons Th(OH)^{\cdot} + H^+$$

$$(ThOH)^{\cdot} + Th^{\cdot+} \rightleftharpoons (ThOH)^+ + Th$$

$$(ThOH)^+ \rightleftharpoons ThO + H^+$$

Scheme 8.232

$$Th^{+\cdot} + H_2O \rightleftharpoons Th(OH_2)^{+\cdot}$$

$$Th(OH_2)^{+\cdot} + H_2O \rightleftharpoons (ThOH)^{\cdot} + H_3O^+$$

$$Th(OH_2)^{+\cdot} + Th(OH)^{\cdot} \longrightarrow Th + H_2O + Th(OH)^+$$

$$Th(OH)^+ + H_2O \xrightarrow{fast} ThO + H_3O^+$$

Scheme 8.233

The anodic oxidation of thianthrene has also been studied by spectroelectrochemistry at optically transparent electrodes (740) and by reflectance spectroscopy (741). The results did not support the disproportionation mechanism (740) but did not permit a choice between the half-regeneration and ECE-mechanism oxidation of (THOH)· at the electrode, a reaction that cannot occur in Shine's experiments. This reaction was further investigated by Blount (742), who reported that the reaction follows third-order dependence on water and proposed the half-regeneration mechanism (Scheme 8.233). The homogeneous electron-transfer step is rate-determining, and the kinetic equation is compatible with experimental data. Besides an inverse first-order dependence, the H_3O^+ observed supports this mechanism.

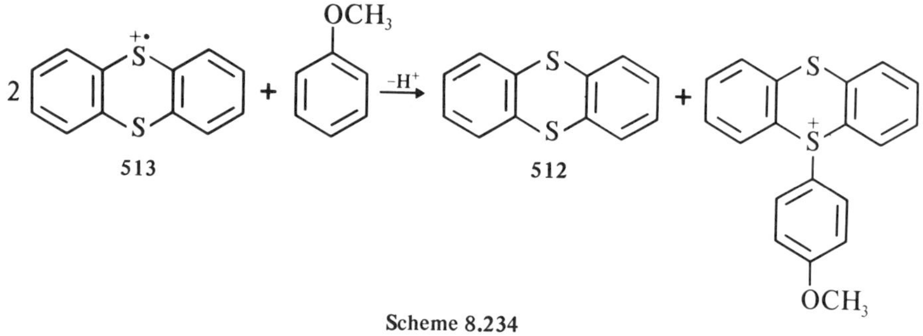

Scheme 8.234

Thianthrene cation radical reacts with anisole as shown in Scheme 8.234. Shine (735) observed second-order kinetics in cation radicals and proposed a disproportionation mechanism (Scheme 8.235). Parker (743) disproved the dispropor-

$$2Th^{+\cdot} \rightleftharpoons Th + Th^{2+}$$

$$Th^{2+} + AnH \longrightarrow ThAn^+ + H^+$$

Scheme 8.235

$$\text{Th}^{\cdot+} + \text{AnH} \rightleftharpoons (\text{Th}-\text{AnH})^{\cdot+}$$

$$(\text{Th}-\text{AnH})^{\cdot+} + \text{Th}^{\cdot+} \rightleftharpoons (\text{Th}-\text{AnH})^{2+} + \text{Th}$$

$$(\text{Th}-\text{AnH})^{2+} \longrightarrow \text{ThAn}^+ + \text{H}^+$$

Scheme 8.236

tionation mechanism and proposed a half-regeneration mechanism (Scheme 8.236) that is supported by the following observations:

1. The observed rate constants are too large compared to those that can be calculated from the disproportionation equilibrium constant (744, 745).
2. At low concentration of $\text{Th}^{\cdot+}$, a first-order rate law is observed, and at higher concentration the second-order rate constant decreases with the initial thianthrene concentration.
3. The rate of reaction is greatly enhanced by the addition of the radical cation of dibenzo p-dioxan, a better oxidant than $\text{Th}^{\cdot+}$, which participates in the electron exchange reaction.

The half-regeneration mechanism was also supported by Geniès (746) on the basis of chronoamperometric and chronoabsorptometric studies. However, in the case of p-dimethoxybenzene, the oxidation potential of which is very close to that of thianthrene, an electron-exchange mechanism takes place (Scheme 8.237).

$$\text{Th}^{\cdot+} + \text{NuH} \rightleftharpoons \text{Th} + \text{NuH}^{\cdot+}$$

$$\text{Th}^{\cdot+} + \text{NuH}^{\cdot+} \rightleftharpoons (\text{ThNuH})^{2+}$$

$$(\text{ThNuH})^{2+} \longrightarrow (\text{ThNu})^+ + \text{H}^+$$

NuH = p-dimethoxybenzene

Scheme 8.237

In the case of phenol as nucleophile, a disproportionation mechanism was proposed by Shine and coworkers (736) but refuted by Parker, who observed (in dichloromethane in the presence of trifluoroacetic acid) (747) the reaction to be second-order in $\text{Th}^{\cdot+}$ inhibited by unoxidized Th and first-order in phenol. The rate of reaction decreased markedly with increasing trifluoroacetic acid concentration. A half-generation mechanism similar to that of anisole has been proposed (747).

The radical cation of thianthrene **513** reacts with 1,1-diphenylethylene to give a vinylsulfonium ion **515**, the parent heterocycle (**512**), and a proton (748a) (Scheme 8.238). The mechanism has been shown to be of the half-regeneration

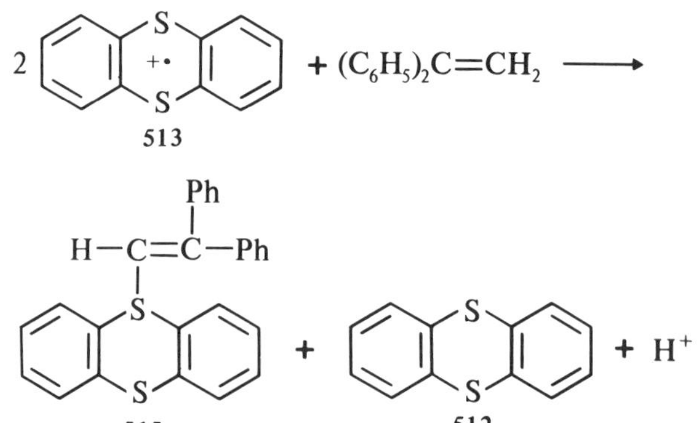

580 Electrochemical Behavior of Heterocyclic Compounds

Scheme 8.238

type by spectroelectrochemistry, and the rate constants have been measured. In liquid sulfur dioxide, the thianthrene radical cation does not react with anisole or water on the coulometric time scale (~1 hr). However, the dication reacts rapidly with anisole (AnH) in an EC reaction ($k \sim 10^8\ M^{-1}\ s^{-1}$), leading first to a (Th-AnH)$^{2+}$ complex and then to the final product (Th-An)$^+$ClO$_4^-$. The reaction of the dication Th^{2+} with water is first-order in water, and the dication ($k_{app} \sim 177\ M^{-1}\ s^{-1}$) leads to thianthrene monoxide (748b).

8.3.1.10. Compounds with Two Nitrogen Atoms and One Phosphorus Atom

Compounds such as **516** are oxidized to dimeric radical cations, which have been characterized by ESR spectroscopy (749) (Scheme 8.239).

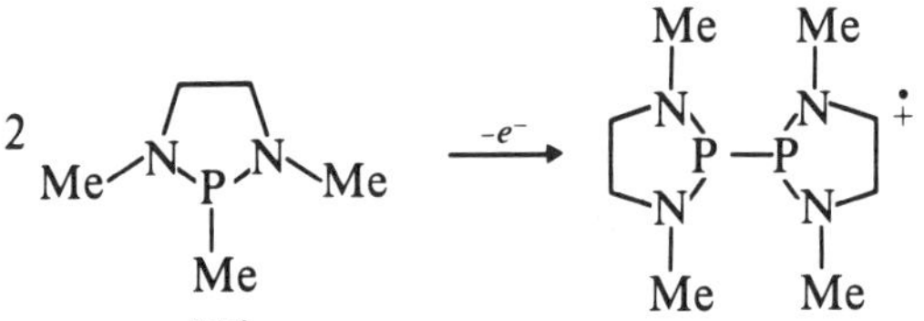

Scheme 8.239

8.3.1.11. Compounds with Three Sulfur Atoms

Trithiapentalenes **517** and related sulfur heterocycles dimerize reversibly in two-electron oxidation (750, 751) (Scheme 8.240).

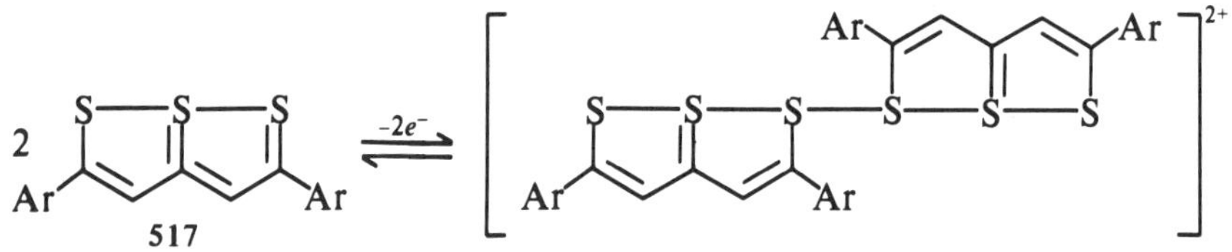

Scheme 8.240

Oxidation of Heterocyclic Compounds   581

8.3.1.12. Compounds with Four Nitrogen Atoms

8.3.1.12.1. Purines. The electrochemical oxidation of purine has been reviewed extensively (669) and compared with their biochemical behavior. The compounds investigated include uric acid **518**, xanthine **519**, theobromine **520**, caffeine **521**, adenine **522**, guanine **523**, and 6-mercaptopurine **524**.

Uric Acid. The voltammetric behavior of this compound has been described in Section 8.1.2.3 (Fig. 8.17). The mechanism proposed in low-phosphate buffers at pH < 5.6 is shown in Scheme 8.241 (752a, b, 753a–d). Between pH 3.7 and 5.7, the products are alloxan, 5-hydroxyhydantoin-5-carboxamide, allantoin, urea, and CO_2. The intermediate obtained on the first two-electron, two-proton oxidation is a diimine **525** (observed as peak I_c in Fig. 8.17), which is very unstable in aqueous solution (half-life 21×10^{-3} s at pH 8) and is rapidly hydrated to an imine alcohol **526** (responsible for peak II_c in Fig. 8.17) involving a first-order reaction (752a, b). Imine alcohol **526** is also unstable but can be observed as a UV-absorbing intermediate by thin layer spectroelectrochemistry on oxidation of uric acid. It is hydrated in a first-order reaction to uric acid 4,5-diol **527a**, which decomposes to the ultimate products. The imine alcohol **527c** can also undergo a ring contraction to **527b**, which leads ultimately to allantoin (753c). At pH > 6 the anion of the diimine **525** is hydrated to the anion of **526**, which undergoes ring contraction to **527b** (753c). The same UV-absorbing species (753d) are obtained on peroxidase oxidation, and the rates of disappearance are also the same. The final products of the electrochemical and enzymatic oxidations are identical, and therefore it has been suggested that the electrochemical and biochemical oxidations proceed through very similar mechanisms.

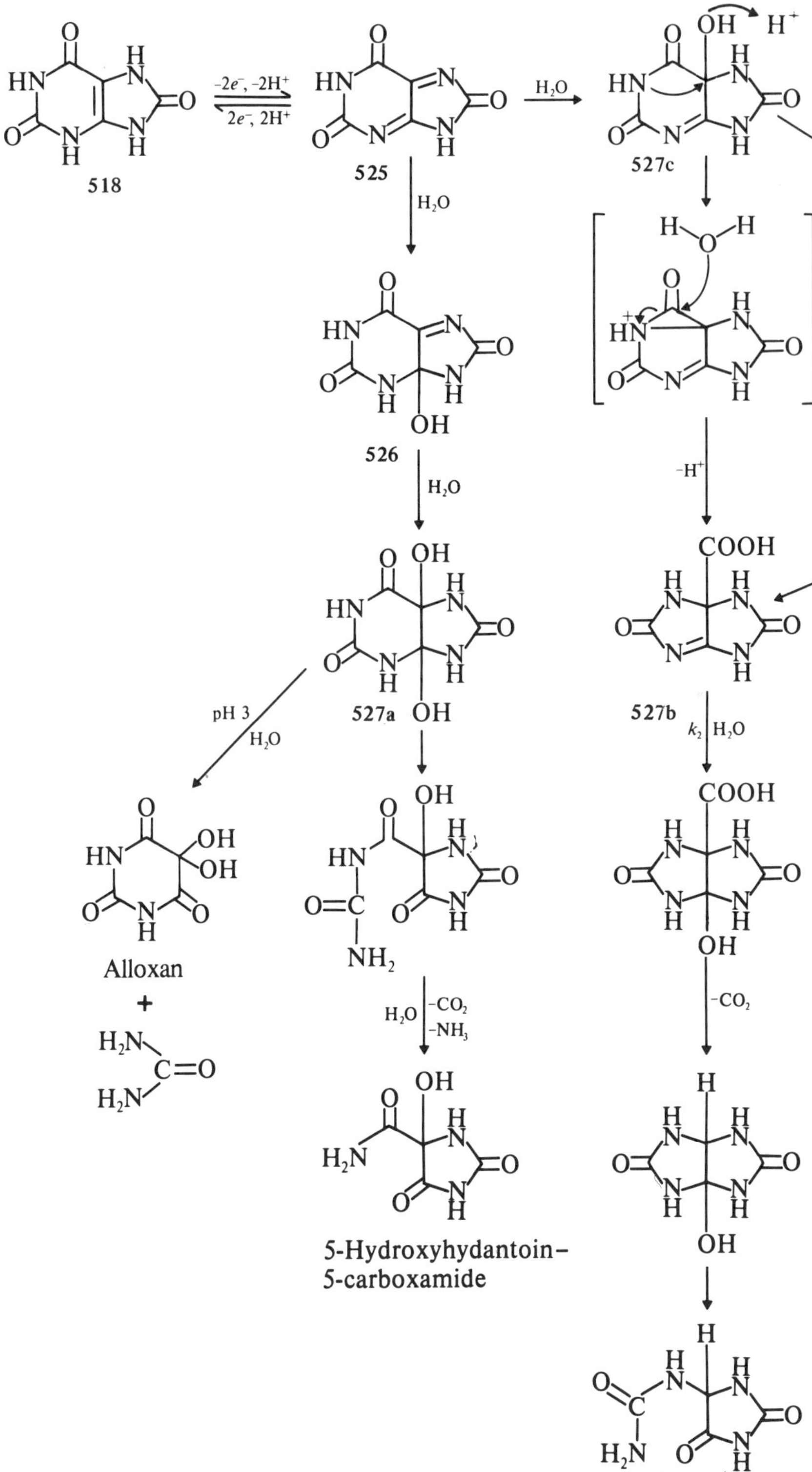

Scheme 8.241

N-Methylated Uric Acids. 1-, 3-, 7-, and 9-methyl; 1,3-, 3,7-, and 7,9-dimethyl; 1,3,7-trimethyl; and 1,3,7,9-tetramethyl uric acids have been investigated (754, 755a, b) by cyclic voltammetry preparative electrolysis and thin layer spectroelectrochemistry. They behave in the same way as uric acid itself, and a two-electron reaction gives an unstable diimine, which is rapidly hydrated to give an imine alcohol. The latter reacts with water to give uric acid 4,5-diol, which fragments to the same products (methylated on the corresponding positions) as uric acid yields. The oxidation of 3,9-dimethyl uric acid in the presence of peroxidase and H_2O_2 gives rise to an intermediate that exhibits spectral, electrochemical, and kinetic properties identical to that of imino alcohol generated electrochemically. It has been reported (756) that the biochemical and electrochemical reactions follow a similar course.

Oxipurinol. Oxipurinol 528 is a structural analog of xanthine and an inhibitor of xanthine oxidase. Its oxidation has been investigated (757) in dilute acetic acid by cyclic voltammetry and controlled potential electrolysis. The mechanism is very complex, and the primary steps are shown in Scheme 8.242. The final oxidation products obtained are 5'-hydroxy-5-carboxy-6,6'-azauracil 536 and 535; uracil 5-carboxylic acid 537 from 530; and alloxan, parabanic acid, and oxaluric acid from 535. Further oxidation of 535 leads to 538.

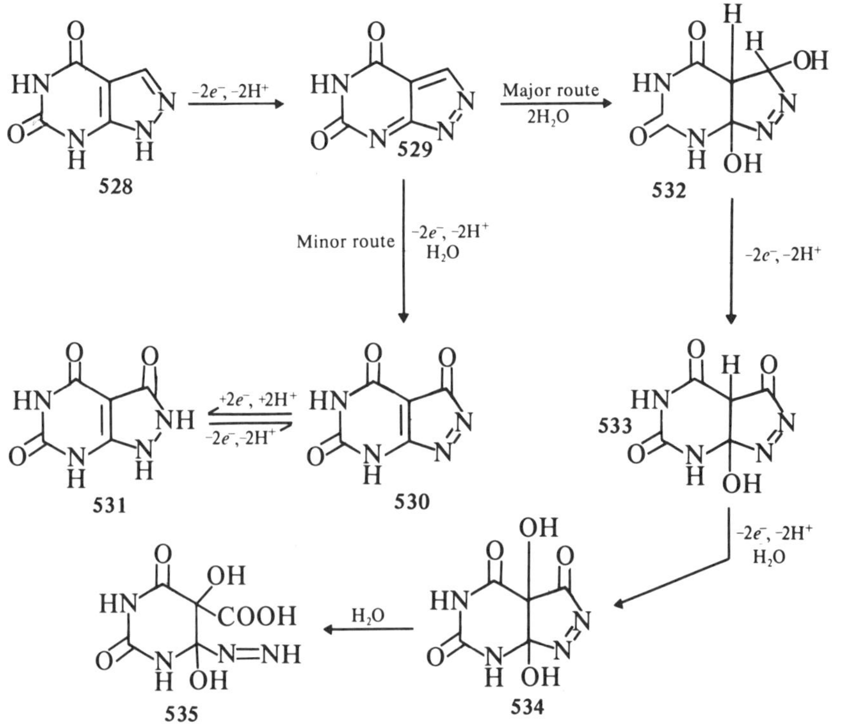

Scheme 8.242

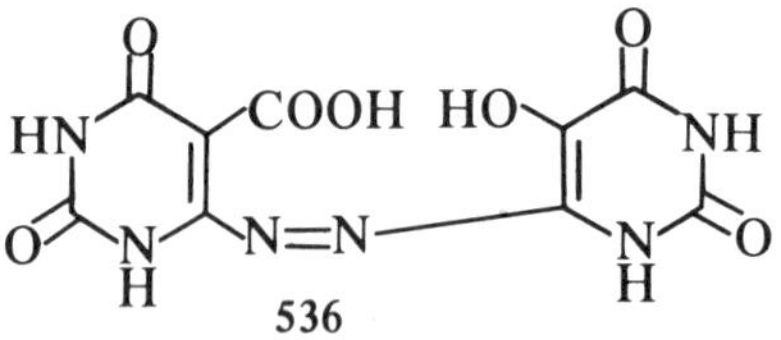

Xanthine. 9-Methylxanthine **539** shows four oxidation peaks in cyclic voltammetry in aqueous media (758). At the first peak, a dimer **540** is obtained via a one-electron, one-proton oxidation (Scheme 8.243). The dimer is further oxidized at the second oxidation peak to **541**, which is unstable and hydrolyzes to 1-methyl allantoin. The third oxidation peak is an adsorption prepeak to the fourth peak. The fourth peak corresponds to a direct four-electron, four-proton oxidation of

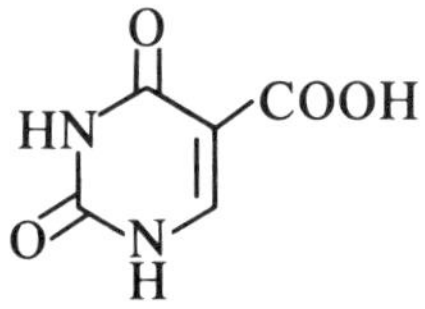

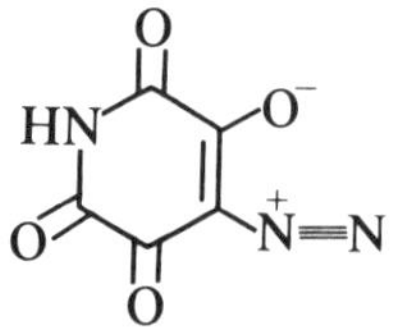

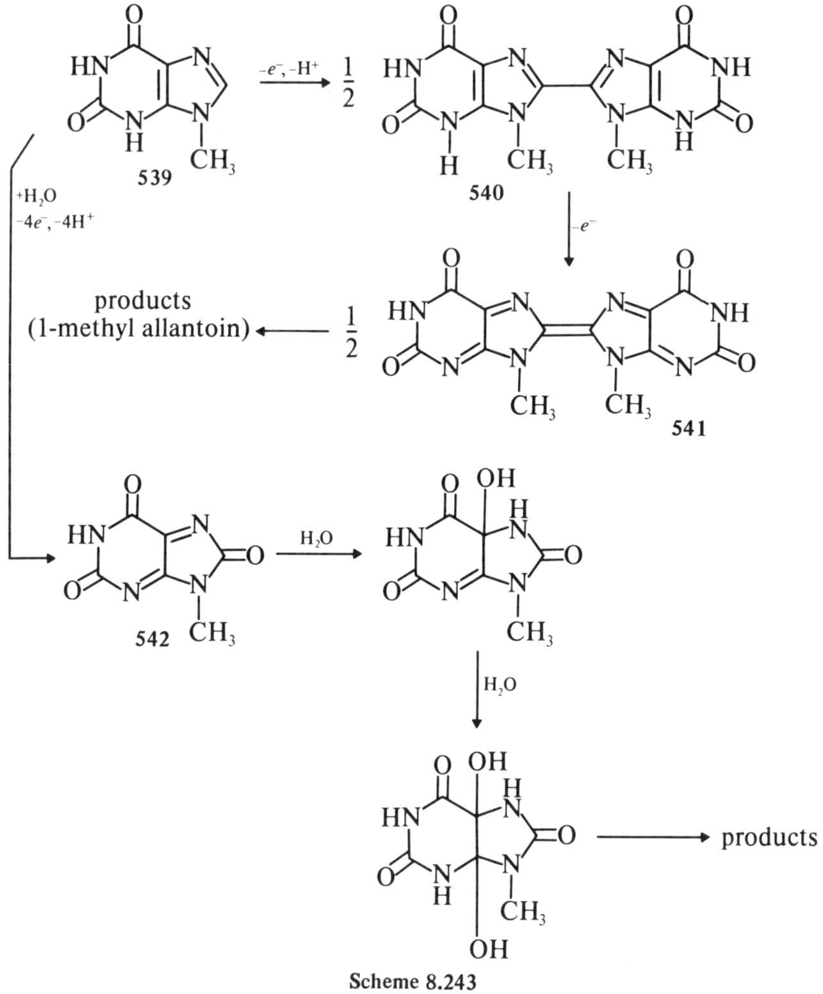

Scheme 8.243

Oxidation of Heterocyclic Compounds 585

539 to an unstable diimine of 9-methyl uric acid 542, which hydrolyzes to a variety of ultimate products (Scheme 8.243). The small amounts of xanthine in xanthosine (the nucleoside 9-β-D-ribofuranoxylxanthine) samples can be determined by linear sweep voltammetry at the pyrolytic graphite electrode (759). Trace amounts of xanthine and xanthosine-5′-monophosphate have been determined in polyriboxanthylic acid by differential pulse voltammetry at the pyrolytic graphite electrode (760).

Hypoxanthine. The first step in the oxidation of hypoxanthine 543 in aqueous media at the pyrolytic graphite electrode is the formation of 6,8-dioxypyrine 544, which is more easily oxidized than 543, and to 6,8-dioxypurine diimine 545 (761) (identified by spectroelectrochemistry). 545 undergoes further oxidation (Scheme 8.244) to 4-amino-4-carboxyimidazole-5-one 546, 5-hydroxyhydantoin-5-carboxamide 547, and 5-imino-2,4-imidazoledione 548.

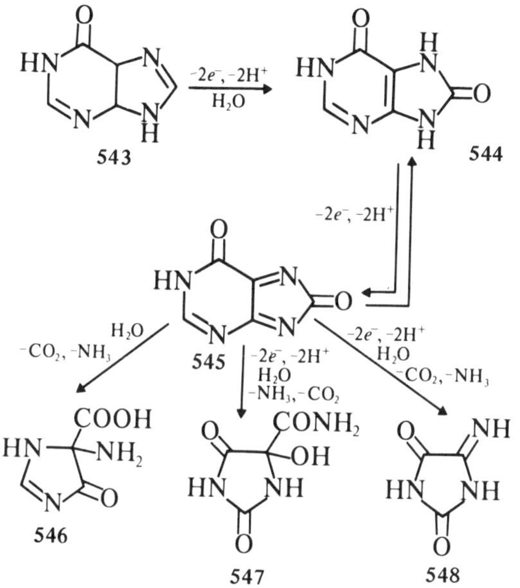

Scheme 8.244

Aminopurines. The oxidation of several aminopurines and their hydroxy derivatives has been investigated at a glassy carbon electrode in $1 M$ H_2SO_4 by cyclic voltammetry, coulometry, and controlled potential electrolysis (762a). The ease of oxidation increases in proportion to the number of amino or keto groups in the molecule. The oxidation consumes 6 F, 4 F, and 2 F per mole for mono-, di-, and trisubstituted purines, respectively. Aminopurines are oxidized first to 8-oxy and then to 2,8- or 6,8-dioxy intermediates (Scheme 8.245). The latter intermediate is further oxidized as uric acid itself in a two-electron process, to give a diimine intermediate that decomposes to the final products parabanic acid, oxaluric acid, urea, ammonia, allantoin, and guanidine.

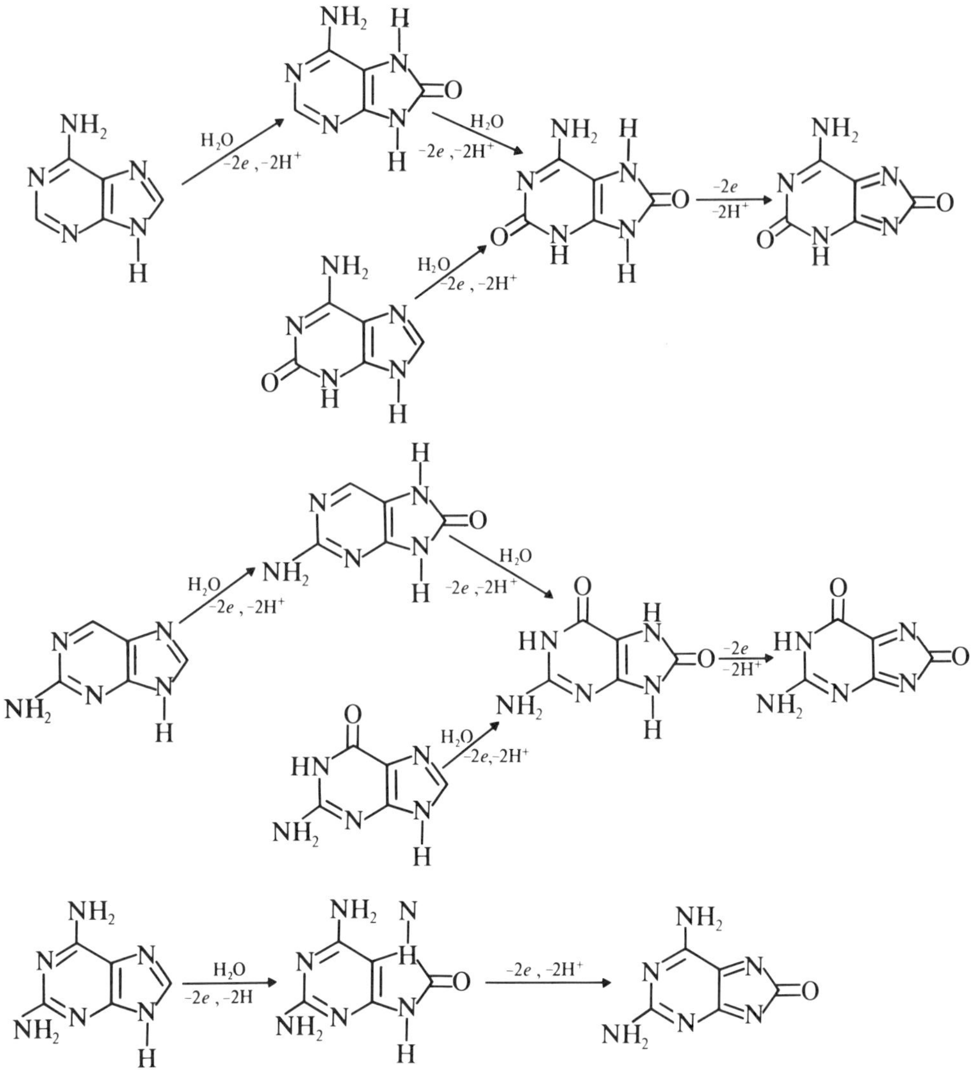

Scheme 8.245

In a more detailed investigation of guanine and 8-oxyguanine (762b), it was confirmed that guanine is initially oxidized in an irreversible two-electron, two-proton step to give 8-oxyguanine, which is immediately further electrooxidized in a two-electron, two-proton process to an unstable quinonoid-diimine. A series of hydration and other followup chemical and electrochemical reactions then occur, leading to the final products 2,5-diimino-4-imidazolone and 5-guanidinohydantoin. 8-Oxyguanine is also oxidized by peroxidase/H_2O_2, and spectral, kinetic, electrochemical, and analytical data indicate that the chemical reaction pathway followed in this process is similar to that observed in the electrochemical reaction (762b).

Adenine could be determined in the presence of adenosine, guanine, and guanosine (763) by voltammetry at a glassy carbon electrode. The oxidation of guanine has been discussed in relation to the photochemical oxidation mechanism (764).

Polynucleotides. The electrochemical oxidation of natural and biosynthetic polynucleotides at the pyrolytic graphite electrode has been studied under differential pulse voltammetric conditions. Denatured DNA, ribosomal RNA, and transfer RNA give two voltammetric peaks. The first (more negative) corresponds to the electrochemical oxidation of guanine residues, whereas the second (more positive) peak corresponds to the electrochemical oxidation of adenine residues (765).

8.3.1.12.2. Pteridines. 6- and 7-Hydroxypteridines **549** (hydrated form) and **550** are oxidized in aqueous media at the pyrolytic graphite electrode to 6,7-dihydroxypteridines **551** (766). 6,7-Dihydroxypteridine is oxidized in a two-electron, two-proton process to a diimine **552** (Scheme 8.246), which rapidly hydrates to a diol **553** and is further oxidized to **554**. Both **553** and **554** decompose to a variety of final products (766).

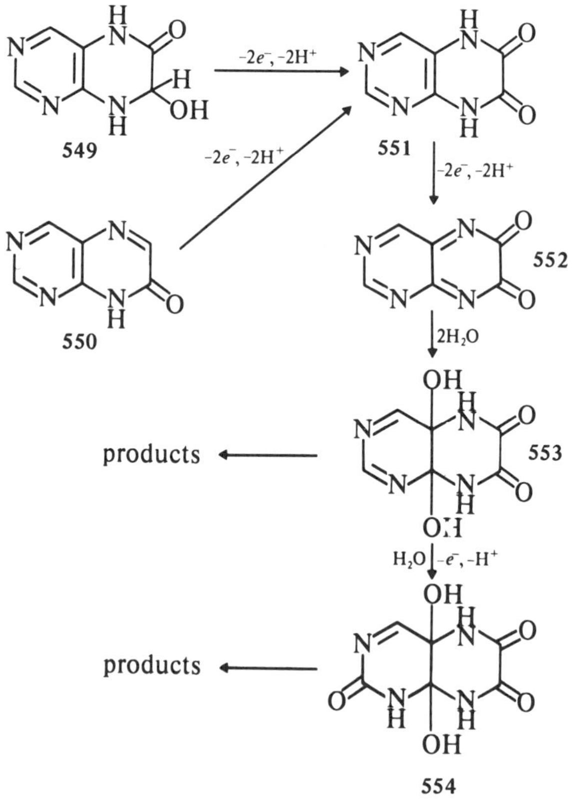

Scheme 8.246

588 Electrochemical Behavior of Heterocyclic Compounds

The electrochemical oxidation of pterins (2-amino-4-ketopteridines) has been the subject of several investigations because of their importance as enzyme cofactors (767a, b, c, d).

6- and 7-Substituted 2-amino-4-hydroxy-5,6,7,8-tetrahydropteridines have been oxidized at the Pt electrode (767a) in a two-electron process, but the oxidation products were not identified. 5,6,7,8-Tetrahydropterin is oxidized in aqueous media at the pyrolytic graphite electrode in an almost reversible two-electron, two-proton process give an unstable quinonoid dihydropterin, which rearranges to 7,8-dihydropterin (767b). 7,8-Dihydropterin is in equilibrium with its covalently hydrated form. The latter species undergoes a quasi-reversible two-electron, two-proton oxidation that leads successively to two quinonoid intermediates, the last one breaking down to an equimolecular mixture of pterin and 7,8-dihydroxanthopterin. The nonhydrated form of 7,8-dihydropterin is irreversibly oxidized to pterin (767b). In aqueous media, at the pyrolytic graphite electrode, 6-methyl-5,6,7,8-tetrahydropterin undergoes a quasi-reversible two-electron, two-proton oxidation, giving an unstable quinonoid-dihydropterin, which undergoes a first-order chemical followup reaction yielding 6-methyl-7,8-dihydropterin. In acidic media, 6-methyl-7,8-dihydropterin is in equilibrium with a covalently hydrated form. Both species are electrochemically oxidized to 6-methylpterins but through different mechanisms (767c). 6,7-Dimethyl-5,6,7,8-tetrahydropterin is oxidized to 6,7-dimethyl-7,8-dihydropterin. Both the hydrated and nonhydrated forms of the latter compound are oxidized to 6,7-dimethylpterins but through different mechanisms (767d). Detailed mechanistic investigations (767b, c, d) were carried out by cyclic voltammetry and thin layer spectroelectrochemistry.

8.3.1.12.3. Porphyrins. The electrochemical behavior of porphyrins has been reviewed (768). Anodic oxidation of porphyrin **555** in alkylnitrile at the Pt electrode take place in two one-electrode steps (769) (Scheme 8.247). The oxidation potentials of **556, 557, 558, 559**, and analogs of tetraphenylporphyrins where two NH groups are replaced by S, Se, and Te have been measured by cyclic voltammetry (770, 771a, b). The oxidation of octaethylisobacteriochlorin **560**, H$_2$(OE$_i$BC), in MeCN and dichloromethane has been reported by cyclic voltammetry, spectroelectrochemistry, and bulk electrolysis. **560** shows two reversible peaks, and on electrolysis an intense blue color is observed, which is assigned to the protonated form of octaethylchlorin **561**, H$_2$(OEC). The proposed mechanism (772) is shown in Scheme 8.248.

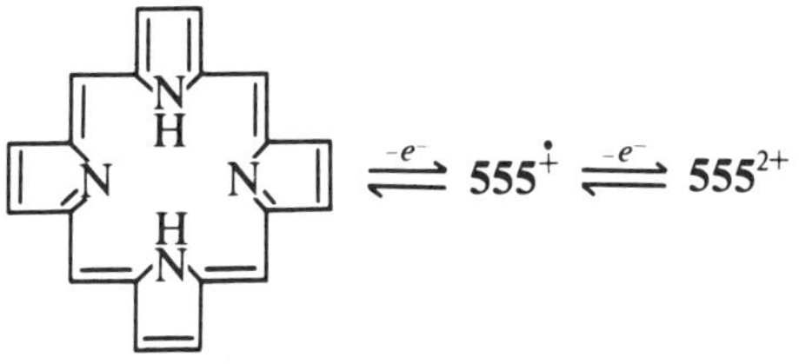

$$\text{555} \underset{}{\overset{-e^-}{\rightleftarrows}} \text{555}^{\dot{+}} \underset{}{\overset{-e^-}{\rightleftarrows}} \text{555}^{2+}$$

555   **Scheme 8.247**

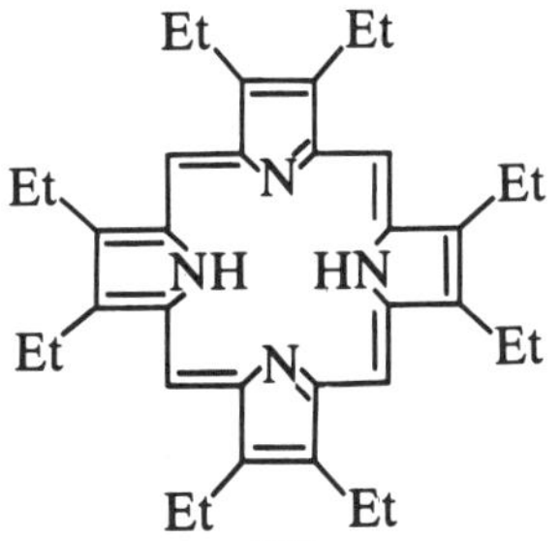

Oxidation of Heterocyclic Compounds   589

556, 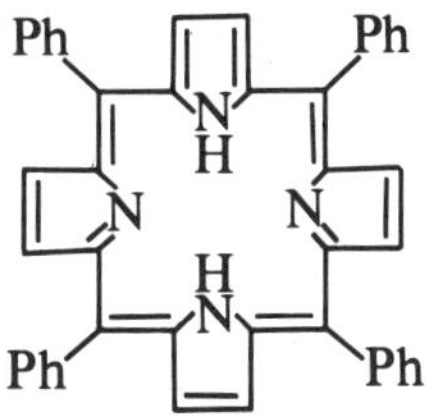 **557**, 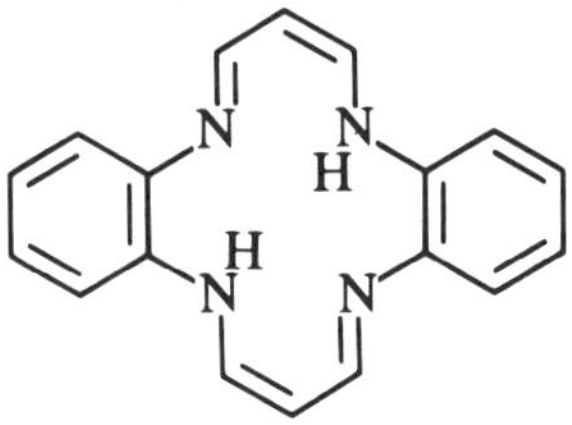 **558**, 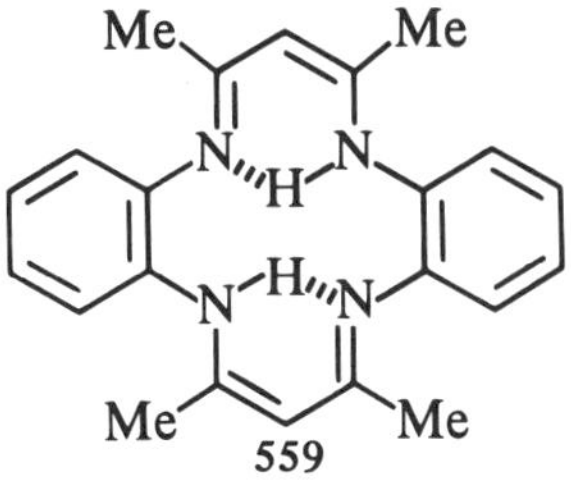 **559**

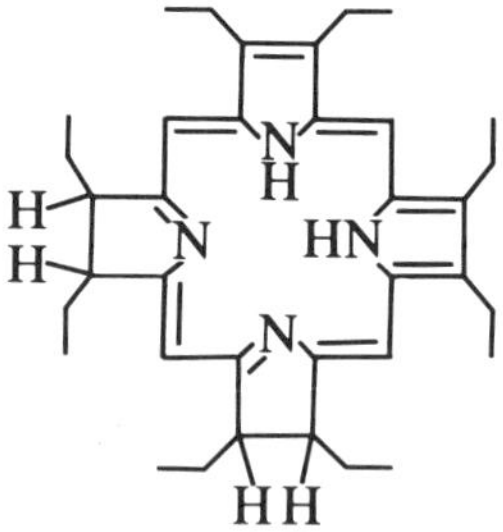

560 H₂(OE$_i$BC)

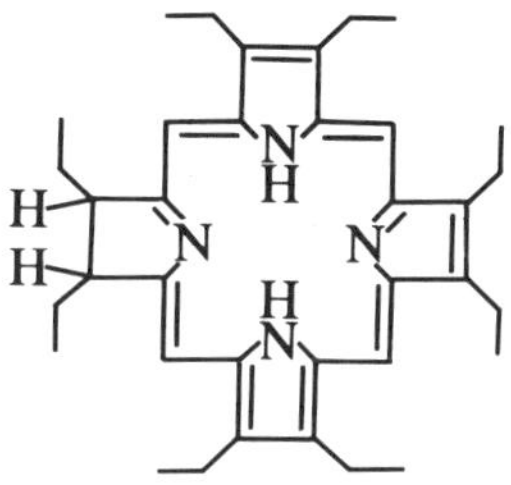

561 H₂(OEC)

$$H_2(OE_iBC) \rightleftharpoons H_2(OE_iBC)^{\dot{+}} + e^-$$

$$2H_2(OE_iBC)^{\dot{+}} \rightleftharpoons H_2(OE_iBC) + H_2(OE_iBC)^{2+}$$

$$H_2(OE_iBC)^{2+} \longrightarrow H_2(OEC) + 2H^+$$

$$H_2(OEC) + H^+ \rightleftharpoons H_3(OEC)^+$$

$$H_2(OE_iBC)^{\dot{+}} \longrightarrow H_3(OEC)^+ + H^+ + e^-$$

Scheme 8.248

590 Electrochemical Behavior of Heterocyclic Compounds

The electrochemical oxidation of Zn octaethylporphyrin in the presence of CN⁻ gives mono-, di-, tri-, and tetracyanooctaethylporphyrin with good selectivity and yields (773).

8.3.1.12.4. Other Compounds with Four Nitrogen Atoms. The synthesis of 2,3,5-triaryltetrazolium salts **563** was achieved by the oxidation of formazan **562**, which can be electrochemically reduced to **562** via an ECCE mechanism (774) (Scheme 8.249). The oxidation and reduction half-wave potentials of **564** have been correlated with orbital levels (775).

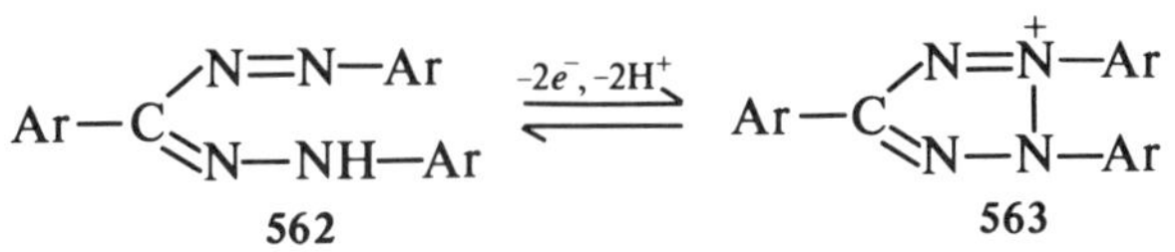

$$Ar-C\begin{matrix}N=N-Ar\\N-NH-Ar\end{matrix} \quad \xrightleftharpoons{-2e^-,-2H^+} \quad Ar-C\begin{matrix}N=\overset{+}{N}-Ar\\N-N-Ar\end{matrix}$$

562 → 563

Scheme 8.249

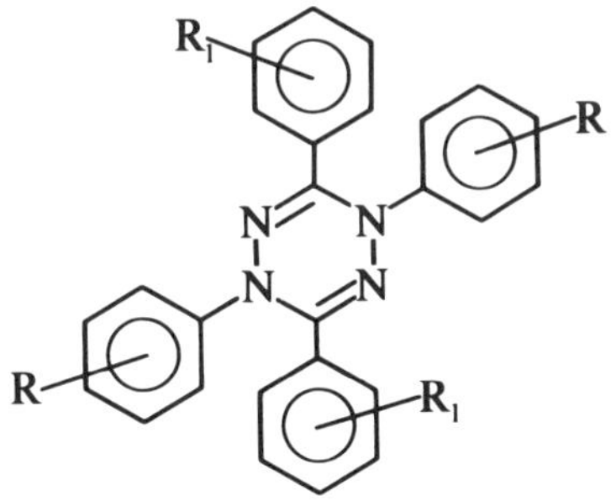

564

8.3.1.13. Compounds with Four Sulfur Atoms

Dehydrotetrathianaphthazarin **565** (a possible component of organic solid-state conductors) shows two reversible peaks on the oxidation in MeCN (776).

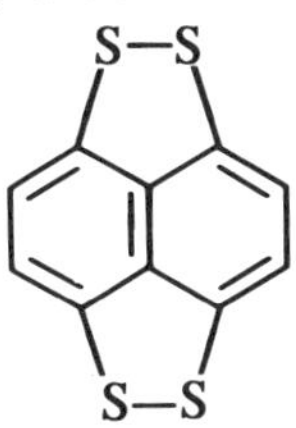

565

8.3.1.14. Compounds with Four Nitrogen and Two Sulfur Atoms

The electrochemical properties of naphthothiadiazine **566** have been investigated (777).

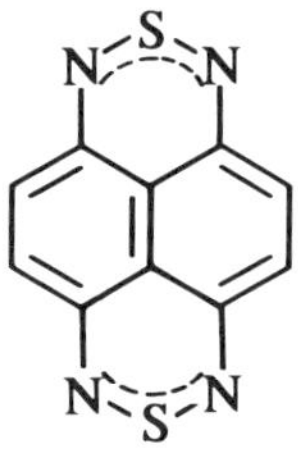

8.3.2. Oxidation of Substituted Heterocycles

8.3.2.1. Substituents with Oxygen Attached to the Ring

The electrochemistry of the compounds with hydroxyl groups has been reviewed (778, 779). Dialuric acid (5-hydroxybarbituric acid) is oxidized to alloxan (780). In aqueous media at pH 6.7, apomorphine is oxidized to oxoapomorphine (781). The proposed mechanism is shown in Scheme 8.250. Intramolecular anodic substitutions are observed during the oxidation of heterocyclic phenols. Pyridyl naphthols 567 (782) lead to intramolecular substitution products and the elimination of t-butyl groups (with 80% efficiency) (Scheme 8.251). Similarly, intramolecular coupling of 2,4,6-tri-t-butyphenol with acetonitrile (783) gives a benzoxazole (Scheme 8.252). Phenolic tetrahydroisoquinolines are oxidized to their C–C or C–O dimers. Optimum conditions involve electrolysis of the sodium salt in MeCN/NEt$_4$ClO$_4$ at the graphite felt anodes. Corypalline 568 leads to 82% 569, which is further oxidized to 570

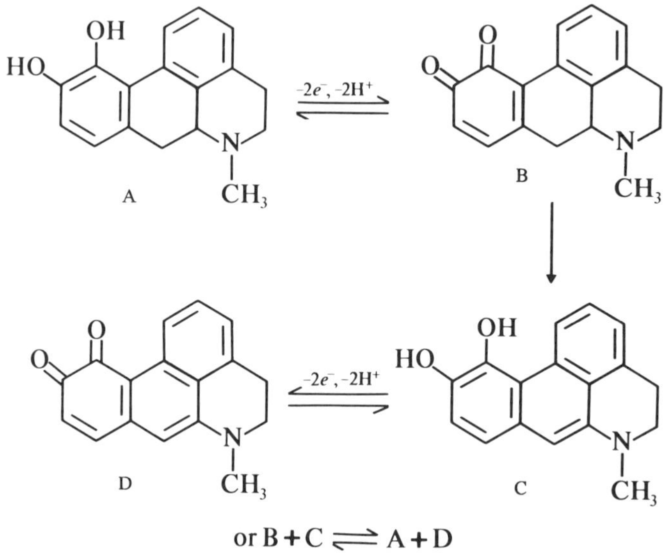

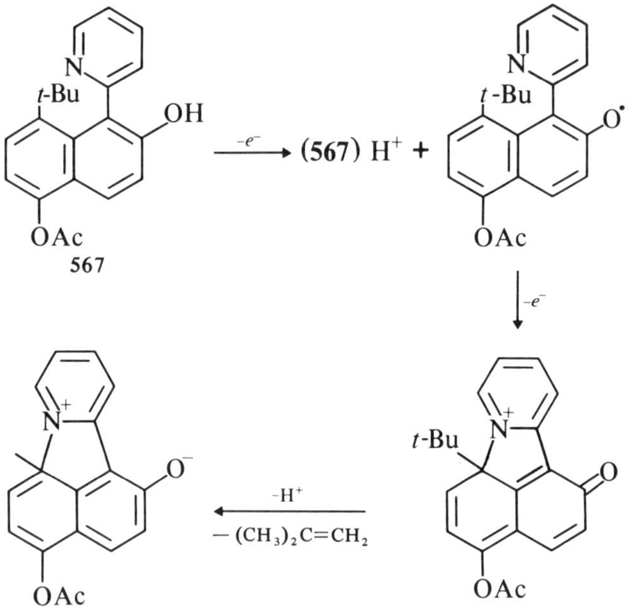

592 Electrochemical Behavior of Heterocyclic Compounds

Scheme 8.251

(784–786) (Scheme 8.253). Electrolysis of **571** under identical conditions leads mainly to **572** (Scheme 8.254). Racemic 1,2-dimethyl-7-hydroxy-6-methoxy-1,2,3,4-tetrahydroisoquinoline **573** is oxidized in wet MeCN/NEt$_4$ClO$_4$ at the graphite felt anode to give a carbon-carbon dimer. Interestingly, only molecules with the same configuration at C-1 coupled with each other to form products (R with R and S with S). Furthermore, only one of the two possible rotational isomers was formed (787, 788) (Scheme 8.255).

Electrochemical oxidation of *N*-carbethoxy *N*-norarmepavine **574** leads to a C–C dimer and also to a C–O dimer with the skeleton of the alkaloid dauricine **575** (789) (Scheme 8.256). The isochromanone **576** is oxidized to spirodienone **577** in dichloromethane trifluoroacetic acid and to **578** in MeCN (790) (Scheme 8.257). The electrochemical oxidation in MeCN/LiBF$_4$ of isochroman **579** and its 9-methyl derivative results in intramolecular coupling, giving **580** and **581** after hydrolysis

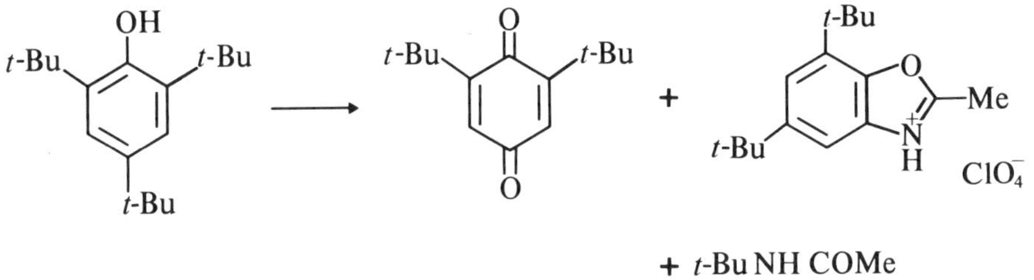

Scheme 8.252

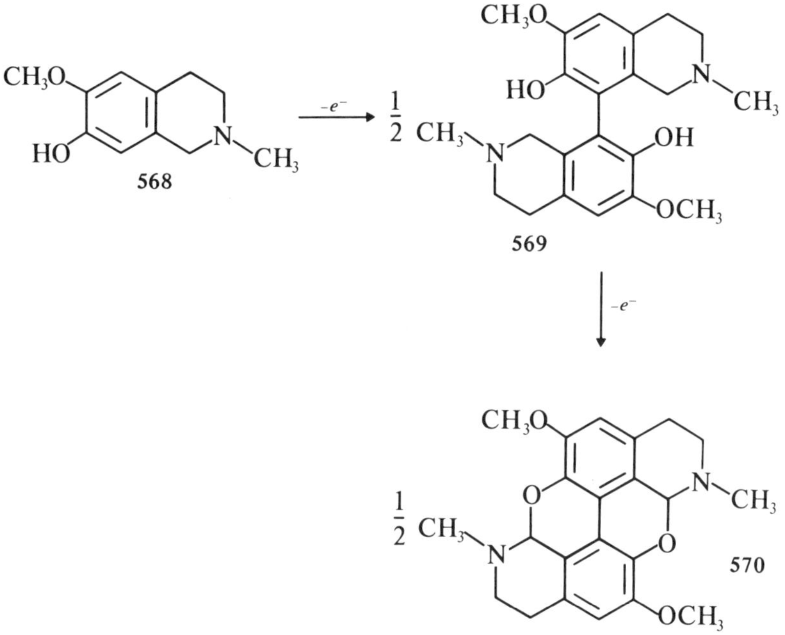

Oxidation of Heterocyclic Compounds 593

Scheme 8.253

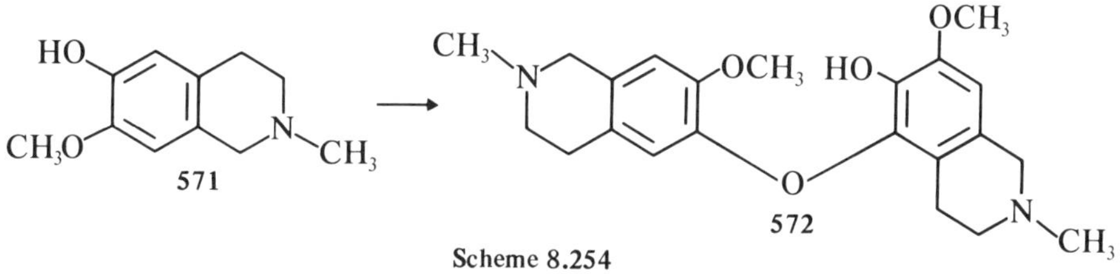

Scheme 8.254

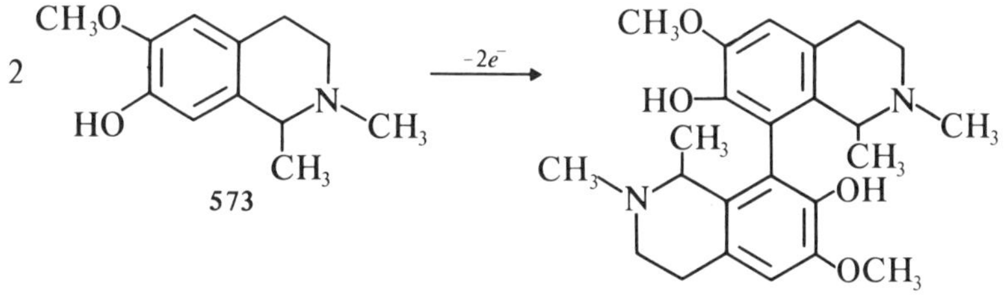

SS rotamer A

Scheme 8.255

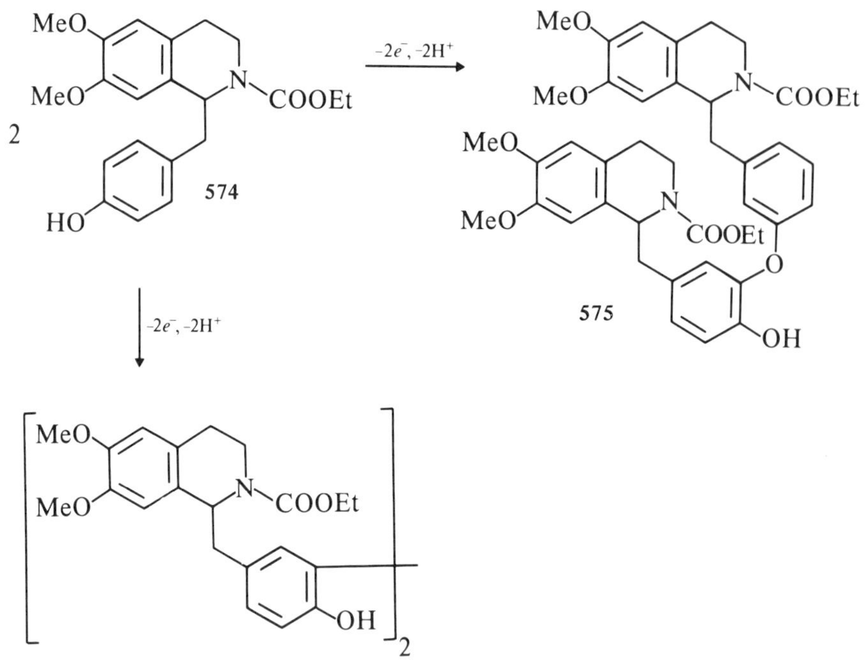

594　Electrochemical Behavior of Heterocyclic Compounds

Scheme 8.256

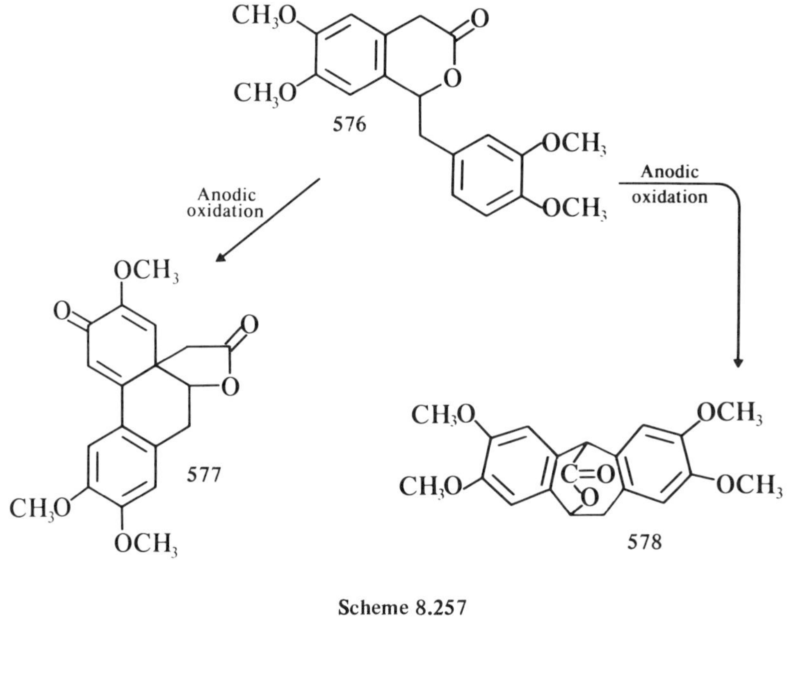

Scheme 8.257

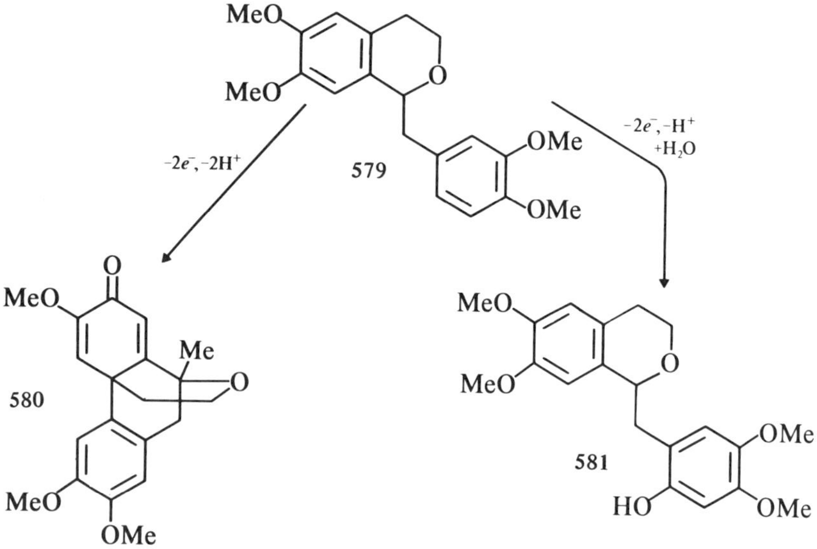

Scheme 8.258

(791) (Scheme 8.258). Similarly, the oxidation of isochroman-3-one **582** in MeCN/NaClO$_4$ gives 55% **583**. The reaction probably involves oxidative cleavage of the 6-methoxy group to a quinone-type intermediate (792) (Scheme 8.259).

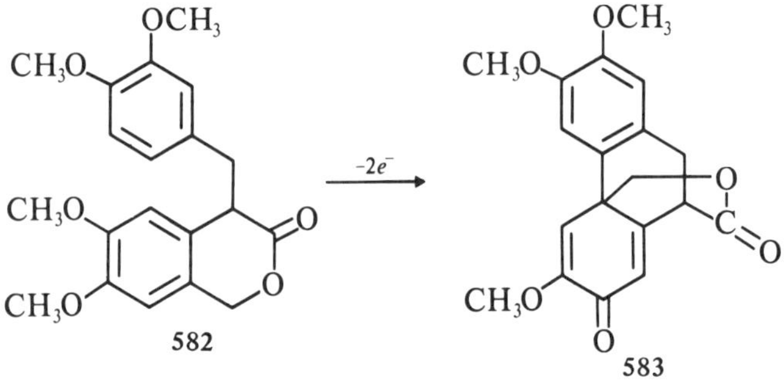

Scheme 8.259

8.3.2.2. Substituents with Nitrogen Attached to the Ring

The electrochemical oxidation of amines and hydrazines has been reviewed (793, 794). 2-Amino-5-ethoxycarbonyl-1,4-methylthiazole **584** is oxidized at the Pt electrode in MeCN/LiClO$_4$ to a radical **585** in a way similar to that of primary amines. **585** dimerizes to the hydrazo compound **586**, which can be formulated as the azine **587**. **585** is oxidized to the nitrene **588**, which affords **589**. Interestingly, **587** is not further oxidized to **589**, probably because of the stabilization due to hydrogen bonding (795) (Scheme 8.260).

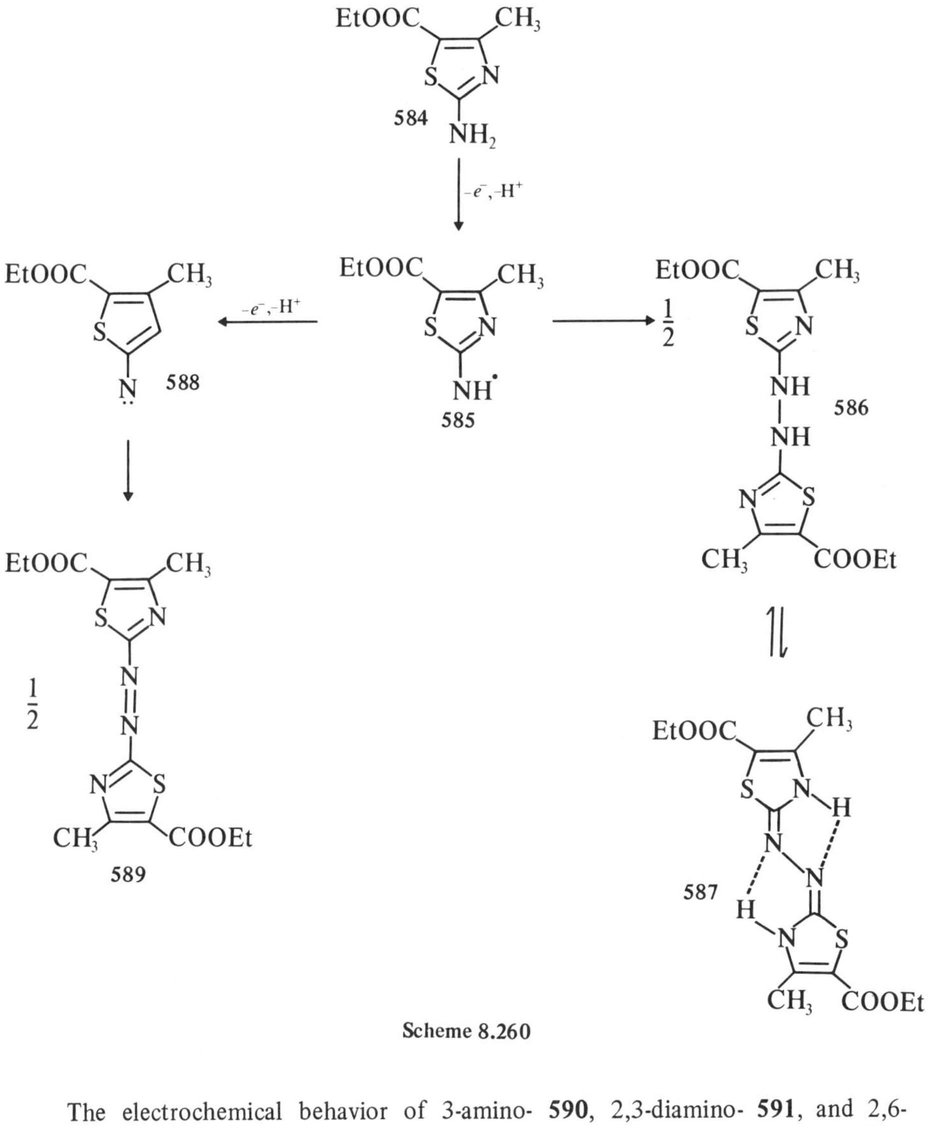

Scheme 8.260

The electrochemical behavior of 3-amino- **590**, 2,3-diamino- **591**, and 2,6-diaminopyridine was investigated in buffered aqueous media. The controlled potential electrolysis of **590** and **591** yielded 3,3'-azopyridine (796). The oxidation of hydrazine **592** in the presence of nucleophiles has been investigated (797) (Scheme 8.261).

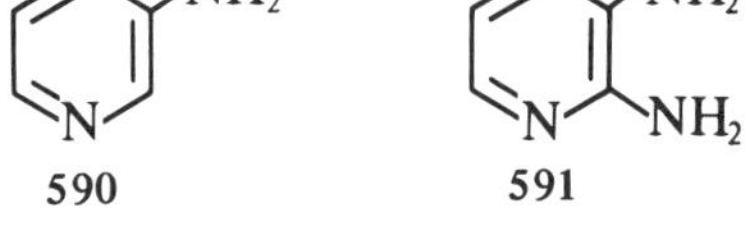

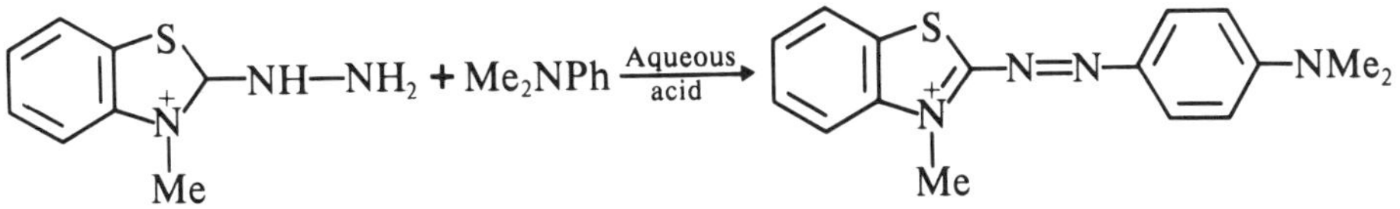

Oxidation of Heterocyclic Compounds 597

[Scheme 8.261 showing conversion of 592 with Me₂NPh in aqueous acid]

592

Scheme 8.261

The two-step reversible oxidation of the compounds such as **593**, **594**, and **595** has been reviewed (720). The controlled potential electrolysis, in MeCN, of **593** gives a stable cation radical, but the dication decomposes to a ketone and an imine (798).

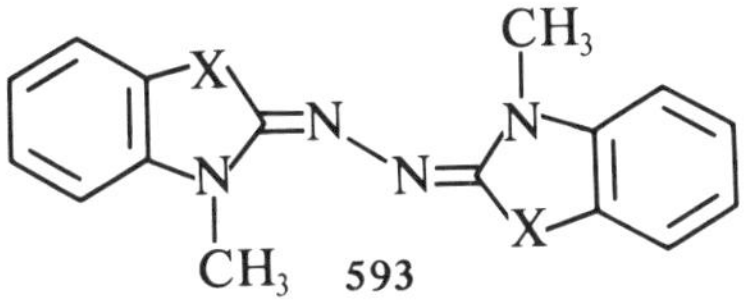

593

X = NCH₃; O; S; Se

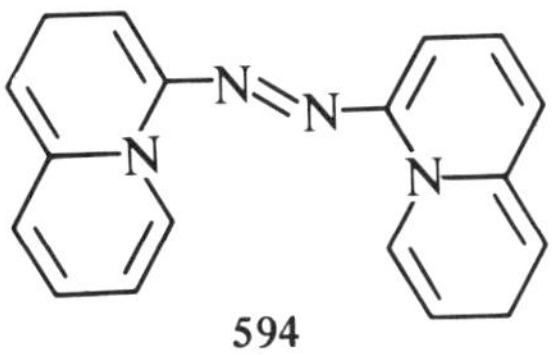

594

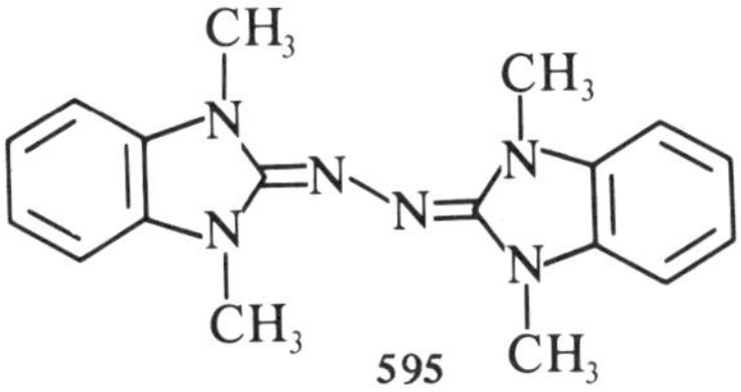

595

8.3.2.3. Substituents with Sulfur Attached to the Ring

The oxidation of sulfur compounds has been reviewed (799). Mercaptans are oxidized to disulfides; for example, 6-thiopurine is oxidized at the pyrolitic graphite electrode to the disulfide (669). The synthesis of **597** is accomplished smoothly by the oxidation of **596** in the presence of various amines at constant current in DMF. The yields are 26 and 98% for dicyclohexylamine and N-propylamine, respectively (800) (Scheme 8.262). At the Hg electrode, the oxidation of mer-

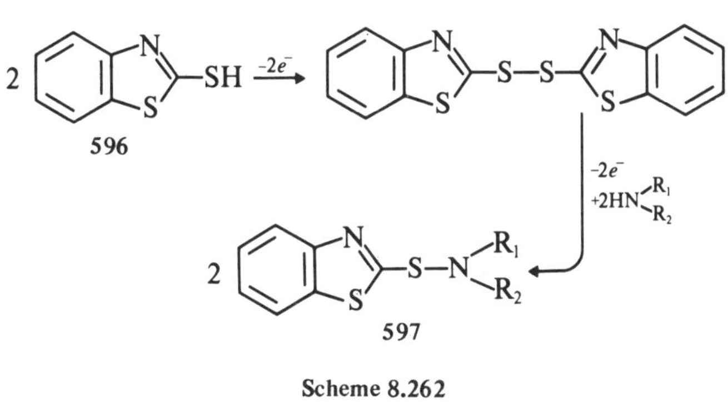

596

597

Scheme 8.262

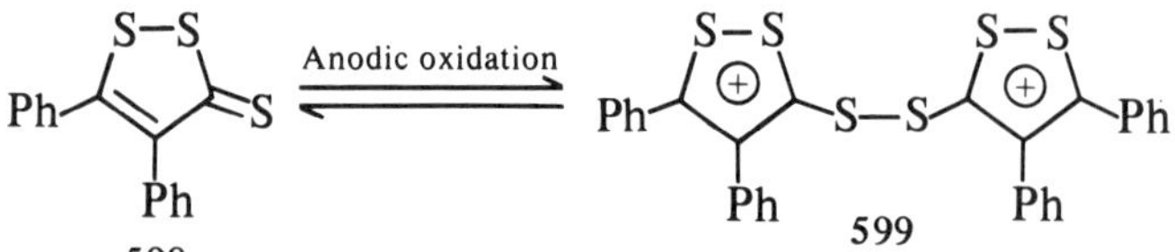

598 Electrochemical Behavior of Heterocyclic Compounds

Scheme 8.263

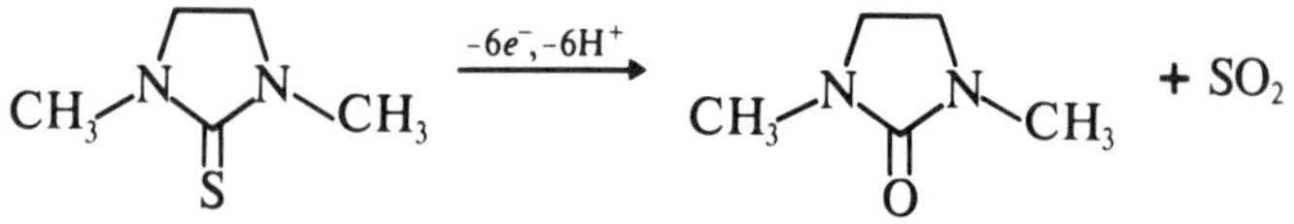

Scheme 8.264

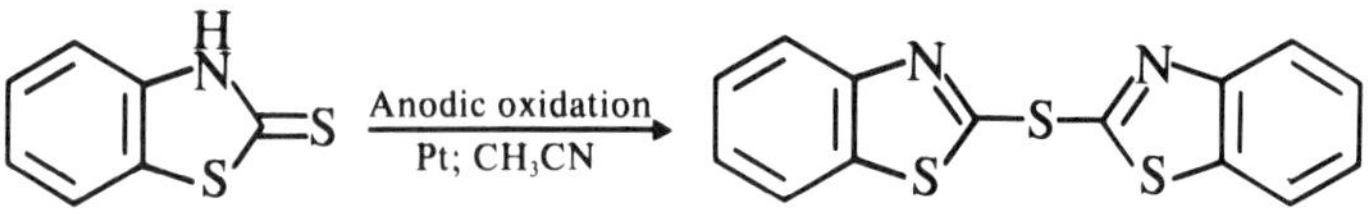

Scheme 8.265

captans leads to mercury derivatives (801). Thiones also lead to mercury derivatives (801), but at the Pt electrode in MeCN they lead to disulfide or disulfide salts **599** as in the case of 1,2-dithiole-3-thiones **598** (802) (Scheme 8.263). In the presence of water they yield the ketone through a complex mechanism involving sulfinic acid (803) (Scheme 8.264). Desulfuration is also observed during the oxidation of 2-mercaptobenzothiazole **602** (804) (Scheme 8.265).

8.3.2.4. Substituents with Carbon Attached to the Ring

8.3.2.4.1. Alkyl Substituents.
The alkyl derivatives of pyridine and pyrazoles are anodically oxidized to carboxylic acids. 3-Picoline forms nicotinic acid on oxidation at the Pt anode (576) in 30% sulfuric acid (60% yield) or at the Pb anode in 7 N sulfuric acid (805). At the Pb anode, isonicotinic acid is obtained from 4-picoline (806).

8.3.2.4.2. Carboxylic Acids and Derivatives as Substituents.
The electrochemical oxidation of carboxylic acids (the Kolbe reaction, its different mechanisms and products) and derivatives have been reviewed (793, 807). The anodic oxidation of 1-azabicyclo[2.2.2]octane-2-carboxylic acid **604** under the Kolbe conditions yields the methoxylated derivative **605** through the intermediate carbocation (808) (Scheme 8.266).

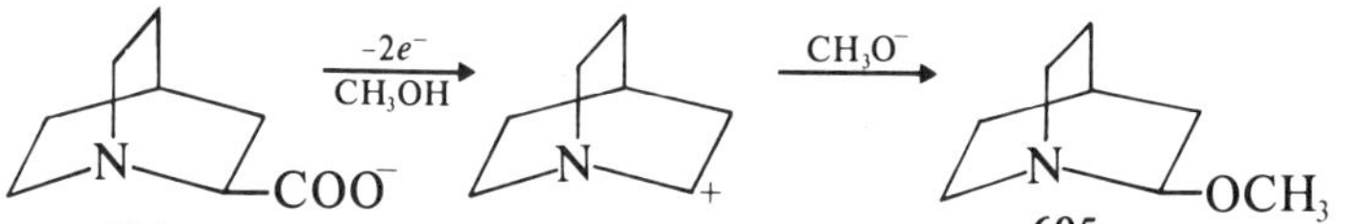

Scheme 8.266

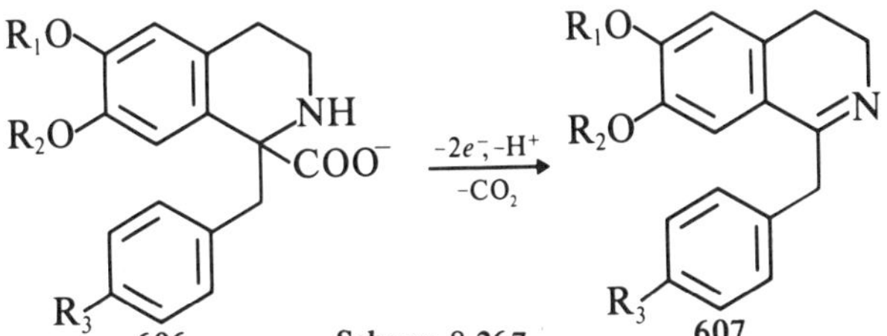

606 Scheme 8.267 607

Tetrahydroisoquinoline-1-carboxylic acids **606** are easily oxidized to 3,4-dihydroisoquinoline **607**. The oxidation potential is about 1.7 V lower than the potential used in the Kolbe electrolysis. This suggests that the oxidation takes place on the aromatic ring with the loss of CO_2 in a "pseudo-Kolbe" reaction (809) (Scheme 8.267). Paraconic acid **608** undergoes a one- or two-electron oxidation (810) (Scheme 8.268).

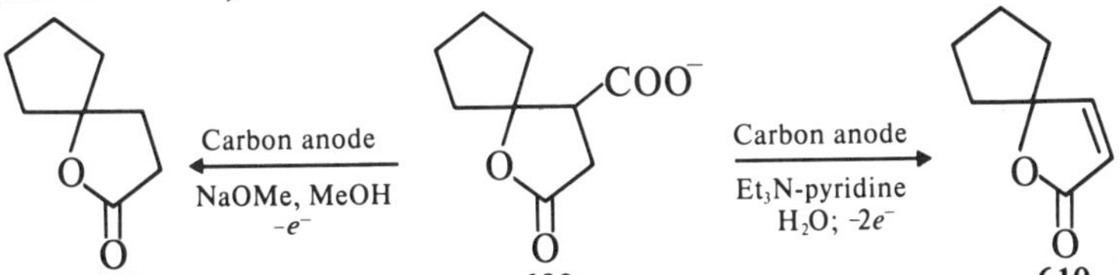

609 608 610

Scheme 8.268

A series of 1,2,3,4-tetrahydrocarboline-1- and 3-carboxylic acids **611** have been electrochemically oxidized in $MeOH/H_2O$ with a graphite felt anode in a divided cell (811) (Scheme 8.269). The anodic oxidation of 2-piperidine carboxylic acid

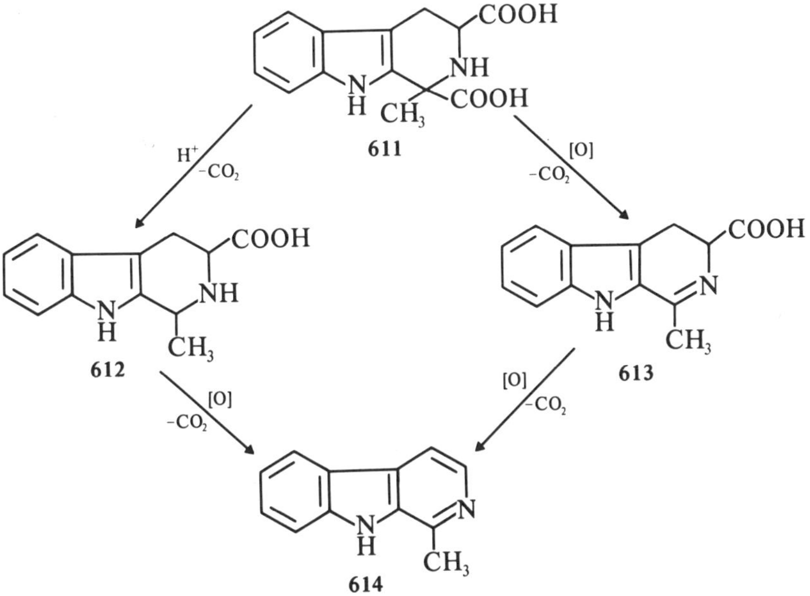

Scheme 8.269

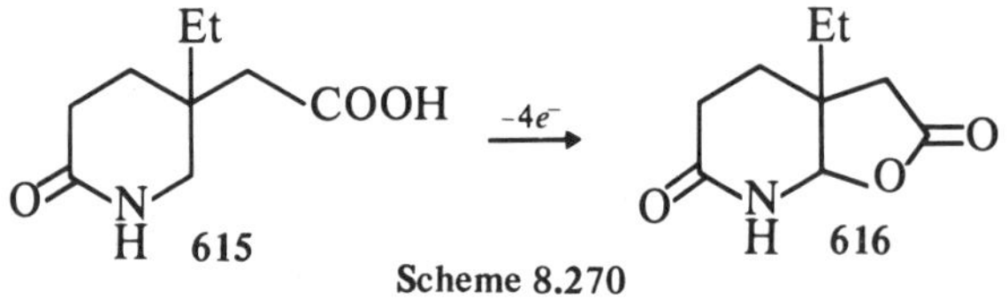

Scheme 8.270

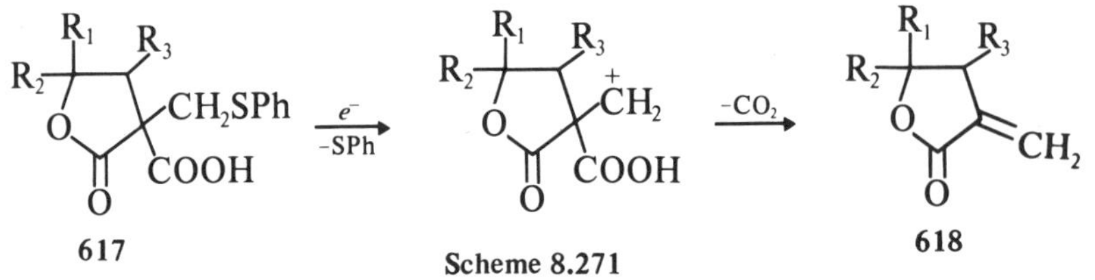

Scheme 8.271

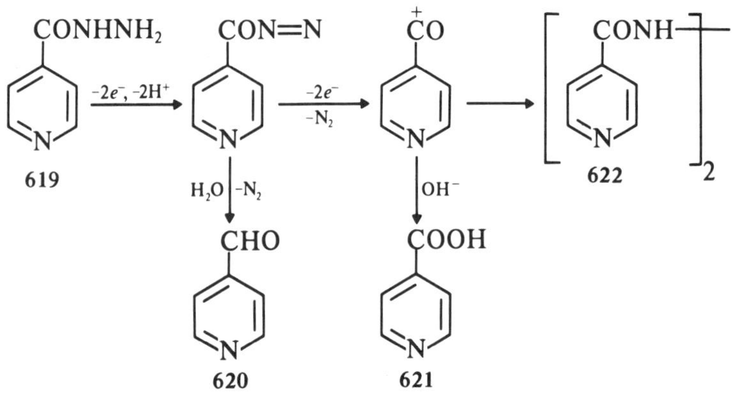

Scheme 8.272

at the Pt electrode in MeCN leads to lactam-lactones. The formation of the lactones takes place stereoselectively at the ring junction. **616** (Scheme 8.270) is a key intermediate in the synthesis of the alkaloid, ± eburnamonine (812).

In an aqueous layer containing triethylamine and LiClO$_4$ covered with an organic extractive layer, with the Pt electrode at constant current, the γ-butyrolactone **617** is electrolyzed to give (92.5%) **618** (813) (Scheme 8.271). Isonicotinic hydrazide **619** can be electrochemically oxidized at pH 11 on the Hg electrode to give 1,2-diisonicotinoylhydrazine **622** by a two-electron process. At pH 13 the oxidation requires 3 F mol^{-1} to give isonicotinic acid (45%) **621** and **622** (55%) (814) (Scheme 8.272).

REFERENCES

1. J. Volke, in A. R. Katritzky, Ed., *Physical Methods in Heterocyclic Chemistry*, Vol. 1, Academic Press, New York, 1963, p. 217.
2. H. Lund, *Adv. Heterocyclic. Chem.*, **12**, 13 (1970).

3. H. Lund, in S. Pataï Ed., *Chemistry of Carbon-Nitrogen Double Bonds*, Interscience, London, 1970, p. 505.
4. (a) G. Dryhurst, *Electrochemistry of Biological Molecules*, Academic Press, New York, 1977. (b) G. Dryhurst, F. Scheller, R. Renneberg, and K. M. Kadish, *Biological Electrochemistry*, Vol. 1, Academic Press, 1983. (c) G. Dryhurst and K. M. Kadish, *Biological Electrochemistry*, Vol. 2, Academic Press, 1983.
5. M. Baizer, *Organic Electrochemistry*, Marcel Dekker, New York, 1973.
6. N. L. Weinberg, *Techniques of Electroorganic Synthesis*, Wiley-Interscience, New York, 1975.
7. A. J. Fry, *Synthetic Organic Electrochemistry*, Harper and Row, New York, 1972.
8. C. K. Mann and K. K. Barnes, *Electrochemical Reactions in Nonaqueous Systems*, Marcel Dekker, New York, 1970.
9. M. Riffi and F. H. Covitz, *Introduction to Organic Electrochemistry*, Marcel Dekker, New York, 1974.
10. S. D. Ross, M. Finkelstein, and E. J. Rudd, *Anodic Oxidation*, Academic Press, New York, 1975.
11. L. Eberson and K. Nyberg, in V. Gold, Ed., *Adv. Phys. Org. Chem.*, Vol. 12, Academic Press, London, 1976, p. 2.
12. A. J. Bard and H. Lund, *Encyclopedia of the Electrochemistry of the Elements*, Vols. 11-14, Marcel Dekker, New York, 1980.
13. Specialist Periodical Reports: *Electrochemistry*, Vols. 1-7, The Chemical Society, London, 198■.
14. L. Meites and P. Zuman, *Handbook Series in Organic Electrochemistry*, CRC Press, Boca Raton, Florida, 1978.
15. *Ascatopics Electrochemistry*, ISI, Philadelphia.
16. C. Bollet, *Analusis*, **5**, 157 (1977); **6**, 54 (1978).
17. C. Bollet, P. Oliva, and M. Caude, *J. Chromatog.*, **149**, 625 (1977).
18. R. Beauchamp, P. Boinet, J. J. Fombont, J. Tacussel, M. Breant, J. Georges, M. Porthault, and O. Vittori, *J. Chromatog.*, **204**, 123 (1981) and references therein.
19. H. Lund and P. Iversen, Chapter 4, p. 165 in ref. 5.
20. D. Danly, Chapter 28, p. 907 in ref. 5.
21. F. Coeuret and A. Storck, *Inf. Chim.*, **210**, 121 (1981).
22. M. M. Baizer, *J. Appl. Electrochem.*, **10**, 285 (1980).
23. R. W. Foreman and R. Veatch (Standard Oil of Ohio), U.S. Patent 3,119,760, Jan. 28 1964.
24. F. Beck and H. Gutke, *Chem. Ing. Tech.*, **41**, 943 (1969).
25. P. M. Robertson, B. Scholder, G. Theis, and N. Ibl, *Chem. Ind.*, **1978**, 459.
26. G. A. Ashworth, P. J. Ayre, and R. E. W. Jansson, *Chem. Ind.*, **1975**, 382.
27. H. Strehlow, in J. J. Lagowski, Ed., *The Chemistry of Non-Aqueous Solvents*, Vol. 1, Academic Press, New York, 1966, p. 129.
28. See ref. 11, pp. 42-44.
29. G. M. McNamee, B. C. Wilett, D. M. La Periere, and D. G. Peters, *J. Amer. Chem. Soc.*, **99**, 1831 (1977).
30. W. F. Caroll, Jr., and D. G. Peters, *J. Org. Chem.*, **43**, 4633 (1978).
31. R. Woods, *Aust. J. Chem.*, **25**, 2329 (1972).
32. J. C. Gressin, D. Michelet, L. Nadjo, and J. M. Saveant, *Nouv. J. Chim.*, **3**, 545 (1979).
33. G. Dryhurst, *Top. Current Chem.*, **34**, 49 (1972).
34. W. E. Van Der Linden, *Anal. Chim. Acta*, **119**, 1 (1980).

35. J. Armand, K. Chekir, N. Ple, G. Queguiner, and M. P Simonnin, *J. Org. Chem.*, **46**, 4754 (1981).
36. G. Manning, V. D. Parker, and R. N. Adams, *J. Amer. Chem. Soc.*, **91**, 4584 (1969).
37. F. M'Halla, J. Pinson, and J. M. Saveant, *J. Electroanal. Chem. Interfacial Electrochem.*, **89**, 347 (1978); F. M'Halla, J. Pinson, and J. M. Saveant, *J. Amer. Chem. Soc.*, **102**, 4120 (1980).
38. P. Zuman and S. Wawzonek, in J. J. Lagowski, Ed., *The Chemistry of Non-Aqueous Solvents*, Vol. 5A, Academic Press, New York, 1978, p. 122.
39. J. J. Lagowski, *Rev. Chim. Min.*, **15**, 1 (1978).
40. J. F. Coetzee, *Pure Appl. Chem.*, **13**, 427 (1967).
41. R. Lindauer and L. M. Mukerjee, *Pure Appl. Chem.*, **23**, 265 (1971).
42. F. Fujinaga and K. Izutzu, *Pure Appl. Chem.*, **27**, 273 (1971).
43. L. Knetch, *Pure Appl. Chem.*, **27**, 281 (1971).
44. T. B. Reddy, *Pure Appl. Chem.*, **25**, 459 (1971).
45. T. B. Hoover, *Pure Appl. Chem.*, **37**, 579 (1974).
46. J. Julliard, *Pure Appl. Chem.*, **49**, 885 (1977).
47. J. E. Coetzee, *Pure Appl. Chem.*, **49**, 213 (1977).
48. I. Mentre, *Ann. Chim. (France)*, **1972**, 333, and references therein.
49. S. D. Ross, M. Fiukelstein, and R. C. Petersen, *J. Org. Chem.*, **35**, 781 (1970).
50. K. Sugino, T. Sekine, and N. Sato, *Electrochem. Technol.*, **1**, 112 (1963).
51. S. D. Ross, M. Finkelstein, and R. C. Petersen, *J. Amer. Chem. Soc.*, **80**, 4139 (1964).
52. J. Badoz-Lambling, A. Thiebault, and P. Oliva, *Electrochim. Acta*, **24**, 1029 (1979).
53. A. Thiebault, C. Mathieu, and P. Oliva, *Compt. Rend. C*, **286**, 417 (1978).
54. A. Jobert-Perol, M. Herlem, F. Bobilliart, and A. Thiebault, *Anal. Lett.*, **10**, 767 (1978).
55. C. Pitti, M. Cerles, A. Thiebault, and M. Herlem, *J. Electroanal. Chem. Interfacial Electrochem.*, **126**, 163 (1981).
56. J. Pinson and J. M. Saveant, *Nouv. J. Chim.*, **5**, 311 (1981).
57. D. Hammerich and V. D. Parker, *J. Amer. Chem. Soc.*, **96**, 5108 (1974).
58. P. Lines, B. S. Jensen, and V. D. Parker, *Acta Chem. Scand., B*, **32**, 510 (1978).
59. See ref. 19, p. 223.
60. K. Bechgaard, V. D. Parker, and C. T. Petersen, *J. Amer. Chem. Soc.*, **95**, 4373 (1973).
61. N. E. Tokel, C. P. Kezthelyi, and A. J. Bard, *J. Amer. Chem. Soc.*, **94**, 4872 (1972).
62. M. Herlem, *Bull. Soc. Chim. France*, **1967**, 1687.
63. M. Herlem, J. Minet, and A. Thiebault, *J. Electroanal. Chem. Interfacial Electrochem.*, **30**, 203 (1971).
64. W. H. Smith and A. J. Bard, *J. Amer. Chem. Soc.*, **97**, 6491 (1975).
65. A. Demortier and A. J. Bard, *J. Amer. Chem. Soc.*, **95**, 3495 (1973).
66. J. M. Saveant and A. Thiebault, *J. Electroanal. Chem. Interfacial Electrochem.*, **89**, 335 (1978).
67. L. A. Constant and D. G. Davis, *Anal. Chem.*, **47**, 2253 (1975).
68. J. M. Saveant and Su Khac Binh, *J. Org. Chem.*, **42**, 1242 (1977).
69. J. Furhop, K. M. Kadhish, and D. G. Davis, *J. Amer. Chem. Soc.*, **95**, 5140 (1973).
70. R. E. Dessy, J. C. Charkoudian, and A. L. Rheingold, *J. Amer. Chem. Soc.*, **94**, 738 (1972).
71. N. R. Armstrong, R. K. Quinn, and N. E. Vanderborgh, *Anal. Chem.*, **46**, 1759 (1974).
72. L. A. Paquette, L. B. Anderson, J. F. Hansen, S. A. Lang, Jr., and H. Berk, *J. Amer. Chem. Soc.*, **94**, 4907 (1972).

73. D. A. Hall, M. Sakuma, and P. J. Elving, *Electrochim. Acta*, **11**, 337 (1966).
74. P. Lemoine and M. Gross, *Compt. Rend. C*, **290**, 231 (1980).
75. See ref. 11, p. 244.
76. A. J. Bard and L. R. Faulkner, *Electrochemical Methods*, John Wiley, New York, 1980.
77. D. E. Smith, *Crit. Rev. Anal. Chem.*, **2**, 247 (1971).
78. D. Garreau and J. M. Saveant, *J. Electroanal. Chem. Interfacial Electrochem.*, **86**, 63 (1978).
79. D. Britz, *J. Electroanal. Chem. Interfacial Electrochem.*, **88**, 309 (1978).
80. P. Delahay, *New Instrumental Methods in Electrochemistry*, Wiley-Interscience, New York, 1954.
81. B. E. Conway, *Theory and Principles of Electrode Processes*, Ronald, New York, 1965.
82. K. J. Vetter, *Electrochemical Kinetics*, Academic Press, New York, 1967.
83. J. O'M. Bockris and A. K. N. Reddy, *Modern Electrochemistry*, Plenum, New York, 1970.
84. T. Erdey-Gruz, *Kinetics of Electrode Processes*, Wiley-Interscience, New York, 1972.
85. H. R. Thirsk, *A Guide to the Study of Electrode Kinetics*, Academic Press, New York, 1972.
86. W. J. Albery, *Electrode Kinetics*, Clarendon, Oxford, 1975.
87. R. N. Adams, *Electrochemistry at Solid Electrodes*, Marcel Dekker, New York, 1969.
88. G. Charlot, J. Badoz-Lambling, and B. Tremillon, *Electrochemical Reactions*, Elsevier, Amsterdam, 1962.
89. B. B. Damaskin, *The Principles of Current Methods for the Study of Electrochemical Reactions*, McGraw-Hill, New York, 1967.
90. Z. Galus, *Fundamentals of Electrochemical Analysis*, Ellis Harwood, Chichester, 1976.
91. D. D. McDonald, *Transient Techniques in Electrochemistry*, Plenum Press, New York, 1977.
92. D. T. Sawyer and J. L. Roberts, Jr., *Experimental Electrochemistry for Chemists*, Wiley-Interscience, New York, 1974.
93. We have used the same symbols as Bard and Faulkner (76) so that the reader can easily refer to their book.
94. D. Pletcher, *Chem. Soc. Rev.*, **4**, 471 (1975).
95. D. H. Evans, *Acc. Chem. Res.*, **10**, 313 (1977).
96. J. M. Saveant, *Acc. Chem. Res.*, **13**, 323 (1980).
97. F. Ammar and J. M. Saveant, *J. Electroanalyt. Chem.*, **47**, 115 (1973).
98. J. Pinson and J. Armand, *Bull. Soc. Chim. France*, 1764 (1971).
99. J. Pinson and J. Armand, *Coll. Czech. Chem. Commun.*, **36**, 585 (1971).
100. For the significance of α, see ref. 76, chap. 3.
101. E. Laviron, *Bull. Soc. Chim. France*, **1961**, 2325.
102. M. P. Strier and J. C. Cavagnol, *J. Amer. Chem. Soc.*, **79**, 4331 (1957).
103. B. B. Damaskin, O. A. Petrii, and V. V. Batrakov, *Adsorption of Organic Compounds on Electrodes*, Plenum Press, New York, 1971.
104. M. Heyrovsky and S. Varicka, *J. Electroanal. Chem. Interfacial Electrochem.*, **36**, 203, 223 (1972).
105. A. C. Riddiford, *Adv. Electrochem. Electrochem. Eng.*, **4**, 47 (1966).
106. F. Opekar and P. Beran, *J. Electroanal. Chem. Interfacial Electrochem.*, **69**, 1 (1976).
107. G. Cauquis, B. Chabaud, and M. Genies, *Bull. Soc. Chim. France*, 583 (1975).
108. J. Armand, K. Chekir, and J. Pinson, *Canad. J. Chem.*, **56**, 1804 (1978).

109. G. Dryhurst, *J. Electrochem. Soc.*, **119**, 1659 (1972).
110. J. L. Owens, H. A. Marsh, and G. Dryhurst, *J. Electroanal. Chem. Interfacial Electrochem.*, **91**, 231 (1978).
111. C. P. Andrieux, *Electrochim. Organ. Tech. Ing., Paris*, **1977**, D960.
112. J. M. Saveant and E. Vianello, *Electrochim. Acta*, **8**, 905 (1963).
113. R. S. Nicholson and I. Shain, *Anal. Chem.*, **36**, 706 (1964).
114. J. M. Saveant and E. Vianello, *Electrochim. Acta*, **12**, 629 (1967).
115. L. Nadjo and J. M. Saveant, *J. Electroanal. Chem. Interfacial Electrochem.*, **30**, 41 (1971).
116. T. Olmstead, R. T. Hamilton, and R. S. Nicholson, *Anal. Chem.*, **41**, 260 (1969).
117. (a) C. P. Andrieux, L. Nadjo, and J. M. Saveant, *J. Electroanal. Chem. Interfacial Electrochem.*, **26**, 147, 223 (1970). (b) C. P. Andrieux, doctoral thesis, Paris, 1969.
118. C. P. Andrieux, L. Nadjo, and J. M. Saveant, *J. Electroanal. Chem. Interfacial Electrochem.*, **42**, 223 (1973).
119. L. Nadjo and J. M. Saveant, *J. Electroanal. Chem. Interfacial Electrochem.*, **44**, 27 (1973).
120. C. P. Andrieux and J. M. Saveant, *J. Electroanal. Chem. Interfacial Electrochem.*, **53**, 165 (1974).
121. C. Amatore, J. Chaussard, J. Pinson, J. M. Saveant, and A. Thiebault, *J. Amer. Chem. Soc.*, **101**, 6012 (1979).
122. R. S. Nicholson and I. Shain, *Anal. Chem.*, **37**, 178 (1965).
123. R. S. Nicholson, J. M. Wilson, and M. L. Olmstead, *Anal. Chem.*, **38**, 542 (1966).
124. M. Mastragostino, L. Nadjo, and J. M. Saveant, *Electrochim. Acta*, **13**, 721 (1966).
125. C. Amatore and J. M. Saveant, *J. Electroanal. Chem. Interfacial Electrochem.*, **85**, 27 (1977); **86**, 227 (1978); **102**, 21 (1979).
126. C. Amatore and J. M. Saveant, *J. Electroanal. Chem. Interfacial Electrochem.*, **86**, 227 (1978).
127. C. Amatore, D. Lexa, and J. M. Saveant, *J. Electroanal. Chem. Interfacial Electrochem.*, **111**, 81 (1980).
128. S. Kwee and H. Lund, *Acta Chem. Scand.*, **25**, 1813 (1972).
129. J. Armand, K. Chekir, and J. Pinson, *Canad. J. Chem.*, **52**, 3971 (1974).
130. A. J. Bard and K. S. V. Santhanam, *Electroanal. Chem.*, **4**, 215 (1970).
131. C. Amatore and J. M. Saveant, *J. Electroanal. Chem. Interfacial Electrochem.*, **123**, 189, 203 (1981).
132. C. Amatore, F. M'Halla, and J. M. Saveant, *J. Electroanal. Chem. Interfacial Electrochem.*, **123**, 219 (1981).
133. C. Amatore, J. Pinson, J. M. Saveant, and A. Thiebault, *J. Electroanal. Chem. Interfacial Electrochem.*, **123**, 231 (1981).
134. T. M. McKinney, *Electroanal. Chem.*, **10**, 97 (1977).
135. M. Genies and P. Campo y Sansano, *J. Electroanal. Chem. Interfacial Electrochem.*, **85**, 351 (1977).
136. G. Bidan and M. Genies, *Nouv. J. Chim.*, **5**, 117 (1981).
137. J. P. Billon, G. Cauquis, and M. Combrisson, *J. Chim. Phys.*, 374 (1964).
138. I. B. Goldberg and A. J. Bard, *J. Phys. Chem.*, **75**, 3281 (1971).
139. R. D. Alloendoerfer, G. A. Martinchek, and S. Bruckenstein, *Anal. Chem.*, **47**, 890 (1975).
140. E. Janzen, *Acc. Chem. Res.*, **4**, 31 (1971).
141. V. E. Norvell and G. Mamantov, *Anal. Chem.*, **49**, 1470 (1977).

References

142. Th. Kuwana and N. Winograd, *Electroanal. Chem.*, **7**, 1 (1974).
143. Th. Kuwana and W. R. Heineman, *Acc. Chem. Res.*, **9**, 241 (1976).
144. M. Genies and A. F. Diaz, *J. Electroanal. Chem. Interfacial Electrochem.*, **98**, 305 (1978).
145. M. K. Hanafey, R. L. Scott, T. H. Ridgway, and C. N. Reilley, *Anal. Chem.*, **50**, 116 (1978).
146. W. N. Hansen, *Adv. Electrochem. Electrochem. Eng.*, **9**, 1 (1973).
147. N. Winograd and T. Kuwana, *J. Electroanal. Chem. Interfacial Electrochem.*, **23**, 333 (1969).
148. N. Winograd and T. Kuwana, *J. Amer. Chem. Soc.*, **93**, 4343 (1971).
149. N. Winograd and T. Kuwana, *J. Amer. Chem. Soc.*, **92**, 224 (1970).
150. J. D. E. McIntyre, *Adv. Electrochem. Electrochem. Eng.*, **9**, 61 (1973).
151. A. Aylmer-Kelly, A. Bewick, P. Cantrill, and A. Tuxford, *Discuss. Faraday Soc.*, **56**, 96 (1973).
152. R. F. Broman, W. R. Heineman, and T. Kuwana, *Discuss. Faraday Soc.*, **56**, 16 (1973).
153. D. L. Jeanmaire, M. R. Suchanski, and R. P. Van Duyne, *J. Amer. Chem. Soc.*, **97**, 1699 (1975).
154. D. L. Jeanmaire and R. P. Van Duyne, *J. Amer. Chem. Soc.*, **98**, 4029 (1976).
155. R. P. Van Duyne, M. R. Suchanski, J. M. Lakovits, A. R. Siedle, K. D. Parks, and T. M. Cotton, *J. Amer. Chem. Soc.*, **101**, 2832 (1979).
156. M. R. Suchanski and R. P. Van Duyne, *J. Amer. Chem. Soc.*, **98**, 252 (1976).
157. D. Barey, M. Balin, P. Boucly, and D. Resibois, unpublished results.
158. T. M. Cotton and R. P. Van Duyne, *Biochem. Biophys. Res. Commun.*, **82**, 424 (1978).
159. R. L. Paul, A. J. McQuillan, P. J. Hendra, and M. Fleischmann, *J. Electroanal. Chem. Interfacial Electrochem.*, **66**, 248 (1975).
160. A. J. McQuillan, P. J. Hendra, and M. Fleischmann, *J. Electroanal. Chem. Interfacial Electrochem.*, **65**, 933 (1975).
161. M. Fleischmann, P. J. Hendra, and A. J. McQuillan, *Chem. Phys. Lett.*, **26**, 163 (1974).
162. J. M. Sequaris, E. Koglin, P. Valenta, and H. W. Nurnberg, unpublished results.
163. M. Dennsted, German Patent 127086, 1901.
164. B. Sakurai, *Bull. Chem. Soc. Japan*, **11**, 374 (1936).
165. O. Carrasco, *Gazzetta, Chim. Ital.*, **38**, 301 (1908).
166. J. Von Braun and W. Sobecki, *Ber.*, **44**, 2158 (1911).
167. J. T. Wrobel and K. M. Pazdro, *Roczn. Chem.*, **41**, 637 (1967).
168. H. Nohe and H. R. Muller, German Patent 2403446, 1975.
169. F. Ender, E. Moisar, K. Schafer, and H. J. Teuber, *Z. Electrochem.*, **63**, 349 (1959).
170. N. P. Shinanskaya, L. A. Pavlova, and V. D. Bezuglyi, *Zh. Obshch. Khim.*, **37**, 974 (1967).
171. S. Hunig, F. Linhart, and D. Scheutzon, *Justus Liebigs Anal. Chem.*, **1975**, 2102.
172. P. C. Tompkins and C. L. A. Schmidt, *J. Biol. Chem.*, **143**, 643 (1942).
173. F. B. Ahrens, *Z. Electrochem.*, **2**, 577 (1895).
174. B. Emmert, *Ber.*, **46**, 1716 (1913).
175. F. J. McNamara, J. W. Nieft, J. F. Ambrose, and E. S. Huyser, *J. Org. Chem.*, **42**, 988 (1977).
176. C. O. Shmakel, K. S. V. Santhanam, and P. J. Elving, *J. Electrochem. Soc.*, **121**, 345 (1974).
177. K. S. V. Santhanam and P. J. Elving, *J. Amer. Chem. Soc.*, **95**, 5482 (1973).

178. H. Lund, *Acta Chem. Scand.*, 17, 2325 (1963).
179. E. L. Colichman and P. A. O'Donovan, *J. Amer. Chem. Soc.*, 76, 3588 (1954) and references therein.
180. M. Naarova and J. Volke, *Coll. Czech. Chem. Commun.*, 38, 2670 (1973).
180a. J. G. Gaudiello, E. Eric, and H. N. Blount, *J. Electroanal. Chem. Interfacial Electrochem.*, 131, 203 (1982).
181. I. Carelli, M. C. Cardinali, A. Casini, and A. Arnone, *J. Org. Chem.*, 41, 3967 (1976).
182. N. Yoshiike, S. Kondo, and M. Furai, *J. Electrochem. Soc.*, 127, 1496 (1980).
183. M. Mohammad, S. V. Sheikh, M. Iqbal, R. Ahmed, M. Razaq, and A. Y. Khan, *J. Electroanal. Chem. Interfacial Electrochem.*, 89, 431 (1978).
184. M. Mohammad, A. Y. Khan, M. Iqbal, and R. Iqbal, *J. Amer. Chem. Soc.*, 100, 7658 (1978).
184a. F. Micheletti-Moracci and S. Tortolla, *Ann. Chim. (Rome)*, 71, 499 (1981).
185. R. Raghavan and R. T. Iwamoto, *J. Electroanal. Chem. Interfacial Electrochem.*, 92, 101 (1978).
186. M. Libert and C. Caullet, *Compt. Rend. C*, 278, 495 (1974).
187. M. K. Polievktov, A. K. Shinkman, and L. N. Morozova, *Khim. Geterosikl. Soedinenii*, 1973, 1067.
188. H. Lund and H. L. Kristensen, *Acta Chem. Scand. Ser. B*, B33, 495 (1979).
189. G. Dryhurst, *Electrochemistry of Biological Molecules*, Academic Press, New York, 1977, p. 533.
190. C. O. Schmakel, K. S. V. Santhanam, and P. J. Elving, *J. Electrochem. Soc.*, 121, 1033 (1974).
191. H. Henschmann, *Studia Biophys.*, 45, 183 (1974).
192. F. Micheletti-Moracci, E. Liberatore, V. Carelli, A. Arnone, I. Carelli, and M. E. Cardinali, *J. Org. Chem.*, 43, 3420 (1978).
193. M. E. Cardinali and F. M. Morracci, *J. Electroanalyt. Chem. Interfacial Electrochem.*, 107, 391 (1980).
193a. H. Jaegfeldt, *Bioelectrochem. Bioenerg.*, 8, 355 (1981).
194. Y. Ohnishi, Y. Kikuchi, and M. Kitami, *Tetrahedron Lett.*, 3005 (1978).
195. K. S. V. Santhanam and P. J. Elving, *J. Amer. Chem. Soc.*, 95, 5482 (1973).
196. H. Ehrard and W. Jaenicke, *J. Electroanal. Chem. Interfacial Electrochem.*, 81, 79, 89 (1977).
197. J. Volke and V. Volkova, *Coll. Czech. Chem. Commun.*, 37, 3686 (1972).
198. V. N. Grachev and S. I. Zhdanov, *Elektrokhim.*, 15, 63 (1979).
199. J. Volke, *Coll. Czech. Chem. Commun.*, 33, 3044 (1968).
200. V. N. Grachev, S. I. Zhdanov, and C. S. Supin, *Elektrokhim.*, 14, 1353 (1978), and references therein.
201. L. Roullier and E. Laviron, *Electrochim. Acta*, 22, 669 (1977) and references therein.
202. S. Hunig and W. Schenk, *Justus Liebigs Anal. Chem.*, 1979, 1523.
202a. A. Deronzier, B. Golland, and M. Vieira, *Nouv. J. Chim.*, 6, 97 (1982).
203. S. Hunig, J. Gross, E. F. Lier, and M. Quast, *Justus Liebigs Anal. Chem.*, 1973, 339.
204. R. Andruzzi, A. Trazza, L. Greci, and L. Marchetti, *J. Electroanal. Chem. Interfacial Electrochem.*, 108, 49 (1980).
205. W. H. Smith and A. J. Bard, *J. Amer. Chem. Soc.*, 97, 6491 (1975).
206. C. Degrand and H. Lund, *Acta Chem. Scand.*, B31, 593 (1977).
207. U. Hess, D. Huhn, and H. Lund, *Acta Chem. Scand.*, B34, 413 (1980).
208. U. Hess, P. Fuchs, E. Jacob, and H. Lund, *Z. Chem.*, 20, 64 (1980).

References 607

209. D. N. Schluter, T. Biegler, E. V. Brown, and H. H. Bauer, *J. Electroanal. Chem. Interfacial Electrochem.*, 75, 545 (1977).
210. D. N. Schluter, T. Biegler, E. V. Brown, and H. H. Bauer, *Electrochim. Acta*, 21, 753 (1976).
211. N. E. Khomutov and V. V. Tsodikov, *Elektrokhim.*, 2, 722 (1966), and references cited therein.
212. J. Bordner and I. W. Elliot, *Cryst. Struct. Commun.*, 3, 689 (1974).
213. S. Kato, J. Nakaya, and E. Imoto, *Denki Kagaku*, 40, 708 (1972).
214. S. Kato, Y. Tanaka, and J. Nakaya, *Denki Kagaku*, 42, 223 (1974).
215. T. Shono, Y. Usui, and H. Hamaguchi, *Tetrahedron Lett.*, 1351 (1980).
216. R. Gottlieb and J. L. Neumeyer, *J. Amer. Chem. Soc.*, 98, 7108 (1976).
217. T. Shono, K. Yoshida, K. Ando, Y. Usui, and H. Hamagushi, *Tetrahedron Lett.*, 4819 (1978).
218. R. C. Kaye and H. I. Stonehill, *J. Chem. Soc.*, 27 (1951).
219. H. Lund, *Elektrodereaktioner i organisk polarografi og voltammetri*, Aarhus Stiftsbogtrykkerie, Aarhus, 1961.
220. V. D. Bezuglyi, M. B. Sidom, V. A. Sidom, V. A. Shapovalov, and A. N. Gaidukevich, *Khim. Geterotsikl. Soedin.*, 1978, 1660.
221. J. G. Frost and J. H. Saylor, *Rec. Trav. Chim.*, 82, 828 (1963).
222. B. S. Jensen, T. Petterson, A. Ronlan, and V. D. Parker, *Acta Chem. Scand., Ser. B*, B30, 773 (1976).
223. A. Zweig, G. Metzler, A. H. Maurer, and B. G. Roberts, *J. Amer. Chem. Soc.*, 88, 2864 (1966).
224. F. Pragst and U. Seydewitz, *J. Prakt. Chem.*, 319, 952 (1977).
225. F. Pragst, P. Ziebig, U. Seydewitz, and G. Driesel, *Electrochim. Acta*, 23, 341 (1980).
226. F. Pragst, M. Janda, and I. Stibor, *Electrochim. Acta*, 25, 779 (1980).
227. M. Vajda and V. Kadri, *Ann. Univ. Sci. Budapest Rolando Entvos Nominature, Sect. Chim.*, 1972, 107.
228. M. M. Evstifeev, F. Kh. Aminova, G. N. Dorofeenko, and F. P. Olekhanovich, *Russ. J. Gen. Chem. (USSR)*, 44, 2225 (1974).
229. G. Le Guillanton, *Bull. Soc. Chim. France*, 1974, 627.
229a. G. Mason, G. Le Guillanton, and J. Simonet, *J. Chem. Soc. Chem. Commun.*, 1982, 571.
230. G. Le Guillanton, *Bull. Soc. Chim. France*, 1974, 1699.
231. A. Lebouc, H. Van Rodijen, O. Riobe, G. Le Guillanton, and J. Delaunay, *Electrochim. Acta*, 24, 1119 (1979).
232. A. J. Marle and L. E. Lyons, *J. Chem. Soc.*, 1950, 1575.
233. R. Patzak and L. Neugebauer, *Monats.*, 82, 662 (1951); 83, 776 (1952).
234. V. S. Griffith and J. B. Westmore, *J. Chem. Soc.*, 1962, 1704.
235. R. N. Gourley, J. Grimshaw, and P. G. Miller, *J. Chem. Soc. (C)*, 1970, 2318.
236. M. K. Polietkov and Z. A. Lomadze, *Zh. Obsch. Khim.*, 47, 1383 (1977).
237. S. G. Mairanovskii, L. I. Kosychenko, and V. P. Litvinov, *Elektrokhim.*, 15, 118 (1979).
238. S. G. Marianovskii, L. I. Kosychenko, and S. Z. Taits, *Izv. Akad. Nauk. SSSR, Ser. Khim.*, 1980, 1382.
239. M. M. Baizer, Ed., *Organic Electrochemistry*, Marcel Dekker, New York, 1973, p. 584.
240. C. Parkanyi, E. L. Khoo, and V. Horak, *J. Heterocycl. Chem.*, 16, 471 (1979).
241. B. Lamm and J. Simonet, *Acta Chem. Scand. B*, 28, 147 (1974).
242. H. Lund, *Coll. Czech. Chem. Commun.*, 31, 4175 (1966).

243. J. P. Kitaev and G. K. Budnikov, *Coll. Czech. Chem. Commun.*, **30**, 4178 (1965).
244. A. Fry, W. E. Britton, R. Wilson, F. D. Greene, and J. G. Pacifici, *J. Org. Chem.*, **38**, 2620 (1973).
245. J. Grimshaw and J. Trocha-Grimshaw, *J. Chem. Soc. Perkin I*, **1973**, 1275.
246. U. Lang and H. Baumgaertel, *J. Electroanal. Chem. Interfacial Electrochem.*, **78**, 133 (1977).
247. H. Lund, *Oester. Chem. Z.*, **68**, 43 (1967).
248. S. Millefiori, *Anal. Chim. (Rome)*, **59**, 15 (1969).
249. L. N. Klatt and R. L. Rouseff, *J. Electroanal. Chem. Interfacial Electrochem.*, **41**, 411 (1973).
250. Ref. 5, p. 590.
251. H. Lund and P. Lunde, *Acta Chem. Scand.*, **21**, 1067 (1967).
252. H. Lund and P. Simonet, *Compt. Rend. C*, **276**, 1387 (1973).
253. H. Lund, *Acta Chem. Scand.*, **21**, 2525 (1967).
254. Ref. 5, p. 592.
255. D. E. Ames, B. Novitt, D. Waite, and H. Lund, *J. Chem. Soc. C*, **1969**, 796.
256. G. Cauquis and M. Genies, *Tetrahedron Lett.*, **1970**, 3403.
257. G. Cauquis and M. Genies, *Tetrahedron Lett.*, **1971**, 3959.
258. S. Millefiori, *J. Heterocycl. Chem.*, **17**, 1541 (1980) and references therein.
259. P. Martigny, H. Lund, and J. Simonet, *Electrochim. Acta*, **21**, 345 (1976).
260. H. Lund and E. T. Jensen, *Acta Chem. Scand.*, **24**, 1867 (1970).
261. H. Lund and E. T. Jensen, *Acta Chem. Scand.*, **25**, 2727 (1971).
262. Ref. 189, p. 186.
263. P. J. Elving, S. J. Race, and J. E. O'Reilly, *J. Amer. Chem. Soc.*, **95**, 647 (1973).
264. J. E. O'Reilly and P. J. Elving, *J. Electroanal. Chem. Interfacial Electrochem.*, **75**, 507 (1977).
264a. B. Czochralska and J. P. Elving, *Electrochim. Acta*, **26**, 1755 (1981).
265. P. Martigny and H. Lund, *Acta Chem. Scand.*, **B33**, 575 (1979).
266. H. Lund, *Acta Chem. Scand.*, **18**, 1984 (1964).
267. S. Kwee and H. Lund, *Acta Chem. Scand.*, **25**, 1813 (1971).
268. J. Volke, D. Dumanovic, and V. Volkova, *Coll. Czech. Chem. Commun.*, **30**, 246 (1965) and references therein.
269. L. N. Klatt and R. L. Rouseff, *J. Amer. Chem. Soc.*, **94**, 7295 (1972).
270. J. Swarcz and F. C. Anson, *J. Electroanal. Chem. Interfacial Electrochem.*, **114**, 117 (1980).
271. H. E. Toma and H. C. Chagas, *Anal. Acad. Brasil Cienc.*, **50**, 487 (1978).
272. J. Armand, P. Bassinet, K. Chekir, J. Pinson, and P. Souchay, *Compt. Rend. C*, **275**, 279 (1972).
273. J. Armand, K. Chekir, and J. Pinson, *Canad. J. Chem.*, **52**, 3971 (1974).
274. J. Pinson and J. Armand, *Bull. Soc. Chim. France*, 1764 (1971).
275. G. Sartori and C. Furlani, *Anal. Chim. (Rome)*, **45**, 251 (1955).
276. M. P. Strier and J. C. Cavagnol, *J. Amer. Chem. Soc.*, **80**, 1565 (1958).
277. J. Pinson and J. Armand, *Coll. Czech. Chem. Commun.*, **36**, 585 (1971).
278. M. Fedoronko and I. Jezo, *Coll. Czech. Chem. Commun.*, **37**, 1781 (1972).
279. J. Armand, K. Chekir, and J. Pinson, *J. Heterocyclic Chem.*, **17**, 1237 (1980).
280. O. N. Nechaeva and A. V. Pushkareva, *Zh. Obshch. Khim.*, **28**, 2693 (1958).
281. L. V. Varyukhina and Z. V. Pushkareva, *Zh. Obshch. Khim.*, **26**, 1740 (1956).

282. L. L. Gordienko, *Electrokhim.*, **1**, 1497 (1965).
283. R. Curti, S. Locchi, and V. Landini, *Ric. Sci.*, **24**, 2053 (1954).
284. D. N. Bailey, D. M. Hercules, and D. K. Poe, *J. Electrochem. Soc.*, **116**, 190 (1969).
285. S. Nakamura and T. Yoshida, *Denki Kagaku*, **39**, 502 (1971).
286. S. Nakamura and T. Yoshida, *Denki Kagaku*, **40**, 714 (1972).
287. D. K. Root, R. O. Pendarvis, and W. H. Smith, *J. Org. Chem.*, **43**, 778 (1978).
288. M. M. Morrison, E. T. Seo, J. K. Howie, and D. T. Sawyer, *J. Amer. Chem. Soc.*, **100**, 207 (1978) and references therein.
289. J. Kulys and A. Malinauskas, *Liet. TSR Mosklu Akad. Dark., Ser. B*, **41** (1979); through *Chem. Abstr.*, **91**, 131180 k.
290. D. Lloyd, A. C. Vincent, and D. J. Walton, *J. Chem. Soc. Perkin Trans.*, **2**, 668 (1980).
291. D. Lloyd, C. Nyns, and A. C. Vincent, *J. Chem. Soc. Perkin Trans.*, **2**, 1441 (1980).
292. M. Van Meersche, G. Germain, and J. P. Declercq, *Acta Cryst. Sect. B*, **1336**, 1418 (1980).
293. H. Lund, *13th Nord. Kemikermoede*, Copenhagen, 1968.
294. R. B. Johns and K. R. Markham, *J. Chem. Soc.*, **1962**, 3712.
295. H. Oelschlager, *Arch. Pharm.*, **296**, 396 (1963).
296. B. Z. Senkowski, M. S. Levin, J. R. Urbigkit, and E. G. Wollish, *Anal. Chem.*, **36**, 1991 (1964).
297. H. Oelschlager, J. Volke, and H. Hoffmann, *Coll. Czech. Chem. Commun.*, **31**, 1264 (1966).
298. H. Oelschlager, J. Volke, H. Hoffmann, and E. Kurek, *Arch. Pharm.*, **300**, 250 (1967).
299. A. V. Bogatskii, S. A. Andronati, V. P. Gultyai, I. Vikhliaev, A. F. Galatin, Z. I. Zhilina, and T. A. Klygul, *Zh. Obshch. Khim.*, **41**, 1358 (1971).
300. H. Oelschlager and H. Hoffmann, *Arch. Pharm.*, **300**, 817 (1967).
301. H. Lund, *Adv. Heterocycl. Chem.*, **12**, 258 (1970).
302. H. Lund, ref. 301, p. 280.
303. K. Butkiewicz, *J. Electroanal. Chem. Interfacial Electrochem.*, **90**, 271 (1978).
304. J. M. Mellor, S. B. Pons, and J. M. A. Stibbard, *J. Chem. Soc. Chem. Commun.*, **1979**, 761.
304a. J. M. Mellor, S. B. Pons, and J. M. A. Stibbard, *J. Chem. Soc. Perkin Trans.*, **1**, 3097 (1981).
305. L. Roullier and E. Laviron, *Electrochim. Acta*, **21**, 421 (1976).
306. L. Roullier and E. Laviron, *Electrochim. Acta*, **23**, 773 (1978).
307. A. V. Lizogub, Z. N. Timofeeva, M. L. Aleksandrova, A. V. El'Tsov, and E. G. Petrova, *J. Gen. Chem. (USSR)*, **43**, 2280 (1973).
308. H. Lund, *Acta Chem. Scand.*, **23**, 563 (1969).
309. H. Lund and L. G. Feoktistov, *Acta Chem. Scand.*, **23**, 3482 (1969).
310. H. Lund, *Acta Chem. Scand.*, **18**, 563 (1964).
311. W. N. Grieg and J. W. Rogers, *J. Electrochem. Soc.*, **117**, 1141 (1970).
312. S. L. Smith, L. D. Cook, and J. W. Rogers, *J. Electrochem. Soc.*, **119**, 1332 (1972).
313. V. D. Bezuglyi, N. P. Shimanskaya, and E. M. Peresleni, *Zh. Obshch. Khim.*, **34**, 3540 (1964).
314. P. E. Iversen and H. Lund, *Acta Chem. Scand.*, **21**, 389 (1967).
315. H. Lund, ref. 5, p. 589.
316. Z. N. Timofeeva, M. V. Petrova, M. Z. Girshovich, and A. V. El'Tzov, *Zh. Obshch. Khim.*, **39**, 54 (1969).

317. S. Roffia and G. Feroci, *J. Electroanal. Chem. Interfacial Electrochem.*, **88**, 169 (1978).
318. O. Manousek, O. Exner, and P. Zuman, *Coll. Czech. Chem. Commun.*, **33**, 4000 (1968).
319. M. Matsui, T. Sawamura, and T. Adachi, *Mem. Coll. Sci. Kyoto Imp. Univ.*, **15A**, 151 (1932); *Chem. Abstr.*, **26**, 5264 (1932).
320. H. Lund, ref. 5, p. 589.
321. G. Piccardi, F. Pergola, M. L. Foresti, and R. Guidelli, *J. Electroanal. Chem. Interfacial Electrochem.*, **84**, 235 (1977).
322. J. H. Stocker, R. M. Jenevein, A. Maguiar, G. W. Prejean, and N. A. Portnoy, *J. Chem. Soc. Chem. Commun.*, **1971**, 1478.
323. R. D. Eeike, R. C. Copenhafer, A. M. Aguiar, M. S. Chatta, and J. C. Williams, *J. Electroanal. Chem. Interfacial Electrochem.*, **42**, 309 (1973).
324. K. Bechgaard, V. D. Parker, and C. T. Pedersen, *J. Amer. Chem. Soc.*, **95**, 4373 (1973).
325. A. Astruc, M. Astruc, D. Goubeau, and G. Pfister-Guillouzo, *Coll. Czech. Chem. Commun.*, **39**, 861 (1974).
326. P. R. Moses and J. Q. Chambers, *J. Amer. Chem. Soc.*, **96**, 945 (1974).
327. R. C. Haddon, F. Wudl, M. L. Kaplan, J. H. Marshall, R. E. Cais, and F. B. Bramwell, *J. Amer. Chem. Soc.*, **100**, 7629 (1978).
328. H. Lund and S. Kwee, *Acta Chim. Scand.*, **22**, 2879 (1968).
329. R. Hazard and A. Tallec, *Bull. Soc. Chim. France*, 433 (1976).
330. H. Lund, ref. 301, p. 283.
331. J. Pinson, J. P. M'Packo, N. Vinot, J. Armand, and P. Bassinet, *Canad. J. Chem.*, **50**, 1581 (1972).
332. G. S. Supin, M. Ya. Fainshraiber, I. A. Mel'Nikova, N. N. Mel'Nikov, and T. N. Motorova, *Zh. Obshch. Khim.*, **47**, 2338 (1977).
333. J. Armand, K. Chekir, and J. Pinson, *Canad. J. Chem.*, **56**, 1804 (1978).
334. C. Bellec, P. Maitte, J. Armand and J. Pinson, *Canad. J. Chem.*, **58**, 2826 (1981).
335. G. L. Smith and J. W. Rogers, *J. Electrochem. Soc.*, **118**, 1089 (1971).
336. R. Schindler, H. Will, and L. Holleck, *Z. Elektrochem.*, **63**, 596 (1959).
337. C. D. Thompson and R. T. Foley, *J. Electrochem. Soc.*, **119**, 177 (1972).
338. Z. I. Fodiman, Z. V. Todres, and E. S. Levin, *Zh. Vses. Khim. Obshch. Mendeleeva*, **19**, 236 (1974).
339. P. Zuman, *Coll. Czech. Chem. Commun.*, **25**, 3245 (1960).
340. V. Sh. Tsvenishvili, V. N. Gaprindashvili, L. A. Tskalobadze, and V. A. Sergeev, *J. Gen. Chem. (USSR)*, **43**, 2114 (1973).
341. L. S. Efros and Z. V. Todres, *Zh. Obshch. Khim.*, **27**, 983 (1957).
342. V. Sh. Tsveniashvili, S. I. Zdanov, and Z. V. Todres, *Z. Anal. Chem.*, **224**, 389 (1967).
343. H. Lund, *Discuss. Faraday Soc.*, **45**, 193 (1968).
344. F. F. Medovschikova and I. Ya. Postovskii, *Zh. Obshch. Khim.*, **24**, 1989 (1954).
345. I. Tabakovic, M. Trkovnik, and Z. Grujic, *J. Chem. Soc. Perkin Trans.*, **2**, 166 (1979).
346. Ref. 301, p. 283.
347. B. Janik and P. J. Elving, *Chem. Rev.*, **68**, 295 (1968).
348. Ref. 189, p. 80.
349. K. S. V. Santhanam and P. J. Elving, *J. Amer. Chem. Soc.*, **96**, 1653 (1974).
350. T. Yao and S. Musha, *Bull. Chem. Soc. Japan*, **47**, 2650 (1974).
351. Z. N. Timofeeva, L. S. Tikhonova, Kh. L. Muravich-Aleksandr, and A. V. El'Tsov, *Russ. J. Gen. Chem. (USSR)*, **44**, 1976 (1974).
352. J. M. Sequaris and J. A. Reynaud, *J. Electroanal. Chem. Interfacial Electrochem.*, **63**, 207 (1975).

References 611

353. J. Komenda and D. Laskafeld, *Coll. Czech. Chem. Commun.*, **27**, 199 (1962).
354. D. L. McAllister and G. Dryhurst, *J. Electroanal. Chem. Interfacial Electrochem.*, **59**, 75 (1975).
354a. R. Raghavan and G. Dryhurst, *J. Electroanal. Chem. Interfacial Electrochem.*, **129**, 181 (1981).
355. J. Armand, K. Chekir, N. Ple, H. Queguiner, and M. P. Simonnin, *J. Org. Chem.*, **46**, 4754 (1981).
356. J. Armand, K. Chekir, and J. Pinson, *Compt. Rend. C*, **284**, 391 (1977).
357. J. Armand, L. Boulares, K. Chekir, and C. Bellec, *Canad. J. Chem.*, **59**, 3237 (1981).
358. P. O. Kosonen and R. Gustafsson, *Finn. Chem. Lett.*, **1977**, 204.
359. J. Armand, L. Boulares, C. Bellec, and J. Pinson, *Canad. J. Chem.*, **60**, 2797 (1982), in press.
360. S. Hünig and H. Pütter, *Chem. Ber.*, **110**, 2532 (1977).
361. Ref. 301, p. 392.
362. D. G. Davis, "Electrochemistry of Porphyrins", in D. Dolphin, Ed., *Porphyrins*, Vol. 5, Academic Press, New York, 1978, p. 127.
363. R. B. Woodward, *J. Pure Appl. Chem.*, **2**, 383 (1961).
364. H. H. Inhoffen, P. Jäger, R. Mahlop, and C. D. Mengler, *Liebigs Ann.*, **704**, 188 (1967) and references therein.
365. G. P. Heiling and G. S. Wilson, *Anal. Chem.*, **43**, 551 (1971).
366. L. Kittler and H. Berg, *J. Electroanal. Chem. Interfacial Electrochem.*, **16**, 251 (1968).
367. J. Elguero, C. Marzin, A. R. Katritzky, and P. Linda, *Adv. Heterocycl. Chem., Suppl. I*, 1976.
368. J. Tafel and M. Stern, *Chem. Ber.*, **33**, 2224 (1900).
369. B. Sakurai, *Bull. Chem. Soc. Japan*, **10**, 311 (1935).
370. A. B. Ershler, T. S. Orekhova, and I. M. Levinson, *Electrokhim.*, **15**, 520 (1979).
371. H. Uchiyama and S. Ozawa, *Jap. Kokai*, 1972, Patent 7227975; *Chem. Abstr.*, **78**, 16032 (1973).
372. J. Tirouflet, M. Rolin, and M. Guyard, *Bull. Soc. Chim. France*, **1956**, 568.
373. A. Ryvolova, *Coll. Czech. Chem. Commun.*, **25**, 420 (1960).
374. A. Ryvolova-Kejharova and P. Zuman, *Coll. Czech. Chem. Commun.*, **36**, 1019 (1971).
375. O. R. Brown, S. Fletcher, and J. A. Harrison, *J. Electroanal. Chem. Interfacial Electrochem.*, **57**, 351 (1974).
376. J. D. Porter, S. Fletcher, and R. G. Barradas, *J. Electrochem. Soc.*, **126**, 1693 (1979).
377. G. E. Hardtmann and H. Ott, *J. Org. Chem.*, **34**, 2244 (1969).
378. R. Andruzzi, A. Trazza, P. Bruni, and L. Greci, *Tetrahedron*, **33**, 665 (1977).
379. M. Ya. Fioshin, F. P. Krysin, I. A. Avrutzkaya, E. V. Zaboroznets, J. G. Tsar'Kova, I. I. Gubenko, and B. M. Kotlyarevskaya, Russ. Patent 433144; *Chem. Abstr.*, **81**, 105969 (1974).
380. G. Dryhurst, B. H. Hansen, and E. B. Hawkins, *J. Electroanal. Chem. Interfacial Electrochem.*, **27**, 375 (1970).
381. V. D. Bezuglyi, M. B. Sidom, and A. N. Gaidukevich, *Zh. Obshch. Khim.*, **48**, 2368 (1978).
382. T. K. Pashkevich, V. A. Shapovalov, V. D. Bezuglyi, I. Ya. Psotovskii, and G. B. Afanas'eva, *Zh. Obshch. Khim.*, **47**, 910 (1977).
383. P. Pflegel, G. Wagner, and O. Manousek, *Z. Chem.*, **6**, 263 (1966).
384. D. H. Miller, *Canad. J. Chem.*, **33**, 1806 (1955).
385. H. Lund, *Acta Chem. Scand.*, **21**, 2525 (1967).

386. H. Lund, *Coll. Czech. Chem. Commun.*, **30**, 4237 (1965).
387. P. Pflegel and G. Wagner, *Pharmazie*, **22**, 147 (1967).
388. A. Daver, *Compt. Rend. C*, **274**, 244 (1972).
389. G. K. Budnikov, *Zh. Obshch. Khim.*, **38**, 2431 (1968).
390. B. Czochralska and D. Shugar, *Biochim. Biophys. Acta*, **281**, 1 (1972).
391. T. Wasa and P. J. Elving, *J. Electroanal. Chem. Interfacial Electrochem.*, **91**, 249 (1978).
392. P. T. Kissinger and C. N. Reilley, *Anal. Chem.*, **42**, 12 (1970).
393. T. E. Cummings and P. J. Elving, *J. Electroanal. Chem. Interfacial Electrochem.*, **94**, 123 (1978).
394. T. E. Cummings and P. J. Elving, *J. Electroanal. Chem. Interfacial Electrochem.*, **102**, 237 (1979).
395. B. H. Hansen and G. Dryhurst, *J. Electrochem. Soc.*, **118**, 1747 (1971).
396. P. Pflegel and G. Wagner, *Pharmazie*, **27**, 24 (1972) and references therein.
397. Y. Armand and L. Boulares, *Compt. Rend. C*, **284**, 13 (1977).
398. J. L. Owens and G. Dryhurst, *Anal. Chim. Acta*, **87**, 37 (1976).
399. P. Pflegel and G. Wagner, *Z. Chem.*, **8**, 179 (1968).
400. W. F. Gum and M. M. Joullie, *J. Org. Chem.*, **32**, 53 (1967).
401. C. Furlani, *Gazzetta*, **85**, 1646 (1955).
402. J. Armand, Y. Armand, and L. Boulares, *Compt. Rend. C*, **286**, 17 (1978).
403. J. Armand, Y. Armand, and L. Boulares, unpublished results.
404. R. Gottlieb and W. Pfleiderer, *Chem. Ber.*, **111**, 1753 (1978).
405. S. Hünig and H. Pütter, *Chem. Ber.*, **110**, 2524 (1977).
406. T. K. Pashkevich, V. A. Shapovalov, V. D. Bezuglyi, I. Ya. Postovskii, and G. B. Afanas'Eva, *Zh. Obshch. Khim.*, **47**, 910 (1977).
407. H. Oelschlaeger and F. I. Senguen, *Arch. Pharm.*, **307**, 909 (1974).
408. M. M. Ellaithy, J. Volke, and J. Hlavaty, *Coll. Czech. Chem. Commun.*, **41**, 3014 (1976), and references therein.
409. M. Falsig and H. Lund, *Acta Chim. Scand.*, **B34**, 591 (1980).
410. H. Lund, ref. 301, p. 256.
411. R. Gottlieb and W. Pfleiderer, *Chem. Ber.*, **111**, 1763 (1978).
412. D. L. Smith and P. J. Elving, *J. Amer. Chem. Soc.*, **84**, 1412 (1962).
413. E. N. Timofeeva, A. V. Lizogub, H. L. Muravich-Alexander, and E. V. Elczov, *Zh. Obshch. Khim.*, **41**, 2539 (1971).
414. P. K. De and G. Dryhurst, *J. Electrochem. Soc.*, **119**, 837 (1972).
415. B. Czochralska, H. Fritsche, and D. Shugar, *Z. Naturforsch.*, **32C**, 488 (1977).
416. Ref. 189, p. 320.
417. D. L. McAllister and G. Dryhurst, *J. Electroanal. Chem. Interfacial Electrochem.*, **47**, 479 (1973).
418. S. Kwee and H. Lund, *Biochim. Biophys. Acta*, **297**, 285 (1973).
419. S. Kwee and H. Lund, *Bioelectrochem. Bioenerg.*, **6**, 441 (1979).
420. Ref. 189, p. 365.
420a. E. J. Nanni, D. T. Sawyer, S. S. Ball, and T. C. Bruice, *J. Amer. Chem. Soc.*, **103**, 2797 (1981).
421. J. P. Stradins and V. T. Glezer, in A. J. Bard, Ed, *Encyclopaedia of Electrochemistry of the Elements*, Vol. 13, Marcel Dekker, New York, 1979, p. 220.
422. A. Foffani and F. Fornasari, *Gazzetta*, **83**, 1051, 1059 (1959).
423. T. Kubota and H. Miyazaki, *Bull. Chem. Soc. Japan*, **35**, 1549 (1962).

424. T. Kubota and H. Miyazaki, *Bull. Chem. Soc. Japan*, 39, 2057 (1966), and references therein.
425. G. Anthoine, J. Nasielski, E. V. Donckt, and N. Vanlantem, *Bull. Soc. Chim. Belg.*, 76, 230 (1967).
426. T. Kubota, K. Nishikita, H. Miyasaki, K. Inatani, and Y. Oishi, *J. Amer. Chem. Soc.*, 90, 5080 (1968).
427. J. L. Sadler and A. J. Bard, *J. Electrochem. Soc.*, 115, 343 (1968).
428. R. Andruzzi, A. Trazza, L. Greci, and L. Marchetti, *J. Electroanal. Chem. Interfacial Electrochem.*, 108, 59 (1980).
429. M. Tachibana, S. Sawaki, and Y. Kawazoe, *Chem. Pharm. Bull.*, 15, 1112 (1967).
430. O. N. Nechaeva and Z. V. Pushkareva, *J. Gen. Chem. (USSR)*, 28, 2721 (1958).
431. L. V. Varyukhina and Z. V. Pushkareva, *J. Gen. Chem. (USSR)*, 26, 1953 (1956).
432. R. Andruzzi, M. E. Cardinali, and A. Trazza, *J. Electroanal. Chem. Interfacial Electrochem.*, 41, 67 (1973).
433. R. Andruzzi, A. Trazza, and P. Bruni, *J. Electroanal. Chem. Interfacial Electrochem.*, 51, 341 (1974).
434. M. B. Neiman, S. G. Mairanovskii, B. M. Kowarskaya, E. G. Rosantsev, and E. G. Gintsberg, *Izv. Akad. Nauk SSSR, Ser. Khim.*, 1964, 1518.
435. I. Suzuki, M. Nakadate, T. Nakashima, and N. Nagasawa, *Tetrahedron Lett.*, 1966, 2899.
436. E. Laviron and T. Lewandowska, *Bull. Soc. Chim. France*, 1970, 3177.
437. J. Volke and S. Beran, *Coll. Czech. Chem. Commun.*, 40, 2232 (1975).
438. T. Okano and K. Ohira, *Yakugaku Zasshi*, 88, 1170 (1968); *Chem. Abstr.*, 70, 16609 (1969).
439. V. M. Kazakova, O. G. Sokol, G. G. Dvoryantseva, I. S. Musatova, and A. S. Elina, *Khim. Geterotsikl. Soedin.*, 1980, 376.
440. T. R. Emerson and C. W. Rees, *J. Chem. Soc.*, 1962, 1923.
441. J. J. Donahue and S. Oliveri-Vigh, *Anal. Chim. Acta*, 63, 415 (1973).
442. C. R. Warner and P. J. Elving, *Coll. Czech. Chem. Commun.*, 30, 4210 (1965).
443. J. Volke, V. Volkova, and H. Oelschlaeger, *Electrochim. Acta*, 25, 1177 (1980).
444. S. Kwee and H. Lund, *Acta Chem. Scand.*, 26, 1195 (1972).
445. D. L. Smith and P. J. Elving, *J. Amer. Chem. Soc.*, 84, 1412 (1962).
446. S. Kwee and H. Lund, *Acta Chem. Scand.*, 25, 1813 (1971).
447. J. W. Webb, B. Janik, and P. J. Elving, *J. Amer. Chem. Soc.*, 95, 991 (1973).
448. A. Darchen and C. Moinet, *J. Chem. Soc. Chem. Commun.*, 487 (1976).
449. A. Darchen and C. Moinet, *J. Electroanal. Chem. Interfacial Electrochem.*, 68, 173 (1976).
450. Ref. 421, Vol. 13, p. 131.
451. J. Armand, Y. Armand, L. Boulares, M. Philoche-Levisalles, and J. Pinson, *Canad. J. Chem.*, 59, 1711 (1981).
452. I. Mazcika and J. Sradins, *Latv. PSR Zinotnu Akad. Vestis*, 85 (1960); *Chem. Abstr.*, 55, 23366b (1961).
453. D. M. Hamel and H. Oelschläger, *J. Electroanal. Chem. Interfacial Electrochem.*, 28, 197 (1970).
454. A. N. Dolgachev, I. A. Avrutskaya, and M. Ya. Fioshin, *Electrokhimiya*, 15, 1882 (1979).
455. M. Ya. Fioshin, I. A. Avrutzkaya, and A. N. Dolgachev, *Khim. Farm. Zh.*, 13, 78 (1979); *Chem. Abstr.*, 92, 64617p (1980).

455a. I. Avrutzkaya and M. Fioshin, *Coll. Czech. Chem. Commun.*, **47**, 196 (1982).
456. L. Holleck and R. Schindler, *Z. Electrochem.*, **60**, 1138 (1956).
457. H. Lund, *Acta Chem. Scand.*, **11**, 990 (1957).
458. E. A. H. Dahmen, D. Vader, and J. D. Van Der Laarse, *Z. Anal. Chem.*, **186**, 161 (1962), and references therein.
459. Ref. 421, Vol. 13, p. 77.
460. B. S. Jensen and V. D. Parker, *J. Chem. Soc. Chem. Commun.*, **1974**, 36.
461. E. Laviron and P. Fournari, *Bull. Soc. Chim. France*, **1966**, 518.
462. E. Laviron, P. Fournari, and J. Greusard, *Bull. Soc. Chim. France*, **1967**, 1255.
463. E. Laviron, P. Fournari, and G. Refalo, *Bull. Soc. Chim. France*, **1969**, 1024.
464. H. Lund and S. K. Sharma, *Acta Chem. Scand.*, **26**, 2324 (1972).
465. Ref. 421, Vol. 13, p. 163.
466. K. G. Boto and F. G. Thomas, *Aust. J. Chem.*, **27**, 1215 (1974).
467. R. Andruzzi, M. E. Cardinali, I. Carelli, and A. Trazza, *J. Electroanal. Chem. Interfacial Electrochem.*, **26**, 211 (1970).
468. Ref. 421, Vol. 12, p. 329.
469. M. Wrona, J. Giziewicz, and D. Shugar, *Nucleic Acid Res.*, **2**, 2209 (1975).
470. M. Wrona, B. Czochralaska, and D. Shugar, *J. Electroanal. Chem. Interfacial Electrochem.*, **68**, 355 (1976).
471. S. Kwee and H. Lund, *Exper. Suppl.*, **18**, 387 (1971).
472. G. Dryhurst, *J. Electrochem. Soc.*, **117**, 1118 (1970).
473. G. Dryhurst, *J. Electrochem. Soc.*, **116**, 1357 (1969).
474. G. Dryhurst, *J. Electroanal. Chem. Interfacial Electrochem.*, **28**, 33 (1970).
475. H. H. Ruettinger, H. Matschiner, and W. Schroth, *J. Prakt. Chem.*, **321**, 274 (1979).
476. M. Wrona and B. Czochralska, *J. Electroanal. Chem. Interfacial Electrochem.*, **48**, 433 (1973).
477. Ref. 5, p. 279.
478. K. Sugino, K. Shirai, T. Sekine, and K. Odo, *J. Electrochem. Soc.*, **104**, 667 (1957).
479. K. Alwair and J. Grimshaw, *J. Chem. Soc. Perkin II*, 1811 (1973).
480. J. Pinson and J. M. Saveant, *J. Amer. Chem. Soc.*, **100**, 1506 (1978).
481. C. Amatore, J. Chaussard, J. Pinson, J. M. Saveant, and A. Thiebault, *J. Amer. Chem. Soc.*, **101**, 6012 (1979).
482. J. Holubek and J. Volke, *Coll. Czech. Chem. Commun.*, **27**, 680 (1962).
483. M. Wrona and B. Czochralska, *Acta Biochim. Polon.*, **17**, 351 (1970).
484. E. Yu. Khmelnistskaya, *Sov. Electrochem.*, **10**, 165 (1974).
485. O. Manousek and P. Zuman, *Coll. Czech. Chem. Commun.*, **29**, 1432 (1964).
486. T. Nonaka, T. Ota and T. Fuchigami, *Bull. Soc. Chem. Japan*, **50**, 2865 (1977).
487. J. Armand and J. Pinson, unpublished results.
488. Ref. 5, p. 347.
489. Ref. 421, Vol. 12, p. 1.
490. J. Volke, *Coll. Czech. Chem. Commun.*, **23**, 1486 (1958).
491. E. Laviron, *Bull. Soc. Chim. France*, **1961**, 2325, and references therein.
492. J. Armand and L. Boulares, *Bull. Soc. Chim. France*, **1975**, 366.
493. J. Armand, L. Boulares, and P. Bassinet, *Compt. Rend. C*, **277**, 695 (1973).
494. R. I. Gelb and L. Meites, *Anal. Chim. Acta*, **26**, 58 (1962).
495. M. E. Cardinali, I. Carelli, and A. Trazza, *J. Electroanal. Chem. Interfacial Electrochem.*, **48**, 277 (1973).

References 615

496. Ref. 5, p. 413.
497. Ref. 421, Vol. 12, p. 261.
498. H. Lund, *Acta Chem. Scand.*, **17**, 972 (1963).
499. P. E. Iversen and H. Lund, *Acta Chem. Scand.*, **21**, 279 (1967).
500. P. E. Iversen and H. Lund, *Acta Chem. Scand.*, **21**, 389 (1967).
501. E. Laviron, *Compt. Rend.*, **250**, 3671 (1960).
502. J. Volke and J. Holubeck, *Coll. Czech. Chem. Commun.*, **28**, 1597 (1963), and references therein.
502a. M. E. Cardinali and I. Carelli, *J. Electroanalyt. Chem. Interfacial Electrochem.*, **125**, 477 (1981).
503. P. Gassman, I. Nishiguchi, and H. Yamamoto, *J. Amer. Chem. Soc.*, **97**, 1600 (1975).
504. R. Kassai and J. Simonet, *Tetrahedron Lett.*, **1980**, 3575.
505. T. Kageyama, S. Sakai, and M. Yokohama, *Nippon Kagaku Kaishi*, 16 (1977); *Chem. Abstr.*, **86**, 129897c (1977).
506. A. Stanienda, *Z. Naturforsch.*, **22b**, 1107 (1967).
507a. K. K. Kanawaza, A. F. Diaz, R. H. Geiss, W. D. Gill, J. F. Kwak, J. A. Logan, J. F. Rabott, and G. B. Street, *J. Chem. Soc. Chem. Commun.*, **1979**, 854.
507b. K. K. Kanawaza, A. F. Diaz, W. D. Gill, P. M. Grant, G. B. Street, G. P. Gardini, and J. F. Kwak, *Synth. Met.*, **1**, 329 (1980).
507c. A. Diaz, *Chem. Scripta.*, **17**, 145 (1981).
508a. A. Watanabe, M. Tanaka, and J. Tanaka, *Bull. Chem. Soc. Japan*, **54**, 2278 (1981).
508b. A. F. Diaz, J. I. Castillo, J. A. Logan, and W. Y. Lee, *J. Electroanal. Chem. Interfacial Chem.*, **129**, 115 (1981).
508c. A. F. Diaz, A. Martinez, and K. Kanazawa, *J. Electroanal. Chem. Interfacial Chem.*, **130**, 181 (1981).
508d. A. F. Diaz, J. Castillo, K. K. Kanazawa, J. A. Logan, M. Salmon, and O. Fajardo, *J. Electroanal. Chem. Interfacial Chem.*, **133**, 233 (1982).
508e. G. B. Street, T. C. Clarke, M. Krounbi, K. Kanawaza, V. Lee, P. Pfluger, J. C. Scott, and G. Weiser, *Mol. Cryst.*, **83**, 1285 (1982).
508f. M. Salmon, A. F. Diaz, A. J. Logan, M. Krounbi, and J. Bargon, *Mol. Cryst.*, **83**, 1297 (1982).
508g. R. A. Bull, F. R. Fan, and A. J. Bard, *J. Electrochem. Soc.*, **129**, 1009 (1982).
509. G. Cauquis and M. Genies, *Bull. Soc. Chim. France*, **1967**, 3220.
510. P. J. Grossi, L. Marchetti, R. Ramasseul, A. Rassat, and D. Serve, *J. Electroanal. Chem. Interfacial Electrochem.*, 87, 353 (1978).
511. N. L. Weinberg and E. A. Brown, *J. Org. Chem.*, **31**, 4054 (1966).
512. M. Libert, C. Caullet, and S. Longchamp, *Bull. Soc. Chim. France*, **1971**, 2367.
513. M. Libert and C. Caullet, *Bull. Soc. Chim. France*, **1971**, 1947.
514. M. Libert, C. Caullet, and J. Huguet, *Bull. Soc. Chim. France*, **1972**, 3639.
515. M. Libert, C. Caullet, and G. Barbey, *Bull. Soc. Chim. France*, **1973**, 536.
516. M. Libert and C. Caullet, *Compt. Rend. C*, **276**, 1073 (1973).
517. M. Libert and C. Caullet, *Bull. Soc. Chim. France*, **1974**, 800.
518. S. Longchamp, C. Caullet, and M. Libert, *Bull. Soc. Chim. France*, **1974**, 353.
519. G. Cauquis, P. J. Grossi, A. Rassat, and D. Serve, *Tetrahedron Lett.*, **1973**, 1863.
520a. K. Yoshida, *J. Amer. Chem. Soc.*, **99**, 6111 (1977).
520b. K. Yoshida, *J. Amer. Chem. Soc.*, **101**, 2116 (1979).
521. L. Eberson, *Acta Chem. Scand. B*, **34**, 747 (1980).

522. H. Horikawa, T. Iwasaki, K. Matsumoto, and M. Miyoshi, *J. Amer. Chem. Soc.*, **43**, 335 (1978).
523. T. Iwasaki, H. Horikawa, K. Matsumoto, and M. Miyoshi, *J. Org. Chem.*, **44**, 1553 (1979).
524. J. A. Guzinski and R. H. Fenton, *J. Chem. Soc. Chem. Commun.*, **1973**, 715.
525a. C. Slifstein and M. Ariel, *J. Electroanal. Chem. Interfacial Electrochem.*, **48**, 447 (1973).
525b. P. Longhi, P. Manitto, D. Monti, T. Mussini, and S. Rondini, *Electrochim. Acta*, **26**, 541 (1981).
526. F. Eivazi, W. M. Lewis, and K. M. Smith, *Tetrahedron Lett.*, **1977**, 3083.
527. F. Eivazi and K. M. Smith, *J. Chem. Soc. Perkin I*, 544 (1979).
528. H. Falk, A. Leodolter, and G. Schade, *Monatsh.*, **109**, 183 (1978).
529. W. R. Turner and P. J. Elving, *Anal. Chem.*, **37**, 467 (1965).
530. Y. Inove, S. Naguse, K. Kodura, H. Baba, and T. Abe, *Bull. Chem. Soc. Japan*, **46**, 2204 (1973).
531. W. J Davis and R. N. Haszeldine, *J. Chem. Soc. Perkin Trans. I*, 1263 (1975).
532. H. Sund in T. P. Singer, Ed., *Biological Oxidations*, John Wiley, New York, 1968, p. 603.
533. K. A. Schellenberg and L. Hellerman, *J. Biol. Chem.*, **231**, 547 (1958).
534. F. H. Westheimer, *Adv. Enzymologia*, **24**, 469 (1962).
535. J. L. Kurz, R. Hutton, and F. H. Westheimer, *J. Amer. Chem. Soc.*, **83**, 584 (1961).
536. E. M. Kosower, A. Teverstein, and A. J. Swallow, *J. Amer. Chem. Soc.*, **95**, 6128 (1973) and references therein.
537. A. Gutman, R. Margalit, and A. Schejter, *Biochemistry*, **7**, 2778 (1968).
538. W. J. Blaedel and R. G. Haas, *Anal. Chem.*, **42**, 918 (1970).
539. J. Klima, A. Kurfurst, J. Kuthan, and J. Volke, *Tetrahedron Lett.*, **1977**, 2725.
540. G. Abou-Elenien, J. Rieser, N. Ismail, and K. Wallenfels, *Z. Naturforsch.*, **36b**, 386 (1981).
541. A. Kufürst, J. Ludvik, P. Rauch, and M. Marek, *Coll. Czech. Chem. Commun.*, **46**, 1141 (1981).
542a. V. Skala, J. Volke, V. Ohanka, and J. Kuthan, *Coll. Czech. Chem. Commun.*, **42**, 292 (1977).
542b. V. Skala, J. Volke, V. Ohanka, and J. Kuthan, Czech., Patent 176,679 through *Chem. Abstr.*, **99**, 128737n (1980).
542c. J. Ludvik, J. Klima, J. Volke, A. Kurfürst, and J. Kuthan, *J. Electroanal. Chem. Interfacial Chem.*, **138**, 131 (1982).
543. J. Stradins, J. Beilis, J. Uldrikjis, G. Duburs, A. E. Sausins, and B. Cekavicius, *Khim. Geterosikl. Soedin.*, **1975**, 1525.
544a. J. Stradins, G. Duburs, J. Beilis, J. Uldrikjis, A. E. Sausins, and B. Cekavicius, *Khim. Geterosikl. Soedin.*, **1975**, 1530.
544b. Ya. V. Ogle, Ya. P. Stradius, Ya. G. Dubur, V. K. Lusis, and V. P. Kadys, *Khim. Getrosikl. Soedin.*, **42**, 292 (1977).
545. Yu. I. Beilis, G. Duburs, J. Uldrikjis, and A. Sausins, *Nov. Polyar. Tezisy. Dokl. Vses. Soveshch. Polyarog. 6th*, 124 (1975) through *Chem. Abstr.*, **86**, 23365n (1977).
546. J. Stradins, J. Beilis, G. Duburs, and T. L. Slonskaya, *Latv. PSR Zinat. Akad. Vestis, Khim. Ser.*, **1978**, 372, through *Chem. Abstr.*, **89**, 154503r (1978).
547. P. Leduc and D. Thevenot, *J. Electroanal. Chem. Interfacial Electrochem.*, **48**, 447 (1973).
548a. J. N. Burnett and A. L. Underwood, *Biochemistry*, **4**, 2060 (1965).

References

548b. H. Jaegfeldt, *J. Electroanal. Chem. Interfacial Chem.*, **110**, 295 (1980).
549. M. D. Smith and C. L. Olson, *Anal. Chem.*, **46**, 1544 (1974).
550. W. J. Blaedel and R. A. Jenkins, *Anal. Chem.*, **47**, 1337 (1975).
551. J. Moiroux and P. J. Elving, *Anal. Chem.*, **50**, 1056 (1978).
552. J. Moiroux and P. J. Elving, *Anal. Chem.*, **51**, 346 (1979).
553. J. Moiroux and P. J. Elving, *J. Amer. Chem. Soc.*, **102**, 6533 (1980).
554. F. Pragst, B. Kaltofen, J. Volke, and J. Kuthan, *J. Electroanal. Chem. Interfacial Electrochem.*, **119**, 301 (1981).
555. P. Klossek and H. Aurich, *Z. Chem.*, **16**, 59 (1976).
556. A. Kitani and K. Sasaki, *Bioelectrochem. Bioenerg.*, **8**, 257 (1981), and references therein.
557. R. M. Kelly and D. J. Kirwan, *Biotechnol. Bioeng.*, **19**, 1215 (1977).
558. A. Torstensson, G. Johansson, M. O. Maanson, P. O. Larsson, and K. Mosbach, *Anal. Lett.*, **13B**, 837 (1980).
559. H. Jaegfeldt, A. Torstensson, and G. Johansson, *Anal. Chim. Acta*, **97**, 221 (1978).
560. K. Sasaki, Z. Kitani, A. Kunai, and H. Miyake, *Bull. Chem. Soc. Japan*, **53**, 3424 (1980).
561a. A. Kitani, H. Y. So, and L. L. Miller, *J. Amer. Chem. Soc.*, **103**, 7636 (1981).
561b. D. Chi-Sing Tse and T. Kuwana, *Anal. Chem.*, **50**, 1315 (1978).
561c. C. Degrand and L. L. Miller, *J. Amer. Chem. Soc.*, **102**, 5728 (1980).
561d. H. Jaegfeldt, A. Torstensson, L. Gorton, and G. Johansson, *Anal. Chem.*, **53**, 1979 (1981).
561e. A. Tortensson and L. Gorton, *J. Electroanal. Chem. Interfacial Electrochem.*, **130**, 199 (1981).
562. R. I. Kaganovitch and B. B. Damaskin, *Sov. Electrochim.*, (*Eng. Transl.*), **4**, 221 (1968).
563. R. G. Barradas, M. C. Giordano, and W. H. Sheffield, *Electrochim. Acta*, **16**, 1235 (1971).
564. M. Okita, T. Wakamatzu, and Y. Bau, *J. Chem. Soc. Chem. Commun.*, **1979**, 749.
565a. S. F. Nelsen, C. R. Kessel, and D. J. Brien, *J. Amer. Chem. Soc.*, **102**, 702 (1980).
565b. S. F. Nelsen, C. R. Kessel, and L. A. Grezzo, and D. J. Steffek, *J. Amer. Chem. Soc.*, **102**, 5482 (1980).
566. S. F. Nelsen, C. R. Kessel, D. J. Brien, and F. Weinhold, *J. Org. Chem.*, **45**, 2116 (1980).
567. V. S. Plashkin, L. N. Pushkina, and S. V. Sokolov, *Zh. Org. Khim.*, **10**, 1215 (1974).
568a. J. C. Nielsen, R. Stotz, G. T. Cheek, and R. F. Nelson, *J. Electroanal. Chem. Interfacial Electrochem.*, **90**, 127 (1978).
568b. G. Tourillon and F. Garnier, *J. Electroanal. Chem. Interfacial Chem.*, **135**, 173 (1982).
569a. S. Torii and T. Yamanaka, Japan Patent 7976576, through *Chem. Abstr.*, **91**, 183177y (1979).
569b. M. Janda, J. Srogl, and P. Holy, *Coll. Czech. Chem. Commun.*, **46**, 3728 (1981).
570. G. T. Cheek and R. F. Nelson, *J. Org. Chem.*, **43**, 1230 (1978).
571. R. Andruzzi and A. Trazza, *J. Electroanal. Chem. Interfacial Electrochem.*, **86**, 201 (1978).
572. R. Andruzzi and A. Romano, *J. Electroanal. Chem. Interfacial Electrochem.*, **90**, 389 (1978).
573. T. E. Young, B. W. Babitt, and L. A. Wolfe, *J. Org. Chem.*, **45**, 2899 (1980).
574. M. Sainsbury, *Heterocycles*, **9**, 1349 (1978).
575. C. Jacubovitz, R. Vallot, L. T. Yu, and J. Reynaud, *Compt. Rend. C*, **290**, 377 (1980).

576. R. Andruzzi, A. Trazza, C. Berti, and L. Greci, *J. Chem. Res.*, **1982**, 178.
577a. M. Kulka, *J. Amer. Chem. Soc.*, **68**, 2472 (1946).
577b. J. C. Cochran and W. F. Little, *J. Org. Chem.*, **26**, 808 (1961).
577c. M. C. Pham, G. Tourillon, P. C. Lacaze, and J. E. Dubois, *J. Electroanal. Chem. Interfacial Electrochem.*, **111**, 385 (1980).
577d. M. C. Pham, J. E. Dubois, and P. C. Lacaze, *J. Electrochem. Soc.*, **130**, 346 (1983).
578. M. Thompson and A. E. Stubley, *Anal. Chim. Acta*, **119**, 179 (1980).
579. I. M. Sosonkin, O. N. Chupakhin, and A. I. Matern, *Zh. Org. Khim.*, **15**, 1976 (1979).
580. G. Cauquis, B. Chabaud, and Y. Gohee, *Tetrahedron Lett.*, **1970**, 2583.
581. N. A. Prikhod'ko, M. Zh. Zhurinov, and M. Ya. Fioshin, *Electrokhimiya*, **14**, 631, 973, 1253 (1978).
582. N. A. Prikohd'ko, M. Zh. Zhurinov, and M. Ya. Fioshin, *Electrokhimiya*, **16**, 1278 (1980).
583. A. Najafi and M. Sainsbury, *Heterocycles*, **6**, 459 (1977).
584. L. L. Miller, F. R. Stermitz, and J. R. Falck, *J. Amer. Chem. Soc.*, **93**, 5941 (1971).
585. L. L. Miller, F. R. Stermitz, and J. F. Falck, *J. Amer. Chem. Soc.*, **95**, 2651 (1973).
586. J. P. Falck, L. L. Miller, and E. R. Stermitz, *Tetrahedron*, **30**, 931 (1974).
587a. L. L. Miller, F. R. Stermitz, J. Y. Becker, and V. Ramachaudran, *J. Amer. Chem. Soc.*, **97**, 2922 (1975).
587b. J. Y. Becker, L. L. Miller, and R. F. Stermitz, *J. Electroanal. Chem. Interfacial Electrochem.*, **68**, 181 (1976).
588. L. Christensen and L. Miller, *J. Org. Chem.*, **46**, 4876 (1981).
589. E. Kotani and S. Tobinaga, *Tetrahedron Lett.*, 4759 (1973).
590a. J. F. Ambrose and R. F. Nelson, *J. Electrochem. Soc.*, **115**, 1159 (1968).
590b. J. E. Dubois, A. Desbene-Monvernay, and P. C. Lacaze, *J. Electroanal. Chem. Interfacial Electrochem.*, **132**, 177 (1982).
590c. A. Desbene-Monvernay, P. C. Lacaze, and J. E. Dubois, *J. Electroanal. Chem. Interfacial Electrochem.*, **129**, 229 (1981).
591. W. Lamm, F. Pragst, and W. Jugelt, *J. Prakt. Chem.*, **317**, 995 (1975).
592. J. M. Bobitt and J. P. Willis, *J. Org. Chem.*, **45**, 1978 (1980).
593a. I. M. Sosonkin, V. A. Subbotin, V. N. Charushin, and O. N. Chupakhin, *Dokl. Akad. Nauk SSSR*, **229**, 888 (1976).
593b. V. N. Charushin, O. N. Chupakhin, E. O. Sidorov, J. Beilis, and I. A. Terent'eva, *Zh. Org. Khim.*, **14**, 140 (1978).
594a. L. Marcoux and R. N. Adams, *J. Electroanal. Chem. Interfacial Electrochem.*, **49**, 111 (1974).
594b. K. Yasukouchi, I. Taniguchi, H. Yamaguchi, and K. Arakawa, *J. Electroanal. Chem. Interfacial Electrochem.*, **121**, 231 (1981).
595a. V. A. Zverev, V. I. Koshutin, and V. B. Eshechenzo, *Electrokhimiya*, **13**, 1382 (1977).
595b. V. A. Zverev, V. I. Koshutin, V. I. Smirnov, V. N. Kulakov, and L. B. Kandyba, Russ. Patent 650990; through *Chem. Abstr.*, **91**, 56360p (1979).
596a. N. Clauson-Kaas, F. Limborg, and K. Glens, *Acta Chem. Scand.*, **6**, 531 (1952).
596b. N. Clauson-Kaas and Z. Tyle, *Acta Chem. Scand.*, **6**, 962 (1952).
597. T. Shono, H. Hamaguchi, and K. Aoki, *Chem. Lett.*, **1977**, 1053.
598. A. J. Baggaley and R. Brettle, *J. Chem. Soc. C*, **1968**, 971.
599. S. D. Ross, M. Finkelstein, and J. J. Vebel, *J. Org. Chem.*, **34**, 1018 (1969).
600. K. E. Kolb and C. L. Wilson, *J. Chem. Soc. Chem. Commun.*, **1966**, 271.

References 619

601. S. Arita, Y. Nagahiro, and T. Takeshita, *Kogyo Kagaku Zasshi*, **72**, 1896 (1969); through *Chem. Abstr.*, **73**, 10122 (1970).
602. K. Yoshida and T. Fueno, *J. Org. Chem.*, **36**, 1523 (1971).
603a. N. Clauson-Kaas, F. Limborg, and P. Dietrich, *Acta Chem. Scand.*, **6**, 545 (1952).
603b. N. Clauson-Kaas, *Acta Chem. Scand.*, **6**, 556 (1952).
603c. J. T. Nielsen, N. Elming, and N. Clauson-Kaas, *Acta Chem. Scand.*, **9**, 9 (1955).
603d. P. Nedenskov, N. Elming, J. T. Nielsen, and N. Clauson-Kaas, *Acta Chem. Scand.*, **9**, 17 (1955).
604. N. Clauson-Kaas, *Acta Chem. Scand.*, **6**, 569 (1952).
605a. I. Stibor, J. Srogl, and M. Janda, *J. Chem. Soc. Chem. Commun.*, **1975**, 397.
605b. M. Janda, J. Srogl, H. Dvorakova, D. Dvorak, and I. Stibor, *Coll. Czech. Chem. Commun.*, **46**, 906 (1981).
606. M. Libert and C. Caullet, *Bull. Soc. Chim. France*, **1974**, 805.
607a. V. I. Mil'Man, V. A. Zverev, V. A. Smirnov, and M. S. Klebanov, *Electrokhimiya*, **14**, 1555 (1978).
607b. V. A. Smirnov and V. A. Zverev, Russ. Patent 535284; through *Chem. Abstr.*, **86**, 89190h (1977).
608. S. Torii, H. Tanaka, H. Ogo, and S. Yamashita, *Bull. Chem. Soc. Japan*, **44**, 1079 (1971).
609. M. Janda, J. Srogl, E. Korblova, and I. Stibor, *Coll. Czech. Chem. Commun.*, **45**, 1361 (1980).
610. H. Tanaka, Y. Kobayashi, and S. Tori, *J. Org. Chem.*, **41**, 3482 (1976).
611. I. A. Markushina and N. V. Shulakovkaya, *Khim. Geterosikl. Soedin.*, **1971**, 1155.
612. A. I. Kirsanova and M. G. Smirnova, *Elektrokhimiya*, **14**, 627 (1978).
613. M. Sugawara, M. Sato, T. Osuda, and Y. Yamamoto, *Denki Kagaku Oyobi Kogyo Butsuri Kagaku*, **42**, 247 (1971); through *Spec. Period. Rep., Electrochem.*, **6**, 63 (1978).
614. M. Sugawara and S. Makoto, *Denki Kagaku Oyobi Kogyo Butsuri Kagaku*, **45**, 198 (1977); through *Chem. Abstr.*, **87**, 52484b (1977).
615. V. V. Berenblit, V. I. Grachev, I. M. Dolgopol'skii, and G. A. Davydov, *Zh. Prikl. Khim.*, **45**, 2360 (1972).
616. R. I. Kruglikova and N. N. Kralinina, *Khim. Geterosikl. Soedin.*, **1972**, 875.
617. N. T. Berberova, A. A. Bumber, M. V. Nekroroshev, B. V. Panov, and O. Yu. Okhlobystin, *Dokl. Akad. Nauk SSSR*, **246**, 108 (1979).
618. F. Pragst and R. Ziebig, *Electrochim. Acta*, **23**, 735 (1978).
619. J. Srogl, M. Janda, I. Stibor, and R. Rozinek, *Synthesis*, **1975**, 717.
620. R. Ziebig and F. Pragst, *Z. Phys. Chem.*, **260**, 795 (1979).
621. M. Janda, *Coll. Czech. Chem. Commun.*, **28**, 2524 (1963).
622. M Janda and J. Radouch, *Coll. Czech. Chem. Commun.*, **32**, 2672 (1967).
623. M. Janda and L. Pavievski, *Coll. Czech. Chem. Commun.*, **32**, 2675 (1967).
624. J. Sprogl, M. Janda, and M. Valentova, *Coll. Czech. Chem. Commun.*, **35**, 148 (1970).
625. M. Nemec, J. Srogl, and M. Janda, *Coll. Czech. Chem. Commun.*, **37**, 3122 (1972).
626. M. Nemec, M. Janda, and J. Srogl, *Coll. Czech. Chem. Commun.*, **38**, 3857 (1973).
627. K. Yoshida, T. Saeki, and T. Fueno, *J. Org. Chem.*, **36**, 3673 (1971).
628. M. Janda, J. Srogl, M. Nemec, and A. Janoussova, *Coll. Czech. Chem. Commun.*, **38**, 1221 (1973).
629. M. Libert and C. Caullet, *Compt. Rend. C*, **278**, 439 (1974).
630. T. Abe, S. Nagase, and H. Baba, *Bull. Chem. Soc. Japan*, **46**, 3845 (1973).

631. J. Srogl, M. Janda, I. Stibor, J. Kos, and V. Vyskocil, *Coll. Czech. Chem. Commun.*, **43**, 2015 (1978).
632. B. L. Chenard and J. S. Swenton, *J. Chem. Soc. Chem. Commun.*, **1979**, 1172.
633. G. Bontempelli, F. Magno, G. A. Mazzochin, and S. Zecchin, *J. Electroanal. Chem. Interfacial Electrochem.*, **43**, 377 (1973).
634. D. S. Houghton and A. A. Humffray, *Electrochim. Acta*, **17**, 2145 (1972).
635. H. Lund, *Coll. Czech. Chem. Commun.*, **31**, 4174 (1966).
636. F. Pragst and I. Schwertfeger, *J. Prakt. Chem.*, **316**, 795 (1974).
637. F. Pragst, *J. Prakt. Chem.*, **315**, 549 (1973).
638. F. Pragst and B. Siefke, *J. Prakt. Chem.*, **316**, 267 (1974).
639. D. B. Baigrie and A. J. Joslin, *J. Chem. Soc. Perkin II*, 77 (1979).
640. P. Corbon, G. Barbey, A. Dupré, and C. Caullet, *Bull. Soc. Chim. France*, **1974**, 768.
641. C. Barbey and C. Caullet, *Compt. Rend. C*, **282**, 911 (1976).
642. F. Pragst and W. Jugelt, *J. Prakt. Chem.*, **316**, 981 (1974).
643. F. Pragst, *Z. Chem.*, **14**, 236 (1974).
644. F. Pragst, K. Köppel, W. Jugelt, and F. G. Weber, *J. Electroanal. Chem. Interfacial Electrochem.*, **60**, 323 (1975).
645. F. Pragst and C. Boeck, *J. Electroanal. Chem. Interfacial Electrochem.*, **61**, 47 (1975).
646. F. Pragst and R. Nastke, *Z. Chem.*, **16**, 487 (1976).
647. G. Barbey and C. Caullet, *Compt. Rend. C*, **280**, 91 (1975).
648. M. Genies, *J. Electroanal. Chem. Interfacial Electrochem.*, **79**, 351 (1977).
649. G. Cauquis, M. Genies, and E. Vieil, *Nouv. J. Chim.*, **1**, 307 (1977).
650. M. Genies and A. F. Diaz, *J. Electroanal. Chem. Interfacial Electrochem.*, **98**, 305 (1979).
651. A. F. Diaz and M. Genies, *Electrochim. Acta*, **26**, 687 (1981).
652. A. F. Diaz, *J. Amer. Chem. Soc.*, **99**, 5838 (1977).
653. A. F. Diaz, *J. Org. Chem.*, **42**, 3949 (1977).
654. D. Vogel and J. Jaenicke, *Ber. Bunsenges. Phys. Chem.*, **75**, 510 (1971).
655. D. Vogel and J. Jaenicke, *Ber. Bunsenges. Phys. Chem.*, **75**, 1302 (1971).
656. H. H. Adam and T. A. Joslin, *J. Electroanal. Chem. Interfacial Electrochem.*, **58**, 393 (1975); **72**, 197 (1976).
657. H. H. Adam, B. D. Baigrie, and T. A. Joslin, *J. Chem. Soc. Perkin II*, 1287 (1977).
658. B. Baigrie and T. A. Joslin, *J. Electroanal. Chem. Interfacial Electrochem.*, **87**, 405 (1978).
659. H. Saya and M. Masui, *J. Chem. Soc. Perkin II*, 1640 (1973).
660. H. Saya and M. Masui, *Chem. Pharm. Bull.*, **24**, 2137 (1976).
661. W. Summerman and H. Baumgaertel, *Coll. Czech. Chem. Commun.*, **36**, 575 (1971).
662. M. Libert and C. Caullet, *Bull. Soc. Chim. France*, **1976**, 345.
663. V. Lang and H. Baumgaertel, *J. Electroanal. Chem. Interfacial Electrochem.*, **78**, 133 (1977).
664. R. Huelnagen and H. Baumgaertel, *J. Electroanal. Chem. Interfacial Electrochem.*, **98**, 119 (1979).
665. V. A. Subbotin, I. M. Sosonkin, N. V. Fedyainov, and V. I. Kumantsov, *Khim. Geterosikl. Soedin.*, **1978**, 516.
666. G. Cauquis and M. Genies, *Tetrahedron Lett.*, **1970**, 3403.
667. G. Cauquis, B. Chabaud, and M. Genies, *Bull. Soc. Chim. France*, **1975**, 679.
668. S. F. Nelsen, C. R. Kessel, and H. B. Brace, *J. Amer. Chem. Soc.*, **101**, 1874 (1979).

669. G. Dryhurst, *Electrochemistry of Biological Molecules*, Academic Press, New York, 1977, p. 533.
670. S. Kato, J. Pinson, and G. Dryhurst, *J. Electroanal. Chem. Interfacial Electrochem.*, **62**, 415 (1975); erratum, *ibid.*, **69**, 444 (1976).
671. S. Kato and G. Dryhurst, *J. Electroanal. Chem. Interfacial Electrochem.*, **80**, 181 (1977).
672. S. Kato, M. Poling, D. Van Der Helm, and G. Dryhurst, *J. Amer. Chem. Soc.*, **96**, 5255 (1974).
673. S. Kato and G. Dryhurst, *J. Electroanal. Chem. Interfacial Electrochem.*, **79**, 931 (1977).
674. S. Kato, B. M. Visinski, and G. Dryhurst, *J. Electroanal. Chem. Interfacial Electrochem.*, **66**, 21 (1975).
675. B. M. Visinski and G. Dryhurst, *J. Electroanal. Chem. Interfacial Electrochem.*, **70**, 199 (1976).
676. J. L. Owens and G. Dryhurst, *J. Electroanal. Chem. Interfacial Electrochem.*, **80**, 171 (1977).
677. H. Meinert and D. Cech, *Z. Chem.*, **12**, 291 (1972).
678. H. Meinert and D. Cech, *Z. Chem.*, **12**, 335 (1972).
679. H. Meinert, D. Cech, G. Berth, P. Langen, and G. Etzold, East German Patent 93561 (1972); through *Chem. Abstr.*, **78**, 143230 (1973).
680. R. F. Nelson, D. W. Leedy, E. T. Seo, and R. N. Adams, *Z. Anal. Chem.*, **224**, 184 (1967).
681. G. Cauquis, J. Cognard, and D. Serve, *Electrochim. Acta*, **20**, 1019 (1975).
682. A. Stuwe and H. Baumgartel, *Ber. Bunsenges, Phys. Chem.*, **78**, 320 (1974).
683. A. Stuwe, M. Weber-Shafer, and H. Baumgartel, *Ber. Bunsenges. Phys. Chem.*, **78**, 309 (1974).
684. L. Simonson and C. K. Mann, *Tetrahedron Lett.*, **1970**, 3303.
685. K. Nyberg and R. Servin, *Acta Chem. Scand. B*, **1976**, 640.
686. K. Nyberg, *Synthesis*, 545 (1976).
687. P. Martigny, H. Lund, and J. Simonet, *Electrochim. Acta*, **21**, 345 (1976).
688. S. F. Nelsen and P. J. Hintz, *J. Amer. Chem. Soc.*, **94**, 7115 (1972).
689. T. M. McKinney and D. H. Geske, *J. Amer. Chem. Soc.*, **87**, 3015 (1965).
690a. J. Moiroux and A. M. Ambruster, *J. Electroanal. Chem. Interfacial Electrochem.*, **114**, 139 (1980).
690b. J. Moiroux and A. Anne, *J. Electroanal. Chem. Interfacial Electrochem.*, **121**, 261 (1981).
691. I. N. Palant, D. M. Kranokutskaya, K. F. Turchin, O. S. Anisimova, I. Y. Vainshtein, and L. N. Yakhoutov, *Khim. Geterosikl. Soedin.*, 1277 (1975).
692. G. Ritzler, F. Peter, and M. Gross, *J. Electroanal. Chem. Interfacial Electrochem.*, **117**, 53 (1981).
693. M. Masui and S. Ozaki, *Chem. Pharm. Bull.*, **27**, 1182 (1979).
694a. P. Kabasakalian and J. McGlotten, *Anal. Chem.*, **31**, 431 (1959).
694b. T. B. Jarbawi and W. R. Heineman, *J. Electroanal. Chem. Interfacial Electrochem.*, **132**, 323 (1982).
695. J. P. Billon, *Bull. Soc. Chim. France*, **1960**, 1784.
696. J. P. Billon, *Bull. Soc. Chim. France*, **1961**, 1923.
697. J. P. Billon, *Ann. Chim. France*, **7**, 183 (1962).
698. J. P. Billon, G. Cauquis, and J. Combrisson, *J. Chim. Phys.*, **61**, 374 (1964).

699. A. L. Tinker and A. J. Bard, *J. Amer. Chem. Soc.*, **101**, 2316 (1979).
700a. E. E. Bancroft, J. E. Pemberton, and A. N. H. Blount, *J. Phys. Chem.*, **84**, 2557 (1980).
700b. K. Yasukouchi, I. Taniguchi, H. Yamaguchi, J. Ayukawa, K. Ohtuka, and Y. Tsuruta, *J. Org. Chem.*, **46**, 1679 (1981).
701. G. Bidan and M. Genies, *Nouv. J. Chim.*, **5**, 177 (1981).
702. C. L. McIntyre and H. N. Blount, *J. Amer. Chem. Soc.*, **101**, 7720 (1979).
703a. M. Genies and M. Thomalla, *Electrochim. Acta*, **26**, 829 (1981).
703b. D. Lardet, E. Laurent, M. Thomalla, and G. Genies, *Nouv. J. Chim.*, **6**, 349 (1982).
704. G. Cauquis, A. Deronzier, D. Serve, and E. Vieil, *J. Electroanal. Chem. Interfacial Electrochem.*, **60**, 205 (1975).
705. G. Cauquis, A. Deronzier, and D. Serve, *J. Electroanal. Chem. Interfacial Electrochem.*, **47**, 193 (1973).
706. G. Cauquis, A. Deronzier, J. L. Lepage, and D. Serve, *Bull. Soc. Chim. France*, **1977**, 295, 303.
707. M. Neptune and R. L. McCreery, *J. Org. Chem.*, **43**, 5007 (1978).
708. M. Neptune, R. L. McCreery, and A. A. Manian, *J. Med. Chem.*, **22**, 196 (1979).
709. T. Iwasita and M. Vielstich, *Electrochim. Acta*, **25**, 703 (1980).
710a. I. I. Krasartsev, *Zh. Org. Khim.*, **17**, 1989 (1981).
710b. R. Borsdorf, R. Herzschuch, and J. Seidler, *Z. Chem.*, **10**, 147 (1970).
711. T. Shono and Y. Matsumara, *J. Amer. Chem. Soc.*, **91**, 2803 (1969).
712. K. Bechgaard and V. D. Parker, *J. Amer. Chem. Soc.*, **94**, 4749 (1972).
713. N. L. Weinberg and B. Belleau, *Tetrahedron*, **29**, 279 (1973).
714a. G. Cauquis and M. Maurey, *Compt. Rend. C*, **266**, 1021 (1968).
714b. G. Cauquis and M. Maurey-Rey, *Bull. Soc. Chim. France*, **1972**, 3588.
715. M. Genies, J. C. Moutet, and G. Reverdy, *Electrochim. Acta*, **26**, 931 (1981).
716. M. Genies and P. Campos y Sansano, *J. Electroanal. Chem. Interfacial Electrochem.*, **85**, 351 (1977).
717. G. Cauquis and M. Maurey-Rey, *Bull. Soc. Chim. France*, **1973**, 2870.
718. G. Cauquis and M. Maurey-Rey, *Bull. Soc. Chim. France*, **1973**, 291.
719. G. Schukat, Le Van Hinh, and E. Faughaenel, *Z. Chem.*, **16**, 360 (1976).
720. S. Hunig and H. Berneth, *Top. Curr. Chem.*, **92**, 1 (1980).
721. J. Q. Chambers, D. C. Green, F. B. Kaufman, E. M. Engler, B. A. Scott, and R. R. Schumaker, *Anal. Chem.*, **49**, 802 (1977).
722. B. F. Kaufman, A. H. Schroeder, E. M. Engler, S. R. Kramers, and J. Q. Chambers, *J. Amer. Chem. Soc.*, **102**, 483 (1980).
723. K. A. Idriss, J. Q. Chambers, and D. C. Green, *J. Electroanal. Chem. Interfacial Electrochem.*, **109**, 341 (1980).
724. Ch. Madec, F. Quentel, and J. Courtot-Coupez, *Anal. Lett. A*, **13**, 33 (1980).
725. F. W. Fung, J. Q. Chambers, and G. Mamantov, *J. Electroanal. Chem. Interfacial Electrochem.*, **47**, 81 (1973).
726. S. Hunig, G. Kiesslich, H. Quast, and D. Schentzow, *Ann.*, **1973**, 310.
727. P. R. Moses, J. Q. Chambers, J. O. Sutherland, and D. R. Williams, *J. Electrochem. Soc.*, **122**, 608 (1975).
728. P. R. Moses, R. M. Harnden, and J. Q. Chambers, *J. Electroanal. Chem. Interfacial Electrochem.*, **84**, 187 (1977).
729. P. M. Harndern, P. R. Moses, and J. Q. Chambers, *J. Chem. Soc. Chem. Commun.*, **1977**, 11.

730. G. S. Wilson, D. D. Swanson, J. T. Klug, R. S. Glass, M. D. Ryan, and W. K. Musker, *J. Amer. Chem. Soc.*, **101**, 1040 (1979).
731. M. D. Ryan, D. D. Swanson, R. S. Glass, and G. S. Wilson, *J. Phys. Chem.*, **85**, 1069 (1981).
732. H. E. Imberger and A. A. Humphrey, *Electrochim. Acta*, **18**, 373 (1973).
733. H. J. Shine and Y. Murata, *J. Amer. Chem. Soc.*, **91**, 1872 (1969).
734. Y. Murata and H. J. Shine, *J. Org. Chem.*, **34**, 3368 (1969).
735. J. Silber and H. J. Shine, *J. Org. Chem.*, **36**, 2923 (1971).
736. K. Kim, V. J. Hull, and H. J. Shine, *J. Org. Chem.*, **39**, 2534 (1974).
737. K. Kim, S. R. Mani, and H. J. Shine, *J. Org. Chem.*, **40**, 3857 (1975).
738. B. K. Blandish, A. G. Padilla, and H. J. Shine, *J. Org. Chem.*, **40**, 2590 (1975).
739. V. D. Parker and L. Eberson, *J. Amer. Chem. Soc.*, **92**, 7488 (1970).
740. R. F. Broman, W. R. Heineman, and T. Kuwana, *Discuss. Faraday Soc.*, **56**, 16 (1973).
741. A. W. B. Aylnaer-Kelly, A. Bewick, P. R. Cantrill, and A. M. Tuxford, *Faraday Discuss.*, **56**, 96 (1973).
742. J. F. Evans and H. N. Blount, *J. Org. Chem.*, **42**, 976 (1977).
743. U. Svanholm, O. Hammerich, and V. D. Parker, *J. Amer. Chem. Soc.*, **97**, 101 (1975).
744. O. Hammerich and V. D. Parker, *Electrochim. Acta*, **18**, 537 (1973).
745. U. Svanholm and V. D. Parker, *J. Chem. Soc. Perkin II*, 1567 (1976).
746. A. Dusserre and M. Genies, *Bull. Soc. Chim. France*, **1979**, 28.
747. U. Svanholm and V. D. Parker, *J. Amer. Chem. Soc.*, **98**, 997 (1976).
748a. M Genies, J. C. Moutet, and G. Reverdy, *Electrochim. Acta*, **26**, 385 (1981).
748b. L. A. Tinker and A. J. Bard, *J. Electroanal. Chem. Interfacial Electrochem.*, **133**, 275 (1982).
749. W. B. Gara and B. P. Roberts, *J. Chem. Soc., Chem. Commun.*, **1975**, 949.
750. C. T. Pedersen and V. D. Parker, *Tetrahedron Lett.*, **1972**, 767.
751. C. T. Pedersen, O. Hammerich, and V. D. Parker, *J. Electroanal. Chem. Interfacial Electrochem.*, **39**, 479 (1972).
752a. J. L. Owens, H. A. Marsh, and G. Dryhurst, *J. Electroanal. Chem. Interfacial Electrochem.*, **91**, 231 (1978).
752b. H. A. Marsh and G. Dryhurst, *J. Electroanal. Chem. Interfacial Electrochem.*, **95**, 81 (1979).
753a. A. Brajter-Toth and G. Dryhurst, *J. Electroanal. Chem. Interfacial Electrochem.*, **122**, 205 (1981).
753b. A. Brajter-Toth, R. N. Goyal, M. Z. Wrona, T. Lacava, N. T. Nguyen, and G. Dryhurst, *J. Electroanal. Chem. Interfacial Electrochem.*, **128**, 413 (1981).
753c. R. N. Goyal, A. Brajter-Toth, and G. Dryhurst, *J. Electroanal. Chem. Interfacial Electrochem.*, **131**, 181 (1982).
753d. R. N. Goyal, A. Brajter-Toth, G. Dryhurst, and N. T. Nguyen, *J. Electroanal. Chem. Interfacial Electrochem.*, **141**, 39 (1982).
754. M. Z. Wrona, J. L. Owens, and G. Dryhurst, *J. Electroanal. Chem. Interfacial Electrochem.*, **105**, 195 (1979).
755a. R. N. Goyal, M. Z. Wrona, and G. Dryhurst, *J. Electroanal. Chem. Interfacial Electrochem.*, **116**, 433 (1980).
755b. R. N. Goyal, A. Brajter-Toth, and G. Dryhurst, *J. Electroanal. Chem. Interfacial Electrochem.*, **133**, 287 (1982).
756. M. Wrona and G. Dryhurst, *Biochem. Biophys. Acta*, **570**, 371 (1979).
757. G. Dryhurst, *J. Electroanal. Chem. Interfacial Electrochem.*, **70**, 171 (1976).

758. M. T. Cleary, J. L. Owens, and G. Dryhurst, *J. Electroanal. Chem. Interfacial Electrochem.*, **123**, 265 (1981).
759. J. L. Owens and G. Dryhurst, *Anal. Chim. Acta*, **89**, 93 (1977).
760. G. Dryhurst and L. G. Karber, *Anal. Chim. Acta*, **100**, 289 (1978).
761. A. C. Conway, R. N. Goyal, and G. Dryhurst, *J. Electroanal. Chem. Interfacial Electrochem.*, **123**, 243 (1981).
762a. T. Yao and S. Musha, *Bull. Chem. Soc. Japan*, **52**, 2307 (1979).
762b. R. N. Goyal and G. Dryhurst, *J. Electroanal. Chem. Interfacial Electrochem.*, **135**, 75 (1982).
763. T. Yao, T. Wasa, and S. Musha, *Bull. Chem. Soc. Japan*, **50**, 2917 (1977).
764. H. Berg, *Bioelectrochem. Bioeng.*, **5**, 347 (1978).
765. V. Brabec and G. Dryhurst, *J. Electroanal. Chem. Interfacial Electrochem.*, **89**, 161 (1978).
766. D. L. McAllister and G. Dryhurst, *J. Electroanal. Chem. Interfacial Electrochem.*, **55**, 69 (1974).
767a. J. Pradac, J. Pradacova, D. Homolka, J. Koryta, J. Weber, K. Slavik, and R. Cihar, *J. Electroanal. Chem. Interfacial Electrochem.*, **74**, 205 (1976).
767b. R. Raghavan and G. Dryhurst, *J. Electroanal. Chem. Interfacial Electrochem.*, **129**, 189 (1981).
767c. L. G. Karber and G. Dryhurst, *J. Electroanal. Chem. Interfacial Electrochem.*, **136**, 271 (1982).
767d. D. Ege-Serpkenci and G. Dryhurst, *J. Electroanal. Chem. Interfacial Electrochem.*, **141**, 175 (1982).
768. D. G. Davis, *Porphyrins*, **5**, 127 (1978).
769. A. Stanienda and G. Biebl, *Z. Phys. Chem.*, **52**, 254 (1967).
770. K. M. Kadish, D. Schaeper, L. A. Bottomley, M. Tsutsui, and R. L. Bobsein, *J. Inorg. Nucl. Chem.*, **42**, 469 (1980).
771a. A. Louati, E. Schaeffer, H. J. Callot, and M. Gross, *Nouv. J. Chim.*, **2**, 163 (1978).
771b. A. Ulman, J. Manassen, F. Frolow, and D. Rabinovich, *Inorg. Chem.*, **20**, 1987 (1981).
772. A. M. Stolzenberg, L. O. Spreer, and R. H. Holm, *J. Amer. Chem. Soc.*, **102**, 364 (1980).
773. H. J. Callot, A. Louati, and M. Gross, *Tetrahedron Lett.*, **1980**, 3281.
774. I. Tabakovic, M. Trkovnik, and Z. Grujic, *J. Chem. Soc. Perkin II*, 166 (1979).
775. V. V. Zverev, B. I. Buzykhin, and V. Kh. Ivanova, *Zh. Obshch. Khim.*, **49**, 1839 (1979).
776. F. Wudl, D. E. Schafer, and B. Miller, *J. Amer. Chem. Soc.*, **98**, 252 (1976).
777. R. C. Haddon, M. L. Kaplan, and J. H. Marshall, *J. Amer. Chem. Soc.*, **100**, 1235 (1978).
778. H. Lund in S. Patai, Ed., *The Chemistry of the Hydroxyl Group*, Wiley-Interscience, New York, 1971, p. 253.
779. V. D. Parker, G. Sundholm, U. Svanholm, A. Ronlan, and O. Hammerich, in A. J. Bard, Ed., *Encyclopedia of Electrochemistry of the Elements*, Vol. 11, Marcel Dekker, New York, 1978, p. 182.
780. W. A. Struck and P. J. Elving, *J. Amer. Chem. Soc.*, **86**, 1229 (1964).
781. H. Y. Cheng, E. Stope, and R. N. Adams, *Anal. Chem.*, **51**, 2243 (1979).
782. G. Popp, *J. Org. Chem.*, **37**, 3058 (1972).
783. G. Popp and N. C. Reitz, *J. Org. Chem.*, **37**, 3646 (1972).
784. G. F. Kirkbright, J. T. Stock, R. D. Pugliese, and J. M. Bobitt, *J. Electrochem. Soc.*, **116**, 219 (1969).

References 625

785. J. M. Bobitt, J. T. Stock, A. Marchand, and K. H. Weisgraber, *Chem. Ind.*, 2127 (1966).
786. J. M. Bobitt, H. Yagi, S. Shibuya, and T. T. Stock, *J. Org. Chem.*, **36**, 3006 (1971).
787. J. M. Bobitt, I. Noguchi, H. Yagi, and K. Weisgraber, *J. Amer. Chem. Soc.*, **93**, 3551 (1971).
788. J. M. Bobitt, I. Noguchi, H. Yagi, and K. Weisgraber, *J. Org. Chem.*, **41**, 845 (1976).
789. J. M. Bobitt and R. C. Hallcher, *J. Chem. Soc. Chem. Commun.*, **1971**, 543.
790. I. W. Elliott, Jr., *J. Org. Chem.*, **42**, 1090 (1977).
791. U. Palmquist, A. Nilsson, T. Petterson, A. Ronlan, and V. D. Parker, *J. Org. Chem.*, **44**, 196 (1979).
792. M. Sainsbury and R. F. Shinazi, *J. Chem. Soc. Chem. Commun.*, **1972**, 718.
793. S. D. Ross, M. Finkelstein, and E. J. Rudd, *Anodic Oxidation*, Academic Press, New York, 1975.
794. U. Eisner and E. Kirowa-Eisner, in A. J. Bard, Ed., *Encyclopedia of the Electrochemistry of the Elements*, Vol. 13, Marcel Dekker, New York, 1979, p. 220.
795. G. Cauquis, H. M. Fahmy, G. Pierre, and M. H. Elnagdi, *J. Heterocycl. Chem.*, **16**, 413 (1979).
796. P. G. Desideri, D. Heimler, and L. Lepri, *J. Electroanal. Chem. Interfacial Electrochem.*, **88**, 407 (1978).
797. G. Henze and E. Keller, *Z. Chem.*, **14**, 238 (1974).
798. J. Janata and M. B. Williams, *J. Phys. Chem.*, **76**, 1178 (1972).
799. J. Q. Chambers, in A. J. Bard, Ed., *Encyclopedia of the Electrochemistry of the Elements*, Vol. 12, Marcel Dekker, New York, 1978.
800. S. Torii, H. Tanaka, and M. Ukida, *J. Org. Chem.*, **43**, 3223 (1978).
801. G. Horn and P. Zuman, *Coll. Czech. Chem. Commun.*, **25**, 3401 (1960).
802. C. T. Pedersen and V. D. Parker, *Tetrahedron Lett.*, **1972**, 771.
803. P. Berstein and M. N. Hull, *J. Electroanal. Chem. Interfacial Electrochem.*, **28**, A1 (1970).
804. J. Q. Chambers, P. R. Moses, R. N. Shelton, and D. L. Coffen, *J. Electroanal. Chem. Interfacial Electrochem.*, **38**, 245 (1972).
805. R. G. Khomyakhov, S. S. Kruglikov, and V. M. Berezovskii, *Zh. Obshch. Khim.*, **28**, 2898 (1958).
806. H. V. Udupa, M. S. Venkatachalapaty, S. Chidambaran, S. S. Karaikudi, and A. Hanamoorthy, Indian Patent 144,132; through *Chem. Abstr.*, **92**, 31033s (1980).
807. L. Eberson and K. Nyberg, in A. J. Bard, Ed., *Encyclopedia of the Electrochemistry of the Elements*, Vol. 12, Marcel Dekker, New York, 1978, p. 261.
808. P. G. Gassman and B. L. Fox, *J. Org. Chem.*, **32**, 480 (1963).
809. J. M. Bobitt and T. Y. Cheng, *J. Org. Chem.*, **41**, 443 (1976).
810. S. Torii, T. Okamoto, and H. Tanaka, *J. Org. Chem.*, **39**, 2486 (1974).
811. J. M. Bobitt and J. P. Willis, *J. Org. Chem.*, **45**, 1978 (1980).
812. K. Irie, M. Okita, T. Wakamatsu, and Y. Ban, *Nouv. J. Chim.*, **4**, 275 (1980).
813. S. Torii, T. Okamoto, and O. Takahide, *J. Org. Chem.*, **43**, 2294 (1978).
814. H. Lund, *Acta Chem. Scand.*, **17**, 1077 (1963).

9 NATURAL AND MAGNETICALLY INDUCED CIRCULAR DICHROISM OF HETEROCYCLIC COMPOUNDS

SVEN E. HARNUNG

Department of Inorganic Chemistry
The H. C. Ørsted Institute
The University of Copenhagen
Universitetsparken, Denmark

ERIK LARSEN

Chemistry Department
The Royal Veterinary and Agricultural University
Thorvaldsensvej, Denmark

9.1. Introduction   628
     9.1.1. Experimental Quantities, 628
     9.1.2. Instrumentation, 629
     9.1.3. Phenomenological Relations, 631
     9.1.4. Units for ORD, CD, and MCD Measurements, 633

9.2. Quantum Mechanical Relations for Circular Dichroism and Magnetically Induced Circular Dichroism   635
     9.2.1. General Relations, 636
     9.2.2. Simple Interpetations of Expressions, 639
         9.2.2.1. Rotatory Strength, 639
         9.2.2.2. Faraday Parameters, 640
     9.2.3. Models Used in the Interpretation of Circular Dichroism, 641
     9.2.4. A Model Used in the Interpretation of Magnetically Induced Circular Dichroism, 642

9.3. Moment Analyses   646
     9.3.1. Dipole Strength, 647
     9.3.2. Rotatory Strength, 649
     9.3.3. Faraday Parameters, 649
     9.3.4. SI Units, 650

9.4. Sample Studies   653
     9.4.1. Analytical Uses of Circular Dichroism, 653
     9.4.2. Analytical Uses of Magnetically Induced Circular Dichrosim, 654

628   Natural and Magnetically Induced Circular Dichroism of Heterocyclic Compounds

        9.4.3. Circular Dichroism of Small Heterocyclic Rings, 655
        9.4.4. Circular Dichroism of Nucleic Acid Monomers and Polymers, 656
        9.4.5. Circular Dichroism of Achiral Heterocyclic Compounds, 657
        9.4.6. Magnetically Induced Circular Dichroism of Isocyclic Polyenes, 658
            9.4.6.1. Benzene and Derivatives, 658
            9.4.6.2. Cyclic Polyenes and Derivatives, 659
        9.4.7. MCD Studies of Heterocyclic Compounds, 660
            9.4.7.1. Azaheterocycles, 660
            9.4.7.2. Inorganic Cyclic π Systems, 660

9.5.   Appendix   661

References   664

9.1. INTRODUCTION

The techniques to be discussed in this chapter offer natural extensions to the measurements of absorption spectra, especially in the visible and the near-ultraviolet regions. Absorption spectra are normally obtained with unpolarized light, and a circular dichroism (CD) spectrum is obtained as the difference in absorption using left and right circularly polarized light. Circular dichroism is absent for samples that are unable to interact differently with the two circularly polarized light beams. All optically active molecules distinguish between left and right, and such molecules have, in principle, circular dichroism at frequencies where they have light absorption. All molecules, even highly symmetric ones, may show an induced circular dichroism when exposed to an external field. Most important is the magnetically induced circular dichroism (MCD), which is measured with the magnetic flux collinear with the light beam.

9.1.1. Experimental Quantities

In ordinary absorption (ABS) spectra it is common to obtain the (decadic) absorbance A as a function of the wavelength λ (in air) of the impinging light. Absorbance is a measure of the internal transmittance T of the sample, $A = \log(1/T)$. For solutions measured on a double-beam spectrophotometer, $1/T$ is replaced by T_0/T, where T_0 is the internal transmittance of the solvent. For a solution obeying Lambert-Beer's relation, $A = \epsilon C l$, where ϵ is the molar (decadic) absorption coefficient, C the substance concentration, and l the path length in the sample. Since the wavelength of the light depends on the refractive index n of the medium, it is common to record ϵ as a function of the energy or the frequency ν of the light. In the general discussions of this chapter the circular frequency ω will be used, thus $\omega = 2\pi\nu$ and $\lambda = 2\pi c/\omega n$, where c is the speed of light in vacuum. We shall need also the absorption index k, and in place of $1/T$ the ratio I_0/I of the intensity of light before and after passing through the sample will be utilized:

$$I = I_0 \times 10^{-A} = I_0 \times 10^{-\epsilon C l} = I_0 e^{-2k(\omega/c)l} = I_0 e^{-\alpha l} \qquad (1)$$

where
$$\alpha = \epsilon C \ln 10 \tag{2}$$
is the napierian absorption coefficient (9).

In CD spectra, the difference $\Delta A = A_L - A_R$ between the absorbances of the left and right circularly polarized light is measured as a function of the wavelength. From equation (1), expression (3) is obtained:

$$\frac{I_L}{I_R} = 10^{-\Delta A} = 10^{-\Delta \epsilon C l} = e^{-2\Delta k(\omega/c)l} = e^{-\Delta \alpha l} \tag{3}$$

where $\Delta \epsilon = \epsilon_L - \epsilon_R$ is the molar (decadic) circular dichroism:

$$\Delta \epsilon = \frac{\Delta \alpha}{C \ln 10} \tag{4}$$

In magnetically induced circular dichroism, circular dichroism is generally measured as a function of the wavelength, the magnetic flux density B, and the temperature. However, in most organic chemical applications it is not necessary to vary the magnetic flux or temperature, and the data are given as the ratio $\Delta \epsilon / B$.

9.1.2. Instrumentation

Measurements of circular dichroism have proliferated since 1960 when commercial dichrographs became available (10). Presently, two companies manufacture automatic recording dichrographs (11, 12) and one manufactures parts for home-build instruments (13). A general block diagram of a fairly recent construction is shown in Figure 9.1.

The source may be a xenon arc lamp, and after passing the monochromator the light must be linearly polarized before it reaches the modulator. The polarization may be achieved by the monochromator (as sketched) or by introducing a Rochon polarizer, for example, in the beam (14). The modulator acts alternately as a quarter-wavelength and a three-quarter-wavelength plate, that changes the linearly polarized light alternately to left and right circularly polarized light. The early dichrographs used a Pockels cell modulator, which makes use of an alternating voltage (50 Hz) across a plate of ammonium dihydrogen phosphate cut perpendicular to the optical

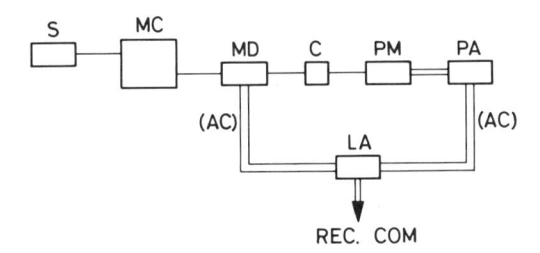

Fig. 9.1. Schematic representation of a popular dichrograph design. S, light source; MC, polarizing monochromator; MD, modulator; C, cell compartment; PM, photomultiplier; PA, preamplifier; LA, lock-in amplifier.

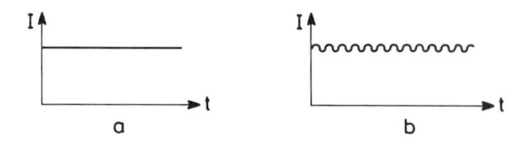

Fig. 9.2. Photomultiplier output at a given wavelength, (*a*) for a sample showing no CD, (*b*) for a CD-active sample.

axis. More recent constructions utilize the piezoelectric effect of, for example, fused quartz in a "photoelastic" modulator. This modulator may operate at a radiofrequency of 10–200 kHz, which is a practical advantage when the signal from the photomultiplier is amplified.

The electric current produced by the light reaching the photomultiplier is constant if no circularly dichroic sample is introduced into the cell compartment. In fact, by varying the voltage over the photomultiplier (Fig. 9.2), the current is also kept constant at different wavelengths where the lamp may emit with different intensity. However, when left and right circularly polarized light is absorbed differently by the sample, the signal from the photomultiplier varies with the frequency of the modulator. The lock-in amplifier selects the ac component that is proportional to ΔA and amplifies it before it is sent to the recorder or the microcomputer.

The use of a microcomputer for data collection is gaining popularity for various reasons. First it should be emphasized that a CD signal is generally very small. Thus, the ac component shown in Figure 9.2*b* is a very small fraction of the total average current. This is a consequence of the circular dichroism being small compared to the absorption. Actually $\Delta \epsilon / \epsilon$ at the maximum of the CD curve for solutions of even the more dichroic substances is only of the order of 0.01. For a given compound a higher CD intensity can be achieved by using a longer cell or a more concentrated solution. This increases the absorption, and for constructions using the principle of a constant average output from the photomultiplier, the voltage over the tube must be increased and the noise level, which varies parallel with the deflection, will also be increased.

A microcomputer can be used to collect a large number of measurements at each wavelength and store their average before proceeding to the next wavelength. It can also store the average of a number of whole curves. Thus it is possible to reach an improved signal-to-noise ratio relative to a CD spectrum displayed directly on a recorder. With a microcomputer it is also possible to make various kinds of curve analyses on experimental CD spectra (see Sec. 9.3).

There has been some interest in detecting the circular dichroism of solids. For a long time it has been an experimentally difficult but standard procedure to measure uniaxial crystals along the optic axis. In this way it is possible to measure the circular dichroism of an oriented ensemble of molecules of known (X-ray diffraction) structure and to compare measurements with theoretical predictions. It is unfortunate that so few optically active compounds crystallize in uniaxial crystals and therefore researchers have attempted to use phase-sensitive spectrometers to detect the circular dichroism of a molecule in a nonuniaxial crystal (15–17).

The circular dichroism of unoriented crystals (powders) has been detected by means of the photoacoustic principle, and now it seems feasible to study spontaneously resolved compounds which in solution would racemize immediately (18–20). Many such systems exist in which upon seeding with a chiral crystal, crystallization of this enantiomeric form continues.

An area of technical development is the IR region. A number of papers concerning vibrational CD spectra have appeared (21–24), and the effect of circularly polarized light in Raman spectroscopy has been considered (25). At the other end of the wavelength scale, the far-UV region, the traditional technique has been extended from ca. 190 nm to ca. 150 nm (26, 27). A completely new source of radiation has been used: synchrotron radiation, which reaches nearly 130 nm (28). The high intensity of the linearly polarized beam from a synchrotron makes this source very attractive, and one can foresee extensive exploitation of this technique.

The magnet used for MCD spectroscopy can be an electromagnet (magnetic flux density 0.5–2 T (1 T = 1 Wb m^{-2})), a permanent magnet (0.5–2 T), or a superconducting magnet (3–6 T). Flux densities as high as 21 T have been reported for MCD measurements on crystals, and even here the signal is proportional to the magnetic flux (29). Thus, for many applications the high cost of running a superconducting magnet is unnecessary. The microcomputer, on the other hand, is very useful in solving the operational difficulties that are introduced by a magnetic field close to an arc lamp and close to the moving electrons of a photomultiplier tube. The difficulties are essentially overcome by obtaining a spectrum with the magnetic flux parallel to the light beam a, a baseline without sample b, a spectrum with the magnetic flux antiparallel to the light beam c, and a baseline with this configuration d. By adding and subtracting the four curves ($a - b - c + d$) the resulting curve is corrected for the influence of the magnet on the dichroism as well as a possible natural circular dichroism.

9.1.3. Phenomenological Relations

A description of the interaction of light waves and matter may start with the solutions of the Maxwell equations. For rapidly varying harmonic electromagnetic fields, the solutions are of the form of traveling waves. We shall specify here the electric field **E** of a plane monochromatic wave:

$$\mathbf{E}(z, t) = E_0 \operatorname{Re}\left\{\mathbf{i} \exp\left[i\omega\left(\frac{\hat{n}z}{c} - t\right)\right]\right\} \tag{5}$$

in terms of the complex refractive index $\hat{n}$,

$$\hat{n} = n + ik \tag{6}$$

where $i = \sqrt{-1}$ and n and k are the refractive and absorption indices, respectively. The wave is polarized along **i**, the unit vector in the direction of the x axis; it enters the medium at the time $t = 0$, with the z coordinate $z = 0$ and amplitude E_0.

Re { } (Im{ }) means that the real (imaginary) part of the expression embraced by { } has to be taken.

The electronic transitions in a molecule contribute separately to the dispersion n as well as to the absorption k. The absorption is confined to a relatively narrow frequency range, whereas the dispersion extends over a great interval of frequencies. It is therefore a practical advantage and in accord with actual experimental conditions that the phenomena may be separated as in equation (6). However, since they have the same physical cause, the two quantities are dependent on one another through an integral transformation, the Kronig-Kramers relation, such that over the whole frequency range the dispersion governs the absorption and vice versa (30). Attempts made in the past to use the relation to transform dispersion data into absorption data have not been successful, mainly because major contributions to the dispersion in the optical ranges have their sources in the far-ultraviolet. Since the commercial availability of dichrographs, the Kronig-Kramers relation has not been utilized in this way.

It can easily be verified that equation (5) expresses correctly the last part of equation (1), because the intensity of the light is proportional to the square of the electric field. This also explains the factor 2 in equation (1) by the desire to have equation (6) in the form given. From a macroscopic point of view the imaginary part of $\hat{n}$ accounts for the energy of the wave being spent on generation of heat in the matter. If the matter is nonabsorbing, $k = 0$, then equation (5) shows directly that the actual speed of the wave is c/n.

In order to describe optical activity, the plane wave is decomposed into a sum of left and right circularly polarized waves:

$$\mathbf{E} = \tfrac{1}{2}(\mathbf{E}_L + \mathbf{E}_R) \tag{7a}$$

with

$$\mathbf{E}_L = E_0 \operatorname{Re}\left\{(\mathbf{i} + i\mathbf{j})\exp\left[i\omega\left(\frac{\hat{n}_L z}{c} - t\right)\right]\right\} \tag{7b}$$

and

$$\mathbf{E}_R = E_0 \operatorname{Re}\left\{(\mathbf{i} - i\mathbf{j})\exp\left[i\omega\left(\frac{\hat{n}_R z}{c} - t\right)\right]\right\} \tag{7c}$$

The cartesian coordinate system is assumed to be right-handed, and $\mathbf{j}$ is a unit vector along the y axis. The complex refractive indices $\hat{n}_L$ and $\hat{n}_R$ for the two waves may be identical, in which case equation (5) is regained. If for some reason the matter in the cell has the property that the difference

$$\Delta\hat{n} = \hat{n}_L - \hat{n}_R = (n_L - n_R) + i(k_L - k_R) = \Delta n + i\Delta k \tag{8}$$

is nonvanishing, then instead of equation (5) equation (7a), the sum of equations (7b) and (7c), yields

$$\mathbf{E} = E_0 \operatorname{Re}\Big\{[(\mathbf{i}\cos\phi - \mathbf{j}\sin\phi)\cosh\psi - i(\mathbf{i}\sin\phi + \mathbf{j}\cos\phi)\sinh\psi] \\ \times \exp\left[i\omega\left(\frac{\hat{n}z}{c} - t\right)\right]\Big\} \tag{9a}$$

with
$$\hat{n} = \tfrac{1}{2}(\hat{n}_L + \hat{n}_R) \tag{9b}$$

$$\phi = \Delta n \left(\frac{\omega}{2c}\right) z \tag{9c}$$

and
$$\psi = \Delta k \left(\frac{\omega}{2c}\right) z \tag{9d}$$

Now, in frequency ranges with no absorption, equation (9d) vanishes and equation (10) is obtained.

$$\mathbf{E} = E_0 \,\mathrm{Re}\left\{(\mathbf{i}\cos\phi - \mathbf{j}\sin\phi)\exp\left[i\omega\left(\frac{nz}{c} - t\right)\right]\right\} \tag{10}$$

Equation (10) represents a plane wave that is rotated clockwise through the angle ϕ when viewed against the z direction. Thus the real part of the $\Delta\hat{n}$ accounts for the ORD through equation (9c).

In ranges of absorption, (9a) represents an elliptically polarized wave whose axes are rotated clockwise through the angle ϕ. By convention the ellipticity is defined as the angle whose tangent is the ratio $\sinh\psi/\cosh\psi$ of the minor amplitude to the major one; that is, the ellipticity is given exactly by $\arctan(\tanh\psi)$. However, as has been pointed out, $\Delta k/k$ is at most of the order of 0.01, and to have an appreciable amount of light transmitted the absorbance cannot greatly exceed 1. Accordingly, ψ is much less than 1, and ψ is a useful approximation for the ellipticity. Combination of equations (9d) and (3) gives

$$4\psi = \Delta\alpha l = \Delta A \ln 10 = \Delta\epsilon Cl \ln 10 \tag{11}$$

which shows that the imaginary part of $\Delta\hat{n}$ accounts for the CD.

The manifestations of optical activity have now been connected to the complex refractive index. The next problem, then, is to express this quantity in terms of more fundamental parameters such as molecular geometry and electronic structure by means of quantum relations, but before discussing this, some comments on the units are desirable.

9.1.4. Units for ORD, CD, and MCD Measurements

Unfortunately, for historical reasons some confusion exists in expressing the experimental results of ORD or CD measurements. Furthermore, since the IUPAC *Manual* (9) makes an unsuccessful attempt to clarify the issue we shall discuss here the definition of the pertinent physical quantities and some ways of representing the results.

Following a historical approach, the basic physical quantities are considered to be the optical rotation ϕ and the ellipticity ψ, which have been obtained by passing a light beam through a homogeneous sample of length l containing the mass m or the amount n of optically active substance in the volume V.

The specific rotation is the rotation per unit of length divided by the mass concentration of optically active substance:

Table 9.1. Representations of Physical Quantities

Quantity	Some Commonly Used Representations of Quantity[a]					
ϕ/l	α	deg dm^{-1}	α^+	rad dm^{-1}	α'	rad cm^{-1}
ψ/l	θ	deg dm^{-1}	θ^+	rad dm^{-1}	θ'	rad cm^{-1}
m/V	ρ	g cm^{-3}	$10^{-2}\tilde{c}$	g cm^{-3}		
$C = n/V$	$\tilde{c}/M$	mol/(100 ml)	C	mol l^{-1}		

[a] Each representation consists of a numerical value times a unit; M is the molar mass, i.e., $m/n = M$ g mol^{-1}; $\tilde{c}$ is the percent concentration, i.e., $m/V = \tilde{c}$ g per 100 ml.

$$[\alpha] = \frac{\phi V}{lm} \tag{12}$$

The molar rotation is the rotation per unit of length divided by the concentration of optically active substance:

$$[M] = \frac{\phi V}{ln} \tag{13}$$

The analogous definitions for the specific and molar ellipticities are:

Specific ellipticity:
$$[\theta] = \frac{\psi V}{lm} \tag{14}$$

Molar ellipticity:
$$[\Theta] = \frac{\psi V}{ln} \tag{15}$$

In Table 9.1 are listed some common representations of the physical quantities that enter the definitions contained in equations (12)–(15). For example, for a neat solvent where the optical rotation α deg (dm^{-1}) has been measured at the sodium D line, one may indicate the specific rotation as

$$[\alpha]_D = \alpha/\rho \text{ deg dm}^{-1} \text{ ml g}^{-1}.$$

The following standards for the molar quantities contained in equations (13) and (15) are found in the older literature.

$$[M] = \frac{\alpha M}{\tilde{c}} \quad \text{deg dm}^{-1} \text{ (100 ml) mol}^{-1} \tag{16}$$

$$[\Theta] = \frac{\theta M}{\tilde{c}} \quad \text{deg dm}^{-1} \text{ (100 ml) mol}^{-1} \tag{17}$$

However, as indicated in previous discussions, the circular dichroism $\Delta\epsilon$ is now experimentally accessible and reflects directly the different absorptions of the two circularly polarized rays. Accordingly, we recommend units as given in equations (18) and (19) to be adopted as measures of the manifestations of optical activity.

$$[M] = \frac{\alpha'}{C} \quad \text{rad l mol}^{-1} \text{ cm}^{-1} \tag{18}$$

Table 9.2. Some Useful Factors of Conversion[a]

$$\frac{\theta}{\theta'} = \frac{1800}{\pi} \qquad \frac{\theta'}{\{\Delta\epsilon\}} = \tfrac{1}{4}C \ln 10$$

$$\frac{C}{\tilde{c}} = \frac{10}{M} \qquad \frac{\alpha'/C}{\alpha M/\tilde{c}} = \frac{\pi}{18 \times 10^3}$$

$$\frac{\{\Delta\epsilon\}}{\theta M/\tilde{c}} = \frac{4\pi}{18 \times 10^3 (\ln 10)} \approx \frac{1}{3298}$$

[a] $\Delta\epsilon$ is the circular dichroism, i.e., $\Delta\epsilon = \{\Delta\epsilon\} 1 \,\mathrm{mol}^{-1}\,\mathrm{cm}^{-1}$.

$$\Delta\epsilon = \{\Delta\epsilon\} \quad 1\,\mathrm{mol}^{-1}\,\mathrm{cm}^{-1} \tag{19}$$

where $\{\Delta\epsilon\}$ denotes the numerical value. It is noted that the units employed are not authorized SI units, but equation (19) is generally accepted and will be used throughout this chapter. Conversion factors between the two sets of units are included in Table 9.2.

Polymer (protein) chemists sometimes use a unit with the concentration of "monomers" or peptide units expressed in decimoles per cubic centimeter and the length is $l = \{l\}$ cm. The molar ellipticity is then given as $[\Theta] = 330\Delta A 109/\tilde{c}\{l\}$ deg cm² dmol⁻¹, where 109 is the average molar mass (g/mol) of the peptide unit. Although the expression is correct, it certainly calls for some standardization. It seems very odd indeed to publish ellipticities from observed CD data, but if, for reasons of continuity or inability to accept new orders of magnitude, ellipticities are used, we recommend the representation

$$[\Theta] = \frac{3298 \Delta A}{C\{l\}} \quad \deg\, l\,\mathrm{m}^{-1}\,\mathrm{mol}^{-1} \tag{20}$$

where C is the molarity of the average peptide unit.

The unit for MCD is the same as for circular dichroism (defined as in natural CD) per unit of magnetic flux density in the direction of the light beam. With this convention the MCD of the first charge-transfer band of $[\mathrm{Fe(CN)_6}]^{3-}$ at $2.39\,\mu\mathrm{m}^{-1}$ is positive, $\Delta\epsilon/B = +3.21\,\mathrm{mol}^{-1}\,\mathrm{cm}^{-1}\,\mathrm{T}^{-1}$ corresponding to a negative $\mathscr{B} + \mathscr{C}/kT$ term. Here the unit T is tesla (Wb m⁻²). The quantities $\mathscr{B}$ and $\mathscr{C}$ are introduced in the following section.

9.2. QUANTUM MECHANICAL RELATIONS FOR CIRCULAR DICHROISM AND MAGNETICALLY INDUCED CIRCULAR DICHROISM

In Condon's review (31) on natural optical activity, the complex refractive index $\Delta\hat{n}$ was expressed in terms of polarizabilities calculated by quantum mechanics.

The same approach has been used by Moffit and Moscowitz (30) for considerations of molecules in solution. The latter authors also introduced analyses of the spectra in terms of moments (see Sec. 9.3) and derived their dependence on the properties of the absorbing molecules.

Magnetic circular dichroism is the molecular equivalent of the longitudinal Zeeman effect in atoms, which has been reviewed by Condon and Shortley (32). MCD theory has been developed by Stephens (33, 34), who followed the classical approach for calculating polarizabilities. He took over the symbols $\mathscr{A}$, $\mathscr{B}$, and $\mathscr{C}$ for the Faraday terms from Serber's work (35) on diatomic molecules. Later, Stephens (36, 37) developed the more direct approach of absorption theory and calculated theoretical expressions for the moments of MCD spectra. The quantum expressions, equations (25)–(30), have been developed by Healy (38) using quantum electrodynamics.

9.2.1. General Relations

In the following description we shall give a survey of quantum relations for CD and MCD. Since it is particularly important here to indicate their uses, we have confined equations to the most common experimental setup, a solution of randomly oriented absorbing molecules in a transparent solvent that is isotropic before the field is applied. Except for a part of Section 9.3, molecular vibrations and librations have not been taken explicitly into account.

The absorption due to an electronic transition from the ground state G (degeneracy d_G, levels g) to the excited state N (degeneracy d_N, levels n) is then given by

$$\frac{\epsilon(N \leftarrow G; \omega)}{\omega} = K_D \mathscr{D}(N \leftarrow G) f^0(\omega) \tag{21}$$

The magnitude of the constant of proportionality K_D depends on the system of units, which is discussed in Section 9.3.4. The line-shape function $f^0(\omega)$ and its derivative $f'(\omega) = (d/d\omega) f^0(\omega)$ are defined to have the properties

$$\int_0^\infty f^0(\omega) d\omega = 1 \quad \text{and} \quad \int_0^\infty f'(\omega) d\omega = 0 \tag{22}$$

$\mathscr{D}$ is the electric dipole strength,

$$\mathscr{D}(N \leftarrow G) = d_G^{-1} \sum_{n, g} \mathscr{D}_{ng} \quad \mathscr{D}_{ng} = \boldsymbol{\mu}^{gn} \cdot \boldsymbol{\mu}^{ng} \tag{23}$$

where matrix elements $\boldsymbol{\mu}^{gn}$ of the electric dipole operator are defined through

$$\boldsymbol{\mu}^{gn} = -e \langle Gg | \sum_i \mathbf{r}_i | Nn \rangle \tag{24}$$

and the analogous expression for $\boldsymbol{\mu}^{ng}$. The state functions may be thought of as Slater determinants, and i runs over the number of electrons. The charge of the electron is $-e$, and the position vector of the ith electron is $\mathbf{r}_i$.

We consider next the equations that describe natural and induced circular dichroism. To the approximations given (38), the two effects are additive. With the static homogeneous magnetic flux density **B** applied parallel to the direction of propagation of the light, the CD is derived to be

$$\frac{\Delta\epsilon(N \leftarrow G; \omega)}{\omega} = -K_F\left[\frac{\mathscr{A}(N \leftarrow G)f'(\omega)}{\hbar} + \left\{\mathscr{B}(N \leftarrow G) + \frac{\mathscr{C}(N \leftarrow G)}{kT}\right\}f^0(\omega)\right]B$$

$$+ K_R\mathscr{R}(N \leftarrow G)f^0(\omega) \quad (25)$$

where k is the Boltzmann constant and T is the absolute temperature. The Faraday parameters $\mathscr{A}$, $\mathscr{B}$, and $\mathscr{C}$ and the rotatory strength $\mathscr{R}$ are given by the expressions

$$\mathscr{A}(N \leftarrow G) = (2d_G)^{-1}\sum_{n,g}\mathscr{A}_{ng} \qquad \mathscr{B}(N \leftarrow G) = d_G^{-1}\sum_{n,g}\mathscr{B}_{ng}$$

$$\mathscr{C}(N \leftarrow G) = (2d_G)^{-1}\sum_{n,g}\mathscr{C}_{ng} \qquad \mathscr{R}(N \leftarrow G) = d_G^{-1}\sum_{n,g}\mathscr{R}_{ng} \quad (26)$$

$$\mathscr{A}_{ng} = \text{Im}\{\boldsymbol{\mu}^{gn} \times \boldsymbol{\mu}^{ng} \cdot (\mathbf{m}^{nn} - \mathbf{m}^{gg})\} \quad (27)$$

$$\mathscr{B}_{ng} = \text{Im}\left\{\sum_k{}' \frac{\boldsymbol{\mu}^{gn} \times \mathbf{m}^{nk} \cdot \boldsymbol{\mu}^{kg}}{\hbar\omega_{NK}} + \sum_k{}' \frac{\boldsymbol{\mu}^{gn} \times \mathbf{m}^{kg} \cdot \boldsymbol{\mu}^{nk}}{\hbar\omega_{GK}}\right\} \quad (28)$$

$$\mathscr{C}_{ng} = \text{Im}\{\boldsymbol{\mu}^{gn} \times \boldsymbol{\mu}^{ng} \cdot \mathbf{m}^{gg}\} \quad (29)$$

$$\mathscr{R}_{ng} = \text{Im}\{\boldsymbol{\mu}^{gn} \cdot \mathbf{m}^{ng}\} \quad (30)$$

In these equations, the matrix elements **m** of the magnetic dipole operator are defined through

$$\mathbf{m}^{gn} = -\frac{e}{2m_e}\langle Gg|\sum_i(\mathbf{l}_i + g_e\mathbf{s}_i)|Nn\rangle \quad (31)$$

with

$$\mathbf{l}_i = -i\hbar\mathbf{r}_i \times \boldsymbol{\nabla}_i \quad (32)$$

For the electron the rest mass is m_e and $\boldsymbol{\nabla}_i = \partial/\partial\mathbf{r}_i$ is the nabla operator, g_e is the Landé g factor for the free electron, with $g_e = 2\mu_e/\mu_B$, where μ_e is the electron magnetic moment, and $\mathbf{s}_i$ is the spin angular momentum. The states $|Kk\rangle$ in the $\mathscr{B}$ term are other states, excluding $|Nn\rangle$ in the first line and $|Gg\rangle$ in the second, respectively, such that the energy differences $E_N - E_K = \hbar\omega_{NK}$ and $E_G - E_K = \hbar\omega_{GK}$ are different from zero. Furthermore, the states are supposed to be diagonal with respect to the magnetic flux, that is, with respect to $\mathbf{m}_z$.

In order to estimate upper limits for the orders of magnitudes to be expected, we appraise the ratio $\Delta\epsilon/\epsilon$. For natural circular dichroism, one has

$$\frac{\Delta\epsilon}{\epsilon} = \frac{K_R\mathscr{R}f^0}{K_D\mathscr{D}f^0} \approx \frac{4|\mathbf{m}|}{c\mu} \approx 10^{-2}$$

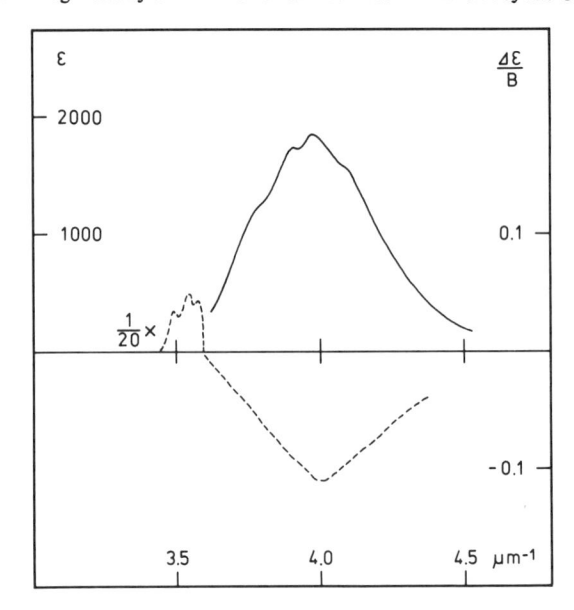

Fig. 9.3. The absorption and MCD (broken curve) spectra of pyridine in heptane. The low-energy part of the MCD has been enhanced 20 times.

using $|m| \approx \hbar$, $\mu \approx 10^{-10}$ m, and the constants K from Section 9.3. Similarly, for magnetically induced circular dichroism,

$$\frac{\Delta\epsilon}{\epsilon} = \frac{K_F \mathscr{A} B f'}{\hbar K_D \mathscr{D} f^0} \approx \frac{2B|m|}{\hbar \Delta\omega} \approx 10^{-1} B_t$$

for a transition with the width $\Delta\omega \approx 0.2\,\mu\text{m}^{-1}$, $|m| \approx \hbar$, and the magnetic flux density B_t in tesla. The relative order of magnitudes for the Faraday terms is (36)

$$\frac{\mathscr{A}f'}{\hbar} : \mathscr{B}f^0 : \frac{\mathscr{C}f^0}{kT} \approx \Delta\omega^{-1} : \Omega^{-1} : (kT)^{-1}$$

where Ω is an estimate of the separation between the considered transition and all other transitions.

Actual MCD intensities of many simple heterocyclic compounds are much less than this estimate. Thus, in the series pyridine, pyridazine, pyrimidine, pyrazine (39) the $\mathscr{B}$ terms due to allowed $\pi^* \leftarrow \pi$ transitions have $(\Delta\epsilon/B)/\epsilon \approx 5 \times 10^{-5}\,\text{T}^{-1}$. In particular, we may consider pyridine in cyclohexane (Fig. 9.3). For the $\pi^* \leftarrow \pi$, $^1B_2 \leftarrow {}^1A_1$ transition at $4.0\,\mu\text{m}^{-1}$, $\epsilon = 2 \times 10^3\,\text{l\,mol}^{-1}\,\text{cm}^{-1}$ and $\Delta\epsilon/B = -0.12$ $\text{l\,mol}^{-1}\,\text{cm}^{-1}\,\text{T}^{-1}$ ($\mathscr{D} \approx e^2 a_0^2$, $\mathscr{B} \approx 5.1 e^2 a_0^2 \mu_B E_h^{-1}$). The lone pair on nitrogen gives rise to the $\pi^* \leftarrow n$ transition at $3.5\,\mu\text{m}^{-1}$, which is difficult to assign in ordinary absorption spectra but is clearly seen in MCD, $\Delta\epsilon/B \approx 3 \times 10^{-3}\,\text{l\,mol}^{-1}\,\text{cm}^{-1}\,\text{T}^{-1}$.

Electrically forbidden, magnetically allowed transitions can be observed only for transitions within the molecular equivalent of a Russell–Saunders term (for atomic transitions: within states having ^{2s+1}L equal), and they are not frequently encountered in organic molecules. Even if they are allowed, they are very weak, and such transitions will not be discussed further.

In order to discuss spin-forbidden transition moments, for example, in singlet-triplet transitions, the spin-orbit coupling contribution to the Hamiltonian must be considered. For the azine series mentioned above, such moments have been calculated in terms of the difference between the dipole moments in the ground and excited states (40). Proper corrections of a crude Hamiltonian to include lower order effects may have "side effects" for which the simple expressions of this section are inadequate (41, 42). However, we consider such topics outside the scope of this chapter.

9.2.2. Simple Interpretations of expressions

9.2.2.1. Rotatory Strength

Natural CD is exhibited only by optically active molecules with absorption bands in the regions investigated. The sign and intensity of a CD band are characteristic for an enantiomer, and important conclusions concerning stereochemistry and electronic structure can be reached from such data, as we shall discuss in Section 9.4. In order to exhibit CD an optically active molecule must have a nonvanishing value of the rotatory strength $\mathscr{R}$, equation (30). Thus it must have an electronic transition during which the charge is linearly promoted (it then has an electric dipole transition moment μ^{ng}) and at the same time rotated perpendicularly to this direction (accounting for the magnetic dipole transition moment $\mathbf{m}^{gn}$). As $\mathscr{R}$ is the observable part of a scalar product, it depends on the angle between the two transition moments, and there may be a very direct and simple connection between the structure of a molecule and the spectral results.

This is particularly obvious for coupled chromophores, where two or more absorbing units are close enough in space (3–10 Å) to allow the individual electronic transition moments to have Coulomb interactions. The coupled chromophore model is elaborated in many places, while here it suffices to postulate that Coulomb interactions lead to ABS and CD spectra, which have patterns that depend on molecular geometry. Thus these spectra can give rather detailed information about the relative orientation of chromophores, for example, between two non-conjugated double bonds. Such a chromophore, with two interacting double bonds, is said to be *inherently chiral*. There are other types of inherently chiral chromophores, for example, disulfides, a twisted π system as in many lactams, or the twisted sulfoxide chromophore of propylenesulfoxide. For all such compounds, CD may give information about the "degree of twist" and its handedness. The measurements do not directly give information about the absolute configuration in terms of the handedness of a carbon atom that may be the originator of the chirality of the compound (2).

In most other cases the connection between the CD spectrum and the absolute configuration is a delicate matter. The reason is that a so-called one-electron transition, in order to be allowed and exhibit an intense absorption, has no rotation of charge during the excitation. Some transitions have a considerable rotation of charge, but without simultaneous linear promotion. These types of transitions of *achiral* chromophores with *chiral* surroundings have induced rotatory strengths, which can often be well understood by means of approximate theoretical models.

9.2.2.2. Faraday Parameters

Magnetically induced circular dichroism is a molecular Zeeman effect measured through the absorption bands and exhibited by all molecules with absorption in the spectral regions investigated. If the linewidth is less than the Zeeman splitting such that the MCD pattern is clearly discernible, then the classification of the spectrum through Faraday parameters may not be useful. This will be the case if for low temperatures only the ground level of a degenerate ground state is populated or if the absorption bands are extremely sharp as they are for the ruby lines in certain solid chromium complexes. For most other cases, in particular for solutes at room temperature, the $\mathscr{A}$, $\mathscr{B}$, $\mathscr{C}$ description is a feasible way of characterizing the spectra.

The $\mathscr{B}$-term contribution arises from the interaction between different energy levels caused by the applied magnetic field and occurs for transitions between degenerate as well as nondegenerate states. The energy differences in the denominators enhance contributions from energy levels lying close to the observed level and correspondingly diminish contributions from more distant levels. As will be discussed in Section 9.2.4, this fact has been used to reduce the rather complicated sum equation (28) to include only a few terms.

On the other hand, the interpretation of the $\mathscr{A}$ and $\mathscr{C}$ parameters is more obvious (8, 34), although these contributions are less general since they are exhibited only by molecules whose ground state or excited state in question is degenerate. They are composed of factors representing the electric dipole strength and the magnetic moment of either the ground state ($\mathscr{A}$ and $\mathscr{C}$ terms) or the excited state ($\mathscr{A}$ terms only). An MCD spectrum in which a $\mathscr{C}$ term dominates is temperature dependent, reflecting the Zeeman splitting of a degenerate ground state and the ensuing Boltzmann distribution between the levels. In most organic molecules the ground state is nondegenerate, and then only $\mathscr{A}$ and $\mathscr{B}$ terms may occur.

To illustrate the interpretation of $\mathscr{A}$ terms, one may consider a transition from the nondegenerate ground state G to an excited state N with electronic degeneracy d_N. In this particular discussion it is assumed that the Zeeman splitting is greater than the linewidth such that one may consider each of the d_N terms of equation (27) independently. The Zeeman energy of a certain level n_1, say, is given by $\hbar\omega_{NG} - \mathbf{m}^{n_1 n_1} \cdot \mathbf{B}$, and it is supposed that the level n_1 has an energy higher than that of level n_2. Generally, selection rules ($\Delta M = \pm 1$ in the case of an atom) allows only MCD in two of the transitions. Now suppose that the transition to level n_1 absorbs left circularly polarized light, which in this experimental setup (8, 43) may be thought of as having an angular momentum of $+1$ (units of $\hbar$); this would correspond to the selection rule $\Delta M = +1$ for an atom (32). Similarly, the transition of level n_2 (of lower energy) is excited by right circularly polarized light. Scanning toward higher frequencies, the Zeeman pattern is first negative and then positive. If the Zeeman effect is less than the linewidth, then the MCD spectrum may be described by equation (25) with a positive $\mathscr{A}$ term, equation (27).

It follows from this dicussion that the $\mathscr{A}$ and $\mathscr{C}$ terms are much easier to interpret than the $\mathscr{B}$ terms. This is due to the closed form of equations (27) and (29), which may lead to compact transparent expressions for these quantities using the technique of irreducible tensorial sets (44–47) for molecules possessing high spatial symmetry and electronic degeneracies.

9.2.3. Models Used in the Interpretation of Circular Dichroism

In recent years, increasing optimism has been expressed about the possibility of a quantitative understanding of the circular dichroism of organic molecules, and many convincing computations have been reported. Those especially interested in the theoretical aspects of optical activity are referred to reference 2. For many practical purposes it is sufficient to be familiar with the exciton coupling model and its extension, the polarizability model.

The molecular exciton or coupled oscillator model considers the simultaneous excitation of two or more chromophores in a molecule. These may be, for example, two conjugated systems bridged by a chain of saturated carbons. In each chromophore an allowed transition from the ground state to the excited state means that an electronic charge is moved along in the direction of the transition dipole moment. If the two chromophores are not coplanar and not perpendicular to one another, then the simultaneous charge propagation in the two chromophores also yields a rotation of charge around the direction of the vectorial sum of the transition dipole moments. The rotation of charge constitutes the magnetic transition dipole moment necessary for producing a rotatory strength. Another transition results with the dipole moment directed along the direction of the vectorial difference between the transition moments of the chromophores. This charge propagation will have a charge rotation along its direction, and it is always the case that the product $R(\pm) = \mu(\pm) \cdot m(\pm)$, for the two resulting transitions being in in phase (+) or out of phase (−) have opposite signs but the same magnitude. The splitting of the two transitions is determined by dipole-dipole coulomb interaction. A vast literature exists on the application of the exciton model to explain CD spectral results, and a book reviewing some authors work has been published (48).

An extension of the exciton model was introduced quite early and has recently gained popularity (1). This model has its application in explaining how a given transition in a chromophore obtains rotatory strength by coupling between the electronic transition dipole moment and a large number of other transition moments asymmetrically arranged around the chromophore and belonging to transitions of much higher energies. The $\pi^* \leftarrow \pi$ transition of an optically active alkene could be an example. The other transitions to interact asymmetrically with this could be, for example, $\sigma^* \leftarrow \sigma$ transitions involving the C–H and C–C bonds. The transition moments may be represented by anisotropic polarizabilities that arise essentially from these transitions. The model is relatively easy to use and is a good approximation when the chromophore is not distorted geometrically.

The above model and others may form the basis for predicting sector rules, among which the octant rule (49) for ketones is the best known. Sector rules have in the past been useful empirical means of correlating the chirooptical properties with structure (50, 51).

9.2.4. A Model Used in the Interpretation of Magnetically Induced Circular Dichroism

It was pointed out above, the interpretation of $\mathscr{A}$ and $\mathscr{C}$ terms is rather direct and presents no genuinely new approach relative to the well-established manipulations of electronic dipole strengths and magnetic moments. Many inorganic complexes have high degeneracies, and their MCD spectra are often dominated by $\mathscr{A}$ and $\mathscr{C}$ terms. This explains the early success of MCD in these fields of chemistry. The $\mathscr{B}$ terms are obviously more difficult to handle, but the understanding of their origins is mandatory if this kind of spectroscopy is to be a working tool for organic chemists. During the last decade, many papers have discussed various aspects of the MCD spectra of organic chromophores. In this section we shall deal in particular with a widely used model for the spectra of the low-energy $\pi^* \leftarrow \pi$ transitions in annulenes and derived compounds, keeping in mind, though, that almost any model that correctly treats molecular symmetry has included the most important aspects of the calculations. The purpose of the present section is to show, by means of qualitative arguments, how $\mathscr{B}$ terms may arise.

The development began with the calculations by Goeppert-Mayer and Sklar on the lower excited levels of benzene by means of molecular orbitals (52). The one-electron orbitals and the orbital energies were found in a π-electron approximation using Hückel's method. Slater determinants for the ground state and for the lowest (degenerate) excited states were investigated by group theoretical methods in order to assess in general terms the splitting due to electronic interaction. Finally, the overlap, coulomb, and exchange integrals were estimated using hydrogen like atomic orbitals. This general scheme of calculation has not changed much since then, although the methods have been refined. Accounts of more recent results may be found in treatises (53, 54) or textbooks (55). The next important step was Platt's perimeter model (56), which was made more quantitative by Moffitt (57). It considers a hypothetical planar cyclic polyene that is regarded as the unperturbed system from which the actual molecule is formed by a perturbation that keeps track of changes of the original properties. In fact, the energy matrix, equation (83), which forms the departure point for many recent discussions of MCD in π systems (58, 59) was given by Moffitt.

Consider first the expression (28) for the $\mathscr{B}$ term and recall that it concerns a transition from the ground state G to an excited state N, where other states K (including G and N) interact through the magnetic field. If we deal with only the lowest transitions, we have for the frequencies

$$\omega_{KG} > \omega_{NG} > |\omega_{KN}|$$

where, for example, $\hbar\omega_{NG}$ is the difference in energy between states N and G. Then the sum, equation (28), can be decomposed into terms of supposedly decreasing order of magnitude:

$$\mathscr{B}(N \leftarrow G) = d_G^{-1} \sum_{n,g} \text{Im} \left\{ \sum_{k \neq g, n} \boldsymbol{\mu}^{gn} \times \mathbf{m}^{nk} \cdot \boldsymbol{\mu}^{kg}/\hbar\omega_{NK} \right.$$

$$+ \sum_{n',g'} (\boldsymbol{\mu}^{gn} \times \mathbf{m}^{ng'} \cdot \boldsymbol{\mu}^{g'g} - \boldsymbol{\mu}^{gn} \times \mathbf{m}^{n'g} \cdot \boldsymbol{\mu}^{nn'})/\hbar\omega_{NG} \quad (33)$$

$$\left. + \sum_{n, k \neq g} \boldsymbol{\mu}^{gn} \times \mathbf{m}^{kg} \cdot \boldsymbol{\mu}^{nk}/\hbar\omega_{GK} \right\}$$

The approximation now consists in breaking off this series after the first term. It has been shown (4, 60) that if a $\mathscr{B}$ term is calculated by means of the truncated series, or equivalently by means of a restricted basic set of functions, then it may not be independent of the origin of the coordinate system unless the symmetry group of the molecule is sufficiently high. However, if the origin is chosen to be at the midpoint between the centers of charge for the ground and excited states, then for most cases the results are reliable and are adequate for the understanding of the basic properties of MCD (61). The problem of origin dependence may be avoided (62) if $\Delta\epsilon/B$ is calculated directly using the eigenfunctions of the system in the presence of the magnetic flux instead of the first-order expansion in B, equation (25).

Next we consider the eigenfunctions for the ground state and the first few excited states, assuming that the parent annulene has an even number of carbon atoms in the perimeter and that the solutions to the one-electron Hamiltonian (a Hamiltonian without terms representing the electronic repulsion) have been found. For this case the molecular orbitals of lowest and highest energy are spatial nondegenerate, while the remaining MOs are double degenerate. For the present example it is further assumed that the molecule has a singlet ground state and possesses more than two π electrons. This implies that there will be four electrons in the orbitals that span the two-dimensional irreducible representation E_n, say, and that the lowest empty orbitals are of symmetry type E_{n+1}. Only the lowest $\pi^* \leftarrow \pi$ transitions will be considered. We shall designate the orbitals ψ_n, $\psi_{\bar{n}}$, ψ_{n+1}, and $\psi_{\overline{n+1}}$, in an obvious notation. The totally symmetric ground state may now be written as a Slater determinant:

$$|^1 A\rangle = |\xi \psi_n^2 \psi_{\bar{n}}^2\rangle \quad (34)$$

where ξ represents the filled orbitals and additional quantum numbers that are necessary for the full characterization of the state but are of no concern here.

The excited singlet states of lowest energy are obtained by transferring an electron from the ψ_n or $\psi_{\bar{n}}$ orbital into the ψ_{n+1} or $\psi_{\overline{n+1}}$ orbital. The energies of the excited states form the point of departure for an interpretation of the MCD spectra. Therefore we give in the appendix (Sec. 9.5) the general fom of the energy matrix together with some important special cases. In the literature the energy eigenfunctions are often given an unintelligible form, but since these functions may be presented exactly and nevertheless in a shorthand way by means of electron creation and annihilation operators, we have used such representations in the following. Details are given in the appendix.

The single excited singlet states and representation of their magnetic moments are, then,

$$|^1B_1\rangle = \Omega(b_{n+1}^+, b_n)|^1A\rangle \qquad m_B e_0$$
$$|^1B_{\bar{1}}\rangle = \Omega(b_{\overline{n+1}}^+, b_{\bar{n}})|^1A\rangle \qquad -m_B e_0$$
$$|^1L_{2n+1}\rangle = \Omega(b_{n+1}^+, b_{\bar{n}})|^1A\rangle \qquad m_L e_0$$
$$|^1L_{\overline{2n+1}}\rangle = \Omega(b_{\overline{n+1}}^+, b_n)|^1A\rangle \qquad -m_L e_0$$
(35)

The operators Ω are functions of electronic creation and annihilation operators and transfer the ground state into excited singlet states. e_0 is a unit vector in the direction of the magnetic flux. Generally, electronic repulsion splits the states; the resulting energies have to be calculated, but the qualitative behavior can be predicted from symmetry. The split pattern depends on the number of carbon atoms in the perimeter and on the actual number of electrons. For example, for a $[4N+2]$-O-annulene [an all-*cis* annulene (63)] with $4N+2$ electrons (integer N) the ^1B states of equation (35) are degenerate and span the irreducible representation $^1E_{1u}$, whereas the two ^1L states are nondegenerate and transform as $^1B_{1u}$ and $^1B_{2u}$. (The actual assignment of the ^1L states depends on the orientation of the coordinate system.) For other annulenes the electronic replusion may split the degenerate ^1B pair. With regard to the magnetic moments, it is enough to say here that the magnitudes m_B and m_L generally are different and that within each pair the vectors are oppositely directed. The transitions to the ^1B states are allowed in electric dipole radiation, with the transition to $^1B_1$ absorbing left circularly polarized light, whereas transtions to the ^1L states are forbidden. Thus, the $^1B \leftarrow {}^1A$ transition carries an $\mathscr{A}$-type MCD which can be evaluated by entering the expressions of the appendix into the spherical (or cartesian) components of equation (27):

$$\mathscr{A}(^1B \leftarrow {}^1A) = -m_B \mathscr{D}(^1B \leftarrow {}^1A) \qquad (36)$$

In this approximation all $\mathscr{B}$ terms vanish.

The introduction of symmetry-lowering perturbations (bridges, aza-substitution etc.) in the parent annulene can be described by an extension of the one-electron Hamiltonian such that the degenerate π orbitals are split. The perturbation may affect either the filled orbitals through an element equivalent to

$$\langle \psi_n | H | \psi_{\bar{n}} \rangle = \hbar \omega_n \qquad (37)$$

or the empty orbitals through an element equivalent to

$$\langle \psi_{n+1} | H | \psi_{\overline{n+1}} \rangle = \hbar \omega_{n+1} \qquad (38)$$

or both. In general the problem is to diagonalize the four-by-four energy matrix, equation (83), constructed with the functions (35), but in some cases only (37) or (38) may be nonvanishing. Suppose now that only $\hbar \omega_n$ is different from zero and that the electronic repulsion between the L states vanishes; then the eigenfunctions are

$$|^1B_i\rangle = \Omega(b^+_{n+1}, b_n \cos \alpha + b_{\bar{n}} \sin \alpha)|^1A\rangle$$
$$|^1B_{\bar{i}}\rangle = \Omega(b^{\pm}_{\overline{n+1}}, b_{\bar{n}} \cos \alpha + b_n \sin \alpha)|^1A\rangle$$
$$|^1L_i\rangle = \Omega(b^+_{n+1}, -b_n \sin \alpha + b_{\bar{n}} \cos \alpha)|^1A\rangle \quad (39)$$
$$|^1L_{\bar{j}}\rangle = \Omega(b^{\pm}_{\overline{n+1}}, -b_{\bar{n}} \sin \alpha + b_n \cos \alpha)|^1A\rangle$$

with $tg\,2\alpha = 2\omega_n/\omega_{BL}$. Similarly, if equation (38) were nonvanishing, then the state $|^1B_1\rangle$ would be mixed into $|^1L_{\overline{2n+1}}\rangle$, and $|^1B_{\bar{1}}\rangle$ into $|^1L_{2n+1}\rangle$, to give states of the form

$$|^1B_k\rangle = \Omega(b^+_{n+1} \cos \alpha + b^{\pm}_{\overline{n+1}} \sin \alpha, b_n)|^1A\rangle \quad (40)$$

The low-energy $\mathscr{B}$ terms of the MCD spectrum can now be calculated by entering the functions (39) into the first term of equation (33) and separating out the electric dipole strength from equation (23). We arrive at the succinct expressions

$$\mathscr{B}(^1B \leftarrow {}^1A) = \frac{2\sin^2\alpha(m_B - m_L)\mathscr{D}(^1B \leftarrow {}^1A)}{\hbar\omega_{BL}}$$
$$\mathscr{B}(^1L \leftarrow {}^1A) = \frac{2\cos^2\alpha(m_B - m_L)\mathscr{D}(^1L \leftarrow {}^1A)}{\hbar\omega_{LB}} \quad (41)$$

for the transitions from the ground state ^1A to the ^1B and ^1L levels, resepctively. It is of interest for the ensuing discussion to enter explicit expressions of the dipole strengths:

$$\mathscr{D}(^1B \leftarrow {}^1A) = 2\cos^2\alpha\,\mathscr{D} \qquad \mathscr{D}(^1L \leftarrow {}^1A) = 2\sin^2\alpha\,\mathscr{D} \quad (42)$$

to give

$$\mathscr{B}(^1B \leftarrow {}^1A) = \frac{\sin^2 2\alpha(m_B - m_L)\mathscr{D}}{\hbar\omega_{BL}}$$
$$\mathscr{B}(^1L \leftarrow {}^1A) = \frac{\sin^2 2\alpha(m_B - m_L)\mathscr{D}}{\hbar\omega_{LB}} \quad (43)$$

in terms of the common constant $\mathscr{D}$. In the present problem there will be $\mathscr{A}$ terms superposed on the $\mathscr{B}$ terms. The theoretical expressions for these are easily derived from the equations (39), (27), and (23). The experimental problem of sorting out the different contributions is discussed in Section 9.3.

We can now summarize some general aspects of models that work within this restricted set of basis orbitals. The $\mathscr{B}$ parameters increase as a function of the genuine semiempirical parameter α whose magnitude may be calculated for specific molecules. In a useful model α is expected to vary in a predictable way through a series of related compounds. The $\mathscr{B}$ terms are due to a mutual borrowing mechanism, and their different signs reflect the trivial relation $\omega_{BL} = -\omega_{LB}$. A main feature of the model is that the magnetic transition moments of equation (33) have been cast into magnetic moments, which may be estimated so that absolute signs of the $\mathscr{B}$ parameters can be predicted.

In order to arrive at the functions (39), the electronic repulsion term c of equation (76) was put to zero. This is not a realistic approximation, and accordingly the model considered is indeed a crude one. Nevertheless such models give an understanding of the origins of the $\mathscr{B}$ parameters in terms that are familiar to organic chemists (64).

One may introduce the commonly used symbols for the energy splittings

$$\Delta E_{\text{HOMO}} = 2\hbar\omega_n \quad \text{and} \quad \Delta E_{\text{LUMO}} = 2\hbar\omega_{n+1} \qquad (44)$$

where the subscripts HOMO and LUMO stand for highest occupied and lowest unoccupied molecular orbitals, respectively. It is an inherent part of standard semiempirical theories that changes in the orbital energies and the ensuing changes in the quantitites of equation (44) can be attributed to inductive effects expressed in terms of coulomb integrals and to resonance effects expressed in terms of bond integrals, both of which are matrix elements of one-electron operators. Thus, in this framework the various details of MCD spectra may be traced back to familiar properties of the substituents that perturb the parent annulene.

We refer to a recent paper (58) that illustrates these aspects. The authors assume a two-parameter model, corresponding to the replacement of α by β in the j functions of equation (39), and connect schematic drawings of the energy levels in a series of benzene derivatives to the MCD pattern.

If all the consequences of a pure $\pi^* \leftarrow \pi$ model are to be exploited, the full matrix, equation (83), must be considered. In this connection it should be pointed out that for each application one must verify that contributions from $\pi^* \leftarrow \sigma$ and $\pi^* \leftarrow n$ transitions may be neglected. The energy levels of special cases (compare the functions (84)) have been discussed (57), and the MCD aspects of the most general cases may be found in references 59 and 65. The latter paper proposes a simple classification of the MCD behavior of the systems discussed in this section. It is based on the relationship between the quantities of equation (44). In the case that ΔE_{HOMO} and ΔE_{LUMO} both are vanishing, the MCD pattern is quite susceptible to the effects of weak perturbations, and for the other extreme case, where both quantities are nonvanishing, their difference governs the MCD pattern, which then is rather insensitive to perturbations. Recently, in a discussion of the MCD of reduced porphyrins (66), these points of view have been given further experimental support.

9.3. MOMENT ANALYSES

In the literature on statistics, distributions are often characterized by their moments (67). Under certain circumstances it may be possible to approximate the distribution by an (asymptotic) series expansion composed of a gaussian followed by terms containing derivatives of gaussians multiplied by linear combinations of the moments or by an expansion onto Hermite's orthogonal polynomials (68). In a similar way a broad absorption band without discernible vibronic structure may be characterized by its moments. In some cases the moments define parameters for a gaussian description of the spectra, but, more important, such moments relate in a significant way the observed data to fundamental constants of the absorbing molecules.

From an experimental point of view the moments of a band-shape function $g(\omega)$ are obtained as follows. The zeroth moment of the function is simply a constant times the area,

$$M = K^{-1} \int g(\omega) d\omega \qquad (45)$$

and plays the role of normalizing the higher moments. The constant K^{-1} is a collection of numbers and physical constants that gives dimension to the physical quantity M. Such constants, which depend on the quantity M, the function g, and the system of units, are discussed in Section 9.3.4. The νth moment with respect to an arbitrary frequency ω_0 is defined through the expression

$$M M_{\omega_0}^{(\nu)} = K^{-1} \int (\omega - \omega_0)^\nu g(\omega) d\omega \qquad (46)$$

with $\nu = 0, 1, 2, \ldots$. The first moment with respect to the zero of frequency,

$$M_0^{(1)} = M^{-1} K^{-1} \int \omega g(\omega) d\omega \qquad (47)$$

locates the band and is termed the mean frequency $\bar{\omega}$. When this frequency is used in equation (46) in place of ω_0, a reduced form of the moments, then called the *central moments*, is obtained. Quite generally, one has $M_{\omega_0}^{(0)} = 1$, $M_{\bar{\omega}}^{(1)} = 0$, and for the second moment $M_{\omega_0}^{(2)} = M_0^{(2)} - 2\omega_0 \bar{\omega} + \omega_0^2$ as a function of ω_0 one finds a minimum for $\omega_0 = \bar{\omega}$,

$$M_{\bar{\omega}}^{(2)} = M_0^{(2)} - \bar{\omega}^2 \qquad (48)$$

This second central moment is a convenient measure of the width of the curve, and one defines $\Delta\omega = \sqrt{M_{\bar{\omega}}^{(2)}}$. It is seen that equation (48) is the variance and $\Delta\omega$ is the standard deviation if the function $g(\omega)$ is considered as being a continuous distribution. The higher central moments are of little importance in spectroscopy; they wear the apt names *skewness* and *excess*, referring to the deviations from a gaussian shape.

In the following we shall discuss the relations between the experimental accessible moments and the corresponding quantum expressions. We consider first ABS and CD spectra and then MCD spectra.

9.3.1. Dipole Strength

As a measure of the total intensity of an absorption band, the dipole strength $\mathscr{D}(N \leftarrow G)$ is defined as the zeroth moment of the function $K_D^{-1} \epsilon(N \leftarrow G; \omega)/\omega$. The integration, equation (45), extends through the complete band arising from the $N \leftarrow G$ transition. By virtue of equation (23) the dipole strength is also given as a sum of contributions of strengths $\mathscr{D}_{ng}$, whose widths are neglected:

$$\mathscr{D}(N \leftarrow G) = K_D^{-1} \int \frac{\epsilon(N \leftarrow G; \omega)}{\omega} d\omega = d_G^{-1} \sum_{n,g} \rho_g \mathscr{D}_{ng} \qquad (49)$$

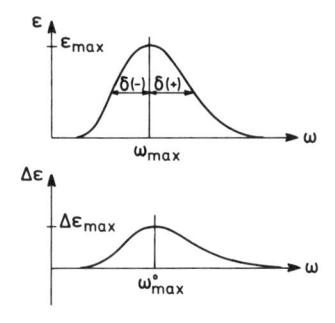

Fig. 9.4. A single absorption band with an associated CD band shown schematically.

We have, consistent with the assumptions, introduced a Boltzmann weighting factor ρ_g governing the appearance of hot bands; this factor will be left out in the expressions to follow.

The theoretical task is to elaborate the right-hand side of equation (49) in some detail, by means of an appropriate quantum model, before it is identified with the left-hand side. The experimental dipole strength embraces all contributions to the intensity (magnetic dipole strength, electric quadrupole strength), and equation (23) is clearly inadequate if the dominating source of intensity is not electric dipolar in nature. In order to show the general line of arguments we shall now restrict the discussion to electric dipole transitions between electronic nondegenerate levels; that is, d_G and d_N are both equal to 1. However, molecular vibrations will be accounted for through the interpretation that g and n denote vibration quantum numbers for the ground state G and the excited state N, respectively. $\mathscr{D}_{ng}$ of equation (49) is then the dipole strength of one narrow vibrational peak in the absorption band; it is located at the frequency

$$\omega_{ng} = \frac{E_{Nn} - E_{Gg}}{\hbar}.$$

For ABS the higher moments are defined by the relation

$$\mathscr{D}(N \leftarrow G)\mathscr{D}_{\omega_0}^{(\nu)}(N \leftarrow G) = K_D^{-1} \int (\omega - \omega_0)^\nu \frac{\epsilon(N \leftarrow G; \omega)}{\omega} d\omega$$

$$= \sum_{n,g} (\omega_{ng} - \omega_0)^\nu \mathscr{D}_{ng} \qquad (50)$$

where the second line expresses the quantum model. From this equation one may obtain explicit expressions for the mean frequency $\bar{\omega}_N$ and the bandwidth $\Delta\omega_N$,

$$\bar{\omega}_N = \mathscr{D}_0^{(1)}(N \leftarrow G) \quad \text{and} \quad \Delta\omega_N = \sqrt{\mathscr{D}_{\bar{\omega}_N}^{(2)}(N \leftarrow G)} \qquad (51)$$

Most complex molecules in solution exhibit absorption bands without much structure. A very common way to characterize such spectra is by stating the position ω_{max} and the intensity ϵ_{max} for the maximum point of each peak. For a more complete description, the half-width given as $\delta(+)$ and $\delta(-)$ may also be recorded (Fig. 9.4). Parameters of this kind in some cases combined with dipole strength, have been used widely in chemistry for empirical assignments of electronic transitions (69). However, the parameters $\mathscr{D}(N \leftarrow G)$, $\bar{\omega}_N$, and $\Delta\omega_N$ have a direct theoretical foundation (30) and may lead to much more structural information about the molecules (70).

9.3.2. Rotatory Strength

The definition of moments for natural CD follows by analogy and does not need many comments. The zeroth moment defines the rotatory strength due to the $N \leftarrow G$ transition:

$$\mathscr{R}(N \leftarrow G) = K_R^{-1} \int \frac{\Delta\epsilon(N \leftarrow G; \omega)}{\omega} d\omega = \sum_{n,g} \mathscr{R}_{ng} \qquad (52)$$

In a similar way the higher moments with respect to ω_0 are given by

$$\mathscr{R}(N \leftarrow G) \mathscr{R}_{\omega_0}^{(\nu)}(N \leftarrow G) = K_R^{-1} \int (\omega - \omega_0)^\nu \frac{\Delta\epsilon(N \leftarrow G; \omega)}{\omega} d\omega$$

$$= \sum_{n,g} (\omega_{ng} - \omega_0)^\nu \mathscr{R}_{ng} \qquad (53)$$

In these equations $\mathscr{R}_{ng}$, the rotatory strength of a narrow vibrational peak, is given by equation (30) with the above-mentioned interpretations. In an obvious notation, the mean frequency for the CD band is $\bar{\omega}_N^0 = \mathscr{R}_0^{(1)}(N \leftarrow G)$, and as a measure of the width one may use

$$\Delta\omega_N^0 = [\mathscr{R}_{\bar{\omega}_N}^{(2)}(N \leftarrow G)]^{1/2}$$

Successful calculations of the moments mentioned here have been made for molecules with transitions between nondegenerate states whose characteristic absorption bands are allowed in magnetic radiation fields (70, 71). Actually, many optically active organic molecules fulfill these conditions. If the transition is electronically allowed, then there is a direct proportionality between $\epsilon(N \leftarrow G; \omega)$ and $\Delta\epsilon(N \leftarrow G; \omega)$ and they have the same mean frequency and bandwidth. In this case the rotatory strength directly reflects the molecular chirality that might be correlated with the sign and magnitude of $\mathscr{R}(N \leftarrow G)$ by a sector rule.

9.3.3. Faraday Parameters

Moment analyses play an important role in the interpretation of MCD spectra of broad bands (8). The reason for this is not only that the moments serve as the connection between the observations and the theoretical expressions, but also that they are used experimentally to separate and characterize the $\mathscr{A}$, $\mathscr{B}$, and $\mathscr{C}$ contributions in a purely phenomenological manner, which of course is supposed to reflect the simple interpretations given in Section 9.2.2.

In the discussion to follow it is stipulated that the natural CD is absent or has been removed during the recording of the observed spectrum. Quite generally, then, the zeroth moment of an MCD band due to the $N \leftarrow G$ transition (electronic degeneracy d_N and d_G, vibrations not explicitly included) is defined as

$$\mathscr{F}(N \leftarrow G) = K_F^{-1} \int \frac{\Delta\epsilon(N \leftarrow G; \omega)}{\omega} d\omega = d_G^{-1} \sum_{n,g} (\mathscr{D}_{ng}^L - \mathscr{D}_{ng}^R) \quad (54)$$

where L and R denote left and right circularly polarized light, respectively. Note that equation (54) is just a restatement of the convention $\Delta\epsilon = \epsilon_L - \epsilon_R$.

The general procedure expressed in equation (46) for obtaining the moments is not directly applicable to MCD spectra because of zeroth moment may vanish, and yet it is valuable to define higher moments. Instead of using the moment, equation (54), as the normalizer for the higher moments, it has been suggested (72) to define the moments in terms of the dipole strength, equation (49), and the mean frequency, equation (51):

$$\mathscr{D}(N \leftarrow G) \mathscr{F}_{\bar{\omega}_N}^{(\nu)}(N \leftarrow G) = K_F^{-1} \int (\omega - \bar{\omega}_N)^\nu \frac{\Delta\epsilon(N \leftarrow G; \omega)}{\omega} d\omega$$

The physical reason for this is that the band shape, as given by the absorption spectrum before application of the magnetic flux, is only slightly changed due to the perturbation. It is often necessary to carry out the corresponding theoretical calculation of the moments (36, 37), in this connection called the rigid shift approximation, and the general approach is applicable also to perturbations caused by other kinds of external fields.

In order to see what one might expect from a moment analysis, we combine equations (54) and (25). Some properties of the ideal shape function $f^0(\omega)$ and its derivative $f'(\omega)$ were given in equation (22); to this we add the definition $\bar{\omega} = \int \omega f^0(\omega) d\omega$ and the integral $\int \omega f'(\omega) d\omega = -1$. Then the moments are

$$\mathscr{F}(N \leftarrow G) = -\left\{\mathscr{B}(N \leftarrow G) + \frac{\mathscr{C}(N \leftarrow G)}{kT}\right\} B \quad (55a)$$

$$\mathscr{F}_{\omega_0}^{(1)}(N \leftarrow G) = \left\{\frac{\mathscr{A}(N \leftarrow G)}{\hbar} - \bar{\omega}\left(\mathscr{B}(N \leftarrow G) + \frac{\mathscr{C}(N \leftarrow G)}{kT}\right)\right.$$
$$\left. + \omega_0\left(\mathscr{B}(N \leftarrow G) + \frac{\mathscr{C}(N \leftarrow G)}{kT}\right)\right\} B \quad (55b)$$

where the constants $\mathscr{A}$, $\mathscr{B}$, and $\mathscr{C}$ are given by equations (26)–(29). Equations (55a) and (55b) show the main feature of this method: First, the zeroth moment yields the $\mathscr{B}$ and $\mathscr{C}$ contributions to the MCD spectrum, and observations at different temperatures are necessary to separate the two terms. Second, if ω_0 is chosen in the vicinity of $\bar{\omega}$, then the $\mathscr{A}$ term is directly obtainable from equation (55b). Stephens (36, 37) has derived quantum expressions for the zeroth and first moments of ABS and MCD spectra under various assumptions.

9.3.4. SI Units

In this field of spectroscopy it has become a well-established practice to present equations using the three-dimensional nonrationalized system of gaussian cgs units.

However, we shall discontinue that practice and follow the recommendations now adopted by several members of the International Council of Scientific Unions (ICSU) (9, 73–75). Thus equations and constants will be expressed using the four-dimensional mksa-Giorgi system of units in its rationalized form together with SI units (74). Similarly, the results of quantum chemistry will be given in terms of selected physical constants (73), which in turn are expressible in SI units. Since we have kept the molar decadic absorption coefficient as given in equation (19), some mixtures of physical constants will not appear in a completely coherent form.

This latter point may be illustrated by an example. Standard derivation (2, 74) of the napierian absorption coefficient $\alpha(\omega)$ in the four-dimensional system of units yields

$$\frac{\alpha(\omega)}{\omega} = \frac{\pi N}{3\hbar c \epsilon_0} e^2 \left| \langle Gg | \sum_i \mathbf{r}_i | Nn \rangle \right|^2 f(\omega) \tag{56}$$

where N is a number of absorbers per unit volume, ϵ_0 is the permittivity of vacuum, and the factor 3 accounts for the random orientation of the molecules. With α as given in equation (2), and Avogadro's number N_A, we have

$$N = C N_A \times 10^3 \, \text{m}^{-3} = C N_A \times 10 \quad \text{m}^{-2} \text{cm}^{-1} \tag{57}$$

and for the transition $N \leftarrow G$ [compare eq. (21)], we obtain

$$\frac{\epsilon(N \leftarrow G; \omega)}{\omega} = \frac{10 \pi N_A}{3\hbar c \epsilon_0 \ln 10} \mathscr{D}(N \leftarrow G) f(\omega) = K_D \mathscr{D}(N \leftarrow G) f(\omega) \tag{58}$$

so that the dipole strength is

$$\mathscr{D}(N \leftarrow G) \approx 1.0221 \times 10^{-61} \int \frac{\epsilon(N \leftarrow G; \omega)}{\omega} d\omega \quad \text{C}^2 \text{m}^2 \tag{59}$$

In a similar way the reciprocals of the constants K_F and K_R in equation (25) are given by

$$K_F^{-1} = \frac{3\hbar c \epsilon_0 \ln 10}{20 \pi N_A} = 5.1105 \times 10^{-62} \quad \text{C}^2 \, \text{m}^2 \, \text{mol} \, \text{l}^{-1} \, \text{cm} \tag{60}$$

and

$$K_R^{-1} = \frac{3\hbar c^2 \epsilon_0 \ln 10}{40 \pi N_A} = 7.6605 \times 10^{-54} \quad \text{C m J T}^{-1} \, \text{mol} \, \text{l}^{-1} \, \text{cm} \tag{61}$$

and the Bohr magneton to be used in this connection, see equation (31), is

$$\mu_B = \frac{e\hbar}{2m_e} \tag{62}$$

For broad, structureless spectra the moments constitute a convenient link between experimental observations and quantum models. In Table 9.3 numerical values of conversion factors between SI units and quantum units (73) have been given with five significant figures, calculated from the CODATA report (75).

To aid in the conversion of data found in the literature, we have given in Table 9.4 a list of some gaussian cgs units and their corresponding quantum SI units. For

Table 9.3. Conversion Factors for Moments

Physical quantity	Relation
Hartree energy[a]	$E_h \stackrel{\wedge}{=} \dfrac{\hbar \times 10^{-6}}{2\pi a_0^2 m_e c} \approx 21.9475\ \mu m^{-1}$
Dipole strength	$e^2 a_0^2 \approx 7.1884 \times 10^{-59}\ C^2\ m^2$
Rotatory strength	$e a_0 \mu_B \approx 7.8630 \times 10^{-53}\ C\ m\ J\ T^{-1}$
Faraday $\mathscr{A}$ and $\mathscr{C}$	$e^2 a_0^2 \mu_B \approx 6.6665 \times 10^{-82}\ C^2\ m^2\ J\ T^{-1}$
Faraday $\mathscr{B}$	$e^2 a_0^2 \mu_B E_h^{-1} \approx 1.5291 \times 10^{-64}\ C^2\ m^2\ T^{-1}$

[a] $E_h = e^2/(4\pi\epsilon_0 a_0)$
The Bohr radius is $a_0 = 4\pi\epsilon_0 \hbar^2/m_e e^2$.
The symbol $\stackrel{\wedge}{=}$ means "corresponds to."

example, in the two systems a rotatory strength is expressible as

$$\mathscr{R}(N \leftarrow G) = \mathscr{R} D\beta = 0.39343\ \mathscr{R} e a_0 \mu_B \tag{63}$$

Finally, we consider the expression for the $\mathscr{B}$ and $\mathscr{A}$ terms for a solution of molecules that possess an electronic nondegenerate ground state. The magnetic flux density is

$$B = B_g\ G \stackrel{\wedge}{=} B_t\ T \tag{64}$$

with

$$B_g = 10^4 B_t \tag{65}$$

The ellipticity per unit magnetic flux density may be given by

$$\frac{[\Theta]}{B} = \frac{\theta M}{\tilde{c} B_g}\quad \deg 1\ m^{-1}\ mol^{-1}\ G^{-1} \tag{66}$$

and analogously, the MCD is

$$\frac{\Delta\epsilon}{B} = \frac{\{\Delta\epsilon\}}{B_t}\quad 1\ mol^{-1}\ cm^{-1}\ T^{-1} \tag{67}$$

Table 9.4. Some CGS Units and Their Corresponding Quantum Units

Physical Quantity	Correspondence between Units
Charge	$1\ esu \stackrel{\wedge}{=} (10c)^{-1}\ A\ m \approx 3.3356 \times 10^{-10}\ C$
Dipole moment[a]	$1\ D \stackrel{\wedge}{=} (10^{21} c)^{-1}\ A\ m^2 \approx 3.3356 \times 10^{-30}\ C\ m$
Dipole strength[b]	$1\ D^2 \stackrel{\wedge}{=} 0.15479 e^2 a_0^2$
Rotatory strength[c]	$1\ D\beta \stackrel{\wedge}{=} 0.39343 e a_0 \mu_B$
Faraday $\mathscr{A}$ and $\mathscr{C}$	$1\ D^2\beta \stackrel{\wedge}{=} 0.15479 e^2 a_0^2 \mu_B$
Faraday $\mathscr{B}$	$1\ D^2\beta/cm^{-1} \stackrel{\wedge}{=} 3.3971 \times 10^4 e^2 a_0^2 \mu_B E_h^{-1}$

[a] See Table 9.3. The Debye unit is $D = 10^{-18}\ esu\ cm$.
[b] The use of physical constants in expressions of results in quantum chemistry is described in reference 73.
[c] The (cgs) Bohr magneton is $\beta = e^s \hbar/2 m_e c \approx 9.274 \times 10^{-21}\ erg\ G^{-1}$

The number $\theta M/\tilde{c} B_g$, which through θ is a function of ω, will be designated by $[\Theta]_m$, and it follows from Table 9.2 and equation (65) that

$$[\Theta]_m = \frac{0.3298\{\Delta\epsilon\}}{B_t} \tag{68}$$

Therefore, in the different systems of units the $\mathscr{B}$ term may be written as

$$\mathscr{B} = \frac{-1}{33.53}\int \frac{[\Theta]_m}{\omega} d\omega \qquad D^2\, \beta(\text{cm}^{-1})$$

$$\triangleq -K_F^{-1}\int \frac{\{\Delta\epsilon\}/B_t}{\omega} d\omega \qquad C^2\, m^2\, T^{-1}$$

$$\approx -334.2\int \frac{\{\Delta\epsilon\}/B_t}{\omega} d\omega \quad \frac{e^2 a_0^2 \mu_B}{E_h} \tag{69}$$

where the first expression has been much used (8, 76, 77). The expressions of the second and third lines are recommended.

In a similar way the $\mathscr{A}$ term may be derived:

$$\mathscr{A} = \hbar K_F^{-1}\int (\omega - \omega_0)\frac{\Delta\epsilon/B}{\omega} d\omega$$

$$\approx 8.0844 \times 10^{-15}\int (\omega - \omega_0)\frac{\{\Delta\epsilon\}/B_t}{\omega} d\omega \qquad e^2 a_0^2 \mu_B \tag{70}$$

in disagreement with the corresponding expressions of references 8 and 77 [the former, however, does not transcribe the quoted source (76) correctly].

9.4. SAMPLE STUDIES

9.4.1. Analytical Uses of Circular Dichroism

The connection between the absolute configuration of a chromophore and the sign of the circular dichroism under the absorption bands is often used in an empirical manner. The absolute configuration of an isolated optically active compound is simply assigned by comparing its CD spectrum with that of a homologous compound of known structure. As always when an empirical correlation is applied, it is important to be cautious. In this case, that means that the compounds to be compared must be closely homologous both in an electronic sense and with respect to relative bulkiness. Some examples of this strategy applied to heterocycles follow.

The structure and absolute configuration of complicated molecules are most often found by the relatively expensive X-ray diffraction technique. This was the case for 8-β-(hydroxymethyl)-podocarpane-13-β-carboxylic acid lactone (78), for example. However, by comparing the CD spectrum of this compound with those of five related compounds the X-ray result was extended at low cost to the other five.

A series of tannins from woody plants have been analyzed, and 26 procyanidins have been isolated. They all show a positive or a negative couplet (two CD bands so positioned under an absorption peak that an exciton coupled pair of transitions is inferred) at 200–220 nm. The signs of these CD bands have been correlated to the absolute configuration of a carbon atom connecting two flavone units (79).

A similar analytic use of CD spectra has related the absolute configuration of the alkaloids $(-)_D$-pelletierine and $(-)_D$-anaferine to that of $(+)_D$-coniine (80), and the stereochemistry of a new benzomorphan has been related to those of known compounds (81). New optically active derivatives of tetrahydro-1,3-oxazine, morpholine, and 1,3-oxazolidine have been studied, and their CD spectra used to establish the absolute configurations (82).

A genuinely analytical application of CD in both a qualitative and quantitative sense has been attempted. Direct analysis of the given alkaloids by means of CD studies was shown to be impractical, since morphine, heroine, codein, and so on have rather similar CD spectra in methanol. However, it seems possible to perform measurements for analysis of the alkaloids in KBr pellets in the same way as for IR spectra or in cholesteric liquid crystalline solvent systems (83, 84).

9.4.2. Analytical Uses of Magnetically Induced Circular Dichroism

One might have thought that a specialized technique like the recording of MCD was of little practical use. However, this is by no means the case (85). The most widespread application of MCD is probably in the analysis of tryptophan in proteins. Tryptophan presented a severe analytical problem because it partially decomposes during standard amino acid analysis. The free amino acid and its peptide derivatives absorb in the 250–300-nm region where phenylalanine and tyrosine also absorb. However, only tryptophan has a strong positive MCD component (at 292 nm), and this is used in the nondestructive analysis for tryptophan. Tyrosine has a negative MCD component, which moves from 276 nm to 294 nm and increases in magnitude with a change in pH from 6 to 11.7. Measurement of MCD spectra of a protein solution at both pH 6 and 11.7 will provide an easy determination of the ratio between tyrosine and tryptophan residues in the protein (86). At pH 7.5 the MCD at 292 nm due to tyrosine in negligible and one can obtain the number of tryptophans rather directly (87).

The rate of oxidation of tryptophan residues by hydrogen peroxide has been followed by means of MCD (88).

The fact that porphyrins have very intense MCD bands has been employed in analytical determinations of porphyrins excreted in urine because of lead poisoning (89).

MCD also finds applications in structural elucidation of natural products. It has been used to distinguish between indole and indoline in the identification of brominated alkaloids of marine origin (90). Recently, MCD data of derivatives of indoles and purines, together with the spectra of amino derivatives of azaindoles, have been published (91).

The tautomerism of pyridinols, pyrimidinols, cytosine, and isocytosine in solution was investigated (92–94) by use of MCD, and good agreement was obtained between the experiments and calculations by the INDO approximation.

9.4.3. Circular Dichroism of Small Heterocyclic Rings

Substitution in thiirane, ethylene sulfoxide, aziridine, etc. may lead to optically active molecules in which the chromophore itself is dissymmetrically distorted, that is, it is inherently *chiral*. It therefore becomes possible to extract more basic informaton from the CD spectra of such molecules in addition to the analytical type of information. For thiirane, theory predicts two rather weak transitions of lowest energies which should both be located largely on the sulfur atom and can be described as $4s \leftarrow 3p$ and $\sigma^* \leftarrow 3p$ transitions. Only the latter carries both electric and magnetic transition moments, and accordingly this transition should have a much more intense CD band than the other one in an optically active derivative of thiiran. This has been verified for the $(+)_D$-t-butyl derivative, and in this way the relative energies of the two alternative transitions are known (95). A semiquantitative agreement has been found (96) between the CD of a number of substituted thiiranes and the rotatory strength computed by means of the dynamic coupling model (97, 98). The perturbing effect of substituents on thiirane can be described by sector rules (96).

In $(+)_D$-2-methyl-1-thiiraneoxide there are two chiral centers, the carbon atom bearing the methyl group and the sulfur atom. The latter chooses its conformation *R* and *S* according to the conformation of lowest energy. It is not surprising that CH$_3$ and O are placed at opposite sides of the plane of the thiirane, thus stabilizing the 1R,1R and 1S,1S forms over the 1R,2S and 1S,2R forms. The out-of-plane angle for the S–O bond is close to 70°, and this value introduces accidentally extensive mixing of orbitals that have different symmetries in the more symmetric ethylene sulfoxide. This mixing in turn produces difficulties for the theoretical calculations (99). The pertinent CD bands are formed below 210 nm, and therefore the experimental situation is also difficult.

A newly studied naturally occurring sulfoxide is sparsomycin. Based on an empirical rule, CD has been applied to determine the absolute configuration of a carbon atom attached to sulfur, and the result has been verified by X-ray diffraction (100).

The rotatory strength of optically active three-membered heterocycles derived from oxirane, aziridine, diaziridine, and oxaziridine has been computed using a Hartree-Fock SCF program and a 6–31 G basis set (101). In general there is fair agreement between the computed and experimental properties. It seems that the calculation of rotatory strength is a sensitive test for the calculational procedure. CD has been utilized in assigning absolute configurations and to test optical purity of acylaziridines (102). In optically active aziridines the nitrogen atom is chiral in general; in 2-substituted pyrrolidines the ring becomes chiral, whereas substitution in piperidine hardly effects the symmetric chair conformation of the six-membered ring (103).

Optically active monocyclic lactams are very important reference compounds to study because of their close resemblance to natural peptides, and they can be made with stereochemical rigidity. The amide chromophore can be varied in a well-defined manner, and the information gained from such studies is important for interpretations of CD results for peptides and proteins. The early CD work was rationalized in terms of a "lactam rule" or a "quadrant rule" (104–106). A series of papers on lactams containing CD measurements, theoretical calculations, and an X-ray structure determination has been published by Blaha (107, 108). The UV absorption of lactams such as (4S)-4-*tert*-butyl-1-aza-2-cyclohexanone is totally dominated by the allowed $\pi^* \leftarrow \pi$ transition, and CD clearly shows the presence of a $\pi^* \leftarrow n$ transition at ca. 230 nm and often several CD bands below 210 nm due to the $\pi^* \leftarrow \pi$ transition. The presence of more than one CD component of the $\pi^* \leftarrow \pi$ transition is a complication that has been explained as due to more than one conformer and possibly to aggregation (109). Such difficulties do not complicate the correlations based on the CD of the $\pi^* \leftarrow n$ transition in the monocyclic β-lactams derived from penicillinates (110).

The CDs associated with the $\pi^* \leftarrow n$ transition of optically active lactones have been studied extensively, and experimental results have been compared with rotatory strengths calculated for various structures (111). The spectral structure relationships have also been noted for chiral α,β-unsaturated γ-lactones, but the many possible contributions to the rotatory strength make predictions somewhat difficult (112).

9.4.4. Circular Dichroism of Nucleic Acids Monomers and Polymers

The absorption bands for nucleosides are similar to those for the free bases. However, the asymmetric character of the ribosyl residue is enough to perturb the $\pi^* \leftarrow n$ and $\pi^* \leftarrow \pi$ transitions of the heterocyclic base, and weak CD bands are observed. Polarizability models are suited for theoretical contemplations over the perturbing effects of sugar residue as a function of relative geometric arrangement between the base and ribosyl unit. The CD seems to vary strongly with the angle of rotation around the bond joining such two groups. Results of conformation energy computations were used to derive a temperature-dependent distribution of adenosine over its conformations. The long-wavelength CD of the $\pi^* \leftarrow \pi$ transitions could be reproduced rather satisfactorily by a weighted sum of the CDs of the predominating conformers. It is important to note that the intrinsic monomer CD may make an important contribution to the CD of polymers (113).

Drastic effects in CD from substitution in the sugar part of adenosine has been observed, but the origins of the effects (steric or electronic) have not been sought (114).

The above-mentioned studies were intended to form a background for investigations of biological materials such as RNA and DNA. In these polymers the CD of the transitions of the bases is more intense per base than it is in a monomer, and it is reasonable to explain this fact by base-base interactions in the polymers. The interaction has been studied in dimeric molecules, for example, in 1,2-di(adenosine-N^6-yl)ethane and 1,4-di(adenosine-N^6-yl)butane. The CD of these dimeric forms of the nucleosides is in certain regions more intense than twice the CD of N^6-ethyladenosine. However, because of the many conformational possibilities encountered and the limited experimental material, it is not possible to draw conclusions about the base-base interaction in this case (115). Similar interactions between two units of formylcamphor condensed with 1,2-ethanediamine, S- and R-1,2-propanediamine, and S,S- and R,R-1,2-cyclohexanediamine have been studied by NMR and CD as a function of temperature (116-118). When the directing power of the bridge between two chromophores is strong, as is the case for 1,2-cyclohexane derivatives, these dimeric molecules show perfect exciton couplets. The base behavior is observed for the dimers formed with less rigid bridges, provided the temperature is so low that the most stable isomer is "frozen out."

Investigations of the biological polymers RNA and DNA have been performed over a long period in order to correlate structural parameters with CD. Considering the difficulties encountered for monomers and dimers of nucleosides, it seems surprising that CD research on the polymers is successful. A prerequisite for successful conclusions is the growing amount of X-ray diffraction results from fibers of RNA and DNA. It is further important to perform reliable computations of the CD of a polymer from guesses of the structural parameters of the double-stranded helixes. The computational side is probably now well developed (119) and it may be possible to find that the polymers have different structures in solution and in "crystals" (fibers).

Due to the pronounced one-dimensionality of DNA it is relatively easy to study molecules partly oriented, for example in a stretched film or by a fast floating solution. This situation has been exploited in a study of denaturation of DNA (120).

9.4.5. Circular Dichroism of Achiral Heterocyclic Compounds

The assignment of absorption bands of heterocyclic molecules is an important task that has taken advantage of many spectroscopic techniques. The use of polarized crystal spectra and other spectra of oriented molecules, MCD, fluorescence, and phosphorescence is widespread, and sometimes CD can also be used when the achiral molecule can be brought into well-defined chiral surroundings. Sometimes this may be achieved by complexation when the heterocycle can act as a ligand toward a metal ion. In this way α,α'-bipyridyle and o-phenanthroline have been studied quite extensively. The internal ligand $\pi^* \leftarrow \pi$ transitions in these ligands couple due to electrostatic interactions (exciton coupling). Only two types of $\pi^* \leftarrow \pi$ transitions are possible. Both types will be located in the plane of the heterocycle and will be polarized along either the long axis (parallel to the N-N direction) or perpendicular to this direction. All the transitions being polarized along the long axis couple in a tris-ligand complex to give two transitions that have both electric and magnetic transition moments. If the complex is resolved into optical antipodes, this coupling shows as two exciton couplets under the absorption band, and thus it proves that this absorption is due to a transition that is long-axis polarized. In practice this technique is often disturbed by overlapping ligand ← metal charge-transfer bands. If a colorless nontransition metal ion like silicon(IV) can be used, such a disturbance does not exist. Similarly one may sometimes choose a transition metal ion that gives a complex with a "window" where the internal ligand bands are undisturbed. Even though many unexpected molecules can be persuaded to act as ligands, this way of assigning transition directions is of limited value. Instead, it has been proposed to let the heterocycle into a chiral hole of a large cyclic polymer of glucose, cyclodextrin. In α-, β-, or γ-cyclodextrin with six, seven, and eight glucose units, respectively, the inner cavity has a diameter of 5–8 Å and apparently accepts many types of guest molecules. The complex is normally soluble and often isolable as crystals suited for X-ray diffraction work (121). The chiral character of glucose is reflected in the chirality of the hole, If an achiral aromatic molecule such as 1-naphthylamine is added to a solution of β-cyclodextrin an inclusion compound is formed and the transitions of 1-naphthylamine are perturbed to produce a measurable CD (122). The CD of the inclusion compound derived from 2-naphthylamine is remarkably different from that of the 1-substituted naphthalene complex, and since similar results have been obtained for other substitutions it has been suggested that the 1-naphthalene derivatives occupy an equatorial position while 2-naphthalene derivatives occupy an axial position (122). It has been possible to account for the experimental CD of the suggested inclusion complexes by means of computations based on the polarizability model. Using the same strategy the electronic transitions of dibenzofuran and dibenzothiophene have recently been investigated and earlier assignments have been confirmed (123). The method seems rather general, but the formation constants of inclusion complexes are small (10^2–10^3) (121) and only very weak CD signals are recorded.

9.4.6. Magnetically Induced Circular Dichroism of Isocyclic Polyenes

In several instances MCD has been applied with success to gain insights into chemical problems that might have been difficult to solve by other means. On the other hand, most of the articles to be cited in the present section and the next serve two main purposes. First, they represent a highly valuable catalog of the MCD of organic molecules in solution at room temperature. The systematic and extensive studies by Michl and coworkers on the MCD of cyclic π-electron systems exemplifies this point of view. Second, the works have shown that semiempirical calculations are able to account for the greater part of the signs and relative intensities of the reported MCD spectra. This is due to the fact that the molecular symmetry, which is an observable quantity, is correctly incorporated in the calculations. It does not imply, however, that the nonobservable quantities ΔE_{HOMO}, ΔE_{LUMO}, or their difference have any independent physical significance; they are mnemonic tools only.

9.4.6.1. Benzene and Derivatives

The lowest spin-allowed transitions in benzene are well described in the π approximation. In order of increasing energy they are transitions from the ground state to $^1L_b\,(^1B_{2u})$, $^1L_a\,(^1B_{1u})$, and $^1B\,(^1E_u)$. Only the last of these is allowed in electric dipole radiation; accordingly the MCD of the two lower excited states is vibronic and has little intensity.

This selection rule does not operate in the unsymmetrical substituted benzene derivatives, which exhibit MCD and have been used in the study of the perturbation of the π system due to each kind of substituent.

Quite early (124) the MCD of the ^1L_b band of 38 benzene derivatives was measured, and it was possible to establish a linear correlation between the MCD peak values of 16 of the compounds and the Hammett σ_{para} constants.

Later, a general method of correlating data from CD and MCD spectra was formulated (61). The observed opposite signed $\mathscr{B}$ terms of o-, p-, and m-directors was accounted for by a variation calculation that included electron donor and acceptor wavefunctions based on occupied and unoccupied substitutent orbitals. The significance of vibronic contributions to the L bands for low symmetry derivatives was pointed out. In dense media (125) the MCD of the ^1L_b band of 27 benzene derivatives was rationalized in a simple perturbation model defined within the π approximation. Calculations in the CNDO approximation have confirmed the results, in particular that $\pi^* \leftarrow n$ transitions in benzaldehyde, say, are of little importance for the MCD spectra (58).

It has been shown (126) that inductive effects of the substituents (in contrast to resonance effects) have little importance for MCD spectra. It has also been stressed that the trends so far observed are related to general theorems on alternant pairing properties so that the trends could be expected for other alternant hydrocarbons and their derivatives.

9.4.6.2. Cyclic Polyenes and Derivatives

Studies of MCD in polycyclic aromatic hydrocarbons (127, 128) indicate a correlation between the polarization of the ^1L_b and ^1L_a transitions and the signs of the $\mathscr{B}$ terms. On the basis of the general properties of wavefunctions and operators in the Parisher-Parr-Pople model, Michl (129) explained the $\mathscr{B}$ terms as the result of a mutual interaction between the ^1L levels and the signs as a function of the direction of the electronic transition dipole moment. He concluded that the sign patterns in MCD of derivatives and heteroanalogs of uncharged alternant hydrocarbons are determined by the nature and location of the substituent or heteroatom in a simply predictable way, while the MCD of derivatives and heteroanalogs of charged alternant hydrocarbons and of nonalternant hydrocarbons (130, 65) are characteristic of the parent chromophore.

The polarized absorption spectra of several hydrocarbons in stretched polymers have been obtained and compared with the MCD results (6).

There is no doubt that the simiempirical results for the $\mathscr{B}$ terms are reliable only for the lower excited states. This has been pointed out by several authors and again confirmed in a study on the MCD of coronene (131). Here, the $\mathscr{A}/\mathscr{D}$ ratio for the $^1E_{1u}$ state as result of an SCF screened potential π-MO CI calculation is in excellent agreement with the observation, whereas the calculated $\mathscr{B}$ value is not.

The general ideas discussed above have been supported by experiments. Thus, MCD has been measured in solutions of the quoted compounds and a selection of their derivatives: naphthalene (132), pyrene (133), azulene (134, 135),

9.4.7. MCD Studies of Heterocyclic Compounds

9.4.7.1. Azaheterocycles

Kaito and coworkers (39) have discussed various approximations in the semi-empirical calculations of MCD of pyridine, pyrimidine, pyrazine, and s-triazine and reported that the CNDO/S-CI method can serve as a quite powerful model for predicting the signs and the order of magnitude of the Faraday $\mathscr{B}$ terms. Electronic repulsion between the single excited states below ca. $8\,\mu m^{-1}$ has been taken into account in the calculations. First, it turns out that the $\pi^* \leftarrow n$ transition observed in the MCD spectra contributes to the $\mathscr{B}$ term of the lowest $\pi^* \leftarrow \pi$ (^1L_b) transition. For pyridazine it is even of the same order of magnitude as the contribution from the third $\pi^* \leftarrow \pi$ transition. Second, a restriction of the series, equation (33), to include the first line only, is a bad approximation for the $\mathscr{B}$ parameter of the $\pi^* \leftarrow n$ transition. Actually, for pyrazine the major contribution comes from the last term of equation (33).

Later calculations (139) confirm the troubles with pyridazine in the simpler approximations.

The MCD spectra of derivatives of pyridine (140, 141), azanaphthalenes (142, 143) and derivatives (144, 145), azaanthracenes and derivatives (146), and azaphenanthrenes (147) in solution have been reported.

9.4.7.2. Inorganic Cyclic π Systems

Many π systems exist which do not contain carbon atoms. Stephens (148) identified the blue color of solutions of sulfur in oleum as being due to the planar cation S_4^{2+}.

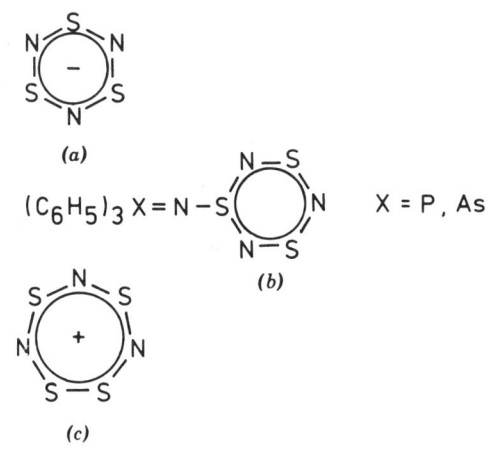

Fig. 9.5. Structure of the inorganic heterocycles *sym*-trithiatriazine (a), a low-symmetry derivative (b), and the tetrathiatriazinium ion (c).

Further, MCD spectra of the dications of the tetrachalcogens S_4^{2+}, Se_4^{2+}, and Te_4^{2+} show an $\mathscr{A}$ term that is explainable within a simple π-electron model.

During the last decade a series of cyclothiazenium ions and their derivatives have been synthesized and investigated by various physicochemical and spectroscopic means (149). The MCD of the anion of the symmetrical trithiatriazine (Fig. 9.5a) shows an $\mathscr{A}$ term that has been assigned to the $E' \leftarrow A_1'(\pi^* \leftarrow \pi^*)$ transition (150). In the simple π-electron picture the one-electron jump corresponding to this transition takes place from the double degenerate $2e''$ orbitals to the nondegenerate $2a_2''$ orbital. In the derivative (Fig. 9.5b) the molecular symmetry is lowered and the MCD pattern (151) is classified as two oppositely signed $\mathscr{B}$ terms.

The tetrathiatriazinium ion (Fig. 9.5c) has an MCD spectrum consisting of four $\mathscr{B}$ terms. which in the π approximation have been assigned (152) to the transitions L_1, L_2, B_1, and B_2 in Platt's notation (56).

9.5. APPENDIX

We consider the special case of an even alternant (64) all-*cis* annulene with $4N+2$ carbon atoms and $4N+2$ π electrons, integer N. In the ground state, the highest filled orbitals ψ_n, $\psi_{\bar{n}}$ span the irreducible representation E_n, and the lowest empty orbitals ψ_{n+1}, $\psi_{\overline{n+1}}$ span E_{n+1}. As explained for equation (34), the ground state may be given as the Slater determinant

$$|^1A\rangle = |\xi\psi_n^2\psi_{\bar{n}}^2\rangle \tag{71}$$

The first excited singlet states are then

$$|^1B_1\rangle = \frac{1}{\sqrt{2}}(|\xi\overset{+}{\psi}_n\psi_{\bar{n}}^2\bar{\psi}_{n+1}\rangle - |\xi\bar{\psi}_n\psi_{\bar{n}}^2\overset{+}{\psi}_{n+1}\rangle)$$

$$|^1B_{\bar{1}}\rangle = \frac{1}{\sqrt{2}}(|\xi\psi_n^2\overset{+}{\psi}_{\bar{n}}\bar{\psi}_{\overline{n+1}}\rangle - |\xi\psi_n^2\bar{\psi}_{\bar{n}}\overset{+}{\psi}_{\overline{n+1}}\rangle) \tag{72}$$

$$|^1L_{2n+1}\rangle = \frac{1}{\sqrt{2}}(|\xi\psi_n^2\overset{+}{\psi}_{\bar{n}}\bar{\psi}_{n+1}\rangle - |\xi\psi_n^2\bar{\psi}_{\bar{n}}\overset{+}{\psi}_{n+1}\rangle)$$

$$|^1L_{\overline{2n+1}}\rangle = \frac{1}{\sqrt{2}}(|\xi\overset{+}{\psi}_n\psi_{\bar{n}}^2\bar{\psi}_{\overline{n+1}}\rangle - |\xi\bar{\psi}_n\psi_{\bar{n}}^2\overset{+}{\psi}_{\overline{n+1}}\rangle)$$

In order to achieve a more transparent notation, we shall introduce the operator notation (153) in place of equation (72)

$$|^1B_1\rangle = \frac{1}{\sqrt{2}}(b_{n+1,\beta}^+ b_{n,\beta} - b_{n+1,\alpha}^+ b_{n,\alpha})|^1A\rangle$$

$$\equiv \Omega(b_{n+1}^+, b_n)|^1A\rangle$$

$$|^1B_{\bar{1}}\rangle = \frac{1}{\sqrt{2}}(b_{\overline{n+1},\beta}^+ b_{\bar{n},\beta} - b_{\overline{n+1},\alpha}^+ b_{\bar{n},\alpha})|^1A\rangle$$

$$\equiv \Omega(b_{\overline{n+1}}^+, b_{\bar{n}})|^1A\rangle \tag{73}$$

$$|^1L_{2n+1}\rangle = \frac{1}{\sqrt{2}}(b^+_{n+1,\beta}b_{\bar{n},\beta} - b^+_{n+1,\alpha}b_{\bar{n},\alpha})|^1A\rangle$$

$$\equiv \Omega(b^+_{n+1}, b_{\bar{n}})|^1A\rangle$$

$$|^1L_{\overline{2n+1}}\rangle = \frac{1}{\sqrt{2}}(b^+_{\overline{n+1},\beta}b_{n,\beta} - b^+_{\overline{n+1},\alpha}b_{n,\alpha})|^1A\rangle$$

$$\equiv \Omega(b^+_{\overline{n+1}}, b_n)|^1A\rangle$$

Here, the annihilation operator b_n, say, destroys an electron from the ψ_n orbital, and the creation operator b^+_{n+1}, say, places an electron in the ψ_{n+1} orbital such that the singlet characters of the states are preserved; α and β refer to the spins. The Ω operators will be used as convenient shorthand expressions.

The magnetic moments of the excited states may be given as

$$\begin{aligned}
m(^1B_1) &= m(\psi_{\bar{n}}) + m(\psi_{n+1}) \equiv m_B e_0 \\
m(^1B_{\bar{1}}) &= m(\psi_n) + m(\psi_{\overline{n+1}}) \equiv -m_B e_0 \\
m(^1L_{\overline{2n+1}}) &= m(\psi_n) + m(\psi_{n+1}) \equiv m_L e_0 \\
m(^1L_{\overline{2n+1}}) &= m(\psi_{\bar{n}}) + m(\psi_{\overline{n+1}}) \equiv -m_L e_0
\end{aligned} \quad (74)$$

Michl (77) uses μ for magnetic moments; thus the relations $\mu^- = m_B$ and $\mu^+ = m_L$ connect his symbols to the present exposition. The electric transition dipole moments are given by

$$\begin{aligned}
\langle^1B_1|-e\sum_i r_i|^1A\rangle &= -e\sqrt{2}\langle\xi\psi_{n+1}|r_i|\xi\psi_n\rangle \equiv \mu e_+ \\
\langle^1B_{\bar{1}}|-e\sum_i r_i|^1A\rangle &= -e\sqrt{2}\langle\xi\psi_{\overline{n+1}}|r_i|\xi\psi_{\bar{n}}\rangle \equiv \mu e_- \\
\langle^1L_{2n+1}|-e\sum_i r_i|^1A\rangle &= 0 \qquad \langle^1L_{\overline{2n+1}}|-e\sum_i r_i|^1A\rangle = 0
\end{aligned} \quad (75)$$

where the vector e_+ (e_-) means that the transition is allowed in left (right) circularly polarized radiation field; e_0 has the direction of the applied magnetic flux, which is the same as that of the wave vector of the light. In terms of cartesian components, the vectors are

$$e_+ = \frac{1}{\sqrt{2}}(-ie_x + e_y) \qquad e_- = \frac{1}{\sqrt{2}}(ie_x + e_y) \qquad e_0 = ie_z$$

In order to set up the energy matrix for the excited states we introduce the following notation:

$\Delta = E(\psi_{n+1}) - E(\psi_n)$ the orbital energy difference between the empty and filled orbitals.

d a diagonal electron repulsion term

c a nondiagonal electron repulsion term, which is taken to be real.

$a = -\langle\psi_{\bar{n}}|\mathcal{H}|\psi_n\rangle$ a (complex) nondiagonal matrix element between the highest occupied orbitals.

$b = \langle \psi_{\overline{n+1}} | \mathcal{H} | \psi_{n+1} \rangle$ a (complex) nondiagonal matrix element between the lowest unoccupied orbitals.

Then the energy matrix is:

$$\begin{array}{c|cccc} & {}^1B_1 & {}^1B_{\overline{1}} & {}^1L_{2n+1} & {}^1L_{\overline{2n+1}} \\ \hline {}^1B_1 & \Delta+d-E & 0 & a & b \\ {}^1B_{\overline{1}} & 0 & \Delta+d-E & \bar{b} & \bar{a} \\ {}^1L_{2n+1} & \bar{a} & b & \Delta-d-E & c \\ {}^1L_{\overline{2n+1}} & \bar{b} & a & c & \Delta-d-E \end{array} = 0 \quad (76)$$

Consider first the special case where b and c both vanish. The problem is then to diagonalize two matrices of the type

$$\begin{vmatrix} \Delta+d-E & a \\ \bar{a} & \Delta-d-E \end{vmatrix} = 0 \quad (77)$$

It is standard matrix calculus (43) to get the eigenfunctions and eigenvalues in terms of the parameter α, given by

$$\tan 2\alpha = \frac{|a|}{d} \quad (78)$$

such that the eigenvalues

$$E = \Delta \pm \frac{d}{\cos 2\alpha} \quad (79)$$

correspond to the eigenfunctions

$$\begin{aligned} |{}^1B_i\rangle &= \cos\alpha\, |{}^1B_1\rangle + \sin\alpha\, |{}^1L_{2n+1}\rangle \\ |{}^1L_i\rangle &= -\sin\alpha\, |{}^1B_1\rangle + \cos\alpha\, |{}^1L_{2n+1}\rangle \end{aligned} \quad (80)$$

We can rewrite this equation by means of the conventions of equation (73) to obtain a transparent form of those eigenfunctions which are connected to an interaction between the orbitals ψ_n, $\psi_{\bar{n}}$:

$$\begin{aligned} |{}^1B_i\rangle &= \Omega(b_{n+1}^+, b_n \cos\alpha + b_{\bar{n}} \sin\alpha)|{}^1A\rangle \\ |{}^1L_i\rangle &= \Omega(b_{n+1}^+, -b_n \sin\alpha + b_{\bar{n}} \cos\alpha)|{}^1A\rangle \end{aligned} \quad (81)$$

In order to discuss the matrix, equation (76), we make a unitary transformation of the functions:

$$\begin{aligned} |X\rangle &= \frac{1}{\sqrt{2}}(|{}^1B_1\rangle + |{}^1B_{\overline{1}}\rangle) \\ |Y\rangle &= \frac{-i}{\sqrt{2}}(|{}^1B_1\rangle - |{}^1B_{\overline{1}}\rangle) \\ |U\rangle &= \frac{1}{\sqrt{2}}(|{}^1L_{2n+1}\rangle + |{}^1L_{\overline{2n+1}}\rangle) \\ |V\rangle &= \frac{-i}{\sqrt{2}}(|{}^1L_{2n+1}\rangle - |{}^1L_{\overline{2n+1}}\rangle) \end{aligned} \quad (82)$$

to give the matrix

$$\begin{array}{c|cccc} & X & Y & U & V \\ \hline X & \Delta + d - E & 0 & \text{Re}\{a+b\} & \text{Im}\{a-b\} \\ Y & 0 & \Delta + d - E & -\text{Im}\{a+b\} & \text{Re}\{a-b\} \\ U & \text{Re}\{a+b\} & -\text{Im}\{a+b\} & \Delta - d + c - E & 0 \\ V & \text{Im}\{a-b\} & \text{Re}\{a-b\} & 0 & \Delta - d - c - E \end{array} = 0 \quad (83)$$

Actually, this matrix (57) is equivalent to that of equation (I) of reference 59. The matrix may be simplified if $a = -b$ or $a = b$. In the latter case the eigenfunctions are

$$\begin{aligned} |X'\rangle &= \cos\tau\,|X\rangle - \sin\tau\,|Y\rangle \\ |Y'\rangle &= \cos\rho\sin\tau\,|X\rangle + \cos\rho\cos\tau\,|Y\rangle - \sin\rho\,|V\rangle \\ |U'\rangle &= |U\rangle \\ |V'\rangle &= \sin\rho\sin\tau\,|X\rangle + \sin\rho\cos\tau\,|Y\rangle + \cos\rho\,|V\rangle \end{aligned} \quad (84)$$

where the parameters ρ, τ can be obtained from an actual diagonalization of the matrix.

REFERENCES

References 1 to 8 are general review articles and books that have been published in recent years. They contain references to earlier works that form the foundation of the subject but have not been used explicitly in this chapter.

1. E. Charney, *The Molecular Basis of Optical Activity*, John Wiley, New York, 1979.
2. Aa. E. Hansen and T. D. Bouman, *Adv. Chem. Phys.*, **44**, 545 (1980).
3. G. Snatzke, *Pure Appl. Chem.*, **51**, 769 (1979).
4. D. J. Caldwell and H. Eyring, *The Theory of Optical Activity*, Wiley-Intersciences, New York, 1971.
5. R. B. Homer, "Optical Rotatory Dispersion, Circular Dichroism, and Magnetic Circular Dichroism," in A. R. Katritzky, Ed., *Physical Methods in Heterocyclic Chemistry*, Vol. 3, Academic Press, New York, 1971, Chap. 7.
6. E. W. Thulstrup, *Linear and Magnetic Circular Dichroism of Planar Organic Molecules* (*Lecture Notes in Chemistry*, Vol. 14), Springer Verlag, Berlin, 1980.
7. P. J. Stephens, *Ann. Rep. Phys. Chem.*, **25**, 201 (1974).
8. P. N. Schatz and A. J. McCaffery, *Quart. Rev.*, **23**, 552 (1969); **24**, 329 (1970).
9. IUPAC, *Manual of Symbols and Terminology for Physicochemical Quantities and Units*, Pergamon, New York, 1979.
10. L. Velluz, M. Legrand, and M. Grosjean, *Optical Circular Dichroism*, Academic Press, New York, 1965.
11. Instruments S.A., Division Jobin Yvon, 16-18 Rue du Canal, F-91169 Longjumeau, France.
12. Japan Spectroscopic Co. Ltd., 2967-5, Ishikawa-Cho Hachioji City, Tokyo 192, Japan.

References

13. Hinds International, Inc., P.O. Box 4192, Portland, Oregon 97208.
14. G. F. Lothian, *Optics and Its Uses*, Van Nostrand-Reinhold, New York, NY, 1980.
15. H. P. Jensen and F. Galsbøl, *Inorg. Chem.*, **16**, 1294 (1977).
16. H. P. Jensen, *Appl. Spectrosc.*, **34**, 360 (1980).
17. H. P. Jensen, J. A. Shellman, and T. Troxell, *Appl. Spectrosc.*, **32**, 192 (1978).
18. R. A. Palmer, J. C. Roark, and J. C. Robinson, *ACS Symp. Ser.*, 1980, 119; *Chem. Abstr.*, **93**, 15942 (1980).
19. J. D. Saxe, T. R. Faulkner, and F. S. Richardson, *J. Appl. Phys.*, **50**, 8204 (1979).
20. J. D. Saxe, T. R. Faulkner, and F. S. Richardson, *Chem. Phys. Lett.*, **68**, 71 (1979).
21. M. Diem, P. J. Gotkin, J. M. Kempfer, and L. A. Nafie, *J. Amer. Chem. Soc.*, **100**, 5644 (1978).
22. C. Marcott, H. A. Harvel, J. Overend, and A. Moscowitz, *J. Amer. Chem. Soc.*, **100**, 7088 (1978).
23. V. J. Heintz and T. A. Keiderling, *J. Amer. Chem. Soc.*, **103**, 2395 (1981).
24. C. N. Su and T. A. Keiderling, *Chem. Phys. Lett.*, **77**, 494 (1981).
25. L. D. Barron, in S. F. Mason, Ed., *Optical Activity and Chiral Discrimination*, Reidel, Dordrecht, 1979.
26. W. C. Johnson, *Ann. Rev. Phys. Chem.*, **29**, 93 (1978).
27. S. Brahms and J. Brahms, *J. Mol. Biol.*, **138**, 149 (1980).
28. P. A. Snyder and E. M. Rowe, *Nucl. Instr. Methods*, **172**, 345 (1980).
29. H. U. Güdel, I. Trabjerg, M. Vala, and C. J. Ballhausen, *Mol. Phys.*, **24**, 1227 (1972).
30. W. Moffitt and A. Moscowitz, *J. Chem. Phys.*, **30**, 648 (1959).
31. E. U. Condon, *Rev. Mod. Phys.*, **9**, 432 (1937).
32. E. U. Condon and G. H. Shortley, *The Theory of Atomic Spectra*, reprint, Cambridge University Press, 1963.
33. P. J. Stephens, "Theoretical Studies of Magneto-optical Phenomena," PhD thesis, part 2, Oxford, 1964.
34. A. D. Buckingham and P. J. Stephens, *Ann. Rev. Phys. Chem.*, **17**, 399 (1966).
35. R. Serber, *Phys. Rev.*, **41**, 489 (1932).
36. P. J. Stephens, *J. Chem. Phys.*, **52**, 3489 (1970).
37. P. J. Stephens, *Adv. Chem. Phys.*, **35**, 197 (1976).
38. W. P. Healey, *J. Chem. Phys.*, **64**, 3111 (1976).
39. A. Kaito, M. Hatano, and A. Tajiri, *J. Amer. Chem. Soc.*, **99**, 5241 (1977).
40. L. Goodman and V. G. Krishna, *Rev. Mod. Phys.*, **35**, 541 (1963).
41. L. L. Lohr, *J. Chem. Phys.*, **45**, 1362 (1966).
42. J. E. Harriman, in E. M. Loebel, Ed., *Theoretical Foundations of Electron Spin Resonance* (*Physical Chemistry*, Vol. 37), Academic Press, New York, 1978.
43. W. Heitler, *The Quantum Theory of Radiation*, 3rd ed., The Clarendon Press, Oxford, 1954.
44. P. A. Dobosh, *Mol. Phys.*, **27**, 689 (1974).
45. S. E. Harnung, *Mol. Phys.*, **26**, 473 (1973).
46. P. A. Dobosh, *Phys. Rev.*, **A5**, 2376 (1972).
47. S. B. Piepho and P. N. Schatz, *Group Theory in Spectroscopy with Applications to Magnetic Circular Dichroism*, Wiley, New York, 1983.
48. N. Harada and K. Nakanishi, *Circular Dichroic Spectroscopy and Exciton Coupling in Organic and Bioorganic Stereochemistry*, University Science Books, Mill Valley, 1983.
49. W. Moffitt, R. B. Woodward, A. Moscowitz, W. Klyne, and C. Djerassi, *J. Amer. Chem. Soc.*, **83**, 4013 (1961).

50. J. A. Shellman, *Acc. Chem. Res.*, **1**, 144 (1968).
51. J. A. Shellman, *J. Chem. Phys.*, **44**, 55 (1966).
52. M. Goeppert-Mayer and A. L. Sklar, *J. Chem. Phys.*, **6**, 645 (1938).
53. R. G. Parr, *Quantum Theory of Molecular Electronic Structure*, W. A. Benjamin, New York, 1963.
54. J. N. Murrell and A. J. Harget, *Semi-empirical Self-consistent-field Molecular-orbital Theory of Molecules*, Wiley-Interscience, New York, 1972.
55. I. N. Levine, *Quantum Chemistry*, 2nd ed., Allyn and Bacon, Boston, 1974.
56. J. R. Platt, *J. Chem. Phys.*, **17**, 484 (1949).
57. W. Moffitt, *J. Chem. Phys.*, **22**, 320 (1954).
58. A. Kaito and M. Hatano, *J. Amer. Chem. Soc.*, **100**, 2034 (1978).
59. J. Michl, *J. Amer. Chem. Soc.*, **100**, 6812 (1978).
60. L. Seamans and A. Moscowitz, *J. Chem. Phys.*, **56**, 1099 (1972).
61. D. J. Caldwell and H. Eyring, *J. Chem. Phys.*, **58**, 1149 (1973).
62. L. Seamans and J. Linderberg, *Mol. Phys.*, **24**, 1393 (1972).
63. J. March, *Advanced Organic Chemistry*, 2nd ed., McGraw-Hill Kogakusha, Tokyo, 1977.
64. M. J. S. Dewar and R. C. Dougherty, *The PMO Theory of Organic Chemistry*, Plenum, New York, 1975.
65. J. Michl, *J. Amer. Chem. Soc.*, **100**, 6819 (1978).
66. J. D. Keegan, A. M. Stolzenberg, Y.-C. Lu, R. E. Linder, G. Barth, A. Moscowitz, E. Bunnenberg, and C. Djerassi, *J. Amer. Chem. Soc.*, **104**, 4305, 4317 (1982).
67. H. Cramér, *Sannolikhetskalkylen och några af dess anvendingar*, Almqvist and Wiksell, Stockholm, 1949.
68. H. Cramér, *Mathematical Methods of Statistics*, Princeton University Press, Princeton, NJ, 1951.
69. R. S. Mulliken, *J. Chem. Phys.*, **7**, 14 (1939).
70. S. E. Harnung, E. C. Ong, and O. E. Weigang, Jr., *J. Chem. Phys.*, **55**, 5711 (1971).
71. O. E. Weigang, Jr. and E. C. Ong, *Tetrahedron*, **30**, 1783 (1974).
72. C. H. Henry, S. E. Schnatterly, and C. P. Slichter, *Phys. Rev.*, **A137**, 583 (1965).
73. *IUPAC Bull.* **49**, "Expression of Results in Quantum Chemistry," Commission I.1. on Physicochemical Symbols, Terminology and Units of the International Union of Pure and Applied Chemistry, 1976.
74. *IUPAP Doc.* **U.I.P.20**, "Symbols, Units and Nomenclature in Physics," Commission for Symbols, Units and Nomenclature in Physics of the International Union of Pure and Applied Physics, *Physica*, **93A**, 1 (1978).
75. *CODATA Bull.* No. 11, "Recommended Consistent Values of the Fundamental Physical Constants, 1973," Committee on Data for Science and Technology of the International Council of Scientific Unions, 1973.
76. P. J. Stephens, *Chem. Phys. Lett.*, **2**, 241 (1968).
77. J. Michl, *J. Amer. Chem. Soc.*, **100**, 6801 (1978), and subsequent papers.
78. A. F. Becham, R. C. Cambie, R. C. Hayward, and B. J. Poppleton, *Aust. J. Chem.*, **32**, 2617 (1979).
79. M. W. Barrett, W. Klyne, P. M. Scopes, A. C. Fletcher, L. J. Porter, and E. Haslam, *J. Chem. Soc. Perkin Trans. I*, 2375 (1979).
80. J. C. Craig, S.-Y. C. Lee, and S. K. Roy, *J. Org. Chem.*, **43**, 347 (1978).
81. N. Yokoyama, P. Almanla, F. B. Block, F. R. Granat, N. Gottfried, R. T. Hill, E. H. McMahan, W. F. Munch, and H. Rachlin, *J. Med. Chem.*, **22**, 537 (1979).
82. W. Klyne, P. M. Scopes, N. Berova, J. Stefanovsky, and B. J. Kurter, *Tetrahedron*, **35**, 2009 (1979).

References

83. J. M. Bowen and N. Purdie, *Anal. Chem.*, **52**, 573 (1980).
84. J. M. Bowen, T. A. Crone, A. O. Hermann, and N. Purdie, *Anal. Chem.*, **52**, 2436 (1980).
85. C. Djerassi, E. Bunnenberg, and D. L. Elder, *Pure Appl. Chem.*, **25**, 51 (1971).
86. G. Barth, E. Bunnenberg, and C. Djerassi, *Anal. Biochem.*, **48**, 471 (1972).
87. B. Holmquist and B. L. Vallee, *Biochem.*, **22**, 4409 (1973).
88. H. Eckstein, G. Barth, R. E. Linder, E. Bunnenberg, and C. Djerassi, *Justus Liebigs Ann. Chem.*, **1974**, 990.
89. S. M. Kalman, G. Barth, R. E. Linder, E. Bunnenberg, and C. Djerassi, *Anal. Biochem.*, **52**, 83 (1973).
90. J. S. Carlé and C. Christoffersen, *J. Amer. Chem. Soc.*, **101**, 4012 (1979).
91. S. L. Wallace and J. Michl, *Tetrahedron*, **36**, 1531 (1980).
92. A. Kaito and M. Hatano, *Bull. Chem. Soc. Japan*, **53**, 3064 (1980).
93. A. Kaito and M. Hatano, *Bull. Chem. Soc. Japan*, **53**, 3069 (1980).
94. A. Kaito, M. Hatano, T. Ueda, and S. Shibuya, *Bull. Chem. Soc. Japan*, **53**, 3073 (1980).
95. G. Bendazzoli, G. Gottarelli, and R. Palmieri, *J. Amer. Chem. Soc.*, **96**, 11 (1974).
96. G. Gottarelli, B. Samori, J. Moretti, and G. Torre, *J. Chem. Soc. Perkin Trans. II*, 1105 (1977).
97. E. G. Höhn and O. E. Weigang, *J. Chem. Phys.*, **48**, 1127 (1968).
98. W. H. Inskeep, D. H. Miles, and H. Eyring, *J. Amer. Chem. Soc.*, **92**, 3866 (1970).
99. G. L. Bendazzoli, P. Palmieri, G. Gottarelli, J. Moretti, and G. Torre, *Gazzetta*, **109**, 19 (1979).
100. H. C. J. Ottenhejm, R. M. J. Liskamp, P. Helquist, J. W. Lauher, and M. S. Shekhani, *J. Amer. Chem. Soc.*, **103**, 1720 (1981).
101. A. Rauk, *J. Amer. Chem. Soc.*, **103**, 1023 (1981).
102. N. Furukawa, T. Yoshimura, M. Ohtsu, T. Akasaka, and S. Oal, *Tetrahedron*, **36**, 73 (1980).
103. B. Ringdahl, W. E. Pereira, and J. C. Craig, *Tetrahedron*, **37**, 1659 (1981).
104. H. Ogura, H. Takayanagi, and K. Furuhata, *Chem. Lett.*, 389 (1973).
105. H. Ogura, H. Takayanagi, K. Kubo, and K. Furuhata, *J. Amer. Chem. Soc.*, **95**, 8056 (1973).
106. B. J. Litman and J. A. Shellman, *J. Phys. Chem.*, **69**, 978 (1965).
107. M. Ticky, P. Malon, I. Fric, and K. Blaha, *Coll. Czech. Chem. Commun.*, **44**, 2653 (1979).
108. K. Blaha and P. Malon, *Acta Univ. Palackianae Olomicencis*, **93**, 81 (1980).
109. M. Tichy, P. Malon, I. Fric, and K. Blaha, *Coll. Czech. Chem. Commun.*, **42**, 3591 (1977).
110. R. Busson, E. Roets, and H. Vanderhaeghe, *J. Org. Chem.*, **43**, 4434 (1978).
111. F. S. Richardson and W. Pitts, *J. Chem. Soc. Perkin Trans. II*, 1276 (1975).
112. H. Kreigh and F. S. Richardson, *J. Chem. Soc. Perkin Trans. II*, 1674 (1976).
113. D. S. Moore, *Biopolymers*, **19**, 1017 (1980).
114. M. J. Robins, S. D. Hawrelak, T. Kanai, J.-M. Seifert, and R. Mengel, *J. Org. Chem.*, **44**, 1317 (1979).
115. J. Zemlicka and J. Owens, *J. Org. Chem.*, **42**, 517 (1977).
116. H. P. Jensen and E. Larsen, *Acta Chem. Scand.*, **A29**, 157 (1975).
117. H. P. Jensen and E. Larsen, *Gazzetta*, **107**, 143 (1977).
118. N. Bernth, E. Larsen, and S. Larsen, *Tetrahedron*, **37**, 2477 (1981).

119. B. B. Johnson, K. S. Dahl, I. Tinoco, V. I. Ivanov, and V. B. Zhurkin, *Biochem.*, **20**, 73 (1981).
120. B. Norden and S. Seth, *Biopolymers*, **18**, 2323 (1979).
121. W. Saenger, *Angew. Chem. Int. Ed.*, **19**, 344 (1980).
122. K. Harata and H. Uedaira, *Bull. Chem. Soc. Japan*, **48**, 375 (1975).
123. H. Yamaguchi and K. Ninomiya, *Spectrochim. Acta*, **37A**, 119 (1981).
124. J. G. Foss and M. E. McCarville, *J. Amer. Chem. Soc.*, **89**, 30 (1967).
125. D. J. Shieh, S. H. Lin, and H. Eyring, *J. Phys. Chem.*, **77**, 1031 (1973).
126. J. Michl and J. Michl, *J. Amer. Chem. Soc.*, **96**, 7887 (1974).
127. J. G. Foss and M. E. McCarville, *J. Chem. Phys.*, **44**, 4350 (1966).
128. J. P. Larkindale and D. J. Simkin, *J. Chem. Phys.*, **55**, 5668 (1971).
129. J. Michl, *Chem. Phys. Lett.*, **39**, 386 (1976).
130. S. M. Warnick and J. Michl, *J. Amer. Chem. Soc.*, **96**, 6280 (1974).
131. H. Yamaguchi and M. Higashi, *Chem. Phys. Lett.*, **68**, 77 (1979).
132. M. R. Whipple, M. Vašák, and J. Michl, *J. Amer. Chem. Soc.*, **100**, 6844 (1978).
133. M. Vašák, M. R. Whipple, A. Berg, and J. Michl, *J. Amer. Chem. Soc.*, **100**, 6872 (1978).
134. W. Gerhartz and J. Michl, *J. Amer. Chem. Soc.*, **100**, 6877 (1978).
135. D. Otteson, C. Jutz, and J. Michl, *J. Amer. Chem. Soc.*, **100**, 6882 (1978).
136. J. W. Kenney III, D. A. Herold, J. Michl, and J. Michl, *J. Amer. Chem. Soc.*, **100**, 6884 (1978).
137. G. P. Dalgaard and J. Michl, *J. Amer. Chem. Soc.*, **100**, 6887 (1978).
138. M. A. Souto, D. Otteson, and J. Michl, *J. Amer. Chem. Soc.*, **100**, 6892 (1978).
139. A. Castellan and J. Michl, *J. Amer. Chem. Soc.*, **100**, 6824 (1978).
140. S. L. Wallace, A. Castellan, D. Müller, and J. Michl, *J. Amer. Chem. Soc.*, **100**, 6828 (1978).
141. I. Jonáš and J. Michl, *J. Amer. Chem. Soc.*, **100**, 6834 (1978).
142. A. Kaito and M. Hatano, *J. Amer. Chem. Soc.*, **100**, 4037 (1978).
143. M. Vašák, M. R. Whipple, and J. Michl, *J. Amer. Chem. Soc.*, **100**, 6838 (1978).
144. M. A. Souto and J. Michl, *J. Amer. Chem. Soc.*, **100**, 6853 (1978).
145. D. Otteson and J. Michl, *J. Amer. Chem. Soc.*, **100**, 6857 (1978).
146. R. P. Steiner and J. Michl, *J. Amer. Chem. Soc.*, **100**, 6861 (1978).
147. M. Vašák, M. R. Whipple, and J. Michl, *J. Amer. Chem. Soc.*, **100**, 6867 (1978).
148. P. J. Stephens, *Chem. Commun.*, 1496 (1969).
149. F. A. Cotton and G. Wilkinson, *Advanced Inorganic Chemistry*, 4th ed., John Wiley, New York, 1980.
150. J. W. Waluk and J. Michl, *Inorg. Chem.*, **20**, 963 (1981).
151. J. W. Waluk, T. Chivers, R. T. Oakley, and J. Michl, *Inorg. Chem.*, **21**, 832 (1982).
152. J. W. Waluk and J. Michl, *Inorg. Chem.*, **21**, 556 (1982).
153. J. Avery, *Creation and Annihilation Operators*, McGraw-Hill, New York, 1976.

INDEX

Absorption (ABS), 628
   intensity correction, Raman spectroscopy, 387–388
Acenaphthylene, 660
Acetic acid, physical properties, 212
Acetone, physical properties, 212
Acetonitrile, 439
   physical properties, 212
Acetylpyridine, 527
   IR data, 71
Achiral heterocyclic compounds, circular dichroism, 657–658
Acoustic ringing, coherent noise, 165–168
Acridan, 269
Acridines:
   derivatives of, 502
   oxidation of, 548
   reduction of, 473–475
Acridone, 502
N-Acyldihydroquinaldonitriles, 80
Adduct formation, 4, 96
Adenine, 267, 581
Adenine-N-oxide, 516
Alcohols, reduction of, 526
Aldehydes, reduction of, 526–528
Alkyl hydrazines, 271
Alkyl substituents, 598
Alkyl substituted pyridines, IR data, 71
Allopurinol, 511
Alloxans, 506
Alloxazines, 269
Aluminum halide complexes, pyridine-N-oxides, 74
Amino derivatives, reduction of, 516–518
Aminopurines, 585–587
Aminopyridines, IR data, 73
5-Aminotetrazoles, amino-imino tautomerism, 49
Anaferine, 654
Anhydrides, IR data, 22–25

Anhydro-1,3-dithiolium-4-oxides, 46
Anion radicals of tetracyanothiophene, 19
[4]Annulene system, 247
Anomalous dispersion, dipole moments, 356
Anthranils, reduction of, 487
Aprotic solvents, 439
   reduction of, 468–469
Aqueous nickel chloride solution, 285
Aqueous solutions, reduction of, 469
Aromatic rings:
   without carbonyl groups, 8–22
   three or four heteroatoms without carbonyl groups, 46–49
Aromatic six-membered rings, 68–111
   diazines, 83–97
   pyridine and related systems, 68–83
   tetrazines, 110–111
   triazines, 107–110
   vibrational spectra of nucleic acid constituents, 97–107
Arsabenzene, 261
   IR data, 75–78
Axial ligand-associated vibrations, 397–404
Axial/transverse method, 389–390
Azaanthracenes, 269, 660
Azabenzenes, 261–263
4-Azafluoren-9-one, 82
Azanaphthalenes, 660
Azaphenanthrenes, 269, 660
Azaphenoxathiins, IR data, 98
Azetidine, 241
   small-ring nitrogen, 338–347
      fused β-lactam systems, 342–347
      β-lactams, 340–342
      simple, 338–340
Azetidine variants, small-ring hydrogen, 347–348
Aziridines, 239, 655
   IR data, 5–6
   oxidation of, 530–531

Aziridines (Continued)
  small-ring nitrogen, 316–327
    electron density "bulging," 318
    fused systems, 321–327
    metal-complexes, 319–321
    simply substituted, 316–319
    stable invertomer, 319
  vibrational assignment, 5–6
Aziridine variants, small-ring nitrogen, 327–338
  azirines, 327–329
  diaziridines, 330–334
  diazirines, 329–330
  oxaziridines, 335–338
  thiadiaziridines, 334
Azirines, small-ring nitrogen, 327–329
  mesomeric form, 328
Azocine, reduction of, 475
Azo derivatives, reduction of, 521
Azulene, 659

Barbituric acid, 504–505, 561
Baseline roll, reduction of, 220–223
"Bench-top" spectrometers, 150–151
Benzomorphan, 654
Benzene, 15, 75, 285, 358
  magnetically induced circular dichroism, 658–659
  physical properties, 212
Benzo[c]cinnoline dioxide, 515
Benzo derivatives, reduction of, 495–496
Benzodiazepines:
  derivatives of, 509
  reduction of, 485–486
Benzodiazocinones, IR data, 32
Benzofuran, 19–20, 266
  oxidation of, 553
Benzo-fused dioxepines, IR data, 115
Benzoic acid, 285
Benzooxalactams, 28
2,1,3-Benzoselenadiazole, 267
1,2,3-Benzothiadiazole, 267
2,1,3-Benzothiadiazole, 267
Benzothiazine derivatives, 58
Benzothiazocinones, IR data, 32
Benzo[b]thiophene, IR data, 19–20
Benzothiophenes, oxidation of, 555
Benzotriazines, reduction of, 491–492
1H-Benzotriazole, 267
Benzotriazoles, reduction of, 491
2,1,3-Benzoxadiazole (and 5-methoxy derivative), 267
Benzoxazine:
  derivatives, 58

  reduction of, 487
Benzoxazocinones, IR data, 32
Benzylcodanine, 546
O-Benzylpallidine, 547
Bicyclic compounds, 60–61
Bicyclic fused aziridine systems, 321
  fusion angle, 322
Bicyclic hydrazines, 271
Bicyclic phosphorane, 273
Bifuran, 20
Biimidazole derivatives, 44
2,2'-Bipyridyl, 472
  IR data, 70–71
Bipyrryls, 255
1,1'-Bisaziridyl, 7
Bismabenzene, 261
2,2'- and 3,3'-Bithienyl, 255
Boron heterocycles, 119–122
Boron trifluoride adducts, 66
Boron trihalide, 60–61
4-tert-Butyl-1-aza-2-cyclohexanone, 656
α-Butyrolacetone, 242
Butyronitrile, 439

Cadmium, 38
Caffeine, 265, 581
Calcium dipicolinate trihydrate, 72
Calculated dipole moments, 358–359
ε-Caprolactam, 28
Carbazoles, 269
  oxidation of, 547–548
  reduction of, 469–470
Carbolines, IR data, 81
Carbon, 436
Carbon disulfide, physical properties, 212
Carbon tetrachloride, 358
  physical properties, 212
Carbonyl stretching vibration, 416
Carboxylic acids, 21, 598–600
  reduction of, 528–530
Cells, electroanalytical experiments, 431–434
Chemical shift phenomenon, 142, 282, 291
Chloroform, physical properties, 212
Chloromethyl diazirine, 330
Chlorophylls, 409–410
Chlorpromazine, 270
2-Chloropyrimidines, 85
Chromone, IR data, 66–68
Cinnoline-N-oxide, 515
Cinnolines:
  oxidation of, 561
  reduction of, 480–481
Cinnolinone, 503–504

Circular dichroism, 627–668
  introduction, 628–635
    experimental quantities, 628–629
    instrumentation, 629–631
    ORD, CD, and MCD measurements, 633–635
    phenomenological relations, 631–633
  moment analyses, 646–653
    dipole strength, 647–649
    Faraday parameters, 649–650
    rotatory strength, 649
    SI units, 650–653
  quantum mechanical relations, 635–646
    general relations, 636–639
    models used, 641–646
    simple interpretation of expressions, 639–641
  sample studies, 653–661
    archiral heterocyclic compounds, 657–658
    analytical uses, 653–654
    MCD of isocyclic polyenes, 658–660
    MCD studies of heterocyclic compounds, 630–631
    nucleic acids monomers and polymers, 656–657
    small heterocyclic rings, 655–656
Clathrates, 15
Coherent anti-Stokes Raman scattering (CARS) technique, 390–392
Coherent noise, 164–169
Coil resonance, coherent noise, 168–169
Collidines, 71
  methyl vibrations, 72
Conformational analysis, dipole moments, 356, 364–365
Coniine, 654
Continuous wave spectrometer, 155
Copper sulfate, 285
Core-expansion correlations, 404–406
Coronene, 659
Corypalline, 591
Coulometry, 463–465
Coumarines, reduction of, 476
Coumarins, diamagnetic susceptibilities, 300
Counter electrodes, 436
α-Crotonolactone, 242
Cyclic dialkyhydrazines, 271
Cyclic hydroxamic acids, 64
Cyclic polyenes, MCD in, 659–660
Cyclic voltammetry, 453–463
1,2-Cycloalkylhydrazines, 277
Cyclodextrin, 658
Cyclohexane, physical properties, 212

Cyclopentane, physical properties, 212
Cytidine, 517
Cytosine, 97–104, 517, 654
  far-infrared frequencies, 102–103

Decoupling methods, reduction of noise, 223–255
3,6-Dehydrooxepin, 263
Depolarization ratio, Raman effect, 387
Derivatives:
  of acridine, 502
  of benzodiazepine, 509
  of imidazolidine, 502
  of isoindole, 501–502
  of phenazine, 509
  of phenoxazine and phenothiazine, 502–503
  of pteridines, 512–513
  of purine, 511–512
  of pyrazine and piperazine, 507
  of pyridopyrazines, 511
  of pyrimidine, 504–506
  of pyrrole, 501
  of quinazoline, 506
  of quinoxaline, 507–508
  of triazines, 510
Deuterium nucleus, 171
t-2,3-Diadamantyl aziridinone, 318
Dialkyl disulfides, 273
Dialuric acid, 591
Diamagnetic susceptibility, heterocyclic compounds, 281–311
  introduction, 282
  measurement, 282–292
    calibration of Gouy tube, 285–286
    experimental technique, 284–285
    Faraday method, 287
    Gouy method, 282–284
    instrumentation, Gouy method, 286–287
    NMR method, 291–292
    optical arrangement, 288–289
    Quincke method, 289–291
    torsion head arrangement, 287–288
  results and discussion, 298–309
    containing nitrogen, 298–300
    containing nitrogen and oxygen, 302
    containing nitrogen and sulfur, 301–302
    containing oxygen, 300–301
    containing sulfur, 301
    list of compounds, 305–309
    in solution, 302–304
  theoretical calculation, 292–298
    Hartree-Fock-Roothann coupled, 298
    Pacault method, 293, 294–295

672 Index

Diamagnetic susceptibility, heterocyclic
  compounds, theoretical calculation
  (*Continued*)
  semiempirical methods, 293–298
Diaphragms, electroanalytical experiment, 434
Diaryl ketones, 19
Diazabasketene, 273
Diazadeltacyclene, 273
1,2-Diazanaphthalene, 268
Diazaphospholanes, 273
Diazepines, reduction of, 485–486
Diazines, IR data, 83
Diaziridines, 655
  "equilateral," 334
  oxidation of, 556
  small-ring nitrogen, 330–334
Diazirines:
  reduction of, 478
  small-ring nitrogen, 329–330
Diazobicyclooctane radical cation, 418
H-Dibenz[*b*,*f*]azepin, 269
Dibenzofuran, 658
Dibenzothiophenes, 20, 658
  oxidation of, 555
1,2-Dichloroethane, physical properties, 212
Dichloromethane, physical properties, 212
Dielectric constant, dipole moments, 357
2,5-Dihydrofuran, 241
1,2-Dihydro-2-hydroxyisophosphinodole-2-
  oxide, 17
2,3-Dihydropyran, 247
Dihydropyridines, oxidation of, 537–541
2,5-Dihydrothiophene, 242
1,4-Dihydroxypyridazino[4,5-*d*] pyridazine, 91
4,6-Dihydroxypyrimidines, 87
2,3-Dihydroxyquinoxaline, 95
Dimensionally stable anodes (DSA), 436
1,2-Dimethoxyethane, 439
2-Dimethylamino-6-hydroxy-9-methyl-purine, 74
cis-2,3-Dimethylaziridine, 5
3,3-Dimethylaziridine, vibrational assignment, 7
3,3-Dimethyldiaziridine, 5
Dimethylformamide (DMF), 439
N,N-Dimethylformamide, physical properties, 212
Dimethylsulfoxide, physical properties, 212
N,N'-Dimethylhydrazines, 271
2,4-Dimethyloxazole, 46
Dimethyl oxadiazole, 302
cis-2,6-Dimethylpeperidine, 49
Dimethylsulfoxide, 439
2,5-Dimethylthiophene, dichloro and dibromo
  derivatives, 17
1,3-Dioxalane, 243, 246

Dioxan, 56, 358
1,4-Dioxan, IR data, 52–55
  physical properties, 212
1,4-Dioxane, 246
Dioxanes, 572
1,2-Dioxetanes, 7
Dioxolane, 571
Dioxolan-2-ones, IR data, 37–38
4,5-Diphenyl-4-imidazoline-2-one, 39
gem-Diphenyl-substituted nydantoin derivatives, 41
Dipicolinic acid, 72
Dipole moments, 355–371
  applications, 359–366
    conformational analysis, 364–365
    electronic structure, 362–364
    structure determination, 360–362
    theoretical calculations, 365–366
  calculated, 358–359
  experimental, 356–358
    determination, 357–358
  introduction, 355–356
Dipole strength, moment analysis, 647–649
Distortion polarization, dipole moments, 361
5,5'-Disubstituted barbituric acids, 299–300
1,4-Dithian, IR data, 52–55
1,3-Dithiane, 246
1,4-Dithia-2,3,5,6,-tetrazine-1,4-dioxide, 59
Dithiatricyclotetradecadiene, 247
1,3-Dithio derivatives, 509–510
1,3-Dithiolane, 243
Dithiole derivatives, IR data, 46
1,3-Dithiole-2-thione, 243, 265
Dithiolylium ions, 490
Double layer, 443
Dropping mercury electrode (DME), 447

Eight-membered lactams, IR data, 31–32
Electroanalytical techniques, 430–468
  experimental procedures, 430–441
    cells, 431–434
    diaphragms, 434
    electronic devices, 440–441
    reference electrodes, 434–435
    solvents, 436–439
    supporting electrolytes, 440
    working and counter electrodes, 436
  organic reactions and, 442–468
    coulometry, 463–465
    cyclic voltammetry, 453–463
    polarography, 447–450
    preparative electrolysis, 465
    rotating disk electrode, 450–453
    spectroscopic methods, 465–468

Electrochemical behavior, 427–625
  experimental procedures, 430–468
  oxidation of compounds, 530–600
    heterocyclic systems, 530–591
    substituted heterocycles, 591–600
  reduction of compounds, 468–530
    heterocyclic systems, 468–500
    substituted heterocycles, 500–530
  symbols, 429
Electrodes, 434–435, 436
Electrolytes, 440
Electromagnets, 151–152
Electron density distribution (EDD), 315
Electronic structure, dipole moments, 362–364
Electron spin resonance (ESR), 465
Enzyme, resonance Raman spectroscopy, 412–413
"Equilateral" diaziridine, 334
Ethanol, physical properties, 212
Ethyl acetate, physical properties, 212
Ethylene carbonate, 37, 243
  IR data, 34–36
tris(Ethylenediamine)nickel(II)thiosulfate, 285
Ethyleneimine quinone, 317–318
Ethylene oxide, vibrational assignment, 4–5, 6
Ethylene sulfoxide, 655
Ethylene trithiocarbonate, IR data, 34–36
Ethyleneurea, IR data, 34–36
Ethyleniminodimethylaluminum trimer, 319
Excitation field, electromagnets, 152
Excitation sources, Raman studies, 383–384
Excited state dipole moments, 363–364
Experimental dipole moments, 356–358
  determination, 357–358
Experimental techniques:
  diamagnetic susceptibility, 284–285
  Raman spectroscopy, 380–397
    depolarization ratio, 387
    discrimination against fluorescence, 390–397
    excitation sources, 383–384
    intensity correction for self-absorption, 387–388
    multichannel detectors, 380–383
    multichannel system, 384–387
    sample-handling techniques, 389–390
Extended X-ray absorption fine structure (EXAF) measurements, 315

Faraday method, 287
  apparatus, 288
Faraday parameters, 640–641
  moment analysis, 649–650
Fast and slow electron transfer, 445

Ferrous ammonium sulfate, 285
Field effect transistors (FETS), 160
Five-membered aromatic rings, vibrational assignments, 11–12
Five-membered heteroaromatic compounds, 248–255
  ionization energies, 250–251
Five-membered lactones, IR data, 33
Five-membered ring heterocyclic compounds, 241–245
  ionization energies, 234–237
Five-membered rings:
  one heteroatom, 8–33
    aromatic rings without carbonyl groups, 8–22
    with carbonyl groups, 22–33
    nonaromatic without carbonyl groups, 8
  two or more heteroatoms, 34–99
    aromatic rings without carbonyl groups, 42–49
      three or four heteroatoms, 46–49
      two heteroatoms, 42–46
    with carbonyl or thiocarbonyl groups, 34–42
Flavins (isoalloxazines), 410–412, 513
Flavoproteins, 410–412
Flavones, IR data, 65–66
Flicker noise, 158
Fluoranthenes, 660
Fluorescence, discrimination against, 390–397
Folic acid, 512
Formamide, physical properties, 212
Formazan, 590
Four-membered cyclic peroxides, 7
Four-membered ring heterocyclic compounds, 241
"Free" aziridine molecule, 316
Free induction decay (FID), 148–150
Free radicals, resonance Raman studies, 418–419
2-Furaldehyde, 21
3-Furaldehyde, 21
Furan derivatives:
  IR data, 20–22
  reduction of, 475
Furans, 15, 248–249
  diamagnetic susceptibilities, 300
  IR data, 8–15
  oxidation of, 549–551
Furazans, reduction of, 494–495
Furfural, 550
α-Furil dioxime, 527
Furoic acids, 550
2- and 3-Furonitrile, 20

Furoxans, 516
    reduction of, 494–495
Furylacryloyl chymotrypsins, 412–413
Fused aziridine systems, 321–327
Fused diazines, IR data, 91–95
N,N-Fused diaziridine, 333
Fused β-lactam systems, 342–347
Fused pyrazines, IR data, 95
Fused pyridazines, IR data, 91
Fused pyrimidines, IR data, 92–94
Fused-ring heterocyclic compounds, 263–270

Glutarimide, IR data, 63
Gold, 436
Gouy method, 282–284
    calibration constant, 285–286
    instrumentation, 286–287
    types of tubes, 286
Guanine, 581
    ring vibrations, 106

Half-wave potential, 488
Halogen derivatives, reduction of, 524–525
Halopyridines, IR data, 71
Hameka method, 293–298
Harney-Miller variable-temperature device, 390
Hartree-Fock-Roothann coupled method, 298
Hemoproteins, Raman spectroscopy, 397–406
Heteroaromatic compounds, 248–263
    azabenzenes, 261–263
    five-membered, 248–255
    pyridine analogs, 261
    pyridine derivatives, 257–260
    six-membered, 255–261
Heterocycles and solvents, diamagnetic studies, 302–303
Heterocyclic compounds:
    circular dichroism, 627–668
        introduction, 628–635
        moment analyses, 646–653
        quantum mechanical relations, 635–646
        sample studies, 653–661
    diamagnetic susceptibility, 281–311
        introduction, 282
        measurement, 282–292
        results and discussion, 298–309
        theoretical calculation, 292–298
    dipole moments, 355–371
        applications, 359–366
        calculated, 358–359
        experimental, 356–358
        introduction, 355–356
    electrochemical behavior:
        experimental procedures, 430–468

oxidation, 530–600
reduction, 468–530
symbols, 429
infrared spectra:
    aromatic six-membered rings, 68–111
    five membered rings:
        one heteroatom, 8–33
        two or more heteroatoms, 34–49
    introduction, 3–4
    metallo-organic and elemento-organic chelate, 114–122
    seven-membered rings, 111–114
    six-membered saturated rings, 49–68
    small rings, 4–7
Raman spectroscopy, 373–425
    applications, 397–419
    experimental techniques, 380–397
    introduction, 374
    theory and principles, 375–380
small-ring nitrogen, 313–354
    azetidine, 338–347
        variants, 347–348
    aziridine, 316–327
        variants, 327–338
    introduction, 314–316
ultraviolet photoelectron spectroscopy, 231–279
    fused-ring, 263–270
    heteroaromatic compounds, 248–263
    introduction, 231–233
    lone-pair interactions, 270–273
    saturated, 233–247
Heterocyclobutanes, 241
Hexachloroacetone, physical properties, 213
Hexahydropyrazolidine, 277
Hexahydropyridazines, 271
Hexamethylphosphoramide, physical properties, 213
Hexamethylphosphorotriamide (HMPA), 439
High performance liquid chromatography (HPLC), 465
High precision Raman difference spectroscopic (RDS) technique, 394–395
Hydantoins, 502
    IR data, 41
Hydrazobenzenes, 277
Hydrogen bonding, 15, 43, 317
Hydrogen chloride, doping studies, 22
Hydrogen-deuterium exchange kinetics, 413–414
C-Hydroxy, reduction of, 500–513
2- and 3-Hydroxybenzo[b]thiophenes, 20
Hydroxycinnolines, 503–504
N-Hydroxy derivatives, 513–516
5-Hydroxyflavone, 66

5-Hydroxyisoflavone, 66
Hydroxynaphthyridines, keto-enol tautomerism, 82
3-Hydroxypiperidine, 50
2-Hydroxypyridines, IR data, 73–74
Hydurilic acid, 562
Hypoxanthine, 265, 511, 585

Imidazole derivatives, reduction of, 478
Imidazoles, 43, 253
  IR data, 43–45
  oxidation of, 560
Imidazolidine, 245
  derivatives of, 502
Imidazolidine-2-thione, 243
Imidazoline derivatives, IR data, 39–40
Imidazoline-5/4-ones, 39
3-Imidazolines, 44
Imidazoline-2-thione, 243
  IR data, 34–36
Imidazo[1,2]pyridine, 44
Imidazopyridinium cations, reduction of, 486–487
Imides, IR data, 22–25
Iminoxyl radicals, 44
1,2- and 1,3-Indanediones, 265
Indolenine-1-oxides, 514
Indoles, 266
  IR data, 15
  oxidation of, 542–544
  reduction of, 469–470
  vibrational assignment, 16
Indolizine, IR data, 15
Industrial cells, 432–434
Infrared spectra, 1–139
  aromatic six-membered rings, 68–111
  five-membered rings:
    one heteroatom, 8–33
    two or more heteroatoms, 34–49
  introduction, 3–4
  metallo-organic and elemento-organic chelate, 114–122
  seven-membered rings, 111–114
  six-membered saturated rings, 49–68
  small rings, 4–7
Inorganic cyclic π systems, MCD studies, 660–661
Instrumentation:
  circular dichroism, 629–633
  used in Gouy method, 286–287
Intensified reticon (SPD) detectors, 380–383
Intensified vidicon (SIT) detectors, 380–383
Intensity correction for self-absorption, 387–388
Iron-histidine stretching frequency, 404

Isatin, 25
  IR data, 15–16
Isoalloxazines, 269, 513
Isoallozazine, 410
Isochromanone, 592
Isocyclic polyenes, MCD studies, 658–660
  benzene and derivatives, 658–659
  cyclic polyenes and derivatives, 659–660
Isocytosine, 654
Isoflavones, IR data, 65–66
Isoindoles:
  derivatives of, 501–502
  IR data, 15
Isophosphindoles, IR data, 17
Isophosphindoline, 17
Isoquinoline, 267
  reduction of, 473–475
Isothiazole, IR data, 45
Isothiazole derivatives, reduction of, 489
Isoxazole, 253, 266

Ketones, reduction of, 526–528
Koopman's theorem, 239
Krypton-ion laser, 383–384

Lactam rule, 656
Lactams, 655
  IR data, 27–31
  β-lactams, 30
  small-ring nitrogen, 340–342
Lactones, IR data, 31, 33
Larmor precession, 143–144
Lasers, 383–384
Lattice, 146
±Laudanosine, 546
Lead, 436
Liquid ammonia, 439
Liquid sulfur dioxide, 439
Local diamagnetic shielding, 147
Local paramagnetic shielding, 147
Lone-pair interactions, 270–273
Low-frequency IR vibrations, 4

Magnetically induced circular dichroism (MCD), 628
  analytical uses, 654
  of isocyclic polyenes, 658–660
  measurements, 633–635
  model used in interpretation, 642–646
Magnetic field, NMR spectrometers, 150–151
  permanent magnets, 150–151
  superconducting magnets, 152–154
Magnetogyric ratio, 143
Maleic anhydride, 243

Maleimide, IR data, 22–25
Manganese, 321
Matrix isolation vibrational spectroscopy, 3, 22
Measurement of diamagnetic susceptibility, 282–292
  experimental technique, 284–285
  Faraday method, 287
  Gouy method, 282–284
    calibration, 285–286
    instrumentation, 286–287
  NMR method, 291–292
  optical arrangement, 288–289
  Quincke method, 289–291
  torsion head arrangement, 287–288
6-Mercaptopurine, 581
8-Mercaptoriboflavin, 411
Mercury, 38, 436
Mercury tetrathiocyanato cobalt, 285
Metal-aziridine complexes, 319–321
Metal chelates, 114–119
Metal ion-nucleotide interactions, 416–418
Metalloporphyrins, Raman spectroscopy, 397–406
Metal oxide semiconductor field effect transistors (MOSFETs), 160
Methanol, physical properties, 213
N-Methylated uric acids, 583
1-Methylbenzotriazole, 267
2-Methylbenzotriazole, 267
Methyl chloride, physical properties, 213
4-Methyl-1,3-dioxolan-2-one, 37
Methylene group, diamagnetic susceptibility, 300–301
Methylene oxalate, IR data, 38
o-Methylflavinantine, 547
3-Methylfuran, 362
4(5)-Methylimidazole, 43
10-Methylphenothiazine radical cation, 418
Methylpyridine, 74
2-Methylpyrimidine, 362
N-Methylquinolinimine, 27
Methyl-substituted diazirines, 329
2-Methyl-1-thiiraneoxide, 655
3-Methylthiophene, 362
Methylviologen, 471
Microchannel plate (MCP), 383
Microwave spectroscopy, 359
Molecular electrostatic potential (MEP), 315
Moment analyses, 646–653
  dipole strength, 647–649
  Faraday parameters, 649–650
  rotatory strength, 649
  SI units, 650–653

Morphinandienones, 547
Morpholine, 654
  IR data, 56–57
Multichannel detector, Raman spectrum, 380–383
Multichannel laser Raman system, 384–387
Multinuclear magnetic resonance, 169–209
  nitrogen, 172–203
  other isotopes of hydrogen, 171–172
  oxygen spectroscopy, 203–209
  sulfur spectroscopy, 209
Multiple isotope labeling techniques, 4

Naphthalene, 285, 659
2-Napthylamine, 658
Naphthyridines, reduction of, 486
Narcotine sulfate, 545
Nicotinamide, 470
Nitro derivatives reduction of, 519–520
2-Nitrofuran, 21
Nitrogen, heterocyclic compounds containing, 298–300
Nitrogen chemical shifts, references for, 216
Nitrogen nuclear magnetic resonance, 172
Nitrogen and oxygen, heterocyclic compounds containing, 302
Nitrogen and sulfur, heterocyclic compounds containing, 301–302
Nitromethane, 215–217
  physical properties, 213
N-Nitropyrazole, 520
Nitropyridines, 518, 520
Nitroso derivatives, reduction of, 518–519
Nitrosopiperidine, 519
Nitrosopyridines, 518
Nitrosouracils, 519
Noise, 158–169
  coherent, 164–169
    acoustic ringing, 165–168
    coil resonance, 168–169
  defined, 158
  random, 158–164
  reduction, 220–225
    baseline roll, 220–223
    decoupling methods, 223–225
Nonaromatic rings without carbonyl groups, 8
Norarmepavine, 592
Normal hydrogen electrode (NHE), 434
Nuclear magnetic resonance (NMR) spectroscopy, 141–229, 291–292
  historical development, 142–147
  multinuclear magnetic resonance, 169–209
    nitrogen, 172–203

Index   677

other isotopes of hydrogen, 171–172
    oxygen spectroscopy, 203–209
    sulfur spectroscopy, 209
noise, 158–169
    coherent, 164–169
    random, 158–164
    reduction of noise, 220–225
        baseline roll, 220–223
        decoupling methods, 223–225
    reference solvents, 210–217
        ^{1}H and ^{13}C, 215
        ^{15}N and ^{14}N, 215–217
    sample temperature, 217–220
    spectrometers, 148–157
        free induction decay, 148–150
        magnetic field, 150–154
        receiver system, 154–157
Nuclear Overhauser effect (NOE), 146–147
Nuclear spin quantum number, 143
Nucleic acid constituents, vibrational spectra, 97–101
    hydrogen-bonding interactions, 107
    observed and calculated in-plane fundamentals, 99–100
    vibrational spectra:
        purines, 105–107
        pyrimidines, 97–105
Nucleic acid derivatives, Raman spectroscopy, 413–418
    hydrogen-deuterium exchange kinetics, 413–414
    metal-ion-nucleotide interactions, 416–418
    ultraviolet resonance effects, 415–416
Nucleic acids monomers, circular dichroism, 656–657

Optical arrangement, 288–289
1,2,5-Oxadiazines, IR data, 59
Oxadiazole:
    IR data, 47
    reduction of, 494
    vibrational assignments, 47
1,2,5-Oxadiazole, 254
1,3,4-Oxadiazole, 254
1,3,4-Oxadiazolin-5-ones, 47
Oxariridines, reduction of, 487
1,4-Oxathian, IR data, 52–55
1,3-Oxazine-4,6-diones, IR data, 63–64
Oxaziridines, 655
    small-ring nitrogen, 335–338
Oxazoles, 253
    IR data, 45–46
    reduction of, 488

1,3-Oxazolidine, 654
Oxazolidine-2-one, IR data, 34–36
Oxazolidine-2-thione, 243
    IR data, 34–36
Oxazoline-2-thione, 243
Oxepine ring systems, 114
Oxetane, 241
Oxidation of heterocyclic compounds:
    heterocyclic systems, 530–590
        four nitrogen atoms, 581–590
        four nitrogen and two sulfur atoms, 590
        four sulfur atoms, 590
        one nitrogen atom, 530–548
        one nitrogen and one oxygen atom, 568
        one nitrogen and one sulfur atom, 568–571
        one oxygen atom, 549–553
        one oxygen and one sulfur atom, 574
        one sulfur atom, 553–556
        three sulfur atoms, 580
        two nitrogen atoms, 556–568
        two nitrogen atoms and one phosphorus atom, 580
        two oxygen atoms, 571–573
        two sulfur atoms, 574–580
    substituted heterocycles, 591–600
        carbon attached to ring, 598–600
        nitrogen attached to ring, 595–597
        oxygen attached to ring, 591–595
        sulfur attached to ring, 597–598
N-Oxides, 513–516
Oxipurinol, 583
Oxirane, 655
Oxo derivatives, reduction of, 500–513
Oxygen, heterocyclic compounds containing, 300–301
Oxygen spectroscopy, 203–209
Oxonides, 273

Pacault method, 293, 294–295
Papaverine, 545
Parabanic acid, 502
Pelletierine, 654
Penems, IR data, 29
Penicillins, 30
cis and trans-Peptide conformation, 28
Permanent magnets, 150–151
α-Peroxylactone, 7
Phenanthridine, 473
o-Phenanthroline, 473
Phenazines:
    derivatives of, 509
    oxidation of, 566–567
    reduction of, 484–485

Phenazinones, 509
Phenomenological relations, circular dichroism, 631–633
Phenothiazine radical cation, 418
Phenothiazines, 270
  derivatives of, 502–503
  derivatives, reduction of, 489
  diamagnetic susceptibilities, 301
  oxidation of, 568–570
Phenoxathine, 574
Phenoxazine radical cation, 418
Phenoxazines, 96, 270
  derivatives of, 502–503
Phenyloxazoles, 46
Phenylquinoline-1-oxides, 514
Phenyl 2-thienyl ketone, 19
Phosphabenzene, 261
  IR data, 75–78
2-Phosphanaphthalene, 267
Phosphirane, 240
Photoelectron spectroscopy, 3, 13
Phthalazines, reduction of, 481
Phthalazinones, 504
Phthalic anhydride, IR data, 26–27
Phthalimide, 26, 265
  IR data, 25
Phthalizinediones, 504
Picoline, 598
2,5- and 2,3-Piperazinediones, IR data, 63
Piperazine, derivatives of, 507
Piperidine:
  derivatives, IR data, 50
  IR data, 49–51
  oxidation of, 541–542
Piperidine nitroxide, 515
Piperimidine, 247
Platinum, 285, 436
Polarography, 447–450
Polycrystalline imidazole, N substitution, 43
Polycyclic fused aziridine (five-, six-, and nine-membered rings), 324
Polycyclic heteroaromatic compounds, IR data, 96–97
Polymers, circular dichroism, 656–657
Polynucleotides, 587
Porphinato core structure, 404
Porphyrin core expansion, 404–405, 406
Porphyrins:
  oxidation of, 588–590
  reduction of, 499–500
Position-sensitive linear detector, 316
Potentiostat, 440
Prazepam, 509
Preparative electrolysis, 465

Procyanidins, 653
Propylene carbonate, 243, 439
Prostaglandin endoperoxide, 273
Protic solvents, 438–439
Protonated imidazole salts, 43
Pseudolaudanine, 546
Pseudorotation, 8, 37
Pteridines:
  derivatives of, 512–513
  oxidation of, 587–588
  reduction of, 497
Pulse laser time-resolved resonance Raman techniques, 406–408
Pulse spectrometers, 156–157
Purines, 267
  derivatives of, 511–512
  nucleic acid constituents, 105–107
  oxidation of, 581–587
  reduction of, 496–497
Pyrans, 247
  oxidation of, 552
Pyrazine-N-oxide, 516
  IR data, 95
Pyrazines, 660
  derivatives of, 507
  IR data, 89–91
  oxidation of, 566–567
  reduction of, 482–483
  vibrational assignment, 90
Pyrazino[2,3-b]pyrazines, reduction of, 498
Pyrazole, 43
  IR data, 42
  reduction of, 478
Pyrazolines, oxidation of, 556–559
Pyrazolinethione, 245
Pyrazolo[1,5a]pyrimidines, 493
Pyrene, 659
Pyridazine-N-oxide, 515
Pyridazines:
  IR data, 83
  oxidation of, 561
  reduction of, 479–480
Pyridazinones, 503
Pyridil, 527
Pyridine, 660
Pyridine-acceptor (metallic) complexes, 75
Pyridine analogs, 261
Pyridine carboxaldehydes, 71
2,6-Pyridinecarboxylic acid, 72
Pyridine derivatives, 257–260
Pyridine derivatives, reduction of, 470–472
Pyridine N-oxide, 514
Pyridines, 75, 439
  IR data, 68–70

oxidation of, 536
physical properties, 213
Pyridinium, 75
Pyridinium cations, 471
Pyridinols, 654
Pyridopyrazines:
  derivatives of, 511
  reduction of, 493
Pyridopyrimidine-4-ones and -2,4-diones, 92
Pyridotriazines, reduction of, 497
Pyrilium derivatives, reduction of, 475
Pyrimidines, 660
  derivatives of, 504–506
  IR data, 83–88
  nucleic acid constituents, 97–105
  oxidation of, 561–565
  reduction of, 481–482
  vibrational assignment, 84
Pyrimidinols, 654
Pyrimidones, 504
Pyrimidotriazole, 500
Pyrones:
  IR data, 62
  reduction of, 476
Pyrrole derivatives, IR data, 15–17
Pyrroles, 43, 248–249
  derivatives of, 501
  IR data, 8–15
  oxidation of, 532–536
  reduction of, 469–470
Pyrrolidines, 541
Pyrrolopyridinium cations, reduction of, 486–487
Pyrylium cation, IR data, 75

Quadrant rule, 656
Quantum-mechanical calculations, dipole moments, 358–359
Quantum mechanical relations, CD and MCD, 635–646
  general relations, 636–639
  magnetically induced, model used, 642–646
  models used, 641–642
  simple interpretation of expressions, 639–641
    Faraday parameters, 640–641
    rotatory strength, 639–640
Quinazoline, 515
  derivatives of, 506
  reduction of, 482
Quinazolinone, 506
Quincke method, 289–291
Quinoline derivatives, IR data, 52
Quinoline-N-oxide, 78

Quinolines, 267
  IR data, 78–81
  oxidation of, 544–547
  reduction of, 473–475
Quinolinic thioanhydride, IR data, 27
Quinolinimide, 265
Quinoxaline derivatives, IR data, 95
Quinoxalinedione, 508
Quinoxalines:
  derivatives of, 507–508
  oxidation of, 566–567
  reduction of, 483
Quinoxalino[2,3-b]quinoxaline, reduction of, 498

Raman hypochromism, 416
Raman-induced Kerr effect spectroscopy (RIKES) technique, 392–393
Raman microprobe analysis, 395–397
Raman spectroscopy, 373–425, 468, 631
  applications, 397–419
    chlorophylls, 409–410
    enzymes, 412
    flavins and flavoproteins, 410–412
    free radicals, 418–419
    metalloporphyrins and hemoproteins, 397–404
    nucleic acid derivatives, 413–418
    spectroscopic ruler for measuring heme central hole, 404–406
    time-resolved resonance studies, 406–408
  experimental techniques, 380–397
    depolarization ratio, 387
    discrimination against fluorescence, 390–397
    excitation sources, 383–384
    intensity correction for self-absorption, 387–388
    multichannel detectors, 380–383
    multichannel system, 384–387
    sample-handling techniques, 389–390
  introduction, 374
  theory and principles, 374–380
    classical, 375–377
    quantum description, 377–380
Random noise, 158
Receiver system, NMR spectrometer, 154–157
  pulse, 156–157
  sweep, 154–156
Reduction of heterocyclic compounds:
  heterocyclic systems, 469–500
    five nitrogen atoms, 500
    four nitrogen atoms, 496–500

Reduction of heterocyclic compounds, heterocyclic systems (*Continued*)
   one nitrogen atom, 469–475
   one nitrogen and one oxygen atom, 487–488
   one nitrogen and one sulfur atom, 488–489
   one sulfur atom, 477
   three nitrogen atoms, 491–493
   two nitrogen atoms, 478–487
   two nitrogen atoms and one oxygen atom, 494–495
   two nitrogen atoms and one sulfur atom, 495–496
   two phosphorus atoms, 489–490
   two sulfur atoms, 490–491
  substituted heterocycles, 500–530
   carbon atom attached to ring, 526–530
   nitrogen atom attached to ring, 516–522
   N-oxides, nitroxides, and N-hydroxy derivatives, 513–516
   oxides atom attached to ring, 500–513
   sulfur atom attached to ring, 522–525
Reduction of noise, 220–225
  baseline roll, 220–223
  decoupling methods, 223–225
Reference electrodes, electroanalytical experiment, 434–435
Reference solvents, NMR spectroscopy, 210–217
  ^{1}H and ^{13}C, 215
  ^{15}N and ^{14}N, 215–217
Relaxation mechanism, 146
Rhodanine, IR data, 38
Rotating disk electrode (RDE), 450–453
Rotating ring disk electrode (RRDE), 453
Rotatory strength, circular dichcroism, 639–640
  moment analysis, 649

Saccharin, IR data, 42
Saturated calomel electrode (SCE), 434
Saturated heterocyclic compounds, 233–247
  five-membered rings, 241–245
  four-membered rings, 241
  six-membered rings, 246–248
  three-membered rings, 238–241
Selenium, 574
Selenodiazole, 495
Selenophene, 14, 248–249
  IR data, 8–15
Selenophthalides, IR data, 27
Seleno and pyrrolo analogs of thieno[2,3-*b*]thiophene, 264
1,4-Selenoxan, IR data, 55–56
Self-absorption, intensity correction, Raman spectroscopy, 387–388

Seven-membered benzolactams, IR data, 31
Seven-membered lactams, IR data, 30
Seven-membered rings, 111–114
Shielding of nucleus, 147
Shot noise, 158, 160
Signal-to-noise ratios, *see* Noise
Silicon heterocycles, 119–122
Silicon photodiode array (SPD), 383
Simple azetidines, small-ring nitrogen, 338–340
Simply substituted aziridines, small-ring nitrogen, 316–319
Sipropyrans, IR data, 51
SI units, 650–653
Six-membered heteroaromatic compounds, 255–261
  ionization energies, 256
Six-membered phosphorus heterocycles, 57–58
Six-membered ring heterocyclic compounds, 246–248
Six-membered saturated rings, 49–68
  with carbonyl groups, 62–68
  one heteroatom, 49–52
  other heterocyclic rings, 57–58
  three or four heteroatoms, 58–62
  two heteroatoms, 52–57
Small heterocyclic rings, circular dichroism, 655–656
Small-ring nitrogen heterocycles, 313–354
  azetidine, 338–347
   fused β-lactam systems, 342–347
   β-lactams, 340–342
   simple, 338–340
   variants, 347–348
  aziridines, 316–327
   electron density "bulging," 318
   fused systems, 321–327
   metal-complexes, 319–321
   simply substituted, 316–319
   stable invertomer, 319
   variants, 327–338
    azirines, 327–329
    diaziridines, 330–334
    diazirines, 329–330
    oxaziridines, 335–338
    thiadiaziridines, 334
  introduction, 314–316
Small rings, infrared spectra, 4–7
  four-membered cyclic peroxides, 7
  three-membered, 4–7
Solid-state IR spectroscopy, 4
Solvent effect, dipole moments, 358
Solvents, electrochemical, 436–439
Sparsomycin, 655
Spectrometers, NMR, 148–157
  free induction decay, 148–150

Index 681

magnetic field, 150–154
  permanent, 150–151
  superconducting, 152–154
receiver system, 154–157
  pulse, 156–157
  sweep, 154–156
Spectroscopic methods, electroanalytical techniques, 465–468
^{14}N Spectroscopy, 173–183
^{15}N Spectroscopy, 183–203
Spin properties of nuclei, 170
Spiro compound, 39
Spiro-fused "aziridinyl" rings, 320
Spiro-fused azirine, 328
Stark effect, 357
Stibabenzene, 261
Structure determination, dipole moments, 360–362
trans-2-Styrylthiophene, 19
Substituted "aniline" diaziridine, 332
N-Substituted α-aziridinyl ketones, 5
C-Substituted diazirede, 333
C,N-Substituted diaziridine, 332
C,N,N-Substituted diaziridine, 333
Substituted diaziridinone, 331
Substituted heterocycles, oxidation of, 591–600
Substituted hydantoins, 41
1-Substituted imidazoles, 253
Substituted isatins, 15
5-Substituted isorhodanines, 38
Substituted pyridines, IR data, 71–75
2-Substituted pyridines, 70
3-Substituted rhodanine, 38
Substituted thiadiazeridine, 334
Substituted thiophenes, IR data, 17–19
Succinic anhydride, 243
Succinimides, 243, 501
Sulfolane, 8, 439
Sulfones, reduction of, 477
Sulfur, heterocyclic compounds containing, 301
Sulfur spectroscopy, 209
Superconducting magnets, 152–154
Surface-enhanced Raman scattering (SERS), 393–394
Sweep spectrometers, 154–156
Sydnones, reduction of, 495
Synchrotron radiation, 315–316

TAILS program, 315
Tannins from woody plants, 653–654
Tellurium, 574
Tellurophene, 14, 248–249
  IR data, 8–15
1,4-Telluroxan, IR data, 55–56

Temperature measurement, NMR spectroscopy, 217–220
Tetraazabinaphthylene, reduction of, 499
Tetrachloroethane, physical properties, 213
Tetrachloro- and tetrabromothiophene, 17
Tetracyanoquinodimethane (TCNQ), 468
Tetrafluoro-1,3-dithietane, 57
Tetrahydro-β-carbolines, 81
Tetrahydrofuran, 241, 242, 439
  oxidation of, 552
  physical properties, 213
  vibrational assignments, 9–10
Tetrahydroisoquinoline, 592
Tetrahydro-1,3-oxazine, 654
Tetrahydropyran, IR data, 49–51
Tetrahydroquinoline, IR data, 51
1,2,3,4-Tetrahydroquinoline-2-carboxylic acid, 51
Tetrahydrotetrathiafulvalene, 245
Tetramethylsilane (TMS), 215
Tetrathiaadamantane, IR data, 61–62
Tetrathiafulvalene, 245, 574
Tetrathiatriazinium ion, 661
s-Tetrazine, 263
Tetrazines, IR data, 110–111
Tetrazole, tautomer equilibrium, 359
1,2,3,4-Tetrazole, 49
Tetrazolium cations, reduction of, 496
Tetroxanes, 273
Theobromine, 265, 581
Theoretical calculation:
  diamagnetic susceptibility, 292–298
    Hartree-Fock-Roothann coupled, 298
    Pacault method, 293, 294–295
    semiempirical methods, 293–298
  dipole moments, 365–366
Therman noise, 158
Thiacyclohexane, IR data, 49–51
Thiadiaziridine, small-ring nitrogen, 334
Thiadiazoles:
  IR data, 47
  vibrational assignments, 47
1,3,4-Thiadiazoles, reduction of, 496
2,1,3-Thiadiazoles, reduction of, 495–496
Thianthrene, 96, 575, 576
6*a*-Thiathiophthene, 264
1,2,3,4-Thiatriazoles, IR data, 48–49
Thiazole, 253–254, 301
  derivatives, reduction of, 488
  IR data, 45, 46
Thiazolidinediones, IR data, 42
Thiazolidine-2-one, IR data, 34–36
Thiazolidine-2-thione, 243
  IR data, 34–36
Thiazoline-2-thione, 243

Index

Thiazolium salts, 45
Thienothiophene, IR data, 19
Thieno[2,3-b]thiophene, 264
Thienylacryloyl chymotrypsins, 412–413
2-Thienyl- and 2-furylcarbonyl compounds, 19
2-Thienylmethacrylates, 19
Thiepienes, IR data, 115
Thietane, 241
Thiirane, 240, 655
Thiirane dioxide, 241
Thioamides, IR data, 41–42
Thiobarbituric acids, 299–300
Thiobenzopropiolactone, 265
Thiochromone, IR data, 66–68
Thiomorpholides, IR data, 49–51
Thionaphthenequinone, IR data, 25
Thiones, 598
Thiophenes, 15
  derivatives:
    IR data, 17–20
    reduction of, 477
    ring vibrations, 18
  IR data, 8–15
  oxidation of, 553–555
Thiophthalic anhydride, IR data, 25
Thiopiperidones, IR data, 49–51
Thiopurine, 597
Thiopyrimidine, 522
1,4-Thioselenan, IR data, 55–56
1,4-Thioxan, IR data, 55–56
Three-membered ring heterocyclic compounds, 238–241
  ionization energies, 232
Three-membered rings, infrared spectra, 4–7
Through-space and through-bond interactions, 270–271
Thymine, 104, 505
  far-infrared frequencies, 102–103
Time-resolved resonance Raman studies, 406–408
Toluene, physical properties, 213
N-p-Tolyldichloromaleimide, 25
Torsion head arrangement, 287–288
Transverse/transverse method, 389–390
Triazaphenothiazine heterocycle, 96
Triazines:
  derivatives of, 510
  IR data, 107–109
  reduction of, 491–492
1,2,4-Triazines, 109–110, 263
s-Triazines, 107–109, 660
1,2,4-Triazole-N-imine derivatives, 46
1,2,4-Triazoles, 254
1,2,3-Triazoles, 46
1,2,3-Triazolinethione, 245

1,3,4-Triazoline-2-thione, 245
Trifluoroacetic acid, physical properties, 213
Trimethylpyridines (collidines), 71
2,3,6-, 2,3,5-, and 2,4,5-Trimethylpyridines, 71
Trioxepines, IR data, 115
1,3,5-Trithiane, IR data, 58
Trithiapentalenes, 580
Tritium, 171–172
Tryptophan, 654
Tyrosine, 654

Ultraviolet photoelectron spectroscopy, 231–279
  fused-ring, 263–270
  heteroaromatic compounds, 248–263
    azabenzenes, 261–263
    five-membered, 248–255
    pyridine analogs, 261
    pyridine derivatives, 257–260
    six-membered, 255–261
  introduction, 231–233
  lone-pair interactions, 270–273
  saturated, 233–247
    five-membered rings, 241–245
    four-membered rings, 241
    six-membered rings, 246–248
    three-membered rings, 238–241
Ultraviolet resonance Raman effects, 415–416
Uracil, 104–105, 505, 561
  far-infrared frequencies, 102–103
Uric acid, 581

δ-Valerolactam, 28
Vapor-phase and low-temperature solid phase spectra, 22
Vector addition, dipole moments, 358, 360
Vinylene carbonate, 243
Visible spectroscopy, 466–467
Volume susceptibility, Gouy tube, 283

Water, 285
  doping studies, 22
  physical properties, 213
Wiedmann's additivity rule, 303

Xanthine, 265, 581, 584–585
  far-infrared frequencies, 102–103
Xenon arc lamp, 629
X-ray diffraction techniques, 314–316
XTAL program, 315

Zeeman effect, 640
Zeroth moment:
  dipole strength, 647
  rotatory strength, 649
Zinc, 38